Semi-Riemannian Geometry

Semi-Riemannian Geometry

The Mathematical Language of General Relativity

STEPHEN C. NEWMAN
University of Alberta
Edmonton, Alberta, Canada

This edition first published 2019
© 2019 John Wiley & Sons, Inc.

The right of Stephen C. Newman to be identified as the author of this work has been asserted in accordance with law.

Registered Office
John Wiley & Sons, Inc., 111 River Street, Hoboken, NJ 07030, USA

Editorial Office
111 River Street, Hoboken, NJ 07030, USA

For details of our global editorial offices, customer services, and more information about Wiley products visit us at www.wiley.com.

Wiley also publishes its books in a variety of electronic formats and by print-on-demand. Some content that appears in standard print versions of this book may not be available in other formats.

Library of Congress Cataloging-in-Publication Data

Names: Newman, Stephen C., 1952- author.

Title: Semi-Riemannian geometry : the mathematical language of general relativity / Stephen C. Newman (University of Alberta, Edmonton, Alberta, Canada).

Description: Hoboken, New Jersey : Wiley, [2019] | Includes bibliographical references and index. |

Identifiers: LCCN 2019011644 (print) | LCCN 2019016822 (ebook) | ISBN 9781119517542 (Adobe PDF) | ISBN 9781119517559 (ePub) | ISBN 9781119517535 (hardcover)

Subjects: LCSH: Semi-Riemannian geometry. | Geometry, Riemannian. | Manifolds (Mathematics) | Geometry, Differential.

Classification: LCC QA671 (ebook) | LCC QA671 .N49 2019 (print) | DDC 516.3/73–dc23

LC record available at https://lccn.loc.gov/2019011644

Cover design: Wiley

Set in 10/12pt Computer Modern by SPi Global, Chennai, India

Printed in the United States of America

V10011687_062619

To Sandra

Contents

Preface

Physics texts on general relativity usually devote several chapters to an overview of semi-Riemannian geometry. Of necessity, the treatment is cursory, covering only the essential elements and typically omitting proofs of theorems. For physics students wanting greater mathematical rigor, there are surprisingly few options. Modern mathematical treatments of semi-Riemannian geometry require grounding in the theory of curves and surfaces, smooth manifolds, and Riemannian geometry. There are numerous books on these topics, several of which are included in Further Reading. Some of them provide a limited amount of material on semi-Riemannian geometry, but there is really only one mathematics text currently available that is devoted to semi-Riemannian geometry and geared toward general relativity, namely, *Semi-Riemannian Geometry: With Applications to Relativity* by Barrett O'Neill. This is a classic, but it is pitched at an advanced level, making it of limited value to the beginner. I wrote the present book with the aim of filling this void in the literature.

There are three parts to the book. Part I and the Appendices present background material on linear algebra, multilinear algebra, abstract algebra, topology, and real analysis. The aim is to make the book as self-contained as possible. Part II discusses aspects of the classical theory of curves and surfaces, but differs from most other expositions in that Lorentz as well as Euclidean signatures are discussed. Part III covers the basics of smooth manifolds, smooth manifolds with boundary, smooth manifolds with a connection, and semi-Riemannian manifolds. It concludes with applications to Lorentz vector spaces, Maxwell's equations, and the Einstein tensor. Not all theorems are provided with a proof, otherwise an already lengthy volume would be even longer.

The manuscript was typed using the WYSIWYG scientific word processor EXP®, and formatted as a camera-ready PDF file using the open-source TEX-LATEX typesetting system MiKTeX, available at `https://miktex.org`. Figure 19.5.1 was prepared using the TEX macro package `diagrams.sty` developed by Paul Taylor. I am indebted to Professor John Lee of the University of Washington for reviewing portions of the manuscript. Any remaining errors or deficiencies are, of course, solely my responsibility.

I am most interested in receiving your comments, which can be emailed to me at `stephen.newman@ualberta.ca`. A list of corrections will be posted on the website `https://sites.ualberta.ca/~sn2/`. Should the email address become unavailable, an alternative will be included with the list of corrections.

On the other hand, if the website becomes inaccessible, the list of corrections will be stored as a public file on Google Drive that can be searched using "Corrections to Semi-Riemannian Geometry by Stephen Newman".

Allow me to close by thanking my wife, Sandra, for her unwavering support and encouragement throughout the writing of the manuscript. It is to her, with love, that this book is dedicated.

Part I

Preliminaries

Differential geometry rests on the twin pillars of linear algebra–multilinear algebra and topology–analysis. Part I of the book provides an overview of selected topics from these areas of mathematics. Most of the linear algebra presented here is likely familiar to the reader, but the same may not be true of the multilinear algebra, with the exception of the material on determinants. Topology and analysis are vast subjects, and only the barest of essentials are touched on here. In order to keep the book to a manageable size, not all theorems are provided with a proof, a remark that also applies to Part II and Part III.

4

Chapter 1

Vector Spaces

1.1   Vector Spaces

The definition of a **vector space** over a field and that of a **subspace** of a vector space are given in Section B.6. Our focus in this book is exclusively on vector spaces over the real numbers (as opposed to the complex numbers or some other field).

Throughout, all vector spaces are over $\mathbb{R}$, the field of real numbers.

For brevity, we will drop the reference to $\mathbb{R}$ whenever possible and write, for example, "linear" instead of "$\mathbb{R}$-linear".

Of particular importance is the vector space $\mathbb{R}^m$, but many other examples of vector spaces will be encountered. It is easily shown that the intersection of any collection of subspaces of a vector space is itself a subspace. The **zero vector** of a vector space is denoted by 0, and the **zero subspace** of a vector space by $\{0\}$. The **zero vector space**, also denoted by $\{0\}$, is the vector space consisting only of the zero vector. We will generally avoid explicit consideration of the zero vector space. Most of the results on vector spaces either apply directly to the zero vector space or can be made applicable with a minor reworking of definitions and proofs. The details are usually left to the reader.

Example 1.1.1. Let V and W be vector spaces. Following Section B.5 and Section B.6, we denote by $\mathrm{Lin}(V, W)$ the vector space of linear maps from V to W, where addition and scalar multiplication are defined as follows: for all maps A, B in $\mathrm{Lin}(V, W)$ and all real numbers c,

$$(A + B)(v) = A(v) + B(v)$$

and

$$(cA)(v) = cA(v)$$

Semi-Riemannian Geometry, First Edition. Stephen C. Newman.
© 2019 John Wiley & Sons, Inc. Published 2019 by John Wiley & Sons, Inc.

for all vectors v in V. The zero element of $\mathrm{Lin}(V, W)$, denoted by 0, is the zero map, that is, the map that sends all vectors in V to the zero vector 0 in W. When $V = W$, we make $\mathrm{Lin}(V, V)$ into a ring by defining multiplication to be composition of maps: for all maps A, B in $\mathrm{Lin}(V, V)$, let

$$A \circ B(v) = A\big(B(v)\big)$$

for all vectors v in V. The identity element of the ring $\mathrm{Lin}(V, V)$ is the identity map on V, denoted by id_V. ◊

A **linear combination** of vectors in a vector space V is defined to be a *finite* sum of the form $a^1 v_1 + \cdots + a^k v_k$, where $a^1, \ldots, a^k$ are real numbers and $v_1, \ldots, v_k$ are vectors in V. The possibility that some (or all) of $a^1, \ldots, a^k$ equal zero is not excluded.

Let us pause here to comment on an aspect of notation. Following the usual convention in differential geometry, we index the scalars and vectors in a linear combination with superscripts and subscripts, respectively. This opens the door to the **Einstein summation convention**, according to which, for example, $a^1 v_1 + \cdots + a^k v_k$ and $\sum_{i=1}^k a^i v_i$ are abbreviated as $a^i v_i$. The logic is that when an expression has a superscript and subscript in common, it is understood that the index is being summed over. Despite the potential advantages of this notation, especially when multiple indices involved, the Einstein summation convention will *not* be adopted here.

Let S be a (nonempty and not necessarily finite) subset of V. The **span of S** is denoted by $\mathrm{span}(S)$ and defined to be the set of linear combinations of vectors in S:

$$\mathrm{span}(S) = \{a^1 v_1 + \cdots + a^k v_k : a^1, \ldots, a^k \in \mathbb{R};$$
$$v_1, \ldots, v_k \in S; k = 1, 2, \ldots\}.$$

For a vector v in V, let us denote

$$\mathrm{span}(\{v\}) = \{av : a \in \mathbb{R}\} \qquad \text{by} \qquad \mathbb{R}v.$$

For example, in $\mathbb{R}^2$, we have

$$\mathrm{span}\big(\{(1, 0), (0, 1)\}\big) = \mathbb{R}^2$$

and

$$\mathrm{span}\big(\{(1, 0)\}\big) = \mathbb{R}(1, 0) = \{(a, 0) \in \mathbb{R}^2 : a \in \mathbb{R}\}.$$

It is easily shown that $\mathrm{span}(S)$ is a subspace of V. In fact, $\mathrm{span}(S)$ is the smallest subspace of V containing S, in the sense that any subspace of V containing S also contains $\mathrm{span}(S)$. When $\mathrm{span}(S) = V$, it is said that **S spans V** or that the **vectors in S span V**, and that each vector in V is in the **span of S**.

We say that **S is linearly independent** or that the **vectors in S are linearly independent** if the only linear combination of *distinct* vectors in S

that equals the zero vector is the one with all coefficients equal to 0. That is, if $v_1, \ldots, v_k$ are distinct vectors in S and $a^1, \ldots, a^k$ are real numbers such that $a^1 v_1 + \cdots + a^k v_k = 0$, then $a^1 = \cdots = a^k = 0$. Evidently, any subset of a linearly independent set is linearly independent. When S is not linearly independent, it is said to be **linearly dependent**. In particular, the zero vector in any vector space is linearly dependent. As further examples, the vectors $(1, 0), (0, 1)$ in $\mathbb{R}^2$ are linearly independent, whereas $(0, 0), (1, 0)$ and $(1, 0), (2, 0)$ are linearly dependent.

The next result shows that when a linearly independent set does not span a vector space, it has a linearly independent **extension**.

Theorem 1.1.2. *Let V be a vector space, let S be a nonempty subset of V such that $\mathrm{span}(S) \neq V$, and let v be a vector in $V \backslash \mathrm{span}(S)$. Then S is linearly independent if and only if $S \cup \{v\}$ is linearly independent.*

Proof. ($\Rightarrow$): Suppose $av + b^1 s_1 + \cdots + b^k s_k = 0$ for distinct vectors $s_1, \ldots, s_k$ in S and real numbers $a, b^1, \ldots, b^k$. Then $a = 0$; for if not, then

$$v = -\left[\left(\frac{b^1}{a}\right) s_1 + \cdots + \left(\frac{b^k}{a}\right) s_k\right],$$

hence v is in $\mathrm{span}(V)$, which is a contradiction. Thus, $b^1 s_1 + \cdots + b^k s_k = 0$, and since S is linearly independent, we have $b^1 = \cdots = b^k = 0$.

($\Leftarrow$): As remarked above, any subset of a linearly independent set is linearly independent. $\qquad \square$

A (not necessarily finite) subset $\mathcal{H}$ of a vector space V is said to be an **unordered basis** for V if it spans V and is linearly independent.

Theorem 1.1.3. *If V is a vector space and $\mathcal{H}$ is an unordered basis for V, then each vector in V can be expressed uniquely (up to order of terms) as a linear combination of vectors in $\mathcal{H}$.*

Proof. Since $\mathcal{H}$ spans V, each vector in V can be expressed as a linear combination of vectors in $\mathcal{H}$. Suppose a vector v in V can be expressed as a linear combination in two ways. Let $h_1, \ldots, h_k$ be the distinct vectors in the linear combinations. Then

$$v = a_1 h_1 + \cdots + a_k h_k \qquad \text{and} \qquad v = b_1 h_1 + \cdots + b_k h_k,$$

for some real numbers $a_1, \ldots, a_k, b_1, \ldots, b_k$, hence

$$(a_1 - b_1) h_1 + \cdots + (a_k - b_k) h_k = 0.$$

Since $\mathcal{H}$ is linearly independent, $a_i - b_i = 0$ for $i = 1, \ldots, k$. $\qquad \square$

Theorem 1.1.4. *Let V be a vector space, and let S and T be nonempty subsets of V, where S is linearly independent, and T is finite and spans V. Then S is finite and $\mathrm{card}(S) \leq \mathrm{card}(T)$, where* card *denotes cardinality.*

Proof. Since S is linearly independent, it does not contain the zero vector. Let $\mathrm{card}(T) = m$ and $T = \{t_1, \ldots, t_m\}$. We proceed in steps. For the first step, let s_1 be a vector in S. Since $V = \mathrm{span}(T)$, s_1 is a linear combination of $t_1, \ldots, t_m$. Because s_1 is not the zero vector, at least one of the coefficients in the linear combination must be nonzero. Renumbering $t_1, \ldots, t_m$ if necessary, suppose it is the coefficient of t_1, and let $S_1 = \{s_1, t_2, \ldots, t_m\}$. Then t_1 can be expressed as a linear combination of the vectors in S_1, hence $V = \mathrm{span}(S_1)$. For the second step, let s_2 be a vector in $S \backslash \{s_1\}$. Since $V = \mathrm{span}(S_1)$, s_2 is a linear combination of $s_1, t_2, \ldots, t_m$. Because s_1, s_2 are linearly independent, at least one of the coefficients of $t_2, \ldots, t_m$ in the linear combination is nonzero. Renumbering $t_2, \ldots, t_m$ if necessary, suppose it is the coefficient of t_2, and let $S_2 = \{s_1, s_2, t_3, \ldots, t_m\}$. Then t_2 can be expressed as a linear combination of the vectors in S_2, hence $V = \mathrm{span}(S_2)$. Proceeding in this way, after $k \leq m$ steps, we have a set $S_k = \{s_1, \ldots, s_k, t_{k+1}, \ldots, t_m\}$, with $V = \mathrm{span}(S_k)$. Then $\mathrm{card}(S) \leq \mathrm{card}(T)$; for if not, at the mth step, we would have $S_m = \{s_1, \ldots, s_m\}$, with $V = \mathrm{span}(S_m)$ and $S \backslash S_m$ nonempty. Then any vector in $S \backslash S_m$ could be expressed as a linear combination of vectors in S_m, which contradicts the assumption that S is linearly independent. $\qquad\square$

We say that a vector space is **finite-dimensional** if it has a finite unordered basis. Finite-dimensional vector spaces have an associated invariant that, as we will see, largely characterizes them.

Theorem 1.1.5. *If V is a finite-dimensional vector space, then every unordered basis for V has the same (finite) number of vectors. This invariant, denoted by $\dim(V)$, is called the **dimension of V**.*

Proof. Let $\mathcal{H}$ and $\mathcal{F}$ be bases for V, with $\mathcal{F}$ finite. By Theorem 1.1.4, $\mathcal{H}$ is finite and $\mathrm{card}(\mathcal{H}) \leq \mathrm{card}(\mathcal{F})$. Then $\mathcal{H}$ is finite, so we use Theorem 1.1.4 again and obtain $\mathrm{card}(\mathcal{F}) \leq \mathrm{card}(\mathcal{H})$. Thus, $\mathrm{card}(\mathcal{H}) = \mathrm{card}(\mathcal{F})$. $\qquad\square$

For completeness, we assign the zero vector space the dimension 0:

$$\dim(\{0\}) = 0.$$

Theorem 1.1.6. *If V is a vector space of dimension m, then:*
(a) *Every subset of V that spans V contains at least m vectors.*
(b) *Every linearly independent subset of V contains at most m vectors.*

Proof. (a): Let $\mathcal{H}$ be an unordered basis for V, and suppose T is a subset of V that spans V. The result is trivial if T is infinite, so assume otherwise. Then Theorem 1.1.4 and Theorem 1.1.5 give $m = \mathrm{card}(\mathcal{H}) \leq \mathrm{card}(T)$.

(b): Suppose S is a linearly independent subset of V. Then Theorem 1.1.4 and Theorem 1.1.5 yield $\mathrm{card}(S) \leq \mathrm{card}(\mathcal{H}) = m$. $\qquad\square$

Theorem 1.1.7. *Let V be a vector space of dimension m, and let U be a subspace of V. Then:*
(a) *U is finite-dimensional and $\dim(U) \leq \dim(V)$.*

(b) *If* $\dim(U) = \dim(V)$, *then* $U = V$.
(c) *If* $\dim(U) < \dim(V)$, *then any unordered basis for* U *can be* **extended** *to an unordered basis for* V. *That is, given an unordered basis* $\{h_1, \ldots, h_k\}$ *for* U, *there are vectors* $h_{k+1}, \ldots, h_m$ *in* V *such that* $\{h_1, \ldots, h_k, h_{k+1}, \ldots, h_m\}$ *is an unordered basis for* V.

Proof. (a): We proceed in steps. For the first step, let u_1 be a vector in U. If $\text{span}(\{u_1\}) = U$, we are done. If not, for the second step, let u_2 be a vector in $U\backslash\text{span}(\{u_1\})$. It follows from Theorem 1.1.2 that u_1, u_2 are linearly independent. If $\text{span}(\{u_1, u_2\}) = U$, we are done, and so on. By Theorem 1.1.6(b), this process ends after $k \leq m$ steps. Then $u_1, \ldots, u_k$ are linearly independent and span U, which is to say that $\{u_1, \ldots, u_k\}$ is an unordered basis for U.

(b): Let $\mathcal{H}$ and $\mathcal{F}$ be bases for U and V, respectively, and suppose $U \neq V$. Since $U = \text{span}(\mathcal{H})$, there is a vector v in $V\backslash\text{span}(\mathcal{H})$. By Theorem 1.1.2, $\mathcal{H} \cup \{v\}$ is linearly independent. We have from Theorem 1.1.5 that

$$\text{card}(\mathcal{H} \cup \{v\}) > \text{card}(\mathcal{H}) = \dim(U) = \dim(V) = \text{card}(\mathcal{F}),$$

which contradicts Theorem 1.1.6(b).

(c): Given the unordered basis $\{h_1, \ldots, h_k\}$ for U, the algorithm described in part (a) can be used to find vectors $h_{k+1}, \ldots, h_m$ in V such that $\{h_1, \ldots, h_k, h_{k+1}, \ldots, h_m\}$ is an unordered basis for V.   □

Throughout the remainder of Part I, unless stated otherwise, all vector spaces are finite-dimensional.

Let V be a vector space, and let $\{h_1, \ldots, h_m\}$ be an unordered basis for V. The m-tuple $(h_1, \ldots, h_m)$ is said to be an **ordered basis** for V, as is any m-tuple derived from $(h_1, \ldots, h_m)$ by permuting $h_1, \ldots, h_m$. For example, $(h_1, h_2, \ldots, h_m)$ and $(h_2, h_1, \ldots, h_m)$ are distinct ordered bases for V.

Example 1.1.8 ($\mathbb{R}^m$). Let e_i be the vector in $\mathbb{R}^m$ defined by

$$e_i = (0, \ldots, 0, 1, 0, \ldots, 0),$$

where 1 is in the ith position and 0s are elsewhere for $i = 1, \ldots, m$. For real numbers $a^1, \ldots, a^m$, we have

$$a^1 e_1 + \cdots + a^m e_m = (a^1, \ldots, a^m),$$

from which it follows that $e_1, \ldots, e_m$ span $\mathbb{R}^m$ and are linearly independent. We refer to $\{e_1, \ldots, e_m\}$ as the **standard unordered basis** for $\mathbb{R}^m$, and to $(e_1, \ldots, e_m)$ as the **standard ordered basis** for $\mathbb{R}^m$. Thus, not surprisingly, $\mathbb{R}^m$ has dimension m.   ◇

Throughout the remainder of Part I, unless stated otherwise, all bases are ordered.

Accordingly, we now refer to $(e_1, \ldots, e_m)$ as the **standard basis** for $\mathbb{R}^m$. Let V and W be vector spaces. A map $A : V \longrightarrow W$ is said to be **linear** if

$$A(cv + w) = cA(v) + A(w)$$

for all vectors v, w in V and all real numbers c. Thus, a linear map respects vector space structure. Suppose A is in fact a linear map. Given a basis $\mathcal{H} = (h_1, \ldots, h_m)$ for V, let us denote

$$\big(A(h_1), \ldots, A(h_m)\big) \qquad \text{by} \qquad A(\mathcal{H}).$$

We say that A is a **linear isomorphism**, and that V and W are **isomorphic**, if A is bijective. To illustrate, let x be an indeterminate, and let

$$\mathbb{P}_m = \{a_0 + a_1 x + \cdots + a_m x^m : a_0, \ldots, a_m \in \mathbb{R}\}$$

be the set of real polynomials of degree at most m. From the properties of polynomials, it is easily shown that $\mathbb{P}_m$ is a vector space of dimension $m+1$, and that the map $A : \mathbb{R}^{m+1} \longrightarrow \mathbb{P}_m$ given by $A(a_0, \ldots, a_m) = a_0 + a_1 x + \cdots + a_m x^m$ for all vectors $(a_0, \ldots, a_m)$ in $\mathbb{R}^{m+1}$ is a linear isomorphism. Following Section B.5, we denote the existence of an isomorphism by $\mathbb{R}^{m+1} \approx \mathbb{P}_m$.

Since a linear isomorphism is a bijective map, it has an inverse map. The next result shows that the inverse of a linear isomorphism is automatically a linear isomorphism.

Theorem 1.1.9. *If V and W are vector spaces and $A : V \longrightarrow W$ is a linear isomorphism, then $A^{-1} : W \longrightarrow V$ is a linear isomorphism.*

Proof. By assumption, A^{-1} is bijective. Let w_1, w_2 be vectors in W, and let c be a real number. Since A is bijective, there are unique vectors v_1, v_2 in V such that $A(v_1) = w_1$ and $A(v_2) = w_2$. Then

$$A^{-1}(cw_1 + w_2) = A^{-1}\big(cA(v_1) + A(v_2)\big) = A^{-1}\big(A(cv_1 + v_2)\big)$$
$$= cv_1 + v_2 = cA^{-1}(w_1) + A^{-1}(w_2). \qquad \square$$

A linear map is completely determined by its values on a basis, as we now show.

Theorem 1.1.10. *Let V and W be vector spaces, let $\mathcal{H} = (h_1, \ldots, h_m)$ be a basis for V, and let $w_1, \ldots, w_m$ be vectors in W. Then there is a unique linear map $A : V \longrightarrow W$ such that $A(\mathcal{H}) = (w_1, \ldots, w_m)$.*

Proof. Uniqueness. Since $\mathcal{H}$ is a basis for V, for each vector v in V, there is a unique m-tuple $(a^1, \ldots, a^m)$ in $\mathbb{R}^m$ such that $v = a^1 h_1 + \cdots + a^m h_m$. Suppose $A : V \longrightarrow W$ is a linear map such that $A(\mathcal{H}) = (w_1, \ldots, w_m)$. Then

$$A(v) = A(a^1 h_1 + \cdots + a^m h_m)$$
$$= a^1 A(h_1) + \cdots + a^m A(h_m) \qquad (1.1.1)$$
$$= a^1 w_1 + \cdots + a^m w_m,$$

from which it follows that A is unique.

Existence. Let us define $A : V \longrightarrow W$ using (1.1.1) for all vectors v in V. The uniqueness of the m-tuple $(a^1, \ldots, a^m)$ ensures that A is well-defined. Clearly, $A(\mathcal{H}) = (w_1, \ldots, w_m)$. Let $u = b^1 h_1 + \cdots + b^m h_m$ be a vector in V, and let c be a real number. Then

$$cv + u = (ca^1 + b^1)h_1 + \cdots + (ca^m + b^m)h_m,$$

hence

$$
\begin{aligned}
A(cv + u) &= (ca^1 + b^1)A(h_1) + \cdots + (ca^m + b^m)A(h_m) \\
&= (ca^1 + b^1)w_1 + \cdots + (ca^m + b^m)w_m \\
&= c(a^1 w_1 + \cdots + a^m w_m) + (b^1 w_1 + \cdots + b^m w_m) \\
&= cA(v) + A(u).
\end{aligned}
$$

Thus, A is linear.   □

From the point of view of linear structure, isomorphic vector spaces are indistinguishable. In fact, it is easily shown using Theorem 1.1.10 that all m-dimensional vector space are isomorphic. More than that, they are all isomorphic to $\mathbb{R}^m$. The isomorphism constructed with the help of Theorem 1.1.10 depends on the choice of bases for the vector spaces. However, we will see an instance in Section 1.2 where an isomorphism can be defined without having to resort to such an arbitrary choice.

Let V and W be vector spaces, and let $A : V \longrightarrow W$ be a linear map. The **kernel of A** is defined by

$$\ker(A) = \{v \in V : A(v) = 0\},$$

and the **image of A** by

$$\mathrm{im}(A) = \{A(v) \in W : v \in V\}.$$

It is easily shown that $\ker(A)$ is a subspace of V, and $\mathrm{im}(A)$ is a subspace of W. The **nullity of A** is defined by

$$\mathrm{null}(A) = \dim\big(\ker(A)\big),$$

and the **rank of A** by

$$\mathrm{rank}(A) = \dim\big(\mathrm{im}(A)\big).$$

The nullity and rank of a linear map satisfy an important identity.

Theorem 1.1.11 (Rank–Nullity Theorem). *If V and W are vector spaces and $A : V \longrightarrow W$ is a linear map, then*

$$\dim(V) = \mathrm{rank}(A) + \mathrm{null}(A). \tag{1.1.2}$$

Proof. By Theorem 1.1.7(c), any basis $(h_1, \ldots, h_k)$ for $\ker(A)$ can be extended to a basis $(h_1, \ldots, h_k, h_{k+1}, \ldots, h_m)$ for V. We claim that $\bigl(A(h_{k+1}), \ldots, A(h_m)\bigr)$ is a basis for $\operatorname{im}(A)$. Let v be a vector in V. Since $\mathcal{H}$ spans V, we have $v = a^1 h_1 + \cdots + a^m h_m$ for some real numbers $a^1, \ldots, a^m$. Then

$$A(v) = a^1 A(h_1) + \cdots + a^k A(h_k) + a^{k+1} A(h_{k+1}) + a^m A(h_m)$$
$$= a^{k+1} A(h_{k+1}) + a^m A(h_m),$$

hence $A(h_{k+1}), \ldots, A(h_m)$ span $\operatorname{im}(A)$. Suppose

$$c^{k+1} A(h_{k+1}) + \cdots + c^m A(h_m) = 0$$

for some real numbers $c^{k+1}, \ldots, c^m$. Then $A(c^{k+1} h_{k+1} + \cdots + c^m h_m) = 0$, so $c^{k+1} h_{k+1} + \cdots + c^m h_m$ is in $\ker(V)$. Since $h_1, \ldots, h_k$ span $\ker(A)$, there are real numbers $b^1, \ldots, b^k$ such that

$$b^1 h_1 + \cdots + b^k h_k = c^{k+1} h_{k+1} + \cdots + c^m h_m,$$

hence

$$b^1 h_1 + \cdots + b^k h_k + (-c^{k+1}) h_{k+1} + \cdots + (-c^m) h_m = 0.$$

From the linear independence of $h_1, \ldots, h_k, h_{k+1}, \ldots, h_m$, we have $c^{k+1} = \cdots = c^m = 0$. Thus, $A(h_{k+1}), \ldots, A(h_m)$ are linearly independent. This proves the claim. It follows that

$$\operatorname{rank}(A) = \dim\bigl(\operatorname{im}(A)\bigr) = m - k = \dim(V) - \dim\bigl(\ker(A)\bigr)$$
$$= \dim(V) - \operatorname{null}(A). \qquad \square$$

As an example of the rank–nullity identity, consider the linear map $A : \mathbb{R}^3 \longrightarrow \mathbb{R}^2$ given by $A(x, y, z) = (x + y, 0)$. Then

$$\ker(A) = \{(x, y, z) \in \mathbb{R}^3 : x + y = 0\}$$

and

$$\operatorname{im}(A) = \{(x, y) \in \mathbb{R}^2 : y = 0\}.$$

In geometric terms, $\ker(A)$ is a plane in $\mathbb{R}^3$ and $\operatorname{im}(A)$ is a line in $\mathbb{R}^2$. Thus, $\operatorname{null}(A) = 2$ and $\operatorname{rank}(A) = 1$, which agrees with Theorem 1.1.11.

In the notation of Theorem 1.1.11, we observe from (1.1.2) that $\operatorname{rank}(A) \leq \dim(V)$. Thus, a linear map at best "preserves" dimension, but never increases it.

Theorem 1.1.12. *If V and W are vector spaces and $A : V \longrightarrow W$ is a linear map, then the following are equivalent:*
(a) $\operatorname{rank}(A) = \dim(V)$.
(b) $\operatorname{null}(A) = 0$.
(c) $\ker(A) = \{0\}$.
(d) *A is injective.*

Proof. (a) $\Leftrightarrow$ (b) $\Leftrightarrow$ (c): By Theorem 1.1.11,

$$\text{rank}(A) = \dim(V)$$
$$\Leftrightarrow \quad \text{null}(A) = 0$$
$$\Leftrightarrow \quad \dim\big(\text{ker}(A)\big) = 0$$
$$\Leftrightarrow \quad \text{ker}(A) = \{0\}.$$

(c) $\Rightarrow$ (d): For vectors v, w in V, we have

$$A(v) = A(w)$$
$$\Leftrightarrow \quad A(v - w) = 0$$
$$\Leftrightarrow \quad v - w \text{ is in } \text{ker}(A)$$
$$\Rightarrow \quad v - w = 0.$$

(d) $\Rightarrow$ (c): Clearly, 0 is in $\text{ker}(V)$. For a vector v in V, we have

$$v \text{ is in } \text{ker}(A)$$
$$\Leftrightarrow \quad A(v) = 0$$
$$\Leftrightarrow \quad A(v) = A(0)$$
$$\Rightarrow \quad v = 0. \qquad \qquad \square$$

Theorem 1.1.13. *Let V and W be vector spaces, let $\mathcal{H}$ be a basis for V, and let $A : V \longrightarrow W$ be a linear map. Then:*
(a) *A is a linear isomorphism if and only if $A(\mathcal{H})$ is a basis for W.*
(b) *If A is a linear isomorphism, then $\dim(V) = \dim(W)$.*

Proof. Let $\mathcal{H} = (h_1, \ldots, h_m)$.
    (a)($\Rightarrow$): Since A is surjective, for each vector w in W, there is a vector v in V such that $A(v) = w$. Let $v = a^1 h_1 + \cdots + a^m h_m$ for some real numbers $a^1, \ldots, a^m$. Then

$$w = A(v) = a^1 A(h_1) + \cdots + a^m A(h_m),$$

so $A(\mathcal{H})$ spans W. Suppose $b^1 A(h_1) + \cdots + b^m A(h_m) = 0$ for some real numbers $b^1, \ldots, b^m$. Then $A(b^1 h_1 + \cdots + b^m h_m) = 0$, hence $b^1 h_1 + \cdots + b^m h_m$ is in $\text{ker}(A)$. Since A is injective, it follows from Theorem 1.1.12 that $b^1 h_1 + \cdots + b^m h_m = 0$, hence $b^1 = \cdots = b^m = 0$. Thus, $A(\mathcal{H})$ is linearly independent.
    (a)($\Leftarrow$): Let w be a vector in W. Since $A(\mathcal{H})$ spans W, we have $w = b^1 A(h_1) + \cdots + b^m A(h_m)$ for some real numbers $b^1, \ldots, b^m$. Then $w = A(b^1 h_1 + \cdots + b^m h_m)$, so A is surjective. Let $v = a^1 h_1 + \cdots + a^m h_m$ be a vector in $\text{ker}(A)$. Then $0 = A(v) = a^1 A(h_1) + \cdots + a^m A(h_m)$. Since $A(\mathcal{H})$ is linearly independent, it follows that $a^1 = \cdots = a^m = 0$, so $v = 0$. Thus, $\text{ker}(A) = \{0\}$. By Theorem 1.1.12, A is injective.
    (b): This follows from part (a). $\qquad \qquad \square$

We pause here to comment on the way proofs are presented when there is an equation or other type of display that stretches over several lines of text. The necessary justification for logical steps in such displays, whether it be equation numbers, theorem numbers, example numbers, and so on, are often provided in brackets at the end of corresponding lines. In order to economize on space, "[Theorem x.y.z]" and "[Example x.y.z]" are abbreviated to "[Th x.y.z]" and "[Ex x.y.z]". The proof of the next result illustrates these conventions.

Theorem 1.1.14. *If V and W are vector spaces of dimension m and $A : V \longrightarrow W$ is a linear map, then the following are equivalent:*
(a) *A is a linear isomorphism.*
(b) *A is injective.*
(c) *A is surjective.*
(d) *$\operatorname{rank}(A) = m$.*

Proof. (a) $\Rightarrow$ (b): This is true by definition.
   (b) $\Leftrightarrow$ (c): By Theorem 1.1.11,

$$\dim(W) = \dim(V) = \operatorname{rank}(A) + \operatorname{null}(A) = \dim\big(\operatorname{im}(A)\big) + \operatorname{null}(A),$$

hence
$$W = \operatorname{im}(A)$$
$$\Leftrightarrow \quad \operatorname{null}(A) = 0 \qquad \text{[Th 1.1.7(b)]}$$
$$\Leftrightarrow \quad A \text{ is injective.} \qquad \text{[Th 1.1.12]}$$

   (c) $\Rightarrow$ (a): Since A is surjective, we have from (b) $\Leftrightarrow$ (c) that A is also injective.
   (d) $\Leftrightarrow$ (b): This follows from Theorem 1.1.12. $\square$

Let V be a vector space, and let $U_1, \ldots, U_k$ be subspaces. The **sum of $U_1, \ldots, U_k$** is denoted by $U_1 + \cdots + U_k$ and defined by

$$U_1 + \cdots + U_k = \{u_1 + \cdots + u_k : u_1 \in U_1, \ldots, u_k \in U_k\}.$$

For example, $\mathbb{R}(1,0) + \mathbb{R}(0,1) = \mathbb{R}^2$. It is easily shown that

$$U_1 + \cdots + U_k = \operatorname{span}(U_1 \cup \cdots \cup U_k),$$

from which it follows that $U_1 + \cdots + U_k$ is the smallest subspace of V containing each of $U_1, \ldots, U_k$, in the sense that any subspace containing each of $U_1, \ldots, U_k$ also contains $U_1 + \cdots + U_k$. We observe that

$$U_1 + \cdots + U_k + \{0\} = U_1 + \cdots + U_k,$$

which shows that adding the zero vector spaces does not change a sum. For vectors $v_1, \ldots, v_k$ in V, we have the following connection between spans and sums:

$$\operatorname{span}(\{v_1, \ldots, v_k\}) = \mathbb{R}v_1 + \cdots + \mathbb{R}v_k.$$

Theorem 1.1.15. *If V is a vector space, and U_1 and U_2 are subspaces of V, then*

$$\dim(U_1 + U_2) = \dim(U_1) + \dim(U_2) - \dim(U_1 \cap U_2).$$

Proof. Let $\mathcal{H} = (h_1, \ldots, h_k)$ be a basis for $U_1 \cap U_2$. By Theorem 1.1.7(c), $\mathcal{H}$ can be extended to a basis $(h_1, \ldots, h_k, f_1, \ldots, f_m)$ for U_1, and also to a basis $(h_1, \ldots, h_k, g_1, \ldots, g_n)$ for U_2. Let

$$\mathcal{V} = \{h_1, \ldots, h_k, f_1, \ldots, f_m, g_1, \ldots, g_n\}.$$

We claim that $\mathcal{V}$ is basis for $U_1 + U_2$. Clearly, $\mathcal{V}$ spans $U_1 + U_2$. To show that $\mathcal{V}$ is linearly independent, suppose

$$(a^1 h_1 + \cdots + a^k h_k) + (b^1 f_1 + \cdots + b^m f_m) + (c^1 g_1 + \cdots + c^n g_n) = 0$$

for some real numbers $a^1, \ldots, a^k, b^1, \ldots, b^m, c^1, \ldots, c^n$. Then

$$c^1 g_1 + \cdots + c^n g_n = -(a^1 h_1 + \cdots + a^k h_k) - (b^1 f_1 + \cdots + b^m f_m). \qquad (1.1.3)$$

Since $h_1, \ldots, h_k, f_1, \ldots, f_m$ are in U_1, so is $c^1 g_1 + \cdots + c^n g_n$, and because $g_1, \ldots, g_n$ are in U_2, so is $c^1 g_1 + \cdots + c^n g_n$. Thus, $c^1 g_1 + \cdots + c^n g_n$ is in $U_1 \cap U_2$, hence

$$c^1 g_1 + \cdots + c^n g_n = d^1 h_1 + \cdots + d^k h_k$$

for some real numbers $d^1, \ldots, d^k$, so

$$(-d^1) h_1 + \cdots + (-d^k) h_k + c^1 g_1 + \cdots + c^n g_n = 0.$$

Since $h_1, \ldots, h_k, g_1, \ldots, g_n$ are linearly independent,

$$c^1 = \cdots = c^n = 0. \qquad (1.1.4)$$

Then (1.1.3) gives

$$(a^1 h_1 + \cdots + a^k h_k) + (b^1 f_1 + \cdots + b^m f_m) = 0.$$

Because $h_1, \ldots, h_k, f_1, \ldots, f_m$ are linearly independent,

$$a^1 = \cdots = a^k = 0 \qquad \text{and} \qquad b^1 = \cdots = b^m = 0. \qquad (1.1.5)$$

It follows from (1.1.4) and (1.1.5) that $\mathcal{V}$ is linearly independent. This proves the claim. By Theorem 1.1.5,

$$\dim(U_1) + \dim(U_2) = (k + m) + (k + n) = k + (k + m + n)$$
$$= \dim(U_1 \cap U_2) + \dim(U_1 + U_2). \qquad \square$$

Let V be a vector space, and let $U_1, \ldots, U_k$ be subspaces of V. We say that the subspace $U_1 + \cdots + U_k$ of V is a **direct sum**, and write

$$U_1 + \cdots + U_k = U_1 \oplus \cdots \oplus U_k,$$

if each vector v in $U_1 + \cdots + U_k$ can be expressed uniquely (up to order of terms) in the form $v = u_1 + \cdots + u_k$ for some vectors u_i in U_i for $i = 1, \ldots, k$. As a matter of notation, writing $V = U_1 \oplus \cdots \oplus U_k$ is shorthand for

$$V = U_1 + \cdots + U_k \qquad \text{and} \qquad U_1 + \cdots + U_k = U_1 \oplus \cdots \oplus U_k.$$

For example, $\mathbb{R}^2 = \mathbb{R}(1,0) \oplus \mathbb{R}(0,1)$.

Theorem 1.1.16. *Let V be a vector space and let $v_1, \ldots, v_k$ be nonzero vectors in V. Then $v_1, \ldots, v_k$ are linearly independent if and only if*

$$\mathbb{R}v_1 + \cdots + \mathbb{R}v_k = \mathbb{R}v_1 \oplus \cdots \oplus \mathbb{R}v_k.$$

Proof. ($\Rightarrow$): If $a^1, \ldots, a^k, b^1, \ldots, b^k$ are real numbers such that

$$a^1 v_1 + \cdots + a^k v_k = b^1 v_1 + \cdots + b^k v_k,$$

then linear independence gives $a^i = b^i$ for $i = 1, \ldots, k$.

($\Leftarrow$): If $a^1, \ldots, a^k$ are real numbers such that

$$a^1 v_1 + \cdots + a^k v_k = 0 = 0 + \cdots + 0 \ [k \text{ terms}],$$

then the uniqueness property of $\mathbb{R}v_1 \oplus \cdots \oplus \mathbb{R}v_k$ gives $a^i v_i = 0$ for $i = 1, \ldots, k$. It follows that each $a^i \neq 0$; for if not, then $v_i = (a^i)^{-1} a^i v_i = 0$ for some i, which is a contradiction. $\qquad\square$

Theorem 1.1.17. *Let V be a vector space, and let U_1, U_2 be subspaces of V. Then $U_1 + U_2 = U_1 \oplus U_2$ if and only if $U_1 \cap U_2 = \{0\}$.*

Proof. ($\Rightarrow$): Evidently, $\{0\} \subseteq U_1 \cap U_2$. Let u be a vector in $U_1 \cap U_2$. Then 0 can be expressed as $0 = u + (-u)$, where u is in U_1 and $-u$ is in U_2. Since $0 = 0 + 0$, it follows from the uniqueness property of $U_1 \oplus U_2$ that $u = 0$. Thus, $U_1 \cap U_2 \subseteq \{0\}$.

($\Leftarrow$): Let u be a vector in $U_1 + U_2$ such that $u = v_1 + v_2$ and $u = w_1 + w_2$ for some vectors v_i, w_i in U_i for $1 = 1, 2$. Then $v_1 - w_1 = w_2 - v_2$ is in $U_1 \cap U_2 = \{0\}$, so $v_i - w_i = 0$ for $i = 1, 2$. $\qquad\square$

Theorem 1.1.18. *If V is a vector space and $U_1, \ldots, U_k$ are subspaces of V such that $U_1 + \cdots + U_k = U_1 \oplus \cdots \oplus U_k$, then*

$$\dim(U_1 + \cdots + U_k) = \dim(U_1) + \cdots + \dim(U_k).$$

Proof. The proof is by induction. For $k = 2$, the result follows from Theorem 1.1.15 and Theorem 1.1.17. Let $k > 2$, and suppose the assertion is true for all indices $< k$. Since $U_1 + \cdots + U_k$ is a direct sum, so are $U_1 + \cdots + U_{k-1}$ and $(U_1 + \cdots + U_{k-1}) + U_k$, hence

$$U_1 + \cdots + U_k = (U_1 + \cdots + U_{k-1}) + U_k$$
$$= (U_1 \oplus \cdots \oplus U_{k-1}) \oplus U_k.$$

Using the induction hypothesis twice gives

$$\dim(U_1 + \cdots + U_k) = \dim(U_1 \oplus \cdots \oplus U_{k-1}) + \dim(U_k)$$
$$= \dim(U_1) + \cdots + \dim(U_{k-1}) + \dim(U_k). \qquad\square$$

Let $V_1, \ldots, V_k$ be vector spaces. Following Section B.5, we make $V_1 \times \cdots \times V_k$ into a vector space, called the **product of $V_1, \ldots, V_k$**, as follows: for all vectors $(v_1, \ldots, v_k)$, $(w_1, \ldots, w_k)$ in $V_1 \times \cdots \times V_k$ and all real numbers c, let

$$(v_1, \ldots, v_k) + (w_1, \ldots, w_k) = (v_1 + w_1, \ldots, v_k + w_k)$$

and

$$c(v_1, \ldots, v_k) = (cv_1, \ldots, cv_k).$$

When $V_1 = \cdots = V_k = V$, we denote

$$V \times \cdots \times V \qquad \text{by} \qquad V^k.$$

We close this section with two definitions that have obvious geometric content. Let V be a vector space. A subset S of V is said to be **star-shaped** if there is a vector v_0 in S such that for all vectors v in S and all real numbers $0 \le t \le 1$, the vector $tv + (1 - t)v_0$ is in S. In that case, we say that S is **star-shaped about v_0**. Since $tv + (1 - t)v_0 = v_0 + t(v - v_0)$, we can think of $\{tv + (1 - t)v_0 : 0 \le t \le 1\}$ as the "line segment" joining v_0 to v. A subset C of V is said to be **cone-shaped** if for all vectors v, v_1, v_2 in C and all real numbers $c > 0$, the vectors cv and $v_1 + v_2$ are in C. It is easily shown that if C is cone-shaped, then it is star-shaped about any vector it contains. For example, the closed cell $\{(x, y) : x, y \in [-1, 1]\}$ in $\mathbb{R}^2$ is star-shaped about $(0, 0)$, but not cone-shaped; whereas the half-plane $\{(x, y) \in \mathbb{R}^2 : x > 0\}$ in $\mathbb{R}^2$ is cone-shaped, hence star-shaped about $(0, 0)$.

1.2   Dual Spaces

In this section, we define the dual (vector) space of a vector space. From this humble beginning, the theory of differential forms will eventually emerge (see Section 15.8).

Let V be a vector space. Following Section B.5 and Section B.6, we denote by $\mathrm{Lin}(V, \mathbb{R})$ the vector space of linear maps from V to $\mathbb{R}$. By definition, for all maps η, ζ in $\mathrm{Lin}(V, \mathbb{R})$ and all real numbers c,

$$(\eta + \zeta)(v) = \eta(v) + \zeta(v)$$

and

$$(c\eta)(v) = c\eta(v)$$

for all vectors v in V. For brevity, we henceforth denote

$$\mathrm{Lin}(V, \mathbb{R}) \qquad \text{by} \qquad V^*.$$

We say that V^* is the **dual (vector) space** of V and refer to each map in V^* as a **covector**. As an example, the map $\eta : \mathbb{R}^2 \longrightarrow \mathbb{R}$ given by $\eta(x, y) = x + y$ is in $(\mathbb{R}^2)^*$. Let us denote

$$(V^*)^* \qquad \text{by} \qquad V^{**}$$

and say that V^{**} is the **double dual** of V.

Theorem 1.2.1. *If V is a vector space and $\mathcal{H} = (h_1, \ldots, h_m)$ is a basis for V, then:*

(a) *There is a unique covector θ^i in V^* such that*

$$\theta^i(h_j) = \delta^i_j$$

for $i, j = 1, \ldots, m$, where δ^i_j is Kronecker's delta.

(b) *$(\theta^1, \ldots, \theta^m)$ is a basis for V^*, called the **dual basis** corresponding to $\mathcal{H}$.*

(c) *$\dim(V^*) = \dim(V)$.*

(d) *For all vectors v in V,*

$$v = \sum_i \theta^i(v) h_i.$$

(e) *For all covectors η in V^*,*

$$\eta = \sum_i \eta(h_i) \theta^i.$$

Proof. (a): This follows from Theorem 1.1.10.

(d): Let $v = \sum_i a^i h_i$. By part (a),

$$\theta^i(v) = \theta^i \left(\sum_j a^j h_j \right) = \sum_j a^j \theta^i(h_j) = a^i.$$

(e): For a vector v in V, we have from part (d) that

$$\eta(v) = \eta \left(\sum_i \theta^i(v) h_i \right) = \sum_i \theta^i(v) \eta(h_i) = \left(\sum_i \eta(h_i) \theta^i \right)(v).$$

Since v was arbitrary, the result follows.

(b): It follows from part (e) that $\theta^1, \ldots, \theta^m$ span V^*. Suppose $\sum_i a^i \theta^i = 0$ for some real numbers $a^1, \ldots, a^m$. By part (a),

$$0 = \left(\sum_j a^j \theta^j \right)(h_i) = \sum_j a^j \theta^j(h_i) = a^i$$

for $1 = 1, \ldots, m$, hence $\theta^1, \ldots, \theta^m$ are linearly independent.

(c): This follows from part (b). $\square$

Theorem 1.2.2. *Let V be a vector space, and let $\iota : V \longrightarrow V^{**}$ be the map defined by*

$$\iota(v)(\eta) = \eta(v) \tag{1.2.1}$$

for all vectors v in V and all covectors η in V^*. Then ι is a linear isomorphism:

$$V \approx V^{**}.$$

Remark. We observe that ι is defined without choosing specific bases for V and V^*. If we denote $\iota(v)$ by v^{**}, then (1.2.1) can be expressed more "symmetrically" as

$$v^{**}(\eta) = \eta(v). \tag{1.2.2}$$

Proof. It is clear that ι is a linear map. Let $(h_1, \ldots, h_m)$ be a basis for V, and let $(\theta^1, \ldots, \theta^m)$ be its dual basis. If v is a vector in V such that $\iota(v) = 0$, then $\eta(v) = 0$ for all covectors η in V^*. In particular, $\theta^i(v) = 0$ for $i = 1, \ldots, m$. It follows from Theorem 1.2.1(d) that $v = 0$, hence $\ker(\iota) = \{0\}$. By Theorem 1.1.12, ι is injective. Using Theorem 1.2.1(c) twice yields

$$\dim(V) = \dim(V^*) = \dim(V^{**}).$$

The result now follows Theorem 1.1.14. $\square$

In view of Theorem 1.2.2, and especially because ι was defined without choosing specific bases for V and V^*, we adopt the following convention:

Throughout, we identify V^{**} with V, and write $V^{**} = V$.

Let v be a vector in V, and let η be a covector in V^*. Having made the identification $V^{**} = V$, we henceforth denote

$$\iota(v) \qquad \text{by} \qquad v.$$

Thus, (1.2.1) and (1.2.2) both become

$$v(\eta) = \eta(v). \tag{1.2.3}$$

In particular, we have

$$h_j(\theta^i) = \theta^i(h_j)$$

for $i, j = 1, \ldots, m$.

1.3   Pullback of Covectors

In Section 1.2, we introduced the dual space of a vector space. Continuing with that theme, we now associate with a given linear map a corresponding linear map between their dual spaces.

Let V and W be vector spaces, and let $A : V \longrightarrow W$ be a linear map. **Pullback by A** is the linear map

$$A^* : W^* \longrightarrow V^*$$

defined by

$$A^*(\eta) = \eta \circ A$$

for all covectors η in W^*; that is,

$$A^*(\eta)(v) = \eta\big(A(v)\big) \tag{1.3.1}$$

for all vectors v in V. We refer to $A^*(\eta)$ as the **pullback of η by A**. Note that the pullback "reverses the order" of vector spaces. Let us denote

$$(A^*)^* \qquad \text{by} \qquad A^{**}.$$

and observe that with the identifications $V^{**} = V$ and $W^{**} = W$, we have

$$A^{**} : V \longrightarrow W.$$

As an example, consider the map $A : \mathbb{R}^3 \longrightarrow \mathbb{R}^2$ defined by $A(x, y, z) = (x + z, y + z)$ for all vectors (x, y, z) in $\mathbb{R}^3$, and let η be the covector in $(\mathbb{R}^2)^*$ given by $\eta(x, y) = x + y$. Then

$$A^*(\eta)(x, y, z) = \eta\big(A(x, y, z)\big) = \eta(x + z, y + z) = x + y + 2z.$$

Pullbacks behave well with respect to basic algebraic structure.

Theorem 1.3.1. *Let U, V, and W be vector spaces, and let $A, B : U \longrightarrow V$ and $C : V \longrightarrow W$ be linear maps. Then:*
(a) $(A + B)^* = A^* + B^*$.
(b) $(C \circ B)^* = B^* \circ C^*$.
(c) $A^{**} = A$.
(d) *If A is a linear isomorphism, then $(A^{-1})^* = (A^*)^{-1}$.*
(e) *A is a linear isomorphism if and only if A^* is a linear isomorphism.*

Proof. (a), (b): Straightforward.
(c): For a vector v in V and a covector η in W^*, we have

$$
\begin{aligned}
A^{**}(v)(\eta) &= v\big(A^*(\eta)\big) && [(1.3.1)] \\
&= A^*(\eta)(v) && [(1.2.3)] \\
&= \eta\big(A(v)\big) && [(1.3.1)] \\
&= A(v)(\eta). && [(1.2.3)]
\end{aligned}
$$

Since v and η were arbitrary, $A^{**} = A$.
(d): By part (b),

$$\mathrm{id}_{V^*} = (\mathrm{id}_V)^* = (A^{-1} \circ A)^* = A^* \circ (A^{-1})^*,$$

from which the result follows.
(e)($\Rightarrow$): Since A is a linear isomorphism, we have from Theorem 1.1.13(b) that $\dim(V) = \dim(W)$, and then from Theorem 1.2.1(c) that $\dim(V^*) = \dim(W^*)$. If η is a covector in W^* such that $A^*(\eta) = 0$, then (1.3.1) gives $\eta\big(A(v)\big) = 0$ for all vectors v in V. Since A is surjective, $\eta(w) = 0$ for all vectors w in W, hence $\eta = 0$. Thus, $\ker(A^*) = \{0\}$. The result now follows from Theorem 1.1.12 and Theorem 1.1.14.
(e)($\Leftarrow$): Since $A^* : W^* \longrightarrow V^*$ is a linear isomorphism, we have from (e)($\Rightarrow$) that so is $A^{**} : V^{**} \longrightarrow W^{**}$. Then part (c) and the identifications $V = V^{**}$ and $W = W^{**}$ give the result. $\square$

1.4   Annihilators

Let V be a vector space, and let U be a subspace of V. The **annihilator of U** in V is denoted by U^0 and defined by

$$
\begin{aligned}
U^0 &= \{\eta \in V^* : \eta(u) = 0 \text{ for all } u \in U\} \\
&= \{\eta \in V^* : U \subseteq \ker(\eta)\}.
\end{aligned}
$$

It is easily shown that U^0 is a subspace of V^*. Let us denote

$$(U^0)^0 \quad \text{by} \quad U^{00}$$

and observe that with the identification $V^{**} = V$, U^{00} is a subspace of V.

Theorem 1.4.1. *If V is a vector space and U is a subspace of V, then*

$$\dim(V) = \dim(U) + \dim(U^0).$$

Proof. If U is the zero subspace, the result is trivial, so assume otherwise. Let $(h_1, \ldots, h_k)$ be a basis for U. Using Theorem 1.1.7(c), we extend $(h_1, \ldots, h_k)$ to a basis $(h_1, \ldots, h_k, h_{k+1}, \ldots, h_m)$ for V. Let $(\theta^1, \ldots, \theta^k, \theta^{k+1}, \ldots, \theta^m)$ be its dual basis, so that $(\theta^{k+1}, \ldots, \theta^m)$ is the dual basis of $(h_{k+1}, \ldots, h_m)$. It follows from $\theta^i(h_1) = \cdots = \theta^i(h_k) = 0$ for $i = k+1, \ldots, m$ that $\theta^{k+1}, \ldots, \theta^m$ are covectors in U^0. We claim that $(\theta^{k+1}, \ldots, \theta^m)$ is a basis for U^0. For a covector η in U^0, we have from Theorem 1.2.1(e) that

$$\eta = \sum_{i=i}^{m} \eta(h_i)\theta^i = \sum_{i=k+1}^{m} \eta(h_i)\theta^i,$$

hence $\theta^{k+1}, \ldots, \theta^m$ span U^0. Since $(\theta^1, \ldots, \theta^m)$ is a basis for V^*, it follows that $\theta^{k+1}, \ldots, \theta^m$ are linearly independent. This proves the claim. By Theorem 1.1.5,

$$\dim(U^0) = m - k = \dim(V) - \dim(U). \qquad \square$$

Theorem 1.4.2. *If V is a vector space and U is a subspace of V, then*

$$U^{00} = U.$$

Proof. If u is in U, then $u(\eta) = \eta(u) = 0$ for all covectors η in U^0, hence u is in U^{00}. Thus, $U \subseteq U^{00}$. We have

$$\begin{aligned}
\dim(U) + \dim(U^0) &= \dim(V) && \text{[Th 1.4.1]} \\
&= \dim(V^*) && \text{[Th 1.2.1(c)]} \\
&= \dim(U^0) + \dim(U^{00}), && \text{[Th 1.4.1]}
\end{aligned}$$

so $\dim(U) = \dim(U^{00})$. The result now follows from Theorem 1.1.7(b). $\qquad \square$

Theorem 1.4.3. *Let V and W be vector spaces, and let $A : V \longrightarrow W$ be a linear map. Then:*
(a) $\operatorname{rank}(A^*) = \operatorname{rank}(A)$.
(b) *If* $\dim(V) = \dim(W)$, *then* $\operatorname{null}(A^*) = \operatorname{null}(A)$.
(c) $\ker(A^*) = \operatorname{im}(A)^0$.
(d) $\operatorname{im}(A^*) = \ker(A)^0$.

Proof. (c): We have

$$\eta \text{ is in } \ker(A^*)$$
$$\Leftrightarrow \quad A^*(\eta) = 0$$
$$\Leftrightarrow \quad A^*(\eta)(v) = 0 \text{ for all } v \text{ in } V$$
$$\Leftrightarrow \quad \eta(A(v)) = 0 \text{ for all } v \text{ in } V \qquad [(1.3.1)]$$
$$\Leftrightarrow \quad \eta \text{ is in } \mathrm{im}(A)^0.$$

(d): It follows from Theorem 1.3.1(c) and part (c) that

$$\ker(A) = \ker(A^{**}) = \mathrm{im}(A^*)^0,$$

and then from Theorem 1.4.2 that

$$\ker(A)^0 = \mathrm{im}(A^*)^{00} = \mathrm{im}(A^*).$$

(a): We have

$$\mathrm{rank}(A^*) = \dim(\mathrm{im}(A^*))$$
$$= \dim(\ker(A)^0) \qquad [\text{part (d)}]$$
$$= \dim(V) - \dim(\ker(A)) \qquad [\text{Th 1.4.1}]$$
$$= \mathrm{rank}(A). \qquad [\text{Th 1.1.11}]$$

(b): We have

$$\mathrm{rank}(A^*) + \mathrm{null}(A^*) = \dim(W^*) \qquad [\text{Th 1.1.11}]$$
$$= \dim(W) \qquad [\text{Th 1.2.1(c)}]$$
$$= \dim(V) \qquad [\text{assumption}]$$
$$= \mathrm{rank}(A) + \mathrm{null}(A). \qquad [\text{Th 1.1.11}]$$

The result now follows from part (a). $\qquad\qquad\qquad\qquad\qquad\qquad \square$

Chapter 2

Matrices and Determinants

In this chapter, we review some of the basic results from the theory of matrices and determinants.

2.1 Matrices

Let us denote by $\mathrm{Mat}_{m \times n}$ the set of $m \times n$ matrices (that is, m rows and n columns) with real entries. When $m = n$, we say that the matrices are **square**. It is easily shown that with the usual matrix addition and scalar multiplication, $\mathrm{Mat}_{m \times n}$ is a vector space, and that with the usual matrix multiplication, $\mathrm{Mat}_{m \times m}$ is a ring.

Let P be a matrix in $\mathrm{Mat}_{m \times n}$, with

$$P = [p_j^i] = \begin{bmatrix} p_1^1 & \cdots & p_n^1 \\ \vdots & \ddots & \vdots \\ p_1^m & \cdots & p_n^m \end{bmatrix}.$$

The **transpose of P** is the matrix P^{T} in $\mathrm{Mat}_{n \times m}$ defined by

$$P^{\mathrm{T}} = [p_i^j] = \begin{bmatrix} p_1^1 & \cdots & p_1^m \\ \vdots & \ddots & \vdots \\ p_n^1 & \cdots & p_n^m \end{bmatrix}.$$

The **row matrices** of P are

$$\begin{bmatrix} p_1^1 & \cdots & p_n^1 \end{bmatrix} \quad \begin{bmatrix} p_1^2 & \cdots & p_n^2 \end{bmatrix} \quad \cdots \quad \begin{bmatrix} p_1^m & \cdots & p_n^m \end{bmatrix},$$

Semi-Riemannian Geometry, First Edition. Stephen C. Newman.
© 2019 John Wiley & Sons, Inc. Published 2019 by John Wiley & Sons, Inc.

and the **column matrices** of P are

$$\begin{bmatrix} p_1^1 \\ \vdots \\ p_1^m \end{bmatrix} \quad \begin{bmatrix} p_2^1 \\ \vdots \\ p_2^m \end{bmatrix} \quad \cdots \quad \begin{bmatrix} p_n^1 \\ \vdots \\ p_n^m \end{bmatrix}.$$

Example 2.1.1. For

$$P = \begin{bmatrix} 1 & 2 & 3 \\ 4 & 5 & 6 \\ 7 & 8 & 9 \end{bmatrix},$$

the transpose is

$$P^{\mathrm{T}} = \begin{bmatrix} 1 & 4 & 7 \\ 2 & 5 & 8 \\ 3 & 6 & 9 \end{bmatrix}$$

and the column matrices are

$$\begin{bmatrix} 1 \\ 4 \\ 7 \end{bmatrix} \quad \begin{bmatrix} 2 \\ 5 \\ 8 \end{bmatrix} \quad \begin{bmatrix} 3 \\ 6 \\ 9 \end{bmatrix}. \qquad\qquad \Diamond$$

Theorem 2.1.2. *If P and Q are matrices in* $\mathrm{Mat}_{l \times m}$ *and* $\mathrm{Mat}_{m \times n}$, *respectively, then:*
(a) $(P^{\mathrm{T}})^{\mathrm{T}} = P$.
(b) $(PQ)^{\mathrm{T}} = Q^{\mathrm{T}} P^{\mathrm{T}}$.

Proof. Straightforward. $\qquad\qquad\qquad\qquad\qquad\qquad\qquad\qquad\qquad\qquad\quad$ $\square$

We say that a matrix $Q = \begin{bmatrix} q_j^i \end{bmatrix}$ in $\mathrm{Mat}_{m \times m}$ is **symmetric** if $Q = Q^{\mathrm{T}}$, and **diagonal** if $q_j^i = 0$ for all $i \neq j$. Evidently, a diagonal matrix is symmetric. Given a vector $(a^1, \ldots, a^m)$ in $\mathbb{R}^m$, the corresponding diagonal matrix is defined by

$$\mathrm{diag}(a^1, \ldots, a^m) = \begin{bmatrix} a^1 & \cdots & 0 \\ \vdots & \ddots & \vdots \\ 0 & \cdots & a^m \end{bmatrix},$$

where all the entries not on the (upper-left to lower-right) diagonal are equal to 0. For example,

$$\mathrm{diag}(1) = \begin{bmatrix} 1 \end{bmatrix} \qquad \mathrm{diag}(1,2) = \begin{bmatrix} 1 & 0 \\ 0 & 2 \end{bmatrix} \qquad \mathrm{diag}(1,2,3) = \begin{bmatrix} 1 & 0 & 0 \\ 0 & 2 & 0 \\ 0 & 0 & 3 \end{bmatrix}.$$

The **zero matrix** in $\mathrm{Mat}_{m \times n}$, denoted by $O_{m \times n}$, is the matrix that has all entries equal to 0. The **identity matrix** in $\mathrm{Mat}_{m \times m}$ is defined by

$$I_m = \mathrm{diag}(1, \ldots, 1),$$

so that, for example,

$$I_1 = \begin{bmatrix} 1 \end{bmatrix} \qquad I_2 = \begin{bmatrix} 1 & 0 \\ 0 & 1 \end{bmatrix} \qquad I_3 = \begin{bmatrix} 1 & 0 & 0 \\ 0 & 1 & 0 \\ 0 & 0 & 1 \end{bmatrix}.$$

We say that a matrix Q in $\mathrm{Mat}_{m \times m}$ is **invertible** if there is a matrix in $\mathrm{Mat}_{m \times m}$, denoted by Q^{-1} and called the **inverse of Q**, such that

$$QQ^{-1} = Q^{-1}Q = I_m.$$

It is easily shown that if the inverse of a matrix exists, then it is unique.

Theorem 2.1.3. *If P and Q are invertible matrices in $\mathrm{Mat}_{m \times m}$, then:*
(a) $(P^{-1})^{-1} = P.$
(b) $(PQ)^{-1} = Q^{-1}P^{-1}.$
(c) $(P^{-1})^{\mathrm{T}} = (P^{\mathrm{T}})^{-1}.$

Proof. (a), (b): Straightforward.
     (c): By Theorem 2.1.2(b),

$$I_m = (I_m)^{\mathrm{T}} = (PP^{-1})^{\mathrm{T}} = (P^{-1})^{\mathrm{T}}P^{\mathrm{T}},$$

from which the result follows.   □

    Multi-index notation, introduced in Appendix A, provides a convenient way to specify **submatrices** of matrices. Let $1 \le r \le m$ and $1 \le s \le n$ be integers, and let $I = (i_1, \ldots, i_r)$ and $J = (j_1, \ldots, j_s)$ be multi-indices in $\mathcal{I}_{r,m}$ and $\mathcal{I}_{s,n}$, respectively. For a matrix $P = \begin{bmatrix} p_j^i \end{bmatrix}$ in $\mathrm{Mat}_{m \times n}$, we denote by

$$P_{(j_1,\ldots,j_s)}^{(i_1,\ldots,i_r)}, \qquad P_J^I, \qquad \begin{bmatrix} p_j^i \end{bmatrix}_{(j_1,\ldots,j_s)}^{(i_1,\ldots,i_r)}, \qquad \text{or} \qquad \begin{bmatrix} p_j^i \end{bmatrix}_J^I$$

the $r \times s$ submatrix of P consisting of the overlap of rows $i_1, i_2, \ldots, i_r$ and columns $j_1, j_2, \ldots, j_s$ (in that order); that is,

$$P_{(j_1,\ldots,j_s)}^{(i_1,\ldots,i_r)} = \begin{bmatrix} p_{j_1}^{i_1} & \cdots & p_{j_s}^{i_1} \\ \vdots & \ddots & \vdots \\ p_{j_1}^{i_r} & \cdots & p_{j_s}^{i_r} \end{bmatrix}.$$

When $r = m$, in which case $(i_1, \ldots, i_m) = (1, \ldots, m)$, we denote

$$P_{(j_1,\ldots,j_s)}^{(1,\ldots,m)} \qquad \text{by} \qquad P_{(j_1,\ldots,j_s)},$$

and when $s = n$, in which case $(j_1, \ldots, j_n) = (1, \ldots, n)$, we denote

$$P_{(1,\ldots,n)}^{(i_1,\ldots,i_r)} \qquad \text{by} \qquad P^{(i_1,\ldots,i_r)}.$$

When $r = 1$, so that $I = (i)$ for some $1 \le i \le m$, we have

$$P^{(i)} = \begin{bmatrix} p_1^i & \cdots & p_n^i \end{bmatrix},$$

which is the ith row matrix of P. Similarly, when $s = 1$, so that $J = (j)$ for some $1 \leq j \leq n$, we have

$$P_{(j)} = \begin{bmatrix} p_j^1 \\ \vdots \\ p_j^m \end{bmatrix},$$

which is the jth column matrix of P.

Example 2.1.4. Continuing with Example 2.1.1, we have

$$P^{(1)} = \begin{bmatrix} 1 & 2 & 3 \end{bmatrix} \qquad P_{(2)} = \begin{bmatrix} 2 \\ 5 \\ 8 \end{bmatrix}$$

$$P^{(1,2)}_{(2,3)} = \begin{bmatrix} 2 & 3 \\ 5 & 6 \end{bmatrix} \qquad P^{(1,2)} = \begin{bmatrix} 1 & 2 & 3 \\ 4 & 5 & 6 \end{bmatrix} \qquad P_{(2,3)} = \begin{bmatrix} 2 & 3 \\ 5 & 6 \\ 8 & 9 \end{bmatrix}. \qquad \Diamond$$

Theorem 2.1.5. *If P, Q, and R are matrices in $\mathrm{Mat}_{k \times l}$, $\mathrm{Mat}_{l \times m}$, and $\mathrm{Mat}_{m \times n}$, respectively, then:*
(a)

$$PQ = \begin{bmatrix} P^{(1)}Q_{(1)} & \cdots & P^{(1)}Q_{(m)} \\ \vdots & \ddots & \vdots \\ P^{(k)}Q_{(1)} & \cdots & P^{(k)}Q_{(m)} \end{bmatrix}.$$

(b)

$$PQR = \begin{bmatrix} P^{(1)}QR_{(1)} & \cdots & P^{(1)}QR_{(n)} \\ \vdots & \ddots & \vdots \\ P^{(k)}QR_{(1)} & \cdots & P^{(k)}QR_{(n)} \end{bmatrix}.$$

Proof. (a): By definition of matrix multiplication, the ij-th entry of PQ is $P^{(i)}Q_{(j)}$.
(b): This follows from part (a) and the observation that $(QR)_{(j)} = QR_{(j)}$. $\qquad \square$

Theorem 2.1.6. *Let P, Q, and R be matrices in $\mathrm{Mat}_{k \times l}$, $\mathrm{Mat}_{l \times m}$, and $\mathrm{Mat}_{m \times n}$, respectively, and let I, J, K, and L be multi-indices in $\mathcal{I}_{r,k}$, $\mathcal{I}_{s,l}$, $\mathcal{I}_{t,m}$, and $\mathcal{I}_{u,n}$, respectively. Then:*
(a) $(P^I)^\mathrm{T} = (P^\mathrm{T})_I$.
(b) $(P_J)^\mathrm{T} = (P^\mathrm{T})^J$.
(c) $(PQ)^I_K = P^I Q_K$.
(d) $(PQR)^I_L = P^I Q R_L$.

Proof. (a), (b), (c): Straightforward.
(d): By part (c),

$$(PQR)^I_L = (PQ)^I R_L = P^I Q R_L. \qquad \square$$

For a matrix $P = \begin{bmatrix} p^i_j \end{bmatrix}$ in $\text{Mat}_{m \times m}$, the **trace of P** is defined by

$$\text{tr}(P) = \sum_i p^i_i.$$

Theorem 2.1.7. *If P and Q are matrices in $\text{Mat}_{m \times m}$, then:*
(a) $\text{tr}(P + Q) = \text{tr}(P) + \text{tr}(Q)$.
(b) $\text{tr}(PQ) = \text{tr}(QP)$.
(c) *If Q is invertible, then $\text{tr}(Q^{-1}PQ) = \text{tr}(P)$.*

Proof. (a): Straightforward.
(b): Let $P = \begin{bmatrix} p^i_j \end{bmatrix}$ and $Q = \begin{bmatrix} q^i_j \end{bmatrix}$. By Theorem 2.1.5(a),

$$\text{tr}(PQ) = \sum_i P^{(i)} Q_{(i)} = \sum_i \left(\sum_j p^i_j q^j_i \right) = \sum_j \left(\sum_i q^j_i p^i_j \right)$$
$$= \sum_j Q^{(j)} P_{(j)} = \text{tr}(QP).$$

(c): This follows from part (b). $\square$

2.2 Matrix Representations

Matrices have many desirable computational properties. For this reason, when computing in vector spaces, it is often convenient to reformulate arguments in terms of matrices. We employ this device often.

Let V be a vector space, let $\mathcal{H} = (h_1, \dots, h_m)$ be a basis for V, and let v be a vector in V, with

$$v = \sum_i a^i h_i.$$

The **matrix representation of v** with respect to $\mathcal{H}$ is denoted by $\begin{bmatrix} v \end{bmatrix}_{\mathcal{H}}$ and defined by

$$\begin{bmatrix} v \end{bmatrix}_{\mathcal{H}} = \begin{bmatrix} a^1 \\ \vdots \\ a^m \end{bmatrix}.$$

We refer to $a^1, \dots, a^m$ as the **components of v** with respect to $\mathcal{H}$. In particular,

$$\begin{bmatrix} h_i \end{bmatrix}_{\mathcal{H}} = \begin{bmatrix} 0 \\ \vdots \\ 1 \\ \vdots \\ 0 \end{bmatrix}, \tag{2.2.1}$$

where 1 is in the ith position and 0s are elsewhere for $i = 1, \dots, m$.

Theorem 2.2.1 (Representation of Vectors). *If V is a vector space of dimension m and $\mathcal{H}$ is a basis for V, then the map*

$$\mathfrak{L}_{\mathcal{H}} : V \longrightarrow \mathrm{Mat}_{m \times 1}$$

defined by

$$\mathfrak{L}_{\mathcal{H}}(v) = [v]_{\mathcal{H}}$$

for all vectors v in V is a linear isomorphism:

$$V \approx \mathrm{Mat}_{m \times 1}.$$

Proof. Straightforward.   □

With V and $\mathcal{H}$ as above, let W be another vector space, and let $\mathcal{F} = (f_1, \ldots, f_n)$ be a basis for W. Let $A : V \longrightarrow W$ be a linear map, with

$$A(h_j) = \sum_i a^i_j f_i, \qquad (2.2.2)$$

so that

$$[A(h_j)]_{\mathcal{F}} = \begin{bmatrix} a^1_j \\ \vdots \\ a^n_j \end{bmatrix}$$

for $j = 1, \ldots, m$. The **matrix representation of A** with respect to $\mathcal{H}$ and $\mathcal{F}$ is denoted by $[A]^{\mathcal{F}}_{\mathcal{H}}$ and defined to be the $n \times m$ matrix

$$[A]^{\mathcal{F}}_{\mathcal{H}} = \begin{bmatrix} a^1_1 & \cdots & a^1_m \\ \vdots & \ddots & \vdots \\ a^n_1 & \cdots & a^n_m \end{bmatrix} = \begin{bmatrix} [A(h_1)]_{\mathcal{F}} & \cdots & [A(h_m)]_{\mathcal{F}} \end{bmatrix}. \qquad (2.2.3)$$

As an example, consider the linear map $A : \mathbb{R}^2 \longrightarrow \mathbb{R}^3$ given by $A(x, y) = (y, 2x, 3x + 4y)$, and let $\mathcal{E}$ and $\mathcal{F}$ be the standard bases for $\mathbb{R}^2$ and $\mathbb{R}^3$, respectively. Then

$$[A]^{\mathcal{F}}_{\mathcal{E}} = \begin{bmatrix} 0 & 1 \\ 2 & 0 \\ 3 & 4 \end{bmatrix}.$$

Theorem 2.2.2. *Let U, V, and W be vector spaces, let $\mathcal{H}$, $\mathcal{F}$, and $\mathcal{G}$ be respective bases, let $A, B : U \longrightarrow V$ and $C : V \longrightarrow W$ be linear maps, and let c be a real number. Then:*

(a) $[A + B]^{\mathcal{F}}_{\mathcal{H}} = [A]^{\mathcal{F}}_{\mathcal{H}} + [B]^{\mathcal{F}}_{\mathcal{H}}$.

(b) $[cA]^{\mathcal{F}}_{\mathcal{H}} = c[A]^{\mathcal{F}}_{\mathcal{H}}$.

(c) $[C \circ A]^{\mathcal{G}}_{\mathcal{H}} = [C]^{\mathcal{G}}_{\mathcal{F}} [A]^{\mathcal{F}}_{\mathcal{H}}$.

Proof. (a), (b): Straightforward.

(c): Let $\mathcal{H} = (h_1, \ldots, h_m)$, $\mathcal{F} = (f_1, \ldots, f_n)$, and $\mathcal{G} = (g_1, \ldots, g_r)$, and let $[A]_{\mathcal{H}}^{\mathcal{F}} = [a_j^i]$ and $[C]_{\mathcal{F}}^{\mathcal{G}} = [c_l^k]$. According to (2.2.2) and (2.2.3),

$$A(h_j) = \sum_i a_j^i f_i \qquad \text{and} \qquad C(f_l) = \sum_k c_l^k g_k.$$

Then

$$C\big(A(h_j)\big) = C\left(\sum_k a_j^k f_k\right) = \sum_k a_j^k C(f_k) = \sum_k a_j^k \left(\sum_i c_k^i g_i\right)$$

$$= \sum_i \left(\sum_k c_k^i a_j^k\right) g_i = \sum_i \left([C]_{\mathcal{F}}^{\mathcal{G}}\right)^{(i)} \left([A]_{\mathcal{H}}^{\mathcal{F}}\right)_{(j)} g_i,$$

from which the result follows.   □

Theorem 2.2.3 (Representation of Linear Maps). *Let V and W be vector spaces of dimensions m and n, respectively, and let $\mathcal{H}$ and $\mathcal{F}$ be respective bases. Define a map*

$$\mathcal{L}_{\mathcal{H}}^{\mathcal{F}} : \mathrm{Lin}(V, W) \longrightarrow \mathrm{Mat}_{n \times m}$$

by

$$\mathcal{L}_{\mathcal{H}}^{\mathcal{F}}(A) = [A]_{\mathcal{H}}^{\mathcal{F}}$$

for all maps A in $\mathrm{Lin}(V, W)$, where $\mathrm{Lin}(V, W)$ is defined in Example 1.1.1. Then:

(a) $\mathcal{L}_{\mathcal{H}}^{\mathcal{F}}$ *is a linear isomorphism with respect to the additive structure of* $\mathrm{Lin}(V, W)$:

$$\mathrm{Lin}(V, W) \approx \mathrm{Mat}_{n \times m}.$$

(b) *If $V = W$, then $\mathcal{L}_{\mathcal{H}}^{\mathcal{F}}$ is a ring isomorphism with respect to the multiplicative structure of* $\mathrm{Lin}(V, V)$.

Remark. We showed in Example 1.1.1 that $\mathrm{Lin}(V, V)$ is both a vector space and a ring, and remarked at the beginning of Secion 2.1 that the same is true of $\mathrm{Mat}_{m \times m}$, so the assertion in part (b) makes sense.

Proof. (a): By parts (a) and (b) of Theorem 2.2.2, $\mathcal{L}_{\mathcal{H}}^{\mathcal{F}}$ is a linear map, and it is easily shown that $\mathcal{L}_{\mathcal{H}}^{\mathcal{F}}$ is injective. We claim that $\mathrm{Mat}_{n \times m}$ and $\mathrm{Lin}(V, W)$ both have dimension mn. Let E_{ij} be the matrix in $\mathrm{Mat}_{n \times m}$ with 1 in the ij-th position and 0s elsewhere for $i = 1, \ldots, n$ and $j = 1, \ldots, m$. It is readily demonstrated that the E_{ij} comprise a basis for $\mathrm{Mat}_{n \times m}$, which therefore has dimension mn. Let $\mathcal{H} = (h_1, \ldots, h_m)$ and $\mathcal{F} = (f_1, \ldots, f_n)$, and using Theorem 1.1.10, define linear maps L_{ij} in $\mathrm{Lin}(V, W)$ by

$$L_{ij}(h_k) = \begin{cases} f_i & \text{if } k = j \\ 0 & \text{if } k \neq j \end{cases}$$

for $i = 1, \ldots, n$ and $j = 1, \ldots, m$. Then

$$\left(\sum_{ij} a^i_j L_{ij}\right)(h_k) = \sum_{ij} a^i_j L_{ij}(h_k) = \sum_i a^i_k f_i \qquad (2.2.4)$$

for all real numbers a^i_j. Let B be a map in $\mathrm{Lin}(V, W)$, with $[B]^{\mathcal{F}}_{\mathcal{H}} = [b^i_j]$. We have from (2.2.4) that

$$\left(\sum_{ij} b^i_j L_{ij}\right)(h_k) = \sum_i b^i_k f_i = B(h_k)$$

for $k = 1, \ldots, m$. By Theorem 1.1.10, $\sum_{ij} b^i_j L_{ij} = B$, so the L_{ij} span $\mathrm{Lin}(V, W)$. Suppose $\sum_{ij} a^i_j L_{ij} = 0$ for some real numbers a^i_j, where 0 denotes the zero map in $\mathrm{Lin}(V, W)$. It follows from (2.2.4) that $\sum_i a^i_k f_i = 0$ for $k = 1, \ldots, m$. Since $\mathcal{F}$ is a basis for W, $a^i_k = 0$ for $i = 1, \ldots, n$ and $k = 1, \ldots, m$, so the L_{ij} are linearly independent. Thus, the L_{ij} comprise a basis for $\mathrm{Lin}(V, W)$, and therefore, $\mathrm{Lin}(V, W)$ has dimension mn. This proves the claim. Suppose $\mathcal{L}^{\mathcal{F}}_{\mathcal{H}}(A) = O_{n \times m}$ for some map A in $\mathrm{Lin}(V, W)$. It follows from Theorem 1.1.10, (2.2.2), and (2.2.3) that A is the zero map, hence $\ker(\mathcal{L}^{\mathcal{F}}_{\mathcal{H}}) = \{0\}$. The result now follows from Theorem 1.1.12 and Theorem 1.1.14.

(b): This follows from Theorem 2.2.2(c) and part (a).   □

Theorem 2.2.4. *Let V and W be vector spaces, let $\mathcal{H}$ and $\mathcal{F}$ be respective bases, let $A : V \longrightarrow W$ be a linear map, and let v be a vector in V. Then*

$$[A(v)]_{\mathcal{F}} = [A]^{\mathcal{F}}_{\mathcal{H}} [v]_{\mathcal{H}}.$$

Proof. Let $\mathcal{H} = (h_1, \ldots, h_m)$ and $v = \sum_i a^i h_i$. It follows from (2.2.1) and (2.2.2) that

$$[A(h_i)]_{\mathcal{F}} = [A]^{\mathcal{F}}_{\mathcal{H}} [h_i]_{\mathcal{H}}.$$

By parts (a) and (b) of Theorem 2.2.2,

$$[A(v)]_{\mathcal{F}} = \left[A\left(\sum_i a^i h_i\right)\right]_{\mathcal{F}} = \left[\sum_i a^i A(h_i)\right]_{\mathcal{F}} = \sum_i a^i [A(h_i)]_{\mathcal{F}}$$

$$= \sum_i a^i \left([A]^{\mathcal{F}}_{\mathcal{H}} [h_i]_{\mathcal{H}}\right) = [A]^{\mathcal{F}}_{\mathcal{H}} \left(\sum_i a^i [h_i]_{\mathcal{H}}\right) = [A]^{\mathcal{F}}_{\mathcal{H}} \begin{bmatrix} a^1 \\ \vdots \\ a^m \end{bmatrix}$$

$$= [A]^{\mathcal{F}}_{\mathcal{H}} [v]_{\mathcal{H}}.$$ □

Let V be a vector space, and let $\mathcal{H}$ and $\mathcal{F}$ be bases for V. Setting $A = \mathrm{id}_V$ in Theorem 2.2.4 yields

$$[v]_{\mathcal{F}} = [\mathrm{id}_V]^{\mathcal{F}}_{\mathcal{H}} [v]_{\mathcal{H}}. \qquad (2.2.5)$$

This shows that $[\mathrm{id}_V]^{\mathcal{F}}_{\mathcal{H}}$ is the matrix that transforms components with respect to $\mathcal{H}$ into components with respect to $\mathcal{F}$. For this reason, $[\mathrm{id}_V]^{\mathcal{F}}_{\mathcal{H}}$ is called the

change of basis matrix from $\mathcal{H}$ to $\mathcal{F}$. Let $\left[\mathrm{id}_V\right]_{\mathcal{H}}^{\mathcal{F}} = [a_j^i]$. Then (2.2.2) and (2.2.3) specialize to

$$h_j = \sum_i a_j^i f_i \tag{2.2.6}$$

for $i = 1, \ldots, m$ and

$$\left[\mathrm{id}_V\right]_{\mathcal{H}}^{\mathcal{F}} = \left[\left[h_1\right]_{\mathcal{F}} \quad \cdots \quad \left[h_m\right]_{\mathcal{F}}\right]. \tag{2.2.7}$$

Theorem 2.2.5. *Let V and W be vector spaces, let $\mathcal{H}$ and $\mathcal{F}$ be respective bases, and let $A : V \longrightarrow W$ be a linear isomorphism. Then:*
(a)

$$[A]_{\mathcal{H}}^{\mathcal{F}} = [\mathrm{id}_W]_{A(\mathcal{H})}^{\mathcal{F}}.$$

(b)

$$[A^{-1}]_{\mathcal{F}}^{\mathcal{H}} = \left([A]_{\mathcal{H}}^{\mathcal{F}}\right)^{-1}.$$

Remark. By Theorem 1.1.13(a), $A(\mathcal{H})$ is a basis for W, so the assertion in part (a) makes sense.

Proof. (a): Let $\mathcal{H} = (h_1, \ldots, h_m)$, so that

$$[A]_{\mathcal{H}}^{\mathcal{F}} = \left[\left[A(h_1)\right]_{\mathcal{F}} \quad \cdots \quad \left[A(h_m)\right]_{\mathcal{F}}\right] \qquad [(2.2.3)]$$
$$= [\mathrm{id}_W]_{A(\mathcal{H})}^{\mathcal{F}}. \qquad [(2.2.7)]$$

(b): By Theorem 2.2.2(c),

$$I_m = [\mathrm{id}_V]_{\mathcal{H}}^{\mathcal{H}} = [A^{-1} \circ A]_{\mathcal{H}}^{\mathcal{H}} = [A^{-1}]_{\mathcal{F}}^{\mathcal{H}} [A]_{\mathcal{H}}^{\mathcal{F}},$$

from which the result follows. $\qquad\square$

Theorem 2.2.6 (Change of Basis). *Let V be a vector space, let $\mathcal{H}$ and $\mathcal{F}$ be bases for V, and let $A : V \longrightarrow V$ be a linear map. Then*

$$[A]_{\mathcal{F}}^{\mathcal{F}} = \left([\mathrm{id}_V]_{\mathcal{F}}^{\mathcal{H}}\right)^{-1} [A]_{\mathcal{H}}^{\mathcal{H}} [\mathrm{id}_V]_{\mathcal{F}}^{\mathcal{H}}.$$

Proof. By Theorem 2.2.2(c),

$$[A]_{\mathcal{F}}^{\mathcal{F}} = [\mathrm{id}_V \circ A \circ \mathrm{id}_V]_{\mathcal{F}}^{\mathcal{F}} = [\mathrm{id}_V]_{\mathcal{H}}^{\mathcal{F}} [A]_{\mathcal{H}}^{\mathcal{H}} [\mathrm{id}_V]_{\mathcal{F}}^{\mathcal{H}}.$$

The result now follows from Theorem 2.2.5(b). $\qquad\square$

Theorem 2.2.7. *Let V be a vector space, let $\mathcal{H}$ and $\widetilde{\mathcal{H}}$ be bases for V, let Θ and $\widetilde{\Theta}$ be the corresponding dual bases, and let $A : V \longrightarrow V$ be a linear map. Then*

$$[A^*]_{\widetilde{\Theta}}^{\Theta} = \left([A]_{\mathcal{H}}^{\widetilde{\mathcal{H}}}\right)^{\mathrm{T}},$$

where A^* is the pullback by A.

Proof. Let $\mathcal{H} = (h_1, \ldots, h_m)$ and $\widetilde{\mathcal{H}} = (\widetilde{h}_1, \ldots, \widetilde{h}_m)$, let $\Theta = (\theta^1, \ldots, \theta^m)$ and $\widetilde{\Theta} = (\widetilde{\theta}^1, \ldots, \widetilde{\theta}^m)$, and let $[A]_{\mathcal{H}}^{\widetilde{\mathcal{H}}} = [a_j^i]$. By (2.2.2) and (2.2.3), $A(h_j) = \sum_i a_j^i \widetilde{h}_i$. Then

$$A^*(\widetilde{\theta}^j) = \sum_i A^*(\widetilde{\theta}^j)(h_i)\theta^i \qquad \text{[Th 1.2.1(e)]}$$

$$= \sum_i \widetilde{\theta}^j(A(h_i))\theta^i \qquad \text{[(1.3.1)]}$$

$$= \sum_i \left[\widetilde{\theta}^j\left(\sum_k a_i^k \widetilde{h}_k\right)\right]\theta^i = \sum_{ik} a_i^k \widetilde{\theta}^j(\widetilde{h}_k)\theta^i$$

$$= \sum_i a_i^j \theta^i. \qquad \text{[Th 1.2.1(a)]}$$

Again by (2.2.2) and (2.2.3), $[A^*]_{\widetilde{\Theta}}^{\Theta} = [a_i^j] = ([a_j^i])^{\mathrm{T}}$. $\qquad\square$

2.3   Rank of Matrices

Consider the n-dimensional vector space $\mathrm{Mat}_{1 \times n}$ of row matrices and the m-dimensional vector space $\mathrm{Mat}_{m \times 1}$ of column matrices. Let P be a matrix in $\mathrm{Mat}_{m \times n}$. The **row rank of P** is defined to be the dimension of the subspace of $\mathrm{Mat}_{1 \times n}$ spanned by the rows of P:

$$\mathrm{rowrank}(P) = \dim(\mathrm{span}\{P^{(1)}, \ldots, P^{(m)}\}).$$

Similarly, the **column rank of P** is defined to be the dimension of the subspace of $\mathrm{Mat}_{m \times 1}$ spanned by the columns of P:

$$\mathrm{colrank}(P) = \dim(\mathrm{span}\{P_{(1)}, \ldots, P_{(n)}\}).$$

To illustrate, for

$$P = \begin{bmatrix} 1 & 0 & 0 \\ 0 & 1 & 0 \end{bmatrix},$$

we have $\mathrm{rowrank}(P) = \mathrm{colrank}(P) = 2$. As shown below, it is not a coincidence that the row rank and column rank of P are equal.

Theorem 2.3.1. *Let V and W be vector spaces, let $\mathcal{H}$ and $\mathcal{F}$ be respective bases, and let $A : V \longrightarrow W$ be a linear map. Then*

$$\mathrm{rank}(A) = \mathrm{colrank}\left([A]_{\mathcal{H}}^{\mathcal{F}}\right).$$

Proof. Let $\mathcal{H} = (h_1, \ldots, h_m)$. It follows from Theorem 2.2.1 and (2.2.3) that

$$\left([A]_{\mathcal{H}}^{\mathcal{F}}\right)_{(j)} = [A(h_j)]_{\mathcal{F}} = \mathfrak{L}_{\mathcal{F}}(A(h_j))$$

for $j = 1, \ldots, m$. Since $\mathfrak{L}_{\mathcal{F}}$ is an isomorphism and $A(h_1), \ldots, A(h_m)$ span the image of A, we have

$$\operatorname{colrank}\left([A]_{\mathcal{H}}^{\mathcal{F}}\right) = \dim\Big(\operatorname{span}\big(\{A(h_1), \ldots, A(h_m)\}\big)\Big) = \dim\big(\operatorname{im}(A)\big)$$
$$= \operatorname{rank}(A). \qquad \square$$

Theorem 2.3.2. *If P is a matrix in $\operatorname{Mat}_{m \times n}$, then*

$$\operatorname{rowrank}(P) = \operatorname{colrank}(P).$$

Proof. Let $\mathcal{E}$ and $\mathcal{F}$ be the standard bases for $\mathbb{R}^n$ and $\mathbb{R}^m$, respectively, and let Ξ and Φ be the corresponding dual bases. Let $P = [p_j^i]$, and let $A : \mathbb{R}^n \longrightarrow \mathbb{R}^m$ be the linear map defined by

$$A(x^1, \ldots, x^n) = \left(\sum_j p_j^1 x^j, \ldots, \sum_j p_j^m x^j\right).$$

Then $[A]_{\mathcal{E}}^{\mathcal{F}} = P$. By Theorem 2.3.1,

$$\operatorname{rank}(A) = \operatorname{colrank}(P).$$

We have from Theorem 2.2.7 that $[A^*]_{\Phi}^{\Xi} = \left([A]_{\mathcal{E}}^{\mathcal{F}}\right)^{\mathrm{T}} = P^{\mathrm{T}}$, so a similar argument gives

$$\operatorname{rank}(A^*) = \operatorname{colrank}(P^{\mathrm{T}}) = \operatorname{rowrank}(P).$$

The result now follows from Theorem 1.4.3(a). $\qquad \square$

In light of Theorem 2.3.2, the common value of the row rank and column rank of P is denoted by $\operatorname{rank}(P)$ and called the **rank of P**. Thus,

$$\operatorname{rank}(P) = \operatorname{rowrank}(P) = \operatorname{colrank}(P). \tag{2.3.1}$$

2.4 Determinant of Matrices

This section presents the basic results on the determinant of matrices.

Consider the m-dimensional vector space $\operatorname{Mat}_{m \times 1}$ of column matrices and the corresponding product vector space $(\operatorname{Mat}_{m \times 1})^m$. We denote by $(E_1, \ldots, E_m)$ the **standard basis** for $\operatorname{Mat}_{m \times 1}$, where E_j has 1 in the jth row and 0s elsewhere for $j = 1, \ldots, m$. Let

$$\Delta : (\operatorname{Mat}_{m \times 1})^m \longrightarrow \mathbb{R}$$

be an arbitrary function, and let σ be a permutation in $\mathcal{S}_m$, the symmetric group on $\{1, 2, \ldots, m\}$. We define a function

$$\sigma(\Delta) : (\operatorname{Mat}_{m \times 1})^m \longrightarrow \mathbb{R}$$

by

$$\sigma(\Delta)(P_1, \ldots, P_m) = \Delta(P_{\sigma(1)}, \ldots, P_{\sigma(m)})$$

for all matrices $P_1, \ldots, P_m$ in $\operatorname{Mat}_{m \times 1}$.

Theorem 2.4.1. *If $\Delta : (\mathrm{Mat}_{m\times 1})^m \longrightarrow \mathbb{R}$ is a function, and σ and ρ are permutations in $\mathcal{S}_m$, then*

$$(\sigma\rho)(\Delta) = \sigma\big(\rho(\Delta)\big).$$

Proof. Setting $P_{\sigma(1)} = Q_1, \ldots, P_{\sigma(m)} = Q_m$, we have

$$\begin{aligned}
(\sigma\rho)(\Delta)(P_1, \ldots, P_m) &= \Delta(P_{(\sigma\rho)(1)}, \ldots, P_{(\sigma\rho)(m)}) = \Delta(P_{\sigma(\rho(1))}, \ldots, P_{\sigma(\rho(m))}) \\
&= \Delta(Q_{\rho(1)}, \ldots, Q_{\rho(m)}) = \rho(\Delta)(Q_1, \ldots, Q_m) \\
&= \rho(\Delta)(P_{\sigma(1)}, \ldots, P_{\sigma(m)}) = \sigma\big(\rho(\Delta)\big)(P_1, \ldots, P_m).
\end{aligned}$$

Since $P_1, \ldots, P_m$ were arbitrary, the result follows.   □

A function $\Delta : (\mathrm{Mat}_{m\times 1})^m \longrightarrow \mathbb{R}$ is said to be **multilinear** if for all matrices $P_1, \ldots, P_m, Q$ in $\mathrm{Mat}_{m\times 1}$ and all real numbers c,

$$\Delta(P_1, \ldots, cP_i + Q, \ldots, P_m) = c\,\Delta(P_1, \ldots, P_i, \ldots, P_m) + \Delta(P_1, \ldots, Q, \ldots, P_m)$$

for $i = 1, \ldots, m$. We say that Δ is **alternating** if for all matrices $P_1, \ldots, P_m$ in $\mathrm{Mat}_{m\times 1}$,

$$\Delta(P_1, \ldots, P_i, \ldots, P_j, \ldots, P_m) = -\Delta(P_1, \ldots, P_j, \ldots, P_i, \ldots, P_m)$$

for all $1 \le i < j \le m$. Equivalently, Δ is alternating if $\tau(\Delta) = -\Delta$ for all transpositions τ in $\mathcal{S}_m$.

Theorem 2.4.2. *If $\Delta : (\mathrm{Mat}_{m\times 1})^m \longrightarrow \mathbb{R}$ is a multilinear function, then the following are equivalent:*
(a) *Δ is alternating.*
(b) *$\sigma(\Delta) = \mathrm{sgn}(\sigma)\,\Delta$ for all permutations σ in $\mathcal{S}_m$.*
(c) *If $P_1, \ldots, P_m$ are matrices in $\mathrm{Mat}_{m\times 1}$ and two (or more) of them are equal, then $\Delta(P_1, \ldots, P_m) = 0$.*
(d) *If $P_1, \ldots, P_m$ are matrices in $\mathrm{Mat}_{m\times 1}$ and $\Delta(P_1, \ldots, P_m) \neq 0$, then $P_1, \ldots, P_m$ are linearly independent.*

Proof. (a) $\Rightarrow$ (b): Let $\sigma = \tau_1 \cdots \tau_k$ be a decomposition of σ into transpositions. By Theorem 2.4.1,

$$\begin{aligned}
\sigma(\Delta) &= (\tau_1 \cdots \tau_k)(\Delta) = (\tau_1 \cdots \tau_{k-1})\big(\tau_k(\Delta)\big) \\
&= (\tau_1 \cdots \tau_{k-1})(-\Delta) = -(\tau_1 \cdots \tau_{k-1})(\Delta).
\end{aligned}$$

Repeating the process $k - 1$ more times gives $\sigma(\Delta) = (-1)^k\Delta$. By Theorem B.2.3, $\mathrm{sgn}(\sigma) = (-1)^k$.

(b) $\Rightarrow$ (a): If τ is a transposition in $\mathcal{S}_m$, then, by Theorem B.2.3, $\tau(\Delta) = \mathrm{sgn}(\tau)\,\Delta = -\Delta$.

(a) $\Rightarrow$ (c): If $P_i = P_j$ for some $1 \le i < j \le m$, then

$$\Delta(P_1, \ldots, P_i, \ldots, P_j, \ldots, P_m) = \Delta(P_1, \ldots, P_j, \ldots, P_i, \ldots, P_m).$$

On the other hand, since Δ is alternating,

$$\Delta(P_1, \ldots, P_j, \ldots, P_i, \ldots, P_m) = -\Delta(P_1, \ldots, P_i, \ldots, P_j, \ldots, P_m).$$

Thus,

$$\Delta(P_1, \ldots, P_i, \ldots, P_j, \ldots, P_m) = -\Delta(P_1, \ldots, P_i, \ldots, P_j, \ldots, P_m),$$

from which the result follows.

(c) $\Rightarrow$ (a): For $1 \leq i < j \leq m$, we have

$$
\begin{aligned}
0 &= \Delta(P_1, \ldots, P_i + P_j, \ldots, P_i + P_j, \ldots, P_m) \\
&= \Delta(P_1, \ldots, P_i, \ldots, P_i, \ldots, P_m) + \Delta(P_1, \ldots, P_i, \ldots, P_j, \ldots, P_m) \\
&\quad + \Delta(P_1, \ldots, P_j, \ldots, P_i, \ldots, P_m) + \Delta(P_1, \ldots, P_j, \ldots, P_j, \ldots, P_m) \\
&= \Delta(P_1, \ldots, P_i, \ldots, P_j, \ldots, P_m) + \Delta(P_1, \ldots, P_j, \ldots, P_i, \ldots, P_m),
\end{aligned}
$$

from which the result follows.

To prove (c) $\Leftrightarrow$ (d), we replace the assertion in part (d) with the following logically equivalent assertion:

(d$'$): If $P_1, \ldots, P_m$ are linearly dependent matrices in $\mathrm{Mat}_{m \times 1}$, then $\Delta(P_1, \ldots, P_m) = 0$.

(c) $\Rightarrow$ (d$'$): Since $P_1, \ldots, P_m$ are linearly dependent, one of them can be expressed as a linear combination of the others, say, $P_1 = \sum_{i=2}^{m} a^i P_i$. Then

$$\Delta(P_1, \ldots, P_m) = \Delta\left(\sum_{i=2}^{m} a^i P_i, P_2, \ldots, P_m\right) = \sum_{i=2}^{m} a^i \Delta(P_i, P_2, \ldots, P_m).$$

Since $\Delta(P_i, P_2, \ldots, P_m)$ has P_i in (at least) two positions, $\Delta(P_i, P_2, \ldots, P_m) = 0$ for $i = 2, \ldots, m$. Thus, $\Delta(P_1, \ldots, P_m) = 0$.

(d$'$) $\Rightarrow$ (c): If two (or more) of $P_1, \ldots, P_m$ are equal, then $P_1, \ldots, P_m$ are linearly dependent, so $\Delta(P_1, \ldots, P_m) = 0$. $\qquad\square$

A function $\Delta : (\mathrm{Mat}_{m \times 1})^m \longrightarrow \mathbb{R}$ is said to be a **determinant function (on $\mathrm{Mat}_{m \times 1}$)** if it is both multilinear and alternating.

Theorem 2.4.3. *Let $\Delta : (\mathrm{Mat}_{m \times 1})^m \longrightarrow \mathbb{R}$ be a determinant function, and let $P_1, \ldots, P_m$ be matrices in $\mathrm{Mat}_{m \times 1}$, with*

$$
P_1 = \begin{bmatrix} p_1^1 \\ \vdots \\ p_1^m \end{bmatrix}, \qquad \cdots \qquad P_m = \begin{bmatrix} p_m^1 \\ \vdots \\ p_m^m \end{bmatrix}.
$$

Then

$$\Delta(P_1, \ldots, P_m) = \Delta(E_1, \ldots, E_m) \sum_{\sigma \in S_m} \mathrm{sgn}(\sigma)\, p_1^{\sigma(1)} \cdots p_m^{\sigma(m)}.$$

Proof. We have

$$P_1 = \sum_{i_1=1}^{m} p_1^{i_1} E_{i_1}, \qquad \cdots \qquad P_m = \sum_{i_m=1}^{m} p_m^{i_m} E_{i_m},$$

hence

$$\Delta(P_1, \ldots, P_m)$$

$$= \Delta\left(\sum_{i_1=1}^{m} p_1^{i_1} E_{i_1}, \ldots, \sum_{i_m=1}^{m} p_m^{i_m} E_{i_m} \right)$$

$$= \sum_{1 \leq i_1, \ldots, i_m \leq m} p_1^{i_1} \cdots p_m^{i_m} \, \Delta(E_{i_1}, \ldots, E_{i_m})$$

$$= \sum_{\substack{1 \leq i_1, \ldots, i_m \leq m \\ i_1, \ldots, i_m \text{ distinct}}} p_1^{i_1} \cdots p_m^{i_m} \, \Delta(E_{i_1}, \ldots, E_{i_m}) \qquad \text{[Th 2.4.2]}$$

$$= \sum_{\sigma \in S_m} p_1^{\sigma(1)} \cdots p_m^{\sigma(m)} \, \Delta(E_{\sigma(1)}, \ldots, E_{\sigma(m)})$$

$$= \sum_{\sigma \in S_m} p_1^{\sigma(1)} \cdots p_m^{\sigma(m)} \, \sigma(\Delta)(E_1, \ldots, E_m)$$

$$= \Delta(E_1, \ldots, E_m) \sum_{\sigma \in S_m} \text{sgn}(\sigma) \, p_1^{\sigma(1)} \cdots p_m^{\sigma(m)}. \qquad \text{[Th 2.4.2]} \qquad \square$$

Theorem 2.4.4 (Existence of Determinant Function). *There is a unique determinant function*

$$\det : (\text{Mat}_{m \times 1})^m \longrightarrow \mathbb{R}$$

on $\text{Mat}_{m \times 1}$ *such that* $\det(E_1, \ldots, E_m) = 1$. *More specifically,*

$$\det(P_1, \ldots, P_m) = \sum_{\sigma \in S_m} \text{sgn}(\sigma) \, p_1^{\sigma(1)} \cdots p_m^{\sigma(m)}$$

$$= \sum_{\sigma \in S_m} \text{sgn}(\sigma) \, p_{\sigma(1)}^1 \cdots p_{\sigma(m)}^m \qquad (2.4.1)$$

for all matrices $P_1, \ldots, P_m$ *in* $\text{Mat}_{m \times 1}$, *where*

$$P_1 = \begin{bmatrix} p_1^1 \\ \vdots \\ p_1^m \end{bmatrix}, \qquad \cdots \qquad P_m = \begin{bmatrix} p_m^1 \\ \vdots \\ p_m^m \end{bmatrix}.$$

Proof. Let us begin by showing that the right-hand sides of (2.4.1) are equal. We have

$$\sum_{\sigma \in S_m} \text{sgn}(\sigma) \, p_1^{\sigma(1)} \cdots p_m^{\sigma(m)} = \sum_{\sigma \in S_m} \text{sgn}(\sigma) \, p_{\sigma^{-1}(1)}^1 \cdots p_{\sigma^{-1}(m)}^m$$

$$= \sum_{\sigma \in S_m} \text{sgn}(\sigma^{-1}) \, p_{\sigma^{-1}(1)}^1 \cdots p_{\sigma^{-1}(m)}^m$$

$$= \sum_{\sigma \in S_m} \text{sgn}(\sigma) \, p_{\sigma(1)}^1 \cdots p_{\sigma(m)}^m,$$

where the second equality follows from Theorem B.2.2(b), and the third equality from the observation that as σ varies over S_m, so does σ^{-1}.

Existence. With $P_1, \ldots, P_m$ as above, define a function $\det : (\mathrm{Mat}_{m \times 1})^m \longrightarrow \mathbb{R}$ using the first equality in (2.4.1). It is easily shown that det is multilinear. For a given permutation ρ in S_m, we have

$$P_{\rho(1)} = \begin{bmatrix} p^1_{\rho(1)} \\ \vdots \\ p^m_{\rho(1)} \end{bmatrix}, \quad \cdots \quad P_{\rho(m)} = \begin{bmatrix} p^1_{\rho(m)} \\ \vdots \\ p^m_{\rho(m)} \end{bmatrix},$$

hence

$$\begin{aligned} \rho(\det)(P_1, \ldots, P_m) &= \det(P_{\rho(1)}, \ldots, P_{\rho(m)}) \\ &= \sum_{\sigma \in S_m} \mathrm{sgn}(\sigma)\, p^{\sigma(1)}_{\rho(1)} \cdots p^{\sigma(m)}_{\rho(m)} \\ &= \sum_{\sigma \in S_m} \mathrm{sgn}(\sigma)\, p^{\sigma\rho^{-1}(1)}_{1} \cdots p^{\sigma\rho^{-1}(m)}_{m} \\ &= \mathrm{sgn}(\rho) \sum_{\sigma \in S_m} \mathrm{sgn}(\sigma\rho^{-1})\, p^{\sigma\rho^{-1}(1)}_{1} \cdots p^{\sigma\rho^{-1}(m)}_{m} \\ &= \mathrm{sgn}(\rho) \det(P_1, \ldots, P_m), \end{aligned}$$

where the fourth equality follows from Theorem B.2.2, and the last equality follows from the observation that as σ varies over S_m, so does $\sigma\rho^{-1}$. Since $P_1, \ldots, P_m$ were arbitrary, $\rho(\det) = \mathrm{sgn}(\rho) \det$. By Theorem 2.4.2, det is alternating. Thus, det is a determinant function. A straightforward computation shows that $\det(E_1, \ldots, E_m) = 1$.

Uniqueness. This follows from Theorem 2.4.3. $\square$

Let P be a matrix in $\mathrm{Mat}_{m \times m}$, and recall that in multi-index notation the column matrices of P are $P_{(1)}, \ldots, P_{(m)}$. Setting

$$P = \begin{bmatrix} P_{(1)} & \cdots & P_{(m)} \end{bmatrix},$$

we henceforth view P as an m-tuple of column matrices. In this way, the vector spaces $\mathrm{Mat}_{m \times m}$ and $(\mathrm{Mat}_{m \times 1})^m$ are identified. Accordingly, we now express the determinant function in Theorem 2.4.4 as

$$\det : \mathrm{Mat}_{m \times m} \longrightarrow \mathbb{R},$$

so that

$$\det(P) = \det(P_{(1)}, \ldots, P_{(m)}),$$

where $\det(P)$ is referred to as the **determinant of P**. In particular, the condition $\det(E_1, \ldots, E_m) = 1$ in Theorem 2.4.4 becomes

$$\det(I_m) = 1. \tag{2.4.2}$$

Theorem 2.4.5. *If P and Q are matrices in $\mathrm{Mat}_{m \times m}$, then:*
(a) $\det(PQ) = \det(P)\det(Q)$.
(b) *If P is invertible, then $\det(P^{-1}) = \det(P)^{-1}$.*
(c) $\det(P^{\mathrm{T}}) = \det(P)$.

Proof. (a): Let us define a function $\Delta : (\mathrm{Mat}_{m \times 1})^m \longrightarrow \mathbb{R}$ by

$$\Delta(R_1, \ldots, R_m) = \det\big(\begin{bmatrix} PR_1 & \cdots & PR_m \end{bmatrix}\big)$$

for all matrices $R_1, \ldots, R_m$ in $\mathrm{Mat}_{m \times 1}$. It is easily shown that Δ is a determinant function. We have

$$\begin{aligned}
\det\big(\begin{bmatrix} PQ_{(1)} & \cdots & PQ_{(m)} \end{bmatrix}\big) &= \Delta(Q_{(1)}, \ldots, Q_{(m)}) \\
&= \Delta(E_1, \ldots, E_m)\det\big(\begin{bmatrix} Q_{(1)} & \cdots & Q_{(m)} \end{bmatrix}\big) \qquad \text{[(2.4.1), Th 2.4.3]} \\
&= \det(P)\det(Q).
\end{aligned}$$

(b): We have from (2.4.1) and part (a) that

$$1 = \det(I_m) = \det(PP^{-1}) = \det(P)\det(P^{-1}),$$

from which the result follows.

(c): Let $P^{\mathrm{T}} = \big[p_j^i\big]^{\mathrm{T}} = \big[q_j^i\big]$. Then

$$\det(P^{\mathrm{T}}) = \sum_{\sigma \in S_m} \mathrm{sgn}(\sigma)\, q_{\sigma(1)}^1 \cdots q_{\sigma(m)}^m = \sum_{\sigma \in S_m} \mathrm{sgn}(\sigma)\, p_1^{\sigma(1)} \cdots p_m^{\sigma(m)}$$
$$= \det(P),$$

where the first and last equalities follow from (2.4.1), and the second equality from the observation that $q_{\sigma(i)}^i = p_i^{\sigma(i)}$ for all permutations σ in S_m. $\qquad \square$

Let P be a matrix in $\mathrm{Mat}_{m \times m}$. The ***ij*-th cofactor of P** is defined by

$$c_j^i = (-1)^{i+j} \det\left(P_{(1,\ldots,\widehat{j},\ldots,m)}^{(1,\ldots,\widehat{i},\ldots,m)}\right),$$

where $\widehat{}$ indicates that an expression is omitted. Thus, $P_{(1,\ldots,\widehat{j},\ldots,m)}^{(1,\ldots,\widehat{i},\ldots,m)}$ is the matrix in $\mathrm{Mat}_{(m-1) \times (m-1)}$ obtained by deleting the ith row and jth column of P. The **adjugate of P** is the matrix in $\mathrm{Mat}_{m \times m}$ defined by

$$\mathrm{adj}(P) = \big[c_j^i\big]^{\mathrm{T}}.$$

Theorem 2.4.6 (Column Expansion of Determinant). *If $P = \big[p_j^i\big]$ is a matrix in $\mathrm{Mat}_{m \times m}$ and $1 \le j \le m$, then*

$$\det(P) = \sum_i c_j^i p_j^i,$$

*which is called the **expansion of $\det(P)$** along the jth column of P.*

Proof. We have

$$P = \begin{bmatrix} P_{(1,\ldots,j-1)} & \sum_i p_j^i E_i & P_{(j+1,\ldots,m)} \end{bmatrix},$$

hence

$$\det(P) = \sum_i p_j^i \det\left(\begin{bmatrix} P_{(1,\ldots,j-1)} & E_i & P_{(j+1,\ldots,m)} \end{bmatrix}\right).$$

We also have

$$\det\left(\begin{bmatrix} P_{(1,\ldots,j-1)} & E_i & P_{(j+1,\ldots,m)} \end{bmatrix}\right)$$

$$= \det\left(\begin{bmatrix} P_{(1,\ldots,j-1)}^{(1,\ldots,i-1)} & O_{(i-1)\times 1} & P_{(j+1,\ldots,m)}^{(1,\ldots,i-1)} \\ P_{(1,\ldots,j-1)}^{(i)} & 1 & P_{(j+1,\ldots,m)}^{(i)} \\ P_{(1,\ldots,j-1)}^{(i+1,\ldots,m)} & O_{(m-i)\times 1} & P_{(j+1,\ldots,m)}^{(i+1,\ldots,m)} \end{bmatrix}\right)$$

$$= (-1)^{(i-1)+(j-1)} \det\left(\begin{bmatrix} 1 & P_{(1,\ldots,j-1)}^{(i)} & P_{(j+1,\ldots,m)}^{(i)} \\ O_{(i-1)\times 1} & P_{(1,\ldots,j-1)}^{(1,\ldots,i-1)} & P_{(j+1,\ldots,m)}^{(1,\ldots,i-1)} \\ O_{(m-i)\times 1} & P_{(1,\ldots,j-1)}^{(i+1,\ldots,m)} & P_{(j+1,\ldots,m)}^{(i+1,\ldots,m)} \end{bmatrix}\right)$$

$$= (-1)^{i+j} \det\left(\begin{bmatrix} 1 & P_{(1,\ldots,\widehat{j},\ldots,m)}^{(i)} \\ O_{(m-1)\times 1} & P_{(1,\ldots,\widehat{j},\ldots,m)}^{(1,\ldots,\widehat{i},\ldots,m)} \end{bmatrix}\right)$$

$$= (-1)^{i+j} \det\left(P_{(1,\ldots,\widehat{j},\ldots,m)}^{(1,\ldots,\widehat{i},\ldots,m)}\right) = c_j^i,$$

where the second equality is obtained by switching $i-1$ adjacent rows, and then switching $j - 1$ adjacent columns, and where the fourth equality follows from (2.4.1). Combining the above identities gives the result.   □

Theorem 2.4.7. *If P is a matrix in* $\mathrm{Mat}_{m\times m}$, *then*

$$P\,\mathrm{adj}(P) = \det(P)\, I_m.$$

Proof. Let $P = \begin{bmatrix} p_j^i \end{bmatrix}$. For given integers $1 \le k, l \le m$, let $Q = \begin{bmatrix} q_j^i \end{bmatrix}$ be the matrix in $\mathrm{Mat}_{m\times m}$ obtained by replacing the kth column of P with the lth column of P; that is,

$$Q = \begin{bmatrix} P_{(1)} & \cdots & P_{(k-1)} & P_{(l)} & P_{(k+1)} & \cdots & P_{(m)} \end{bmatrix}.$$

Let $\mathrm{adj}(P) = \begin{bmatrix} c_j^i \end{bmatrix}^{\mathrm{T}}$ and $\mathrm{adj}(Q) = \begin{bmatrix} d_j^i \end{bmatrix}^{\mathrm{T}}$. By Theorem 2.4.6,

$$\sum_i c_k^i p_k^i = \det(P).$$

If $k \neq l$, then two columns of Q are equal and

$$Q^{(1,\ldots,\widehat{i},\ldots,m)}_{(1,\ldots,\widehat{k},\ldots,m)} = P^{(1,\ldots,\widehat{i},\ldots,m)}_{(1,\ldots,\widehat{k},\ldots,m)}.$$

It follows from Theorem 2.4.2 and Theorem 2.4.6 that

$$\sum_i c^i_k p^i_l = \sum_i d^i_k q^i_k = \det(Q) = 0.$$

Thus,

$$\sum_i c^i_k p^i_l = \det(P)\, \delta_{kl}$$

for $k, l = 1, \ldots, m$, where δ_{kl} is Kronecker's delta. Then $\left[c^i_j\right]\left[p^i_j\right]^{\mathrm{T}} = \det(P)I_m$. Taking transposes gives the result. $\square$

Theorem 2.4.8. *If P is a matrix in $\mathrm{Mat}_{m \times m}$, then the following are equivalent:*
(a) *P is invertible.*
(b) $\det(P) \neq 0$.
(c) $\mathrm{rank}(P) = m$.
  If any of the above equivalent conditions is satisfied, then

$$P^{-1} = \frac{1}{\det(P)}\, \mathrm{adj}(P).$$

Proof. (a) $\Rightarrow$ (b): By Theorem 2.4.5(b), $\det(P)\det(P^{-1}) = 1$, hence $\det(P) \neq 0$.

  (b) $\Rightarrow$ (c): Since $\det\!\left(\begin{bmatrix} P_{(1)} & \cdots & P_{(m)} \end{bmatrix}\right) = \det(P) \neq 0$, it follows from Theorem 2.4.2 that $P_{(1)}, \ldots, P_{(m)}$ are linearly independent, so

$$\mathrm{rank}(P) = \mathrm{rank}\!\left(\begin{bmatrix} P_{(1)} & \cdots & P_{(m)} \end{bmatrix}\right) = m.$$

  (c) $\Rightarrow$ (a): Since $\mathrm{rank}(P) = m$, $\mathbb{P} = (P_{(1)}, \ldots, P_{(m)})$ is a basis for $\mathrm{Mat}_{m \times 1}$. Let $Q = \begin{bmatrix} [E_1]_{\mathbb{P}} & \cdots & [E_m]_{\mathbb{P}} \end{bmatrix}$. By definition, $E_j = P[E_j]_{\mathbb{P}}$ for $j = 1, \ldots, m$, hence

$$I_m = \begin{bmatrix} E_1 & \cdots & E_m \end{bmatrix} = PQ,$$

so $Q = P^{-1}$.

  The final assertion follows from Theorem 2.4.7. $\square$

Theorem 2.4.9. *The matrices $P_1, \ldots, P_m$ in $\mathrm{Mat}_{m \times 1}$ are linearly independent if and only if $\det\!\left(\begin{bmatrix} P_1 & \cdots & P_m \end{bmatrix}\right) \neq 0$.*

Proof. We have

$$P_1, \ldots, P_m \text{ are linearly independent}$$
$$\Leftrightarrow \quad \mathrm{rank}\!\left(\begin{bmatrix} P_1 & \cdots & P_m \end{bmatrix}\right) = m$$
$$\Leftrightarrow \quad \det\!\left(\begin{bmatrix} P_1 & \cdots & P_m \end{bmatrix}\right) \neq 0. \qquad \text{[Th 2.4.8]} \qquad \square$$

The next result is not usually included in an overview of determinants, but it will prove invaluable later on.

Theorem 2.4.10 (Cauchy–Binet Identity). *Let* $1 \le k \le m$ *be integers.*
(a) *If P and Q are matrices in* $\mathrm{Mat}_{m \times k}$, *then*

$$\det(P^{\mathrm{T}}Q) = \sum_{I \in \mathcal{I}_{k,m}} \det(P^I) \det(Q^I).$$

(b) *If R and S are matrices in* $\mathrm{Mat}_{k \times m}$, *then*

$$\det(RS^{\mathrm{T}}) = \sum_{J \in \mathcal{I}_{k,m}} \det(R_J) \det(S_J).$$

Proof. (a): Let $Q = \left[q_j^i\right]$, so that

$$Q = \begin{bmatrix} Q_{(1)} & \cdots & Q_{(k)} \end{bmatrix}, \qquad Q_{(j)} = \begin{bmatrix} q_j^1 & \cdots & q_j^m \end{bmatrix}^{\mathrm{T}},$$

and

$$P^{\mathrm{T}}Q_{(j)} = \sum_i q_j^i (P^{\mathrm{T}})_{(i)}$$

for $j = 1, \dots, k$. We have

$$\det(P^{\mathrm{T}}Q)$$
$$= \det\left(\begin{bmatrix} P^{\mathrm{T}}Q_{(1)} & \cdots & P^{\mathrm{T}}Q_{(k)} \end{bmatrix}\right)$$
$$= \det\left(\begin{bmatrix} \sum_{i_1} q_1^{i_1}(P^{\mathrm{T}})_{(i_1)} & \cdots & \sum_{i_k} q_k^{i_k}(P^{\mathrm{T}})_{(i_k)} \end{bmatrix}\right)$$
$$= \sum_{1 \le i_1, \dots, i_k \le m} q_1^{i_1} \cdots q_k^{i_k} \det\left(\begin{bmatrix} (P^{\mathrm{T}})_{(i_1)} & \cdots & (P^{\mathrm{T}})_{(i_k)} \end{bmatrix}\right)$$
$$= \sum_{\substack{1 \le i_1, \dots, i_k \le m \\ i_1, \dots, i_k \text{ distinct}}} q_1^{i_1} \cdots q_k^{i_k} \det\left(\begin{bmatrix} (P^{\mathrm{T}})_{(i_1)} & \cdots & (P^{\mathrm{T}})_{(i_k)} \end{bmatrix}\right)$$
$$= \sum_{1 \le i_1 < \cdots < i_k \le m} \left(\sum_{\sigma \in \mathcal{S}_k} q_1^{i_{\sigma(1)}} \cdots q_k^{i_{\sigma(k)}} \det\left(\begin{bmatrix} (P^{\mathrm{T}})_{(i_{\sigma(1)})} & \cdots & (P^{\mathrm{T}})_{(i_{\sigma(k)})} \end{bmatrix}\right) \right),$$

where the third equality follows from Theorem 2.4.2. The term in large parentheses in the last row of the preceding display can be expressed as

$$\sum_{\sigma \in \mathcal{S}_k} q_1^{i_{\sigma(1)}} \cdots q_k^{i_{\sigma(k)}} \det\left(\begin{bmatrix} (P^{\mathrm{T}})_{(i_{\sigma(1)})} & \cdots & (P^{\mathrm{T}})_{(i_{\sigma(k)})} \end{bmatrix}\right)$$
$$= \sum_{\sigma \in \mathcal{S}_k} q_1^{i_{\sigma(1)}} \cdots q_k^{i_{\sigma(k)}} \operatorname{sgn}(\sigma) \det\left(\begin{bmatrix} (P^{\mathrm{T}})_{(i_1)} & \cdots & (P^{\mathrm{T}})_{(i_k)} \end{bmatrix}\right)$$
$$= \det\left(\begin{bmatrix} (P^{\mathrm{T}})_{(i_1)} & \cdots & (P^{\mathrm{T}})_{(i_k)} \end{bmatrix}\right) \left(\sum_{\sigma \in \mathcal{S}_k} \operatorname{sgn}(\sigma) q_1^{i_{\sigma(1)}} \cdots q_k^{i_{\sigma(k)}} \right)$$
$$= \det\left((P^{\mathrm{T}})_{(i_1, \dots, i_k)}\right) \det(Q^{(i_1, \dots, i_k)}),$$

where the first equality follows from Theorem 2.4.2, and the third equality from (2.4.1). Then

$$\det(P^T Q) = \sum_{1 \le i_1 < \cdots < i_k \le m} \det\big((P^T)_{(i_1,\ldots,i_k)}\big) \det\big(Q^{(i_1,\ldots,i_k)}\big)$$

$$= \sum_{I \in \mathcal{I}_{k,m}} \det\big((P^T)_I\big) \det(Q^I)$$

$$= \sum_{I \in \mathcal{I}_{k,m}} \det\big((P^I)^T\big) \det(Q^I) \qquad\qquad \text{[Th 2.1.6(a)]}$$

$$= \sum_{I \in \mathcal{I}_{k,m}} \det(P^I) \det(Q^I). \qquad\qquad \text{[Th 2.4.5(c)]}$$

(b): We have

$$\det(RS^T) = \det\big((R^T)^T S^T\big) \qquad\qquad \text{[Th 2.1.2(a)]}$$

$$= \sum_{J \in \mathcal{I}_{k,m}} \det\big((R^T)^J\big) \det\big((S^T)^J\big) \qquad \text{[part (a)]}$$

$$= \sum_{J \in \mathcal{I}_{k,m}} \det\big((R_J)^T\big) \det\big((S_J)^T\big) \qquad \text{[Th 2.1.6(b)]}$$

$$= \sum_{J \in \mathcal{I}_{k,m}} \det(R_J) \det(S_J). \qquad\qquad \text{[Th 2.4.5(c)]} \qquad \square$$

Example 2.4.11 (Classical Cauchy–Binet Identity). Let P and Q be matrices in $\mathrm{Mat}_{m \times 2}$, with

$$P = \begin{bmatrix} a^1 & b^1 \\ \vdots & \vdots \\ a^m & b^m \end{bmatrix} \quad \text{and} \quad Q = \begin{bmatrix} c^1 & d^1 \\ \vdots & \vdots \\ c^m & d^m \end{bmatrix}.$$

Then

$$\det(P^T Q) = \det\left(\begin{bmatrix} \sum_i a^i c^i & \sum_i a^i d^i \\ \sum_i b^i c^i & \sum_i b^i d^i \end{bmatrix}\right)$$

$$= \left(\sum_i a^i c^i\right)\left(\sum_i b^i d^i\right) - \left(\sum_i a^i d^i\right)\left(\sum_i b^i c^i\right)$$

and

$$\sum_{I \in \mathcal{I}_{2,m}} \det(P^I) \det(Q^I) = \sum_{1 \le i < j \le m} \det(P^{(i,j)}) \det(Q^{(i,j)})$$

$$= \sum_{1 \le i < j \le m} (a^i b^j - a^j b^i)(c^i d^j - c^j d^i).$$

By Theorem 2.4.10(a),

$$\left(\sum_i a^i c^i\right)\left(\sum_i b^i d^i\right) - \left(\sum_i a^i d^i\right)\left(\sum_i b^i c^i\right)$$

$$= \sum_{1 \le i < j \le m} (a^i b^j - a^j b^i)(c^i d^j - c^j d^i),$$

which is the classical version of the Cauchy–Binet identity. $\diamond$

Theorem 2.4.12. *If V is a vector space, and $\mathcal{H}$ and $\mathcal{F}$ are bases for V, then*

$$\det\left(\left[\mathrm{id}_V\right]_{\mathcal{H}}^{\mathcal{F}}\right) \neq 0$$

Proof. We have

$$1 = \det\left(\left[\mathrm{id}_V\right]_{\mathcal{H}}^{\mathcal{H}}\right) \qquad\qquad [(2.4.2)]$$

$$= \det\left(\left[\mathrm{id}_V\right]_{\mathcal{F}}^{\mathcal{H}}\right)\det\left(\left[\mathrm{id}_V\right]_{\mathcal{H}}^{\mathcal{F}}\right), \qquad [\text{Th } 2.2.2\text{(c), Th } 2.4.5\text{(a)}]$$

from which the result follows. $\qquad\square$

It was remarked in connection with Theorem 2.2.3 that $\mathrm{Mat}_{m\times m}$ is a ring under the usual operations of matrix addition and matrix multiplication, hence it is a group under matrix addition. But $\mathrm{Mat}_{m\times m}$ is clearly not a group under matrix multiplication because not all matrices have an inverse. However, $\mathrm{Mat}_{m\times m}$ contains a number of **matrix groups**. The largest of these is the **general linear group**, consisting of all invertible matrices:

$$\mathrm{GL}(m) = \{P \in \mathrm{Mat}_{m\times m} : \det(P) \neq 0\},$$

where the characterization using determinants follows from Theorem 2.4.8.

We say that a matrix P in $\mathrm{Mat}_{m\times m}$ is **orthogonal** if $P^{\mathrm{T}}P = I_m$. If so, then $P^{\mathrm{T}} = P^{-1}$, hence $PP^{\mathrm{T}} = PP^{-1} = I_m$. Thus, $P^{\mathrm{T}}P = I_m$ if and only if $PP^{\mathrm{T}} = I_m$. The **orthogonal group** is the subgroup of $\mathrm{GL}(m)$ consisting of all orthogonal matrices:

$$\mathrm{O}(m) = \{P \in \mathrm{GL}(m) : P^{\mathrm{T}}P = I_m\}.$$

For a matrix P in $\mathrm{O}(m)$, we have

$$1 = \det(I_m) = \det(P^{\mathrm{T}}P) = \det(P^{\mathrm{T}})\det(P) = \det(P)^2,$$

hence $\det(P) = \pm 1$. The **special orthogonal group** is the subgroup of $\mathrm{O}(m)$ consisting of all matrices with determinant equal to 1:

$$\mathrm{SO}(m) = \{P \in \mathrm{O}(m) : \det(P) = 1\}.$$

2.5 Trace and Determinant of Linear Maps

The trace and determinant of a matrix were defined in Section 2.1 and Section 2.4, respectively. We now extend these concepts to linear maps.

Let V be a vector space, let $\mathcal{H}$ be a basis for V, and let $A : V \longrightarrow V$ be a linear map. The **trace of A** is defined by

$$\mathrm{tr}(A) = \mathrm{tr}\left(\left[A\right]_{\mathcal{H}}^{\mathcal{H}}\right), \tag{2.5.1}$$

and the **determinant of A** by

$$\det(A) = \det\left(\left[A\right]_{\mathcal{H}}^{\mathcal{H}}\right). \tag{2.5.2}$$

Theorem 2.5.1. *With the above setup,* $\mathrm{tr}(A)$ *and* $\det(A)$ *are independent of the choice of basis for* V.

Proof. Let $\mathcal{F}$ be another basis for V. By Theorem 2.2.6,

$$[A]_{\mathcal{H}}^{\mathcal{H}} = \left([\mathrm{id}_V]_{\mathcal{H}}^{\mathcal{F}} \right)^{-1} [A]_{\mathcal{F}}^{\mathcal{F}} [\mathrm{id}_V]_{\mathcal{H}}^{\mathcal{F}}.$$

We have from Theorem 2.1.7(c) that

$$\mathrm{tr}\left([A]_{\mathcal{H}}^{\mathcal{H}} \right) = \mathrm{tr}\left([A]_{\mathcal{F}}^{\mathcal{F}} \right),$$

and from Theorem 2.4.5 that

$$\det\left([A]_{\mathcal{H}}^{\mathcal{H}} \right) = \det\left([A]_{\mathcal{F}}^{\mathcal{F}} \right). \qquad \square$$

Theorem 2.5.2. *If* V *is a vector space and* $A, B : V \longrightarrow V$ *are linear maps, then:*
(a) $\det(\mathrm{id}_V) = 1$.
(b) $\det(A \circ B) = \det(A) \det(B)$.
(c) *If* A *is a linear isomorphism, then* $\det(A^{-1}) = \det(A)^{-1}$.
(d) $\det(A^*) = \det(A)$.

Proof. (a): This follows from Theorem 2.2.3(b) and (2.4.2).
    (b): This follows from Theorem 2.2.3(b) and Theorem 2.4.5(a).
    (c): This follows from Theorem 2.2.3(b) and Theorem 2.4.5(b).
    (d): This follows from Theorem 2.2.7 and Theorem 2.4.5(c). $\square$

Theorem 2.5.3. *Let* V *be a vector space, and let* $A : V \longrightarrow V$ *be a linear map. Then* A *is a linear isomorphism if and only if* $\det(A) \neq 0$.

Proof. This follows from Theorem 2.2.3(b) and Theorem 2.4.8. $\square$

Chapter 3

Bilinear Functions

The inner product is an example of a bilinear function that likely is familiar to the reader. In this book, we are more interested in a less restrictive type of bilinear function called a scalar product, but we will not ignore the inner product altogether. In order to decide whether a bilinear function is a scalar product, we need to determine whether it is symmetric and nondegenerate. These properties of bilinear functions will be the focus of the present chapter.

3.1 Bilinear Functions

Let V be a vector space. A function

$$\mathfrak{b} : V \times V \longrightarrow \mathbb{R}$$

is said to be **bilinear (on V)** if it is linear in both arguments; that is,

$$\mathfrak{b}(cu + v, w) = c\,\mathfrak{b}(u, w) + \mathfrak{b}(v, w)$$

and

$$\mathfrak{b}(u, cv + w) = c\,\mathfrak{b}(u, v) + \mathfrak{b}(u, w)$$

for all vectors u, v, w in V and all real numbers c . In the literature, a bilinear function is sometimes called a **quadratic form**. We often denote

$$\mathfrak{b}(\cdot, \cdot) \qquad \text{by} \qquad \langle \cdot, \cdot \rangle,$$

writing $\langle v, w \rangle$ in place of $\mathfrak{b}(v, w)$.

Semi-Riemannian Geometry, First Edition. Stephen C. Newman.
© 2019 John Wiley & Sons, Inc. Published 2019 by John Wiley & Sons, Inc.

We say that $\flat$ is:

symmetric	if $\langle v, w \rangle = \langle w, v \rangle$ for all v, w in V.
alternating	if $\langle v, w \rangle = -\langle w, v \rangle$ for all v, w in V.
nondegenerate	if for all v in V, $\langle v, w \rangle = 0$ for all w in V implies $v = 0$.
degenerate	if $\flat$ is not nondegenerate.

positive definite	if $\langle v, v \rangle > 0$ for all nonzero v in V.
negative definite	if $\langle v, v \rangle < 0$ for all nonzero v in V.
definite	if $\flat$ is either positive definite or negative definite.
indefinite	if $\flat$ is not definite.

positive semidefinite	if $\langle v, v \rangle \geq 0$ for all v in V.
negative semidefinite	if $\langle v, v \rangle \leq 0$ for all v in V.
semidefinite	if $\flat$ is either positive semidefinite or negative semidefinite.

Let U be a subspace of V. For brevity, we denote the restriction

$$\flat|_{U \times U} \quad \text{by} \quad \flat|_U.$$

In fact, we often drop "$|_U$" from the notation altogether and say, for example, that $\flat$ is symmetric on U if $\flat|_U$ is symmetric. Evidently, $\flat$ is bilinear on U, and if $\flat$ is symmetric (resp., alternating, definite, semidefinite) on V, then $\flat$ is symmetric (resp., alternating, definite, semidefinite) on U. The situation is more complicated with nondegeneracy. If $\flat$ is nondegenerate on V, it does not necessarily follow that $\flat$ is nondegenerate on U. We will explore this phenomenon later on.

Example 3.1.1. The following table gives examples of bilinear functions on $\mathbb{R}^3$. Observe that a bilinear function can be either symmetric or not, and independent of that, nondegenerate or not.

$\langle (x^1, x^2, x^3), (y^1, y^2, y^3) \rangle$	Symmetric	Definite	Semidef.	Nondegen.
$x^1y^1 + x^2y^2 + x^3y^3$	yes	yes	yes	yes
$x^1y^1 + x^2y^2$	yes	no	yes	no
$x^1y^1 + x^2y^2 - x^3y^3$	yes	no	no	yes
$x^1y^1 - x^3y^3$	yes	no	no	no
$x^1y^1 + x^2y^3 - x^3y^2$	no	no	yes	yes
$x^1y^1 + x^2y^3$	no	no	no	no

$\Diamond$

Borrowing terminology from the special theory of relativity, we say that (with respect to $\mathfrak{b}$) a vector v in V is:

spacelike	if $v = 0$ or $\langle v, v \rangle > 0$.
timelike	if $\langle v, v \rangle < 0$.
lightlike	if $v \neq 0$ and $\langle v, v \rangle = 0$.

The zero vector is usually considered to be spacelike (as is the case here), but in some of the literature it is taken to be lightlike. Evidently, each vector is either spacelike, timelike, or lightlike, and this results in a corresponding partition of V. (There are several standard but differing uses of the term "partition" found at various places in this book. The present instance refers to the definition of partition given in Appendix A.)

Let $\mathfrak{b}$ be a bilinear function on a vector space V, and let $\mathcal{H} = (h_1, \ldots, h_m)$ be a basis for V. The **matrix of $\mathfrak{b}$** with respect to $\mathcal{H}$ is denoted by $\mathfrak{b}_{\mathcal{H}}$ and defined by

$$
\mathfrak{b}_{\mathcal{H}} = \begin{bmatrix} \mathfrak{b}_{11} & \cdots & \mathfrak{b}_{1m} \\ \vdots & \ddots & \vdots \\ \mathfrak{b}_{m1} & \cdots & \mathfrak{b}_{mm} \end{bmatrix},
$$

where

$$
\mathfrak{b}_{ij} = \langle h_i, h_j \rangle
$$

for $i, j = 1, \ldots, m$.

The next two results illustrate how matrix representations can be used to facilitate computations involving bilinear functions.

Theorem 3.1.2. *Let V be a vector space of dimension m, let $\mathfrak{b}$ be a bilinear function on V, and let $\mathcal{H}$ be a basis for V. Then:*
(a) *For all vectors v, w in V,*

$$
\langle v, w \rangle = \left([v]_{\mathcal{H}} \right)^{\mathrm{T}} \mathfrak{b}_{\mathcal{H}} \, [w]_{\mathcal{H}}.
$$

(b) *For all integers $1 \leq k \leq m$ and all vectors $v_1, \ldots, v_k, w_1, \ldots, w_k$ in V,*

$$
\begin{bmatrix} \langle v_1, w_1 \rangle & \cdots & \langle v_1, w_k \rangle \\ \vdots & \ddots & \vdots \\ \langle v_k, w_1 \rangle & \cdots & \langle v_k, w_k \rangle \end{bmatrix} = \begin{bmatrix} [v_1]_{\mathcal{H}} & \cdots & [v_k]_{\mathcal{H}} \end{bmatrix}^{\mathrm{T}} \mathfrak{b}_{\mathcal{H}} \begin{bmatrix} [w_1]_{\mathcal{H}} & \cdots & [w_k]_{\mathcal{H}} \end{bmatrix}.
$$

Proof. (a): Let $\mathcal{H} = (h_1, \ldots, h_m)$, and let $v = \sum_i a^i h_i$ and $w = \sum_j b^j h_j$. Then

$$
\left([v]_{\mathcal{H}} \right)^{\mathrm{T}} = \begin{bmatrix} a^1 & \cdots & a^m \end{bmatrix} \qquad \text{and} \qquad [w]_{\mathcal{H}} = \begin{bmatrix} b^1 & \cdots & b^m \end{bmatrix}^{\mathrm{T}},
$$

hence

$$
\langle v, w \rangle = \left\langle \sum_i a^i h_i, \sum_j b^j h_j \right\rangle = \sum_{ij} a^i b^j \langle h_i, h_j \rangle = \sum_{ij} a^i b^j \mathfrak{b}_{ij}
$$

$$
= \left([v]_{\mathcal{H}} \right)^{\mathrm{T}} \mathfrak{b}_{\mathcal{H}} \, [w]_{\mathcal{H}}.
$$

(b): It follows from Theorem 2.1.5(b) and part (a) that

$$\left[[v_1]_\mathcal{H} \quad \cdots \quad [v_k]_\mathcal{H}\right]^\mathrm{T} \mathfrak{b}_\mathcal{H} \left[[w_1]_\mathcal{H} \quad \cdots \quad [w_k]_\mathcal{H}\right]$$
$$= \left[\left([v_i]_\mathcal{H}\right)^\mathrm{T} \mathfrak{b}_\mathcal{H} [w_j]_\mathcal{H}\right] = [\langle v_i, w_j \rangle]. \qquad \square$$

Theorem 3.1.3 (Change of Basis). *Let V be a vector space, let $\mathfrak{b}$ be a bilinear function on V, and let $\mathcal{H}$ and $\mathcal{F}$ be bases for V. Then*

$$\mathfrak{b}_\mathcal{F} = \left([\mathrm{id}_V]_\mathcal{F}^\mathcal{H}\right)^\mathrm{T} \mathfrak{b}_\mathcal{H} [\mathrm{id}_V]_\mathcal{F}^\mathcal{H}.$$

Proof. We have

$$\left([\mathrm{id}_V]_\mathcal{F}^\mathcal{H}\right)^\mathrm{T} \mathfrak{b}_\mathcal{H} [\mathrm{id}_V]_\mathcal{F}^\mathcal{H}$$
$$= \left[[f_1]_\mathcal{H} \quad \cdots \quad [f_m]_\mathcal{H}\right]^\mathrm{T} \mathfrak{b}_\mathcal{H} \left[[f_1]_\mathcal{H} \quad \cdots \quad [f_m]_\mathcal{H}\right] \qquad [(2.2.7)]$$
$$= [\langle f_i, f_j \rangle] \qquad\qquad\qquad\qquad\qquad\qquad\qquad\qquad [\text{Th 3.1.2(b)}]$$
$$= \mathfrak{b}_\mathcal{F}. \qquad\qquad\qquad\qquad\qquad\qquad\qquad\qquad\qquad\qquad\quad \square$$

We close this section with a few definitions that will be useful later on. Let V be a vector space, and let $\mathfrak{b}$ be a bilinear function on V. The **quadratic function** corresponding to $\mathfrak{b}$ is the function

$$\mathfrak{q} : V \longrightarrow \mathbb{R}$$

defined by

$$\mathfrak{q}(v) = \langle v, v \rangle$$

for all vectors v in V. It is easily shown that

$$\mathfrak{q}(v + w) - \mathfrak{q}(v) - \mathfrak{q}(w) = \mathfrak{b}(v, w) + \mathfrak{b}(w, v) \qquad (3.1.1)$$

for all vectors v, w in V.

Several of the earlier definitions can be expressed in terms of $\mathfrak{q}$. Specifically, $\mathfrak{b}$ is:

positive definite	if $\mathfrak{q}(v) > 0$ for all nonzero v in V.
negative definite	if $\mathfrak{q}(v) < 0$ for all nonzero v in V.
positive semidefinite	if $\mathfrak{q}(v) \geq 0$ for all v in V.
negative semidefinite	if $\mathfrak{q}(v) \leq 0$ for all v in V.
spacelike	if $v = 0$ or $\mathfrak{q}(v) > 0$.
timelike	if $\mathfrak{q}(v) < 0$.
lightlike	if $v \neq 0$ and $\mathfrak{q}(v) = 0$.

We say that v is a **unit vector** if $\langle v, v \rangle = \pm 1$, and that v is **orthogonal** to w if $\langle v, w \rangle = 0$. In particular, the zero vector is orthogonal to every vector in V, and a lightlike vector is orthogonal to itself. Unless $\mathfrak{b}$ is symmetric, $\langle w, v \rangle$ may not equal $\langle v, w \rangle$. Thus, even when v is orthogonal to w, it does not necessarily follow that w is orthogonal to v.

Let U be a subspace of V. The **perp of U (in V with respect to $\mathfrak{b}$)** is denoted by $U^{\perp}$ and defined to be the set of vectors in V that are orthogonal (perpendicular) to every vector in U:

$$U^{\perp} = \{v \in V : \langle v, u \rangle = 0 \text{ for all } u \in U\}.$$

It is easily shown that $U^{\perp}$ is a subspace of V. Let us denote

$$(U^{\perp})^{\perp} \qquad \text{by} \qquad U^{\perp\perp}.$$

The **light set of V** is denoted by Λ and defined to be the set of lightlike vectors in V:

$$\Lambda = \{v \in V : \langle v, v \rangle = 0\} \backslash \{0\}.$$

3.2   Symmetric Bilinear Functions

Let V be a vector space, and let $\mathfrak{b}$ be a bilinear function on V. Recall from Section 3.1 that $\mathfrak{b}$ is said to be symmetric on V if $\langle v, w \rangle = \langle w, v \rangle$ for all vectors v, w in V.

Theorem 3.2.1. *If V is a vector space and $\mathfrak{b}$ is a bilinear function on V, then the following are equivalent:*
(a) *$\mathfrak{b}$ is symmetric (as a bilinear function).*
(b) *$\mathfrak{b}_{\mathcal{H}}$ is symmetric (as a matrix) for some basis $\mathcal{H}$ for V.*
(c) *$\mathfrak{b}_{\mathcal{H}}$ is symmetric (as a matrix) for every basis $\mathcal{H}$ for V.*

Proof. (a) $\Rightarrow$ (c): Straightforward.
   (c) $\Rightarrow$ (b): Straightforward.
   (b) $\Rightarrow$ (a): We have

$$\begin{aligned}
\langle v, w \rangle &= \left([v]_{\mathcal{H}}\right)^{\mathrm{T}} \mathfrak{b}_{\mathcal{H}} \, [w]_{\mathcal{H}} && \text{[Th 3.1.2(a)]}\\
&= \left([v]_{\mathcal{H}}\right)^{\mathrm{T}} \mathfrak{b}_{\mathcal{H}}^{\mathrm{T}} \, [w]_{\mathcal{H}} \\
&= \left(\left([w]_{\mathcal{H}}\right)^{\mathrm{T}} \mathfrak{b}_{\mathcal{H}} \, [v]_{\mathcal{H}}\right)^{\mathrm{T}} && \text{[Th 2.1.2(b)]}\\
&= \langle w, v \rangle^{\mathrm{T}} && \text{[Th 3.1.2(a)]}\\
&= \langle w, v \rangle. && \square
\end{aligned}$$

Theorem 3.2.2. *Let V be a vector space, let $\mathfrak{b}$ be a bilinear function on V, and let $\mathfrak{q}$ be the corresponding quadratic function. Then $\mathfrak{b}$ is symmetric if and only if*

$$\mathfrak{b}(v, w) = \frac{\mathfrak{q}(v + w) - \mathfrak{q}(v) - \mathfrak{q}(w)}{2} \qquad \text{(polarization identity)}$$

for all vectors v, w in V.

Proof. ($\Rightarrow$): This follows from (3.1.1).

$\quad$ ($\Leftarrow$): Straightforward. $\qquad\qquad\qquad\qquad\qquad\qquad\qquad\qquad\square$

Theorem 3.2.2 shows that provided $\mathfrak{b}$ is symmetric, $\mathfrak{b}$ and $\mathfrak{q}$ determine each other completely.

Theorem 3.2.3. *Let V be a vector space, let $\mathfrak{b}$ be a bilinear function on V, and let $\mathfrak{q}$ be the corresponding quadratic function. Then the following are equivalent:*
(a) $\mathfrak{b}$ *is alternating.*
(b) $\mathfrak{q} = 0$.
(c) V *does not have a unit vector.*

Proof. (a) $\Rightarrow$ (b): Since $\langle v, w \rangle = -\langle w, v \rangle$ for all vectors v, w in V, setting $w = v$ gives $\mathfrak{q}(v) = -\mathfrak{q}(v)$, hence $\mathfrak{q}(v) = 0$, for all vectors v in V.

$\quad$ (b) $\Rightarrow$ (a): This follows from (3.1.1).

$\quad$ (b) $\Leftrightarrow$ (c): We prove the logically equivalent assertion: $\mathfrak{q} \neq 0$ if and only if V has a unit vector. If $\mathfrak{q} \neq 0$, then there is a (nonzero) vector v in V such that $\mathfrak{q}(v) \neq 0$, hence $v/\sqrt{|\mathfrak{q}(v)|}$ is a unit vector. The converse is equally straightforward. $\qquad\qquad\qquad\qquad\qquad\qquad\qquad\qquad\square$

We are especially interested in vector spaces on which there is a nonzero symmetric bilinear function. The reason is that such vector spaces have a unit vector, as we now show.

Theorem 3.2.4. *Let V be a vector space, let $\mathfrak{b}$ be a symmetric bilinear function on V, and let $\mathfrak{q}$ be the corresponding quadratic function. Then:*
(a) $\mathfrak{b} = 0$ *if and only if $\mathfrak{q} = 0$.*
(b) *If $\mathfrak{b} \neq 0$, then V has a unit vector.*

Proof. (a)($\Rightarrow$): Straightforward.

$\quad$ (a)($\Leftarrow$): This follows from Theorem 3.2.2.

$\quad$ (b): By part (a), $\mathfrak{q} \neq 0$. The result now follows from Theorem 3.2.3. $\quad\square$

The light set of a vector space is a rather mysterious entity. The next result provides a characterization when the bilinear function satisfies certain properties.

Theorem 3.2.5. *If V is a vector space and $\mathfrak{b}$ is a symmetric semidefinite bilinear function on V, then*

$$\Lambda = V^{\perp} \setminus \{0\}.$$

Remark. The preceding identity is equivalent to saying that for every nonzero vector v in V, $\langle v, v \rangle = 0$ if and only if $\langle v, w \rangle = 0$ for all vectors w in V.

Proof. Clearly, $V^{\perp} \setminus \{0\} \subseteq \Lambda$. To prove the reverse inclusion, let $\mathfrak{q}$ be the quadratic function corresponding to $\mathfrak{b}$, and let v and w be vectors in Λ and V, respectively. Consider the polynomial $p(t) : \mathbb{R} \longrightarrow \mathbb{R}$ defined by

$$p(t) = \mathfrak{q}(tv + w) = 2\langle v, w \rangle t + \langle w, w \rangle.$$

We have $\langle v, w \rangle = 0$; for if not, then $p(t)$ takes both positive and negative values, which contradicts the semidefiniteness of $\mathfrak{b}$. Since w was arbitrary, v is in $V^\perp \backslash \{0\}$. Thus, $\Lambda \subseteq V^\perp \backslash \{0\}$. $\qquad\square$

The next result is just one of the many versions of the Cauchy–Schwarz inequality that arise in several areas of mathematics. We will see another, likely more familiar, variation in Section 4.6.

Theorem 3.2.6 (Cauchy–Schwarz Inequality). *If V is a vector space and $\mathfrak{b}$ is a symmetric semidefinite bilinear function on V, then*

$$\langle v, w \rangle^2 \le \langle v, v \rangle \langle w, w \rangle \tag{3.2.1}$$

for all vectors v, w in V.

Proof. The result is trivial if either $v = 0$ or $w = 0$, so suppose $v, w \ne 0$. Let $\mathfrak{q}$ be the quadratic function corresponding to $\mathfrak{b}$, and consider the polynomial $p(t) : \mathbb{R} \longrightarrow \mathbb{R}$ defined by

$$p(t) = \mathfrak{q}(tv + w) = \langle v, v \rangle t^2 + 2 \langle v, w \rangle t + \langle w, w \rangle.$$

If either $\langle v, v \rangle$ or $\langle w, w \rangle = 0$, that is, if either v or w is in Λ, then, by Theorem 3.2.5, $\langle v, w \rangle = 0$, so (3.2.1) is satisfied. Now suppose $\langle v, v \rangle, \langle w, w \rangle \ne 0$. According to the quadratic formula, $p(t)$ has the two (not necessarily distinct) roots

$$-2 \langle v, w \rangle \pm \frac{\sqrt{\langle v, w \rangle^2 - \langle v, v \rangle \langle w, w \rangle}}{\langle v, v \rangle}.$$

By assumption, $\mathfrak{b}$ is semidefinite. Suppose $\mathfrak{b}$ is positive semidefinite, in which case $p(t) \ge 0$. If $p(t) > 0$ for all t in $\mathbb{R}$, then the graph of $p(t)$ versus t does not intersect the t-axis. It follows that $p(t)$ has two nonreal complex roots, which is equivalent to $\langle v, w \rangle^2 - \langle v, v \rangle \langle w, w \rangle < 0$. Thus, (3.2.1) is satisfied. If $p(t_0) = 0$ for some t_0 in $\mathbb{R}$, then the graph of $p(t)$ versus t intersects the t-axis at the single point t_0. It follows that t_0 is a repeated root, which is equivalent to $\langle v, w \rangle^2 - \langle v, v \rangle \langle w, w \rangle = 0$. Once again, (3.2.1) is satisfied. This proves the positive semidefinite case. If $\mathfrak{b}$ is negative semidefinite, then $-\mathfrak{b}$ is positive semidefinite, and we again arrive at (3.2.1). $\qquad\square$

3.3 Flat Maps and Sharp Maps

Readers acquainted with tensor analysis, especially in the context of theoretical physics, may have some familiarity with computations involving "lowering" and "raising" indices. This section and Section 4.5 develop these ideas using flat maps and sharp maps.

Let V be a vector space, and let $\mathfrak{b}$ be a bilinear function on V. The **flat map** corresponding to $\mathfrak{b}$ is the linear map

$$\mathsf{F} : V \longrightarrow V^*$$

defined by

$$F(v)(w) = \langle v, w \rangle$$

for all vectors v, w in V. We usually denote

$$F(v) \qquad \text{by} \qquad v^\mathsf{F},$$

so that

$$v^\mathsf{F}(w) = \langle v, w \rangle. \tag{3.3.1}$$

Theorem 3.3.1. *Let V be a vector space, and let $\mathfrak{b}$ be a bilinear function on V. Let $\mathcal{H} = (h_1, \ldots, h_m)$ be a basis for V, let $\Theta = (\theta^1, \ldots, \theta^m)$ be its dual basis, and let $\mathfrak{b}_{\mathcal{H}} = [\mathfrak{b}_{ij}]$. Then:*
(a)

$$h_j^\mathsf{F} = \sum_i \mathfrak{b}_{ji} \theta^i$$

   for $j = 1, \ldots, m$.
(b)

$$[\mathsf{F}]_{\mathcal{H}}^{\Theta} = \mathfrak{b}_{\mathcal{H}}^{\mathsf{T}}.$$

Proof. By Theorem 1.2.1(e),

$$h_j^\mathsf{F} = \sum_i h_j^\mathsf{F}(h_i)\, \theta^i = \sum_i \langle h_j, h_i \rangle \theta^i = \sum_i \mathfrak{b}_{ji} \theta^i.$$

Then (2.2.2) and (2.2.3) give $[\mathsf{F}]_{\mathcal{H}}^{\Theta} = [\mathfrak{b}_{ji}] = [\mathfrak{b}_{ij}]^{\mathsf{T}}$. $\square$

Theorem 3.3.2 (Lowering an Index). *Let V be a vector space, let $\mathfrak{b}$ be a bilinear function on V, let $\mathcal{H} = (h_1, \ldots, h_m)$ be a basis for V, and let $\Theta = (\theta^1, \ldots, \theta^m)$ be its dual basis. Let $\mathfrak{b}_{\mathcal{H}} = [\mathfrak{b}_{ij}]$, and let v be a vector in V, with*

$$v = \sum_i a^i h_i. \tag{3.3.2}$$

Then

$$v^\mathsf{F} = \sum_i a_i \theta^i, \tag{3.3.3}$$

where

$$a_i = \sum_j \mathfrak{b}_{ji} a^j \tag{3.3.4}$$

for $i = 1, \ldots, m$. Expressed more succinctly:

$$[v^\mathsf{F}]_{\Theta} = \mathfrak{b}_{\mathcal{H}}^{\mathsf{T}} [v]_{\mathcal{H}}.$$

Proof. For a vector w in V, Theorem 1.2.1(d) gives $w = \sum_j \theta^j(w)h_j$. Then

$$v^{\mathsf{F}}(w) = \langle v, w \rangle = \left\langle \sum_i a^i h_i, \sum_j \theta^j(w)h_j \right\rangle = \sum_{ij} a^i \theta^j(w)\langle h_i, h_j \rangle$$

$$= \sum_{ij} a^i \theta^j(w)\mathfrak{b}_{ij} = \left(\sum_{ij} \mathfrak{b}_{ji} a^j \theta^i \right)(w)$$

$$= \left(\sum_i \left[\sum_j \mathfrak{b}_{ji} a^j \right] \theta^i \right)(w) = \left(\sum_i a_i \theta^i \right)(w).$$

Since w was arbitrary, $v^{\mathsf{F}} = \sum_i a_i \theta^i$. To prove the succinct version directly:

$$[v^{\mathsf{F}}]_{\Theta} = [\mathsf{F}(v)]_{\Theta}$$

$$= [\mathsf{F}]^{\Theta}_{\mathcal{H}} [v]_{\mathcal{H}} \qquad [\text{Th 2.2.4}]$$

$$= \mathfrak{b}^{\mathsf{T}}_{\mathcal{H}} [v]_{\mathcal{H}}. \qquad [\text{Th 3.3.1(b)}] \qquad \square$$

We see from (3.3.2)–(3.3.4) why taking the flat of v to obtain v^{F} is classically referred to as **lowering an index** by $\mathfrak{b}$.

Let V be a vector space, and let $\mathfrak{b}$ be a bilinear function on V. Recall from Section 3.1 that $\mathfrak{b}$ is said to be nondegenerate on V if for all vectors v in V, $\langle v, w \rangle = 0$ for all vectors w in V implies $v = 0$.

Theorem 3.3.3. *If V is a vector space and $\mathfrak{b}$ is a bilinear function on V, then the following are equivalent:*
(a) *$\mathfrak{b}$ is nondegenerate on V.*
(b) *$V^{\perp} = \{0\}$.*
(c) *F is a linear isomorphism.*
(d) *The rows (hence columns) of $\mathfrak{b}_{\mathcal{H}}$ are linearly independent for some (hence every) basis $\mathcal{H}$ for V.*
(e) *$\mathfrak{b}_{\mathcal{H}}$ is invertible for some (hence every) basis $\mathcal{H}$ for V.*
(f) *$\det(\mathfrak{b}_{\mathcal{H}}) \neq 0$ for some (hence every) basis $\mathcal{H}$ for V.*

Proof. (a) $\Leftrightarrow$ (b): Straightforward.
   (b) $\Leftrightarrow$ (c): We have

$$V^{\perp} = \{v \in V : v^{\mathsf{F}}(w) = 0 \text{ for all } w \in V\}$$

$$= \{v \in V : v^{\mathsf{F}} = 0\} = \ker(\mathsf{F}).$$

The result now follows from Theorem 1.1.12, Theorem 1.1.14, and Theorem 1.2.1(c).
   (c) $\Leftrightarrow$ (d): Let $\dim(V) = m$, and let Θ be the dual basis corresponding to $\mathcal{H}$. Then

$$\text{rank}(\mathsf{F}) = \text{rank}\left([\mathsf{F}]^{\Theta}_{\mathcal{H}} \right) \qquad [\text{Th 2.3.1, (2.3.1)}]$$

$$= \text{rank}(\mathfrak{b}^{\mathsf{T}}_{\mathcal{H}}) \qquad [\text{Th 3.3.1(b)}]$$

$$= \text{rank}(\mathfrak{b}_{\mathcal{H}}), \qquad [\text{Th 2.3.2}]$$

hence

F is a linear isomorphism

$\quad\Leftrightarrow\quad \text{rank}(\mathsf{F}) = m$ [Th 1.1.14, Th 1.2.1(c)]

$\quad\Leftrightarrow\quad \text{rank}(\mathfrak{b}_{\mathcal{H}}) = m$

$\quad\Leftrightarrow\quad$ statement of part (d). [Th 2.3.2]

(d) $\Leftrightarrow$ (e) $\Leftrightarrow$ (f): The result for "some basis" follows from Theorem 2.3.2, Theorem 2.4.8, and Theorem 2.4.9. The extension to "every basis" can be shown using Theorem 3.1.3. □

Theorem 3.3.4. *If V is a vector space and $\mathfrak{b}$ is a nondegenerate bilinear function on V, then $\mathfrak{b} \neq 0$.*

Proof. Let v be a nonzero vector in V. Since $\mathfrak{b}$ is nondegenerate, there is a (nonzero) vector w in V such that $\langle v, w \rangle \neq 0$. Thus, $\mathfrak{b}$ is nonzero. □

Let V be a vector space, and let $\mathfrak{b}$ be a nondegenerate bilinear function on V. By Theorem 3.3.3, F is a linear isomorphism. Its inverse, also a linear isomorphism according to Theorem 1.1.9, is denoted by

$$\mathsf{S} : V^* \longrightarrow V$$

and called the **sharp map** corresponding to $\mathfrak{b}$. Thus,

$$\mathsf{S} = \mathsf{F}^{-1}. \tag{3.3.5}$$

We usually denote

$$\mathsf{S}(\eta) \qquad \text{by} \qquad \eta^{\mathsf{S}}$$

for all covectors η in V^*. In much of the mathematical literature, F and S are represented by the flat and sharp music symbols, $\flat$ and $\sharp$. For our purposes the present notation is more convenient.

Let us observe that with the identification $V^{**} = V$, the pullback by $\mathsf{F} : V \longrightarrow V^*$ and the pullback by $\mathsf{S} : V^* \longrightarrow V$ are expressed as

$$\mathsf{F}^* : V \longrightarrow V^* \qquad \text{and} \qquad \mathsf{S} : V^* \longrightarrow V.$$

Theorem 3.3.5. *If V is a vector space and $\mathfrak{b}$ is a symmetric nondegenerate bilinear function on V, then:*
(a) $\mathsf{F}^* = \mathsf{F}$.
(b) $\mathsf{S}^* = \mathsf{S}$.

Remark. Beginning in Section 4.1, we will refer to a symmetric nondegenerate bilinear function as a *scalar product*.

Proof. (a): For vectors v and w in V, we have

$$\begin{aligned}
\mathsf{F}^*(v)(w) &= v\big(\mathsf{F}(w)\big) && [(1.3.1)] \\
&= \mathsf{F}(w)(v) && [(1.2.3)] \\
&= \langle w, v \rangle = \langle v, w \rangle \\
&= \mathsf{F}(v)(w). && [(1.2.3)]
\end{aligned}$$

Since v and w were arbitrary, $\mathsf{F}^* = \mathsf{F}$.

(b): The proof is similar to that given for part (a). $\square$

Chapter 4

Scalar Product Spaces

4.1 Scalar Product Spaces

We now begin the study of a type of vector space that will occupy us, in one way or another, for the rest of the book.

Let V be a vector space, and let $\mathfrak{g} : V \times V \longrightarrow \mathbb{R}$ be a bilinear function. We say that $\mathfrak{g}$ is a **scalar product** on V, and that the pair $(V, \mathfrak{g})$ is a **scalar product space**, if $\mathfrak{g}$ is symmetric and nondegenerate on V. Recall from Section 3.1 that $\mathfrak{g}$ is symmetric on V if $\langle v, w \rangle = \langle w, v \rangle$ for all vectors v, w in V; and that $\mathfrak{g}$ is nondegenerate on V if for all vectors v in V, $\langle v, w \rangle = 0$ for all vectors w in V implies $v = 0$.

Suppose $(V, \mathfrak{g})$ is in fact a scalar product space. It follows from the symmetry of $\mathfrak{g}$ that for all bases $\mathcal{H}$ for V, $\mathfrak{g}_{\mathcal{H}}$ is a symmetric matrix. The **norm** corresponding to $\mathfrak{g}$ is the function

$$\|\cdot\| : V \longrightarrow \mathbb{R}$$

defined by

$$\|v\| = \sqrt{|\langle v, v \rangle|} = \sqrt{|\mathfrak{q}(v)|} \qquad (4.1.1)$$

for all vectors v in V, where $\mathfrak{q}$ is the quadratic function corresponding to $\mathfrak{g}$. Taking the absolute value in (4.1.1) is necessary because a scalar product can have negative values. We refer to $\|v\|$ as the **norm of v**. Observe that $\|v\| = 0$ if and only if $v = 0$ or v is lightlike; and $\|v\| = 1$ if and only if v is a unit vector. Recall from Theorem 3.3.3 that since $\mathfrak{g}$ is nondegenerate on V, the flat map F is a linear isomorphism, so its inverse, the sharp map S, exists and is also a linear isomorphism.

Example 4.1.1 ($\mathbb{R}^m_\nu$). For an integer $0 \leq \nu \leq m$, let

$$\mathfrak{s} : \mathbb{R}^m \times \mathbb{R}^m \longrightarrow \mathbb{R}$$

Semi-Riemannian Geometry, First Edition. Stephen C. Newman.
© 2019 John Wiley & Sons, Inc. Published 2019 by John Wiley & Sons, Inc.

be the function given by

$$s\big((x^1,\ldots,x^m),(y^1,\ldots,y^m)\big)$$
$$= \langle(x^1,\ldots,x^m),(y^1,\ldots,y^m)\rangle$$
$$= \begin{cases} \sum_{i=1}^m x^i y^i & \text{if } \nu = 0 \\ \sum_{i=1}^{m-\nu} x^i y^i - \sum_{i=m-\nu+1}^m x^i y^i & \text{if } 1 \le \nu \le m-1 \\ -\sum_{i=1}^m x^i y^i & \text{if } \nu = m. \end{cases}$$

Clearly, s is bilinear and symmetric. Let $(e_1,\ldots,e_m)$ be the standard basis for $\mathbb{R}^m$, and suppose $v = (a^1,\ldots,a^m)$ is a vector in $\mathbb{R}^m$ such that $\langle v, w \rangle = 0$ for all vectors w in $\mathbb{R}^m$. Then, in particular, $0 = \langle v, e_i \rangle = \pm a^i$ for each i, hence $v = 0$. Thus, s is nondegenerate. This shows that s is a scalar product on $\mathbb{R}^m$. The scalar product space $(\mathbb{R}^m, s)$ is called **semi-Euclidean (m, ν)-space.** We usually denote $(\mathbb{R}^m, s)$ by $\mathbb{R}^m_\nu$, and write

$$\mathbb{R}^m_\nu = (\mathbb{R}^m, s).$$

Let us observe that the presence or absence of a scalar product on $\mathbb{R}^m$ has nothing to do with the existence of a basis. Accordingly, the standard basis for $\mathbb{R}^m_\nu$ is defined to be the standard basis for $\mathbb{R}^m$.   ◇

Theorem 4.1.2. *If $(V, \mathfrak{g})$ is a scalar product space and U is a subspace of V, then:*
(a) $\dim(U^\perp) = \dim(U^0)$.
(b) $\dim(V) = \dim(U) + \dim(U^\perp)$.
(c) $U^{\perp\perp} = U$.

Proof. (a): We have

$$U^\perp = \{v \in V : \langle v, u \rangle = 0 \text{ for all } u \in U\} = \{v \in V : \mathsf{F}(v) \in U^0\}$$
$$= \mathsf{F}^{-1}(U^0) = \mathsf{S}(U^0) = \mathrm{im}(\mathsf{S}|_{U^0}),$$

hence

$$\dim(U^\perp) = \mathrm{rank}(\mathsf{S}|_{U^0}).$$

Since $\mathsf{S} : V^* \longrightarrow V$ is a linear isomorphism, so is $\mathsf{S}|_{U^0} : U^0 \longrightarrow \mathsf{S}(U^0)$. Then Theorem 1.1.14 gives

$$\mathrm{rank}(\mathsf{S}|_{U^0}) = \dim(U^0).$$

The result follows.

(b): This follows from Theorem 1.4.1 and part (a).

(c): If u is a vector in U, then $\langle u, v \rangle = 0$ for all vectors v in $U^\perp$, hence u is in $U^{\perp\perp}$. Thus, $U \subseteq U^{\perp\perp}$. Using part (b) twice yields

$$\dim(U) + \dim(U^\perp) = \dim(V) = \dim(U^\perp) + \dim(U^{\perp\perp}),$$

so $\dim(U) = \dim(U^{\perp\perp})$. The result now follows from Theorem 1.1.7(b).   □

Theorem 4.1.3. *If $(V, \mathfrak{g})$ is a scalar product space and U is a subspace of V, then the following are equivalent:*
(a) $\mathfrak{g}$ *is nondegenerate on* U.
(b) $U \cap U^\perp = \{0\}$.
(c) $V = U \oplus U^\perp$.
(d) $\mathfrak{g}$ *is nondegenerate on* $U^\perp$.

Proof. (a) $\Leftrightarrow$ (b): We have

> $\mathfrak{g}$ is nondegenerate on U
>
> $\Leftrightarrow$ v is in U, and $\langle v, w \rangle = 0$ for all w in U implies $v = 0$
>
> $\Leftrightarrow$ v is in U, and v is in $U^\perp$ implies $v = 0$
>
> $\Leftrightarrow$ $U \cap U^\perp = \{0\}$.

(b) $\Rightarrow$ (c): It follows from

$$\begin{aligned}
\dim(V) &= \dim(U) + \dim(U^\perp) & \text{[Th 4.1.2(b)]} \\
&= \dim(U + U^\perp) + \dim(U \cap U^\perp) & \text{[Th 1.1.15]}
\end{aligned}$$

and $U \cap U^\perp = \{0\}$ that $\dim(V) = \dim(U + U^\perp)$, and then from Theorem 1.1.7(b) that $V = U + U^\perp$. By Theorem 1.1.17, $U + U^\perp = U \oplus U^\perp$, hence $V = U \oplus U^\perp$.
(c) $\Rightarrow$ (b): This follows from Theorem 1.1.17.
(c) $\Leftrightarrow$ (d): This follows from (a) $\Leftrightarrow$ (c) and Theorem 4.1.2(c). $\square$

Continuing with the notation of Theorem 4.1.3, suppose $\mathfrak{g}$ is nondegenerate on U, in which case $V = U \oplus U^\perp$. In this setting only, we refer to $U^\perp$ (previously called the perp of U in V with respect to $\mathfrak{g}$) as the **orthogonal complement of U (in V with respect to $\mathfrak{g}$)**. It follows from Theorem 4.1.2(c) that U is the orthogonal complement of $U^\perp$ in V with respect to $\mathfrak{g}$. Thus, U and $U^\perp$ are orthogonal complements of each other in V with respect to $\mathfrak{g}$. To illustrate, suppose $\nu \neq 0$. It is easily shown that the subspaces $\mathbb{R}^{m-\nu} \times \{0\}^\nu$ and $\{0\}^{m-\nu} \times \mathbb{R}^\nu$ of $\mathbb{R}^m_\nu$ are orthogonal complements of each other in $\mathbb{R}^m_\nu$ with respect to $\mathfrak{s}$. Then Theorem 4.1.3 gives

$$\mathbb{R}^m_\nu = (\mathbb{R}^{m-\nu} \times \{0\}^\nu) \oplus (\{0\}^{m-\nu} \times \mathbb{R}^\nu). \tag{4.1.2}$$

Making the obvious identifications and abusing notation slightly, this can be expressed succinctly as $\mathbb{R}^m_\nu = \mathbb{R}^{m-\nu} \oplus \mathbb{R}^\nu$.

Theorem 4.1.4. *If $(V, \mathfrak{g})$ is a scalar product space and U is a subspace of V on which $\mathfrak{g}$ is nondegenerate, then each vector v in V can be expressed uniquely in the form*

$$v = \mathcal{P}_U(v) + \mathcal{P}_{U^\perp}(v),$$

where $\mathcal{P}_U(v)$ is a vector in U and $\mathcal{P}_{U^\perp}(v)$ is a vector in $U^\perp$.

Proof. This follows from Theorem 4.1.3. $\square$

In the notation of Theorem 4.1.4, the **orthogonal projection** on U is the linear map

$$\mathcal{P}_U : V \longrightarrow U$$

defined by the assignment

$$v \longmapsto \mathcal{P}_U(v).$$

More generally, for a subset S of V, let us define

$$\mathcal{P}_U(S) = \{\mathcal{P}_U(s) : s \in S\}.$$

When S consists of a single vector, say, v, we denote

$$\mathcal{P}_U(\{v\}) \qquad \text{by} \qquad \mathcal{P}_U(v).$$

Theorem 4.1.5. *If $(V, \mathfrak{g})$ is a scalar product space and v, w are vectors in V such that $\langle v, v \rangle \neq 0$, then*

$$\mathcal{P}_{\mathbb{R}v}(w) = \frac{\langle v, w \rangle}{\langle v, v \rangle}\, v.$$

Proof. Since $\langle v, v \rangle \neq 0$, $\mathfrak{g}$ is nondegenerate on $\mathbb{R}v$, so Theorem 4.1.4 applies. Thus, $\mathcal{P}_{\mathbb{R}v}(w) = cv$ and $w = cv + \mathcal{P}_{(\mathbb{R}v)^\perp}(w)$ for some real number c. Then

$$\langle v, w \rangle = \langle v, cv + \mathcal{P}_{(\mathbb{R}v)^\perp}(w) \rangle = c\langle v, v \rangle,$$

so $c = \langle v, w \rangle / \langle v, v \rangle$. $\qquad\qquad\qquad\qquad\qquad\qquad\qquad\qquad\qquad\square$

An observation related to (4.1.2) is that $\mathfrak{s}$ is positive definite on $\mathbb{R}^{m-\nu} \times \{0\}^\nu$ and negative definite on $\{0\}^{m-\nu} \times \mathbb{R}^\nu$. The next result shows that this is a specific instance of a more general phenomenon.

Theorem 4.1.6 (Sylvester's Law of Inertia). *If $(V, \mathfrak{g})$ is a scalar product space, then:*
(a) *There are subspaces V^+ and V^- of V such $V = V^+ \oplus V^-$, with $\mathfrak{g}$ positive definite on V^+ and negative definite on V^-.*
(b) *The dimensions $\dim(V^+)$ and $\dim(V^-)$ are uniquely determined by $\mathfrak{g}$ in the following sense: if W^+ and W^- are subspaces of V such that $V = W^+ \oplus W^-$, with $\mathfrak{g}$ positive definite on W^+ and negative definite on W^-, then $\dim(V^+) = \dim(W^+)$ and $\dim(V^-) = \dim(W^-)$.*

Proof. (a): If $\mathfrak{g} = 0$, the result is trivial, so assume otherwise. By Theorem 3.2.4(b), there is a nonzero vector v in V such that either $\langle v, v \rangle > 0$ or $\langle v, v \rangle < 0$. Suppose $\langle v, v \rangle > 0$; the argument when $\langle v, v \rangle < 0$ is similar. Then $\mathbb{R}v$ is a nonzero subspace of V on which $\mathfrak{g}$ is positive definite. Let V^+ be a (not necessarily unique) subspace of V that has maximal dimension among subspaces on which $\mathfrak{g}$ is positive definite, and let $V^- = (V^+)^\perp$. Then $\mathfrak{g}$ is nondegenerate on V^+, so Theorem 4.1.3 gives

$$V = V^+ \oplus V^- \tag{4.1.3}$$

and

$$V^+ \cap V^- = \{0\}. \tag{4.1.4}$$

We claim that $\mathfrak{g}$ is negative semidefinite on V^-. Suppose, for a contradiction, that there is a nonzero vector v^- in V^- such that $\langle v^-, v^- \rangle > 0$, and consider the subspace $V^+ + \mathbb{R}v^-$ of V. Since $\mathbb{R}v^- \subseteq V^-$, it follows from (4.1.3) that

$$V^+ + \mathbb{R}v^- = V^+ \oplus \mathbb{R}v^-. \tag{4.1.5}$$

Let $v^+ + av^-$ be a vector in $V^+ + \mathbb{R}v^-$. Then

$$\langle v^+ + av^-, v^+ + av^- \rangle = \langle v^+, v^+ \rangle + a^2 \langle v^-, v^- \rangle \geq 0,$$

with

$$\langle v^+ + av^-, v^+ + av^- \rangle = 0$$
$$\Leftrightarrow \quad \langle v^+, v^+ \rangle = 0 \text{ and } a^2 \langle v^-, v^- \rangle = 0$$
$$\Leftrightarrow \quad v^+ = 0 \text{ and } a = 0$$
$$\Leftrightarrow \quad v^+ + av^- = 0. \qquad\qquad [(4.1.5)]$$

This shows that $\mathfrak{g}$ is positive definite on $V^+ + \mathbb{R}v^-$. It follows from the maximality property of V^+ that $V^+ + \mathbb{R}v^- = V^+$, so v^- is in V^+, which contradicts (4.1.4). This proves the claim.

We now claim that $\mathfrak{g}$ is negative definite on V^-. Let w^- be a vector in V^- such that $\langle w^-, w^- \rangle = 0$. Since $\mathfrak{g}$ is negative semidefinite on V^-, Theorem 3.2.6 applies. Thus, $\langle w^-, v^- \rangle^2 \leq \langle w^-, w^- \rangle \langle v^-, v^- \rangle = 0$ for all vectors v^- in V^-, hence $\langle w^-, v^- \rangle = 0$ for all vectors v^- in V^-. Since w^- is in $V^- = (V^+)^\perp$, $\langle w^-, v^+ \rangle = 0$ for all vectors v^+ in V^+. We have from (4.1.3) that $V = V^+ + V^-$, so $\langle w^-, v \rangle = 0$ for all vectors v in V. Since $\mathfrak{g}$ is nondegenerate, $w^- = 0$. Thus, $\mathfrak{g}$ is negative definite on V^-. This proves the claim.

(b): Suppose there is a nonzero vector v in $V^+ \cap W^-$. By assumption, $\mathfrak{g}$ is positive definite on V^+, so $\langle v, v \rangle > 0$. Also by assumption, $\mathfrak{g}$ is negative definite on W^-, hence $\langle v, v \rangle < 0$, which is a contradiction. Thus, $V^+ \cap W^- = \{0\}$. We have from Theorem 1.1.17 that $V^+ + W^- = V^+ \oplus W^-$, and then from Theorem 1.1.18 that

$$\dim(V^+) + \dim(W^-) = \dim(V^+ + W^-) \leq \dim(V).$$

By Theorem 1.1.18 and (4.1.3),

$$\dim(V^+) + \dim(V^-) = \dim(V).$$

It follows that $\dim(W^-) \leq \dim(V^-)$. Similarly, since $\mathfrak{g}$ is positive definite on W^+ and negative definite on V^-, we have $\dim(V^-) \leq \dim(W^-)$. Thus, $\dim(V^-) = \dim(W^-)$. A corresponding argument shows that $\dim(V^+) = \dim(W^+)$. $\qquad\square$

4.2   Orthonormal Bases

Let $(V, \mathfrak{g})$ be a scalar product space of dimension m, and let v, w be vectors in V. Recall from Section 3.1 that v is said to be orthogonal to w if $\langle v, w \rangle = 0$. Since $\mathfrak{g}$ is symmetric, v is orthogonal to w if and only if w is orthogonal to v. For an integer $k \geq 2$, a k-tuple $(v_1, \ldots, v_k)$ of distinct *nonzero* vectors in V is said to be an **orthogonal k-tuple** if v_i and v_j are orthogonal for $i \neq j = 1, \ldots, k$. This allows the possibility that v_i is orthogonal to itself; that is, v_i might be a lightlike vector for one or more $1 \leq i \leq m$. If an orthogonal k-tuple $(v_1, \ldots, v_k)$ consists of unit vectors, it is said to be an **orthonormal k-tuple**, in which case lightlike vectors are excluded. When an orthogonal m-tuple is a basis for V, it is called an **orthogonal basis**. It follows from Theorem 4.2.3 that an orthonormal m-tuple is necessarily a basis for V, which is called an **orthonormal basis**. Thus, if $(e_1, \ldots, e_m)$ is an orthonormal basis for V, then, by definition, each of $e_1, \ldots, e_m$ is a unit vector, so $\langle e_i, e_j \rangle = \pm \delta_{ij}$ for $i, j = 1, \ldots, m$, where δ_{ij} is Kronecker's delta. In particular, the standard basis for $\mathbb{R}^m_\nu$ is orthonormal.

Orthonormal bases are computationally convenient. It turns out that a scalar product space always has one, as we now show.

Theorem 4.2.1. *Every nonzero scalar product space contains a unit vector.*

Proof. This follows from Theorem 3.2.4(b) and Theorem 3.3.4. □

Theorem 4.2.2. *If $(V, \mathfrak{g})$ is a scalar product space of dimension m and $1 \leq k \leq m - 1$ is an integer, then any orthonormal k-tuple of vectors in V can be extended to an orthonormal $(k + 1)$-tuple of vectors in V.*

Proof. Let $\mathcal{E} = (e_1, \ldots, e_k)$ be an orthonormal k-tuple of vectors in V. Then $\mathcal{E}$ is an orthonormal basis for the subspace $U = \mathbb{R}e_1 + \cdots + \mathbb{R}e_k$ of V. By Theorem 1.1.16, $U = \mathbb{R}e_1 \oplus \cdots \oplus \mathbb{R}e_k$. Since $(\mathfrak{g}|_U)_\mathcal{E}$, the matrix of $\mathfrak{g}|_U$ with respect to $\mathcal{E}$, is a diagonal matrix with each diagonal entry either 1 or -1, $(\mathfrak{g}|_U)_\mathcal{E}$ is invertible. It follows from Theorem 3.3.3 that $\mathfrak{g}$ is nondegenerate on U, and then from Theorem 4.1.3 that $\mathfrak{g}$ is nondegenerate on $U^\perp$. Thus, $(U^\perp, \mathfrak{g}|_{U^\perp})$ is a nonzero scalar product space. By Theorem 4.2.1, $U^\perp$ contains a unit vector e_{k+1}. Then $(e_1, \ldots, e_{k+1})$ is the desired orthonormal $(k + 1)$-tuple of vectors in V. □

Theorem 4.2.3. *If $(V, \mathfrak{g})$ is a scalar product space and $(e_1, \ldots, e_k)$ is an orthonormal k-tuple of vectors in V, then $e_1, \ldots, e_k$ are linearly independent.*

Proof. If $\sum_i a_i e_i = 0$, then

$$0 = \left\langle e_i, \sum_j a_j e_j \right\rangle = \sum_j a_j \langle e_i, e_j \rangle = a_i \langle e_i, e_i \rangle = \pm a_i,$$

hence $a_i = 0$ for $i = 1, \ldots, m$. □

Theorem 4.2.4. *Every nonzero scalar product space has an orthonormal basis.*

Proof. Let $(V, \mathfrak{g})$ be a scalar product space of dimension $m \geq 1$. By Theorem 4.2.1, V contains a unit vector. Applying Theorem 4.2.2 inductively, we construct an orthonormal m-tuple of vectors in V. It follows from Theorem 4.2.3 that this is an orthonormal basis for V.   □

It is convenient to restrict attention to orthonormal bases satisfying a certain condition.

Throughout, all orthonormal bases $(e_1, \ldots, e_m)$ are assumed ordered such that any $+1$s among $\langle e_1, e_1 \rangle, \ldots, \langle e_m, e_m \rangle$ precede any -1s.

To illustrate, let $\mathcal{E} = (e_1, \ldots, e_{m-2}, e_{m-1}, e_m)$ be the standard basis for $\mathbb{R}_1^m$, and let $\mathcal{F} = (e_1, \ldots, e_{m-2}, e_m, e_{m-1})$. Then $\mathcal{E}$ and $\mathcal{F}$ are both orthonormal bases for $\mathbb{R}_1^m$. Since $\langle e_1, e_1 \rangle = \cdots = \langle e_{m-1}, e_{m-1} \rangle = 1$ and $\langle e_m, e_m \rangle = -1$, we see that $\mathcal{E}$ satisfies the preceding convention but $\mathcal{F}$ does not.

Let $(V, \mathfrak{g})$ be a scalar product space, and let $V = V^+ \oplus V^-$ be a direct sum of the type given by Theorem 4.1.6(a). The **index of $\mathfrak{g}$**, also called the **index of V**, is defined by

$$\mathrm{ind}(\mathfrak{g}) = \dim(V^-).$$

We often denote

$$\mathrm{ind}(\mathfrak{g}) \qquad \text{by} \qquad \nu.$$

According to Theorem 4.1.6(b), $\mathrm{ind}(\mathfrak{g})$ is independent of the choice of direct sum. Taking $\mathbb{R}_\nu^m = (\mathbb{R}^m, \mathfrak{s})$ as an example, we see from (4.1.2) that $\mathrm{ind}(\mathfrak{s}) = \nu$.

Let $\mathcal{E} = (e_1, \ldots, e_m)$ be an orthonormal basis for V. The **signature of $\mathfrak{g}$**, also called the **signature of V**, is defined to be the m-tuple

$$(\langle e_1, e_1 \rangle, \ldots, \langle e_m, e_m \rangle).$$

In light of the above convention on orthonormal bases, the signature of $\mathfrak{g}$ has $+1$s in the first $m - \nu$ positions and -1s in the remaining ν positions. Let us denote

$$\langle e_i, e_i \rangle \qquad \text{by} \qquad \varepsilon_i$$

for $i = 1, \ldots, m$. Then

$$\langle e_i, e_j \rangle = \varepsilon_i \delta_{ij} \tag{4.2.1}$$

for $i, j = 1, \ldots, m$, where δ_{ij} is Kronecker's delta. In this notation, the signature of $\mathfrak{g}$ is

$$(\varepsilon_1, \ldots, \varepsilon_m).$$

We note that

$$\mathfrak{g}_{\mathcal{E}} = \mathrm{diag}(\varepsilon_1, \ldots, \varepsilon_m), \tag{4.2.2}$$

hence

$$\det(\mathfrak{g}_{\mathcal{E}}) = \varepsilon_1 \cdots \varepsilon_m = (-1)^\nu. \tag{4.2.3}$$

Here is generalization of the above identities.

Theorem 4.2.5. *Let $(V, \mathfrak{g})$ be a scalar product space with signature $(\varepsilon_1, \ldots, \varepsilon_m)$ and index ν. Let $\mathcal{E}$ and $\mathcal{H}$ be bases for V, with $\mathcal{E}$ orthonormal, and let $A : V \longrightarrow V$ be the linear isomorphism defined by $A(\mathcal{E}) = \mathcal{H}$. Then*

$$\det(\mathfrak{g}_{\mathcal{H}}) = \varepsilon_1 \cdots \varepsilon_m \det(A)^2 = (-1)^\nu \det(A)^2.$$

Proof. We have

$$\mathfrak{g}_{\mathcal{H}} = \left([\mathrm{id}_V]_{A(\mathcal{E})}^{\mathcal{E}} \right)^{\mathrm{T}} \mathfrak{g}_{\mathcal{E}} \, [\mathrm{id}_V]_{A(\mathcal{E})}^{\mathcal{E}} \qquad \text{[Th 3.1.3]}$$

$$= \left([A]_{\mathcal{E}}^{\mathcal{E}} \right)^{\mathrm{T}} \mathfrak{g}_{\mathcal{E}} \, [A]_{\mathcal{E}}^{\mathcal{E}}, \qquad \text{[Th 2.2.5(a)]}$$

hence $\det(\mathfrak{g}_{\mathcal{H}}) = \det(\mathfrak{g}_{\mathcal{E}}) \det(A)^2$. The result now follows from (4.2.3). $\qquad \square$

Theorem 4.2.6. *Let $(V, \mathfrak{g})$ be a scalar product space, and let U be a subspace of V. If $\mathfrak{g}$ is nondegenerate on U, then*

$$\mathrm{ind}(\mathfrak{g}) = \mathrm{ind}(\mathfrak{g}|_U) + \mathrm{ind}(\mathfrak{g}|_{U^\perp}).$$

Remark. By Theorem 4.1.3, $\mathfrak{g}$ is nondegenerate on $U^\perp$. Thus, $(U, \mathfrak{g}|_U)$ and $(U^\perp, \mathfrak{g}|_{U^\perp})$ are both scalar product spaces, so the assertion makes sense.

Proof. Let $\dim(V) = m$ and $\dim(U) = k$. We have from Theorem 4.1.3 that $V = U \oplus U^\perp$, and then from Theorem 1.1.18 that $\dim(U^\perp) = m - k$. Let $(f_1, \ldots, f_k)$ and $(f_{k+1}, \ldots, f_m)$ be orthonormal bases for U and $U^\perp$, respectively, and let $e_1, \ldots, e_m$ be a reordering of $f_1, \ldots, f_m$ such that $(e_1, \ldots, e_m)$ is an orthonormal basis for V satisfying the above convention on the ordering of signatures. Evidently, the number of -1s in $(e_1, \ldots, e_m)$ equals the number of -1s in $(f_1, \ldots, f_k)$ plus the number of -1s in $(f_{k+1}, \ldots, f_m)$. The result follows. $\qquad \square$

Theorem 4.2.7. *Let $(V, \mathfrak{g})$ be a scalar product space with signature $(\varepsilon_1, \ldots, \varepsilon_m)$, let $(e_1, \ldots, e_m)$ be an orthonormal basis for V, and let v be a vector in V. Then*

$$v = \sum_i \varepsilon_i \langle v, e_i \rangle e_i.$$

Proof. Let $v = \sum_j a^j e_j$. Then

$$\langle v, e_i \rangle = \left\langle \sum_j a^j e_j, e_i \right\rangle = \sum_j a^j \langle e_j, e_i \rangle = \varepsilon_i a^i,$$

hence $a^i = \varepsilon_i \langle v, e_i \rangle$ for $i = 1, \ldots, m$. $\qquad \square$

Theorem 4.2.8. *Let $(V, \mathfrak{g})$ be a scalar product space with signature $(\varepsilon_1, \ldots, \varepsilon_m)$, let $(e_1, \ldots, e_m)$ be an orthonormal basis for V, and let v, w be vectors in V, with $v = \sum_i a^i e_i$ and $w = \sum_j b^j e_j$. Then*

$$\langle v, w \rangle = \sum_i \varepsilon_i a^i b^i.$$

Proof. We have

$$\langle v, w \rangle = \left\langle \sum_i a^i e_i, \sum_j b^j e_j \right\rangle = \sum_{ij} a^i b^j \langle e_i, e_j \rangle = \sum_i \varepsilon_i a^i b^i. \qquad \square$$

4.3 Adjoints

We previously defined what it means for a matrix to be symmetric. In this section, we introduce a type of linear map that under certain circumstances has a related symmetry property (see Theorem 4.6.10).

Let $(V, \mathfrak{g})$ and $(W, \mathfrak{h})$ be scalar product spaces, and let $A : V \longrightarrow W$ be a linear map. We define a map

$$A^\dagger : W \longrightarrow V,$$

called the **adjoint of A**, by

$$A^\dagger = \mathsf{S} \circ A^* \circ \mathfrak{F}, \qquad (4.3.1)$$

where S and $\mathfrak{F}$ are the sharp map and flat map corresponding to $\mathfrak{g}$ and $\mathfrak{h}$, respectively, and A^* is the pullback by A. Let us denote

$$(A^\dagger)^\dagger \qquad \text{by} \qquad A^{\dagger\dagger}.$$

Theorem 4.3.1. *With the above setup:*
(a) $A^\dagger$ is the unique linear map such that

$$\langle A(v), w \rangle = \langle v, A^\dagger(w) \rangle$$

for all vectors v in V and w in W.
(b) If $V = W$, then $\det(A^\dagger) = \det(A)$.

Proof. (a): We have

$$A^\dagger = \mathsf{S} \circ A^* \circ \mathfrak{F}$$
$$\Leftrightarrow \quad A^* \circ \mathfrak{F} = \mathsf{F} \circ A^\dagger$$
$$\Leftrightarrow \quad (A^* \circ \mathfrak{F})(w) = (\mathsf{F} \circ A^\dagger)(w) \text{ for all } w \text{ in } W$$
$$\Leftrightarrow \quad A^*(\mathfrak{F}(w))(v) = \mathsf{F}(A^\dagger(w))(v) \text{ for all } v \text{ in } V \text{ and } w \text{ in } W$$
$$\Leftrightarrow \quad \mathfrak{F}(w)(A(v)) = \mathsf{F}(A^\dagger(w))(v) \text{ for all } v \text{ in } V \text{ and } w \text{ in } W$$
$$\Leftrightarrow \quad \langle w, A(v) \rangle = \langle A^\dagger(w), v \rangle \text{ for all } v \text{ in } V \text{ and } w \text{ in } W$$
$$\Leftrightarrow \quad \langle A(v), w \rangle = \langle v, A^\dagger(w) \rangle \text{ for all } v \text{ in } V \text{ and } w \text{ in } W.$$

This shows that $A^\dagger$ has the desired property, and at the same time that it is uniquely determined by this property.

(b): With $\mathfrak{F} = \mathsf{F}$, we have $\mathsf{S} = \mathsf{F}^{-1}$, so the result follows from Theorem 2.4.5 and (4.3.1). $\qquad \square$

Adjoints behave well with respect to basic algebraic structure.

Theorem 4.3.2. *If $(V, \mathfrak{g})$ is a scalar product space and $A, B : V \longrightarrow V$ are linear maps, then:*
(a) $(A + B)^\dagger = A^\dagger + B^\dagger$.
(b) $(A \circ B)^\dagger = B^\dagger \circ A^\dagger$.
(c) $A^{\dagger\dagger} = A$.
(d) *If A is a linear isomorphism, then $(A^{-1})^\dagger = (A^\dagger)^{-1}$.*
(e) *A is a linear isomorphism if and only if $A^\dagger$ is a linear isomorphism.*

Proof. We have $A^\dagger = \mathsf{S} \circ A^* \circ \mathsf{F}$ and $B^\dagger = \mathsf{S} \circ B^* \circ \mathsf{F}$.
   (a): By Theorem 1.3.1(a),

$$(A + B)^\dagger = \mathsf{S} \circ (A + B)^* \circ \mathsf{F} = \mathsf{S} \circ (A^* + B^*) \circ \mathsf{F}$$
$$= (\mathsf{S} \circ A^* \circ \mathsf{F}) + (\mathsf{S} \circ B^* \circ \mathsf{F}) = A^\dagger + B^\dagger.$$

   (b): By Theorem 1.3.1(b),

$$(A \circ B)^\dagger = \mathsf{S} \circ (A \circ B)^* \circ \mathsf{F} = \mathsf{S} \circ (B^* \circ A^*) \circ \mathsf{F}$$
$$= (\mathsf{S} \circ B^* \circ \mathsf{F}) \circ (\mathsf{S} \circ A^* \circ \mathsf{F}) = B^\dagger \circ A^\dagger.$$

   (c): By Theorem 1.3.1(c) and Theorem 3.3.5,

$$A^{\dagger\dagger} = \mathsf{S} \circ (A^\dagger)^* \circ \mathsf{F} = \mathsf{S} \circ (\mathsf{S} \circ A^* \circ \mathsf{F})^* \circ \mathsf{F}$$
$$= \mathsf{S} \circ (\mathsf{F}^* \circ A^{**} \circ \mathsf{S}^*) \circ \mathsf{F} = A.$$

   (d): By Theorem 1.3.1(d),

$$(A^{-1})^\dagger = \mathsf{S} \circ (A^{-1})^* \circ \mathsf{F} = \mathsf{F}^{-1} \circ (A^*)^{-1} \circ \mathsf{S}^{-1}$$
$$= (\mathsf{S} \circ A^* \circ \mathsf{F})^{-1} = (A^\dagger)^{-1}.$$

   (e): Since $A^\dagger = \mathsf{S} \circ A^* \circ \mathsf{F}$, and F and S are isomorphisms, the result follows.
                                                                                                         $\square$

For a scalar product space $(V, \mathfrak{g})$, we say that a linear map $A : V \longrightarrow V$ is **self-adjoint** if $A = A^\dagger$. In view of Theorem 4.3.1(a), this condition is equivalent to

$$\langle A(v), w \rangle = \langle v, A(w) \rangle$$

for all vectors v, w in V.

Theorem 4.3.3. *Let $(V, \mathfrak{g})$ be a scalar product space, let $A : V \longrightarrow V$ be a linear map, and let $\mathcal{H}$ be a basis for V. Then*

$$[A^\dagger]_{\mathcal{H}}^{\mathcal{H}} = \mathfrak{g}_{\mathcal{H}}^{-1} \left([A]_{\mathcal{H}}^{\mathcal{H}} \right)^{\mathrm{T}} \mathfrak{g}_{\mathcal{H}}.$$

Remark. It follows from Theorem 3.3.3 that the inverse of $\mathfrak{g}_{\mathcal{H}}$ exists, so the assertion makes sense.

Proof. Let $\mathcal{H} = (h_1, \ldots, h_m)$, and let $[A] = [A]_{\mathcal{H}}^{\mathcal{H}} = [a_j^i]$ and $[A^\dagger] = [A^\dagger]_{\mathcal{H}}^{\mathcal{H}} = [b_j^i]$. We have

$$[\langle A(h_i), h_j \rangle] = [\langle h_i, A^\dagger(h_j) \rangle].$$

By Theorem 2.1.6(b), (2.2.2), and (2.2.3),

$$\langle A(h_i), h_j \rangle = \left\langle \sum_k a_i^k h_k, h_j \right\rangle = \sum_k a_i^k \langle h_k, h_j \rangle = \sum_k a_i^k \mathfrak{g}_{kj}$$

$$= ([A]_{(i)})^{\mathrm{T}} (\mathfrak{g}_{\mathcal{H}})_{(j)} = ([A]^{\mathrm{T}})^{(i)} (\mathfrak{g}_{\mathcal{H}})_{(j)},$$

hence

$$[\langle A(h_i), h_j \rangle] = [A]^{\mathrm{T}} \mathfrak{g}_{\mathcal{H}}.$$

On the other hand,

$$\langle h_i, A^\dagger(h_j) \rangle = \left\langle h_i, \sum_k b_j^k h_k \right\rangle = \sum_k b_j^k \langle h_i, h_k \rangle = \sum_k \mathfrak{g}_{ik} b_j^k$$

$$= (\mathfrak{g}_{\mathcal{H}})^{(i)} [A^\dagger]_{(j)},$$

so

$$[\langle h_i, A^\dagger(h_j) \rangle] = \mathfrak{g}_{\mathcal{H}} [A^\dagger].$$

Combining the above identities gives $[A]^{\mathrm{T}} \mathfrak{g}_{\mathcal{H}} = \mathfrak{g}_{\mathcal{H}} [A^\dagger]$, from which the result follows. ◻

Theorem 4.3.4. *Let $(V, \mathfrak{g})$ and $(W, \mathfrak{h})$ be scalar product spaces, and let $A : V \longrightarrow W$ be a linear map. Then:*
(a) $\mathrm{rank}(A^\dagger) = \mathrm{rank}(A)$.
(b) *If* $\dim(V) = \dim(W)$, *then* $\mathrm{null}(A^\dagger) = \mathrm{null}(A)$.
(c) $\ker(A^\dagger) = \mathrm{im}(A)^\perp$.
(d) $\mathrm{im}(A^\dagger) = \ker(A)^\perp$.

Proof. The proof emulates that of Theorem 1.4.3.
  (c): We have

$$w \text{ is in } \ker(A^\dagger)$$
$$\Leftrightarrow \quad A^\dagger(w) = 0$$
$$\Leftrightarrow \quad \langle A^\dagger(w), v \rangle = 0 \text{ for all } v \text{ in } V \qquad [\mathfrak{h} \text{ is nondegenerate}]$$
$$\Leftrightarrow \quad \langle w, A(v) \rangle = 0 \text{ for all } v \text{ in } V \qquad [\text{Th } 4.3.1(a)]$$
$$\Leftrightarrow \quad w \text{ is in } \mathrm{im}(A)^\perp.$$

  (d): It follows from Theorem 4.3.2(c) and part (c) that

$$\ker(A) = \ker(A^{\dagger\dagger}) = \mathrm{im}(A^\dagger)^\perp,$$

and then from Theorem 4.1.2(c) that

$$\ker(A)^\perp = \mathrm{im}(A^\dagger)^{\perp\perp} = \mathrm{im}(A^\dagger).$$

(a): We have

$$
\begin{aligned}
\operatorname{rank}(A^\dagger) &= \dim\bigl(\operatorname{im}(A^\dagger)\bigr) \\
&= \dim\bigl(\ker(A)^\perp\bigr) & &\text{[part (d)]} \\
&= \dim(V) - \dim\bigl(\ker(A)\bigr) & &\text{[Th 4.1.2(b)]} \\
&= \operatorname{rank}(A). & &\text{[Th 1.1.11]}
\end{aligned}
$$

(b): We have

$$
\begin{aligned}
\operatorname{rank}(A^\dagger) + \operatorname{null}(A^\dagger) &= \dim(W) & &\text{[Th 1.1.11]} \\
&= \dim(V) & &\text{[assumption]} \\
&= \operatorname{rank}(A) + \operatorname{null}(A). & &\text{[Th 1.1.11]}
\end{aligned}
$$

The result now follows from part (a). $\square$

Let $(V, \mathfrak{g})$ be a scalar product space, let $A : V \longrightarrow V$ be a linear map, and let U be a subspace of V on which $\mathfrak{g}$ is nondegenerate. The left-hand side of the following table summarizes selected results on annihilators and pullbacks, while the right-hand side of the table does the same for perps and adjoints. The correspondence between annihilators and perps on the one hand, and pullbacks and adjoints on the other, is evident and demonstrates the similar roles played by U^0 and $U^\perp$.

Annihilators and pullbacks	Perps and adjoints
$\dim(V) = \dim(U) + \dim(U^0)$	$\dim(V) = \dim(U) + \dim(U^\perp)$
$U^{00} = U$	$U^{\perp\perp} = U$
$A^{**} = A$	$A^{\dagger\dagger} = A$
$\operatorname{rank}(A^*) = \operatorname{rank}(A)$	$\operatorname{rank}(A^\dagger) = \operatorname{rank}(A)$
$\operatorname{null}(A^*) = \operatorname{null}(A)$	$\operatorname{null}(A^\dagger) = \operatorname{null}(A)$
$\ker(A^*) = \operatorname{im}(A)^0$	$\ker(A^\dagger) = \operatorname{im}(A)^\perp$
$\operatorname{im}(A^*) = \ker(A)^0$	$\operatorname{im}(A^\dagger) = \ker(A)^\perp$

4.4   Linear Isometries

A linear map is one that respects the linear structure of a vector space. In a scalar product space, there is the additional feature of a scalar product. This section introduces a type of linear map that also respects scalar products.

Let $(V, \mathfrak{g})$ and $(W, \mathfrak{h})$ be scalar product spaces, and let $A : V \longrightarrow W$ be a linear map. We say that A is a **linear isometry** (or **orthogonal transformation**) if

$$
\langle A(v), A(w) \rangle = \langle v, w \rangle
$$

for all vectors v, w in V.

Theorem 4.4.1. *If $(V, \mathfrak{g})$ and $(W, \mathfrak{h})$ are scalar product spaces with the same dimension and $A : V \longrightarrow W$ is a linear isometry, then:*
(a) *A is a linear isomorphism.*
(b) *A maps every orthonormal basis for V to an orthonormal basis for W.*

Proof. (a): Let v be a vector in $\ker(A)$. Then $\langle v, w \rangle = \langle A(v), A(w) \rangle = 0$ for all vectors w in V. Since $\mathfrak{g}$ is nondegenerate, $v = 0$. Thus, $\ker(A) = \{0\}$. By Theorem 1.1.12 and Theorem 1.1.14, A is a linear isomorphism.

  (b): Let $\mathcal{E} = (e_1, \ldots, e_m)$ be an orthonormal basis for V. By Theorem 1.1.13(a) and part (a), $A(\mathcal{E})$ is basis for V. Since A is a linear isometry, $\langle A(e_i), A(e_j) \rangle = \langle e_i, e_j \rangle$ for $i, j = 1, \ldots, m$. Thus, $A(\mathcal{E})$ is orthonormal.   □

Theorem 4.4.2. *Let $(V, \mathfrak{g})$ be a scalar product space, and let $A : V \longrightarrow V$ be a linear map. Then the following are equivalent:*
(a) *A is a linear isometry.*
(b) *A is a linear isomorphism and $A^\dagger = A^{-1}$.*
(c) *For some (hence every) basis $\mathcal{H}$ for V,*

$$\mathfrak{g}_{\mathcal{H}} = \left([A]_{\mathcal{H}}^{\mathcal{H}} \right)^{\mathrm{T}} \mathfrak{g}_{\mathcal{H}} \, [A]_{\mathcal{H}}^{\mathcal{H}} .$$

Proof. (a) $\Rightarrow$ (b): By Theorem 4.4.1(a), A is a linear isomorphism. Let v, w be vectors in V. Then

$$\langle v, w \rangle = \langle A(v), A(w) \rangle = \langle A^\dagger(A(v)), w \rangle,$$

hence

$$0 = \langle A^\dagger(A(v)) - v, w \rangle = \langle ((A^\dagger \circ A) - \mathrm{id}_V)(v), w \rangle.$$

Since $\mathfrak{g}$ is nondegenerate and w was arbitrary, $((A^\dagger \circ A) - \mathrm{id}_V)(v) = 0$ for all v in V. That is, $(A^\dagger \circ A) - \mathrm{id}_V = 0$, hence $A^\dagger \circ A = \mathrm{id}_V$, so $A^\dagger = A^{-1}$.
  (b) $\Rightarrow$ (c): Since $A^\dagger = A^{-1}$, by Theorem 2.2.5(b),

$$[A^\dagger]_{\mathcal{H}}^{\mathcal{H}} = [A^{-1}]_{\mathcal{H}}^{\mathcal{H}} = \left([A]_{\mathcal{H}}^{\mathcal{H}} \right)^{-1} .$$

The result now follows from Theorem 4.3.3.
  (c) $\Rightarrow$ (a): For vectors v, w in V, we have

$$
\begin{aligned}
\langle A(v), A(w) \rangle &= \left([A(v)]_{\mathcal{H}} \right)^{\mathrm{T}} \mathfrak{g}_{\mathcal{H}} \, [A(w)]_{\mathcal{H}} && \text{[Th 3.1.2(a)]} \\
&= \left([A]_{\mathcal{H}}^{\mathcal{H}} [v]_{\mathcal{H}} \right)^{\mathrm{T}} \mathfrak{g}_{\mathcal{H}} \, [A]_{\mathcal{H}}^{\mathcal{H}} [w]_{\mathcal{H}} && \text{[Th 2.2.4]} \\
&= ([v]_{\mathcal{H}})^{\mathrm{T}} \left(\left([A]_{\mathcal{H}}^{\mathcal{H}} \right)^{\mathrm{T}} \mathfrak{g}_{\mathcal{H}} \, [A]_{\mathcal{H}}^{\mathcal{H}} \right) [w]_{\mathcal{H}} && \\
&= ([v]_{\mathcal{H}})^{\mathrm{T}} \mathfrak{g}_{\mathcal{H}} \, [w]_{\mathcal{H}} && \\
&= \langle v, w \rangle. && \text{[Th 3.1.2(a)]}   □
\end{aligned}
$$

Theorem 4.4.3. *If $(V, \mathfrak{g})$ is a scalar product space and $A : V \longrightarrow V$ is a linear isometry, then $\det(A) = \pm 1$.*

Proof. Let $\mathcal{H}$ be a basis for V. We have from Theorem 4.4.2 that $\det(\mathfrak{g}_{\mathcal{H}}) = \det(\mathfrak{g}_{\mathcal{H}}) \det(A)^2$, and from Theorem 3.3.3 that $\det(\mathfrak{g}_{\mathcal{H}}) \neq 0$. Thus, $\det(A)^2 = 1$, from which the result follows.   □

Linear isometries are of special interest because they "preserve norms", as the next result shows.

Theorem 4.4.4. *Let $(V, \mathfrak{g})$ be a scalar product space, and let $A : V \longrightarrow V$ be a linear map. Then:*
(a) *If A is a linear isometry, then $\|A(v)\| = \|v\|$ for all vectors v in V.*
(b) *If $\|A(v)\| = \|v\|$ for all vectors v in V, and A maps spacelike (resp., timelike, lightlike) vectors to spacelike (resp., timelike, lightlike) vectors, then A is a linear isometry.*

Proof. (a): We have

$$\|A(v)\| = \sqrt{|\langle A(v), A(v)\rangle|} = \sqrt{|\langle v, v\rangle|} = \|v\|.$$

(b): Since $\|A(v)\| = \|v\|$ is equivalent to $|\langle A(v), A(v)\rangle| = |\langle v, v\rangle|$, the assumption regarding the way A maps vectors yields $\langle A(v), A(v)\rangle = \langle v, v\rangle$.   □

Theorem 4.4.5. *Let $(V, \mathfrak{g})$ be a scalar product space, and let $A : V \longrightarrow V$ be a linear map. Then A is a linear isometry if and only if $A^\dagger$ is a linear isometry.*

Proof. ($\Rightarrow$): By Theorem 4.4.2, A is a linear isomorphism and $A^\dagger = A^{-1}$. For vectors v, w in V, we then have

$$\begin{aligned}
\langle A^\dagger(v), A^\dagger(w)\rangle &= \langle v, A^{\dagger\dagger}(A^\dagger(w))\rangle && \text{[Th 4.3.1(a)]} \\
&= \langle v, A \circ A^\dagger(w)\rangle && \text{[Th 4.3.2(c)]} \\
&= \langle v, w\rangle.
\end{aligned}$$

($\Leftarrow$): This follows from Theorem 4.3.2(c) and ($\Rightarrow$).   □

Let $(V, \mathfrak{g})$ be a scalar product space of dimension m and index ν, let $\mathcal{E}$ be an orthonormal basis for V, and let $A : V \longrightarrow V$ be a linear isometry. To simplify notation, we denote $[A]_{\mathcal{E}}^{\mathcal{E}}$ by $[A]$ and adopt similar abbreviations for other matrices. Consider the partition of $[A]$ given by

$$[A] = \begin{bmatrix} [A]_1^1 & [A]_2^1 \\ [A]_1^2 & [A]_2^2 \end{bmatrix}, \tag{4.4.1}$$

where the submatrices have the following dimensions:

$$\begin{array}{ll}
[A]_1^1 & (m - \nu) \times (m - \nu) \\
[A]_2^1 & (m - \nu) \times \nu \\
[A]_1^2 & \nu \times (m - \nu) \\
[A]_2^2 & \nu \times \nu.
\end{array}$$

Since $\mathfrak{g}_\mathcal{E} = \mathfrak{g}_\mathcal{E}^{-1}$, we have from Theorem 4.4.2 that

$$[A]^{-1} = \mathfrak{g}_\mathcal{E} \, [A]^\mathrm{T} \, \mathfrak{g}_\mathcal{E},$$

hence

$$[A]^{-1} = \begin{bmatrix} ([A]_1^1)^\mathrm{T} & -([A]_1^2)^\mathrm{T} \\ -([A]_2^1)^\mathrm{T} & ([A]_2^2)^\mathrm{T} \end{bmatrix}. \tag{4.4.2}$$

Theorem 4.4.6. *With the above setup:*
(a)
$$[A]_1^1([A]_1^1)^\mathrm{T} = I_{m-\nu} + [A]_2^1([A]_2^1)^\mathrm{T}.$$

(b)
$$[A]_1^1([A]_1^2)^\mathrm{T} = [A]_2^1([A]_2^2)^\mathrm{T}.$$

(c)
$$[A]_2^2([A]_2^2)^\mathrm{T} = I_\nu + [A]_1^2([A]_1^2)^\mathrm{T}.$$

(d)
$$([A]_1^1)^\mathrm{T}[A]_1^1 = I_{m-\nu} + ([A]_1^2)^\mathrm{T}[A]_1^2.$$

(e)
$$([A]_1^1)^\mathrm{T}[A]_2^1 = ([A]_1^2)^\mathrm{T}[A]_2^2.$$

(f)
$$([A]_2^1)^\mathrm{T}[A]_2^2 = I_\nu + ([A]_2^1)^\mathrm{T}[A]_2^1.$$

Proof. Since $[A][A]^{-1} = I_m$, we have

$$\begin{bmatrix} [A]_1^1 & [A]_2^1 \\ [A]_1^2 & [A]_2^2 \end{bmatrix} \begin{bmatrix} ([A]_1^1)^\mathrm{T} & -([A]_1^2)^\mathrm{T} \\ -([A]_2^1)^\mathrm{T} & ([A]_2^2)^\mathrm{T} \end{bmatrix} = \begin{bmatrix} I_{m-\nu} & O_{(m-\nu)\times\nu} \\ O_{\nu\times(m-\nu)} & I_\nu \end{bmatrix},$$

hence

$$[A]_1^1([A]_1^1)^\mathrm{T} - [A]_2^1([A]_2^1)^\mathrm{T} = I_{m-\nu}$$

$$-[A]_1^1([A]_1^2)^\mathrm{T} + [A]_2^1([A]_2^2)^\mathrm{T} = O_{(m-\nu)\times\nu}$$

$$[A]_1^2([A]_1^1)^\mathrm{T} - [A]_2^2([A]_2^1)^\mathrm{T} = O_{\nu\times(m-\nu)} \tag{4.4.3}$$

$$-[A]_1^2([A]_1^2)^\mathrm{T} + [A]_2^2([A]_2^2)^\mathrm{T} = I_\nu.$$

Since $[A]^{-1}[A] = I_m$, we have

$$\begin{bmatrix} ([A]_1^1)^{\mathrm{T}} & -([A]_1^2)^{\mathrm{T}} \\ -([A]_2^1)^{\mathrm{T}} & ([A]_2^2)^{\mathrm{T}} \end{bmatrix} \begin{bmatrix} [A]_1^1 & [A]_2^1 \\ [A]_1^2 & [A]_2^2 \end{bmatrix} = \begin{bmatrix} I_{m-\nu} & O_{(m-\nu)\times\nu} \\ O_{\nu\times(m-\nu)} & I_\nu \end{bmatrix},$$

hence

$$([A]_1^1)^{\mathrm{T}} [A]_1^1 - ([A]_1^2)^{\mathrm{T}} [A]_1^2 = I_{m-\nu}$$

$$([A]_1^1)^{\mathrm{T}} [A]_2^1 - ([A]_1^2)^{\mathrm{T}} [A]_2^2 = O_{(m-\nu)\times\nu}$$

(4.4.4)

$$-([A]_2^1)^{\mathrm{T}} [A]_1^1 + ([A]_2^2)^{\mathrm{T}} [A]_1^2 = O_{\nu\times(m-\nu)}$$

$$-([A]_2^1)^{\mathrm{T}} [A]_2^1 + ([A]_2^2)^{\mathrm{T}} [A]_2^2 = I_\nu.$$

(a): This follows from the first identity of (4.4.3).
(b): This follows from either the second or third identity of (4.4.3).
(c): This follows from the fourth identity of (4.4.3).
(d): This follows from the first identity of (4.4.4).
(e): This follows from either the second or third identity of (4.4.4).
(f): This follows from the fourth identity of (4.4.4).   □

4.5   Dual Scalar Product Spaces

In this section, we show that a scalar product on a vector space induces a corresponding scalar product on its dual space.

Let $(V, \mathfrak{g})$ be a scalar product space, let V^* be its dual space, and let F and S be the flat map and sharp map corresponding to $\mathfrak{g}$. We define a function

$$\mathfrak{g}^* = \langle\cdot,\cdot\rangle^* : V^* \times V^* \longrightarrow \mathbb{R}$$

by

$$\langle\eta, \zeta\rangle^* = \langle\eta^{\mathsf{S}}, \zeta^{\mathsf{S}}\rangle \tag{4.5.1}$$

for all covectors η, ζ in V^*. According to (3.3.1), $\langle\eta^{\mathsf{S}}, \zeta^{\mathsf{S}}\rangle = (\eta^{\mathsf{S}})^{\mathsf{F}}(\zeta^{\mathsf{S}}) = \eta(\zeta^{\mathsf{S}})$, hence

$$\langle\eta, \zeta\rangle^* = \eta(\zeta^{\mathsf{S}}). \tag{4.5.2}$$

Theorem 4.5.1. *With the above setup, $(V^*, \mathfrak{g}^*)$ is a scalar product space, called the **dual scalar product space** corresponding to $(V, \mathfrak{g})$.*

Proof. Clearly, $\mathfrak{g}^*$ is bilinear and symmetric. Suppose η is a covector in V^* such that $\langle\eta, \zeta\rangle^* = 0$, that is, $\langle\eta^{\mathsf{S}}, \zeta^{\mathsf{S}}\rangle = 0$, for all covectors ζ in V^*. Since S is surjective, as ζ varies over V^*, ζ^{S} varies over V. Because $\mathfrak{g}$ is nondegenerate, $\eta^{\mathsf{S}} = 0$; and since S is injective, by Theorem 1.1.12, $\eta = 0$. Thus, $\mathfrak{g}^*$ is nondegenerate.   □

Let $\mathcal{F}$ and $\mathcal{S}$ be the flat map and sharp map corresponding to $\mathfrak{g}^*$. We have from the identification $V^{**} = V$ that

$$\mathcal{F} : V^* \longrightarrow V \qquad \text{and} \qquad \mathcal{S} : V \longrightarrow V^*.$$

Perhaps the more obvious choice of notation would be F^* instead of $\mathcal{F}$, and S^* instead of $\mathcal{S}$. However, looking back at Theorem 3.3.5, this would confuse what we now denote by $\mathcal{F}$ and $\mathcal{S}$ with the pullback by F and the pullback by S.

Theorem 4.5.2. *If $(V, \mathfrak{g})$ is a scalar product space, and F and S are the flat map and sharp map corresponding to $\mathfrak{g}$, then:*
(a) $\mathcal{F} = \mathsf{S}$.
(b) $\mathcal{S} = \mathsf{F}$.

Proof. (a): For covectors η, ζ in V^*, we have

$$\begin{aligned}
\eta^{\mathcal{F}}(\zeta) &= \langle \eta, \zeta \rangle^* & [(3.3.1)] \\
&= \langle \eta^{\mathsf{S}}, \zeta^{\mathsf{S}} \rangle & [(4.5.1)] \\
&= \langle \zeta^{\mathsf{S}}, \eta^{\mathsf{S}} \rangle & \\
&= (\zeta^{\mathsf{S}})^{\mathsf{F}}(\eta^{\mathsf{S}}) & [(3.3.1)] \\
&= \zeta(\eta^{\mathsf{S}}) & \\
&= \eta^{\mathsf{S}}(\zeta). & [(1.2.3)]
\end{aligned}$$

Since η and ζ were arbitrary, $\mathcal{F} = \mathsf{S}$.
   (b): We have

$$\begin{aligned}
\mathcal{S} &= \mathcal{F}^{-1} & [(3.3.5)] \\
&= \mathsf{S}^{-1} & [\text{part (a)}] \\
&= \mathsf{F}. & [(3.3.5)]
\end{aligned}$$ $\qquad\square$

Theorem 4.5.3. *Let $(V, \mathfrak{g})$ be a scalar product space, let $\mathcal{H} = (h_1, \ldots, h_m)$ be a basis for V, let $\Theta = (\theta^1, \ldots, \theta^m)$ be its dual basis, and let $\mathfrak{g}_{\mathcal{H}}^{-1} = [\mathfrak{g}^{ij}]$. Then:*
(a)

$$\theta^{j\mathsf{S}} = \sum_i \mathfrak{g}^{ij} h_i$$

   for $i = 1, \ldots, m$.
(b)

$$[\mathsf{S}]_{\Theta}^{\mathcal{H}} = \mathfrak{g}_{\Theta}^* = \mathfrak{g}_{\mathcal{H}}^{-1}.$$

Proof. (b): Since a scalar product is symmetric, the same is true of its matrix with respect to a given basis. Accordingly, the transpose in Theorem 3.3.1(b) can be dropped. We then have

$$\mathfrak{g}_\Theta^* = [\mathcal{F}]_\Theta^{\mathcal{H}} \qquad \text{[Th 3.3.1(b)]}$$

$$= [\mathsf{S}]_\Theta^{\mathcal{H}} \qquad \text{[Th 4.5.2(a)]}$$

$$= [\mathsf{F}^{-1}]_\Theta^{\mathcal{H}}$$

$$= \left([\mathsf{F}]_{\mathcal{H}}^\Theta\right)^{-1} \qquad \text{[Th 2.2.5(b)]}$$

$$= \mathfrak{g}_{\mathcal{H}}^{-1}. \qquad \text{[Th 3.3.1(b)]}$$

(a): This follows from (2.2.2), (2.2.3), and part (b). $\qquad\qquad\square$

Theorem 4.5.4 (Raising an Index). *Let $(V, \mathfrak{g})$ be a scalar product space, let $\mathcal{H} = (h_1, \ldots, h_m)$ be a basis for V, and let $\Theta = (\theta^1, \ldots, \theta^m)$ be its dual basis. Let $\mathfrak{g}_{\mathcal{H}}^{-1} = [\mathfrak{g}^{ij}]$, and let η be a covector in V^*, with*

$$\eta = \sum_i a_i \theta^i. \tag{4.5.3}$$

Then

$$\eta^\mathsf{S} = \sum_i a^i h_i,$$

where

$$a^i = \sum_j \mathfrak{g}^{ij} a_j \tag{4.5.4}$$

for $i = 1, \ldots, m$. Expressed more succinctly:

$$[\eta^\mathsf{S}]_{\mathcal{H}} = \mathfrak{g}_{\mathcal{H}}^{-1} [\eta]_\Theta.$$

Proof. This follows from Theorem 3.3.2 and Theorem 4.5.2(a), but it is instructive to work through the details separately. Let ζ be a covector in V^*, and observe that $\mathfrak{g}_\Theta^* = [\langle \theta^i, \theta^j \rangle^*]$. Then

$$\eta^\mathsf{S}(\zeta) = \eta^\mathcal{F}(\zeta) \qquad \text{[Th 4.5.2(a)]}$$

$$= \langle \eta, \zeta \rangle^* \qquad \text{[(3.3.1)]}$$

$$= \langle \zeta, \eta \rangle^*$$

$$= \left\langle \sum_i \zeta(h_i)\theta^i, \sum_j a_j\theta^j \right\rangle^* \qquad \text{[Th 1.2.1(e)]}$$

$$= \sum_{ij} \zeta(h_i)a_j \langle \theta^i, \theta^j \rangle^*$$

$$= \sum_{ij} \mathfrak{g}^{ij} a_j h_i(\zeta) \qquad \text{[(1.2.3), Th 4.5.3(b)]}$$

$$= \left(\sum_i \left[\sum_j \mathfrak{g}^{ij} a_j\right] h_i\right)(\zeta)$$

$$= \left(\sum_i a^i h_i\right)(\zeta).$$

Since ζ was arbitrary, $\eta^S = \sum_i a^i h_i$. To prove the succinct version directly:

$$\left[\eta^S\right]_{\mathcal{H}} = \left[S(\eta)\right]_{\mathcal{H}}$$
$$= \left[S\right]_{\Theta}^{\mathcal{H}} \left[\eta\right]_{\Theta} \qquad \text{[Th 2.2.4]}$$
$$= \mathfrak{g}_{\mathcal{H}}^{-1} \left[\eta\right]_{\Theta}. \qquad \text{[Th 4.5.3]} \qquad \square$$

We see from (4.5.3) and (4.5.4) why taking the sharp of η to obtain η^S is classically referred to as **raising an index** by $\mathfrak{g}$. The flat map F and sharp map S therefore lower and raise indices, respectively. In the mathematical literature, these maps are colorfully referred to as the **musical isomorphisms**.

A scalar product space and its dual scalar product space are isomorphic under the flat map and sharp map. The remaining results of this section show that there is also a close connection between other familiar structures.

Theorem 4.5.5. *Let $(V, \mathfrak{g})$ be a scalar product space with signature $(\varepsilon_1, \ldots, \varepsilon_m)$, let $(e_1, \ldots, e_m)$ be an orthonormal basis for V, and let $(\xi^1, \ldots, \xi^m)$ be its dual basis. Then, for $i = 1, \ldots, m$:*
(a) $e_i^F = \varepsilon_i \xi^i$.
(b) $\xi^{iS} = \varepsilon_i e_i$.

Proof. (a): This follows from Theorem 3.3.1(a).
   (b): This follows from Theorem 4.5.3(a) or from part (a). $\square$

Theorem 4.5.6. *Let $(V, \mathfrak{g})$ be a scalar product space, let $\mathcal{E}$ be an orthonormal basis for V, and let Ξ be its dual basis. Then:*
(a) *Ξ is an orthonormal basis for $(V^*, \mathfrak{g}^*)$.*
(b) *$(V, \mathfrak{g})$ and $(V^*, \mathfrak{g}^*)$ have the same signature.*

Proof. Let V have signature $(\varepsilon_1, \ldots, \varepsilon_m)$, and let $\mathcal{E} = (e_1, \ldots, e_m)$ and $\Xi = (\xi^1, \ldots, \xi^m)$. By (4.5.1) and Theorem 4.5.5(b),

$$\langle \xi^i, \xi^j \rangle^* = \langle \xi^{iS}, \xi^{jS} \rangle = \varepsilon_i \varepsilon_j \langle e_i, e_j \rangle = \begin{cases} \varepsilon_i & \text{if } i = j \\ 0 & \text{if } i \neq j, \end{cases}$$

hence $\langle \xi^i, \xi^j \rangle^* = \langle e_i, e_j \rangle$ for $i, j = 1, \ldots, m$. The result follows. $\square$

4.6   Inner Product Spaces

Having discussed in some detail the features of scalar product spaces, we now specialize to the type of scalar product space that is usually the focus in introductory linear algebra.

Let V be a vector space, and let $\mathfrak{g}$ be a bilinear function on V. We say that $\mathfrak{g}$ is an **inner product** on V, and that the pair $(V, \mathfrak{g})$ is an **inner product space**, if $\mathfrak{g}$ is symmetric and positive definite on V. Recall from Section 3.1 that $\mathfrak{g}$ is symmetric on V if $\langle v, w \rangle = \langle w, v \rangle$ for all vectors v, w in V, and that $\mathfrak{g}$ is positive definite on V if $\langle v, v \rangle > 0$ for all nonzero vectors v in V.

Suppose $(V, \mathfrak{g})$ is in fact an inner product space. If v is a vector in V such that $\langle v, w \rangle = 0$ for all vectors w in V, then $\langle v, v \rangle = 0$, hence $v = 0$. Thus, $\mathfrak{g}$ is nondegenerate on V, demonstrating that $(V, \mathfrak{g})$ is automatically a scalar product space. In the present context, the **norm** corresponding to $\mathfrak{g}$, given by (4.1.1), simplifies to the function

$$\|\cdot\| : V \longrightarrow \mathbb{R}$$

defined by

$$\|v\| = \sqrt{\langle v, v \rangle} = \sqrt{\mathfrak{q}(v)} \tag{4.6.1}$$

for all vectors v in V. Since an inner product space is a type of scalar product space, the definitions of orthogonal, orthogonal k-tuple, orthogonal basis, orthonormal, orthonormal k-tuple, and orthonormal basis given in Section 4.2 also apply here. Let $(e_1, \ldots, e_m)$ be an orthonormal basis for V. Since $\mathfrak{g}$ is an inner product, $\langle e_i, e_i \rangle = 1$ for $i = 1, \ldots, m$. Thus, $\mathrm{ind}(\mathfrak{g}) = 0$ and the signature of $\mathfrak{g}$ is an m-tuple of $+1$s.

The fact that an inner product is positive definite distinguishes it from a scalar product space in ways that are fundamental. For example, in an inner product space, only the zero vector is orthogonal to itself; or equivalently, only the zero vector has zero norm. Perhaps most important of all, in an inner product space, the inner product is positive definite (hence nondegenerate) on every subspace. Consequently, pairing any subspace of an inner product with the corresponding restriction of the inner product yields an inner product space.

Example 4.6.1 ($\mathbb{R}_0^m$). Continuing with Example 4.1.1, let $\nu = 0$. In this case, we denote $\mathfrak{s}$ by $\mathfrak{e}$. Then

$$\mathfrak{e} : \mathbb{R}^m \times \mathbb{R}^m \longrightarrow \mathbb{R}$$

is given by

$$\mathfrak{e}\big((x^1, \ldots, x^m), (y^1, \ldots, y^m)\big) = x^1 y^1 + \cdots + x^m y^m.$$

We refer to

$$\mathbb{R}_0^m = (\mathbb{R}^m, \mathfrak{e})$$

as **Euclidean m-space** and to $\mathfrak{e}$ as the **Euclidean inner product**. It is easily shown that $\mathfrak{e}$ is in fact an inner product on $\mathbb{R}^m$. Following Example 4.1.1, the standard basis for $\mathbb{R}_0^m$ is simply the standard basis for $\mathbb{R}^m$. The norm corresponding to $\mathfrak{e}$ is given by

$$\big\|(x^1, \ldots, x^m)\big\| = \sqrt{(x^1)^2 + \cdots + (x^m)^2}$$

and called the **Euclidean norm**. The reader is likely familiar with $\|\cdot\|$ as a measure of "length" in $\mathbb{R}^m$, a notion to be explored further in Section 9.3.   ◇

Theorem 4.6.2. *Let $(V, \mathfrak{g})$ be a scalar product space of index ν. Then $\mathfrak{g}$ is an inner product if and only if $\nu = 0$.*

Proof. ($\Rightarrow$): This follows from above remarks.

($\Leftarrow$): Let $(e_1, \ldots, e_m)$ be an orthonormal basis for V, and let v be a vector in V, with $v = \sum_i a^i e_i$. Since $\nu = 0$,

$$\langle v, v \rangle = \left\langle \sum_i a^i e_i, \sum_j a^j e_j \right\rangle = \sum_{ij} \langle e_i, e_j \rangle a^i a^j = \sum_i (a^i)^2,$$

hence $\langle v, v \rangle \geq 0$, with equality if and only if each $a^i = 0$, that is, if and only if $v = 0$. $\square$

Theorem 4.6.3. *If $(V, \mathfrak{g})$ is an inner product space and $\mathcal{H}$ is a basis for V, then $\det(\mathfrak{g}_\mathcal{H}) > 0$.*

Proof. We have from Theorem 4.2.5 and Theorem 4.6.2 that $\det(\mathfrak{g}_\mathcal{H}) = \det(A)^2 \geq 0$. By Theorem 3.3.3, $\det(\mathfrak{g}_\mathcal{H}) \neq 0$. $\square$

We encountered a version of the Cauchy–Schwarz inequality in Theorem 3.2.6, the proof of which was somewhat involved. For an inner product space, we get a stronger result with much less effort.

Theorem 4.6.4 (Cauchy–Schwarz Inequality). *If $(V, \mathfrak{g})$ is an inner product space and v, w are vectors in V, then*

$$\langle v, w \rangle^2 \leq \langle v, v \rangle \langle w, w \rangle, \tag{4.6.2}$$

with equality if and only if v, w are linearly dependent.

Proof. The result is trivial if either $v = 0$ or $w = 0$, so assume $v, w \neq 0$. Since

$$\langle w, w \rangle (\langle v, v \rangle \langle w, w \rangle - \langle v, w \rangle^2)$$
$$= \langle \langle w, w \rangle v - \langle v, w \rangle w, \langle w, w \rangle v - \langle v, w \rangle w \rangle \geq 0, \tag{4.6.3}$$

we have

$$\langle v, v \rangle \langle w, w \rangle - \langle v, w \rangle^2 \geq 0,$$

from which (4.6.2) follows. If equality holds in (4.6.2), then equality holds in (4.6.3), that is,

$$\langle \langle w, w \rangle v - \langle v, w \rangle w, \langle w, w \rangle v - \langle v, w \rangle w \rangle = 0,$$

hence $\langle w, w \rangle v - \langle v, w \rangle w = 0$, so v, w are linearly dependent. Conversely, if $w = cv$ for some nonzero real number c, then

$$\langle v, v \rangle \langle w, w \rangle = \langle v, w/c \rangle \langle cv, w \rangle = \langle v, w \rangle^2. \qquad \square$$

Theorem 4.6.5 (Triangle Inequality). *If $(V, \mathfrak{g})$ is an inner product space and v, w are vectors in V, then*

$$\|v + w\| \leq \|v\| + \|w\|,$$

with equality if and only if v, w are linearly dependent.

Proof. By Theorem 4.6.4,

$$\langle v, w \rangle \leq |\langle v, w \rangle| = \sqrt{\langle v, w \rangle^2} \leq \sqrt{\langle v, v \rangle}\sqrt{\langle w, w \rangle} = \|v\|\,\|w\|,$$

hence

$$\|v + w\|^2 = \langle v + w, v + w \rangle = \langle v, v \rangle + 2\langle v, w \rangle + \langle w, w \rangle$$
$$\leq \|v\|^2 + 2\|v\|\,\|w\| + \|w\|^2 = (\|v\| + \|w\|)^2,$$

so $\|v + w\| \leq \|v\| + \|w\|$. There is equality if and only if $\langle v, w \rangle = \|v\|\,\|w\|$, which is equivalent to $\langle v, w \rangle \geq 0$ and $\langle v, w \rangle^2 = \langle v, v \rangle\langle w, w \rangle$, which in turn, by Theorem 4.6.4, is equivalent to v, w being linearly dependent. $\qquad\square$

In geometric terms, the triangle inequality has the following intuitive interpretation: in an inner product space, it is shorter to walk diagonally across a rectangular field than to "go around the corner". As we will see in Section 4.8, this commonplace observation depends crucially on the properties of the inner product.

Example 4.6.6 ($\mathbb{R}_0^m$). For vectors $(x^1, \ldots, x^m)$ and $(y^1, \ldots, y^m)$ in $\mathbb{R}_0^m$, the Cauchy–Schwarz inequality and triangle inequality become

$$\left(\sum_i x^i y^i\right)^2 \leq \left(\sum_i (x^i)^2\right)\left(\sum_i (y^i)^2\right)$$

and

$$\sqrt{\sum_i (x^i + y^i)^2} \leq \sqrt{\sum_i (x^i)^2} + \sqrt{\sum_i (y^i)^2},$$

respectively. $\qquad\qquad\diamond$

Theorem 4.6.7 (Pythagora's Theorem). *If $(V, \mathfrak{g})$ is an inner product space and U is a subspace of V, then*

$$\|v\|^2 = \|\mathcal{P}_U(v)\|^2 + \|\mathcal{P}_{U^\perp}(v)\|^2,$$

where $\mathcal{P}_U$ and $\mathcal{P}_{U^\perp}$ are the orthogonal projection maps defined in Section 4.1.

Proof. By Theorem 4.1.4,

$$\|v\|^2 = \langle \mathcal{P}_U(v) + \mathcal{P}_{U^\perp}(v), \mathcal{P}_U(v) + \mathcal{P}_{U^\perp}(v) \rangle$$
$$= \langle \mathcal{P}_U(v), \mathcal{P}_U(v) \rangle + \langle \mathcal{P}_{U^\perp}(v), \mathcal{P}_{U^\perp}(v) \rangle$$
$$= \|\mathcal{P}_U(v)\|^2 + \|\mathcal{P}_{U^\perp}(v)\|^2. \qquad\square$$

The next result is the inner product space counterpart of Theorem 4.2.3.

Theorem 4.6.8. *If $(V, \mathfrak{g})$ is an inner product space and $(e_1, \ldots, e_k)$ is an orthogonal (but not necessarily orthonormal) k-tuple of vectors in V, then $e_1, \ldots, e_k$ are linearly independent.*

Proof. If $\sum_i a_i e_i = 0$, then

$$0 = \left\langle e_i, \sum_j a_j e_j \right\rangle = \sum_j a_j \langle e_i, e_j \rangle = a_i \langle e_i, e_i \rangle.$$

By definition, e_i is nonzero, hence $\langle e_i, e_i \rangle \neq 0$, so $a_i = 0$ for $i = 1, \ldots, m$.   □

Theorem 4.2.4 guarantees the existence of an orthonormal basis for every scalar product space, but does not provide an explicit mechanism for constructing one. In an inner product space, there is a step-by-step algorithm that produces the desired basis.

Theorem 4.6.9 (Gram–Schmidt Orthogonalization Process). *Let $(V, \mathfrak{g})$ be an inner product space, and let $\mathcal{H} = (h_1, \ldots, h_m)$ be a basis for V. Computing in a sequential manner, let*

$$f_1 = h_1$$

$$f_2 = h_2 - \frac{\langle h_2, f_1 \rangle}{\langle f_1, f_1 \rangle} f_1$$

$$f_3 = h_3 - \frac{\langle h_3, f_1 \rangle}{\langle f_1, f_1 \rangle} f_1 - \frac{\langle h_3, f_2 \rangle}{\langle f_2, f_2 \rangle} f_2$$

$$\vdots$$

$$f_m = h_m - \frac{\langle h_m, f_1 \rangle}{\langle f_1, f_1 \rangle} f_1 - \frac{\langle h_m, f_2 \rangle}{\langle f_2, f_2 \rangle} f_2 - \cdots - \frac{\langle h_m, f_{m-1} \rangle}{\langle f_{m-1}, f_{m-1} \rangle} f_{m-1}.$$

Then:
(a) $\mathcal{F} = (f_1, \ldots, f_m)$ is an orthogonal basis for V.
(b) $\mathcal{E} = (f_1 / \|f_1\|, \ldots, f_m / \|f_m\|)$ is an orthonormal basis for V.

Proof. The first three steps of the sequential process are as follows, where we make repeated use of Theorem 4.1.4.

Step 1. Let $f_1 = h_1$. By definition, f_1 is nonzero.

Step 2. Let $f_2 = \mathcal{P}_{(\mathbb{R}f_1)^\perp}(h_2)$. Then $h_2 = a^1 f_1 + f_2$ for some real number a^1, hence

$$\langle h_2, f_1 \rangle = \langle a^1 f_1 + f_2, f_1 \rangle = a^1 \langle f_1, f_1 \rangle,$$

so

$$a^1 = \frac{\langle h_2, f_1 \rangle}{\langle f_1, f_1 \rangle}.$$

Thus,

$$f_2 = h_2 - \frac{\langle h_2, f_1 \rangle}{\langle f_1, f_1 \rangle} f_1.$$

Since h_2 is not in $\mathbb{R}f_1 = \mathbb{R}h_1$, it follows that f_2 is nonzero.

Step 3. Let $f_3 = \mathcal{P}_{(\mathbb{R}f_1 \oplus \mathbb{R}f_2)^\perp}(h_3)$. Then $h_3 = b^1 f_1 + b^2 f_2 + f_3$ for some real numbers b^1, b^2, hence

$$\langle h_3, f_1 \rangle = \langle b^1 f_1 + b^2 f_2 + f_3, f_1 \rangle = b^1 \langle f_1, f_1 \rangle$$

and

$$\langle h_3, f_2 \rangle = \langle b^1 f_1 + b^2 f_2 + f_3, f_2 \rangle = b^2 \langle f_2, f_2 \rangle,$$

so

$$b^1 = \frac{\langle h_3, f_1 \rangle}{\langle f_1, f_1 \rangle} \qquad \text{and} \qquad b^2 = \frac{\langle h_3, f_2 \rangle}{\langle f_2, f_2 \rangle}.$$

Thus,

$$f_3 = h_3 - \frac{\langle h_3, f_1 \rangle}{\langle f_1, f_1 \rangle} f_1 - \frac{\langle h_3, f_2 \rangle}{\langle f_2, f_2 \rangle} f_2.$$

Since h_3 is not in $\mathbb{R} f_1 \oplus \mathbb{R} f_2 = \mathbb{R} h_1 \oplus \mathbb{R} h_2$, it follows that f_3 is nonzero.

Proceeding in this way, after m steps we arrive at $\mathcal{F} = (f_1, \ldots, f_m)$. It is clear that, by construction, $\mathcal{F}$ is an orthogonal basis for V, and $\mathcal{E}$ is an orthonormal basis for V. $\square$

With the aid of Theorem 4.6.9, we are now able to provide an alternative proof of Theorem 4.2.4. Let $(V, \mathfrak{g})$ be a scalar product space of dimension m and index $\nu \neq 0$. It follows from Theorem 4.1.6(a) that there are subspaces V^+ and V^- of V such that $V = V^+ \oplus V^-$, with $\mathfrak{g}$ an inner product on V^+, and $-\mathfrak{g}$ an inner product on V^-. By Theorem 4.6.9, V^+ has a basis $(e_1, \ldots, e_{m-\nu})$ that is orthonormal with respect to the inner product $\mathfrak{g}|_{V^+}$. Similarly, V^- has a basis $(e_{m-\nu+1}, \ldots, e_m)$ that is orthonormal with respect to the inner product $-\mathfrak{g}|_{V^-}$. Then $(e_1, \ldots, e_{m-\nu}, e_{m-\nu+1}, \ldots, e_m)$ is basis for V that is orthonormal with respect to the scalar product $\mathfrak{g}$ (except that it needs to be reordered to conform with the convention on signs as they appear in signatures).

It was remarked in Section 4.3 that adjoints of linear maps have a certain symmetry property that is akin to the symmetry of a matrix. Part (b) of the next result justifies that observation.

Theorem 4.6.10. *Let $(V, \mathfrak{g})$ be an inner product space, let $A : V \longrightarrow V$ be a linear map, and let $\mathcal{E}$ be an orthonormal basis for V. Then:*
(a) $\left[A^\dagger \right]_{\mathcal{E}}^{\mathcal{E}} = \left([A]_{\mathcal{E}}^{\mathcal{E}} \right)^{\mathrm{T}}.$
(b) *If A is self-adjoint, then $[A]_{\mathcal{E}}^{\mathcal{E}}$ is a symmetric matrix.*

Proof. (a): Since $\mathfrak{g}_{\mathcal{E}}$ is the identity matrix, the result follows from Theorem 4.3.3.
    (b): This follows from part (a). $\square$

Theorem 4.6.11. *If $(V, \mathfrak{g})$ and $(W, \mathfrak{h})$ are inner product spaces with the same dimension and $A : V \longrightarrow W$ is a linear map, then the following are equivalent:*
(a) *A is a linear isometry.*
(b) *A maps some orthonormal basis for V to an orthonormal basis for W.*
(c) *A maps every orthonormal basis for V to an orthonormal basis for W.*

Proof. (a) $\Rightarrow$ (c): This follows from Theorem 4.4.1(b).
    (c) $\Rightarrow$ (b): Straightforward.
    (b) $\Rightarrow$ (a): Let $\mathcal{E} = (e_1, \ldots, e_m)$ be an orthonormal basis for V such that $A(\mathcal{E})$ is an orthonormal basis for W. Since V and W are inner product spaces,

their signatures are both $(+1, \ldots, +1)$. Let v, w be vectors in V, with $v = \sum_i a^i e_i$ and $w = \sum_j b^j e_j$, hence $A(v) = \sum_i a^i A(e_i)$ and $A(w) = \sum_j b^j A(e_j)$. By Theorem 4.2.8,

$$\langle A(v), A(w) \rangle = \sum_i a^i b^i = \langle v, w \rangle. \qquad \square$$

Theorem 4.6.12. *If $(V, \mathfrak{g})$ is an inner product space and $A : V \longrightarrow V$ is a linear map, then the following are equivalent:*
(a) *A is a linear isometry.*
(b) *A maps some orthonormal basis for V to an orthonormal basis for V.*
(c) *A maps every orthonormal basis for V to an orthonormal basis for V.*
(d) *$\|A(v)\| = \|v\|$ for all vectors v in V.*
(e) *For every orthonormal basis $\mathcal{E}$ for V, the matrix $[A]_{\mathcal{E}}^{\mathcal{E}}$ is in $O(m)$, the orthogonal group defined in Section 2.4.*

Proof. (a) $\Leftrightarrow$ (b) $\Leftrightarrow$ (c): This is just Theorem 4.6.11 with $W = V$.
(a) $\Leftrightarrow$ (d): This follows from Theorem 4.4.4.
(c) $\Leftrightarrow$ (e): Let $\mathcal{E} = (e_1, \ldots, e_m)$, and observe that $\mathfrak{g}_{\mathcal{E}} = I_m$. Then

$$
\begin{aligned}
\langle A(e_i), A(e_j) \rangle &= \big([A(e_i)]_{\mathcal{E}}\big)^{\mathrm{T}} [A(e_j)]_{\mathcal{E}} && \text{[Th 3.1.2(a)]} \\
&= \big([A]_{\mathcal{E}}^{\mathcal{E}} [e_i]_{\mathcal{E}}\big)^{\mathrm{T}} \big([A]_{\mathcal{E}}^{\mathcal{E}} [e_j]_{\mathcal{E}}\big) && \text{[Th 2.2.4]} \\
&= \big(([A]_{\mathcal{E}}^{\mathcal{E}})_{(i)}\big)^{\mathrm{T}} \big([A]_{\mathcal{E}}^{\mathcal{E}}\big)_{(j)} \\
&= \big(([A]_{\mathcal{E}}^{\mathcal{E}})^{\mathrm{T}}\big)^{(i)} \big([A]_{\mathcal{E}}^{\mathcal{E}}\big)_{(j)}, && \text{[Th 2.1.6(b)]}
\end{aligned}
$$

hence

$$[\langle A(e_i), A(e_j) \rangle] = \big([A]_{\mathcal{E}}^{\mathcal{E}}\big)^{\mathrm{T}} [A]_{\mathcal{E}}^{\mathcal{E}}.$$

It follows that

$$
\begin{aligned}
A(\mathcal{E}) \text{ is orthonormal} & \\
\Leftrightarrow \quad [\langle A(e_i), A(e_j) \rangle] &= I_m \\
\Leftrightarrow \quad \big([A]_{\mathcal{E}}^{\mathcal{E}}\big)^{\mathrm{T}} [A]_{\mathcal{E}}^{\mathcal{E}} &= I_m \\
\Leftrightarrow \quad [A]_{\mathcal{E}}^{\mathcal{E}} \text{ is in } O(m). && \square
\end{aligned}
$$

4.7   Eigenvalues and Eigenvectors

Let P be a matrix in $\mathrm{Mat}_{m \times m}$, and let x be an "indeterminate". The **characteristic polynomial of P** is defined by

$$\mathrm{char}_P(x) = \det(x I_m - P).$$

Theorem 4.7.1. *If P and Q are matrices in $\mathrm{Mat}_{m \times m}$, with Q invertible, then*

$$\mathrm{char}_{Q^{-1} P Q}(x) = \mathrm{char}_P(x).$$

Proof. We have
$$xI_m - Q^{-1}PQ = Q^{-1}(xI_m - P)Q,$$
hence $\det(xI_m - Q^{-1}PQ) = \det(xI_m - P)$.   □

Let V be a vector space of dimension m, and let $A : V \longrightarrow V$ be a linear map. We say that a real number κ is an **eigenvalue of A** if there is a *nonzero* vector v in V such that $A(v) = \kappa v$, or equivalently, $(\kappa \operatorname{id}_V - A)(v) = 0$. In that case, v is said to be an **eigenvector of A** corresponding to κ. Since $A(cv) = c\kappa v$ for all real numbers c, if there is one eigenvector of A corresponding to κ, then there are infinitely many.

Let $\mathcal{H}$ be a basis for V. The **characteristic polynomial of A** is denoted by $\operatorname{char}_A(x)$ and defined to be the characteristic polynomial of $[A]_{\mathcal{H}}^{\mathcal{H}}$:

$$\operatorname{char}_A(x) = \det\left(xI_m - [A]_{\mathcal{H}}^{\mathcal{H}}\right).$$

We observe that

$$\operatorname{char}_A(0) = (-1)^m \det(A). \tag{4.7.1}$$

Theorem 4.7.2. *With the above setup, $\operatorname{char}_A(x)$ is independent of the choice of basis.*

Proof. Let $\mathcal{F}$ be another basis for V. By Theorem 2.2.6,

$$
\begin{aligned}
xI_m - [A]_{\mathcal{F}}^{\mathcal{F}} &= [x\operatorname{id}_V - A]_{\mathcal{F}}^{\mathcal{F}} \\
&= \left([\operatorname{id}_V]_{\mathcal{F}}^{\mathcal{H}}\right)^{-1} [x\operatorname{id}_V - A]_{\mathcal{H}}^{\mathcal{H}} [\operatorname{id}_V]_{\mathcal{F}}^{\mathcal{H}}.
\end{aligned}
$$

The result now follows from Theorem 4.7.1.   □

Theorem 4.7.3. *Let V be a vector space, let $A : V \longrightarrow V$ be a linear map, and let κ be a real number. Then the following are equivalent:*
(a) κ is an eigenvalue of A.
(b) $\det(\kappa \operatorname{id}_V - A) = 0$.
(c) $\operatorname{char}_A(\kappa) = 0$.

Proof. Let $\mathcal{H}$ be a basis for V. We have

$$
\begin{aligned}
&\kappa \text{ is an eigenvalue of } A \\
\Leftrightarrow\quad & (\kappa \operatorname{id}_V - A)(v) = 0 \text{ for some nonzero vector } v \text{ in } V \\
\Leftrightarrow\quad & \ker(\kappa \operatorname{id}_V - A) \neq \{0\} \\
\Leftrightarrow\quad & \kappa \operatorname{id}_V - A \text{ is not a linear isomorphism} \\
\Leftrightarrow\quad & \det(\kappa \operatorname{id}_V - A) = 0 \\
\Leftrightarrow\quad & \det\left([\kappa \operatorname{id}_V - A]_{\mathcal{H}}^{\mathcal{H}}\right) = 0 \\
\Leftrightarrow\quad & \det\left(\kappa I_m - [A]_{\mathcal{H}}^{\mathcal{H}}\right) = 0 \\
\Leftrightarrow\quad & \operatorname{char}_A(\kappa) = 0,
\end{aligned}
$$

where the third equivalence follows from Theorem 1.1.12 and Theorem 1.1.14, and the fourth equivalence from Theorem 2.5.3. □

Let V be a vector space of dimension m, and let $A : V \longrightarrow V$ be a linear map. It is easily shown that $\mathrm{char}_A(x)$ is a polynomial of degree m with real coefficients. By the fundamental theorem of algebra, $\mathrm{char}_A(x)$ has m (not necessarily distinct) roots in the field of complex numbers. Without further assumptions, there is no guarantee that any of them are real. In what follows, we focus on the case $m = 2$.

Theorem 4.7.4. *Let $(V, \mathfrak{g})$ be a 2-dimensional inner product space, and let $A : V \longrightarrow V$ be a self-adjoint linear map. Then A has two real eigenvalues, and they are equal if and only if $A = c\,\mathrm{id}_V$ for some real number c.*

Proof. Let $\mathcal{E}$ be an orthonormal basis for V. By Theorem 4.6.10(b), $[A]_{\mathcal{E}}^{\mathcal{E}}$ is symmetric, so

$$[A]_{\mathcal{E}}^{\mathcal{E}} = \begin{bmatrix} a & b \\ b & c \end{bmatrix}$$

for some real numbers a, b, and c. Then

$$\mathrm{char}_A(x) = \det\left(\begin{bmatrix} x - a & -b \\ -b & x - c \end{bmatrix} \right) = x^2 - (a + c)x + ac - b^2.$$

From the quadratic formula, the two roots of $\mathrm{char}_A(x)$ are

$$\frac{a + c \pm \sqrt{(a - c)^2 + 4b^2}}{2}.$$

Since $(a - c)^2 + 4b^2 \geq 0$, both roots are real. They are equal if and only if $(a-c)^2 + 4b^2 = 0$, which is equivalent to $a = c$ and $b = 0$; that is, $[A]_{\mathcal{E}}^{\mathcal{E}} = cI_2$. □

Theorem 4.7.5. *Let $(V, \mathfrak{g})$ be a 2-dimensional inner product space, let $A : V \longrightarrow V$ be a self-adjoint linear map, and let κ_1, κ_2 be the (real but not necessarily distinct) eigenvalues of A. Then*

$$\det(A) = \kappa_1 \kappa_2.$$

Proof. Since κ_1 and κ_2 are roots of $\mathrm{char}_A(x)$, we have from the theory of polynomials that

$$\mathrm{char}_A(x) = (x - \kappa_1)(x - \kappa_2),$$

hence $\mathrm{char}_A(0) = \kappa_1 \kappa_2$. On the other hand, (4.7.1) gives $\mathrm{char}_A(0) = \det(A)$. □

Theorem 4.7.6 (Euler's Rotation Theorem). *Let $(V, \mathfrak{g})$ be an inner product space of odd dimension, and let $A : V \longrightarrow V$ be a linear isometry. If $\det(A) = 1$, then 1 is an eigenvalue of A.*

Remark. Recall from Theorem 4.4.3 that $\det(A) = \pm 1$, so the assertion makes sense.

Proof. Let $\dim(V) = m$, and let $\mathcal{E}$ be an orthonormal basis for V. By Theorem 4.6.12, the matrix $[A]_{\mathcal{E}}^{\mathcal{E}}$ is in $\mathrm{O}(m)$, so $\left([A]_{\mathcal{E}}^{\mathcal{E}}\right)^{\mathrm{T}}[A]_{\mathcal{E}}^{\mathcal{E}} = I_m$. Then

$$\left([\mathrm{id}_V - A]_{\mathcal{E}}^{\mathcal{E}}\right)^{\mathrm{T}} = \left(I_m - [A]_{\mathcal{E}}^{\mathcal{E}}\right)^{\mathrm{T}} = \left(-[A]_{\mathcal{E}}^{\mathcal{E}}\right)^{\mathrm{T}}\left(I_m - [A]_{\mathcal{E}}^{\mathcal{E}}\right)$$

$$= \left(-[A]_{\mathcal{E}}^{\mathcal{E}}\right)^{\mathrm{T}}[\mathrm{id}_V - A]_{\mathcal{E}}^{\mathcal{E}},$$

hence

$$\det(\mathrm{id}_V - A) = (-1)^m \det(A)\det(\mathrm{id}_V - A) = -\det(\mathrm{id}_V - A),$$

so $\det(\mathrm{id}_V - A) = 0$. By Theorem 4.7.3, $\kappa = 1$ is an eigenvalue of A. $\square$

According to Theorem 4.6.12, a linear isometry on an inner product space "preserves norms", and as we will see in Section 9.3, this means that it also "preserves distance". Such a linear isometry is said to produce a "rigid motion" about the origin. Theorem 4.7.10 has an interpretation that fits nicely with these observations. Suppose 1 is an eigenvalue of A, with corresponding eigenvector v. Then $A(cv) = cv$ for all real numbers c, so that A fixes the subspace $\mathbb{R}v$. Thus, the action of A is to rigidly rotate V about the origin, with $\mathbb{R}v$ as the "axis of rotation".

4.8   Lorentz Vector Spaces

Lorentz vector spaces play a central role in the special theory of relativity and, by extension, in general relativity. The geometry of Lorentz vector spaces is considerably more complicated than that of inner product spaces. As we will see, when translated into physical terms, the implications can be surprising to our Euclidean sensibilities.

Let $(V, \mathfrak{g})$ be a nonzero scalar product space. We say that $\mathfrak{g}$ is a **Lorentz scalar product**, and that the pair $(V, \mathfrak{g})$ is a **Lorentz vector space**, if $\mathrm{ind}(\mathfrak{g}) = 1$. Based on our convention on signs, the signature of $\mathfrak{g}$ is therefore $(+1, \ldots, +1, -1)$. Recall from Section 3.1 that a vector v in V is:

spacelike	if $v = 0$ or $\langle v, v \rangle > 0$.
timelike	if $\langle v, v \rangle < 0$.
lightlike	if $v \neq 0$ and $\langle v, v \rangle = 0$.

Thus, by definition, $e_1, \ldots, e_{m-1}$ are spacelike and e_m is timelike. The convention on signs adopted here is not the only one appearing in the literature. For instance, some authors take the signature to be $(-1, +1, \ldots, +1)$. The choice of one approach over another can lead to differences in the number and location of negative signs in expressions, but does not affect substance.

Let U be a nonzero subspace of V. If $\mathfrak{g}$ is nondegenerate on U, then according to Theorem 4.2.6,

$$\mathrm{ind}(\mathfrak{g}|_U) + \mathrm{ind}(\mathfrak{g}|_{U^\perp}) = 1.$$

Thus, $\text{ind}(\mathfrak{g}|_U)$ equals 0 or 1. We say that U is:

spacelike	if $\mathfrak{g}$ is nondegenerate on U and $\text{ind}(\mathfrak{g}|_U) = 0$;
	that is, $\mathfrak{g}$ is an inner product on U.
timelike	if $\mathfrak{g}$ is nondegenerate on U and $\text{ind}(\mathfrak{g}|_U) = 1$; (4.8.1)
	that is, $\mathfrak{g}$ is a Lorentz scalar product on U.
lightlike	if $\mathfrak{g}$ is degenerate on U.

Evidently, every subspace of V is either spacelike, timelike, or lightlike. By convention, the zero subspace of V is regarded as spacelike. We see from (4.8.1) that in contrast to the situation with an inner product, a Lorentz scalar product on V may be degenerate on U and therefore not a scalar product (not to mention a Lorentz scalar product) on U. Recall from Section 3.1 that the light set of V is

$$\Lambda = \{v \in V : \langle v, v \rangle = 0\} \backslash \{0\}.$$

In the present context, we refer to Λ as the **light cone** of V, for reasons that will become apparent shortly.

Example 4.8.1 ($\mathbb{R}_1^m$). Continuing with Example 4.1.1, let $\nu = 1$. In this case, we denote $\mathfrak{s}$ by $\mathfrak{m}$. Then

$$\mathfrak{m} : \mathbb{R}^m \times \mathbb{R}^m \longrightarrow \mathbb{R}$$

is the function given by

$$\mathfrak{m}\big((x^1, \ldots, x^m), (y^1, \ldots, y^m)\big) = \sum_{i=1}^{m-1} x^i y^i - x^m y^m.$$

We refer to

$$\mathbb{R}_1^m = (\mathbb{R}^m, \mathfrak{m})$$

as **Minkowski m-space** and to $\mathfrak{m}$ as the **Minkowski scalar product**. The standard basis for $\mathbb{R}_1^m$ is the same as the standard basis for $\mathbb{R}_0^m$ (and $\mathbb{R}^m$). The norm corresponding to $\mathfrak{m}$ is given by

$$\big\|(x^1, \ldots, x^m)\big\| = \sqrt{\left| \sum_{i=1}^{m-1} (x^i)^2 - (x^m)^2 \right|},$$

and the light cone of $\mathbb{R}_1^m$ is denoted by Λ_m. In $\mathbb{R}_1^3$, for example, we have

$$\|(x, y, z)\| = \sqrt{|x^2 + y^2 - z^2|}$$

and

$$\Lambda_3 = \{(x, y, z) \in \mathbb{R}_1^3 : x^2 + y^2 - z^2 = 0\} \backslash \{(0, 0, 0)\}.$$

See Figure 4.8.1. In geometric terms, Λ_3 is the union of a pair of cones with their vertices removed. A bit of analytic geometry shows that the set of timelike

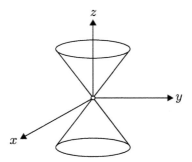

Figure 4.8.1. Light cone: Diagram for Example 4.8.1

vectors in $\mathbb{R}_1^3$ is the "inside" of Λ_3, and the set of spacelike vectors in $\mathbb{R}_1^3$ is the "outside" of Λ_3.

Any 2-dimensional subspace of $\mathbb{R}_1^3$ can be represented by a plane Π through the origin. This gives precisely three possibilities, as depicted in Figure 4.8.2, where dashed lines indicate intersections. Π_1 does not intersect Λ_3: it is space-like and contains only spacelike vectors. Π_2 intersects Λ_3 in a pair of lines (minus the origin): it is timelike and contains spacelike, timelike, and lightlike vectors. Π_3 intersects Λ_3 in a single line (minus the origin): it is lightlike and contains spacelike and lightlike vectors, but no timelike vectors. We observe that according to this analysis, every 2-dimensional subspace of $\mathbb{R}_1^3$ contains spacelike vectors. $\diamond$

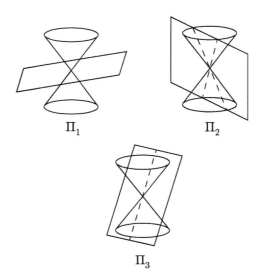

Figure 4.8.2. Diagram for Example 4.8.1

In the following series of results, we show that the characteristics of 2-dimensional subspaces of $\mathbb{R}_1^3$ delineated in Example 4.8.1 hold more generally.

Theorem 4.8.2. *Let $(V, \mathfrak{g})$ be a Lorentz vector space, and let v be a vector in V. Then v is a spacelike (resp., timelike, lightlike) vector if and only if $\mathbb{R}v$ is a spacelike (resp., timelike, lightlike) subspace.*

Proof. Each vector in $\mathbb{R}v$ is of the form cv for some real number c, so we have $\langle cv, cv \rangle = c^2 \langle v, v \rangle$. The result follows. □

Theorem 4.8.3. *Let $(V, \mathfrak{g})$ be a Lorentz vector space, and let U be a subspace of V. Then U is spacelike if and only if it consists entirely of spacelike vectors.*

Proof. We have

U is spacelike

$\quad \Leftrightarrow \quad \mathfrak{g}$ is an inner product on U $\qquad\qquad\qquad$ [(4.8.1)]

$\quad \Leftrightarrow \quad \mathfrak{g}$ is an inner product on $\mathbb{R}u$ for all u in U

$\quad \Leftrightarrow \quad \mathbb{R}u$ is spacelike for all u in U $\qquad\qquad$ [(4.8.1)]

$\quad \Leftrightarrow \quad u$ is spacelike for all u in U. $\qquad\qquad$ [Th 4.8.2] □

Theorem 4.8.4. *Let $(V, \mathfrak{g})$ be a Lorentz vector space, and let U be a subspace of V. Then:*
(a) *U is spacelike if and only if $U^\perp$ timelike.*
(b) *U is timelike if and only if $U^\perp$ spacelike.*
(c) *U is lightlike if and only if $U^\perp$ is lightlike.*

Proof. (a)($\Rightarrow$): Since U is spacelike, $\mathfrak{g}$ is nondegenerate on U and $\mathrm{ind}(\mathfrak{g}|_U) = 0$. It follows from Theorem 4.1.3 that $\mathfrak{g}$ is nondegenerate on $U^\perp$, and then from Theorem 4.2.6 that $1 = \mathrm{ind}(\mathfrak{g}) = \mathrm{ind}(\mathfrak{g}|_{U^\perp})$. Thus, $U^\perp$ is timelike.
 (b)($\Rightarrow$): The proof is similar to that given for (a)($\Rightarrow$).
 (c)($\Rightarrow$): We observe that $U^\perp$ is not timelike; for if it were, then, by Theorem 4.1.2(c) and (b)($\Rightarrow$), U would be spacelike, which contradicts the assumption on U. On the other hand, $U^\perp$ is not spacelike; for if it were, then, by Theorem 4.1.2(c) and (a)($\Rightarrow$), U would be timelike, which again contradicts the assumption on U. Thus, $U^\perp$ is lightlike.
 (a)($\Leftarrow$): This follows from Theorem 4.1.2(c) and (b)($\Rightarrow$).
 (b)($\Leftarrow$): This follows from Theorem 4.1.2(c) and (a)($\Rightarrow$).
 (c)($\Leftarrow$): This follows from Theorem 4.1.2(c) and (c)($\Rightarrow$). □

Theorem 4.8.5. *If $(V, \mathfrak{g})$ is a Lorentz vector space and v is a timelike vector in V, then $V = (\mathbb{R}v)^\perp \oplus \mathbb{R}v$, where $(\mathbb{R}v)^\perp$ is spacelike and $\mathbb{R}v$ is timelike. Thus, each vector w in V can be expressed uniquely in the form*

$$w = \overline{w} + cv,$$

where $\overline{w}$ is a (spacelike) vector in $(\mathbb{R}v)^\perp$ and c is a real number.

Proof. Since v is timelike, it follows from Theorem 4.8.2 that $\mathbb{R}v$ is timelike, and then from Theorem 4.8.4(b) that $(\mathbb{R}v)^\perp$ is spacelike. Thus, $\mathfrak{g}$ is nondegenerate on $(\mathbb{R}v)^\perp$. By Theorem 4.1.2(c) and Theorem 4.1.3, $V = (\mathbb{R}v)^\perp \oplus \mathbb{R}v$.   □

In the notation of Theorem 4.8.5, since $(\mathbb{R}v)^\perp$ is spacelike and $\mathbb{R}v$ is timelike, it follows that $\mathfrak{g}$ is positive definite on $(\mathbb{R}v)^\perp$ and negative definite on $\mathbb{R}v$. Thus, $V = (\mathbb{R}v)^\perp \oplus \mathbb{R}v$ is a direct sum of the type given by Theorem 4.1.6(a).

Theorem 4.8.6. *Let $(V, \mathfrak{g})$ be a Lorentz vector space, and let v, w be lightlike vectors in V. Then v and w are orthogonal if and only if they are linearly dependent.*

Proof. ($\Rightarrow$): Let $(e_1, \ldots, e_m)$ be an orthonormal basis for V. Then e_m is timelike, so we have from Theorem 4.8.5 that v and w can be expressed as

$$v = \overline{v} + ae_m \qquad \text{and} \qquad w = \overline{w} + be_m, \tag{4.8.2}$$

where $\overline{v}, \overline{w}$ are (spacelike) vectors in $(\mathbb{R}e_m)^\perp$ and a, b are real numbers. Then

$$0 = \langle v, v \rangle = \langle \overline{v}, \overline{v} \rangle - a^2$$
$$0 = \langle w, w \rangle = \langle \overline{w}, \overline{w} \rangle - b^2$$
$$0 = \langle v, w \rangle = \langle \overline{v}, \overline{w} \rangle - ab,$$

hence

$$a^2 = \langle \overline{v}, \overline{v} \rangle \qquad b^2 = \langle \overline{w}, \overline{w} \rangle \qquad ab = \langle \overline{v}, \overline{w} \rangle, \tag{4.8.3}$$

so

$$\langle \overline{v}, \overline{w} \rangle^2 = \langle \overline{v}, \overline{v} \rangle \langle \overline{w}, \overline{w} \rangle. \tag{4.8.4}$$

Since e_m is timelike, we have from Theorem 4.8.2 that $\mathbb{R}e_m$ is timelike, and then from Theorem 4.8.4(b) that $(\mathbb{R}e_m)^\perp$ is spacelike. Thus, $\mathfrak{g}$ is an inner product on $(\mathbb{R}e_m)^\perp$. It follows from Theorem 4.6.4 and (4.8.4) that

$$\overline{v} = c\overline{w} \tag{4.8.5}$$

for some real number c. Then (4.8.5) and the second and third identities in (4.8.3) give $ab = b^2 c$. We have $b \neq 0$; for if not, then $w = \overline{w}$, where w is lightlike and $\overline{w}$ is spacelike, which is a contradiction. Thus,

$$a = bc. \tag{4.8.6}$$

Combining (4.8.2), (4.8.5), and (4.8.6) yields $v = cw$.

($\Leftarrow$): If v, w are linearly dependent, then $w = cv$ for some real number c, hence $\langle v, w \rangle = c\langle v, v \rangle = 0$.   □

Theorem 4.8.7. *If $(V, \mathfrak{g})$ is a Lorentz vector space and U is a subspace of V of dimension $m \geq 2$, then the following are equivalent:*
(a) *U is timelike.*
(b) *U contains two linearly independent lightlike vectors.*

(c) *U contains a timelike vector.*

Proof. (a) $\Rightarrow$ (b): By assumption, U is timelike, so $\mathfrak{g}$ is a (Lorentz) scalar product on U. By Theorem 4.2.4, there is an orthonormal basis $(e_1, \ldots, e_k)$ for U. Then e_1 is a spacelike unit vector, e_k is a timelike unit vector, and e_1 and e_k are orthogonal. It follows that

$$\langle e_1 + e_k, e_1 + e_k \rangle = 0 \qquad \text{and} \qquad \langle e_1 - e_k, e_1 - e_k \rangle = 0,$$

so $e_1 + e_k$ and $e_1 - e_k$ are lightlike. Suppose

$$0 = a(e_1 + e_k) + b(e_1 - e_k) = (a + b)e_1 + (a - b)e_k$$

for some real numbers a, b. Since e_1, e_k are linearly independent, we have $a = b = 0$. Thus, $e_1 + e_k, e_1 - e_k$ are linearly independent.

(b) $\Rightarrow$ (c): By assumption, U contains two linearly independent lightlike vectors v and w. Then

$$\langle v + w, v + w \rangle = 2\langle v, w \rangle \qquad \text{and} \qquad \langle v - w, v - w \rangle = -2\langle v, w \rangle.$$

It follows from Theorem 4.8.6 that $\langle v, w \rangle \neq 0$, so either $v + w$ or $v - w$ is timelike.

(c) $\Rightarrow$ (a): By assumption, U contains a timelike vector v. It follows from Theorem 4.8.2 that $\mathbb{R}v$ is timelike, and then from Theorem 4.8.4(b) that $(\mathbb{R}v)^\perp$ is spacelike. Since $\mathbb{R}v \subseteq U$, we have $U^\perp \subseteq (\mathbb{R}v)^\perp$, and then from Theorem 4.8.3 that $U^\perp$ is spacelike. By Theorem 4.1.2(c) and Theorem 4.8.4(a), $U = U^{\perp\perp}$ is timelike. $\qquad\square$

Theorem 4.8.8. *If $(V, \mathfrak{g})$ is a Lorentz vector space and U is a subspace of V, then the following are equivalent:*
(a) *U is lightlike.*
(b) *U contains a lightlike vector but not a timelike vector.*
(c) *$U \cap \Lambda = \mathbb{R}v \backslash \{0\}$ for some lightlike vector v in U.*

Proof. If V is 1-dimensional, the result is trivial, so assume $\dim(V) \geq 2$.

(a) $\Rightarrow$ (b): By assumption, U is lightlike. Then $\mathfrak{g}$ is degenerate on U, so there is a nonzero vector v in U such that $\langle v, w \rangle = 0$ for all vectors w in U. In particular, $\langle v, v \rangle = 0$, hence v is lightlike. Furthermore, U does not contain a timelike vector; for if it did, then, by Theorem 4.8.7, U would be timelike, which contradicts the assumption on U.

(b) $\Rightarrow$ (c): By assumption, U contains a lightlike vector v. Then $U \cap \Lambda$ does not contain a lightlike vector that is linearly independent of v; for if it did, then, by Theorem 4.8.7, U would contain a timelike vector, which contradicts the assumption on U. Thus, $U \cap \Lambda = \mathbb{R}v \backslash \{0\}$.

(c) $\Rightarrow$ (a): By assumption, $U \cap \Lambda = \mathbb{R}v \backslash \{0\}$ for some lightlike vector v in U. We have from Theorem 4.8.3 that U is not spacelike. Since $U \cap \Lambda$ is contained in the 1-dimensional subspace $\mathbb{R}v$ of V, it does not contain two linearly independent (lightlike) vectors. By Theorem 4.8.7, U is not timelike. Therefore, U is lightlike. $\qquad\square$

Theorem 4.8.9. *If $(V, \mathfrak{g})$ is a Lorentz vector space and U is a subspace of V of dimension $m \geq 2$, then U contains a nonzero spacelike vector.*

Proof. There are three cases to consider.

Case 1. U is spacelike.

By Theorem 4.8.3, any nonzero vector in U is spacelike.

Case 2. U is timelike.

Let $\dim(V) = m$. According to Theorem 4.8.7, U contains a timelike vector v. Since a timelike vector is nonzero, $\dim(\mathbb{R}v) = 1$, so Theorem 4.1.2(b) gives $\dim\big((\mathbb{R}v)^{\perp}\big) = m - 1$. By assumption, $\dim(U) \geq 2$. We have from Theorem 1.1.15 that

$$
\begin{aligned}
m &\geq \dim\big((\mathbb{R}v)^{\perp} + U\big) \\
&= \dim\big((\mathbb{R}v)^{\perp}\big) + \dim(U) - \dim\big((\mathbb{R}v)^{\perp} \cap U\big) \\
&\geq m + 1 - \dim\big((\mathbb{R}v)^{\perp} \cap U\big),
\end{aligned}
$$

hence $\dim\big((\mathbb{R}v)^{\perp} \cap U\big) \geq 1$. Thus, $(\mathbb{R}v)^{\perp} \cap U$ is not the zero subspace. It follows from Theorem 4.8.2 that $\mathbb{R}v$ is timelike, and then from Theorem 4.8.4(b) that $(\mathbb{R}v)^{\perp}$ is spacelike. By Theorem 4.8.3, any nonzero vector in $(\mathbb{R}v)^{\perp} \cap U$ is spacelike.

Case 3. U is lightlike.

By Theorem 4.8.8, U contains a lightlike vector v. Let w be a vector in $U \backslash \mathbb{R}v$. There are three cases to consider: w is either spacelike, timelike, or lightlike. If w is spacelike, then we are done. Suppose w is not spacelike. By Theorem 4.8.8, w cannot be timelike, so it is lightlike. It follows from Theorem 1.1.2 that v, w are linearly independent, and then from Theorem 4.8.6 that they are not orthogonal. Thus,

$$
\langle v + w, v + w \rangle = 2\langle v, w \rangle \neq 0,
$$

so the vector $v + w$ in U is nonzero and not lightlike. By Theorem 4.8.8, $v + w$ cannot be timelike, hence it is spacelike. $\square$

The following table summarizes some of the above findings on Lorentz vector spaces. As can be seen, the observations for 2-dimensional subspaces of $\mathbb{R}_1^3$ made in Example 4.8.1 hold for Lorentz vector spaces in general.

Type of subspace (dimension ≥ 2)	Types of vectors in subspace (excluding zero vector)		
	Spacelike	Timelike	Lightlike
Spacelike	yes	no	no
Timelike	yes	yes	yes
Lightlike	yes	no	yes

Looking again at Figure 4.8.2, we confirm using Theorem 4.8.3, Theorem 4.8.7, and Theorem 4.8.8, respectively, that Π_1 is spacelike, Π_2 is timelike, and Π_3 is lightlike.

We conclude this section with yet another version of the Cauchy–Schwarz inequality. It is labeled "reversed" because of the direction of the inequality. This is a reminder of how differently behaved a Lorentz vector space is compared to an inner product space.

Theorem 4.8.10 (Reversed Cauchy–Schwarz Inequality). *If $(V, \mathfrak{g})$ is a Lorentz vector space and v, w are timelike vectors, then*

$$\langle v, w \rangle^2 \geq \langle v, v \rangle \langle w, w \rangle,$$

with equality if and only if v, w are linearly dependent.

Proof. Since v is timelike, it follows from Theorem 4.8.2 that $\mathbb{R}v$ is timelike, and then from Theorem 4.8.4(b) that $(\mathbb{R}v)^\perp$ is spacelike. Thus, $\mathfrak{g}$ is an inner product on $(\mathbb{R}v)^\perp$. By Theorem 4.8.5, $w = \overline{w} + cv$, where $\overline{w}$ is a (spacelike) vector in $(\mathbb{R}v)^\perp$ and c is a real number. Then

$$\langle v, w \rangle = c \langle v, v \rangle \qquad \text{and} \qquad \langle w, w \rangle = \langle \overline{w}, \overline{w} \rangle + c^2 \langle v, v \rangle,$$

hence
$$\langle v, w \rangle^2 = c^2 \langle v, v \rangle^2 = \langle v, v \rangle (\langle w, w \rangle - \langle \overline{w}, \overline{w} \rangle)$$
$$= \langle v, v \rangle \langle w, w \rangle - \langle v, v \rangle \langle \overline{w}, \overline{w} \rangle.$$

Since v and w are timelike, $\langle v, v \rangle \langle w, w \rangle > 0$, and because v is timelike and $\overline{w}$ is spacelike, $\langle v, v \rangle \langle \overline{w}, \overline{w} \rangle \leq 0$. It follows that $\langle v, w \rangle^2 \geq \langle v, v \rangle \langle w, w \rangle$, with equality if and only if $\langle \overline{w}, \overline{w} \rangle = 0$. Since $\mathfrak{g}$ is an inner product on $(\mathbb{R}v)^\perp$, $\langle \overline{w}, \overline{w} \rangle = 0$ if and only if $\overline{w} = 0$, which is equivalent to $w = cv$, which in turn is equivalent to v and w being linearly dependent. $\qquad\qquad\square$

4.9   Time Cones

Let $(V, \mathfrak{g})$ be a Lorentz vector space. The **time cone** of V is denoted by $\mathcal{T}$ and defined to be the set of timelike vectors in V:

$$\mathcal{T} = \{v \in V : \langle v, v \rangle < 0\}.$$

In $\mathbb{R}_1^m$, the time cone is denoted by $\mathcal{T}_m$. For example,

$$\mathcal{T}_3 = \{(x, y, z) \in \mathbb{R}_1^3 : x^2 + y^2 - z^2 < 0\}.$$

Recall from Example 4.8.1 that the light cone of $\mathbb{R}_1^3$ is

$$\Lambda_3 = \{(x, y, z) \in \mathbb{R}_1^3 : x^2 + y^2 - z^2 = 0\} \backslash \{(0, 0, 0)\}.$$

Thus, $\mathcal{T}_3$ is "inside" Λ_3.

Theorem 4.9.1. *Let $(V, \mathfrak{g})$ be a Lorentz vector space. If v is a timelike vector in V and w is a timelike or lightlike vector in V, then $\langle v, w \rangle \neq 0$.*

Proof. Since v is timelike, it follows from Theorem 4.8.2 that $\mathbb{R}v$ is timelike, and then from Theorem 4.8.4(b) that $(\mathbb{R}v)^{\perp}$ is spacelike. Since w is not spacelike, by Theorem 4.8.3, it is not in $(\mathbb{R}v)^{\perp}$, hence $\langle v, w \rangle \neq 0$. $\qquad\square$

Let $(V, \mathfrak{g})$ be a Lorentz vector space, and let v, w be timelike vectors in V. It follows from Theorem 4.9.1 that either $\langle v, w \rangle < 0$ or $\langle v, w \rangle > 0$. Let us write

$$v \sim w \qquad \text{if} \qquad \langle v, w \rangle < 0.$$

Theorem 4.9.2. *With the above setup, $\sim$ is an equivalence relation on $\mathcal{T}$.*

Proof. Clearly, $\sim$ is reflexive and symmetric. It remains to show that $\sim$ is transitive; that is, if u, v, w are vectors in $\mathcal{T}$ such that $\langle u, v \rangle < 0$ and $\langle u, w \rangle < 0$, then $\langle v, w \rangle < 0$. Suppose without loss of generality that u is a (timelike) unit vector. By Theorem 4.8.5,

$$v = \overline{v} + au \qquad \text{and} \qquad w = \overline{w} + bu,$$

where $\overline{v}, \overline{w}$ are (spacelike) vectors in $(\mathbb{R}u)^{\perp}$ and a, b are real numbers. It follows that

$$\langle u, v \rangle = -a \qquad \langle u, w \rangle = -b \tag{4.9.1}$$

$$\langle v, w \rangle = \langle \overline{v}, \overline{w} \rangle - ab \tag{4.9.2}$$

$$\langle v, v \rangle = \langle \overline{v}, \overline{v} \rangle - a^2 \qquad \langle w, w \rangle = \langle \overline{w}, \overline{w} \rangle - b^2. \tag{4.9.3}$$

By assumption, $\langle v, v \rangle < 0$ and $\langle w, w \rangle < 0$. Since $\overline{v}$ and $\overline{w}$ are spacelike, we have $\langle \overline{v}, \overline{v} \rangle \geq 0$ and $\langle \overline{w}, \overline{w} \rangle \geq 0$. It follows from (4.9.3) that

$$0 \leq \langle \overline{v}, \overline{v} \rangle < a^2 \qquad \text{and} \qquad 0 \leq \langle \overline{w}, \overline{w} \rangle < b^2,$$

hence

$$(ab)^2 > \langle \overline{v}, \overline{v} \rangle \langle \overline{w}, \overline{w} \rangle \geq 0. \tag{4.9.4}$$

By assumption, $\langle u, v \rangle < 0$ and $\langle u, w \rangle < 0$, so $\langle u, v \rangle \langle u, w \rangle > 0$. Then (4.9.1) yields

$$ab > 0. \tag{4.9.5}$$

Since u is timelike, we have from Theorem 4.8.2 that $\mathbb{R}u$ is timelike, and then from Theorem 4.8.4(b) that $(\mathbb{R}u)^{\perp}$ is spacelike. Thus, $\mathfrak{g}$ is an inner product on $(\mathbb{R}u)^{\perp}$. By Theorem 4.6.4,

$$\langle \overline{v}, \overline{v} \rangle \langle \overline{w}, \overline{w} \rangle \geq \langle \overline{v}, \overline{w} \rangle^2. \tag{4.9.6}$$

It follows from (4.9.4) and (4.9.6) that $(ab)^2 > \langle \overline{v}, \overline{w} \rangle^2$, and then from (4.9.5) that

$$ab > |\langle \overline{v}, \overline{w} \rangle|. \tag{4.9.7}$$

Finally, (4.9.2) and (4.9.7) give

$$\langle v, w \rangle = \langle \overline{v}, \overline{w} \rangle - ab \leq |\langle \overline{v}, \overline{w} \rangle| - ab < 0. \qquad\square$$

Let $(V, \mathfrak{g})$ be a Lorentz vector space, and let $\mathcal{T}$ be its time cone. It is clear that $\sim$ partitions $\mathcal{T}$ into two equivalence classes, which we denote by $\mathcal{T}^+$ and $\mathcal{T}^-$. We arbitrarily label $\mathcal{T}^+$ the **future time cone**, and $\mathcal{T}^-$ the **past time cone**. Vectors in $\mathcal{T}^+$ are said to be **future-directed**, while those in $\mathcal{T}^-$ are called **past-directed**. In $\mathbb{R}_1^m$, the future time cone and past time cone are denoted by $\mathcal{T}_m^+$ and $\mathcal{T}_m^-$, respectively.

Continuing with the above example, we observe that $(0,0,1)$ is a vector in $\mathcal{T}_3$. For a vector (x, y, z) in $\mathcal{T}_3$, we have

$$(x, y, z) \sim (0, 0, 1)$$
$$\Leftrightarrow \quad \langle (x, y, z), (0, 0, 1) \rangle < 0$$
$$\Leftrightarrow \quad -z < 0$$
$$\Leftrightarrow \quad z > 0.$$

Similarly, $(x, y, z) \sim (0, 0, -1)$ if and only if $z < 0$. Thus, $\mathcal{T}_3^+$ is either

$$\{(x, y, z) \in \mathcal{T}_3 : z > 0\} \qquad \text{or} \qquad \{(x, y, z) \in \mathcal{T}_3 : z < 0\},$$

but which one is a matter of choice.

Part (a) of the next result shows that time cones are aptly named.

Theorem 4.9.3. *If $(V, \mathfrak{g})$ is a Lorentz vector space, then:*
(a) $\mathcal{T}^+$ *and* $\mathcal{T}^-$ *are cone-shaped.*
(b) $\mathcal{T}^- = -\mathcal{T}^+$.

Proof. (a): Let v, w be vectors in $\mathcal{T}^+$, and let $c > 0$ be a real number. By definition, $\langle v, v \rangle, \langle w, w \rangle, \langle v, w \rangle < 0$. We have $\langle cv, cv \rangle < 0$ and $\langle cv, v \rangle < 0$, hence cv is in $\mathcal{T}$ and $cv \sim v$. By Theorem 4.9.2, cv is in $\mathcal{T}^+$. We also have $\langle v + w, v + w \rangle < 0$ and $\langle v + w, v \rangle < 0$, so $v + w$ is in $\mathcal{T}$ and $v + w \sim v$. Again by Theorem 4.9.2, $v + w$ is in $\mathcal{T}^+$. Thus, $\mathcal{T}^+$ is cone-shaped. A similar argument shows that so is $\mathcal{T}^-$.

(b): Let v be a vector in $\mathcal{T}^+$. Then $\langle -v, -v \rangle < 0$, hence $-v$ is in $\mathcal{T}$. Furthermore, $-v$ is not in $\mathcal{T}^+$; for if it were, then, since v is in $\mathcal{T}^+$, we would have $v \sim -v$, that is, $\langle v, -v \rangle < 0$, hence $\langle v, v \rangle > 0$, which contradicts the assumption on v. It follows that $-v$ is in $\mathcal{T}^-$, so v is in $-\mathcal{T}^-$. Then $\mathcal{T}^+ \subseteq -\mathcal{T}^-$, hence $-\mathcal{T}^+ \subseteq \mathcal{T}^-$. A similar argument shows that $\mathcal{T}^- \subseteq -\mathcal{T}^+$. Thus, $\mathcal{T}^- = -\mathcal{T}^+$. $\qquad\square$

Let $(V, \mathfrak{g})$ be a Lorentz vector space, let v, w be timelike vectors in V, and let u be a lightlike vector in V. It follows from Theorem 4.9.1 that $\langle u, v \rangle, \langle u, w \rangle \neq 0$, so $\langle u, v \rangle$ and $\langle u, w \rangle$ either have the same sign or opposite signs.

Theorem 4.9.4. *With the above setup, $\langle u, v \rangle$ and $\langle u, w \rangle$ have the same sign if and only if $v \sim w$.*

Proof. ($\Leftarrow$): By assumption, $v \sim w$, so either v, w are in $\mathcal{T}^+$ or v, w are in $\mathcal{T}^-$. Assume the former and suppose, for a contradiction, that $\langle u, v \rangle$ and $\langle u, w \rangle$ have opposite signs, say, $\langle u, v \rangle < 0$ and $\langle u, w \rangle > 0$. Observe that $-\langle u, v \rangle / \langle u, w \rangle > 0$, and let $\widetilde{w} = -(\langle u, v \rangle / \langle u, w \rangle)w$. By Theorem 4.9.3(a), $v + \widetilde{w}$ is in $\mathcal{T}^+$. It follows from Theorem 4.8.2 that $\mathbb{R}(v + \widetilde{w})$ is timelike, and then from Theorem 4.8.4(b) that $\mathbb{R}(v + \widetilde{w})^\perp$ is spacelike. Since $\langle u, v \rangle = -\langle u, \widetilde{w} \rangle$, hence $\langle u, v + \widetilde{w} \rangle = 0$, we see that u is in $\mathbb{R}(v + \widetilde{w})^\perp$. By Theorem 4.8.3, u is spacelike, which contradicts the assumption on u. Thus, $\langle u, v \rangle$ and $\langle u, w \rangle$ have the same sign. The proof when v, w are in $\mathcal{T}^-$ is similar.

($\Rightarrow$): We prove the logically equivalent assertion: if $v \not\sim w$, then $\langle u, v \rangle$ and $\langle u, w \rangle$ have opposite signs. Since $v \not\sim w$, either v is in $\mathcal{T}^+$ and w is in $\mathcal{T}^-$, or v is in $\mathcal{T}^-$ and w is in $\mathcal{T}^+$. Suppose the former. By Theorem 4.9.3(b), $-w$ is in $\mathcal{T}^+$, hence $v \sim -w$. It follows from ($\Leftarrow$) that $\langle u, v \rangle$ and $\langle u, -w \rangle$ have the same sign, hence $\langle u, v \rangle$ and $\langle u, w \rangle$ have opposite signs. The proof when v is in $\mathcal{T}^-$ and w is in $\mathcal{T}^+$ is similar. $\qquad\square$

Theorem 4.9.4 provides a mechanism for partitioning a light cone in a way that is consistent with the time structure of the time cone. Let $(V, \mathfrak{g})$ be a Lorentz vector space, and let u be a lightlike vector in V. We say that u is **future-directed** if $\langle u, v \rangle < 0$ for some (hence every) vector v in $\mathcal{T}^+$, and **past-directed** if $\langle u, w \rangle < 0$ for some (hence every) vector w in $\mathcal{T}^-$. The **future light cone** is denoted by Λ^+ and defined to be the set of future-directed lightlike vectors:

$$\Lambda^+ = \{u \in \Lambda : \langle u, v \rangle < 0 \text{ for some (hence every) } v \in \mathcal{T}^+\}.$$

Similarly, the **past light cone** is denoted by Λ^- and defined to be the set of past-directed lightlike vectors:

$$\Lambda^- = \{u \in \Lambda : \langle u, w \rangle < 0 \text{ for some (hence every) } w \in \mathcal{T}^-\}.$$

The future and past light cones in $\mathbb{R}_1^m$ are denoted by Λ_m^+ and Λ_m^-, respectively.

Theorem 4.9.5. *Let $(V, \mathfrak{g})$ be a Lorentz vector space, let $A : V \longrightarrow V$ be a linear isometry, and let v be a vector in V. Then v is spacelike (resp., timelike, lightlike) if and only if $A(v)$ is spacelike (resp., timelike, lightlike).*

Proof. Straightforward. $\qquad\square$

Theorem 4.9.6. *Let $(V, \mathfrak{g})$ be a Lorentz vector space, and let $A : V \longrightarrow V$ be a linear isometry. Then either $A(\mathcal{T}^+) = \mathcal{T}^+$ (equivalently, $A(\mathcal{T}^-) = \mathcal{T}^-$), in which case A is said to be **orthochronous**, or $A(\mathcal{T}^+) = \mathcal{T}^-$ (equivalently, $A(\mathcal{T}^-) = \mathcal{T}^+$), in which case A is said to be **nonorthochronous**.*

Proof. It follows from Theorem 4.9.5 that $A(\mathcal{T}) = \mathcal{T}$.

Let v, w be vectors in $\mathcal{T}^+$. Then $\langle A(v), A(w) \rangle = \langle v, w \rangle < 0$, so either $A(v), A(w)$ are in $\mathcal{T}^+$ or $A(v), A(w)$ are $\mathcal{T}^-$. Since v, w were arbitrary, it follows that either $A(\mathcal{T}^+) \subseteq \mathcal{T}^+$ or $A(\mathcal{T}^+) \subseteq \mathcal{T}^-$. Suppose $A(\mathcal{T}^+) \subseteq \mathcal{T}^+$. Since A^{-1} is a linear isometry, the preceding argument shows that either $A^{-1}(\mathcal{T}^+) \subseteq \mathcal{T}^+$ or $A^{-1}(\mathcal{T}^+) \subseteq \mathcal{T}^-$, or equivalently, either $\mathcal{T}^+ \subseteq A(\mathcal{T}^+)$ or $\mathcal{T}^+ \subseteq A(\mathcal{T}^-)$. If $\mathcal{T}^+ \subseteq A(\mathcal{T}^+)$, then, since $A(\mathcal{T}^+) \subseteq \mathcal{T}^+$, we have $A(\mathcal{T}^+) = \mathcal{T}^+$. On the other hand, if $\mathcal{T}^+ \subseteq A(\mathcal{T}^-)$, then Theorem 4.9.3(b) gives $\mathcal{T}^+ \subseteq -A(\mathcal{T}^+)$, hence $\mathcal{T}^- = -\mathcal{T}^+ \subseteq A(\mathcal{T}^+) \subseteq \mathcal{T}^+$, which is contradiction. Thus, $A(\mathcal{T}^+) = \mathcal{T}^+$. Now suppose $A(\mathcal{T}^+) \subseteq \mathcal{T}^-$. An argument similar to that just given shows that $A(\mathcal{T}^+) = \mathcal{T}^-$.

Let v, w be vectors in $\mathcal{T}^-$. An argument similar to that just given shows that either $A(\mathcal{T}^-) = \mathcal{T}^-$ or $A(\mathcal{T}^-) = \mathcal{T}^+$. $\qquad\square$

Let $(V, \mathfrak{g})$ be a Lorentz vector space, and let $(e_1, \ldots, e_m)$ be an orthonormal basis for V. Since e_m is timelike, it is in either $\mathcal{T}^+$ or $\mathcal{T}^-$. The labeling of time cones in V is arbitrary, so we are free to adopt the following convention.

Given an orthonormal basis $(e_1, \ldots, e_m)$ in a Lorentz vector space, $\mathcal{T}^+$ is the time cone containing e_m.

Thus,

$$\mathcal{T}^+ = \{v \in \mathcal{T} : \langle v, e_m \rangle < 0\} \qquad (4.9.8)$$

and

$$\Lambda^+ = \{u \in \Lambda : \langle u, e_m \rangle < 0\},$$

where we note that e_m is in $\mathcal{T}^+$. It needs to be emphasized that $\mathcal{T}^+$ and Λ^+ depend on the choice of orthonormal basis. In particular, if we decide to use the orthonormal basis $(e_1, \ldots, -e_m)$ instead of $(e_1, \ldots, e_m)$, then "future" and "past" are reversed.

Theorem 4.9.7. *With the above setup, let v be a vector in $\mathcal{T}$, with $v = \sum_i a^i e_i$. Then v is in $\mathcal{T}^+$ if and only if $a^m > 0$.*

Proof. We have

$$\langle v, e_m \rangle = \left\langle \sum_i a^i e_i, e_m \right\rangle = \langle e_m, e_m \rangle a^m = -a^m,$$

from which the result follows. $\qquad\square$

Continuing with the earlier example, let us choose the standard basis (e_1, e_2, e_3) for $\mathbb{R}_1^3$. According to the above convention, $e_3 = (0, 0, 1)$ is a future-directed vector, so

$$\mathcal{T}_3^+ = \{(x, y, z) \in \mathcal{T}_3 : z > 0\} \qquad \text{and} \qquad \Lambda_3^+ = \{(x, y, z) \in \Lambda_3 : z > 0\}.$$

We close this section with what is perhaps the most striking illustration of the divide that separates the geometry of Lorentz vector spaces from that of inner product spaces. We observed in connection with Theorem 4.6.5 that in an inner product space, it is shorter to walk diagonally across a rectangular field than to go around the corner. The next result shows that, by contrast, in a Lorentz vector space, the opposite is true when the rectangular field is determined by equivalent timelike vectors.

Theorem 4.9.8 (Reversed Triangle Inequality). *If $(V, \mathfrak{g})$ is a Lorentz vector space and v, w are timelike vectors, with $v \sim w$, then*

$$\|v + w\| \geq \|v\| + \|w\|,$$

with equality if and only if v, w are linearly dependent.

Proof. It follows from $\langle v, v \rangle < 0$ and $\langle w, w \rangle < 0$ that

$$\|v\|^2 = -\langle v, v \rangle \qquad \text{and} \qquad \|w\|^2 = -\langle w, w \rangle, \tag{4.9.9}$$

and then from Theorem 4.8.10 that

$$\|v\|^2 \|w\|^2 = \langle v, v \rangle \langle w, w \rangle \leq \langle v, w \rangle^2.$$

Since $v \sim w$, we have $\langle v, w \rangle < 0$, hence

$$\|v\| \|w\| \leq -\langle v, w \rangle. \tag{4.9.10}$$

Also, since $v \sim w$, either v, w are in $\mathcal{T}^+$ or v, w are in $\mathcal{T}^-$. Suppose the former. By Theorem 4.9.3(a), $v + w$ is in $\mathcal{T}^+$, hence $\langle v + w, v + w \rangle < 0$, so

$$\|v + w\|^2 = -\langle v + w, v + w \rangle. \tag{4.9.11}$$

Then

$$
\begin{aligned}
(\|v\| + \|w\|)^2 &= \|v\|^2 + 2\|v\| \|w\| + \|w\|^2 \\
&\leq -\langle v, v \rangle - 2\langle v, w \rangle - \langle w, w \rangle && [(4.9.9), (4.9.10)] \\
&= -\langle v + w, v + w \rangle \\
&= \|v + w\|^2, && [(4.9.11)]
\end{aligned}
$$

hence $\|v\| + \|w\| \leq \|v + w\|$, with equality if and only if $\|v\| \|w\| = -\langle v, w \rangle$. We have from (4.9.9) that the preceding identity is equivalent to $\langle v, v \rangle \langle w, w \rangle = \langle v, w \rangle^2$, which in turn, by Theorem 4.8.10, is equivalent to v, w being linearly dependent. The proof when v, w are in $\mathcal{T}^-$ is similar. $\qquad \square$

Chapter 5

Tensors on Vector Spaces

5.1 Tensors

Let V be a vector space, and let $r, s \geq 1$ be integers. Following Section B.5 and Section B.6, we denote by $\mathrm{Mult}(V^{*r} \times V^s, \mathbb{R})$ the vector space of $\mathbb{R}$-multilinear functions from $V^{*r} \times V^s$ to $\mathbb{R}$, where addition and scalar multiplication are defined as follows: for all functions $\mathcal{A}, \mathcal{B}$ in $\mathrm{Mult}(V^{*r} \times V^s, \mathbb{R})$ and all real numbers c,

$$(\mathcal{A} + \mathcal{B})(\eta^1, \ldots, \eta^r, v_1, \ldots, v_s)$$
$$= \mathcal{A}(\eta^1, \ldots, \eta^r, v_1, \ldots, v_s) + \mathcal{B}(\eta^1, \ldots, \eta^r, v_1, \ldots, v_s)$$

and

$$(c\mathcal{A})(\eta^1, \ldots, \eta^r, v_1, \ldots, v_s) = c\mathcal{A}(\eta^1, \ldots, \eta^r, v_1, \ldots, v_s)$$

for all covectors $\eta^1, \ldots, \eta^r$ in V^* and all vectors $v_1, \ldots, v_s$ in V. For brevity, let us denote

$$\mathrm{Mult}(V^{*r} \times V^s, \mathbb{R}) \qquad \text{by} \qquad \mathcal{T}_s^r(V).$$

We refer to a function $\mathcal{A}$ in $\mathcal{T}_s^r(V)$ as an **(r, s)-tensor**, an **r-contravariant-s-covariant tensor**, or simply a **tensor** on V, and we define the **rank of $\mathcal{A}$** to be (r, s). When $s = 0$, $\mathcal{A}$ is said to be an **r-contravariant tensor** or just a **contravariant tensor**, and when $r = 0$, $\mathcal{A}$ is said to be an **s-covariant tensor** or simply a **covariant tensor**. In Section 1.2, we used the term covector to describe what we now call a 1-covariant tensor. Note that the sum of tensors is defined only when they have the same rank. The zero element of $\mathcal{T}_s^r(V)$, called the **zero tensor** and denoted by 0, is the tensor that sends all elements of $V^{*r} \times V^s$ to the real number 0. A tensor in $\mathcal{T}_s^r(V)$ is said to be **nonzero** if it is not the zero tensor. Let us observe that

$$\mathcal{T}_1^0(V) = V^* \qquad \text{and} \qquad \mathcal{T}_0^1(V) = V^{**} = V. \tag{5.1.1}$$

Semi-Riemannian Geometry, First Edition. Stephen C. Newman.
© 2019 John Wiley & Sons, Inc. Published 2019 by John Wiley & Sons, Inc.

For completeness, we define
$$\mathcal{T}_0^0(V) = \mathbb{R}. \tag{5.1.2}$$

It is important not to confuse the above vector space operations with the multilinear property. The latter means that a tensor is linear in each of its arguments. That is, for all covectors $\eta^1, \ldots, \eta^r, \zeta$ in V^*, all vectors $v_1, \ldots, v_s, w$ in V, and all real numbers c, we have

$$\mathcal{A}(\eta^1, \ldots, c\eta^i + \zeta, \ldots, \eta^r, v_1, \ldots, v_s)$$
$$= c\,\mathcal{A}(\eta^1, \ldots, \eta^i, \ldots, \eta^r, v_1, \ldots, v_s) + \mathcal{A}(\eta^1, \ldots, \zeta, \ldots, \eta^r, v_1, \ldots, v_s)$$

for $i = 1, \ldots, r$, and

$$\mathcal{A}(\eta^1, \ldots, \eta^r, v_1, \ldots, cv_j + w, \ldots, v_s)$$
$$= c\,\mathcal{A}(\eta^1, \ldots, \eta^r, v_1, \ldots, v_j, \ldots, v_s) + \mathcal{A}(\eta^1, \ldots, \eta^r, v_1, \ldots, w, \ldots, v_s)$$

for $j = 1, \ldots, s$.

Example 5.1.1. The classic example of a tensor is the determinant function det, which is an m-covariant tensor on $\mathrm{Mat}_{m \times 1}$. Another obvious example is a bilinear function $\mathfrak{b}$ on a vector space V, which is a 2-covariant tensor on V. Here are two more examples of tensors. For given vectors v, w in V, the function $\mathcal{A} : V^{*2} \longrightarrow \mathbb{R}$ defined by $\mathcal{A}(\eta, \zeta) = \eta(v)\zeta(w)$ for all covector η, ζ in V^* is a 2-contravariant tensor on V^*. Similarly, for given covectors η, ζ in V^*, the function $\mathcal{B} : V^2 \longrightarrow \mathbb{R}$ defined by $\mathcal{B}(v, w) = \eta(v)\zeta(w)$ for all vector v, w in V is a 2-covariant tensor on V. $\diamond$

Computations involving tensors rely in a crucial way on the identification $V^{**} = V$, which is included as part of (5.1.1). We repeat here the remarks made at the close of Section 1.2: for a vector v in V and a covector η in V^*, the element of V^{**} corresponding to v is also denoted by v, so that, by definition,

$$v(\eta) = \eta(v). \tag{5.1.3}$$

Thus, $\mathcal{B}$ in Example 5.1.1 can be expressed as $\mathcal{B}(v, w) = v(\eta)w(\zeta)$. Despite (5.1.3), we usually prefer the notation $\eta(v)$ to $v(\eta)$.

We now define a type of multiplication of tensors. **Tensor product** is the family of linear maps

$$\otimes \colon \mathcal{T}_s^r(V) \times \mathcal{T}_{s'}^{r'}(V) \longrightarrow \mathcal{T}_{s+s'}^{r+r'}(V)$$

defined for $r, s, r', s' \geq 0$ by

$$(\mathcal{A} \otimes \mathcal{B})(\eta^1, \ldots, \eta^{r+r'}, v_1, \ldots, v_{s+s'})$$
$$= \mathcal{A}(\eta^1, \ldots, \eta^r, v_1, \ldots, v_s)\,\mathcal{B}(\eta^{r+1}, \ldots, \eta^{r+r'}, v_{s+1}, \ldots, v_{s+s'}) \tag{5.1.4}$$

for all tensors $\mathcal{A}$ in $\mathcal{T}_s^r(V)$ and $\mathcal{B}$ in $\mathcal{T}_{s'}^{r'}(V)$, all covectors $\eta^1, \ldots, \eta^{r+r'}$ in V^*, and all vectors $v_1, \ldots, v_{s+s'}$ in V. We refer to $\mathcal{A} \otimes \mathcal{B}$ as the **tensor product**

of $\mathcal{A}$ **and** $\mathcal{B}$. Unlike with addition, the product of tensors does not require $\mathcal{A}$ and $\mathcal{B}$ to have the same rank. Usually, $\mathcal{A} \otimes \mathcal{B} \neq \mathcal{B} \otimes \mathcal{A}$, so the tensor product is generally not commutative.

The next result gives the basic algebraic properties of tensors.

Theorem 5.1.2. *Let V be a vector space, let $\mathcal{A}, \mathcal{A}_1, \mathcal{A}_2$ and $\mathcal{B}, \mathcal{B}_1, \mathcal{B}_2$ and C be tensors in $\mathcal{T}_s^r(V)$ and $\mathcal{T}_{s'}^{r'}(V)$ and $\mathcal{T}_{s''}^{r''}(V)$, respectively, and let c be a real number. Then:*
(a) $(\mathcal{A}_1 + \mathcal{A}_2) \otimes \mathcal{B} = \mathcal{A}_1 \otimes \mathcal{B} + \mathcal{A}_2 \otimes \mathcal{B}$.
(b) $\mathcal{A} \otimes (\mathcal{B}_1 + \mathcal{B}_2) = \mathcal{A} \otimes \mathcal{B}_1 + \mathcal{A} \otimes \mathcal{B}_2$.
(c) $(c\mathcal{A}) \otimes \mathcal{B} = c(\mathcal{A} \otimes \mathcal{B}) = \mathcal{A} \otimes (c\mathcal{B})$.
(d) $(\mathcal{A} \otimes \mathcal{B}) \otimes C = \mathcal{A} \otimes (\mathcal{B} \otimes C)$.

Proof. Straightforward. □

In view of Theorem 5.1.2(d), we drop parentheses around tensor products and, for example, denote $(\mathcal{A} \otimes \mathcal{B}) \otimes C$ and $\mathcal{A} \otimes (\mathcal{B} \otimes C)$ by $\mathcal{A} \otimes \mathcal{B} \otimes C$, with corresponding notation for tensor products of more than three terms. By forming the tensor product of collections of vectors and covectors, we can construct tensors of arbitrary rank. To illustrate, from vectors $w_1, \ldots, w_r$ in V and covectors $\zeta^1, \ldots, \zeta^s$ in V^*, we obtain the tensor $w_1 \otimes \cdots \otimes w_r \otimes \zeta^1 \otimes \cdots \otimes \zeta^s$ in $\mathcal{T}_s^r(V)$. Then (5.1.4) gives

$$w_1 \otimes \cdots \otimes w_r \otimes \zeta^1 \otimes \cdots \otimes \zeta^s (\eta^1, \ldots, \eta^r, v_1, \ldots, v_s)$$
$$= w_1(\eta^1) \cdots w_r(\eta^r) \zeta^1(v_1) \cdots \zeta^s(v_s)$$

for all all covectors $\eta^1, \ldots, \eta^r$ in V^* and all vectors $v_1, \ldots, v_s$ in V.

Let V be a vector space, let $\mathcal{A}$ be a tensor in $\mathcal{T}_s^r(V)$, let $\mathcal{H} = (h_1, \ldots, h_m)$ be a basis for V, and let $\Theta = (\theta^1, \ldots, \theta^m)$ be its dual basis. The **components of** $\mathcal{A}$ with respect to $\mathcal{H}$ are defined by

$$\mathcal{A}_{j_1 \ldots j_s}^{i_1 \ldots i_r} = \mathcal{A}(\theta^{i_1}, \ldots, \theta^{i_r}, h_{j_1}, \ldots, h_{j_s})$$

for all $1 \leq i_1, \ldots, i_r \leq m$ and $1 \leq j_1, \ldots, j_s \leq m$.

Theorem 5.1.3 (Basis for $\mathcal{T}_s^r(V)$). *With the above setup:*
(a)

$$\mathcal{A} = \sum_{\substack{1 \leq i_1, \ldots, i_r \leq m \\ 1 \leq j_1, \ldots, j_s \leq m}} \mathcal{A}_{j_1 \ldots j_s}^{i_1 \ldots i_r} h_{i_1} \otimes \cdots \otimes h_{i_r} \otimes \theta^{j_1} \otimes \cdots \otimes \theta^{j_s}. \qquad (5.1.5)$$

(b)

$$\{h_{i_1} \otimes \cdots \otimes h_{i_r} \otimes \theta^{j_1} \otimes \cdots \otimes \theta^{j_s} \in \mathcal{T}_s^r(V) : 1 \leq i_1, \ldots, i_r \leq m;$$
$$1 \leq j_1, \ldots, j_s \leq m\}$$

is an unordered basis for $\mathcal{T}_s^r(V)$.
(c) *$\mathcal{T}_s^r(V)$ has dimension m^{r+s}.*

Proof. (a), (b): In view of preceding remarks, each $h_{i_1} \otimes \cdots \otimes h_{i_r} \otimes \theta^{j_1} \otimes \cdots \otimes \theta^{j_s}$ is in $\mathcal{T}_s^r(V)$. Let $\widetilde{\mathcal{A}}$ be the tensor in $\mathcal{T}_s^r(V)$ defined by

$$\widetilde{\mathcal{A}} = \sum_{\substack{1 \le i_1,\ldots,i_r \le m \\ 1 \le j_1,\ldots,j_s \le m}} \mathcal{A}_{j_1 \ldots j_s}^{i_1 \ldots i_r} h_{i_1} \otimes \cdots \otimes h_{i_r} \otimes \theta^{j_1} \otimes \cdots \otimes \theta^{j_s}.$$

Then

$$\widetilde{\mathcal{A}}(\theta^{i_1},\ldots,\theta^{i_r},h_{j_1},\ldots,h_{j_s})$$

$$= \sum_{\substack{1 \le k_1,\ldots,k_r \le m \\ 1 \le l_1,\ldots,l_s \le m}} \mathcal{A}_{l_1 \ldots l_s}^{k_1 \ldots k_r} h_{k_1} \otimes \cdots \otimes h_{k_r} \otimes \theta^{l_1} \otimes \cdots \otimes \theta^{l_s}$$

$$(\theta^{i_1},\ldots,\theta^{i_r},h_{j_1},\ldots,h_{j_s})$$

$$= \sum_{\substack{1 \le k_1,\ldots,k_r \le m \\ 1 \le l_1,\ldots,l_s \le m}} \mathcal{A}_{l_1 \ldots l_s}^{k_1 \ldots k_r} h_{k_1}(\theta^{i_1}) \cdots h_{k_r}(\theta^{i_r}) \theta^{l_1}(h_{j_1}) \cdots \theta^{l_s}(h_{j_s})$$

$$= \sum_{\substack{1 \le k_1,\ldots,k_r \le m \\ 1 \le l_1,\ldots,l_s \le m}} \mathcal{A}_{l_1 \ldots l_s}^{k_1 \ldots k_r} \delta_{k_1}^{i_1} \cdots \delta_{k_r}^{i_r} \delta_{j_1}^{l_1} \cdots \delta_{j_s}^{l_s}$$

$$= \mathcal{A}_{j_1 \ldots j_s}^{i_1 \ldots i_r} = \mathcal{A}(\theta^{i_1},\ldots,\theta^{i_r},h_{j_1},\ldots,h_{j_s}).$$

Since $\mathcal{H}$ and Θ are bases for V and V^*, respectively, and $\mathcal{A}$ and $\mathcal{B}$ are multilinear, it follows that $\mathcal{A} = \widetilde{\mathcal{A}}$. Thus, the $h_{i_1} \otimes \cdots \otimes h_{i_r} \otimes \theta^{j_1} \otimes \cdots \otimes \theta^{j_s}$ span $\mathcal{T}_s^r(V)$. If

$$\sum_{\substack{1 \le i_1,\ldots,i_r \le m \\ 1 \le j_1,\ldots,j_s \le m}} a_{j_1 \ldots j_s}^{i_1 \ldots i_r} h_{i_1} \otimes \cdots \otimes h_{i_r} \otimes \theta^{j_1} \otimes \cdots \otimes \theta^{j_s} = 0$$

for some real numbers $a_{j_1 \ldots j_s}^{i_1 \ldots i_r}$, then applying both sides of the preceding identity to $(\theta^{k_1},\ldots,\theta^{k_r},h_{l_1},\ldots,h_{l_s})$ yields $a_{l_1 \ldots l_s}^{k_1 \ldots k_r} = 0$. Thus, the $h_{i_1} \otimes \cdots \otimes h_{i_r} \otimes \theta^{j_1} \otimes \cdots \otimes \theta^{j_s}$ are linearly independent.

(c): This follows from part (a). $\qquad\square$

We see from Theorem 5.1.3 that a tensor is completely determined by its components with respect to a given basis. In fact, this is the classical way of defining a tensor. Choosing another basis generally yields different values for the components. The next result shows how to convert components when changing from one basis to another.

Theorem 5.1.4 (Change of Basis). *Let V be a vector space, let $\mathcal{H}$ and $\widetilde{\mathcal{H}}$ be bases for V, and let $[\mathrm{id}_V]_{\widetilde{\mathcal{H}}}^{\mathcal{H}} = [a_j^i]$ and $\left([\mathrm{id}_V]_{\widetilde{\mathcal{H}}}^{\mathcal{H}}\right)^{-1} = [b_j^i]$. Let $\mathcal{A}$ be a tensor in $\mathcal{T}_s^r(V)$, and let $\mathcal{A}_{j_1 \ldots j_s}^{i_1 \ldots i_r}$ and $\widetilde{\mathcal{A}}_{j_1 \ldots j_s}^{i_1 \ldots i_r}$ be the components of $\mathcal{A}$ with respect to $\mathcal{H}$ and $\widetilde{\mathcal{H}}$, respectively. Then*

$$\widetilde{\mathcal{A}}_{j_1 \ldots j_s}^{i_1 \ldots i_r} = \sum_{\substack{1 \le k_1,\ldots,k_r \le m \\ 1 \le l_1,\ldots,l_s \le m}} b_{k_1}^{i_1} \cdots b_{k_r}^{i_r} a_{j_1}^{l_1} \cdots a_{j_s}^{l_s} \mathcal{A}_{l_1 \ldots l_s}^{k_1 \ldots k_r}.$$

Proof. Let $\mathcal{H} = (h_1, \ldots, h_m)$ and $\widetilde{\mathcal{H}} = (\widetilde{h}_1, \ldots, \widetilde{h}_m)$, and let $\Theta = (\theta^1, \ldots, \theta^m)$ and $\widetilde{\Theta} = (\widetilde{\theta}^1, \ldots, \widetilde{\theta}^m)$ be the corresponding dual bases. We have

$$
\begin{aligned}
\left[\mathrm{id}_{V^*}\right]_{\widetilde{\Theta}}^{\Theta} &= \left(\left[\mathrm{id}_{V^*}\right]_{\Theta}^{\widetilde{\Theta}}\right)^{-1} && \text{[Th 2.2.5(b)]} \\
&= \left(\left[(\mathrm{id}_V)^*\right]_{\Theta}^{\widetilde{\Theta}}\right)^{-1} \\
&= \left(\left(\left[\mathrm{id}_V\right]_{\widetilde{\mathcal{H}}}^{\mathcal{H}}\right)^{\mathrm{T}}\right)^{-1} && \text{[Th 2.2.7]} \\
&= \left(\left(\left[\mathrm{id}_V\right]_{\widetilde{\mathcal{H}}}^{\mathcal{H}}\right)^{-1}\right)^{\mathrm{T}} && \text{[Th 2.1.3(c)]} \\
&= \left[b_j^i\right]^{\mathrm{T}}.
\end{aligned}
$$

Using (2.2.6) and (2.2.7) gives

$$
\begin{aligned}
\widetilde{\mathcal{A}}_{j_1 \ldots j_s}^{i_1 \ldots i_r} &= \mathcal{A}(\widetilde{\theta}^{i_1}, \ldots, \widetilde{\theta}^{i_r}, \widetilde{h}_{j_1}, \ldots, \widetilde{h}_{j_s}) \\
&= \mathcal{A}\left(\sum_{k_1} b_{k_1}^{i_1} \theta^{k_1}, \ldots, \sum_{k_r} b_{k_r}^{i_r} \theta^{k_r}, \sum_{l_1} a_{j_1}^{l_1} h_{l_1}, \ldots, \sum_{l_s} a_{j_s}^{l_s} h_{l_s}\right) \\
&= \sum_{\substack{1 \le k_1, \ldots, k_r \le m \\ 1 \le l_1, \ldots, l_s \le m}} b_{k_1}^{i_1} \cdots b_{k_r}^{i_r} a_{j_1}^{l_1} \cdots a_{j_s}^{l_s} \, \mathcal{A}(\theta^{k_1}, \ldots, \theta^{k_r}, h_{l_1}, \ldots, h_{l_s}) \\
&= \sum_{\substack{1 \le k_1, \ldots, k_r \le m \\ 1 \le l_1, \ldots, l_s \le m}} b_{k_1}^{i_1} \cdots b_{k_r}^{i_r} a_{j_1}^{l_1} \cdots a_{j_s}^{l_s} \, \mathcal{A}_{l_1 \ldots l_s}^{k_1 \ldots k_r}. \qquad \square
\end{aligned}
$$

As the next result shows, the components of the product of tensors are given by the product of the respective components.

Theorem 5.1.5 (Product of Tensors). *Let V be a vector space, let $\mathcal{A}$ and $\mathcal{B}$ be tensors in $T_s^r(V)$ and $T_{s'}^{r'}(V)$, respectively, let $\mathcal{H} = (h_1, \ldots, h_m)$ be a basis for V, and let $(\theta^1, \ldots, \theta^m)$ be its dual basis. Let $\mathcal{A}_{j_1 \ldots j_s}^{i_1 \ldots i_r}$ and $\mathcal{B}_{l_1 \ldots l_{s'}}^{k_1 \ldots k_{r'}}$ be the components of $\mathcal{A}$ and $\mathcal{B}$ with respect to $\mathcal{H}$, respectively, so that*

$$
\mathcal{A} = \sum_{\substack{1 \le i_1, \ldots, i_r \le m \\ 1 \le j_1, \ldots, j_s \le m}} \mathcal{A}_{j_1 \ldots j_s}^{i_1 \ldots i_r} h_{i_1} \otimes \cdots \otimes h_{i_r} \otimes \theta^{j_1} \otimes \cdots \otimes \theta^{j_s}
$$

and

$$
\mathcal{B} = \sum_{\substack{1 \le k_1, \ldots, k_{r'} \le m \\ 1 \le l_1, \ldots, l_{s'} \le m}} \mathcal{B}_{l_1 \ldots l_{s'}}^{k_1 \ldots k_{r'}} h_{k_1} \otimes \cdots \otimes h_{k_{r'}} \otimes \theta^{l_1} \otimes \cdots \otimes \theta^{l_{s'}}.
$$

Then $\mathcal{A} \otimes \mathcal{B}$ has the components $\mathcal{A}_{j_1 \ldots j_s}^{i_1 \ldots i_r} \mathcal{B}_{l_1 \ldots l_{s'}}^{k_1 \ldots k_{r'}}$ with respect to $\mathcal{H}$, so that

$$
\begin{aligned}
\mathcal{A} \otimes \mathcal{B} = \sum_{\substack{1 \le i_1, \ldots, i_r, k_1, \ldots, k_{r'} \le m \\ 1 \le j_1, \ldots, j_s, l_1, \ldots, l_{s'} \le m}} &\mathcal{A}_{j_1 \ldots j_s}^{i_1 \ldots i_r} \mathcal{B}_{l_1 \ldots l_{s'}}^{k_1 \ldots k_{r'}} h_{i_1} \otimes \cdots \otimes h_{i_r} \otimes h_{k_1} \otimes \cdots \otimes h_{k_{r'}} \\
&\otimes \theta^{j_1} \otimes \cdots \otimes \theta^{j_s} \otimes \theta^{l_1} \otimes \cdots \otimes \theta^{l_{s'}}.
\end{aligned}
$$

Proof. It follows from Theorem 5.1.2 that

$$
\mathcal{A} \otimes \mathcal{B} = \sum_{\substack{1 \leq i_1, \ldots, i_r, k_1, \ldots, k_{r'} \leq m \\ 1 \leq j_1, \ldots, j_s, l_1, \ldots, l_{s'} \leq m}} \mathcal{A}^{i_1 \ldots i_r}_{j_1 \ldots j_s} \mathcal{B}^{k_1 \ldots k_{r'}}_{l_1 \ldots l_{s'}} (h_{i_1} \otimes \cdots \otimes h_{i_r} \otimes \theta^{j_1} \otimes \cdots \otimes \theta^{j_s})
$$

$$
\otimes (h_{k_1} \otimes \cdots \otimes h_{k_{r'}} \otimes \theta^{l_1} \otimes \cdots \otimes \theta^{l_{s'}}).
$$

We need to check that

$$
(h_{i_1} \otimes \cdots \otimes h_{i_r} \otimes \theta^{j_1} \otimes \cdots \otimes \theta^{j_s}) \otimes (h_{k_1} \otimes \cdots \otimes h_{k_{r'}} \otimes \theta^{l_1} \otimes \cdots \otimes \theta^{l_{s'}})
$$

$$
= (h_{i_1} \otimes \cdots \otimes h_{i_r} \otimes h_{k_1} \otimes \cdots \otimes h_{k_{r'}}) \otimes (\theta^{j_1} \otimes \cdots \otimes \theta^{j_s} \otimes \theta^{l_1} \otimes \cdots \otimes \theta^{l_{s'}}).
$$

Let $\eta^1, \ldots, \eta^{r+r'}$ be covectors in V^*, and let $v_1, \ldots, v_{s+s'}$ be vectors in V. We have from (5.1.4) that applying the left-hand side of the above expression to $(\eta^1, \ldots, \eta^{r+r'}, v_1, \ldots, v_{s+s'})$ gives

$$
[h_{i_1}(\eta^1) \cdots h_{i_r}(\eta^r)][\theta^{j_1}(v_1) \cdots \theta^{j_s}(v_s)]
$$

$$
\cdot [h_{k_1}(\eta^{r+1}) \cdots h_{k_{r'}}(\eta^{r+r'})][\theta^{l_1}(v_{s+1}) \cdots \theta^{l_{s'}}(v_{s+s'})],
$$

while applying the right-hand side gives

$$
[h_{i_1}(\eta^1) \cdots h_{i_r}(\eta^r)][h_{k_1}(\eta^{r+1}) \cdots h_{k_{r'}}(\eta^{r+r'})]
$$

$$
\cdot [\theta^{j_1}(v_1) \cdots \theta^{j_s}(v_s)][\theta^{l_1}(v_{s+1}) \cdots \theta^{l_{s'}}(v_{s+s'})].
$$

The result follows.   □

Theorem 5.1.5 shows that the product of tensors, each of which is expressed in the standard format given by (5.1.5), can in turn be expressed in that same format.

Example 5.1.6 (Kronecker's Delta). Let V be a vector space, let $\mathcal{H} = (h_1, \ldots, h_m)$ be a basis for V, and let $(\theta^1, \ldots, \theta^m)$ be its dual basis. As remarked above, in the classical literature a tensor is specified by giving its components with respect to a chosen basis for V. From that perspective, perhaps the most basic of all tensors in $\mathcal{T}^1_1(V)$ is the one with components δ^i_j given by Kronecker's delta. Accordingly, we consider the $(1,1)$-tensor

$$
\delta = \sum_{ij} \delta^i_j h_i \otimes \theta^j.
$$

Let η be a covector in V^*, and let v be a vector in V, with $\eta = \sum_k a_k \theta^k$ and $v = \sum_l b^l h_l$. Then

$$
\begin{aligned}
\delta(\eta, v) &= \left(\sum_{ij} \delta^i_j h_i \otimes \theta^j \right) \left(\sum_k a_k \theta^k, \sum_l b^l h_l \right) \\
&= \sum_{ij} \delta^i_j \left[h_i \left(\sum_k a_k \theta^k \right) \theta^j \left(\sum_l b^l h_l \right) \right] \\
&= \sum_{ij} \delta^i_j \left[\left(\sum_k a_k h_i(\theta^k) \right) \left(\sum_l b^l \theta^j (h_l) \right) \right] \\
&= \sum_{ij} \delta^i_j a_i b^j = \sum_i a_i b^i
\end{aligned}
$$

and

$$
\begin{aligned}
\eta(v) &= \left(\sum_i a_i \theta^i \right) \left(\sum_j b^j h_j \right) = \sum_i a_i \left[\theta^i \left(\sum_j b^j h_j \right) \right] \\
&= \sum_{ij} a_i b^j \theta^i (h_j) = \sum_i a_i b^i.
\end{aligned}
$$

Thus, $\delta(\eta, v) = \eta(v)$. $\Diamond$

5.2   Pullback of Covariant Tensors

In this brief section, we generalize to tensors the definition of pullback by a linear map given in Section 1.3. Let V and W be vector spaces, and let $A : V \longrightarrow W$ be a linear map. **Pullback by A (for covariant tensors)** is the family of linear maps

$$
A^* : T^0_s(W) \longrightarrow T^0_s(V)
$$

defined for $s \geq 1$ by

$$
A^*(\mathcal{B})(v_1, \ldots, v_s) = \mathcal{B}\big(A(v_1), \ldots, A(v_s)\big) \tag{5.2.1}
$$

for all tensors $\mathcal{B}$ in $T^0_s(W)$ and all vectors $v_1, \ldots, v_s$ in V. We refer to $A^*(\mathcal{B})$ as the **pullback of $\mathcal{B}$ by A**. When $s = 1$, the earlier definition is recovered. As an example, let $(V, \mathfrak{g})$ be a scalar product space, and let $A : V \longrightarrow V$ be a linear map. Then the tensor $A^*(\mathfrak{g})$ in $T^0_2(V)$ is given by $A^*(\mathfrak{g})(v, w) = \langle A(v), A(w) \rangle$ for all vectors v, w in V.

Pullbacks of covariant tensors behave well with respect to basic algebraic structure.

Theorem 5.2.1. *Let $U, V,$ and W be vector spaces, let $A : U \longrightarrow V$ and $B : V \longrightarrow W$ be linear maps, and let $\mathcal{A}, \mathcal{B}$ and $\mathcal{C}$ be tensors in $T^0_s(V)$ and $T^0_{s'}(V)$, respectively. Then:*
(a) $A^*(\mathcal{A} + \mathcal{B}) = A^*(\mathcal{A}) + A^*(\mathcal{B})$.
(b) $A^*(\mathcal{A} \otimes \mathcal{C}) = A^*(\mathcal{A}) \otimes A^*(\mathcal{C})$.

(c) $(B \circ A)^* = A^* \circ B^*$.

Proof. Straightforward.   □

5.3   Representation of Tensors

Let V be a vector space, let $\mathcal{H} = (h_1, \ldots, h_m)$ be a basis for V, let $(\theta^1, \ldots, \theta^m)$ be its dual basis, and let $s \geq 1$ be an integer. Following Section B.5 and Section B.6, we denote by $\mathrm{Mult}(V^s, V)$ the vector space of $\mathbb{R}$-multilinear maps from V^s to V. Let us define a map

$$\mathfrak{R}_s : \mathrm{Mult}(V^s, V) \longrightarrow \mathcal{T}^1_s(V),$$

called the **representation map**, by

$$\mathfrak{R}_s(\Psi)(\eta, v_1, \ldots, v_s) = \eta\big(\Psi(v_1, \ldots, v_s)\big) \tag{5.3.1}$$

for all maps Ψ in $\mathrm{Mult}(V^s, V)$, all covectors η in V^*, and all vectors $v_1, \ldots, v_s$ in V. When $s = 1$, we denote

$$\mathfrak{R}_1 \qquad \text{by} \qquad \mathfrak{R}.$$

Theorem 5.3.1 (Representation of Tensors). *With the above setup:*
(a) $\mathfrak{R}_s$ *is a linear isomorphism:*

$$\mathcal{T}^1_s(V) \approx \mathrm{Mult}(V^s, V).$$

(b) *For all maps* Ψ *in* $\mathrm{Mult}(V^s, V)$, *the components of the tensor* $\mathfrak{R}_s(\Psi)$ *in* $\mathcal{T}^1_s(V)$ *with respect to* $\mathcal{H}$ *are*

$$\mathfrak{R}_s(\Psi)^i_{j_1 \ldots j_s} = \theta^i\big(\Psi(h_{j_1}, \ldots, h_{j_s})\big).$$

(c) *For all tensors* $\mathcal{A}$ *in* $\mathcal{T}^1_s(V)$,

$$\mathfrak{R}_s^{-1}(\mathcal{A})(h_{j_1}, \ldots, h_{j_s}) = \sum_i \mathcal{A}^i_{j_1 \ldots j_s} h_i,$$

where the $\mathcal{A}^i_{j_1 \ldots j_s}$ *are the components of* $\mathcal{A}$ *with respect to* $\mathcal{H}$.

Proof. (a): It is easily shown that $\mathfrak{R}_s$ is a linear map. If $\mathfrak{R}_s(\Psi) = 0$, then $\eta\big(\Psi(v_1, \ldots, v_s)\big) = 0$ for all covectors η in V^* and vectors $v_1, \ldots, v_s$ in V. In particular, each $\theta^i\big(\Psi(v_1, \ldots, v_s)\big) = 0$. Because $\Psi(v_1, \ldots, v_s)$ is a vector in V, by Theorem 1.2.1(d),

$$\Psi(v_1, \ldots, v_s) = \sum_i \theta^i\big(\Psi(v_1, \ldots, v_s)\big) h_i,$$

hence $\Psi(v_1, \ldots, v_s) = 0$. Since $v_1, \ldots, v_s$ were arbitrary, $\Psi = 0$. Thus, $\ker(\mathfrak{R}_s) = \{0\}$. By Theorem 5.1.3(c), $\mathcal{T}^1_s(V)$ has dimension m^{s+1}, and it can be shown that $\text{Mult}(V^s, V)$ has the same dimension. It follows from Theorem 1.1.12 and Theorem 1.1.14 that $\mathfrak{R}_s$ is a linear isomorphism.

(b): We have from (5.3.1) that

$$\mathfrak{R}_s(\Psi)^i_{j_1 \ldots j_s} = \mathfrak{R}_s(\Psi)(\theta^i, h_{j_1}, \ldots, h_{j_s}) = \theta^i\big(\Psi(h_{j_1}, \ldots, h_{j_s})\big).$$

(c): Since $\mathfrak{R}_s^{-1}(\mathcal{A})(h_{j_1}, \ldots, h_{j_s})$ is a vector in V, by Theorem 1.2.1(d),

$$\mathfrak{R}_s^{-1}(\mathcal{A})(h_{j_1}, \ldots, h_{j_s}) = \sum_i \theta^i\big(\mathfrak{R}_s^{-1}(\mathcal{A})(h_{j_1}, \ldots, h_{j_s})\big)h_i.$$

We have from (5.3.1) and part (a) that

$$\begin{aligned}
\mathcal{A}^i_{j_1 \ldots j_s} &= \mathcal{A}(\theta^i, h_{j_1}, \ldots, h_{j_s}) = \mathfrak{R}_s\big(\mathfrak{R}_s^{-1}(\mathcal{A})\big)(\theta^i, h_{j_1}, \ldots, h_{j_s}) \\
&= \theta^i\big(\mathfrak{R}_s^{-1}(\mathcal{A})(h_{j_1}, \ldots, h_{j_s})\big).
\end{aligned}$$

The result follows.   □

It is instructive to specialize Theorem 5.3.1 to the case $s = 1$, which for us is the situation of most interest. We then have $\text{Mult}(V, V) = \text{Lin}(V, V)$ and

$$\mathfrak{R} : \text{Lin}(V, V) \longrightarrow \mathcal{T}^1_1(V)$$

defined by

$$\mathfrak{R}(B)(\eta, v) = \eta\big(B(v)\big) = B^*(\eta)(v) \tag{5.3.2}$$

for all maps B in $\text{Lin}(V, V)$, all covectors η in V^*, and all vectors v in V.

Theorem 5.3.2. *With the above setup:*

(a) *$\mathfrak{R}$ is a linear isomorphism:*

$$\mathcal{T}^1_1(V) \approx \text{Lin}(V, V).$$

(b) *For all maps B in $\text{Lin}(V, V)$, the components of the tensor $\mathfrak{R}(B)$ in $\mathcal{T}^1_1(V)$ with respect to $\mathcal{H}$ are*

$$\mathfrak{R}(B)^i_j = \theta^i\big(B(h_j)\big).$$

(c) *For all tensors $\mathcal{A}$ in $\mathcal{T}^1_1(V)$,*

$$\mathfrak{R}^{-1}(\mathcal{A})(h_j) = \sum_i \mathcal{A}^i_j h_i,$$

hence

$$\big[\mathfrak{R}^{-1}(\mathcal{A})\big]^{\mathcal{H}}_{\mathcal{H}} = \big[\mathcal{A}^i_j\big],$$

where the $\mathcal{A}^i_j$ are the components of $\mathcal{A}$ with respect to $\mathcal{H}$.

Proof. This follows from Theorem 5.3.1, except for the second identity in part (c), which comes from (2.2.2) and (2.2.3).   □

Theorem 5.3.2(c) shows that each tensor in $\mathcal{T}^1_1(V)$ has a representation as a matrix. Continuing with Example 5.1.6, the tensor δ in $\mathcal{T}^1_1(V)$ has the components δ^i_j with respect to $\mathcal{H}$. By Theorem 5.3.2(c),

$$\big[\mathfrak{R}^{-1}(\delta)\big]^{\mathcal{H}}_{\mathcal{H}} = \big[\delta^i_j\big] = I_m,$$

so, not surprisingly, $\mathfrak{R}^{-1}(\delta) = \text{id}_V$.

5.4   Contraction of Tensors

Contraction of tensors is closely related to the trace of matrices. Like dual vector spaces, it is another of those deceptively simple algebraic constructions that has far-reaching applications.

Theorem 5.4.1. *Let V be a vector space, and define a linear function*

$$\mathsf{C}_1^1 : \mathcal{T}_1^1(V) \longrightarrow \mathbb{R},$$

called $(1,1)$-contraction, by

$$\mathsf{C}_1^1 = \mathrm{tr} \circ \mathfrak{R}^{-1},$$

where tr *is trace and* $\mathfrak{R}$ *is given by (5.3.2). Then* C_1^1 *is the unique linear function such that*

$$\mathsf{C}_1^1(v \otimes \eta) = v(\eta) \tag{5.4.1}$$

for all vectors v in V and all covectors η in V^*.

Proof. Let $\mathcal{H} = (h_1, \ldots, h_m)$ be a basis for V, and let $(\theta^1, \ldots, \theta^m)$ be its dual basis.

Existence. By Theorem 5.3.2(c),

$$\left[\mathfrak{R}^{-1}(v \otimes \eta)\right]_{\mathcal{H}}^{\mathcal{H}} = \left[(v \otimes \eta)_j^i\right]. \tag{5.4.2}$$

Since

$$(v \otimes \eta)_j^i = (v \otimes \eta)(\theta^i, h_j) = v(\theta^i)\eta(h_j),$$

it follows from Theorem 1.2.1(d) and (1.2.3) that

$$\mathrm{tr}\left(\left[(v \otimes \eta)_j^i\right]\right) = \sum_i v(\theta^i)\eta(h_i) = \eta\left(\sum_i \theta^i(v)h_i\right) = \eta(v). \tag{5.4.3}$$

Thus,

$$
\begin{aligned}
\mathrm{tr}\left(\mathfrak{R}^{-1}(v \otimes \eta)\right) &= \mathrm{tr}\left(\left[\mathfrak{R}^{-1}(v \otimes \eta)\right]_{\mathcal{H}}^{\mathcal{H}}\right) && [(2.5.1)] \\
&= \mathrm{tr}\left(\left[(v \otimes \eta)_j^i\right]\right) && [(5.4.2)] \\
&= \eta(v) && [(5.4.3)] \\
&= v(\eta), && [(1.2.3)]
\end{aligned}
$$

which shows that $\mathrm{tr} \circ \mathfrak{R}^{-1}$ satisfies (5.4.1).

Uniqueness. Suppose $\ell : \mathcal{T}_1^1(V) \longrightarrow \mathbb{R}$ is a linear function satisfying (5.4.1), let $\mathcal{A}$ be a tensor in $\mathcal{T}_1^1(V)$, and let $\mathcal{A}_j^i$ be the components of $\mathcal{A}$ with respect to

$\mathcal{H}$. Then

$$
\ell(\mathcal{A}) = \ell\left(\sum_{ij} \mathcal{A}^i_j h_i \otimes \theta^j\right) = \sum_{ij} \mathcal{A}^i_j \ell(h_i \otimes \theta^j)
$$

$$
= \sum_{ij} \mathcal{A}^i_j h_i(\theta^j) \qquad\qquad [(5.4.1)]
$$

$$
= \sum_i \mathcal{A}^i_i = \mathrm{tr}\left([\mathcal{A}^i_j]\right)
$$

$$
= \mathrm{tr}\left([\mathfrak{R}^{-1}(\mathcal{A})]^{\mathcal{H}}_{\mathcal{H}}\right) \qquad\qquad [\text{Th 5.3.2(c)}]
$$

$$
= \mathrm{tr}\left(\mathfrak{R}^{-1}(\mathcal{A})\right). \qquad\qquad [(2.5.1)]
$$

Since $\mathcal{A}$ was arbitrary, $\ell = \mathrm{tr} \circ \mathfrak{R}^{-1}$.   □

Theorem 5.4.2. *Let V be a vector space, let $\mathcal{H}$ be a basis for V, let $\mathcal{A}$ be a tensor in $T^1_1(V)$, and let $\mathcal{A}^i_j$ be the components of $\mathcal{A}$ with respect to $\mathcal{H}$. Then*

$$
\mathsf{C}^1_1(\mathcal{A}) = \sum_i \mathcal{A}^i_i.
$$

Proof. This was shown as part of the proof of Theorem 5.4.1.   □

Theorem 5.4.2 explains why in the literature "contraction" is sometimes called "trace". The next result is a generalization of Theorem 5.4.1.

Theorem 5.4.3. *If V is a vector space, and $1 \le k \le r$ and $1 \le l \le s$ are integers, then there is a unique linear map*

$$
\mathsf{C}^k_l : T^r_s(V) \longrightarrow T^{r-1}_{s-1}(V),
$$

called (k, l)-contraction, such that

$$
\mathsf{C}^k_l(v_1 \otimes \cdots \otimes v_r \otimes \eta^1 \otimes \cdots \otimes \eta^s)
$$

$$
= v_k(\eta^l)\, v_1 \otimes \cdots \otimes \widehat{v_k} \otimes \cdots \otimes v_r \otimes \eta^1 \otimes \cdots \otimes \widehat{\eta^l} \otimes \cdots \otimes \eta^s
$$

for all vectors $v_1, \ldots, v_r$ in V and all covectors $\eta^1, \ldots, \eta^s$ in V^*, where $\widehat{}$ indicates that an expression is omitted.

Remark. In light of (5.1.2), when $r = s = 1$, we recover (5.4.1).

Proof. Existence. Let $\mathcal{A}$ be a tensor in $T^r_s(V)$, let $\zeta^1, \ldots, \zeta^{k-1}, \zeta^{k+1}, \ldots, \zeta^r$ be covectors in V^*, and let $w_1, \ldots, w_{l-1}, w_{l+1}, \ldots, w_s$ be vectors in V. We temporarily treat the preceding covectors and vectors as fixed and define a tensor $\tilde{\mathcal{A}}$ in $T^1_1(V)$ by the assignment

$$
(\zeta, w) \longmapsto \mathcal{A}(\zeta^1, \ldots, \zeta^{k-1}, \zeta, \zeta^{k+1}, \ldots, \zeta^r, w_1, \ldots, w_{l-1}, w, w_{l+1}, \ldots, w_s)
$$

for all covectors ζ in V^* and all vectors w in V. Using Theorem 5.4.1, we obtain the corresponding $(1, 1)$-contraction $\mathsf{C}^1_1(\tilde{\mathcal{A}})$. Allowing $\zeta^1, \ldots, \zeta^{k-1}, \zeta^{k+1}, \ldots, \zeta^r$

and $w_1, \ldots, w_{l-1}, w_{l+1}, \ldots, w_s$ to vary defines a tensor $\mathsf{C}_l^k(\mathcal{A})$ in $\mathcal{T}_{s-1}^{r-1}(V)$. It follows from

$$
\begin{aligned}
&v_1 \otimes \cdots \otimes v_r \otimes \eta^1 \otimes \cdots \otimes \eta^s \\
&\quad (\zeta^1, \ldots, \zeta^{k-1}, \zeta, \zeta^{k+1}, \ldots, \zeta^r, w_1, \ldots, w_{l-1}, w, w_{l+1}, \ldots, w_s) \\
&= [v_1(\zeta^1) \cdots v_{k-1}(\zeta^{k-1})] \, v_k(\zeta) \, [v_{k+1}(\zeta^{k+1}) \cdots v_r(\zeta^r)] \\
&\quad \cdot [\eta^1(w_1) \cdots \eta^{l-1}(w_{l-1})] \, \eta^l(w) \, [\eta^{l+1}(w_{l+1}) \cdots \eta^s(w_s)] \\
&= \big([v_1(\zeta^1) \cdots v_{k-1}(\zeta^{k-1})] \, [v_{k+1}(\zeta^{k+1}) \cdots v_r(\zeta^r)] \\
&\quad \cdot [\eta^1(w_1) \cdots \eta^{l-1}(w_{l-1})] \, [\eta^{l+1}(w_{l+1}) \cdots \eta^s(w_s)] \, (v_k \otimes \eta^l) \big) (\zeta, w)
\end{aligned}
$$

and Theorem 5.4.1 that

$$
\begin{aligned}
&\mathsf{C}_l^k(v_1 \otimes \cdots \otimes v_r \otimes \eta^1 \otimes \cdots \otimes \eta^s) \\
&\quad (\zeta^1, \ldots, \zeta^{k-1}, \zeta^{k+1}, \ldots, \zeta^r, w_1, \ldots, w_{l-1}, w_{l+1}, \ldots, w_s) \\
&= [v_1(\zeta^1) \cdots v_{k-1}(\zeta^{k-1})] \, [v_{k+1}(\zeta^{k+1}) \cdots v_r(\zeta^r)] \\
&\quad \cdot [\eta^1(w_1) \cdots \eta^{l-1}(w_{l-1})] \, [\eta^{l+1}(w_{l+1}) \cdots \eta^s(w_s)] \, \mathsf{C}_1^1(v_k \otimes \eta^l) \\
&= [v_1(\zeta^1) \cdots v_{k-1}(\zeta^{k-1})] \, [v_{k+1}(\zeta^{k+1}) \cdots v_r(\zeta^r)] \\
&\quad \cdot [\eta^1(w_1) \cdots \eta^{l-1}(w_{l-1})] \, [\eta^{l+1}(w_{l+1}) \cdots \eta^s(w_s)] \, v_k(\eta^l) \\
&= v_k(\eta^l) \, v_1 \otimes \cdots \otimes \widehat{v_k} \otimes \cdots \otimes v_r \otimes \eta^1 \otimes \cdots \otimes \widehat{\eta^l} \otimes \cdots \otimes \eta^s \\
&\quad (\zeta^1, \ldots, \zeta^{k-1}, \zeta^{k+1}, \ldots, \zeta^r, w_1, \ldots, w_{l-1}, w_{l+1}, \ldots, w_s).
\end{aligned}
$$

Since $\zeta^1, \ldots, \zeta^{k-1}, \zeta^{k+1}, \ldots, \zeta^r, w_1, \ldots, w_{l-1}, w_{l+1}, \ldots, w_s$ were arbitrary, the result follows.

Uniqueness. This follows from the uniqueness property of C_1^1 described in Theorem 5.4.1. $\qquad\square$

We say that (k, l) is the **rank** of a (k, l)-contraction. The composite of contractions (of various ranks) is called simply a **contraction**. We sometimes refer to the type of contractions defined here as **ordinary**, to distinguish them from a variant to be described in Section 6.5.

Example 5.4.4. Let V be a vector space, let $\mathcal{H} = (h_1, \ldots, h_m)$ be a basis for V, and let $(\theta^1, \ldots, \theta^m)$ be its dual basis. Let $\mathcal{A}$ be a tensor in $\mathcal{T}_2^2(V)$, and let $\mathcal{A}_{kl}^{ij}$ be the components of $\mathcal{A}$ with respect to $\mathcal{H}$, so that

$$
\mathcal{A} = \sum_{ijkl} \mathcal{A}_{kl}^{ij} \, h_i \otimes h_j \otimes \theta^k \otimes \theta^l.
$$

Then the tensor $\mathsf{C}_2^1(\mathcal{A})$ in $\mathcal{T}_1^1(V)$ is given by

$$\mathsf{C}_2^1(\mathcal{A}) = \sum_{ijkl} \mathcal{A}_{kl}^{ij} \, \mathsf{C}_2^1(h_i \otimes h_j \otimes \theta^k \otimes \theta^l)$$

$$= \sum_{ijkl} \mathcal{A}_{kl}^{ij} \, h_i(\theta^l) \, h_j \otimes \theta^k = \sum_{jkp} \mathcal{A}_{kp}^{pj} h_j \otimes \theta^k$$

$$= \sum_{jk} \left(\sum_p \mathcal{A}_{kp}^{pj} \right) h_j \otimes \theta^k = \sum_{ij} \left(\sum_p \mathcal{A}_{jp}^{pi} \right) h_i \otimes \theta^j.$$

Thus, $\mathsf{C}_2^1(\mathcal{A})$ has the components $\mathsf{C}_2^1(\mathcal{A})_j^i = \sum_p \mathcal{A}_{jp}^{pi}$ with respect to $\mathcal{H}$. $\Diamond$

Theorem 5.4.5. *Let V be a vector space, let $\mathcal{H}$ be a basis for V, and let $1 \le k \le r$ and $1 \le l \le s$ be integers. Let $\mathcal{A}$ be a tensor in $\mathcal{T}_s^r(V)$ with the components $\mathcal{A}_{j_1 \ldots j_s}^{i_1 \ldots i_r}$ with respect to $\mathcal{H}$. Then the tensor $\mathsf{C}_l^k(\mathcal{A})$ in $\mathcal{T}_{s-1}^{r-1}(V)$ has the components*

$$\mathsf{C}_l^k(\mathcal{A})_{j_1 \ldots j_{s-1}}^{i_1 \ldots i_{r-1}} = \sum_p \mathcal{A}_{j_1 \ldots j_{l-1} \, p \, j_l \ldots j_{s-1}}^{i_1 \ldots i_{k-1} \, p \, i_k \ldots i_{r-1}}$$

with respect to $\mathcal{H}$.

Proof. The proof is an elaboration of Example 5.4.4. Let $\mathcal{H} = (h_1, \ldots, h_m)$, and let $(\theta^1, \ldots, \theta^m)$ be its dual basis, so that

$$\mathcal{A} = \sum_{\substack{1 \le i_1, \ldots, i_r \le m \\ 1 \le j_1, \ldots, j_s \le m}} \mathcal{A}_{j_1 \ldots j_s}^{i_1 \ldots i_r} h_{i_1} \otimes \cdots \otimes h_{i_r} \otimes \theta^{j_1} \otimes \cdots \otimes \theta^{j_s}.$$

Then

$$\mathsf{C}_l^k(\mathcal{A}) = \sum_{\substack{1 \le i_1, \ldots, i_r \le m \\ 1 \le j_1, \ldots, j_s \le m}} \mathcal{A}_{j_1 \ldots j_s}^{i_1 \ldots i_r} \, \mathsf{C}_l^k(h_{i_1} \otimes \cdots \otimes h_{i_r} \otimes \theta^{j_1} \otimes \cdots \otimes \theta^{j_s})$$

$$= \sum_{\substack{1 \le i_1, \ldots, i_r \le m \\ 1 \le j_1, \ldots, j_s \le m}} \mathcal{A}_{j_1 \ldots j_s}^{i_1 \ldots i_r} h_{i_k}(\theta^{j_l})$$

$$\cdot \, h_{i_1} \otimes \cdots \otimes \widehat{h_{i_k}} \otimes \cdots \otimes h_{i_r} \otimes \theta^{j_1} \otimes \cdots \otimes \widehat{\theta^{j_l}} \otimes \cdots \otimes \theta^{j_s}$$

$$= \sum_{\substack{1 \le i_1, \ldots, \widehat{i_k}, \ldots i_r \le m \\ 1 \le j_1, \ldots, \widehat{j_l}, \ldots, j_s \le m}} \left(\sum_p \mathcal{A}_{j_1 \ldots j_{l-1} \, p \, j_{l+1} \ldots j_s}^{i_1 \ldots i_{k-1} \, p \, i_{k+1} \ldots i_r} \right)$$

$$\cdot \, h_{i_1} \otimes \cdots \otimes \widehat{h_{i_k}} \otimes \cdots \otimes h_{i_r} \otimes \theta^{j_1} \otimes \cdots \otimes \widehat{\theta^{j_l}} \otimes \cdots \otimes \theta^{j_s},$$

where $\frown$ indicates that an expression is omitted. Relabeling indices $(i_{k+1}, \ldots, i_r)$ as $(i_k, \ldots, i_{r-1})$, and $(j_{l+1}, \ldots, j_s)$ as $(j_l, \ldots, j_{s-1})$, gives

$$\mathsf{C}_l^k(\mathcal{A}) = \sum_{\substack{1 \le i_1, \ldots i_{r-1} \le m \\ 1 \le j_1, \ldots, j_{s-1} \le m}} \left(\sum_p \mathcal{A}_{j_1 \ldots j_{l-1} \, p \, j_l \ldots j_{s-1}}^{i_1 \ldots i_{k-1} \, p \, i_k \ldots i_{r-1}} \right)$$

$$\cdot \, h_{i_1} \otimes \cdots \otimes h_{i_{r-1}} \otimes \theta^{j_1} \otimes \cdots \otimes \theta^{j_{s-1}},$$

from which the result follows.   □

Example 5.4.6. Let V be a vector space, let $\mathcal{A}$ be a tensor in $\mathcal{T}_1^1(V)$, let η be a covector in V^*, and let v be a vector in V. We claim that

$$\mathcal{A}(\eta, v) = \mathsf{C}_1^1 \circ \mathsf{C}_2^2(v \otimes \mathcal{A} \otimes \eta).$$

Let $\mathcal{H} = (h_1, \ldots, h_m)$ be a basis for V, let $(\theta^1, \ldots, \theta^m)$ be its dual basis, and, in local coordinates, let

$$\mathcal{A} = \sum_{ij} \mathcal{A}_j^i h_i \otimes \theta^j, \qquad \eta = \sum_k a_k \theta^k, \qquad \text{and} \qquad v = \sum_l b^l h_l.$$

Then

$$
\begin{aligned}
\mathcal{A}(\eta, v) &= \left(\sum_{ij} \mathcal{A}_j^i h_i \otimes \theta^j \right) \left(\sum_k a_k \theta^k, \sum_l b^l h_l \right) \\
&= \sum_{ij} \mathcal{A}_j^i \left[h_i \left(\sum_k a_k \theta^k \right) \theta^j \left(\sum_l b^l h_l \right) \right] \\
&= \sum_{ij} \mathcal{A}_j^i \left[\left(\sum_k a_k h_i(\theta^k) \right) \left(\sum_l b^l \theta^j(h_l) \right) \right] \\
&= \sum_{ij} \mathcal{A}_j^i a_i b^j
\end{aligned}
\tag{5.4.4}
$$

and

$$
\begin{aligned}
v \otimes \mathcal{A} \otimes \eta &= \left(\sum_l b^l h_l \right) \otimes \left(\sum_{ij} \mathcal{A}_j^i h_i \otimes \theta^j \right) \otimes \left(\sum_k a_k \theta^k \right) \\
&= \sum_{ijkl} b^l \mathcal{A}_j^i a_k \, h_l \otimes h_i \otimes \theta^j \otimes \theta^k,
\end{aligned}
$$

hence

$$
\begin{aligned}
\mathsf{C}_2^2(v \otimes \mathcal{A} \otimes \eta) &= \sum_{ijkl} b^l \mathcal{A}_j^i a_k \mathsf{C}_2^2(h_l \otimes h_i \otimes \theta^j \otimes \theta^k) \\
&= \sum_{ijkl} b^l \mathcal{A}_j^i a_k h_i(\theta^k) h_l \otimes \theta^j = \sum_{ijl} b^l \mathcal{A}_j^i a_i h_l \otimes \theta^j,
\end{aligned}
$$

so

$$
\begin{aligned}
\mathsf{C}_1^1 \circ \mathsf{C}_2^2(v \otimes \mathcal{A} \otimes \eta) &= \sum_{ijl} b^l \mathcal{A}_j^i a_i \mathsf{C}_1^1(h_l \otimes \theta^j) \\
&= \sum_{ijl} b^l \mathcal{A}_j^i a_i \, h_l(\theta^j) = \sum_{ij} a_i b^j \mathcal{A}_j^i.
\end{aligned}
\tag{5.4.5}
$$

The result follows from (5.4.4) and (5.4.5).   ◇

Theorem 5.4.7. *Let V be a vector space, let $\mathcal{A}$ be a tensor in $\mathcal{T}_s^r(V)$, let $\eta^1, \ldots, \eta^r$ be covectors in V^*, and let $v_1, \ldots, v_s$ be vectors in V. Then there is a contraction C on V such that*

$$\mathcal{A}(\eta^1, \ldots, \eta^r, v_1, \ldots, v_s) = \mathsf{C}(v_1 \otimes \cdots \otimes v_s \otimes \mathcal{A} \otimes \eta^1 \otimes \cdots \otimes \eta^r).$$

Proof. The proof is an elaboration of Example 5.4.6. Let $\mathcal{H} = (h_1, \ldots, h_m)$ be a basis for V, let $(\theta^1, \ldots, \theta^m)$ be its dual basis, and, in local coordinates, let

$$\mathcal{A} = \sum_{\substack{1 \le i_1, \ldots, i_r \le m \\ 1 \le j_1, \ldots, j_s \le m}} \mathcal{A}^{i_1 \ldots i_r}_{j_1 \ldots j_s} h_{i_1} \otimes \cdots \otimes h_{i_r} \otimes \theta^{j_1} \otimes \cdots \otimes \theta^{j_s},$$

$$\eta^1 = \sum_{k_1} a_{1k_1} \theta^{k_1}, \ldots, \eta^r = \sum_{k_r} a_{rk_r} \theta^{k_r}$$

and

$$v_1 = \sum_{l_1} b^{1l_1} h_{l_1}, \ldots, v_s = \sum_{l_s} b^{sl_s} h_{l_s}.$$

Then

$$\mathcal{A}(\eta^1, \ldots, \eta^r, v_1, \ldots, v_s)$$

$$= \left(\sum_{\substack{1 \le i_1, \ldots, i_r \le m \\ 1 \le j_1, \ldots, j_s \le m}} \mathcal{A}^{i_1 \ldots i_r}_{j_1 \ldots j_s} h_{i_1} \otimes \cdots \otimes h_{i_r} \otimes \theta^{j_1} \otimes \cdots \otimes \theta^{j_s} \right)$$

$$\left(\sum_{k_1} a_{1k_1} \theta^{k_1}, \ldots, \sum_{k_r} a_{rk_r} \theta^{k_r}, \sum_{l_1} b^{1l_1} h_{l_1}, \ldots, \sum_{l_s} b^{sl_s} h_{l_s} \right)$$

$$= \sum_{\substack{1 \le i_1, \ldots, i_r \le m \\ 1 \le j_1, \ldots, j_s \le m}} \mathcal{A}^{i_1 \ldots i_r}_{j_1 \ldots j_s} \left[h_{i_1} \left(\sum_{k_1} a_{1k_1} \theta^{k_1} \right) \cdots h_{i_r} \left(\sum_{k_r} a_{rk_r} \theta^{k_r} \right) \right.$$

$$\left. \cdot \theta^{j_1} \left(\sum_{l_1} b^{1l_1} h_{l_1} \right) \cdots \theta^{j_s} \left(\sum_{l_s} b^{sl_s} h_{l_s} \right) \right] \qquad (5.4.6)$$

$$= \sum_{\substack{1 \le i_1, \ldots, i_r \le m \\ 1 \le j_1, \ldots, j_s \le m}} \mathcal{A}^{i_1 \ldots i_r}_{j_1 \ldots j_s} \left[\left(\sum_{k_1} a_{1k_1} h_{i_1}(\theta^{k_1}) \right) \cdots \left(\sum_{k_r} a_{rk_r} h_{i_r}(\theta^{k_r}) \right) \right.$$

$$\left. \cdot \left(\sum_{l_1} b^{1l_1} \theta^{j_1}(h_{l_1}) \right) \cdots \left(\sum_{l_s} b^{sl_s} \theta^{j_s}(h_{l_s}) \right) \right]$$

$$= \sum_{\substack{1 \le i_1, \ldots, i_r \le m \\ 1 \le j_1, \ldots, j_s \le m}} \mathcal{A}^{i_1 \ldots i_r}_{j_1 \ldots j_s} a_{1i_1} \cdots a_{ri_r} b^{1j_1} \cdots b^{sj_s}$$

and

$$v_1 \otimes \cdots \otimes v_s \otimes \mathcal{A} \otimes \eta^1 \otimes \cdots \otimes \eta^r$$

$$= \left(\sum_{l_1} b^{1l_1} h_{l_1} \right) \otimes \cdots \otimes \left(\sum_{l_s} b^{sl_s} h_{l_s} \right)$$

$$\otimes \sum_{\substack{1 \le i_1, \ldots, i_r \le m \\ 1 \le j_1, \ldots, j_s \le m}} \mathcal{A}^{i_1 \ldots i_r}_{j_1 \ldots j_s} h_{i_1} \otimes \cdots \otimes h_{i_r} \otimes \theta^{j_1} \otimes \cdots \otimes \theta^{j_s}$$

$$\otimes \left(\sum_{k_1} a_{1k_1} \theta^{k_1} \right) \otimes \cdots \otimes \left(\sum_{k_r} a_{rk_r} \theta^{k_r} \right)$$

$$= \sum_{\substack{1 \le i_1,\ldots,i_r,k_1,\ldots,k_r \le m \\ 1 \le j_1,\ldots,j_s,l_1,\ldots,l_s \le m}} b^{1l_1} \cdots b^{sl_s} \mathcal{A}^{i_1\ldots i_r}_{j_1\ldots j_s} a_{1k_1} \cdots a_{rk_r}$$

$$\cdot (h_{l_1} \otimes \cdots \otimes h_{l_s}) \otimes (h_{i_1} \otimes \cdots \otimes h_{i_r} \otimes \theta^{j_1} \otimes \cdots \otimes \theta^{j_s})$$
$$\otimes (\theta^{k_1} \otimes \cdots \otimes \theta^{k_r}).$$

To complete the proof, it is now a matter of defining a sequence of (k,l)-contractions to apply to $v_1 \otimes \cdots \otimes v_s \otimes \mathcal{A} \otimes \eta^1 \otimes \cdots \otimes \eta^r$ so that

h_{i_1} is paired with θ^{k_1} to give $h_{i_1}(\theta^{k_1})$, then

$\vdots$

h_{i_r} is paired with θ^{k_r} to give $h_{i_r}(\theta^{k_r})$, then

θ^{j_1} is paired with h_{l_1} to give $\theta^{j_1}(h_{l_1})$, then

$\vdots$

θ^{j_s} is paired with h_{l_s} to give $\theta^{j_s}(h_{l_s})$.

Denoting the composition of these contractions by C, we obtain

$$\mathsf{C}(v_1 \otimes \cdots \otimes v_s \otimes \mathcal{A} \otimes \eta^1 \otimes \cdots \otimes \eta^r)$$
$$= \sum_{\substack{1 \le i_1,\ldots,i_r \le m \\ 1 \le j_1,\ldots,j_s \le m}} b^{1j_1} \cdots b^{sj_s} \mathcal{A}^{i_1\ldots i_r}_{j_1\ldots j_s} a_{1i_1} \cdots a_{ri_r}. \tag{5.4.7}$$

The result now follows from (5.4.6) and (5.4.7). □

Chapter 6

Tensors on Scalar Product Spaces

6.1 Contraction of Tensors

In Section 5.4, we discussed contraction of tensors over vector spaces. In the setting of scalar product spaces, there is more to say.

Theorem 6.1.1 (Orthonormal Basis Expression for Ordinary Contraction). *Let $(V, \mathfrak{g})$ be a scalar product space with signature $(\varepsilon_1, \ldots, \varepsilon_m)$, let $(e_1, \ldots, e_m)$ be an orthonormal basis for V, and let $\mathcal{A}$ be a tensor in $\mathcal{T}^1_s(V)$. Then:*

(a) *For integers $s \geq 2$ and $1 \leq l \leq s-1$, the tensor $\mathsf{C}^1_l(\mathcal{A})$ in $\mathcal{T}^0_{s-1}(V)$ is given by*

$$\mathsf{C}^1_l(\mathcal{A})(v_1, \ldots, v_{s-1}) = \sum_i \varepsilon_i \langle \mathfrak{R}_s^{-1}(\mathcal{A})(v_1, \ldots, v_{l-1}, e_i, v_l, \ldots, v_{s-1}), e_i \rangle$$

for all vectors $v_1, \ldots, v_{s-1}$ in V, where $\mathfrak{R}_s$ is given by (5.3.1).

(b) *For $s = 1$, the real number $\mathsf{C}^1_1(\mathcal{A})$ is given by*

$$\mathsf{C}^1_1(\mathcal{A}) = \sum_i \varepsilon_i \langle \mathfrak{R}^{-1}(\mathcal{A})(e_i), e_i \rangle,$$

where $\mathfrak{R}$ is given by (5.3.2).

Proof. (a): By Theorem 5.4.5, the components of $\mathsf{C}^1_l(\mathcal{A})$ with respect to $(e_1, \ldots, e_m)$ are

$$\mathsf{C}^1_l(\mathcal{A})(e_{j_1}, \ldots, e_{j_{s-1}}) = \mathsf{C}^1_l(\mathcal{A})_{j_1 \ldots j_{s-1}} = \sum_p \mathcal{A}^p_{j_1 \ldots j_{l-1} \, p \, j_l \ldots j_{s-1}}. \tag{6.1.1}$$

Semi-Riemannian Geometry, First Edition. Stephen C. Newman.
© 2019 John Wiley & Sons, Inc. Published 2019 by John Wiley & Sons, Inc.

We have from Theorem 5.3.1(c) that

$$\mathfrak{R}_s^{-1}(\mathcal{A})(e_{j_1},\ldots,e_{j_{l-1}},e_p,e_{j_l},\ldots,e_{j_{s-1}}) = \sum_q \mathcal{A}^q_{j_1\ldots j_{l-1}\, p\, j_l\ldots j_{s-1}} e_q,$$

hence

$$\langle \mathfrak{R}_s^{-1}(\mathcal{A})(e_{j_1},\ldots,e_{j_{l-1}},e_p,e_{j_l},\ldots,e_{j_{s-1}}), e_p \rangle = \left\langle \sum_q \mathcal{A}^q_{j_1\ldots j_{l-1}\, p\, j_l\ldots j_{s-1}} e_q, e_p \right\rangle$$

$$= \sum_q \mathcal{A}^q_{j_1\ldots j_{l-1}\, p\, j_l\ldots j_{s-1}} \langle e_q, e_p \rangle$$

$$= \varepsilon_p \mathcal{A}^p_{j_1\ldots j_{l-1}\, p\, j_l\ldots j_{s-1}},$$

so

$$\mathcal{A}^p_{j_1\ldots j_{l-1}\, p\, j_l\ldots j_{s-1}} = \varepsilon_p \langle \mathfrak{R}_s^{-1}(\mathcal{A})(e_{j_1},\ldots,e_{j_{l-1}},e_p,e_{j_l},\ldots,e_{j_{s-1}}), e_p \rangle. \qquad (6.1.2)$$

Then (6.1.1) and (6.1.2) yield

$$\mathsf{C}^1_l(\mathcal{A})(e_{j_1},\ldots,e_{j_{s-1}}) = \sum_p \varepsilon_p \langle \mathfrak{R}_s^{-1}(\mathcal{A})(e_{j_1},\ldots,e_{j_{l-1}},e_p,e_{j_l},\ldots,e_{j_{s-1}}), e_p \rangle.$$

The result for arbitrary vectors $v_1,\ldots,v_{s-1}$ in V follows from the multilinearity of $\mathsf{C}^1_l(\mathcal{A})$ and $\mathfrak{R}_s^{-1}(\mathcal{A})$.

(b): We have

$$\sum_i \varepsilon_i \langle \mathfrak{R}^{-1}(\mathcal{A})(e_i), e_i \rangle = \sum_i \varepsilon_i \left\langle \sum_j \mathcal{A}^j_i e_j, e_i \right\rangle \qquad \text{[Th 5.3.2(c)]}$$

$$= \sum_{ij} \varepsilon_i \langle e_j, e_i \rangle \mathcal{A}^j_i = \sum_i \mathcal{A}^i_i$$

$$= \mathsf{C}^1_1(\mathcal{A}). \qquad \text{[Th 5.4.2]} \qquad \square$$

6.2   Flat Maps

Let $(V,\mathfrak{g})$ be a scalar product space, let $\mathcal{H} = (h_1,\ldots,h_m)$ be a basis for V, and let $(\theta^1,\ldots,\theta^m)$ be its dual basis. Recall from (5.1.1) that $\mathcal{T}^1_0(V) = V$ and $\mathcal{T}^0_1(V) = V^*$. The flat map of Section 3.3 can therefore be expressed as

$$\mathsf{F} : \mathcal{T}^1_0(V) \longrightarrow \mathcal{T}^0_1(V).$$

More generally, let $1 \le k \le r$ and $1 \le l \le s+1$ be integers (so that $r \ge 1$ and $s \ge 0$). The **(k,l)-flat map** is denoted by

$$\mathsf{F}^k_l : \mathcal{T}^r_s(V) \longrightarrow \mathcal{T}^{r-1}_{s+1}(V)$$

and defined for all tensors $\mathcal{A}$ in $\mathcal{T}^r_s(V)$, all covectors $\eta^1,\ldots,\eta^{r-1}$ in V^*, and all vectors $v_1,\ldots,v_{s+1}$ in V as follows, where $\widehat{}$ indicates that an expression is omitted:

[F1] For $r = 1$ (so that $k = 1$):

$$\mathsf{F}_l^1(\mathcal{A})(v_1, \ldots, v_{s+1}) = \mathcal{A}(v_l^\mathsf{F}, v_1, \ldots, \widehat{v_l}, \ldots, v_{s+1}).$$

[F2] For $r \geq 2$ and $1 \leq k \leq r - 1$:

$$\mathsf{F}_l^k(\mathcal{A})(\eta^1, \ldots, \eta^{r-1}, v_1, \ldots, v_{s+1})$$
$$= \mathcal{A}(\eta^1, \ldots, \eta^{k-1}, v_l^\mathsf{F}, \eta^k, \ldots, \eta^{r-1}, v_1, \ldots, \widehat{v_l}, \ldots, v_{s+1}).$$

[F3] For $r \geq 2$ and $k = r$:

$$\mathsf{F}_l^r(\mathcal{A})(\eta^1, \ldots, \eta^{r-1}, v_1, \ldots, v_{s+1})$$
$$= \mathcal{A}(\eta^1, \ldots, \eta^{r-1}, v_l^\mathsf{F}, v_1, \ldots, \widehat{v_l}, \ldots, v_{s+1}).$$

The computational procedure is most transparent for [F2]: extract the vector in the lth covariant position, flat it, and insert the result in the kth contravariant position.

Example 6.2.1. To illustrate [F1], let $r = 1$ (so that $k = 1$) and $s = 0$ (so that $l = 1$), and let w be a vector in $\mathcal{T}_0^1(V) = V$. Then the covector $\mathsf{F}_1^1(w)$ in $\mathcal{T}_1^0(V) = V^*$ is given by

$$\mathsf{F}_1^1(w)(v) = w(v^\mathsf{F})$$

for all vectors v in V. We have from (1.2.3) and (3.3.1) that

$$w(v^\mathsf{F}) = v^\mathsf{F}(w) = \langle v, w \rangle = \langle w, v \rangle = w^\mathsf{F}(v) = \mathsf{F}(w)(v),$$

hence $\mathsf{F}_1^1(w)(v) = \mathsf{F}(w)(v)$. Since v and w were arbitrary,

$$\mathsf{F}_1^1 = \mathsf{F}.$$

To illustrate [F2], let $r = 2$ (so that $k = 1$) and $s = 0$ (so that $l = 1$), and let $\mathcal{A}$ be a tensor in $\mathcal{T}_0^2(V)$. Then the tensor $\mathsf{F}_1^1(\mathcal{A})$ in $\mathcal{T}_1^1(V)$ is given by

$$\mathsf{F}_1^1(\mathcal{A})(\eta, v) = \mathcal{A}(v^\mathsf{F}, \eta).$$

To illustrate [F3], let $r = 2$ (so that $k = 2$) and $s = 0$ (so that $l = 1$), and let $\mathcal{A}$ be a tensor in $\mathcal{T}_0^2(V)$. Then the tensor $\mathsf{F}_1^2(\mathcal{A})$ in $\mathcal{T}_1^1(V)$ is given by

$$\mathsf{F}_1^2(\mathcal{A})(\eta, v) = \mathcal{A}(\eta, v^\mathsf{F}). \qquad \diamond$$

Example 6.2.2. Let $\mathcal{A}$ be a tensor in $\mathcal{T}_1^1(V)$. For $r = s = 1$, there are flat maps for $k = 1$ and $l = 1, 2$. For $k = 1$ and $l = 1$, the tensor $\mathsf{F}_1^1(\mathcal{A})$ in $\mathcal{T}_2^0(V)$ is given by

$$\mathsf{F}_1^1(\mathcal{A})(v_1, v_2) = \mathcal{A}(v_1^\mathsf{F}, v_2).$$

For $k = 1$ and $l = 2$, the tensor $\mathsf{F}_2^1(\mathcal{A})$ in $\mathcal{T}_2^0(V)$ is given by

$$\mathsf{F}_2^1(\mathcal{A})(v_1, v_2) = \mathcal{A}(v_2^\mathsf{F}, v_1). \qquad \diamond$$

Example 6.2.3. For a tensor $\mathcal{A}$ in $\mathcal{T}_2^2(V)$, the tensor $\mathsf{F}_1^2(\mathcal{A})$ in $\mathcal{T}_3^1(V)$ is given by

$$\mathsf{F}_1^2(\mathcal{A})(\eta^1, v_1, v_2, v_3) = \mathcal{A}(\eta^1, v_1^{\mathsf{F}}, v_2, v_3).$$

By Theorem 3.3.1(a), $h_j^{\mathsf{F}} = \sum_p \mathfrak{g}_{jp}\theta^p$, so $\mathsf{F}_1^2(\mathcal{A})$ has the components

$$\mathsf{F}_1^2(\mathcal{A})_{jkl}^i = \mathsf{F}_1^2(\mathcal{A})(\theta^i, h_j, h_k, h_l) = \mathcal{A}(\theta^i, h_j^{\mathsf{F}}, h_k, h_l)$$

$$= \mathcal{A}\left(\theta^i, \sum_p \mathfrak{g}_{jp}\theta^p, h_k, h_l\right) = \sum_p \mathfrak{g}_{jp}\mathcal{A}(\theta^i, \theta^p, h_k, h_l)$$

$$= \sum_p \mathfrak{g}_{jp}\mathcal{A}_{kl}^{ip}$$

with respect to $\mathcal{H}$. $\diamond$

Theorem 6.2.4. *Let* $(V, \mathfrak{g})$ *be a scalar product space, let* $\mathcal{H}$ *be a basis for* V, *and let* $\mathfrak{g}_{\mathcal{H}} = [\mathfrak{g}_{ij}]$. *Let* $\mathcal{A}$ *be a tensor in* $\mathcal{T}_s^r(V)$, *and let* $1 \le k \le r$ *and* $1 \le l \le s+1$ *be integers. Then the tensor* $\mathsf{F}_l^k(\mathcal{A})$ *in* $\mathcal{T}_{s+1}^{r-1}(V)$ *has the components*

$$\mathsf{F}_l^k(\mathcal{A})_{j_1\dots j_{s+1}}^{i_1\dots i_{r-1}} = \sum_p \mathfrak{g}_{j_l p} \mathcal{A}_{j_1\dots \widehat{j_l}\dots j_{s+1}}^{i_1\dots i_{k-1}\, p\, i_k\dots i_{r-1}}$$

with respect to $\mathcal{H}$.

Proof. The proof is an elaboration of Example 6.2.3. We consider only the case [F2]. The tensor $\mathsf{F}_l^k(\mathcal{A})$ in $\mathcal{T}_{s+1}^{r-1}(V)$ is given by

$$\mathsf{F}_l^k(\mathcal{A})(\eta^1, \dots, \eta^{r-1}, v_1, \dots, v_{s+1})$$
$$= \mathcal{A}(\eta^1, \dots, \eta^{k-1}, v_l^{\mathsf{F}}, \eta^k, \dots, \eta^{r-1}, v_1, \dots, \widehat{v_l}, \dots, v_{s+1})$$

for all covectors $\eta^1, \dots, \eta^{r-1}$ in V^* and all vectors $v_1, \dots, v_{s+1}$ in V. Let $\mathcal{H} = (h_1, \dots, h_m)$, and let $(\theta^1, \dots, \theta^m)$ be its dual basis. By Theorem 3.3.1(a), $h_j^{\mathsf{F}} = \sum_p \mathfrak{g}_{jp}\theta^p$, so $\mathsf{F}_l^k(\mathcal{A})$ has the components

$$\mathsf{F}_l^k(\mathcal{A})_{j_1\dots j_{s+1}}^{i_1\dots i_{r-1}}$$
$$= \mathsf{F}_l^k(\mathcal{A})(\theta^{i_1}, \dots, \theta^{i_k}, \dots, \theta^{i_{r-1}}, h_{j_1}, \dots, h_{j_l}, \dots, h_{j_{s+1}})$$
$$= \mathcal{A}(\theta^{i_1}, \dots, \theta^{i_{k-1}}, h_{j_l}^{\mathsf{F}}, \theta^{i_k}, \dots, \theta^{i_{r-1}}, h_{j_1}, \dots, h_{j_{l-1}}, h_{j_{l+1}}, \dots, h_{j_{s+1}})$$
$$= \mathcal{A}\left(\theta^{i_1}, \dots, \theta^{i_{k-1}}, \sum_p \mathfrak{g}_{j_l p}\theta^p, \theta^{i_k}, \dots, \theta^{i_{r-1}}, h_{j_1}, \dots, h_{j_{l-1}}, h_{j_{l+1}}, \dots, h_{j_{s+1}}\right)$$
$$= \sum_p \mathfrak{g}_{j_l p}\mathcal{A}(\theta^{i_1}, \dots, \theta^{i_{k-1}}, \theta^p, \theta^{i_k}, \dots, \theta^{i_{r-1}}, h_{j_1}, \dots, h_{j_{l-1}}, h_{j_{l+1}}, \dots, h_{j_{s+1}})$$
$$= \sum_p \mathfrak{g}_{j_l p}\mathcal{A}_{j_1\dots \widehat{j_l}\dots j_{s+1}}^{i_1\dots i_{k-1}\, p\, i_k\dots i_{r-1}}$$

with respect to $\mathcal{H}$. $\square$

Theorem 6.2.5. *Let $(V, \mathfrak{g})$ be a scalar product space, let $\mathcal{H}$ be a basis for V, and let $\mathfrak{g}_{\mathcal{H}} = [\mathfrak{g}_{ij}]$. Let $\mathcal{A}$ be a tensor in $\mathcal{T}^r_s(V)$, and let $r \geq 2$ and $s \geq 0$ be integers. Then:*

(a) *For integers $1 \leq k < l \leq r$ and $1 \leq n \leq s$, the components of the tensor $\mathsf{C}^k_n \circ \mathsf{F}^l_n(\mathcal{A})$ in $\mathcal{T}^{r-2}_s(V)$ with respect to $\mathcal{H}$ are*

$$\mathsf{C}^k_n \circ \mathsf{F}^l_n(\mathcal{A})^{i_1 \ldots i_{r-2}}_{j_1 \ldots j_s} = \sum_{pq} \mathfrak{g}_{pq} \mathcal{A}^{i_1 \ldots i_{k-1}\, p\, i_k \ldots i_{l-2}\, q\, i_{l-1} \ldots i_{r-2}}_{j_1 \ldots j_s}. \tag{6.2.1}$$

(b) *For integers $1 \leq k \leq r$ and $1 \leq n \leq s$, the components of the tensor $\mathsf{C}^k_n \circ \mathsf{F}^k_n(\mathcal{A})$ in $\mathcal{T}^{r-2}_s(V)$ with respect to $\mathcal{H}$ are*

$$
\begin{aligned}
\mathsf{C}^k_n \circ \mathsf{F}^k_n(\mathcal{A})^{i_1 \ldots i_{r-2}}_{j_1 \ldots j_s} &= \mathsf{C}^k_n \circ \mathsf{F}^{k+1}_n(\mathcal{A})^{i_1 \ldots i_{r-2}}_{j_1 \ldots j_s} \\
&= \sum_{pq} \mathfrak{g}_{pq} \mathcal{A}^{i_1 \ldots i_{k-1}\, p\, q\, i_k \ldots i_{r-2}}_{j_1 \ldots j_s}.
\end{aligned} \tag{6.2.2}
$$

Proof. By Theorem 6.2.4, the components of the tensor $\mathsf{F}^l_n(\mathcal{A})$ in $\mathcal{T}^{r-1}_{s+1}(V)$ with respect to $\mathcal{H}$ are

$$\mathsf{F}^l_n(\mathcal{A})^{i_1 \ldots i_{r-1}}_{j_1 \ldots j_{s+1}} = \sum_q \mathfrak{g}_{j_n q} \mathcal{A}^{i_1 \ldots i_{l-1}\, q\, i_l \ldots i_{r-1}}_{j_1 \ldots \widehat{j_n} \ldots j_{s+1}}. \tag{6.2.3}$$

(a): We have

$$
\begin{aligned}
\mathsf{C}^k_n &\circ \mathsf{F}^l_n(\mathcal{A})^{i_1 \ldots \widehat{i_k} \ldots i_{r-1}}_{j_1 \ldots \widehat{j_n} \ldots j_{s+1}} \\
&= \sum_p \mathsf{F}^l_n(\mathcal{A})^{i_1 \ldots i_{k-1}\, p\, i_{k+1} \ldots i_{r-1}}_{j_1 \ldots j_{n-1}\, p\, j_{n+1} \ldots j_{s+1}} && \text{[Th 5.4.5]} \\
&= \sum_p \left(\sum_q \mathfrak{g}_{pq} \mathcal{A}^{i_1 \ldots i_{k-1}\, p\, i_{k+1} \ldots i_{l-1}\, q\, i_l \ldots i_{r-1}}_{j_1 \ldots \widehat{j_n} \ldots j_{s+1}} \right) && \text{[(6.2.3)]} \\
&= \sum_{pq} \mathfrak{g}_{pq} \mathcal{A}^{i_1 \ldots i_{k-1}\, p\, i_{k+1} \ldots i_{l-1}\, q\, i_l \ldots i_{r-1}}_{j_1 \ldots \widehat{j_n} \ldots j_{s+1}}.
\end{aligned}
$$

Relabeling indices $(i_{k+1}, \ldots, i_{l-1}, q, i_l, \ldots, i_{r-1})$ as $(i_k, \ldots, i_{l-2}, q, i_{l-1}, \ldots, i_{r-2})$, and $(j_{n+1}, \ldots, j_{s+1})$ as $(j_n, \ldots, j_s)$, gives the result.

(b): We have

$$
\begin{aligned}
\mathsf{C}^k_n \circ \mathsf{F}^k_n(\mathcal{A})^{i_1 \ldots \widehat{i_k} \ldots i_{r-1}}_{j_1 \ldots \widehat{j_n} \ldots j_{s+1}} &= \sum_p \mathsf{F}^k_n(\mathcal{A})^{i_1 \ldots i_{k-1}\, p\, i_{k+1} \ldots i_{r-1}}_{j_1 \ldots j_{n-1}\, p\, j_{n+1} \ldots j_{s+1}} && \text{[Th 5.4.5]} \\
&= \sum_p \left(\sum_q \mathfrak{g}_{pq} \mathcal{A}^{i_1 \ldots i_{k-1}\, q\, p\, i_{k+1} \ldots i_{r-1}}_{j_1 \ldots \widehat{j_n} \ldots j_{s+1}} \right) && \text{[(6.2.3)]} \\
&= \sum_{pq} \mathfrak{g}_{qp} \mathcal{A}^{i_1 \ldots i_{k-1}\, q\, p\, i_{k+1} \ldots i_{r-1}}_{j_1 \ldots \widehat{j_n} \ldots j_{s+1}} \\
&= \sum_{pq} \mathfrak{g}_{pq} \mathcal{A}^{i_1 \ldots i_{k-1}\, p\, q\, i_{k+1} \ldots i_{r-1}}_{j_1 \ldots \widehat{j_n} \ldots j_{s+1}}
\end{aligned}
$$

and

$$C_n^k \circ F_n^{k+1}(\mathcal{A})_{j_1 \dots \hat{j_n} \dots j_{s+1}}^{i_1 \dots \hat{i_k} \dots i_{r-1}} = \sum_p F_n^{k+1}(\mathcal{A})_{j_1 \dots j_{n-1}\, p\, j_{n+1} \dots j_{s+1}}^{i_1 \dots i_{k-1}\, p\, i_{k+1} \dots i_{r-1}} \qquad \text{[Th 5.4.5]}$$

$$= \sum_p \left(\sum_q \mathfrak{g}_{pq} \mathcal{A}_{j_1 \dots \hat{j_n} \dots j_{s+1}}^{i_1 \dots i_{k-1}\, p\, q\, i_{k+1} \dots i_{r-1}} \right) \qquad \text{[(6.2.3)]}$$

$$= \sum_{pq} \mathfrak{g}_{pq} \mathcal{A}_{j_1 \dots \hat{j_n} \dots j_{s+1}}^{i_1 \dots i_{k-1}\, p\, q\, i_{k+1} \dots i_{r-1}},$$

which gives the first equality. Relabeling indices $(i_{k+1}, \dots, i_{r-1})$ as $(i_k, \dots, i_{r-2})$, and $(j_{n+1}, \dots, j_{s+1})$ as $(j_n, \dots, j_s)$, gives the second equality. $\qquad\square$

We observe that p and q in (6.2.1) are in the kth and lth contravariant positions of $\mathcal{A}$, respectively, and that p and q in (6.2.2) are in the kth and $(k+1)$th contravariant positions of $\mathcal{A}$, respectively. We also note that the right-hand side of (6.2.1) is independent of n. This makes sense because F_n^l lowers the lth contravariant index to the nth covariant position, and then C_n^k contracts over the kth contravariant index and the nth covariant index. In a manner of speaking, the nth covariant position is simply a temporary location for the lowered lth contravariant index to reside prior to its being contracted with the kth contravariant index.

Theorem 6.2.6. *Let $(V, \mathfrak{g})$ be a scalar product space, let $\mathcal{A}$ be a tensor in $\mathcal{T}_s^r(V)$, and let $1 \le k \le r$ be an integer. Then*

$$F_1^k(\mathcal{A}) = C_1^k(\mathfrak{g} \otimes \mathcal{A}).$$

Proof. All the components to follow are computed with respect to a basis $\mathcal{H}$ for V. By Theorem 6.2.4, the tensor $F_1^k(\mathcal{A})$ in $\mathcal{T}_{s+1}^{r-1}(V)$ has the components

$$F_1^k(\mathcal{A})_{j_1 \dots j_{s+1}}^{i_1 \dots i_{r-1}} = \sum_p \mathfrak{g}_{j_1 p} \mathcal{A}_{j_2 \dots j_{s+1}}^{i_1 \dots i_{k-1}\, p\, i_k \dots i_{r-1}}. \qquad (6.2.4)$$

It follows from Theorem 5.1.5 that the tensor $\mathfrak{g} \otimes \mathcal{A}$ in $\mathcal{T}_{s+2}^r(V)$ has the components

$$(\mathfrak{g} \otimes \mathcal{A})_{j_1 \dots j_{s+2}}^{i_1 \dots i_r} = \mathfrak{g}_{j_1 j_2} \mathcal{A}_{j_3 \dots j_{s+1}}^{i_1 \dots i_r},$$

and then from Theorem 5.4.5 that the tensor $C_1^k(\mathfrak{g} \otimes \mathcal{A})$ in $\mathcal{T}_{s+1}^{r-1}(V)$ has the components

$$C_1^k(\mathfrak{g} \otimes \mathcal{A})_{j_2 \dots j_{s+2}}^{i_1 \dots \hat{i_k} \dots i_r} = \sum_p \mathfrak{g}_{pj_2} \mathcal{A}_{j_3 \dots j_{s+2}}^{i_1 \dots i_{k-1}\, p\, i_{k+1} \dots i_r}.$$

Relabeling $(i_{k+1}, \dots, i_r)$ as $(i_k, \dots, i_{r-1})$, and $(j_2, \dots, j_{s+2})$ as $(j_1, \dots, j_{s+1})$, gives

$$C_1^k(\mathfrak{g} \otimes \mathcal{A})_{j_1 \dots j_{s+1}}^{i_1 \dots i_{r-1}} = \sum_p \mathfrak{g}_{pj_1} \mathcal{A}_{j_2 \dots j_{s+1}}^{i_1 \dots i_{k-1}\, p\, i_k \dots i_{r-1}}. \qquad (6.2.5)$$

The result now follows from (6.2.4) and (6.2.5). $\qquad\square$

6.3 Sharp Maps

Let $(V, \mathfrak{g})$ be a scalar product space, let $\mathcal{H} = (h_1, \ldots, h_m)$ be a basis for V, and let $(\theta^1, \ldots, \theta^m)$ be its dual basis. Recall from (5.1.1) that $\mathcal{T}_1^0(V) = V$ and $\mathcal{T}_1^0(V) = V^*$. The sharp map in Section 3.3 can therefore be expressed as

$$\mathsf{S} : \mathcal{T}_1^0(V) \longrightarrow \mathcal{T}_0^1(V).$$

More generally, let $1 \leq k \leq r+1$ and $1 \leq l \leq s$ be integers (so that $r \geq 0$ and $s \geq 1$). The **(k, l)-sharp map** is denoted by

$$\mathsf{S}_l^k : \mathcal{T}_s^r(V) \longrightarrow \mathcal{T}_{s-1}^{r+1}(V)$$

and defined for all tensors $\mathcal{A}$ in $\mathcal{T}_s^r(V)$, all covectors $\eta^1, \ldots, \eta^{r+1}$ in V^*, and all vectors $v_1, \ldots, v_{s-1}$ in V as follows, where $\widehat{}$ indicates that an expression is omitted:

[S1] For $s = 1$ (so that $l = 1$):

$$\mathsf{S}_1^k(\mathcal{A})(\eta^1, \ldots, \eta^{r+1}) = \mathcal{A}(\eta^1, \ldots, \widehat{\eta^k}, \ldots, \eta^{r+1}, \eta^{k\mathsf{S}}).$$

[S2] For $s \geq 2$ and $1 \leq l \leq s-1$:

$$\mathsf{S}_l^k(\mathcal{A})(\eta^1, \ldots, \eta^{r+1}, v_1, \ldots v_{s-1})$$
$$= \mathcal{A}(\eta^1, \ldots, \widehat{\eta^k}, \ldots, \eta^{r+1}, v_1, \ldots, v_{l-1}, \eta^{k\mathsf{S}}, v_l, \ldots, v_{s-1}).$$

[S3] For $s \geq 2$ and $l = s$:

$$\mathsf{S}_s^k(\mathcal{A})(\eta^1, \ldots, \eta^{r+1}, v_1, \ldots, v_{s-1})$$
$$= \mathcal{A}(\eta^1, \ldots, \widehat{\eta^k}, \ldots, \eta^{r+1}, v_1, \ldots, v_{s-1}, \eta^{k\mathsf{S}}).$$

The computational procedure is most transparent for [S2]: extract the covector in the kth contravariant position, sharp it, and insert the result in the lth covariant position.

Example 6.3.1. To illustrate [S1], let $r = 0$ (so that $k = 1$) and $s = 1$ (so that $l = 1$), and let ζ be a covector in $\mathcal{T}_1^0(V) = V^*$. Then the tensor $\mathsf{S}_1^1(\zeta)$ in $\mathcal{T}_0^1(V) = V$ is given by

$$\mathsf{S}_1^1(\zeta)(\eta) = \zeta(\eta^\mathsf{S})$$

for all covectors η in V^*. We have from (1.2.3) and (4.5.2) that

$$\zeta(\eta^\mathsf{S}) = \langle \zeta, \eta \rangle^* = \langle \eta, \zeta \rangle^* = \eta(\zeta^\mathsf{S}) = \zeta^\mathsf{S}(\eta) = \mathsf{S}(\zeta)(\eta),$$

hence $\mathsf{S}_1^1(\zeta)(\eta) = \mathsf{S}(\zeta)(\eta)$. Since η and ζ were arbitrary,

$$\mathsf{S}_1^1 = \mathsf{S}.$$

To illustrate [S2], let $r = 0$ (so that $k = 1$) and $s = 2$ (so that $l = 1$), and let $\mathcal{A}$ be a tensor in $\mathcal{T}_2^0(V)$. Then the tensor $\mathsf{S}_1^1(\mathcal{A})$ in $\mathcal{T}_1^1(V)$ is given by

$$\mathsf{S}_1^1(\mathcal{A})(\eta, v) = \mathcal{A}(\eta^\mathsf{S}, v).$$

To illustrate [S3], let $r = 0$ (so that $k = 1$) and $s = 2$ (so that $l = 2$), and let $\mathcal{A}$ be a tensor in $\mathcal{T}_2^0(V)$. Then the tensor $\mathsf{S}_2^1(\mathcal{A})$ in $\mathcal{T}_1^1(V)$ is given by

$$\mathsf{S}_2^1(\mathcal{A})(\eta, v) = \mathcal{A}(v, \eta^\mathsf{S}). \qquad \Diamond$$

Example 6.3.2. Let $\mathcal{A}$ be a tensor in $\mathcal{T}_1^1(V)$. For $r = s = 1$, there are sharp maps for $k = 1, 2$ and $l = 1$. For $k = 1$ and $l = 1$, the tensor $\mathsf{S}_1^1(\mathcal{A})$ in $\mathcal{T}_0^2(V)$ is given by

$$\mathsf{S}_1^1(\mathcal{A})(\eta^1, \eta^2) = \mathcal{A}(\eta^2, \eta^{1\mathsf{S}}).$$

For $k = 2$ and $l = 1$, the tensor $\mathsf{S}_1^2(\mathcal{A})$ in $\mathcal{T}_0^2(V)$ is given by

$$\mathsf{S}_1^2(\mathcal{A})(\eta^1, \eta^2) = \mathcal{A}(\eta^1, \eta^{2\mathsf{S}}). \qquad \Diamond$$

Example 6.3.3. For a tensor $\mathcal{B}$ in $\mathcal{T}_3^1(V)$, the tensor $\mathsf{S}_1^2(\mathcal{B})$ in $\mathcal{T}_2^2(V)$ is given by

$$\mathsf{S}_1^2(\mathcal{B})(\eta^1, \eta^2, v_1, v_2) = \mathcal{B}(\eta^1, \eta^{2\mathsf{S}}, v_1, v_2).$$

By Theorem 4.5.3(a), $\theta^{j\mathsf{S}} = \sum_p \mathfrak{g}^{jp} h_p$, so $\mathsf{S}_1^2(\mathcal{B})$ has the components

$$\begin{aligned}
\mathsf{S}_1^2(\mathcal{B})_{kl}^{ij} &= \mathsf{S}_1^2(\mathcal{B})(\theta^i, \theta^j, h_k, h_l) = \mathcal{B}(\theta^i, \theta^{j\mathsf{S}}, h_k, h_l) \\
&= \mathcal{B}\left(\theta^i, \sum_p \mathfrak{g}^{jp} h_p, h_k, h_l\right) = \sum_p \mathfrak{g}^{jp} \mathcal{B}(\theta^i, h_p, h_k, h_l) \\
&= \sum_p \mathfrak{g}^{jp} \mathcal{B}_{pkl}^i
\end{aligned} \qquad (6.3.1)$$

with respect to $\mathcal{H}$. $\Diamond$

Theorem 6.3.4. *Let* $(V, \mathfrak{g})$ *be a scalar product space, let* $\mathcal{H}$ *be a basis for* V, *and let* $\mathfrak{g}_{\mathcal{H}}^{-1} = [\mathfrak{g}^{ij}]$. *Let* $\mathcal{A}$ *be a tensor in* $\mathcal{T}_s^r(V)$, *and let* $1 \le k \le r + 1$ *and* $1 \le l \le s$ *be integers. Then the tensor* $\mathsf{S}_l^k(\mathcal{A})$ *in* $\mathcal{T}_{s-1}^{r+1}(V)$ *has the components*

$$\mathsf{S}_l^k(\mathcal{A})_{j_1 \dots j_{s-1}}^{i_1 \dots i_{r+1}} = \sum_p \mathfrak{g}^{i_k p} \mathcal{A}_{j_1 \dots j_{l-1} \, p \, j_l \dots j_{s-1}}^{i_1 \dots \widehat{i_k} \dots i_{r+1}}$$

with respect to $\mathcal{H}$.

Proof. The proof is an elaboration of Example 6.3.3. We consider only the case [S2]. The tensor $\mathsf{S}_l^k(\mathcal{A})$ in $\mathcal{T}_{s-1}^{r+1}(V)$ is given by

$$\begin{aligned}
&\mathsf{S}_l^k(\mathcal{A})(\eta^1, \dots, \eta^{r+1}, v_1, \dots, v_{s-1}) \\
&\quad = \mathcal{A}(\eta^1, \dots, \widehat{\eta^k}, \dots, \eta^{r+1}, v_1, \dots, v_{l-1}, \eta^{k\mathsf{S}}, v_{l+1}, \dots, v_{s-1})
\end{aligned}$$

for all covectors $\eta^1, \dots, \eta^{r-1}$ in V^* and all vectors $v_1, \dots, v_{s+1}$ in V. Let $\mathcal{H} = (h_1, \dots, h_m)$, and let $(\theta^1, \dots, \theta^m)$ be its dual basis. By Theorem 4.5.3(a), $\theta^{i\mathsf{S}} =$

$\sum_p \mathfrak{g}^{ip} h_p$, so $\mathsf{S}_l^k(\mathcal{A})$ has the components

$$
\begin{aligned}
\mathsf{S}_l^k(\mathcal{A})_{j_1\ldots j_{s-1}}^{i_1\ldots i_{r+1}}
&= \mathsf{S}_l^k(\mathcal{A})(\theta^{i_1},\ldots,\theta^{i_k},\ldots,\theta^{i_{r+1}},h_{j_1},\ldots,h_{j_l},\ldots,h_{j_{s-1}}) \\
&= \mathcal{A}(\theta^{i_1},\ldots,\theta^{i_{k-1}},\theta^{i_{k+1}},\ldots,\theta^{i_{r+1}},h_{j_1},\ldots,h_{j_{l-1}},\theta^{i_k}\mathsf{S},h_{j_l},\ldots,h_{j_{s-1}}) \\
&= \mathcal{A}\left(\theta^{i_1},\ldots,\theta^{i_{k-1}},\theta^{i_{k+1}},\ldots,\theta^{i_{r+1}},h_{j_1},\ldots,h_{j_{l-1}},\sum_p \mathfrak{g}^{i_k p}h_p,h_{j_l},\ldots,h_{j_{s-1}}\right) \\
&= \sum_p \mathfrak{g}^{i_k p}\mathcal{A}(\theta^{i_1},\ldots,\theta^{i_{k-1}},\theta^{i_{k+1}},\ldots,\theta^{i_{r+1}},h_{j_1},\ldots,h_{j_{l-1}},h_p,h_{j_l},\ldots,h_{j_{s-1}}) \\
&= \sum_p \mathfrak{g}^{i_k p}\mathcal{A}_{j_1\ldots j_{l-1}\,p\,j_l\ldots j_{s-1}}^{i_1\ldots \widehat{i_k}\ldots i_{r+1}}
\end{aligned}
$$

with respect to $\mathcal{H}$. $\qquad\square$

Theorem 6.3.5. *Let $(V,\mathfrak{g})$ be a scalar product space, let $\mathcal{H}$ be a basis for V, and let $\mathfrak{g}_{\mathcal{H}}^{-1} = \left[\mathfrak{g}^{ij}\right]$. Let $\mathcal{A}$ be a tensor in $T_s^r(V)$, and let $r \geq 0$ and $s \geq 2$ be integers. Then:*

(a) *For integers $1 \leq k < l \leq s$ and $1 \leq n \leq r$, the components of the tensor $\mathsf{C}_k^n \circ \mathsf{S}_l^n(\mathcal{A})$ in $T_{s-2}^r(V)$ with respect to $\mathcal{H}$ are*

$$
\mathsf{C}_k^n \circ \mathsf{S}_l^n(\mathcal{A})_{j_1\ldots j_{s-2}}^{i_1\ldots i_r} = \sum_{pq} \mathfrak{g}^{pq}\mathcal{A}_{j_1\ldots j_{k-1}\,p\,j_k\ldots j_{l-2}\,q\,j_{l-1}\ldots j_{s-2}}^{i_1\ldots i_r}. \tag{6.3.2}
$$

(b) *For integers $1 \leq k \leq s$ and $1 \leq n \leq r$, the components of the tensor $\mathsf{C}_k^n \circ \mathsf{S}_l^n(\mathcal{A})$ in $T_{s-2}^r(V)$ with respect to $\mathcal{H}$ are*

$$
\begin{aligned}
\mathsf{C}_k^n \circ \mathsf{S}_k^n(\mathcal{A})_{j_1\ldots j_{s-2}}^{i_1\ldots i_r}
&= \mathsf{C}_k^n \circ \mathsf{S}_{k+1}^n(\mathcal{A})_{j_1\ldots j_{s-2}}^{i_1\ldots i_r} \\
&= \sum_{pq} \mathfrak{g}^{pq}\mathcal{A}_{j_1\ldots j_{k-1}\,p\,q\,j_k\ldots j_{s-2}}^{i_1\ldots i_r}.
\end{aligned} \tag{6.3.3}
$$

Proof. By Theorem 6.3.4, the components of the tensor $\mathsf{S}_l^n(\mathcal{A})$ in $T_{s-1}^{r+1}(V)$ with respect to $\mathcal{H}$ are

$$
\mathsf{S}_l^n(\mathcal{A})_{j_1\ldots j_{s-1}}^{i_1\ldots i_{r+1}} = \sum_q \mathfrak{g}^{i_n q}\mathcal{A}_{j_1\ldots j_{l-1}\,q\,j_l\ldots j_{s-1}}^{i_1\ldots \widehat{i_n}\ldots i_{r+1}}. \tag{6.3.4}
$$

(a): We have

$$
\begin{aligned}
\mathsf{C}_k^n \circ \mathsf{S}_l^n(\mathcal{A})_{j_1\ldots \widehat{j_l}\ldots j_{s-1}}^{i_1\ldots \widehat{i_n}\ldots i_{r+1}}
&= \sum_p \mathsf{S}_l^n(\mathcal{A})_{j_1\ldots j_{k-1}\,p\,j_{k+1}\ldots j_{s-1}}^{i_1\ldots i_{n-1}\,p\,i_{n+1}\ldots i_{r+1}} && \text{[Th 5.4.5]} \\
&= \sum_p \left(\sum_q \mathfrak{g}^{pq}\mathcal{A}_{j_1\ldots j_{k-1}\,p\,j_{k+1}\ldots j_{l-1}\,q\,j_l\ldots j_{s-1}}^{i_1\ldots \widehat{i_n}\ldots i_{r+1}}\right) && \text{[(6.3.4)]} \\
&= \sum_{pq} \mathfrak{g}^{pq}\mathcal{A}_{j_1\ldots j_{k-1}\,p\,j_{k+1}\ldots j_{l-1}\,q\,j_l\ldots j_{s-1}}^{i_1\ldots \widehat{i_n}\ldots i_{r+1}}.
\end{aligned}
$$

Relabeling indices $(i_{n+1}, \ldots, i_{r+1})$ as $(i_n, \ldots, i_r)$, and $(j_{k+1}, \ldots, j_{l-1}, q, j_l, \ldots,$ $j_{s-1})$ as $(j_k, \ldots, j_{l-2}, q, j_{l-1}, \ldots, j_{s-2})$, gives the result.

(b): We have

$$\mathsf{C}_k^n \circ \mathsf{S}_k^n(\mathcal{A})_{j_1\ldots\widehat{j}_k\ldots j_{s-1}}^{i_1\ldots\widehat{i}_n\ldots i_{r+1}} = \sum_p \mathsf{S}_k^n(\mathcal{A})_{j_1\ldots j_{k-1}\,p\,j_{k+1}\ldots j_{s-1}}^{i_1\ldots i_{n-1}\,p\,i_{n+1}\ldots i_{r+1}} \qquad \text{[Th 5.4.5]}$$

$$= \sum_p \left(\sum_q \mathfrak{g}^{pq}\mathcal{A}_{j_1\ldots j_{k-1}\,q\,p\,j_{k+1}\ldots j_{s-1}}^{i_1\ldots\widehat{i}_n\ldots i_{r+1}} \right) \qquad \text{[(6.3.4)]}$$

$$= \sum_{pq} \mathfrak{g}^{qp}\mathcal{A}_{j_1\ldots j_{k-1}\,q\,p\,j_{k+1}\ldots j_{s-1}}^{i_1\ldots\widehat{i}_n\ldots i_{r+1}}$$

$$= \sum_{pq} \mathfrak{g}^{pq}\mathcal{A}_{j_1\ldots j_{k-1}\,p\,q\,j_{k+1}\ldots j_{s-1}}^{i_1\ldots\widehat{i}_n\ldots i_{r+1}}$$

and

$$\mathsf{C}_k^n \circ \mathsf{S}_{k+1}^n(\mathcal{A})_{j_1\ldots\widehat{j}_k\ldots j_{s-1}}^{i_1\ldots\widehat{i}_n\ldots i_{r+1}} = \sum_p \mathsf{S}_{k+1}^n(\mathcal{A})_{j_1\ldots j_{k-1}\,p\,j_{k+1}\ldots j_{s-1}}^{i_1\ldots i_{n-1}\,p\,i_{n+1}\ldots i_{r+1}} \qquad \text{[Th 5.4.5]}$$

$$= \sum_p \left(\sum_q \mathfrak{g}^{pq}\mathcal{A}_{j_1\ldots j_{k-1}\,p\,q\,j_{k+1}\ldots j_{s-1}}^{i_1\ldots\widehat{i}_n\ldots i_{r+1}} \right) \qquad \text{[(6.3.4)]}$$

$$= \sum_{pq} \mathfrak{g}^{pq}\mathcal{A}_{j_1\ldots j_{k-1}\,p\,q\,j_{k+1}\ldots j_{s-1}}^{i_1\ldots\widehat{i}_n\ldots i_{r+1}},$$

which gives the first equality. Relabeling indices $(i_{n+1}, \ldots, i_{r+1})$ as $(i_n, \ldots, i_r)$, and $(j_{k+1}, \ldots, j_{s-1})$ as $(j_k, \ldots, j_{s-2})$, gives the second equality. $\qquad \square$

We observe that p and q in (6.3.2) are in the kth and lth covariant positions of $\mathcal{A}$, respectively, and that p and q in (6.3.3) are in the kth and $(k+1)$th covariant positions of $\mathcal{A}$, respectively. We also note that the right hand side of (6.3.2) is independent of n. See the corresponding remarks following Theorem 6.2.5.

Example 6.3.6. For a tensor $\mathcal{A}$ in $\mathcal{T}_2^2(V)$, we have from Example 6.2.3 that

$$\mathsf{F}_1^2(\mathcal{A})_{pkl}^i = \sum_q \mathfrak{g}_{pq}\mathcal{A}_{kl}^{iq}.$$

Substituting $\mathcal{B} = \mathsf{F}_1^2(\mathcal{A})$, as given by the above expression, into (6.3.1) yields

$$\mathsf{S}_1^2 \circ \mathsf{F}_1^2(\mathcal{A})_{kl}^{ij} = \sum_p \mathfrak{g}^{jp}\mathsf{F}_1^2(\mathcal{A})_{pkl}^i = \sum_p \mathfrak{g}^{jp}\left(\sum_q \mathfrak{g}_{pq}\mathcal{A}_{kl}^{iq} \right)$$

$$= \sum_q \left(\sum_p \mathfrak{g}^{jp}\mathfrak{g}_{pq} \right) \mathcal{A}_{kl}^{iq} = \sum_q \delta_q^j \mathcal{A}_{kl}^{iq} = \mathcal{A}_{kl}^{ij}.$$

Thus, $\mathsf{S}_1^2 \circ \mathsf{F}_1^2(\mathcal{A}) = \mathcal{A}$. Likewise, $\mathsf{F}_1^2 \circ \mathsf{S}_1^2(\mathcal{A}) = \mathcal{A}$. Since $\mathcal{A}$ was arbitrary, it follows from Theorem A.4 that F_1^2 and S_1^2 are inverses of each other, and are therefore linear isomorphisms. $\qquad \diamond$

Theorem 6.3.7. *If $(V, \mathfrak{g})$ is a scalar product space, and $1 \leq k \leq r$ and $1 \leq l \leq s$ are integers, then F_l^k and S_l^k are linear isomorphisms that are inverses.*

Proof. The proof is an elaboration of Example 6.3.6. Let $\mathcal{H}$ be a basis for V, let $\mathcal{A}$ and $\mathcal{B}$ be tensors in $\mathcal{T}_s^r(V)$ and $\mathcal{T}_{s+1}^{r-1}(V)$, respectively, and let $\mathfrak{g}_{\mathcal{H}} = [\mathfrak{g}_{ij}]$. By Theorem 6.2.4, $\mathsf{F}_l^k(\mathcal{A})$ is a tensor in $\mathcal{T}_{s+1}^{r-1}(V)$ that has the components

$$\mathsf{F}_l^k(\mathcal{A})^{i_1 \dots i_{r-1}}_{j_1 \dots j_{s+1}} = \sum_q \mathfrak{g}_{j_l q} \mathcal{A}^{i_1 \dots i_{k-1} \, q \, i_k \dots i_{r-1}}_{j_1 \dots \widehat{j_l} \dots j_{s+1}}$$

with respect to $\mathcal{H}$. Relabeling indices $(i_k, \dots, i_{r-1})$ as $(i_{k+1}, \dots, i_r)$, and $(j_l, \dots, j_{s+1})$ as $(p, j_l, \dots, j_s)$, yields

$$\mathsf{F}_l^k(\mathcal{A})^{i_1 \dots \widehat{i_k} \dots i_r}_{j_1 \dots j_{l-1} \, p \, j_l \dots j_s} = \sum_q \mathfrak{g}_{pq} \mathcal{A}^{i_1 \dots i_{k-1} \, q \, i_{k+1} \dots i_r}_{j_1 \dots j_s}. \tag{6.3.5}$$

By Theorem 6.3.4, the tensor $\mathsf{S}_l^k(\mathcal{B})$ in $\mathcal{T}_s^r(V)$ has the components

$$\mathsf{S}_l^k(\mathcal{B})^{i_1 \dots i_r}_{j_1 \dots j_s} = \sum_p \mathfrak{g}^{i_k p} \mathcal{B}^{i_1 \dots \widehat{i_k} \dots i_r}_{j_1 \dots j_{l-1} \, p \, j_l \dots j_s} \tag{6.3.6}$$

with respect to $\mathcal{H}$. Substituting $\mathcal{B} = \mathsf{F}_l^k(\mathcal{A})$, as given by (6.3.5), into (6.3.6) gives

$$\begin{aligned}
\mathsf{S}_l^k \circ \mathsf{F}_l^k(\mathcal{A})^{i_1 \dots i_r}_{j_1 \dots j_s} &= \sum_p \mathfrak{g}^{i_k p} \mathsf{F}_l^k(\mathcal{A})^{i_1 \dots \widehat{i_k} \dots i_r}_{j_1 \dots j_{l-1} \, p \, j_l \dots j_s} \\
&= \sum_p \mathfrak{g}^{i_k p} \left(\sum_q \mathfrak{g}_{pq} \mathcal{A}^{i_1 \dots i_{k-1} \, q \, i_{k+1} \dots i_r}_{j_1 \dots j_s} \right) \\
&= \sum_q \left(\sum_p \mathfrak{g}^{i_k p} \mathfrak{g}_{pq} \right) \mathcal{A}^{i_1 \dots i_{k-1} \, q \, i_{k+1} \dots i_r}_{j_1 \dots j_s} \\
&= \sum_q \delta_q^{i_k} \mathcal{A}^{i_1 \dots i_{k-1} \, q \, i_{k+1} \dots i_r}_{j_1 \dots j_s} \\
&= \mathcal{A}^{i_1 \dots i_{k-1} \, i_k \, i_{k+1} \dots i_r}_{j_1 \dots j_s} = \mathcal{A}^{i_1 \dots i_r}_{j_1 \dots j_s}.
\end{aligned}$$

The rest of the proof uses the argument presented in Example 6.3.6. $\qquad\square$

6.4   Representation of Tensors

Section 5.3 was devoted to the topic of representation of tensors over vector spaces. We now extend that discussion to tensors over scalar product spaces.

Let $(V, \mathfrak{g})$ be a scalar product space, let $\mathcal{H} = (h_1, \dots, h_m)$ be a basis for V, let $(\theta^1, \dots \theta^m)$ be its dual basis, and let $\mathfrak{g}_{\mathcal{H}} = [\mathfrak{g}_{ij}]$. Following Section B.5 and Section B.6, we denote by $\mathrm{Mult}(V^s, V)$ the vector space of $\mathbb{R}$-multilinear maps from V^s to V. Let us define a map

$$\mathfrak{S}_s : \mathrm{Mult}(V^s, V) \longrightarrow \mathcal{T}_{s+1}^0(V),$$

called the **scalar product map**, by

$$\mathfrak{S}_s(\Psi)(v_1, \ldots, v_{s+1}) = \langle v_1, \Psi(v_2, \ldots, v_{s+1}) \rangle \tag{6.4.1}$$

for all maps Ψ in $\mathrm{Mult}(V^s, V)$ and all vectors $v_1, \ldots, v_{s+1}$ in V. When $s = 1$, we denote

$$\mathfrak{S}_1 \qquad \text{by} \qquad \mathfrak{S}.$$

Theorem 6.4.1 (Representation of Tensors). *With the above setup:*
(a) $\mathfrak{S}_s$ *is a linear isomorphism:*

$$\mathcal{T}^0_{s+1}(V) \approx \mathrm{Mult}(V^s, V).$$

(b) *For all maps* Ψ *in* $\mathrm{Mult}(V^s, V)$, *the components of the tensor* $\mathfrak{S}_s(\Psi)$ *in* $\mathcal{T}^0_{s+1}(V)$ *with respect to* $\mathcal{H}$ *are*

$$\mathfrak{S}_s(\Psi)_{j_1 \ldots j_{s+1}} = \sum_i \mathfrak{g}_{ij_1} \theta^i \big(\Psi(h_{j_2}, \ldots, h_{j_{s+1}}) \big).$$

(c) *For all tensors* $\mathcal{A}$ *in* $\mathcal{T}^0_{s+1}(V)$,

$$\mathfrak{S}_s^{-1}(\mathcal{A})(h_{j_1}, \ldots, h_{j_s}) = \sum_i \mathsf{S}_1^1(\mathcal{A})^i_{j_1 \ldots j_s} h_i,$$

where S_1^1 *is defined in Section 6.3.*
(d)

$$\mathfrak{R}_s = \mathsf{S}_1^1 \circ \mathfrak{S}_s,$$

where $\mathfrak{R}_s$ *is given by (5.3.1).*

Proof. (a): It is easily shown that $\mathfrak{S}_s$ is a linear map. If $\mathfrak{S}_s(\Psi) = 0$, then $\langle \Psi(v_2, \ldots, v_{s+1}), v_1 \rangle = 0$ for all vectors $v_1, \ldots, v_{s+1}$ in V. Since $\mathfrak{g}$ is nondegenerate, we have $\Psi(v_2, \ldots, v_{s+1}) = 0$, and because $v_2, \ldots, v_{s+1}$ were arbitrary, $\Psi = 0$. Thus, $\ker(\mathfrak{S}_s) = \{0\}$. By Theorem 5.1.3(c), $\mathcal{T}^0_{s+1}(V)$ has dimension m^{s+1}, and it can be shown that $\mathrm{Mult}(V^s, V)$ has the same dimension. It follows from Theorem 1.1.12 and Theorem 1.1.14 that $\mathfrak{S}_s$ is a linear isomorphism.

(b): Since $\Psi(h_{j_2}, \ldots, h_{j_{s+1}})$ is a vector in V, we have from Theorem 1.2.1(d) that

$$\Psi(h_{j_2}, \ldots, h_{j_{s+1}}) = \sum_i \theta^i \big(\Psi(h_{j_2}, \ldots, h_{j_{s+1}}) \big) h_i,$$

and then from (6.4.1) that

$$\mathfrak{S}_s(\Psi)_{j_1 \ldots j_{s+1}} = \mathfrak{S}_s(\Psi)(h_{j_1}, \ldots, h_{j_{s+1}}) = \langle h_{j_1}, \Psi(h_{j_2}, \ldots, h_{j_{s+1}}) \rangle$$

$$= \Big\langle h_{j_1}, \sum_i \theta^i \big(\Psi(h_{j_2}, \ldots, h_{j_{s+1}}) \big) h_i \Big\rangle$$

$$= \sum_i \mathfrak{g}_{ij_1} \theta^i \big(\Psi(h_{j_2}, \ldots, h_{j_{s+1}}) \big).$$

(c):   Since $\mathfrak{S}_s^{-1}(\mathcal{A})$ is a map in $\mathrm{Mult}(V^s, V)$, $\mathfrak{S}_s^{-1}(\mathcal{A})(h_{j_2}, \ldots, h_{j_{s+1}})$ is a vector in V. We have from Theorem 1.2.1(d) that

$$\mathfrak{S}_s^{-1}(\mathcal{A})(h_{j_2}, \ldots, h_{j_{s+1}}) = \sum_i \theta^i \big(\mathfrak{S}_s^{-1}(\mathcal{A})(h_{j_2}, \ldots, h_{j_{s+1}})\big) h_i. \qquad (6.4.2)$$

Then

$$
\begin{aligned}
\mathcal{A}_{j_1 \ldots j_{s+1}} &= \mathcal{A}(h_{j_1}, \ldots, h_{j_{s+1}}) \\
&= \mathfrak{S}_s\big(\mathfrak{S}_s^{-1}(\mathcal{A})\big)(h_{j_1}, \ldots, h_{j_{s+1}}) & \text{[part(a)]} \\
&= \langle h_{j_1}, \mathfrak{S}_s^{-1}(\mathcal{A})(h_{j_2}, \ldots, h_{j_{s+1}}) \rangle & \text{[(6.4.1)]} \\
&= \Big\langle h_{j_1}, \sum_i \theta^i \big(\mathfrak{S}_s^{-1}(\mathcal{A})(h_{j_2}, \ldots, h_{j_{s+1}})\big) h_i \Big\rangle & \text{[(6.4.2)]} \\
&= \sum_i \mathfrak{g}_{ij_1} \theta^i \big(\mathfrak{S}_s^{-1}(\mathcal{A})(h_{j_2}, \ldots, h_{j_{s+1}})\big),
\end{aligned}
$$

hence

$$
\begin{aligned}
\sum_{j_1} \mathfrak{g}^{ij_1} \mathcal{A}_{j_1 \ldots j_{s+1}} &= \sum_{j_1} \mathfrak{g}^{ij_1} \Big(\sum_k \mathfrak{g}_{kj_1} \theta^k \big(\mathfrak{S}_s^{-1}(\mathcal{A})(h_{j_2}, \ldots, h_{j_{s+1}})\big)\Big) \\
&= \sum_k \Big(\sum_{j_1} \mathfrak{g}^{ij_1} \mathfrak{g}_{j_1 k}\Big) \theta^k \big(\mathfrak{S}_s^{-1}(\mathcal{A})(h_{j_2}, \ldots, h_{j_{s+1}})\big) \\
&= \sum_k \delta_k^i \theta^k \big(\mathfrak{S}_s^{-1}(\mathcal{A})(h_{j_2}, \ldots, h_{j_{s+1}})\big) \\
&= \theta^i \big(\mathfrak{S}_s^{-1}(\mathcal{A})(h_{j_2}, \ldots, h_{j_{s+1}})\big),
\end{aligned}
$$

so

$$\sum_{ij_1} \mathfrak{g}^{ij_1} \mathcal{A}_{j_1 \ldots j_{s+1}} h_i = \sum_i \theta^i \big(\mathfrak{S}_s^{-1}(\mathcal{A})(h_{j_2}, \ldots, h_{j_{s+1}})\big) h_i. \qquad (6.4.3)$$

By Theorem 6.3.4,

$$\mathsf{S}_1^1(\mathcal{A})_{j_2 \ldots j_{s+1}}^i = \sum_{j_1} \mathfrak{g}^{ij_1} \mathcal{A}_{j_1 \ldots j_{s+1}},$$

hence

$$\sum_i \mathsf{S}_1^1(\mathcal{A})_{j_2 \ldots j_{s+1}}^i h_i = \sum_{ij_1} \mathfrak{g}^{ij_1} \mathcal{A}_{j_1 \ldots j_{s+1}} h_i. \qquad (6.4.4)$$

Combining (6.4.2)–(6.4.4) yields

$$\mathfrak{S}_s^{-1}(\mathcal{A})(h_{j_2}, \ldots, h_{j_{s+1}}) = \sum_i \mathsf{S}_1^1(\mathcal{A})_{j_2 \ldots j_{s+1}}^i h_i,$$

and relabeling indices $(j_2, \ldots, j_{s+1})$ as $(j_1, \ldots, j_s)$ gives the result.

(d): Let Ψ be a multilinear function in $\mathrm{Mult}(V^s, V)$, let η be a covector in V^*, and let $v_1, \ldots, v_s$ be vectors in V. Then

$$
\begin{aligned}
\mathfrak{R}_s(\Psi)(\eta, v_1, \ldots, v_s) &= \eta\big(\Psi(v_1, \ldots, v_s)\big) && [(5.3.1)] \\
&= (\eta^{\mathsf{S}})^{\mathsf{F}}\big(\Psi(v_1, \ldots, v_s)\big) && [(3.3.5)] \\
&= \langle \eta^{\mathsf{S}}, \Psi(v_1, \ldots, v_s)\rangle && [(3.3.1)] \\
&= \mathfrak{S}_s(\Psi)(\eta^{\mathsf{S}}, v_1, \ldots, v_s) && [(6.4.1)] \\
&= \mathsf{S}_1^1\big(\mathfrak{S}_s(\Psi)\big)(\eta, v_1, \ldots, v_s). && [[\mathrm{S2}] \text{ in } \S 6.3]
\end{aligned}
$$

Since Ψ, η, and $v_1, \ldots, v_s$ were arbitrary, the result follows. $\square$

Theorem 6.4.2. *With the above setup:*

$$
\mathcal{T}^0_{s+1}(V) \approx \mathrm{Mult}(V^s, V) \approx \mathcal{T}^1_s(V).
$$

Proof. This follows from Theorem 5.3.1(a) and Theorem 6.4.1(a). $\square$

It is instructive to specialize Theorem 6.4.1 to the case $s = 1$. We then have $\mathrm{Mult}(V, V) = \mathrm{Lin}(V, V)$ and

$$
\mathfrak{S} : \mathrm{Lin}(V, V) \longrightarrow \mathcal{T}^0_2(V)
$$

defined by

$$
\mathfrak{S}(B)(v, w) = \langle v, B(w)\rangle \tag{6.4.5}
$$

for all maps B in $\mathrm{Lin}(V, V)$ and all vectors v, w in V. For example,

$$
\mathfrak{S}(\mathrm{id}_V)(v, w) = \langle v, w\rangle = \mathfrak{g}(v, w),
$$

so $\mathfrak{S}(\mathrm{id}_V) = \mathfrak{g}$.

Theorem 6.4.3. *With the above setup:*
(a) $\mathfrak{S}$ *is a linear isomorphism:*

$$
\mathcal{T}^0_2(V) \approx \mathrm{Lin}(V, V).
$$

(b) *For all maps B in $\mathrm{Lin}(V, V)$, the components of the tensor $\mathfrak{S}(B)$ in $\mathcal{T}^0_2(V)$ with respect to $\mathcal{H}$ are*

$$
\mathfrak{S}(B)_{ij} = \sum_k \mathfrak{g}_{ik}\theta^k\big(B(h_j)\big).
$$

(c) *For all tensors $\mathcal{A}$ in $\mathcal{T}^0_2(V)$,*

$$
\big[\mathfrak{S}^{-1}(\mathcal{A})\big]^{\mathcal{H}}_{\mathcal{H}} = \big[\mathfrak{g}^{ij}\big]\big[\mathcal{A}_{ij}\big],
$$

where $\big[\mathfrak{g}^{ij}\big] = \mathfrak{g}_{\mathcal{H}}^{-1}$ and the $\mathcal{A}_{ij}$ are the components of $\mathcal{A}$ with respect to $\mathcal{H}$.

(d)
$$\mathfrak{R} = \mathsf{S}_1^1 \circ \mathfrak{S},$$

where $\mathfrak{R}$ is given by (5.3.2).

Proof. (a), (b), (d): These follow from the corresponding parts of Theorem 6.4.1.

(c): By Theorem 6.4.1(c),

$$\mathfrak{S}^{-1}(\mathcal{A})(h_j) = \sum_i \mathsf{S}_1^1(\mathcal{A})_j^i h_i,$$

so that (2.2.2) and (2.2.3) give

$$\left[\mathfrak{S}^{-1}(\mathcal{A})\right]_{\mathcal{H}}^{\mathcal{H}} = \left[\mathsf{S}_1^1(\mathcal{A})_j^i\right].$$

From Theorem 6.3.4,

$$\mathsf{S}_1^1(\mathcal{A})_j^i = \sum_p \mathfrak{g}^{ip} \mathcal{A}_{pj},$$

hence

$$\left[\mathsf{S}_1^1(\mathcal{A})_j^i\right] = \left[\mathfrak{g}^{ij}\right]\left[\mathcal{A}_{ij}\right].$$

The result follows.   □

6.5   Metric Contraction of Tensors

For tensors over a vector space, we are restricted to contracting over one contravariant index and one covariant index. However, for tensors over a scalar product space, the flat map and sharp map make it possible to contract over two contravariant or two covariant indices, yielding what are referred to as **metric contractions**.

Let $(V, \mathfrak{g})$ be a scalar product space. The following definitions are motivated by Theorem 6.2.5 and Theorem 6.3.5.

For integers $r \geq 2$, $s \geq 0$, and $1 \leq k < l \leq r$, the **(k, l)-contravariant metric contraction** is the linear map

$$\mathsf{C}^{kl} : \mathcal{T}_s^r(V) \longrightarrow \mathcal{T}_s^{r-2}(V)$$

defined by

$$\mathsf{C}^{kl} = \mathsf{C}_1^k \circ \mathsf{F}_1^l. \tag{6.5.1}$$

We have from Theorem 6.2.5(b) that

$$\mathsf{C}^{k,k+1} = \mathsf{C}_1^k \circ \mathsf{F}_1^k. \tag{6.5.2}$$

For integers $r \geq 0$, $s \geq 2$, and $1 \leq k < l \leq s$, the **(k, l)-covariant metric contraction** is the linear map

$$\mathsf{C}_{kl} : \mathcal{T}_s^r(V) \longrightarrow \mathcal{T}_{s-2}^r(V)$$

defined by

$$C_{kl} = C_k^1 \circ S_l^1. \tag{6.5.3}$$

We have from Theorem 6.3.5(b) that

$$C_{k,k+1} = C_k^1 \circ S_k^1. \tag{6.5.4}$$

In order to avoid confusion between metric contractions and the contractions defined in Section 5.4, we recall that the latter are sometimes referred to as **ordinary contractions**.

Theorem 6.5.1 (Orthonormal Basis Expression for Metric Contraction). *Let* $(V, \mathfrak{g})$ *be a scalar product space with signature* $(\varepsilon_1, \ldots, \varepsilon_m)$, *let* $(e_1, \ldots, e_m)$ *be an orthonormal basis for* V, *and let* $\mathcal{A}$ *be a tensor in* $\mathcal{T}_s^0(V)$. *Then:*

(a) *For integers* $s \geq 3$ *and* $1 \leq k < l \leq s$, *the tensor* $C_{kl}(\mathcal{A})$ *in* $\mathcal{T}_{s-2}^0(V)$ *is given by*

$$
\begin{aligned}
C_{kl}&(\mathcal{A})(v_1, \ldots, v_{s-2}) \\
&= \sum_i \varepsilon_i \mathcal{A}(v_1, \ldots, v_{k-1}, e_i, v_k, \ldots, v_{l-2}, e_i, v_{l-1}, \ldots, v_{s-2})
\end{aligned} \tag{6.5.5}
$$

for all vectors $v_1, \ldots, v_{s-2}$ *in* V.

(b) *For* $s = 2$, *the real number* $C_{12}(\mathcal{A})$ *is given by*

$$C_{12}(\mathcal{A}) = \sum_i \varepsilon_i \mathcal{A}(e_i, e_i).$$

Proof. (a): Let $\mathcal{E} = (e_1, \ldots, e_m)$ and $\mathfrak{g}_{\mathcal{E}} = [\mathfrak{g}_{ij}]$, and observe that $\mathfrak{g}^{ij} = \mathfrak{g}_{ij} = \varepsilon_i \delta_{ij}$. We have from Theorem 6.3.5(a) and (6.5.3) that $C_{kl}(\mathcal{A})$ has the components

$$
\begin{aligned}
C_{kl}(\mathcal{A})_{j_1 \ldots j_{s-2}} &= \sum_{pq} \mathfrak{g}^{pq} \mathcal{A}_{j_1 \ldots j_{k-1} \, p \, j_k \ldots j_{l-2} \, q \, j_{l-1} \ldots j_{s-2}} \\
&= \sum_p \varepsilon_p \mathcal{A}_{j_1 \ldots j_{k-1} \, p \, j_k \ldots j_{l-2} \, p \, j_{l-1} \ldots j_{s-2}} \\
&= \sum_p \varepsilon_p \mathcal{A}(e_{j_1}, \ldots, e_{j_{k-1}}, e_p, e_{j_k} \ldots, e_{j_{l-2}}, e_p, e_{j_{l-1}}, \ldots, e_{j_{s-2}})
\end{aligned}
$$

with respect to $\mathcal{H}$. The result for arbitrary vectors $v_1, \ldots, v_{s-2}$ in V follows from the multilinearity of $\mathcal{A}$ and $C_{kl}(\mathcal{A})$.

(b): We have

$$
\begin{aligned}
\mathsf{C}_{12}(\mathcal{A}) &= \mathsf{C}_1^1 \circ \mathsf{S}_1^1(\mathcal{A}) & [(6.5.4)] \\
&= \sum_i \varepsilon_i \langle \mathfrak{R}^{-1} \circ \mathsf{S}_1^1(\mathcal{A})(e_i), e_i \rangle & [\text{Th } 6.1.1(b)] \\
&= \sum_i \varepsilon_i \langle e_i, \mathfrak{S}^{-1}(\mathcal{A})(e_i) \rangle & [\text{Th } 6.4.3(d)] \\
&= \sum_i \varepsilon_i \mathfrak{S}(\mathfrak{S}^{-1}(\mathcal{A}))(e_i, e_i) & [(6.4.5)] \\
&= \sum_i \varepsilon_i \mathcal{A}(e_i, e_i). & \square
\end{aligned}
$$

Observe that in (6.5.5) the e_i appear in the kth and lth positions.

We saw in Theorem 5.4.1 that "contraction" and "trace" are related. Here is another manifestation of that same phenomenon.

Theorem 6.5.2. *If $(V, \mathfrak{g})$ is a scalar product space, then*

$$
\mathsf{C}_{12} = \operatorname{tr} \circ \mathfrak{S}^{-1},
$$

where tr *is trace and* $\mathfrak{S}$ *is given by (6.4.5).*

Proof. We have

$$
\begin{aligned}
\mathsf{C}_{12} &= \mathsf{C}_1^1 \circ \mathsf{S}_1^1 & [(6.5.4)] \\
&= (\operatorname{tr} \circ \mathfrak{R}^{-1}) \circ (\mathfrak{R} \circ \mathfrak{S}^{-1}) & [\text{Th } 5.4.1, \text{Th } 6.4.3(d)] \\
&= \operatorname{tr} \circ \mathfrak{S}^{-1}. & \square
\end{aligned}
$$

6.6   Symmetries of $(0,4)$-Tensors

Let V be a vector space, and let $\mathcal{A}$ be a tensor in $T_4^0(V)$. Consider the following symmetries that $\mathcal{A}$ might satisfy for all vectors v_1, v_2, v_3, v_4 in V:

[S1] $\mathcal{A}(v_1, v_2, v_3, v_4) = -\mathcal{A}(v_2, v_1, v_3, v_4)$.
[S2] $\mathcal{A}(v_1, v_2, v_3, v_4) = -\mathcal{A}(v_1, v_2, v_4, v_3)$.
[S3] $\mathcal{A}(v_1, v_2, v_3, v_4) = \mathcal{A}(v_3, v_4, v_1, v_2)$.
[S4] $\mathcal{A}(v_1, v_2, v_3, v_4) + \mathcal{A}(v_2, v_3, v_1, v_4) + \mathcal{A}(v_3, v_1, v_2, v_4) = 0$.

Observe that the left-hand side of [S4] is obtained by cyclically permuting v_1, v_2, v_3 while leaving v_4 in place. It follows from [S1] that

$$
\mathcal{A}(v_1, v_1, v_2, v_3) = 0, \tag{6.6.1}
$$

and from [S2] that

$$
\mathcal{A}(v_1, v_2, v_3, v_3) = 0 \tag{6.6.2}
$$

for all vectors v_1, v_2, v_3 in V.

The next result shows that [S1]–[S4] are not independent.

Theorem 6.6.1. *Let V be a vector space, and let $\mathcal{A}$ be a tensor in $T_4^0(V)$. If $\mathcal{A}$ satisfies* [S1], [S2], *and* [S4], *then it satisfies* [S3].

Proof. Using [S4] four times gives

$$\mathcal{A}(v_3, v_1, v_4, v_2) + \mathcal{A}(v_1, v_4, v_3, v_2) + \mathcal{A}(v_4, v_3, v_1, v_2) = 0$$
$$\mathcal{A}(v_1, v_4, v_2, v_3) + \mathcal{A}(v_4, v_2, v_1, v_3) + \mathcal{A}(v_2, v_1, v_4, v_3) = 0$$
$$\mathcal{A}(v_4, v_2, v_3, v_1) + \mathcal{A}(v_2, v_3, v_4, v_1) + \mathcal{A}(v_3, v_4, v_2, v_1) = 0$$
$$\mathcal{A}(v_2, v_3, v_1, v_4) + \mathcal{A}(v_3, v_1, v_2, v_4) + \mathcal{A}(v_1, v_2, v_3, v_4) = 0.$$

$$(6.6.3)$$

Applying [S1] and [S2] to the second and third columns of (6.6.3) gives

$$\mathcal{A}(v_3, v_1, v_4, v_2) - \mathcal{A}(v_1, v_4, v_2, v_3) - \mathcal{A}(v_3, v_4, v_1, v_2) = 0$$
$$\mathcal{A}(v_1, v_4, v_2, v_3) - \mathcal{A}(v_4, v_2, v_3, v_1) + \mathcal{A}(v_1, v_2, v_3, v_4) = 0$$
$$\mathcal{A}(v_4, v_2, v_3, v_1) - \mathcal{A}(v_2, v_3, v_1, v_4) - \mathcal{A}(v_3, v_4, v_1, v_2) = 0$$
$$\mathcal{A}(v_2, v_3, v_1, v_4) - \mathcal{A}(v_3, v_1, v_4, v_2) + \mathcal{A}(v_1, v_2, v_3, v_4) = 0.$$

$$(6.6.4)$$

Summing both sides of (6.6.4) yields

$$2\mathcal{A}(v_1, v_2, v_3, v_4) - 2\mathcal{A}(v_3, v_4, v_1, v_2) = 0,$$

from which the result follows.   □

Let V be a vector space, and let $\mathcal{S}(V)$ be the set of tensors in $T_4^0(V)$ that satisfy [S1] and [S2]. It is easily shown that $\mathcal{S}(V)$ is a subspace of $T_4^0(V)$. Let $\mathcal{A}$ be a tensor in $T_4^0(V)$, and consider the function $\overline{\mathcal{A}}$ defined by

$$\overline{\mathcal{A}}(v_1, v_2) = \mathcal{A}(v_1, v_2, v_2, v_1)$$

for all vectors v_1, v_2 in V. Observe that if $\mathcal{A}$ is in $\mathcal{S}(V)$, then

$$\overline{\mathcal{A}}(v_1, v_2) = \mathcal{A}(v_1, v_2, v_2, v_1) = \mathcal{A}(v_2, v_1, v_1, v_2) = \overline{\mathcal{A}}(v_2, v_1),$$

so $\overline{\mathcal{A}}$ is symmetric.

Theorem 6.6.2. *Let V be a 2-dimensional vector space, and let $\mathcal{A}$ be a tensor in $\mathcal{S}(V)$. Then the following are equivalent:*
(a) $\mathcal{A}$ *is the zero tensor.*
(b) $\overline{\mathcal{A}}(h_1, h_2) = 0$ *for some basis* (h_1, h_2) *for* V.
(c) $\overline{\mathcal{A}}(h_1, h_2) = 0$ *for every basis* (h_1, h_2) *for* V.

Proof. Let $\mathcal{H} = (h_1, h_2)$ be a basis for V. For the moment, we make no further assumptions about $\mathcal{A}$ and $\mathcal{H}$. Let v_1, v_2, v_3, v_4 be vectors in V, with

$$[v_1]_{\mathcal{H}} = \begin{bmatrix} a^1 \\ a^2 \end{bmatrix} \qquad [v_2]_{\mathcal{H}} = \begin{bmatrix} b^1 \\ b^2 \end{bmatrix} \qquad [v_3]_{\mathcal{H}} = \begin{bmatrix} c^1 \\ c^2 \end{bmatrix} \qquad [v_4]_{\mathcal{H}} = \begin{bmatrix} d^1 \\ d^2 \end{bmatrix}.$$

It follows from [S1] and (6.6.1) that

$$
\begin{aligned}
\mathcal{A}(v_1, v_2, v_3, v_4) &= \mathcal{A}(a^1 h_1 + a^2 h_2, b^1 h_1 + b^2 h_2, v_3, v_4) \\
&= a^1 b^1 \mathcal{A}(h_1, h_1, v_3, v_4) + a^1 b^2 \mathcal{A}(h_1, h_2, v_3, v_4) \\
&\quad + a^2 b^1 \mathcal{A}(h_2, h_1, v_3, v_4) + a^2 b^2 \mathcal{A}(h_2, h_2, v_3, v_4) \\
&= (a^1 b^2 - a^2 b^1) \mathcal{A}(h_1, h_2, v_3, v_4),
\end{aligned}
$$

and from [S2] and (6.6.2) that

$$
\begin{aligned}
\mathcal{A}(h_1, h_2, v_3, v_4) &= \mathcal{A}(h_1, h_2, c^1 h_1 + c^2 h_2, d^1 h_1 + d^2 h_2) \\
&= c^1 d^1 \mathcal{A}(h_1, h_2, h_1, h_1) + c^1 d^2 \mathcal{A}(h_1, h_2, h_1, h_2) \\
&\quad + c^2 d^1 \mathcal{A}(h_1, h_2, h_2, h_1) + c^2 d^2 \mathcal{A}(h_1, h_2, h_2, h_2) \\
&= -(c^1 d^2 - c^2 d^1) \mathcal{A}(h_1, h_2, h_2, h_1) \\
&= -(c^1 d^2 - c^2 d^1) \overline{\mathcal{A}}(h_1, h_2).
\end{aligned}
$$

Thus,

$$
\mathcal{A}(v_1, v_2, v_3, v_4) = -(a^1 b^2 - a^2 b^1)(c^1 d^2 - c^2 d^1) \overline{\mathcal{A}}(h_1, h_2). \tag{6.6.5}
$$

(a) $\Rightarrow$ (b): Since $\mathcal{A}$ is the zero tensor,

$$
\overline{\mathcal{A}}(h_1, h_2) = \mathcal{A}(h_1, h_2, h_2, h_1) = 0.
$$

(b) $\Rightarrow$ (a): Since $\overline{\mathcal{A}}(h_1, h_2) = 0$, it follows from (6.6.5) that $\mathcal{A}(v_1, v_2, v_3, v_4) = 0$. Since v_1, v_2, v_3, v_4 were arbitrary, $\mathcal{A}$ is the zero tensor.

(b) $\Leftrightarrow$ (c): Let (f_1, f_2) be another basis for V. Setting $v_1 = v_4 = f_1$ and $v_2 = v_3 = f_2$, we find that $a^i = d^i$ and $b^i = c^i$ for $i = 1, 2$. Then (6.6.5) gives

$$
\overline{\mathcal{A}}(f_1, f_2) = (a^1 b^2 - a^2 b^1)^2 \overline{\mathcal{A}}(h_1, h_2). \tag{6.6.6}
$$

From (2.2.6), (2.2.7), and Theorem 2.4.12,

$$
0 \neq \det\left([\mathrm{id}_V]_{\mathcal{F}}^{\mathcal{H}} \right) = \det\left(\begin{bmatrix} a^1 & b^1 \\ a^2 & b^2 \end{bmatrix} \right) = a^1 b^2 - a^2 b^1.
$$

The result follows. $\square$

Theorem 6.6.3. *Let $(V, \mathfrak{g})$ be a 2-dimensional scalar product space, and let $\mathcal{D}$ be the function defined by*

$$
\mathcal{D}(v_1, v_2, v_3, v_4) = \det\left(\begin{bmatrix} \langle v_1, v_3 \rangle & \langle v_1, v_4 \rangle \\ \langle v_2, v_3 \rangle & \langle v_2, v_4 \rangle \end{bmatrix} \right) \tag{6.6.7}
$$

for all vectors v_1, v_2, v_3, v_4 in V. Then:
(a) $\mathcal{D}$ is a nonzero tensor in $\mathcal{S}(V)$, and $\overline{\mathcal{D}}$ is given by

$$
\overline{\mathcal{D}}(v_1, v_2) = -\det\left(\begin{bmatrix} \langle v_1, v_1 \rangle & \langle v_1, v_2 \rangle \\ \langle v_2, v_1 \rangle & \langle v_2, v_2 \rangle \end{bmatrix} \right)
$$

for all vectors v_1, v_2 in V.

(b) $\mathcal{S}(V)$ *is a 1-dimensional subspace of* $T_4^0(V)$ *that is spanned by* $\mathcal{D}$.

Proof. (a): That $\mathcal{D}$ is a tensor in $\mathcal{S}(V)$ follows from the properties of determinants, or perhaps more easily from the observation that

$$\mathcal{D}(v_1, v_2, v_3, v_4) = \langle v_1, v_3 \rangle \langle v_2, v_4 \rangle - \langle v_2, v_3 \rangle \langle v_1, v_4 \rangle.$$

Let (e_1, e_2) be an orthonormal basis for V. Then

$$\mathcal{D}(e_1, e_2, e_2, e_1) = -\langle e_1, e_1 \rangle \langle e_2, e_2 \rangle \neq 0,$$

hence $\mathcal{D}$ is nonzero.

(b): Let $\mathcal{A}$ be a tensor in $\mathcal{S}(V)$. We have from (6.6.5) that

$$\mathcal{A}(v_1, v_2, v_3, v_4) = -(a^1 b^2 - a^2 b^1)(c^1 d^2 - c^2 d^1)\,\overline{\mathcal{A}}(h_1, h_2)$$

and

$$\mathcal{D}(v_1, v_2, v_3, v_4) = -(a^1 b^2 - a^2 b^1)(c^1 d^2 - c^2 d^1)\,\overline{\mathcal{D}}(h_1, h_2),$$

hence

$$\mathcal{A}(v_1, v_2, v_3, v_4) = \frac{\overline{\mathcal{A}}(h_1, h_2)}{\overline{\mathcal{D}}(h_1, h_2)}\,\mathcal{D}(v_1, v_2, v_3, v_4),$$

where Theorem 6.6.2 and part (a) ensure that the denominator is nonzero. Since v_1, v_2, v_3, v_4 were arbitrary,

$$\mathcal{A} = \frac{\overline{\mathcal{A}}(h_1, h_2)}{\overline{\mathcal{D}}(h_1, h_2)}\,\mathcal{D},$$

so $\mathcal{D}$ spans $\mathcal{S}(V)$. $\square$

Theorem 6.6.4. *Let* $(V, \mathfrak{g})$ *be a 2-dimensional scalar product space, let* $\mathcal{A}$ *be a tensor in* $\mathcal{S}(V)$, *and let* (h_1, h_2) *and* (f_1, f_2) *be bases for* V. *Then*

$$\frac{\overline{\mathcal{A}}(h_1, h_2)}{\overline{\mathcal{D}}(h_1, h_2)} = \frac{\overline{\mathcal{A}}(f_1, f_2)}{\overline{\mathcal{D}}(f_1, f_2)}.$$

Proof. Arguing as in the proof of part (b) of Theorem 6.6.3, but with (6.6.6) in place of (6.6.5), gives the result. $\square$

Chapter 7

Multicovectors

It was remarked in Section 5.1 that the determinant function is the classic example of a multilinear function. In addition to its multilinearity, the determinant function has another characteristic feature—it is alternating. This chapter is devoted to an examination of tensors that have a corresponding property.

7.1   Multicovectors

Let V be a vector space of dimension m, and let $s \geq 1$ be an integer. Following Section B.2, we denote by $\mathcal{S}_s$ the group of permutations on $\{1, \ldots, s\}$. For each permutation σ in $\mathcal{S}_s$, consider the linear map

$$\sigma : \mathcal{T}^0_s(V) \longrightarrow \mathcal{T}^0_s(V)$$

defined by

$$\sigma(\mathcal{A})(v_1, \ldots, v_s) = \mathcal{A}(v_{\sigma(1)}, \ldots, v_{\sigma(s)}) \qquad (7.1.1)$$

for all tensors $\mathcal{A}$ in $\mathcal{T}^0_s(V)$ and all vectors $v_1, \ldots, v_s$ in V. By saying that σ is linear, we mean that for all tensors $\mathcal{A}, \mathcal{B}$ in $\mathcal{T}^0_s(V)$ and all real numbers c,

$$\sigma(c\mathcal{A} + \mathcal{B})(v_1, \ldots, v_s) = c\mathcal{A}(v_{\sigma(1)}, \ldots, v_{\sigma(s)}) + \mathcal{B}(v_{\sigma(1)}, \ldots, v_{\sigma(s)}).$$

There is potential confusion arising from (7.1.1), as a simple example illustrates. Let $s = 3$, let $\mathcal{A}$ be a tensor in $\mathcal{T}^0_3(V)$, and consider the permutation $\sigma = (1\ 2\ 3)$ in $\mathcal{S}_3$. According to (7.1.1),

$$\sigma(\mathcal{A})(v_1, v_2, v_3) = \mathcal{A}(v_{\sigma(1)}, v_{\sigma(2)}, v_{\sigma(3)}) = \mathcal{A}(v_2, v_3, v_1)$$

for all vectors v_1, v_2, v_3 in V. To be consistent, it seems that $\sigma(\mathcal{A})(v_1, v_3, v_2)$ should be interpreted as $\mathcal{A}(v_{\sigma(1)}, v_{\sigma(3)}, v_{\sigma(2)}) = \mathcal{A}(v_2, v_1, v_3)$, but this is incorrect. The issue is that the indices in (v_1, v_3, v_2) are not sequential, which is

Semi-Riemannian Geometry, First Edition. Stephen C. Newman.
© 2019 John Wiley & Sons, Inc. Published 2019 by John Wiley & Sons, Inc.

implicit in the way (7.1.1) is presented. Setting $(w_1, w_2, w_3) = (v_1, v_3, v_2)$, we have from (7.1.1) that

$$\sigma(\mathcal{A})(v_1, v_3, v_2) = \sigma(\mathcal{A})(w_1, w_2, w_3) = \mathcal{A}(w_{\sigma(1)}, w_{\sigma(2)}, w_{\sigma(3)})$$
$$= \mathcal{A}(w_2, w_3, w_1) = \mathcal{A}(v_3, v_2, v_1).$$

In most of the computations that follow, the indices are sequential, thereby avoiding this issue.

Theorem 7.1.1. *Let V be a vector space, let σ, ρ be permutations in $\mathcal{S}_s$, and let $\mathcal{A}$ be a tensor in $\mathcal{T}_s^0(V)$. Then*

$$(\sigma\rho)(\mathcal{A}) = \sigma\big(\rho(\mathcal{A})\big).$$

Proof. Let $v_1, \ldots, v_s$ be vectors in V, and let $(w_1, \ldots, w_s) = (v_{\sigma(1)}, \ldots, v_{\sigma(s)})$. Then

$$(\sigma\rho)(\mathcal{A})(v_1, \ldots, v_s) = \mathcal{A}(v_{(\sigma\rho)(1)}, \ldots, v_{(\sigma\rho)(s)}) = \mathcal{A}(v_{\sigma(\rho(1))}, \ldots, v_{\sigma(\rho(s))})$$
$$= \mathcal{A}(w_{\rho(1)}, \ldots, w_{\rho(s)}) = \rho(\mathcal{A})(w_1, \ldots, w_s)$$
$$= \rho(\mathcal{A})(v_{\sigma(1)}, \ldots, v_{\sigma(s)}) = \sigma\big(\rho(\mathcal{A})\big)(v_1, \ldots, v_s).$$

Since $v_1, \ldots, v_s$ were arbitrary, the result follows. $\square$

We say that a tensor $\mathcal{A}$ in $\mathcal{T}_s^0(V)$ is **symmetric** if $\sigma(\mathcal{A}) = \mathcal{A}$ for all permutations σ in $\mathcal{S}_s$; that is,

$$\mathcal{A}(v_{\sigma(1)}, \ldots, v_{\sigma(s)}) = \mathcal{A}(v_1, \ldots, v_s)$$

for all vectors $v_1, \ldots, v_s$ in V. We denote the set of symmetric tensors in $\mathcal{T}_s^0(V)$ by $\Sigma_s(V)$. It is easily shown that $\Sigma_s(V)$ is subspace of $\mathcal{T}_s^0(V)$.

A tensor $\mathcal{B}$ in $\mathcal{T}_s^0(V)$ is said to be **alternating** if $\tau(\mathcal{B}) = -\mathcal{B}$ for all transpositions τ in $\mathcal{S}_s$, or equivalently, if

$$\mathcal{B}(v_1, \ldots, v_j, \ldots, v_i, \ldots, v_s) = -\mathcal{B}(v_1, \ldots, v_i, \ldots, v_j, \ldots, v_s)$$

for all vectors $v_1, \ldots, v_s$ in V for all $1 \leq i < j \leq s$. The set of alternating tensors in $\mathcal{T}_s^0(V)$ is denoted by $\Lambda^s(V)$. It is readily demonstrated that $\Lambda^s(V)$ is a subspace of $\mathcal{T}_s^0(V)$. An element of $\Lambda^s(V)$ is called an **s-covector** or **multicovector**. When $s = 1$, the alternating criterion is vacuous, and a 1-covector is simply a covector. Thus,

$$\Lambda^1(V) = V^*. \tag{7.1.2}$$

Since $\Lambda^s(V)$ is a subspace of $\mathcal{T}_s^0(V)$, the **zero multicovector** in $\Lambda^s(V)$ is precisely the zero tensor in $\mathcal{T}_s^0(V)$. A multicovector in $\Lambda^s(V)$ is **nonzero** when it is nonzero as a tensor in $\mathcal{T}_s^0(V)$. To be consistent with (5.1.2), we define

$$\Lambda^0(V) = \mathbb{R}. \tag{7.1.3}$$

With these definitions, the determinant function $\det : (\mathrm{Mat}_{m \times 1})^m \longrightarrow \mathbb{R}$ is seen to be a multicovector in $\Lambda^m(\mathrm{Mat}_{m \times 1})$.

There are several equivalent ways to characterize multicovectors in $\Lambda^s(V)$.

Theorem 7.1.2. *Let V be a vector space and let η be a tensor in $\mathcal{T}_s^0(V)$. Then the following are equivalent:*

(a) *η is a multicovector in $\Lambda^s(V)$.*
(b) *$\sigma(\eta) = \operatorname{sgn}(\sigma)\,\eta$ for all permutations σ in $\mathcal{S}_s$.*
(c) *If $v_1, \ldots, v_s$ are vectors in V and two (or more) of them are equal, then $\eta(v_1, \ldots, v_s) = 0$.*
(d) *If $v_1, \ldots, v_s$ are vectors in V and $\eta(v_1, \ldots, v_s) \neq 0$, then $v_1, \ldots, v_s$ are linearly independent.*

Proof. The proof is similar to that of Theorem 2.4.2.

(a) $\Rightarrow$ (b): Let $\sigma = \tau_1 \cdots \tau_k$ be a decomposition of σ into transpositions. By Theorem 7.1.1,

$$\sigma(\eta) = (\tau_1 \cdots \tau_k)(\eta) = (\tau_1 \cdots \tau_{k-1})(\tau_k(\eta))$$
$$= (\tau_1 \cdots \tau_{k-1})(-\eta) = -(\tau_1 \cdots \tau_{k-1})(\eta).$$

Repeating the process $k - 1$ more times gives $\sigma(\eta) = (-1)^k \eta$. By Theorem B.2.3, $\operatorname{sgn}(\sigma) = (-1)^k$.

(b) $\Rightarrow$ (a): If τ is a transposition in $\mathcal{S}_s$, then, by Theorem B.2.3, $\tau(\eta) = \operatorname{sgn}(\tau)\,\eta = -\eta$.

(a) $\Rightarrow$ (c): If $v_i = v_j$ for some $1 \leq i < j \leq s$, then

$$\eta(v_1, \ldots, v_i, \ldots, v_j, \ldots, v_s) = \eta(v_1, \ldots, v_j, \ldots, v_i, \ldots, v_s).$$

On the other hand, since η is alternating,

$$\eta(v_1, \ldots, v_j, \ldots, v_i, \ldots, v_s) = -\eta(v_1, \ldots, v_i, \ldots, v_j, \ldots, v_s).$$

Thus,

$$\eta(v_1, \ldots, v_i, \ldots, v_j, \ldots, v_s) = -\eta(v_1, \ldots, v_i, \ldots, v_j, \ldots, v_s),$$

from which the result follows.

(c) $\Rightarrow$ (a): For all $1 \leq i < j \leq s$, we have

$$0 = \eta(v_1, \ldots, v_i + v_j, \ldots, v_i + v_j, \ldots, v_s)$$
$$= \eta(v_1, \ldots, v_i, \ldots, v_i, \ldots, v_s) + \eta(v_1, \ldots, v_i, \ldots, v_j, \ldots, v_s)$$
$$+ \eta(v_1, \ldots, v_j, \ldots, v_i, \ldots, v_s) + \eta(v_1, \ldots, v_j, \ldots, v_j, \ldots, v_s)$$
$$= \eta(v_1, \ldots, v_i, \ldots, v_j, \ldots, v_s) + \eta(v_1, \ldots, v_j, \ldots, v_i, \ldots, v_s),$$

from which the result follows.

To prove (c) $\Leftrightarrow$ (d), we replace the assertion in part (d) with the following logically equivalent assertion:

(d'): If $v_1, \ldots, v_s$ are linearly dependent vectors in V, then $\eta(v_1, \ldots, v_s) = 0$.

(c) $\Rightarrow$ (d'): Since $v_1, \ldots, v_s$ are linearly dependent, one of them can be expressed as a linear combination of the others, say, $v_1 = \sum_{i=2}^s a^i v_i$. Then

$$\eta(v_1, \ldots, v_s) = \eta\left(\sum_{i=2}^s a^i v_i, v_2, \ldots, v_s\right) = \sum_{i=2}^s a^i \eta(v_i, v_2, \ldots, v_s).$$

Since $\eta(v_i, v_2, \ldots, v_s)$ has v_i in (at least) two positions, $\eta(v_i, v_2, \ldots, v_s) = 0$ for $i = 2, \ldots, s$. Thus, $\eta(v_1, \ldots, v_s) = 0$.

(d') $\Rightarrow$ (c): If two (or more) of $v_1, \ldots, v_s$ are equal, then they are linearly dependent, so $\eta(v_1, \ldots, v_s) = 0$. $\qquad\qquad\qquad\qquad\qquad\qquad\qquad\qquad\square$

We now introduce a way of associating a multicovector to a given tensor. Let V be a vector space. **Alternating map** is the family of linear maps

$$\mathrm{Alt} : \mathcal{T}_s^0(V) \longrightarrow \mathcal{T}_s^0(V)$$

defined for $s \geq 0$ by

$$\mathrm{Alt}(\mathcal{A}) = \frac{1}{s!} \sum_{\sigma \in \mathcal{S}_s} \mathrm{sgn}(\sigma)\, \sigma(\mathcal{A}) \qquad\qquad (7.1.4)$$

for all tensors $\mathcal{A}$ in $\mathcal{T}_s^0(V)$.

Theorem 7.1.3. *Let V be a vector space, let $\mathcal{A}$ be a tensor in $\mathcal{T}_s^0(V)$, let σ be a permutation in $\mathcal{S}_s$, and let η be a multicovector in $\Lambda^s(V)$. Then:*
(a) $\sigma\big(\mathrm{Alt}(\mathcal{A})\big) = \mathrm{sgn}(\sigma)\,\mathrm{Alt}(\mathcal{A}) = \mathrm{Alt}\big(\sigma(\mathcal{A})\big)$.
(b) $\mathrm{Alt}(\mathcal{A})$ *is a multicovector in* $\Lambda^s(V)$.
(c) $\mathrm{Alt}(\eta) = \eta$.
(d) $\mathrm{Alt}\big(\mathrm{Alt}(\mathcal{A})\big) = \mathrm{Alt}(\mathcal{A})$.
(e) $\mathrm{Alt}\big(\mathcal{T}_s^0(V)\big) = \Lambda^s(V)$.

Proof. (a): We have

$$\sigma\big(\mathrm{Alt}(\mathcal{A})\big) = \sigma\left(\frac{1}{s!} \sum_{\rho \in \mathcal{S}_s} \mathrm{sgn}(\rho)\,\rho(\mathcal{A})\right) \qquad [(7.1.4)]$$

$$= \frac{1}{s!} \sum_{\rho \in \mathcal{S}_s} \mathrm{sgn}(\rho)\, \sigma\big(\rho(\mathcal{A})\big)$$

$$= \frac{1}{s!} \sum_{\rho \in \mathcal{S}_s} \mathrm{sgn}(\rho)\, (\sigma\rho)(\mathcal{A}) \qquad [\text{Th 7.1.1}]$$

$$= \mathrm{sgn}(\sigma)\, \frac{1}{s!} \sum_{\rho \in \mathcal{S}_s} \mathrm{sgn}(\sigma\rho)\, (\sigma\rho)(\mathcal{A}) \qquad [\text{Th B.2.2}]$$

$$= \mathrm{sgn}(\sigma)\, \mathrm{Alt}(\mathcal{A}), \qquad [(7.1.4)]$$

where the last equality follows from the observation that as ρ varies over $\mathcal{S}_s$, so does $\sigma\rho$. We also have

$$\mathrm{Alt}\big(\sigma(\mathcal{A})\big) = \frac{1}{s!} \sum_{\rho \in \mathcal{S}_s} \mathrm{sgn}(\rho)\, \rho\big(\sigma(\mathcal{A})\big) \qquad [(7.1.4)]$$

$$= \mathrm{sgn}(\sigma)\, \frac{1}{s!} \sum_{\rho \in \mathcal{S}_s} \mathrm{sgn}(\rho\sigma)\, (\rho\sigma)(\mathcal{A}) \qquad [\text{Th 7.1.1, Th B.2.2}]$$

$$= \mathrm{sgn}(\sigma)\, \mathrm{Alt}(\mathcal{A}),$$

where the last equality is justified as above.

(b): For all transpositions τ in $\mathcal{S}_s$, it follows from part (a) that $\tau\big(\mathrm{Alt}(\mathcal{A})\big) = -\mathrm{Alt}(\mathcal{A})$, so $\mathrm{Alt}(\mathcal{A})$ is alternating.

(c): We have

$$\mathrm{Alt}(\eta) = \frac{1}{s!} \sum_{\sigma \in \mathcal{S}_s} \mathrm{sgn}(\sigma)\,\sigma(\eta) \qquad [(7.1.4)]$$

$$= \frac{1}{s!} \sum_{\sigma \in \mathcal{S}_s} \eta \qquad [\text{Th } 7.1.2(b)]$$

$$= \eta. \qquad [\mathrm{card}(\mathcal{S}_s) = s!]$$

(d): This follows from parts (b) and (c).

(e): We have from part (b) that $\mathrm{Alt}\big(\mathcal{T}_s^0(V)\big) \subseteq \Lambda^s(V)$. Since $\Lambda^s(V) \subseteq \mathcal{T}_s^0(V)$, by part (c),

$$\Lambda^s(V) = \mathrm{Alt}\big(\Lambda^s(V)\big) \subseteq \mathrm{Alt}\big(\mathcal{T}_s^0(V)\big). \qquad \square$$

In view of Theorem 7.1.3(b), we can replace the map in (7.1.4) with

$$\mathrm{Alt} : \mathcal{T}_s^0(V) \longrightarrow \Lambda^s(V).$$

7.2   Wedge Products

In Section 5.1, we introduced a type of multiplication of tensors called the tensor product. Our next task is to define a corresponding operation for multicovectors.

Let V be a vector space. **Wedge product** is the family of linear maps

$$\wedge : \Lambda^s(V) \times \Lambda^{s'}(V) \longrightarrow \Lambda^{s+s'}(V)$$

defined for $s, s' \geq 0$ by

$$\eta \wedge \zeta = \frac{(s+s')!}{s!\,s'!}\,\mathrm{Alt}(\eta \otimes \zeta) \tag{7.2.1}$$

for all multicovectors η in $\Lambda^s(V)$ and ζ in $\Lambda^{s'}(V)$. That is,

$$(\eta \wedge \zeta)(v_1, \ldots, v_{s+s'})$$
$$= \frac{(s+s')!}{s!\,s'!} \left[\frac{1}{(s+s')!} \sum_{\sigma \in \mathcal{S}_{s+s'}} \mathrm{sgn}(\sigma)\,\sigma(\eta \otimes \zeta)(v_{\sigma(1)}, \ldots, v_{\sigma(s+s')}) \right]$$
$$= \frac{1}{s!\,s'!} \sum_{\sigma \in \mathcal{S}_{s+s'}} \mathrm{sgn}(\sigma)\,\eta(v_{\sigma(1)}, \ldots, v_{\sigma(s)})\,\zeta(v_{\sigma(s+1)}, \ldots, v_{\sigma(s+s')})$$

for all vectors $v_1, \ldots, v_{s+s'}$ in V, where the first equality follows from (7.2.1), and the second equality from (5.1.4) and (7.1.1).

Example 7.2.1. Let η, ζ be covectors in $\Lambda^1(V) = V^*$, and let v_1, v_2 be vectors in V. With $\mathcal{S}_2 = \{\mathrm{id}, (1\,2)\}$, we have

$$(\eta \wedge \zeta)(v_1, v_2) = \frac{1}{1!1!} \sum_{\sigma \in \mathcal{S}_2} \mathrm{sgn}(\sigma)\, \eta(v_{\sigma(1)})\, \zeta(v_{\sigma(2)})$$

$$= \eta(v_1)\zeta(v_2) - \eta(v_2)\zeta(v_1).$$

Now let η be a multicovector in $\Lambda^2(V)$, let ζ be a covector in $\Lambda^1(V) = V^*$, and let v_1, v_2 be vectors in V. With

$$\mathcal{S}_3 = \{\mathrm{id}, (1\,2), (1\,3), (2\,3), (1\,2\,3), (1\,3\,2)\},$$

we have

$$(\eta \wedge \zeta)(v_1, v_2, v_3) = \frac{1}{2!1!} \sum_{\sigma \in \mathcal{S}_3} \mathrm{sgn}(\sigma)\, \eta(v_{\sigma(1)}, v_{\sigma(2)})\, \zeta(v_{\sigma(3)})$$

$$= [\eta(v_1, v_2)\zeta(v_3) - \eta(v_2, v_1)\zeta(v_3) - \eta(v_3, v_2)\zeta(v_1)$$
$$\quad - \eta(v_1, v_3)\zeta(v_2) + \eta(v_2, v_3)\zeta(v_1) + \eta(v_3, v_1)\zeta(v_2)]/2$$
$$= [\eta(v_1, v_2)\zeta(v_3) + \eta(v_1, v_2)\zeta(v_3) + \eta(v_2, v_3)\zeta(v_1)$$
$$\quad + \eta(v_3, v_1)\zeta(v_2) + \eta(v_2, v_3)\zeta(v_1) + \eta(v_3, v_1)\zeta(v_2)]/2$$
$$= \eta(v_1, v_2)\zeta(v_3) + \eta(v_2, v_3)\zeta(v_1) + \eta(v_3, v_1)\zeta(v_2). \qquad \Diamond$$

Wedge products behave well with respect to basic algebraic structure.

Theorem 7.2.2. *Let V be a vector space, let η, η^1, η^2 and ζ, ζ^1, ζ^2 be multicovectors in $\Lambda^s(V)$ and $\Lambda^{s'}(V)$, respectively, and let c be a real number. Then:*
(a) $(\eta^1 + \eta^2) \wedge \zeta = \eta^1 \wedge \zeta + \eta^2 \wedge \zeta$.
(b) $\eta \wedge (\zeta^1 + \zeta^2) = \eta \wedge \zeta^1 + \eta \wedge \zeta^2$.
(c) $(c\eta) \wedge \zeta = c(\eta \wedge \zeta) = \eta \wedge (c\zeta)$.

Proof. Straightforward. $\qquad\qquad\qquad\qquad\qquad\qquad\qquad\qquad\qquad\qquad\square$

Theorem 7.2.3. *If V is a vector space, and η and ζ are multicovectors in $\Lambda^s(V)$ and $\Lambda^{s'}(V)$, respectively, then:*
(a) $\eta \wedge \zeta = (-1)^{ss'} \zeta \wedge \eta$.
(b) *If $s = s' = 1$, then $\eta \wedge \zeta = -\zeta \wedge \eta$.*
(c) *If $s = 1$, then $\eta \wedge \eta = 0$.*

Proof. (a): Let $v_1, \ldots, v_{s+s'}$ be vectors in V, and define a permutation σ in $\mathcal{S}_{s+s'}$ by

$$\sigma = \begin{pmatrix} 1 & \cdots & s & s+1 & \cdots & s+s' \\ s'+1 & \cdots & s'+s & 1 & \cdots & s' \end{pmatrix}.$$

We have

$$\sigma(\eta \otimes \zeta)(v_1, \ldots, v_{s+s'}) = \eta \otimes \zeta(v_{\sigma(1)}, \ldots, v_{\sigma(s)}, v_{\sigma(s+1)}, \ldots, v_{\sigma(s+s')})$$
$$= \eta \otimes \zeta(v_{s'+1}, \ldots, v_{s'+s}, v_1, \ldots, v_{s'})$$
$$= \eta(v_{s'+1}, \ldots, v_{s'+s})\, \zeta(v_1, \ldots, v_{s'})$$
$$= \zeta(v_1, \ldots, v_{s'})\, \eta(v_{s'+1}, \ldots, v_{s'+s})$$
$$= \zeta \otimes \eta(v_1, \ldots, v_{s'}, v_{s'+1}, \ldots, v_{s'+s}).$$

Since $v_1, \ldots, v_{s+s'}$ were arbitrary, $\sigma(\eta \otimes \zeta) = \zeta \otimes \eta$. Then

$$
\begin{aligned}
\mathrm{Alt}(\zeta \otimes \eta) &= \mathrm{Alt}\big(\sigma(\eta \otimes \zeta)\big) \\
&= \mathrm{sgn}(\sigma)\, \mathrm{Alt}(\eta \otimes \zeta) && \text{[Th 7.1.3(a)]} \\
&= (-1)^{ss'} \mathrm{Alt}(\eta \otimes \zeta), && \text{[Th B.2.4]}
\end{aligned}
$$

hence

$$
\eta \wedge \zeta = \frac{(s+s')!}{s!s'!}\, \mathrm{Alt}(\eta \otimes \zeta) = (-1)^{ss'} \frac{(s'+s)!}{s'!s!}\, \mathrm{Alt}(\zeta \otimes \eta)
$$
$$
= (-1)^{ss'} \zeta \wedge \eta.
$$

(b): This follows from part (a).
(c): This follows from part (b). $\qquad\qquad\qquad\qquad\qquad\qquad\qquad\square$

Theorem 7.2.4. *If V is a vector space, and $\mathcal{A}$ and $\mathcal{B}$ are multicovectors in $\Lambda^s(V)$ and $\Lambda^{s'}(V)$, respectively, then*

$$
\mathrm{Alt}\big(\mathrm{Alt}(\mathcal{A}) \otimes \mathcal{B}\big) = \mathrm{Alt}(\mathcal{A} \otimes \mathcal{B}) = \mathrm{Alt}\big(\mathcal{A} \otimes \mathrm{Alt}(\mathcal{B})\big).
$$

Proof. Let us define a map $\iota : \mathcal{S}_s \longrightarrow \mathcal{S}_{s+s'}$ as follows. For each permutation σ in $\mathcal{S}_s$, let $\iota(\sigma)$ be the permutation in $\mathcal{S}_{s+s'}$ that restricts to σ on $\{1, \ldots, s\}$ and fixes $s+1, \ldots, s+s'$; that is,

$$
\iota(\sigma) = \begin{pmatrix} 1 & \cdots & s & s+1 & \cdots & s+s' \\ \sigma(1) & \cdots & \sigma(s) & s+1 & \cdots & s+s' \end{pmatrix}.
$$

Evidently, ι is a group isomorphism between $\mathcal{S}_s$ and the subgroup $\iota(\mathcal{S}_s)$ of $\mathcal{S}_{s+s'}$. It is easily shown that

$$
\mathrm{sgn}\big(\iota(\sigma)\big) = \mathrm{sgn}(\sigma) \qquad \text{and} \qquad \iota(\sigma)(\mathcal{A} \otimes \mathcal{B}) = \sigma(\mathcal{A}) \otimes \mathcal{B}. \qquad (7.2.2)
$$

Then

$$
\begin{aligned}
\mathrm{Alt}\big(\mathrm{Alt}(\mathcal{A}) \otimes \mathcal{B}\big) &= \mathrm{Alt}\left(\left[\frac{1}{s!} \sum_{\sigma \in \mathcal{S}_s} \mathrm{sgn}(\sigma)\, \sigma(\mathcal{A})\right] \otimes \mathcal{B}\right) && \text{[(7.1.4)]} \\
&= \mathrm{Alt}\left(\frac{1}{s!} \sum_{\sigma \in \mathcal{S}_s} \mathrm{sgn}(\sigma)\, \big(\sigma(\mathcal{A}) \otimes \mathcal{B}\big)\right) && \text{[Th 5.1.2(a)]} \\
&= \frac{1}{s!} \sum_{\sigma \in \mathcal{S}_s} \mathrm{sgn}(\sigma)\, \mathrm{Alt}\big(\sigma(\mathcal{A}) \otimes \mathcal{B}\big) \\
&= \frac{1}{s!} \sum_{\sigma \in \mathcal{S}_s} \mathrm{sgn}\big(\iota(\sigma)\big)\, \mathrm{Alt}\big(\iota(\sigma)(\mathcal{A} \otimes \mathcal{B})\big) && \text{[(7.2.2)]} \\
&= \frac{1}{s!} \sum_{\sigma \in \mathcal{S}_s} \mathrm{Alt}(\mathcal{A} \otimes \mathcal{B}) && \text{[Th 7.1.3(a)]} \\
&= \mathrm{Alt}(\mathcal{A} \otimes \mathcal{B}), && [\mathrm{card}(\mathcal{S}_s) = s!]
\end{aligned}
$$

which proves the first equality. The proof of the second equality is similar. $\quad\square$

Any operation that purports to be a type of "multiplication" should be associative. The wedge product meets this requirement.

Theorem 7.2.5 (Associativity of Wedge Product). *Let V be a vector space, and let η and ζ and ξ be multicovectors in $\Lambda^s(V)$ and $\Lambda^{s'}(V)$ and $\Lambda^{s''}(V)$, respectively. Then*

$$(\eta \wedge \zeta) \wedge \xi = \frac{(s + s' + s'')!}{s!s'!s''!} \, \mathrm{Alt}(\eta \otimes \zeta \otimes \xi) = \eta \wedge (\zeta \wedge \xi).$$

Proof. We have

$$
\begin{aligned}
(\eta \wedge \zeta) \wedge \xi &= \frac{[(s + s') + s'']!}{(s + s')!s''!} \, \mathrm{Alt}\big((\eta \wedge \zeta) \otimes \xi\big) && [(7.2.1)] \\
&= \frac{(s + s' + s'')!}{(s + s')!s''!} \, \mathrm{Alt}\left(\left[\frac{(s + s')!}{s!s'!} \, \mathrm{Alt}(\eta \otimes \zeta)\right] \otimes \xi\right) && [(7.2.1)] \\
&= \frac{(s + s' + s'')!}{s!s'!s''!} \, \mathrm{Alt}\big(\mathrm{Alt}(\eta \otimes \zeta) \otimes \xi\big) \\
&= \frac{(s + s' + s'')!}{s!s'!s''!} \, \mathrm{Alt}(\eta \otimes \zeta \otimes \xi), && [\text{Th } 7.2.4]
\end{aligned}
$$

which proves the first equality. The proof of the second equality is similar. $\quad\square$

In light of the associativity of the wedge product, we drop parentheses and, for example, denote $(\eta \wedge \zeta) \wedge \xi$ and $\eta \wedge (\zeta \wedge \xi)$ by $\eta \wedge \zeta \wedge \xi$, with corresponding notation for wedge products of more than three terms.

Theorem 7.2.6. *Let V be a vector space, and let $\eta^1, \ldots, \eta^s$ be covectors in V^*. If $\eta^i = \eta^j$ for some integers $1 \le i < j \le s$, then $\eta^1 \wedge \cdots \wedge \eta^s = 0$.*

Proof. We have

$$
\begin{aligned}
&\eta^1 \wedge \cdots \wedge \eta^i \wedge \cdots \wedge \eta^j \wedge \cdots \wedge \eta^k \\
&\quad = (-1)^{(i-1)+(j-2)}(\eta^i \wedge \eta^j) \wedge (\eta^1 \wedge \cdots \wedge \widehat{\eta^i} \wedge \cdots \wedge \widehat{\eta^j} \wedge \cdots \wedge \eta^k) = 0,
\end{aligned}
$$

where the first equality follows from repeated applications of Theorem 7.2.3(b), and the second equality from Theorem 7.2.3(c), and where $\widehat{}$ indicates that an expression is omitted. $\quad\square$

The next result is a generalization of Theorem 7.2.5.

Theorem 7.2.7. *If V is a vector space and η^i is a multicovector in $\Lambda^{s_i}(V)$ for $i = 1, \ldots, k$, then*

$$\eta^1 \wedge \cdots \wedge \eta^k = \frac{(s_1 + \cdots + s_k)!}{s_1! \cdots s_k!} \, \mathrm{Alt}(\eta^1 \otimes \cdots \otimes \eta^k).$$

Proof. The result is trivial for $k = 1$. For $k = 2$, the result is simply the definition of Alt. For $k = 3$, the result is given by Theorem 7.2.5. For $k \ge 4$, the proof is by induction. Let $k \geq 4$, and suppose the assertion is true for all indices $< k$. Then

$$(\eta^1 \wedge \eta^2) \wedge \eta^3 \wedge \cdots \wedge \eta^k$$

$$= \frac{[(s_1 + s_2) + s_3 + \cdots + s_k]!}{(s_1 + s_2)! s_3! \cdots s_k!} \operatorname{Alt}\big((\eta^1 \wedge \eta^2) \otimes \eta^3 \otimes \cdots \otimes \eta^k\big)$$

$$= \frac{(s_1 + s_2 + s_3 + \cdots + s_k)!}{(s_1 + s_2)! s_3! \cdots s_k!} \operatorname{Alt}\left(\frac{(s_1 + s_2)!}{s_1! s_2!} \operatorname{Alt}(\eta^1 \otimes \eta^2) \otimes (\eta^3 \otimes \cdots \otimes \eta^k)\right)$$

$$= \frac{(s_1 + \cdots + s_k)!}{s_1! \cdots s_k!} \operatorname{Alt}\big(\operatorname{Alt}(\eta^1 \otimes \eta^2) \otimes (\eta^3 \otimes \cdots \otimes \eta^k)\big)$$

$$= \frac{(s_1 + \cdots + s_k)!}{s_1! \cdots s_k!} \operatorname{Alt}(\eta^1 \otimes \cdots \otimes \eta^k),$$

where the first equality follows from the induction hypothesis, the second equality from (7.2.1), and the last equality from Theorem 7.2.4. $\qquad\square$

The next result shows that wedge products and determinants are closely related, which is not so surprising.

Theorem 7.2.8. *Let V be a vector space, let $\eta^1, \ldots, \eta^s$ be covectors in V^*, and let $v_1, \ldots, v_s$ be vectors in V. Then*

$$\eta^1 \wedge \cdots \wedge \eta^s(v_1, \ldots, v_s) = \det\left(\begin{bmatrix} \eta^1(v_1) & \cdots & \eta^1(v_s) \\ \vdots & \ddots & \vdots \\ \eta^s(v_1) & \cdots & \eta^s(v_s) \end{bmatrix}\right).$$

Proof. We have

$$\eta^1 \wedge \cdots \wedge \eta^s(v_1, \ldots, v_s)$$

$$= s! \operatorname{Alt}(\eta^1 \otimes \cdots \otimes \eta^s)(v_1, \ldots, v_s) \qquad\qquad\text{[Th 7.2.7]}$$

$$= \sum_{\sigma \in \mathcal{S}_s} \operatorname{sgn}(\sigma)\, \sigma(\eta^1 \otimes \cdots \otimes \eta^s)(v_1, \ldots, v_s) \qquad\text{[(7.1.4)]}$$

$$= \sum_{\sigma \in \mathcal{S}_s} \operatorname{sgn}(\sigma)\, \eta^1 \otimes \cdots \otimes \eta^s(v_{\sigma(1)}, \ldots, v_{\sigma(s)}) \qquad\text{[(7.1.1)]}$$

$$= \sum_{\sigma \in \mathcal{S}_s} \operatorname{sgn}(\sigma)\, \eta^1(v_{\sigma(1)}) \cdots \eta^s(v_{\sigma(s)})$$

$$= \det\big([\eta^i(v_j)]\big). \qquad\qquad\qquad\qquad\qquad\text{[(2.4.1)]} \qquad\square$$

Theorem 7.2.9. *Let V be a vector space, let $(h_1, \ldots, h_m)$ be a basis for V, and let $(\theta^1, \ldots, \theta^m)$ be its dual basis. Then*

$$\theta^{j_1} \wedge \cdots \wedge \theta^{j_s}(h_{i_1}, \ldots, h_{i_s}) = \begin{cases} 1 & \text{if } (i_1, \ldots, i_s) = (j_1, \ldots, j_s) \\ 0 & \text{if } (i_1, \ldots, i_s) \neq (j_1, \ldots, j_s) \end{cases}$$

for all $1 \leq i_1 < \cdots < i_s \leq m$ and $1 \leq j_1 < \cdots < j_s \leq m$.

Proof. By Theorem 7.2.8,

$$\theta^{j_1} \wedge \cdots \wedge \theta^{j_s}(h_{i_1}, \ldots, h_{i_s}) = \det\left(\left[\theta^i(h_j)\right]_{(i_1,\ldots,i_s)}^{(j_1,\ldots,j_s)}\right) = \det\left((I_m)_{(i_1,\ldots,i_s)}^{(j_1,\ldots,j_s)}\right).$$

If $(i_1, \ldots, i_s) = (j_1, \ldots, j_s)$, then $(I_m)_{(i_1,\ldots,i_s)}^{(j_1,\ldots,j_s)} = I_s$, and if $(i_1, \ldots, i_s) \neq (j_1, \ldots, j_s)$, then $(I_m)_{(i_1,\ldots,i_s)}^{(j_1,\ldots,j_s)}$ has at least one row (and at least one column) of 0s. The result follows. $\square$

Theorem 7.2.10. *Let V be a vector space, let $\eta^1, \ldots, \eta^s$ be covectors in V^*, and let σ be a permutation in $\mathcal{S}_s$. Then*

$$\eta^{\sigma(1)} \wedge \cdots \wedge \eta^{\sigma(s)} = \operatorname{sgn}(\sigma)\, \eta^1 \wedge \cdots \wedge \eta^s.$$

Proof. For vectors $v_1, \ldots, v_s$ in V, we have

$$
\begin{aligned}
\eta^{\sigma(1)} &\wedge \cdots \wedge \eta^{\sigma(s)}(v_1, \ldots, v_s) \\
&= \det\left(\left[\eta^{\sigma(i)}(v_j)\right]\right) && \text{[Th 7.2.8]} \\
&= \det\left(\left[\eta^{\sigma(j)}(v_i)\right]\right) && \text{[Th 2.4.5(c)]} \\
&= \operatorname{sgn}(\sigma) \det\left(\left[\eta^j(v_i)\right]\right) && \text{[Th 2.4.2]} \\
&= \operatorname{sgn}(\sigma) \det\left(\left[\eta^i(v_j)\right]\right) && \text{[Th 2.4.5(c)]} \\
&= \operatorname{sgn}(\sigma)\, \eta^1 \wedge \cdots \wedge \eta^s(v_1, \ldots, v_s), && \text{[Th 7.2.8]}
\end{aligned}
$$

where $\left[\eta^{\sigma(j)}(v_i)\right] = \left(\left[\eta^{\sigma(i)}(v_j)\right]\right)^{\mathrm{T}}$. Since $v_1, \ldots, v_s$ were arbitrary, the result follows. $\square$

Theorem 7.2.11. *Let V be a vector space, let $\mathcal{H} = (h_1, \ldots, h_m)$ and $\mathcal{F}$ be bases for V, and let $(\varphi^1, \ldots, \varphi^m)$ be the dual basis corresponding to $\mathcal{F}$. Then*

$$\varphi^1 \wedge \cdots \wedge \varphi^m(h_1, \ldots, h_m) = \det\left(\left[\operatorname{id}_V\right]_{\mathcal{H}}^{\mathcal{F}}\right).$$

Proof. By Theorem 1.2.1(d), $h_j = \sum_i \varphi^i(h_j) f_i$ for $j = 1, \ldots, m$. Then (2.2.6) and (2.2.7) give $\left[\operatorname{id}_V\right]_{\mathcal{H}}^{\mathcal{F}} = \left[\varphi^i(h_j)\right]$. The result now follows from Theorem 7.2.8. $\square$

Theorem 7.2.12 (Basis for $\Lambda^s(V)$). *Let V be a vector space, let $(h_1, \ldots, h_m)$ be a basis for V, and let $(\theta^1, \ldots, \theta^m)$ be its dual basis. Then:*
(a) *If $1 \leq s \leq m$, then*

$$\{\theta^{i_1} \wedge \cdots \wedge \theta^{i_s} : 1 \leq i_1 < \cdots < i_s \leq m\}$$

is an unordered basis for $\Lambda^s(V)$.
(b) *If $s > m$, then $\Lambda^s(V) = \{0\}$.*
(c) *$\Lambda^s(V)$ has dimension $\binom{m}{s}$.*
(d) *Any nonzero multicovector in $\Lambda^m(V)$ spans $\Lambda^m(V)$.*

(e) $(\eta^1, \ldots, \eta^m)$ *is a basis for* $\Lambda^{m-1}(V)$*, where*

$$\eta^i = (-1)^{i-1}\theta^1 \wedge \cdots \wedge \widehat{\theta^i} \wedge \cdots \wedge \theta^m$$

for $i = 1, \ldots, m$*, and where* $\widehat{\ }$ *indicates that an expression is omitted.*

Proof. Let

$$\Omega_s = \{\theta^{i_i} \otimes \cdots \otimes \theta^{i_s} : 1 \leq i_1, \ldots, i_s \leq m\}$$

and

$$\Theta_s = \{\theta^{i_1} \wedge \cdots \wedge \theta^{i_s} : 1 \leq i_1 < \cdots < i_s \leq m\}.$$

Then

$$\begin{aligned}
\text{span}(\Theta_s) &= \text{span}\big(\text{Alt}(\Omega_s)\big) && [\text{Th 7.2.7}] \\
&= \text{Alt}\big(\text{span}(\Omega_s)\big) \\
&= \text{Alt}\big(\mathcal{T}_s^0(V)\big) && [\text{Th 5.1.3(b)}] \\
&= \Lambda^s(V), && [\text{Th 7.1.3(e)}]
\end{aligned}$$

so Θ_s spans $\Lambda^s(V)$ for all $s \geq 1$.

(a): If

$$\sum_{1 \leq i_1 < \cdots < i_s \leq m} a_{i_1 \ldots i_s}\theta^{i_1} \wedge \cdots \wedge \theta^{i_s} = 0$$

for some real numbers $a_{i_1 \ldots i_s}$, then evaluating both sides of the preceding identity at $(h_{j_1}, \ldots, h_{j_s})$ and using Theorem 7.2.9 yields $a_{j_1 \ldots j_s} = 0$ for all $1 \leq j_1 < \cdots < j_s \leq m$. Thus, Θ_s is linearly independent.

(b): Since $s > m$, each element of Θ_s contains a repeated covector, so the result follows from Theorem 7.2.6.

(c): This follows from parts (a) and (b).

(d): By part (c), $\Lambda^m(V)$ is 1-dimensional.

(e): For $s = m - 1$, we have $\Theta_{m-1} = \{(-1)^{i-1}\eta^i : i = 1, \ldots, m\}$, so the result follows from part (a). □

Part (a) of the next result is a generalization of Theorem 1.2.1(e).

Theorem 7.2.13. *Let* V *be a vector space, let* η *be a multicovector in* $\Lambda^s(V)$*, let* $(h_1, \ldots, h_m)$ *be a basis for* V*, and let* $(\theta^1, \ldots, \theta^m)$ *be its dual basis. Then:*
(a)

$$\eta = \sum_{1 \leq i_1 < \cdots < i_s \leq m} \eta(h_{i_1}, \ldots, h_{i_s})\, \theta^{i_1} \wedge \cdots \wedge \theta^{i_s}.$$

(b) *If* $s = m$*, then*

$$\eta = \eta(h_1, \ldots, h_m)\, \theta^1 \wedge \cdots \wedge \theta^m.$$

Proof. (a): By Theorem 7.2.12(a),

$$\eta = \sum_{1 \leq i_1 < \cdots < i_s \leq m} a_{i_1 \ldots i_s}\theta^{i_1} \wedge \cdots \wedge \theta^{i_s}$$

for some real numbers $a_{i_1 \ldots i_s}$. Evaluating both sides of the preceding identity at $(h_{j_1}, \ldots, h_{j_s})$ and using Theorem 7.2.9 yields $a_{j_1 \ldots j_s} = \eta(h_{j_1}, \ldots, h_{j_s})$ for all $1 \leq j_1 < \cdots < j_s \leq m$.

(b): This follows from part (a). □

7.3   Pullback of Multicovectors

The pullback of covariant tensors was briefly considered in Section 5.2. The corresponding theory for multicovectors is far richer.

Before proceeding, we pause to consider multi-index notation, which was discussed in Section 2.1 in the context of matrices. Let V be a vector space, let $(h_1, \ldots, h_m)$ be a basis for V, and let $(\theta^1, \ldots, \theta^m)$ be its dual basis. For an integer $1 \leq s \leq m$, let $I = (i_1, \ldots, i_s)$ be a multi-index in $\mathcal{I}_{s,m}$, and let us denote

$$(h_{i_1}, \ldots, h_{i_s}) \qquad \text{by} \qquad h_I$$

and

$$\theta^{i_1} \wedge \cdots \wedge \theta^{i_s} \qquad \text{by} \qquad \theta^I.$$

In this notation, the unordered basis for $\Lambda^s(V)$ in Theorem 7.2.12(a) can be expressed concisely as $\{\theta^I : I \in \mathcal{I}_{s,m}\}$, and the identity in Theorem 7.2.13(a) becomes

$$\eta = \sum_{I \in \mathcal{I}_{s,m}} \eta(h_I)\theta^I.$$

If we order $\mathcal{I}_{s,m}$ in some fashion, for example, lexicographically, then the basis $\{\theta^I : I \in \mathcal{I}_{s,m}\}$ of Theorem 7.2.12(a) can be similarly ordered, yielding an ordered basis for $\Lambda^s(V)$, which we denote by

$$(\theta^I : I \in \mathcal{I}_{s,m}).$$

Kronecker's delta can be generalized to the multi-index setting. Let $J = (j_1, \ldots, j_s)$ be another multi-index in $\mathcal{I}_{s,m}$, and define

$$\delta_J^I = \begin{cases} 1 & \text{if } I = J \\ 0 & \text{if } I \neq J. \end{cases}$$

Then the conditional identity in Theorem 7.2.9 is simply $\theta^J(h_I) = \delta_I^J$. As discussed in Appendix A, the complement of the multi-index (i) in $\mathcal{I}_{1,m}$ is

$$(i)^c = (1, \ldots, \widehat{i}, \ldots, m) = (1, \ldots, i-1, i+1, \ldots, m)$$

for $i = 1, \ldots, m$. In this notation, the multicovectors comprising the basis in Theorem 7.2.12(e) can be expressed as $\eta^i = (-1)^{i-1}\theta^{(i)^c}$. We will find multi-index notation of great utility in what follows.

Let V and W be vector spaces, and let $A : V \longrightarrow W$ be a linear map. **Pullback by A (for multicovectors)** is the family of linear maps

$$A^* : \Lambda^s(W) \longrightarrow \Lambda^s(V)$$

defined for $s \geq 1$ by

$$A^*(\eta)(v_1, \ldots, v_s) = \eta(A(v_1), \ldots, A(v_s)) \tag{7.3.1}$$

for all multicovectors η in $\Lambda^s(W)$ and all vectors $v_1, \ldots, v_s$ in V. It follows from the multilinearity and alternating properties of η and the linearity of A that $A^*(\eta)$ is in $\Lambda^s(V)$, so the definition makes sense. We refer to $A^*(\eta)$ as the **pullback of η by A**.

Theorem 7.3.1. *Let V and W be vector spaces, let $A : V \longrightarrow W$ be a linear map, and let η, ζ and ξ be multicovectors in $\Lambda^s(W)$ and $\Lambda^{s'}(W)$, respectively. Then:*
(a) $A^*(\eta + \zeta) = A^*(\eta) + A^*(\zeta)$.
(b) $A^*(\eta \wedge \xi) = A^*(\eta) \wedge A^*(\xi)$.

Proof. (a): Straightforward.
  (b): We begin with a preliminary result. Let $\zeta^1, \ldots, \zeta^{s+s'}$ be (not necessarily distinct) covectors in $\Lambda^{s+s'}(W)$. For vectors $v_1, \ldots, v_{s+s'}$ in W, we have

$$
\begin{aligned}
A^*(\zeta^1 \wedge \cdots \wedge \zeta^{s+s'})&(v_1, \ldots, v_{s+s'}) \\
&= \zeta^1 \wedge \cdots \wedge \zeta^{s+s'}\big(A(v_1), \ldots, A(v_{s+s'})\big) \\
&= \det\big([\zeta^i(A(v_j))]\big) && \text{[Th 7.2.8]} \\
&= \det\big([A^*(\zeta^i)(v_j)]\big) \\
&= A^*(\zeta^1) \wedge \cdots \wedge A^*(\zeta^{s+s'})(v_1, \ldots, v_{s+s'}). && \text{[Th 7.2.8]}
\end{aligned}
$$

Since $v_1, \ldots, v_{s+s'}$ were arbitrary,

$$A^*(\zeta^1 \wedge \cdots \wedge \zeta^{s+s'}) = A^*(\zeta^1) \wedge \cdots \wedge A^*(\zeta^{s+s'}). \tag{7.3.2}$$

Now for the main proof. Let $(\varphi^1, \ldots, \varphi^n)$ be the dual basis corresponding to a basis for W. By Theorem 7.2.12(a), η and ξ can be expressed as

$$\eta = \sum_{I \in \mathcal{I}_{s,n}} a_I \varphi^I \qquad \text{and} \qquad \xi = \sum_{J \in \mathcal{I}_{s',n}} b_J \varphi^J.$$

Then

$$\eta \wedge \xi = \left(\sum_{I \in \mathcal{I}_{s,n}} a_I \varphi^I \right) \wedge \left(\sum_{J \in \mathcal{I}_{s',n}} b_J \varphi^J \right) = \sum_{I \in \mathcal{I}_{s,n},\, J \in \mathcal{I}_{s',n}} a_I b_J \varphi^I \wedge \varphi^J,$$

hence

$$
\begin{aligned}
A^*(\eta \wedge \xi) &= \sum_{I \in \mathcal{I}_{s,n},\, J \in \mathcal{I}_{s',n}} a_I b_J A^*(\varphi^I \wedge \varphi^J) \\
&= \sum_{I \in \mathcal{I}_{s,n},\, J \in \mathcal{I}_{s',n}} a_I b_J A^*(\varphi^I) \wedge A^*(\varphi^J) \\
&= \left(\sum_{I \in \mathcal{I}_{s,n}} a_I A^*(\varphi^I) \right) \wedge \left(\sum_{J \in \mathcal{I}_{s',n}} b_J A^*(\varphi^J) \right) \\
&= A^*(\eta) \wedge A^*(\xi),
\end{aligned}
$$

where the second equality follows from (7.3.2). $\qquad\square$

Theorem 7.3.2 (Pullback of Multicovector). *Let* V *and* W *be vector spaces of dimensions* m *and* n, *respectively, let* $\mathcal{H}$ *and* $\mathcal{F}$ *be respective bases, and let* $(\theta^1, \ldots, \theta^m)$ *and* $(\varphi^1, \ldots, \varphi^n)$ *be the corresponding dual bases. Let* $A : V \longrightarrow W$ *be a linear map, and let* I *be a multi-index in* $\mathcal{I}_{s,n}$, *where* $s \leq \min(m, n)$. *Then*

$$A^*(\varphi^I) = \sum_{J \varepsilon \mathcal{I}_{s,m}} \det\left(\left([A]^{\mathcal{F}}_{\mathcal{H}}\right)^I_J\right) \theta^J.$$

Proof. Let $\mathcal{H} = (h_1, \ldots, h_m)$. By Theorem 7.2.13(a),

$$A^*(\varphi^I) = \sum_{J \varepsilon \mathcal{I}_{s,m}} A^*(\varphi^I)(h_J)\, \theta^J. \tag{7.3.3}$$

Let $[A]^{\mathcal{F}}_{\mathcal{H}} = [a^i_j]$. Then (2.2.2) and (2.2.3) give

$$\varphi^i\big(A(h_j)\big) = \varphi^i\left(\sum_k a^k_j f_k\right) = \sum_i a^k_j \varphi^i(f_k) = a^i_j,$$

hence

$$[A]^{\mathcal{F}}_{\mathcal{H}} = \left[\varphi^i\big(A(h_j)\big)\right]. \tag{7.3.4}$$

Let $I = (i_1, \ldots, i_s)$ and $J = (j_1, \ldots, j_s)$. We have

$$
\begin{aligned}
A^*(\varphi^I)(h_J) &= A^*(\varphi^{i_1} \wedge \cdots \wedge \varphi^{i_s})(h_{j_1}, \ldots, h_{j_s}) \\
&= A^*(\varphi^{i_1}) \wedge \cdots \wedge A^*(\varphi^{i_s})(h_{j_1}, \ldots, h_{j_s}) \qquad \text{[Th 7.3.1(b)]} \\
&= \det\left(\left[A^*(\varphi^i)(h_j)\right]^{(i_1,\ldots,i_s)}_{(j_1,\ldots,j_s)}\right) \qquad \text{[Th 7.2.8]} \\
&= \det\left(\left[\varphi^i\big(A(h_j)\big)\right]^{(i_1,\ldots,i_s)}_{(j_1,\ldots,j_s)}\right) \\
&= \det\left(\left([A]^{\mathcal{F}}_{\mathcal{H}}\right)^I_J\right). \qquad \text{[(7.3.4)]}
\end{aligned}
\tag{7.3.5}
$$

Substituting (7.3.5) into (7.3.3) gives the result.   □

Special cases of Theorem 7.3.2 provide a number of useful identities.

Theorem 7.3.3. *Let* V *and* W *be vector spaces of dimension* m, *let* $\mathcal{H} = (h_1, \ldots, h_m)$ *and* $\mathcal{F} = (f_1, \ldots, f_m)$ *be respective bases, and let* $(\theta^1, \ldots, \theta^m)$ *and* $(\varphi^1, \ldots, \varphi^m)$ *be the corresponding dual bases. Let* $A : V \longrightarrow W$ *be a linear map, and let* η *be a multicovector in* $\Lambda^m(W)$. *Then:*
(a)

$$A^*(\varphi^1 \wedge \cdots \wedge \varphi^m) = \det\left([A]^{\mathcal{F}}_{\mathcal{H}}\right) \theta^1 \wedge \cdots \wedge \theta^m.$$

(b)

$$A^*(\eta)(h_1, \ldots, h_m) = \det\left([A]^{\mathcal{F}}_{\mathcal{H}}\right) \eta(f_1, \ldots, f_m).$$

Proof. (a): Since $\mathcal{I}_{m,m} = \{(1, \ldots, m)\}$, this follows from Theorem 7.3.2.

(b): It follows from Theorem 7.2.13(b) that

$$\eta = \eta(f_1, \ldots, f_m)\, \varphi^1 \wedge \cdots \wedge \varphi^m,$$

and then from part (a) that

$$\begin{aligned}
A^*(\eta) &= \eta(f_1, \ldots, f_m)\, A^*(\varphi^1 \wedge \cdots \wedge \varphi^m) \\
&= \eta(f_1, \ldots, f_m) \det\left([A]_{\mathcal{H}}^{\mathcal{F}}\right) \theta^1 \wedge \cdots \wedge \theta^m.
\end{aligned} \qquad (7.3.6)$$

Applying both sides of the preceding identity to $(h_1, \ldots, h_m)$ and using Theorem 7.2.9 gives the result. $\qquad\square$

Theorem 7.3.4. *Let V be a vector space of dimension m, let $A : V \longrightarrow V$ be a linear map, let η be a multicovector in $\Lambda^m(V)$, and let $v_1, \ldots, v_m$ be vectors in V. Then:*

(a) $A^*(\eta) = \det(A)\, \eta$.

(b) $\eta(A(v_1), \ldots, A(v_m)) = \det(A)\, \eta(v_1, \ldots, v_m)$.

Proof. (a): In the notation of Theorem 7.3.3, we have

$$\begin{aligned}
A^*(\eta) &= \det\left([A]_{\mathcal{H}}^{\mathcal{H}}\right) \eta(h_1, \ldots, h_m)\, \theta^1 \wedge \cdots \wedge \theta^m && [(7.3.6)] \\
&= \det(A)\, \eta. && [\text{Th } 7.2.13(\text{b})]
\end{aligned}$$

(b): By part (a),

$$\eta(A(v_1), \ldots, A(v_m)) = A^*(\eta)(v_1, \ldots, v_m) = \det(A)\, \eta(v_1, \ldots, v_m). \qquad\square$$

The next result is reminiscent of Theorem 2.4.9.

Theorem 7.3.5. *Let V be a vector space of dimension m, let η be a nonzero multicovector in $\Lambda^m(V)$, and let $v_1, \ldots, v_m$ be vectors in V. Then $v_1, \ldots, v_m$ are linearly independent if and only if $\eta(v_1, \ldots, v_m) \neq 0$.*

Proof. Let $\mathcal{H} = (h_1, \ldots, h_m)$ be a basis for V, and let $(\theta^1, \ldots, \theta^m)$ be its dual basis. Let $\mathcal{V} = (v_1, \ldots, v_m)$, and let $A : V \longrightarrow V$ be the linear map defined by $A(\mathcal{H}) = \mathcal{V}$. Since η is nonzero, we have from Theorem 7.2.13(b) that

$$\eta(h_1, \ldots, h_m) \neq 0, \qquad (7.3.7)$$

and from Theorem 7.3.4(b) that

$$\eta(v_1, \ldots, v_m) = \det(A)\, \eta(h_1, \ldots, h_m). \qquad (7.3.8)$$

Then

$$\begin{aligned}
&v_1, \ldots, v_m \text{ are linearly independent} \\
\Leftrightarrow\quad &\mathcal{V} \text{ is a basis for } V \\
\Leftrightarrow\quad &A \text{ is a linear isomorphism} && [\text{Th } 1.1.13(\text{a})] \\
\Leftrightarrow\quad &\det(A) \neq 0 && [\text{Th } 2.5.3] \\
\Leftrightarrow\quad &\eta(v_1, \ldots, v_m) \neq 0. && [(7.3.7),\ (7.3.8)]
\end{aligned}$$

We note in passing that ($\Leftarrow$) follows from Theorem 7.1.2. $\qquad\square$

Theorem 7.3.6 (Change of Basis). *Let V be a vector space, let $\mathcal{H}$ and $\mathcal{F}$ be bases for V, let $(\theta^1, \ldots, \theta^m)$ and $(\varphi^1, \ldots, \varphi^m)$ be the corresponding dual bases, and let $(\theta^I : I \in \mathcal{I}_{s,m})$ and $(\varphi^J : J \in \mathcal{I}_{s,m})$ be the corresponding bases for $\Lambda^s(V)$. Let η be a multicovector in $\Lambda^s(V)$, with*

$$\eta = \sum_{I \in \mathcal{I}_{s,m}} a_I \theta^I \qquad and \qquad \eta = \sum_{J \in \mathcal{I}_{s,m}} b_J \varphi^J.$$

Then

$$a_I = \sum_{J \in \mathcal{I}_{s,m}} b_J \det\left(\left(([\mathrm{id}_V]^{\mathcal{F}}_{\mathcal{H}})^J_I\right)\right).$$

Proof. Setting $W = V$ and $A = \mathrm{id}_V$ in Theorem 7.3.2 yields

$$\varphi^J = \sum_{I \in \mathcal{I}_{s,m}} \det\left(\left(([\mathrm{id}_V]^{\mathcal{F}}_{\mathcal{H}})^J_I\right)\right)\theta^I,$$

hence

$$\sum_{I \in \mathcal{I}_{s,m}} a_I \theta^I = \sum_{J \in \mathcal{I}_{s,m}} b_J \varphi^J$$

$$= \sum_{J \in \mathcal{I}_{s,m}} b_J \left[\sum_{I \in \mathcal{I}_{s,m}} \det\left(\left(([\mathrm{id}_V]^{\mathcal{F}}_{\mathcal{H}})^J_I\right)\right)\theta^I\right]$$

$$= \sum_{I \in \mathcal{I}_{s,m}} \left[\sum_{J \in \mathcal{I}_{s,m}} b_J \det\left(\left(([\mathrm{id}_V]^{\mathcal{F}}_{\mathcal{H}})^J_I\right)\right)\right]\theta^I,$$

from which the result follows. $\qquad\qquad\qquad\qquad\qquad\qquad\qquad\qquad\qquad\square$

7.4   Interior Multiplication

Let V be a vector space of dimension m, and let v be a vector in V. **Interior multiplication by v** is the family of linear maps

$$i_v : \Lambda^s(V) \longrightarrow \Lambda^{s-1}(V)$$

defined for $s \geq 2$ by

$$i_v(\eta)(v_1, \ldots, v_{s-1}) = \eta(v, w_1, \ldots, w_{s-1})$$

for all multicovectors η in $\Lambda^s(V)$ and all vectors $w_1, \ldots, w_{s-1}$ in V. Since any $m+1$ vectors in V are linearly dependent, it follows from Theorem 7.3.5 that i_v is the zero map when $s > m$. Recalling from (7.1.2) and (7.1.3) that $\Lambda^1(V) = V^*$ and $\Lambda^0(V) = \mathbb{R}$, we extend the preceding definition to $s = 1$ as follows:

$$i_v : \Lambda^1(V) \longrightarrow \Lambda^0(V)$$

is given by

$$i_v(\eta) = \eta(v)$$

for all covectors η in $\Lambda^1(V)$. For $s = 0$, we trivially define $i_v = 0$. Let us denote

$$i_v \circ i_v \qquad \text{by} \qquad i_v^2.$$

Not surprisingly, interior multiplication behaves well with respect to basic algebraic structure, but its handling of wedge products is more complex.

Theorem 7.4.1. *Let V be a vector space, let v, w be vectors in V, let η, ζ be multicovectors in $\Lambda^s(V)$, and let c be a real number. Then:*
(a) $i_{v+w}(\eta) = i_v(\eta) + i_w(\eta)$.
(b) $i_{cv}(\eta) = c\, i_v(\eta)$.
(c) $i_v(\eta + \zeta) = i_v(\eta) + i_v(\zeta)$.
(d) $i_v(c\eta) = c\, i_v(\eta)$.

Proof. Straightforward.   □

Theorem 7.4.2. *Let V be a vector space, let v be a vector in V, and let $\eta^1, \eta^2, \ldots, \eta^s$ be covectors in V^*. Then*

$$i_v(\eta^1 \wedge \cdots \wedge \eta^s) = \sum_{j=1}^{s}(-1)^{j-1}\, \eta^j(v)\, \eta^1 \wedge \cdots \wedge \widehat{\eta^j} \wedge \cdots \wedge \eta^s,$$

where $\widehat{}$ indicates that an expression is omitted.

Proof. For vectors $w_1, \ldots, w_{s-1}$ in V, we have

$$
\begin{aligned}
i_v(\eta^1 &\wedge \cdots \wedge \eta^s)(w_1, \ldots, w_{s-1}) \\
&= (\eta^1 \wedge \cdots \wedge \eta^s)(v, w_1, \ldots, w_{s-1}) \\
&= \det\left(\begin{bmatrix} \eta^1(v) & \eta^1(w_1) & \cdots & \eta^1(w_{s-1}) \\ \vdots & \vdots & \ddots & \vdots \\ \eta^s(v) & \eta^s(w_1) & \cdots & \eta^s(w_{s-1}) \end{bmatrix}\right) \\
&= \sum_{j=1}^{s}(-1)^{j-1}\, \eta^j(v)\, \det\left(\begin{bmatrix} \eta^1(w_1) & \cdots & \eta^1(w_{s-1}) \\ \vdots & \cdots & \vdots \\ \eta^s(w_1) & \cdots & \eta^s(w_{s-1}) \end{bmatrix}^{(j)^c}\right) \\
&= \sum_{j=1}^{s}(-1)^{j-1}\, \eta^j(v)\, \eta^1 \wedge \cdots \wedge \widehat{\eta^j} \wedge \cdots \wedge \eta^s(w_1, \ldots, w_{s-1}),
\end{aligned}
$$

where the second equality follows from Theorem 7.2.8, and the third equality follows from expanding the determinant along the first column. Since $w_1, \ldots, w_{s-1}$ were arbitrary, the result follows.   □

The next result shows that interior multiplication satisfies a novel product rule.

Theorem 7.4.3. *Let V be a vector space, let v be a vector in V, and let η and ζ be multicovectors in $\Lambda^s(V)$ and $\Lambda^{s'}(V)$, respectively. Then*

$$i_v(\eta \wedge \zeta) = i_v(\eta) \wedge \zeta + (-1)^s \eta \wedge i_v(\zeta).$$

Proof. For reasons exhibited by the proof of Theorem 7.3.1(b), it suffices to consider multicovectors of the form $\eta = \xi^1 \wedge \cdots \wedge \xi^s$ and $\zeta = \xi^{s+1} \wedge \cdots \wedge \xi^{s+s'}$, where ξ^j is a covector in V^* for $j = 1, 2, \ldots, s + s'$. By Theorem 7.4.2,

$i_v(\eta \wedge \zeta)$

$= i_v(\xi^1 \wedge \cdots \wedge \xi^s \wedge \xi^{s+1} \wedge \cdots \wedge \xi^{s+s'})$

$= \displaystyle\sum_{j=1}^{s} (-1)^{j-1} \xi^j(v) \, (\xi^1 \wedge \cdots \wedge \widehat{\xi^j} \wedge \cdots \wedge \xi^s) \wedge (\xi^{s+1} \wedge \cdots \wedge \xi^{s+s'})$

$\quad + \displaystyle\sum_{j=s+1}^{s+s'} (-1)^{j-1} \xi^j(v) \, (\xi^1 \wedge \cdots \wedge \xi^s) \wedge (\xi^{s+1} \wedge \cdots \wedge \widehat{\xi^j} \wedge \cdots \wedge \xi^{s+s'})$

$= \left(\displaystyle\sum_{j=1}^{s} (-1)^{j-1} \xi^j(v) \, \xi^1 \wedge \cdots \wedge \widehat{\xi^j} \wedge \cdots \wedge \xi^s \right) \wedge \xi^{s+1} \wedge \cdots \wedge \xi^{s+s'}$

$\quad + (-1)^s \xi^1 \wedge \cdots \wedge \xi^s \wedge \left(\displaystyle\sum_{j=s+1}^{s+s'} (-1)^{j-(s+1)} \xi^j(v) \, \xi^{s+1} \wedge \cdots \wedge \widehat{\xi^j} \wedge \cdots \wedge \xi^{s+s'} \right)$

$= i_v(\eta) \wedge \zeta + (-1)^s \eta \wedge i_v(\zeta).$

$\qquad\qquad\qquad\qquad\qquad\qquad\qquad\qquad\qquad\qquad\qquad\qquad\qquad\qquad\qquad\qquad \square$

7.5   Multicovector Scalar Product Spaces

In Section 4.5, we showed how to construct the dual of a scalar product space. Building on that foundation, we now generalize to multicovectors.

Let $(V, \mathfrak{g})$ be a scalar product space, let $\mathcal{H}$ be a basis for V, and let $\Theta = (\theta^1, \ldots, \theta^m)$ be its dual basis. By Theorem 7.2.12(a), $(\theta^I : I \in \mathcal{I}_{s,m})$ is a basis for $\Lambda^s(V)$. We define a bilinear function

$$\mathfrak{g}^\Lambda = \langle \cdot, \cdot \rangle^\Lambda : \Lambda^s(V) \times \Lambda^s(V) \longrightarrow \mathbb{R}$$

as follows. For multicovectors η, ζ in $\Lambda^s(V)$, with

$$\eta = \sum_{I \in \mathcal{I}_{s,m}} a_I \theta^I \qquad \text{and} \qquad \zeta = \sum_{J \in \mathcal{I}_{s,m}} b_J \theta^J, \tag{7.5.1}$$

let

$$\langle \eta, \zeta \rangle^\Lambda = \sum_{I, J \in \mathcal{I}_{s,m}} \det\!\left((\mathfrak{g}_\Theta^*)_J^I \right) a_I b_J, \tag{7.5.2}$$

where $\mathfrak{g}^*$ is the scalar product on V^* and

$$
\mathfrak{g}_\Theta^* = \begin{bmatrix} \langle\theta^1,\theta^1\rangle^* & \cdots & \langle\theta^1,\theta^m\rangle^* \\ \vdots & \ddots & \vdots \\ \langle\theta^m,\theta^1\rangle^* & \cdots & \langle\theta^m,\theta^m\rangle^* \end{bmatrix}
$$

is the matrix of $\mathfrak{g}^*$ with respect to Θ.

It would appear that the definition of $\mathfrak{g}^\Lambda$ is dependent on the choice of basis for V. Remarkably, this turns out not to be the case.

Theorem 7.5.1. *With the above setup, $\mathfrak{g}^\Lambda$ is independent of the choice of basis for V.*

Proof. Let $\mathcal{F}$ be another basis for V, let $\Phi = (\varphi^1,\ldots,\varphi^m)$ be its dual basis, and let $(\varphi^J : J \in \mathcal{I}_{s,m})$ be the corresponding basis for $\Lambda^s(V)$. Let

$$
\eta = \sum_{I\in\mathcal{I}_{s,m}} c_I\varphi^I \qquad \text{and} \qquad \zeta = \sum_{J\in\mathcal{I}_{s,m}} d_J\varphi^J.
$$

In the notation of (7.5.1) and (7.5.2), we need to show that

$$
\sum_{I,J\in\mathcal{I}_{s,m}} \det\big((\mathfrak{g}_\Theta^*)_J^I\big) a_I b_J = \sum_{I,J\in\mathcal{I}_{s,m}} \det\big((\mathfrak{g}_\Phi^*)_J^I\big) c_I d_J. \tag{7.5.3}
$$

Let

$$
P = \big[\mathrm{id}_V\big]_{\mathcal{F}}^{\mathcal{H}} \qquad \text{and} \qquad Q = \mathfrak{g}_\Phi^*. \tag{7.5.4}
$$

We begin with a few preliminary computations. Since Q is symmetric, we have from Theorem 2.1.6 that

$$
P^L(Q^I)^{\mathrm{T}} = P^L(Q^{\mathrm{T}})_I = P^L Q_I = (PQ)_I^L \tag{7.5.5}
$$

and

$$
P^K\big((PQ)^L\big)^{\mathrm{T}} = P^K\big((PQ)^{\mathrm{T}}\big)_L = (PQ^{\mathrm{T}}P^{\mathrm{T}})_L^K = (PQP^{\mathrm{T}})_L^K. \tag{7.5.6}
$$

We also have

$$
\begin{aligned}
PQP^{\mathrm{T}} \\
&= \big[\mathrm{id}_V\big]_{\mathcal{F}}^{\mathcal{H}} \mathfrak{g}_\Phi^* \Big(\big[\mathrm{id}_V\big]_{\mathcal{F}}^{\mathcal{H}}\Big)^{\mathrm{T}} \\
&= \Big(\big[\mathrm{id}_V\big]_{\mathcal{H}}^{\mathcal{F}}\Big)^{-1} \mathfrak{g}_{\mathcal{F}}^{-1} \Big(\big(\big[\mathrm{id}_V\big]_{\mathcal{H}}^{\mathcal{F}}\big)^{-1}\Big)^{\mathrm{T}} && \text{[Th 2.2.5(b), Th 4.5.3(b)]} \\
&= \Big(\big(\big[\mathrm{id}_V\big]_{\mathcal{H}}^{\mathcal{F}}\big)^{\mathrm{T}} \mathfrak{g}_{\mathcal{F}} \big[\mathrm{id}_V\big]_{\mathcal{H}}^{\mathcal{F}}\Big)^{-1} && \text{[Th 2.1.3]} \\
&= \mathfrak{g}_{\mathcal{H}}^{-1} && \text{[Th 3.1.3]} \\
&= \mathfrak{g}_\Theta^*, && \text{[Th 4.5.3(b)]}
\end{aligned} \tag{7.5.7}
$$

and from Theorem 7.3.6 that

$$c_I = \sum_{K \in \mathcal{I}_{s,m}} a_K \det(P_I^K) \quad \text{and} \quad d_J = \sum_{L \in \mathcal{I}_{s,m}} b_L \det(P_J^L). \tag{7.5.8}$$

Using (7.5.4) and (7.5.8), the right-hand side of (7.5.3) can be expressed as

$$\sum_{I,J \in \mathcal{I}_{s,m}} \det((\mathfrak{g}_\Phi^*)_J^I) c_I d_J$$

$$= \sum_{I,J \in \mathcal{I}_{s,m}} \left[\det(Q_J^I) \left(\sum_{K \in \mathcal{I}_{s,m}} a_K \det(P_I^K) \right) \left(\sum_{L \in \mathcal{I}_{s,m}} b_L \det(P_J^L) \right) \right] \tag{7.5.9}$$

$$= \sum_{K,L \in \mathcal{I}_{s,m}} \left[\left(\sum_{I,J \in \mathcal{I}_{s,m}} \det(P_I^K) \det(P_J^L) \det(Q_J^I) \right) a_K b_L \right].$$

The term in large parentheses in the last line of (7.5.9) can be expressed as

$$\sum_{I,J \in \mathcal{I}_{s,m}} \det(P_I^K) \det(P_J^L) \det(Q_J^I)$$

$$= \sum_{I \in \mathcal{I}_{s,m}} \left[\det(P_I^K) \left(\sum_{J \in \mathcal{I}_{s,m}} \det(P_J^L) \det(Q_J^I) \right) \right]$$

$$= \sum_{I \in \mathcal{I}_{s,m}} \left(\det(P_I^K) \det(P^L (Q^I)^{\mathrm{T}}) \right) \qquad \text{[Th 2.4.10(b)]}$$

$$= \sum_{I \in \mathcal{I}_{s,m}} \left(\det(P_I^K) \det((PQ)_I^L) \right) \qquad \text{[(7.5.5)]} \tag{7.5.10}$$

$$= \det\left(P^K ((PQ)^L)^{\mathrm{T}} \right) \qquad \text{[Th 2.4.10(b)]}$$

$$= \det\left((PQP^{\mathrm{T}})_L^K \right) \qquad \text{[(7.5.6)]}$$

$$= \det((\mathfrak{g}_\Theta^*)_L^K). \qquad \text{[(7.5.7)]}$$

Then (7.5.9) and (7.5.10) give

$$\sum_{I,J \in \mathcal{I}_{s,m}} \det((\mathfrak{g}_\Phi^*)_J^I) c_I d_J = \sum_{K,L \in \mathcal{I}_{s,m}} \det((\mathfrak{g}_\Theta^*)_L^K) a_K b_L,$$

which is equivalent to (7.5.3).   □

Theorem 7.5.2. *Let* $(V, \mathfrak{g})$ *be a scalar product space of signature* $(\varepsilon_1, \ldots, \varepsilon_m)$, *let* $\mathcal{E}$ *be an orthonormal basis for* V, *let* $(\xi^1, \ldots, \xi^m)$ *be its dual basis, and let* $I = (i_1, \ldots, i_s)$ *and* J *be multi-indices in* $\mathcal{I}_{s,m}$. *Then*

$$\langle \xi^I, \xi^J \rangle^\Lambda = \begin{cases} \varepsilon_{i_1} \cdots \varepsilon_{i_s} & \text{if } I = J \\ 0 & \text{if } I \neq J, \end{cases}$$

hence each ξ^I *is a unit vector.*

Proof. Let $\Xi = (\xi^1, \ldots, \xi^m)$. Since $\mathfrak{g}_{\mathcal{E}} = \mathrm{diag}(\varepsilon_1, \ldots, \varepsilon_m)$, we have from Theorem 4.5.3(b) that

$$\mathfrak{g}_{\Xi}^* = \mathfrak{g}_{\mathcal{E}}^{-1} = \mathfrak{g}_{\mathcal{E}} = \mathrm{diag}(\varepsilon_1, \ldots, \varepsilon_m),$$

and then from (7.5.2) that

$$\langle \xi^I, \xi^J \rangle^\wedge = \det\big(\mathrm{diag}(\varepsilon_1, \ldots, \varepsilon_m)_J^I\big).$$

If $I = J$, then $\mathrm{diag}(\varepsilon_1, \ldots, \varepsilon_m)_J^I = \mathrm{diag}(\varepsilon_{i_1}, \ldots, \varepsilon_{i_s})$. On the other hand, if $I \neq J$, then $\mathrm{diag}(\varepsilon_1, \ldots, \varepsilon_m)_J^I$ has at least one row (and at least one column) of 0s. The result follows. $\qquad\square$

We are now in a position to define the multicovector counterpart of the dual scalar product space described in Section 4.5.

Theorem 7.5.3. *Let $(V, \mathfrak{g})$ be a scalar product space, let $\mathcal{E}$ be an orthonormal basis for V, and let $(\xi^1, \ldots, \xi^m)$ be its dual basis. Then:*
(a) *$\big(\Lambda^s(V), \mathfrak{g}^\wedge\big)$ is a scalar product space, called the **multicovector scalar product space** corresponding to $(V, \mathfrak{g})$.*
(b) *$(\xi^I : I \in \mathcal{I}_{s,m})$ is an orthonormal basis for $\Lambda^s(V)$.*
(c) *If $(V, \mathfrak{g})$ is an inner product space, then so is $\big(\Lambda^s(V), \mathfrak{g}^\wedge\big)$.*

Proof. (a): We need to show that $\mathfrak{g}^\wedge$ is a scalar product. By construction, $\mathfrak{g}^\wedge$ is bilinear, and clearly it is symmetric. According to Theorem 7.2.12(a), $(\xi^I : I \in \mathcal{I}_{s,m})$ is a basis for $\Lambda^s(V)$. Suppose η is a multicovector in $\Lambda^s(V)$ such that $\langle \eta, \zeta \rangle^\wedge = 0$ for all multicovectors ζ in $\Lambda^s(V)$, and let

$$\eta = \sum_{I \in \mathcal{I}_{s,m}} a_I \xi^I. \tag{7.5.11}$$

Then Theorem 7.5.2 gives

$$0 = \langle \eta, \xi^I \rangle^\wedge = \Big\langle \sum_{J \in \mathcal{I}_{s,m}} a_J \xi^J, \xi^I \Big\rangle^\wedge = \sum_{J \in \mathcal{I}_{s,m}} a_J \langle \xi^J, \xi^I \rangle^\wedge = \pm a_I,$$

hence $a_I = 0$ for all multi-indices I in $\mathcal{I}_{s,m}$, so $\eta = 0$. Thus, $\mathfrak{g}^\wedge$ is nondegenerate.

(b): It was just observed that $(\xi^I : I \in \mathcal{I}_{s,m})$ is a basis for $\Lambda^s(V)$. By Theorem 7.5.2, it is orthonormal.

(c): Since $\mathfrak{g}$ is an inner product, we have from Theorem 7.5.2 that

$$\langle \xi^I, \xi^J \rangle^\wedge = \begin{cases} 1 & \text{if } I = J \\ 0 & \text{if } I \neq J. \end{cases}$$

With η as in (7.5.11),

$$\langle \eta, \eta \rangle^\wedge = \Big\langle \sum_{I \in \mathcal{I}_{s,m}} a_I \xi^I, \sum_{J \in \mathcal{I}_{s,m}} a_J \xi^J \Big\rangle^\wedge = \sum_{I, J \in \mathcal{I}_{s,m}} \langle \xi^I, \xi^J \rangle^\wedge a_I a_J$$

$$= \sum_{I \in \mathcal{I}_{s,m}} (a_I)^2 \geq 0,$$

and there is equality if and only if each $a_I = 0$, in which case $\eta = 0$. $\qquad\square$

Now that we have the scalar product space $\left(\Lambda^s(V), \mathfrak{g}^\Lambda\right)$, the construction in Section 4.5 yields the corresponding dual scalar product space $\left(\Lambda^s(V)^*, \mathfrak{g}^{\Lambda*}\right)$ and the associated flat map and sharp map:

$$\mathsf{F}^\Lambda : \Lambda^s(V) \longrightarrow \Lambda^s(V)^* \qquad \text{and} \qquad \mathsf{S}^\Lambda : \Lambda^s(V)^* \longrightarrow \Lambda^s(V).$$

By definition,

$$\eta^{\mathsf{F}^\Lambda}(\zeta) = \langle \eta, \zeta \rangle^\Lambda \tag{7.5.12}$$

for all multicovectors η, ζ in $\Lambda^s(V)$.

Chapter 8

Orientation

Orientation is concerned with "sidedness", a concept that is intuitively obvious but surprisingly difficult to formulate in a mathematically rigorous fashion. In Section 8.1, we make a few observations based on concrete examples, and then present a preliminary definition of orientation for $\mathbb{R}^m$. In Section 8.2, these ideas are developed into a computational framework for arbitrary vector spaces using our recently acquired knowledge of multicovectors.

8.1 Orientation of $\mathbb{R}^m$

Let $\mathcal{H} = (h_1, \ldots, h_m)$ be a basis for $\mathbb{R}^m$. Corresponding to each point in $\mathbb{R}^m$ is an m-tuple of real numbers consisting of the components of the point with respect to $\mathcal{H}$. For $m = 1, 2, 3$, we can plot this m-tuple on a rectangular coordinate system, with axes labeled $h_1, \ldots, h_m$. We use the geometry of this approach to motivate a definition of orientation.

Consider the bases $\mathcal{E} = (e_1, e_2)$, $(e_2, -e_1)$, and (e_2, e_1) for $\mathbb{R}^2$, where $\mathcal{E}$ is the standard basis. In Figures 8.1.1(a)–(c), these bases are depicted as the axes of rectangular coordinate systems. The configuration in Figure 8.1.1(a), which we call the standard configuration for $\mathbb{R}^2$, is said to have a **counterclockwise orientation** because the 90° rotation taking e_1 to e_2 is in a counterclockwise direction. The configuration in Figure 8.1.1(b) is also said to have a counterclockwise orientation because its axes can be rigidly rotated about the origin to give the same configuration as in Figure 8.1.1(a). On the other hand, the configuration in Figure 8.1.1(c) is said to have a **clockwise orientation** because its axes cannot be rigidly rotated in such a fashion (or equivalently, because the 90° rotation taking e_1 to e_2 is in a clockwise direction). These observations are summarized in the first three columns of Table 8.1.1.

Now consider the bases $\mathcal{E} = (e_1, e_2, e_3)$, $(e_2, -e_1, e_3)$, and (e_2, e_1, e_3) for $\mathbb{R}^3$, where $\mathcal{E}$ is the standard basis. In Figures 8.1.2(a)–(c), these bases are depicted

Semi-Riemannian Geometry, First Edition. Stephen C. Newman.
© 2019 John Wiley & Sons, Inc. Published 2019 by John Wiley & Sons, Inc.

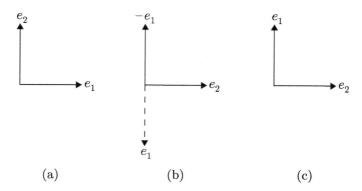

Figure 8.1.1. Orientation in $\mathbb{R}^2$

Figure	$\mathcal{H}$	Orientation	$[\mathrm{id}_{\mathbb{R}^2}]_{\mathcal{E}}^{\mathcal{H}}$	$\det\left([\mathrm{id}_{\mathbb{R}^2}]_{\mathcal{E}}^{\mathcal{H}}\right)$
8.1.1(a)	(e_1, e_2)	counterclockwise	$\begin{bmatrix} 1 & 0 \\ 0 & 1 \end{bmatrix}$	1
8.1.1(b)	$(e_2, -e_1)$	counterclockwise	$\begin{bmatrix} 0 & 1 \\ -1 & 0 \end{bmatrix}$	1
8.1.1(c)	(e_2, e_1)	clockwise	$\begin{bmatrix} 0 & 1 \\ 1 & 0 \end{bmatrix}$	-1

Table 8.1.1. Orientation in $\mathbb{R}^2$

as the axes of rectangular coordinate systems. The configuration in Figure 8.1.2(a), which we call the standard configuration for $\mathbb{R}^3$, is said to have a **right-handed orientation** because you can grasp the positive e_3-axis in your right hand, with your thumb pointing along the positive e_3-axis and your fingers curled around the e_3-axis in a counterclockwise direction with respect to the e_1e_2-plane. The configuration in Figure 8.1.2(b) is also said to have a right-handed orientation because its axes can be rigidly rotated to give the same configuration as in Figure 8.1.2(a). On the other hand, the configuration in Figure 8.1.2(c) is said to have a **left-handed orientation** because it cannot be rigidly rotated in such a fashion (or equivalently, because you can grasp the positive e_3-axis in your left hand, with your thumb pointing along the positive e_3-axis and your fingers curled around the e_3-axis in a clockwise direction with respect to the e_1e_2-plane). These observations are summarized in the first three columns of Table 8.1.2.

In the tables, we also give the change of basis matrix that takes the standard basis $\mathcal{E}$ to the basis $\mathcal{H}$, as well as the value of its determinant. The crucial observation is that for these examples, orientation is preserved when (and only

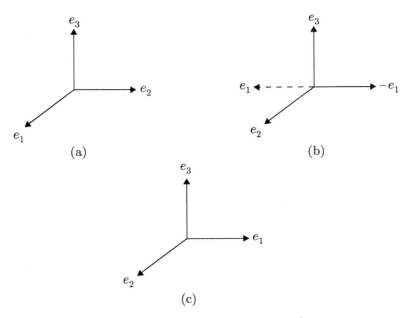

Figure 8.1.2. Orientation in $\mathbb{R}^3$

Figure	$\mathcal{H}$	Orientation	$\left[\mathrm{id}_{\mathbb{R}^3}\right]_{\mathcal{E}}^{\mathcal{H}}$	$\det\left(\left[\mathrm{id}_{\mathbb{R}^3}\right]_{\mathcal{E}}^{\mathcal{H}}\right)$
8.1.2(a)	(e_1, e_2, e_3)	right-handed	$\begin{bmatrix} 1 & 0 & 0 \\ 0 & 1 & 0 \\ 0 & 0 & 1 \end{bmatrix}$	1
8.1.2(b)	$(e_2, -e_1, e_3)$	right-handed	$\begin{bmatrix} 0 & 1 & 0 \\ -1 & 0 & 0 \\ 0 & 0 & 1 \end{bmatrix}$	1
8.1.2(c)	(e_2, e_1, e_3)	left-handed	$\begin{bmatrix} 0 & 1 & 0 \\ 1 & 0 & 0 \\ 0 & 0 & 1 \end{bmatrix}$	-1

Table 8.1.2. Orientation in $\mathbb{R}^3$

when) the change of basis matrix has a positive determinant. This provides a rationale for the following general definition of orientation in $\mathbb{R}^m$.

Let $\mathcal{E}$ and $\mathcal{H}$ be bases for $\mathbb{R}^m$, where $\mathcal{E}$ is the standard basis. It follows from Theorem 2.4.12 that $\det\left(\left[\mathrm{id}_{\mathbb{R}^m}\right]_{\mathcal{E}}^{\mathcal{H}}\right) \neq 0$. We say that $\mathcal{H}$ has the **standard**

orientation if

$$\det\left(\left[\mathrm{id}_{\mathbb{R}^m}\right]_{\mathcal{E}}^{\mathcal{H}}\right) > 0. \tag{8.1.1}$$

In particular, $\mathcal{E}$ has the standard orientation.

8.2   Orientation of Vector Spaces

In Section 8.1, we defined what it means for a basis for $\mathbb{R}^m$ to have a certain orientation by comparing it with the standard basis. This approach to orientation works in $\mathbb{R}^m$ because there is an obvious choice of reference basis, something that may not be the case for an arbitrary vector space.

Let V be a vector space of dimension m, and let $\mathcal{H}$ and $\mathcal{F}$ be bases for V. It follows from Theorem 2.4.12 that $\det\left(\left[\mathrm{id}_V\right]_{\mathcal{H}}^{\mathcal{F}}\right) \neq 0$. We say that $\mathcal{H}$ and $\mathcal{F}$ are **consistent**, and write $\mathcal{H} \sim \mathcal{F}$, if

$$\det\left(\left[\mathrm{id}_V\right]_{\mathcal{H}}^{\mathcal{F}}\right) > 0.$$

Using Theorem 2.2.2(c) and Theorem 2.2.5(b), it is easily shown that $\sim$ is an equivalence relation on the set of bases for V, and that there are precisely two equivalence classes. Each equivalence class is said to be an **orientation of** V. The equivalence class containing a given basis $\mathcal{H}$ is denoted by $[\mathcal{H}]$. Let $\mathcal{H} = (h_1, h_2, \ldots, h_m)$, and let

$$-\mathcal{H} = (-h_1, h_2, \ldots, h_m). \tag{8.2.1}$$

Clearly, $-\mathcal{H}$ is a basis for V and $\det\left(\left[\mathrm{id}_V\right]_{\mathcal{H}}^{-\mathcal{H}}\right) = -1$, so $\mathcal{H}$ and $-\mathcal{H}$ are not consistent. We say that the orientation $[-\mathcal{H}]$ is the **opposite of** $[\mathcal{H}]$. Thus, for any basis $\mathcal{H}$ for V, the orientations of V are $[\mathcal{H}]$ and $[-\mathcal{H}]$. For example, the orientations of $\mathbb{R}^m$ are $[\mathcal{E}]$ and $[-\mathcal{E}]$, where $\mathcal{E}$ is the standard basis for $\mathbb{R}^m$. We refer to $[\mathcal{E}]$ as the **standard orientation of** $\mathbb{R}^m$. According to this terminology, $\mathcal{E}$ is a basis *in* the standard orientation, which is somewhat different from Section 8.1 where $\mathcal{E}$ was said to *have* the standard orientation. It is convenient to adopt this language more generally: for any vector space, a basis *in* an orientation is said to *have* that orientation.

Once an orientation $\mathfrak{O}$ of V has been chosen, the pair $(V, \mathfrak{O})$ is called an **oriented vector space**, and V is said to have the orientation $\mathfrak{O}$. The alternative orientation is denoted by $-\mathfrak{O}$ and called the **opposite of** $\mathfrak{O}$. If a basis for V is in $\mathfrak{O}$, it is said to be **positively oriented (with respect to $\mathfrak{O}$)**. We will often find it convenient to say, for instance, that "$\mathcal{H}$ is a basis for V that is positively oriented with respect to $\mathfrak{O}$" rather than the more concise "$\mathcal{H}$ is in $\mathfrak{O}$". Although convention or circumstances tend to influence the orientation assigned to a given vector space, it needs to be emphasized that ultimately the decision is a matter of choice. There is nothing intrinsic to $\mathfrak{O}$ or $-\mathfrak{O}$ that makes one preferable to the other as an orientation of V. Having said that, in the case of $\mathbb{R}^m$ we adopt the following convention.

Throughout, $\mathbb{R}^m$ is assumed to have the standard orientation.

We now present an equivalent approach to orientation using multicovectors. Let V be a vector space of dimension m. In the present context, a nonzero multicovector in $\Lambda^m(V)$ is called an **orientation multicovector** on V. As an example, let $(h_1, \ldots, h_m)$ be a basis for V, and let $(\theta^1, \ldots, \theta^m)$ be its dual basis. By Theorem 7.2.9, $\theta^1 \wedge \cdots \wedge \theta^m (h_1, \ldots, h_m) = 1$, so $\theta^1 \wedge \cdots \wedge \theta^m$ is an orientation multicovector on V. More generally, we have the following result.

Theorem 8.2.1. *Let V be a vector space of dimension m, and let ϖ be a multicovector in $\Lambda^m(V)$. Then ϖ is an orientation multicovector if and only if $\varpi(h_1, \ldots, h_m) \neq 0$ for some (hence every) basis $(h_1, \ldots, h_m)$ for V.*

Proof. Let $\mathcal{H} = (h_1, \ldots, h_m)$ be a basis for V, and let $(\theta^1, \ldots, \theta^m)$ be its dual basis. We have from Theorem 7.2.13(b) that

$$\varpi = \varpi(h_1, \ldots, h_m)\, \theta^1 \wedge \cdots \wedge \theta^m.$$

Thus, ϖ is nonzero if and only if $\varpi(h_1, \ldots, h_m) \neq 0$. Let $\mathcal{F} = (f_1, \ldots, f_m)$ be another basis for V. By Theorem 7.3.3(b),

$$\varpi(h_1, \ldots, h_m) = \det\left([\mathrm{id}_V]_{\mathcal{H}}^{\mathcal{F}}\right) \varpi(f_1, \ldots, f_m),$$

and by Theorem 2.4.12, $\det\left([\mathrm{id}_V]_{\mathcal{H}}^{\mathcal{F}}\right) \neq 0$. Thus, $\varpi(h_1, \ldots, h_m) \neq 0$ if and only if $\varpi(f_1, \ldots, f_m) \neq 0$. The result follows. $\square$

Theorem 8.2.2. *Let V be a vector space, let $\mathcal{H} = (h_1, \ldots, h_m)$ and $\mathcal{F} = (f_1, \ldots, f_m)$ be bases for V, and let $(\theta^1, \ldots, \theta^m)$ be the dual basis corresponding to $\mathcal{H}$. Then the following are equivalent:*
(a) *$\mathcal{H}$ and $\mathcal{F}$ are consistent.*
(b) *$\theta^1 \wedge \cdots \wedge \theta^m (f_1, \ldots, f_m) > 0$.*
(c) *$\varpi(h_1, \ldots, h_m)$ and $\varpi(f_1, \ldots, f_m)$ have the same sign for some (hence every) orientation multicovector ϖ on V.*

Remark. By Theorem 8.2.1, we have $\varpi(h_1, \ldots, h_m), \varpi(f_1, \ldots, f_m) \neq 0$, so the assertion in part (c) makes sense.

Proof. (a) $\Leftrightarrow$ (b): By Theorem 7.2.9 and Theorem 7.3.3(b),

$$\det\left([\mathrm{id}_V]_{\mathcal{H}}^{\mathcal{F}}\right) \theta^1 \wedge \cdots \wedge \theta^m (f_1, \ldots, f_m) = \theta^1 \wedge \cdots \wedge \theta^m (h_1, \ldots, h_m) = 1.$$

The result follows.
(a) $\Leftrightarrow$ (c): By Theorem 7.3.3(b),

$$\varpi(h_1, \ldots, h_m) = \det\left([\mathrm{id}_V]_{\mathcal{H}}^{\mathcal{F}}\right) \varpi(f_1, \ldots, f_m),$$

and by Theorem 2.4.12, $\det\left([\mathrm{id}_V]_{\mathcal{H}}^{\mathcal{F}}\right) \neq 0$. The result follows. $\square$

Theorem 8.2.3. *Let V be a vector space, let ϖ be an orientation multicovector on V, and let $\mathfrak{O}$ be the set of bases $(h_1, \ldots, h_m)$ for V such that $\varpi(h_1, \ldots, h_m) > 0$. Then $\mathfrak{O}$ is an orientation of V, called the **orientation induced by ϖ**.*

Proof. Let $\mathcal{H} = (h_1, \ldots, h_m)$ and $\mathcal{F} = (f_1, \ldots, f_m)$ be bases for V, with $\varpi(h_1, \ldots, h_m) > 0$. Then

$\quad \mathcal{F}$ is in $\mathfrak{O}$

$\qquad \Leftrightarrow \quad \varpi(f_1, \ldots, f_m) > 0$

$\qquad \Leftrightarrow \quad \varpi(h_1, \ldots, h_m)$ and $\varpi(f_1, \ldots, f_m)$ have the same sign

$\qquad \Leftrightarrow \quad \mathcal{H}$ and $\mathcal{F}$ are consistent,

where the last equivalence follows from Theorem 8.2.2. Thus, $\mathfrak{O}$ is the set of bases for V that are consistent with $\mathcal{H}$; that is, $\mathfrak{O} = [\mathcal{H}]$. $\qquad\square$

Example 8.2.4 ($\mathbb{R}^m$). Let $\mathcal{E} = (e_1, \ldots, e_m)$ be the standard basis for $\mathbb{R}^m$, let $(\xi^1, \ldots, \xi^m)$ be its dual basis, and let $\Omega = \xi^1 \wedge \cdots \wedge \xi^m$. It follows from Theorem 7.2.9 that $\Omega(e_1, \ldots, e_m) = 1$, so Ω is an orientation multicovector on $\mathbb{R}^m$, and $\mathcal{E}$ is in the orientation of $\mathbb{R}^m$ induced by Ω. Thus, Ω induces the standard orientation of $\mathbb{R}^m$. $\qquad\diamond$

Let V be a vector space, and let ϖ and ϑ be orientation multicovectors on V. By Theorem 7.2.12(d) and Theorem 8.2.1, ϖ spans $\Lambda^m(M)$ (as does ϑ). It follows that $\varpi = c\vartheta$ for some nonzero real number c. We say that ϖ and ϑ are **consistent**, and write $\varpi \sim \vartheta$, if $c > 0$. It is easily shown that $\sim$ is an equivalence relation on the set of orientation multicovectors on V, and that there are precisely two equivalence classes. The equivalence class containing ϖ is denoted by $[\varpi]$. By definition, $[\varpi]$ comprises all multicovectors on V that are consistent with ϖ. Evidently, ϖ and $-\varpi$ are not consistent. Thus, for any orientation multicovector ϖ on V, the equivalence classes of orientation multicovectors are $[\varpi]$ and $[-\varpi]$.

Let $\mathcal{H}$ be a basis for V. It is clear that orientation multicovectors in the same equivalence class induce the same orientation, and that orientation multicovectors in different equivalence classes induce opposite orientations. We therefore have a bijective map

$$\iota : \{[\varpi], [-\varpi]\} \longrightarrow \{[\mathcal{H}], -[\mathcal{H}]\}$$

defined by assigning $[\varpi]$ and $[-\varpi]$ to the orientations induced by ϖ and $-\varpi$, respectively. This shows that we are free to specify orientations using either (equivalence classes of) bases for V or (equivalence classes of) orientation multicovectors on V. For purposes of computation, the latter approach is generally more convenient.

Theorem 8.2.5 (Induced Orientation of Subspace). *Let $(V, \mathfrak{O})$ be an oriented vector space of dimension $m \geq 2$, let U be an $(m-1)$-dimensional subspace of V, and let v be a vector in $V \backslash U$. Then:*

(a) *There is a unique orientation $\mathfrak{O}_U$ of U, called the **orientation induced by** $\boldsymbol{v}$, such that $(h_1, \ldots, h_{m-1})$ is a basis for U that is positively oriented with respect to $\mathfrak{O}_U$ if and only if $(v, h_1, \ldots, h_{m-1})$ is a basis for V that is positively oriented with respect to $\mathfrak{O}$.*

(b) *If ϖ is an orientation multicovector on V that induces $\mathfrak{O}$, then $i_v(\varpi)|_U$ is an orientation multicovector on U that induces $\mathfrak{O}_U$.*

(c) *$\mathfrak{O}_U$ is independent of the choice of orientation multicovector on V that induces $\mathfrak{O}$.*

Proof. (a), (b): Let $(h_1, \ldots, h_{m-1})$ be an $(m-1)$-tuple of vectors in U. Since v is in $V \backslash U$, by Theorem 1.1.2, $(h_1, \ldots, h_{m-1})$ is a basis for U if and only if $(v, h_1, \ldots, h_{m-1})$ is a basis for V. Suppose $(h_1, \ldots, h_{m-1})$ is in fact a basis for U. It is clear that $i_v(\varpi)|_U$ is a multicovector in $\Lambda^{m-1}(U)$. By definition,

$$i_v(\varpi)|_U(h_1, \ldots, h_{m-1}) = \varpi(v, h_1, \ldots, h_{m-1}). \tag{8.2.2}$$

We have from Theorem 8.2.1 that $\varpi(v, h_1, \ldots, h_{m-1}) \neq 0$, and then from (8.2.2) that $i_v(\varpi)|_U(h_1, \ldots, h_{m-1}) \neq 0$. This shows that $i_v(\varpi)|_U$ is nonzero, and is therefore an orientation multicovector on U. Let $\mathfrak{O}_U$ be the orientation of U induced by $i_v(\varpi)|_U$. It follows from (8.2.2) and Theorem 8.2.3 that $(h_1, \ldots, h_{m-1})$ is positively oriented with respect to $\mathfrak{O}_U$ if and only if $(v, h_1, \ldots, h_{m-1})$ is positively oriented with respect to $\mathfrak{O}$. Thus, $\mathfrak{O}_U$ has the desired property. Uniqueness follows from the (almost tautological) observation that an orientation of a vector space is determined by the bases it contains.

(c): This follows from part (a). $\square$

A remark is that, in contrast to part (c) of Theorem 8.2.5, the orientation induced on U is not independent of the choice of vector in $V \backslash U$. In particular, if v induces $\mathfrak{O}_U$, then $-v$ induces $-\mathfrak{O}_U$.

Theorem 8.2.6. *Let $(V, \mathfrak{O})$ and $(\widetilde{V}, \widetilde{\mathfrak{O}})$ be oriented vector spaces, let $A : V \longrightarrow \widetilde{V}$ be a linear isomorphism, and let $A(\mathfrak{O}) = \{A(\mathcal{H}) : \mathcal{H} \in \mathfrak{O}\}$. Then either $A(\mathfrak{O}) = \widetilde{\mathfrak{O}}$, in which case A is said to be **orientation-preserving**, or $A(\mathfrak{O}) = -\widetilde{\mathfrak{O}}$, in which case A is said to be **orientation-reversing**.*

Proof. Let $\mathcal{H}$ and $\mathcal{F}$ be bases in $\mathfrak{O}$, and observe that, by Theorem 1.1.13(a), $A(\mathcal{H})$ and $A(\mathcal{F})$ are bases for $\widetilde{V}$. Let $[\mathrm{id}_V]_{\mathcal{H}}^{\mathcal{F}} = [a_j^i]$. We have from (2.2.6) and (2.2.7) that $h_j = \sum_i a_j^i f_i$, hence $A(h_j) = \sum_i a_j^i A(f_i)$, so $[\mathrm{id}_V]_{A(\mathcal{H})}^{A(\mathcal{F})} = [\mathrm{id}_V]_{\mathcal{H}}^{\mathcal{F}}$. By definition, $\mathcal{H}$ and $\mathcal{F}$ are consistent, and therefore, so are $A(\mathcal{H})$ and $A(\mathcal{F})$. Thus, either $A(\mathcal{H})$ and $A(\mathcal{F})$ are in $\widetilde{\mathfrak{O}}$, or $A(\mathcal{H})$ and $A(\mathcal{F})$ are in $-\widetilde{\mathfrak{O}}$. Since $\mathcal{H}$ and $\mathcal{F}$ were arbitrary, either $A(\mathfrak{O}) \subseteq \widetilde{\mathfrak{O}}$ or $A(\mathfrak{O}) \subseteq -\widetilde{\mathfrak{O}}$. By Theorem 1.1.9, A^{-1} is a linear isomorphism, so a similar argument shows that either $A^{-1}(\widetilde{\mathfrak{O}}) \subseteq \mathfrak{O}$ or $A^{-1}(\widetilde{\mathfrak{O}}) \subseteq -\mathfrak{O}$, hence either $\widetilde{\mathfrak{O}} \subseteq A(\mathfrak{O})$ or $\widetilde{\mathfrak{O}} \subseteq A(-\mathfrak{O})$. It is easily shown using (8.2.1) that $\mathcal{A}(-\mathcal{H}) = -\mathcal{A}(\mathcal{H})$ for any basis $\mathcal{H}$ in $\mathfrak{O}$, so either $\widetilde{\mathfrak{O}} \subseteq A(\mathfrak{O})$ or $-\widetilde{\mathfrak{O}} \subseteq A(\mathfrak{O})$. The result follows. $\square$

Theorem 8.2.7. *If $(V, \mathfrak{O})$ and $(\widetilde{V}, \widetilde{\mathfrak{O}})$ are oriented vector spaces and $A : V \longrightarrow \widetilde{V}$ is a linear isomorphism, then the following are equivalent:*

(a) *A is orientation-preserving.*

(b) *$\det\left([A]_{\mathcal{H}}^{\widetilde{\mathcal{H}}}\right) > 0$ for some (hence every) basis $\mathcal{H}$ in $\mathfrak{O}$ and some (hence every) basis $\widetilde{\mathcal{H}}$ in $\widetilde{\mathfrak{O}}$.*

(c) *$A^*(\widetilde{\varpi})$ is an orientation multicovector on V that induces $\mathfrak{O}$ for some (hence every) orientation multicovector $\widetilde{\varpi}$ on $\widetilde{V}$ that induces $\widetilde{\mathfrak{O}}$.*

Remark. It is easily shown that since A is a linear isomorphism, $A^*(\widetilde{\varpi})$ is an orientation multicovector on V, so the assertion in part (c) makes sense.

Proof. (a) $\Leftrightarrow$ (b): We have

A is orientation-preserving

$\Leftrightarrow$ $A(\mathcal{H})$ and $\widetilde{\mathcal{H}}$ are consistent for some (hence every) $\mathcal{H}$ in $\mathfrak{O}$ and $\widetilde{\mathcal{H}}$ in $\widetilde{\mathfrak{O}}$

$\Leftrightarrow$ $\det\left(\left[\mathrm{id}_{\widetilde{V}}\right]_{A(\mathcal{H})}^{\widetilde{\mathcal{H}}}\right) > 0$ for some (hence every) $\mathcal{H}$ in $\mathfrak{O}$ and $\widetilde{\mathcal{H}}$ in $\widetilde{\mathfrak{O}}$

$\Leftrightarrow$ $\det\left([A]_{\mathcal{H}}^{\widetilde{\mathcal{H}}}\right) > 0$ for some (hence every) $\mathcal{H}$ in $\mathfrak{O}$ and $\widetilde{\mathcal{H}}$ in $\widetilde{\mathfrak{O}}$,

where the first equivalence follows from Theorem 8.2.6, and the last equivalence from Theorem 2.2.5(a).

(b) $\Leftrightarrow$ (c): Let $\mathcal{H} = (h_1, \ldots, h_m)$ and $\widetilde{\mathcal{H}} = (\widetilde{h}_1, \ldots, \widetilde{h}_m)$ be bases in $\mathfrak{O}$ and $\widetilde{\mathfrak{O}}$, respectively, and let $\widetilde{\varpi}$ be an orientation multicovector that induces $\widetilde{\mathfrak{O}}$. We have from Theorem 7.3.3(b) that

$$A^*(\widetilde{\varpi})(h_1, \ldots, h_m) = \det\left([A]_{\mathcal{H}}^{\widetilde{\mathcal{H}}}\right) \widetilde{\varpi}(\widetilde{h}_1, \ldots, \widetilde{h}_m).$$

By Theorem 8.2.3, $\widetilde{\varpi}(\widetilde{h}_1, \ldots, \widetilde{h}_m) > 0$, hence

$A^*(\widetilde{\varpi})$ induces $\mathfrak{O}$

$\Leftrightarrow$ $A^*(\widetilde{\varpi})(h_1, \ldots, h_m) > 0$ for some (hence every) $\mathcal{H}$ in $\mathfrak{O}$

$\Leftrightarrow$ $\det\left([A]_{\mathcal{H}}^{\widetilde{\mathcal{H}}}\right) \widetilde{\varpi}(\widetilde{h}_1, \ldots, \widetilde{h}_m) > 0$ for some (hence every) $\mathcal{H}$ in $\mathfrak{O}$ and $\widetilde{\mathcal{H}}$ in $\widetilde{\mathfrak{O}}$

$\Leftrightarrow$ $\det\left([A]_{\mathcal{H}}^{\widetilde{\mathcal{H}}}\right) > 0$ for some (hence every) $\mathcal{H}$ in $\mathfrak{O}$ and $\widetilde{\mathcal{H}}$ in $\widetilde{\mathfrak{O}}$,

where the first equivalence follows from Theorem 8.2.3. $\square$

Theorem 8.2.8. *Let $(V, \mathfrak{O})$ be an oriented vector space, and let $A : V \longrightarrow V$ be a linear isomorphism. Then A is orientation-preserving if and only if $\det(A) > 0$.*

Proof. Setting $(V, \mathfrak{O}) = (\widetilde{V}, \widetilde{\mathfrak{O}})$ and $\mathcal{H} = \widetilde{\mathcal{H}}$ in Theorem 8.2.7 gives the result. $\square$

8.3   Orientation of Scalar Product Spaces

In Section 8.2, we considered orientation of vector spaces. We now expand our coverage to include scalar product spaces.

Theorem 8.3.1. *Let $(V, \mathfrak{g})$ be a scalar product space of index ν, let $\mathcal{E} = (e_1, \ldots, e_m)$ and $\mathcal{H} = (h_1, \ldots, h_m)$ be bases for V, with $\mathcal{E}$ orthonormal, and let $(\xi^1, \ldots, \xi^m)$ and $(\theta^1, \ldots, \theta^m)$ be the corresponding dual bases. Then:*
(a)

$$\det(\mathfrak{g}_{\mathcal{H}}) = (-1)^{\nu} \left(\det \left([\mathrm{id}_V]_{\mathcal{H}}^{\mathcal{E}} \right) \right)^2.$$

(b)

$$\xi^1 \wedge \cdots \wedge \xi^m = \pm \sqrt{|\det(\mathfrak{g}_{\mathcal{H}})|} \, \theta^1 \wedge \cdots \wedge \theta^m,$$

where the positive (negative) sign is chosen if $\mathcal{E}$ and $\mathcal{H}$ are consistent (not consistent).
(c) *If $\mathcal{H}$ is orthonormal, then*

$$\xi^1 \wedge \cdots \wedge \xi^m = \pm \theta^1 \wedge \cdots \wedge \theta^m,$$

where the signs are chosen as in part (b).

Proof. (a): We have from Theorem 3.1.3 that

$$\mathfrak{g}_{\mathcal{H}} = \left([\mathrm{id}_V]_{\mathcal{H}}^{\mathcal{E}} \right)^{\mathrm{T}} \mathfrak{g}_{\mathcal{E}} \, [\mathrm{id}_V]_{\mathcal{H}}^{\mathcal{E}},$$

and from (4.2.3) that $\det(\mathfrak{g}_{\mathcal{E}}) = (-1)^{\nu}$. The result follows.
   (b): By Theorem 7.3.3(a),

$$\xi^1 \wedge \cdots \wedge \xi^m = \det \left([\mathrm{id}_V]_{\mathcal{H}}^{\mathcal{E}} \right) \theta^1 \wedge \cdots \wedge \theta^m,$$

and by part (a), $|\det(\mathfrak{g}_{\mathcal{H}})| = \left(\det([\mathrm{id}_V]_{\mathcal{H}}^{\mathcal{E}}) \right)^2$, hence

$$\det \left([\mathrm{id}_V]_{\mathcal{H}}^{\mathcal{E}} \right) = \pm \sqrt{|\det(\mathfrak{g}_{\mathcal{H}})|},$$

where the positive (negative) sign is chosen if $\mathcal{H}$ and $\mathcal{E}$ are consistent (not consistent). The result follows.
   (c): This follows from (4.2.3) and part (b).   □

Theorem 8.3.2. *Let $(V, \mathfrak{g})$ be a scalar product space, let $\mathcal{E} = (e_1, \ldots, e_m)$ and $\mathcal{F} = (f_1, \ldots, f_m)$ be orthonormal bases for V, and let ϖ be an orientation multicovector on V. Then*

$$\varpi(e_1, \ldots, e_m) = \pm \varpi(f_1, \ldots, f_m),$$

where the positive (negative) sign is chosen if $\mathcal{E}$ and $\mathcal{F}$ are consistent (not consistent).

Proof. Let $(\xi^1, \ldots, \xi^m)$ and $(\varphi^1, \ldots, \varphi^m)$ be the dual bases corresponding to $\mathcal{E}$ and $\mathcal{F}$, respectively. We have

$$\varpi(e_1, \ldots, e_m)\, \xi^1 \wedge \cdots \wedge \xi^m$$
$$= \varpi(f_1, \ldots, f_m)\, \varphi^1 \wedge \cdots \wedge \varphi^m \qquad [\text{Th 7.2.13(b)}]$$
$$= \pm \varpi(f_1, \ldots, f_m)\, \xi^1 \wedge \cdots \wedge \zeta^m, \qquad [\text{Th 8.3.1(c)}]$$

where the positive (negative) sign is chosen if $\mathcal{E}$ and $\mathcal{F}$ are consistent (not consistent). Then Theorem 7.2.9 gives the result. $\qquad\square$

Theorem 8.3.3. *Let $(V, \mathfrak{g})$ be a scalar product space, let $\mathcal{E} = (e_1, \ldots, e_m)$ and $\mathcal{F}$ be orthonormal bases for V, and let $(\varphi^1, \ldots, \varphi^m)$ be the dual basis corresponding to $\mathcal{F}$. Then:*

$$\det\left(\left[\mathrm{id}_V \right]_{\mathcal{E}}^{\mathcal{F}} \right) = \varphi^1 \wedge \cdots \wedge \varphi^m(e_1, \ldots, e_m) = \pm 1,$$

where the positive (negative) sign is chosen if $\mathcal{E}$ and $\mathcal{F}$ are consistent (not consistent).

Proof. The first equality follows from Theorem 7.2.11. Setting $\varpi = \varphi^1 \wedge \cdots \wedge \varphi^m$ in Theorem 8.3.2 and using Theorem 7.2.9 gives the second equality. $\qquad\square$

Theorem 8.3.4 (Existence of Volume Multicovector). *If $(V, \mathfrak{g}, \mathfrak{O})$ is an oriented scalar product space, then there is a unique orientation multicovector Ω_V on V, called the **volume multicovector**, such that: (i) Ω_V induces $\mathfrak{O}$, and (ii) if $(e_1, \ldots, e_m)$ is an orthonormal basis for V that is positively oriented with respect to $\mathfrak{O}$, then $\Omega_V(e_1, \ldots, e_m) = 1$. In fact, if $(f_1, \ldots, f_m)$ is any orthonormal basis for V that is positively oriented with respect to $\mathfrak{O}$ and $(\varphi^1, \ldots, \varphi^m)$ is its dual basis, then $\Omega_V = \varphi^1 \wedge \cdots \wedge \varphi^m$.*

Proof. Existence. By Theorem 7.2.9, $\varphi^1 \wedge \cdots \wedge \varphi^m(f_1, \ldots, f_m) = 1$, so $\varphi^1 \wedge \cdots \wedge \varphi^m$ is an orientation multicovector on V. Then $\varphi^1 \wedge \cdots \wedge \varphi^m$ induces an orientation of V, of which there are two possibilities, namely, $\mathfrak{O}$ and $-\mathfrak{O}$. Since $(f_1, \ldots, f_m)$ is positively oriented with respect to $\mathfrak{O}$ and $\varphi^1 \wedge \cdots \wedge \varphi^m(f_1, \ldots, f_m) > 0$, by Theorem 8.2.3, $\varphi^1 \wedge \cdots \wedge \varphi^m$ induces $\mathfrak{O}$. Since $(e_1, \ldots, e_m)$ and $(f_1, \ldots, f_m)$ are positively oriented with respect to $\mathfrak{O}$, they are consistent. It follows from Theorem 8.3.3 that $\varphi^1 \wedge \cdots \wedge \varphi^m(e_1, \ldots, e_m) = 1$.

   Uniqueness. Suppose ϖ is an orientation multicovector on V such that $\varpi(f_1, \ldots, f_m) = 1$. It follows from Theorem 7.2.13(b) that

$$\varpi = \varpi(f_1, \ldots, f_m)\, \varphi^1 \wedge \cdots \wedge \varphi^m = \varphi^1 \wedge \cdots \wedge \varphi^m. \qquad\square$$

Theorem 8.3.5 (Expression for Volume Multicovector). *Let $(V, \mathfrak{g}, \mathfrak{O})$ be an oriented scalar product space, let $\mathcal{H}$ be a basis for V that is positively oriented with respect to $\mathfrak{O}$, and let $(\theta^1, \ldots, \theta^m)$ be its dual basis. Then*

$$\Omega_V = \sqrt{|\det(\mathfrak{g}_{\mathcal{H}})|}\; \theta^1 \wedge \cdots \wedge \theta^m.$$

Proof. Let $(e_1, \ldots, e_m)$ be an orthonormal basis for V that is positively oriented with respect to $\mathfrak{O}$, and let $(\xi^1, \ldots, \xi^m)$ be its dual basis. Since both $(e_1, \ldots, e_m)$ and $\mathcal{H}$ are positively oriented, they are consistent. The result now follows from Theorem 8.3.1(b) and Theorem 8.3.4. $\square$

Let $(V, \mathfrak{g}, \mathfrak{O})$ be an oriented scalar product space of dimension $m \geq 2$, and let Ω_V be its volume multicovector. By Theorem 8.3.4, Ω_V induces $\mathfrak{O}$. Let U be a subspace of V of dimension $m - 1$ on which $\mathfrak{g}$ is nondegenerate. We have from Theorem 4.1.2(b) that $U^\perp$ is a 1-dimensional subspace of V, and from Theorem 4.1.3 that $V = U \oplus U^\perp$. Let u be a unit vector in $U^\perp$, so that $\mathbb{R}u = U^\perp$, hence $V = U \oplus \mathbb{R}u$. Since $\mathfrak{g}$ is nondegenerate on U, $\mathfrak{g}|_U$ is a scalar product on U. By parts (a) and (b) of Theorem 8.2.5, $i_u(\Omega_V)|_U$ is an orientation multicovector on U that induces a certain orientation of U denoted by $\mathfrak{O}_U$. Thus, $(U, \mathfrak{g}|_U, \mathfrak{O}_U)$ is an oriented scalar product space of dimension $m - 1$. Let Ω_U be its volume multicovector. The question arises as whether the orientation multicovectors $i_u(\Omega_V)|_U$ and Ω_U are related. As the next result shows, they are one and the same.

Theorem 8.3.6. *With the above setup:*
(a)
$$i_u(\Omega_V)|_U = \Omega_U.$$

(b) *More generally, if v is a vector in V and ϵ is the sign of $\langle u, u \rangle$, then*

$$i_v(\Omega_V)|_U = \epsilon \langle u, v \rangle \Omega_U.$$

Proof. (a): Let $(e_1, \ldots, e_{m-1})$ be an orthonormal basis for U that is positively oriented with respect to $\mathfrak{O}_U$. We have from Theorem 8.2.5(a) that $(u, e_1, \ldots, e_{m-1})$ is an orthonormal basis for V that is positively oriented with respect to $\mathfrak{O}$, and then from Theorem 8.3.4 (applied to V) that $\Omega_V(u, e_1, \ldots, e_{m-1}) = 1$. By definition,

$$i_u(\Omega_V)|_U(e_1, \ldots, e_{m-1}) = \Omega_V(u, e_1, \ldots, e_{m-1}),$$

so $i_u(\Omega_V)|_U(e_1, \ldots, e_{m-1}) = 1$. The result now follows from the uniqueness of Ω_U, as given by Theorem 8.3.4 (applied to U).

(b): Since $V = U \oplus \mathbb{R}u$, we have $v = (v - \bar{v}) + \bar{v}$, where $v - \bar{v}$ and $\bar{v}$ are (unique) vectors in U and $\mathbb{R}u$, respectively. By Theorem 7.4.1(a),

$$i_v(\Omega_V)|_U = i_{v - \bar{v}}(\Omega_V)|_U + i_{\bar{v}}(\Omega_V)|_U. \qquad (8.3.1)$$

We seek alternative expressions for the terms on the right-hand side of the preceding identity. For the first term, let $w_1, \ldots, w_{m-1}$ be vectors in U. Since $v - \bar{v}$ is a vector in U, which has dimension $m - 1$, it follows that $v - \bar{v}, w_1, \ldots, w_{m-1}$ are linearly dependent. Then Theorem 7.3.5 gives

$$i_{v - \bar{v}}(\Omega_V)|_U(w_1, \ldots, w_{m-1}) = \Omega_V(v - \bar{v}, w_1, \ldots, w_{m-1}) = 0.$$

Since $w_1, \ldots, w_{m-1}$ were arbitrary,

$$i_{v-\bar{v}}(\Omega_V)|_U = 0. \tag{8.3.2}$$

For the second term, we have from Theorem 4.1.5 that

$$\bar{v} = \mathcal{P}_{\mathbb{R}u}(v) = \epsilon\langle v, u\rangle u,$$

and then from Theorem 7.4.1(b) and part (a) that

$$i_{\bar{v}}(\Omega_V)|_U = \epsilon\langle v, u\rangle i_u(\Omega_V)|_U = \epsilon\langle v, u\rangle \Omega_U. \tag{8.3.3}$$

Substituting (8.3.2) and (8.3.3) into (8.3.1) gives the result. $\square$

8.4   Vector Products

In this section, we generalize the well-known vector product in $\mathbb{R}^3$ to an arbitrary scalar product space.

Let $(V, \mathfrak{g})$ be a scalar product space with signature $(\varepsilon_1, \ldots, \varepsilon_m)$, let $\mathcal{E} = (e_1, \ldots, e_m)$ be an orthonormal basis for V, and let $v_1, \ldots, v_{m-1}$ be vectors in V. The **vector product** (or **cross product**) of $v_1, \ldots, v_{m-1}$ (in that order) with respect to $\mathcal{E}$ is defined by

$$v_1 \times \cdots \times v_{m-1} = \sum_{i=1}^{m} \varepsilon_i(-1)^{i-1}\det\left(\left[[v_1]_{\mathcal{E}} \quad \cdots \quad [v_{m-1}]_{\mathcal{E}}\right]^{(i)^c}\right)e_i, \tag{8.4.1}$$

where $(i)^c$ is the multi-index

$$(i)^c = (1, \ldots, \widehat{i}, \ldots, m).$$

Clearly, the vector product of given vectors depends on the order in which they are taken and the choice of orthonormal basis.

In this chapter, the vector product $\times$ on V is computed with respect to a given orthonormal basis $\mathcal{E}$ for V.

Throughout, the vector product $\times$ on $\mathbb{R}^m_\nu$ is computed with respect to the standard basis for $\mathbb{R}^m_\nu$.

We obtain a computationally convenient alternative to (8.4.1) as follows. Let

$$v_j = \sum_i a^i_j e_i$$

for $j = 1, \ldots, m-1$, so that from (2.2.3),

$$\left[[v_1]_{\mathcal{E}} \quad \cdots \quad [v_{m-1}]_{\mathcal{E}}\right] = \begin{bmatrix} a^1_1 & \cdots & a^1_{m-1} \\ \vdots & \ddots & \vdots \\ a^m_1 & \cdots & a^m_{m-1} \end{bmatrix}. \tag{8.4.2}$$

Then (8.4.1) can be expressed as the formal identity

$$v_1 \times \cdots \times v_{m-1} = \det\left(\begin{bmatrix} \varepsilon_1 e_1 & a_1^1 & \cdots & a_{m-1}^1 \\ \vdots & \vdots & \ddots & \vdots \\ \varepsilon_m e_m & a_1^m & \cdots & a_{m-1}^m \end{bmatrix}\right), \tag{8.4.3}$$

provided we expand the "determinant" along the first column.

When $m = 2$, even though the notation for the left-hand sides of (8.4.1) and (8.4.3) simplifies to a single vector, the right-hand sides can still be computed. Let $v = a^1 e_1 + a^2 e_2$. Then the "vector product" of v is

$$\det\left(\begin{bmatrix} \varepsilon_1 e_1 & a^1 \\ \varepsilon_2 e_2 & a^2 \end{bmatrix}\right) = \varepsilon_1 a^2 e_1 - \varepsilon_2 a^1 e_2.$$

When the context is meaningful, this quantity has the same properties as a vector product when $m \geq 3$.

Example 8.4.1 (Vector Product on $\mathbb{R}_0^2$ and $\mathbb{R}_0^3$). Let (e_1, e_2) be the standard basis for $\mathbb{R}_0^2$, and let (a^1, a^2) be a vector in $\mathbb{R}_0^2$. The "vector product" of (a^1, a^2) is $(a^2, -a^1)$.

Now let (e_1, e_2, e_3) be the standard basis for $\mathbb{R}_0^3$, and let $v = (a^1, a^2, a^3)$ and $w = (b^1, b^2, b^3)$ be vectors in $\mathbb{R}_0^3$. Then

$$[v]_{\mathcal{E}} = \begin{bmatrix} a^1 \\ a^2 \\ a^3 \end{bmatrix} \qquad \text{and} \qquad [w]_{\mathcal{E}} = \begin{bmatrix} b^1 \\ b^2 \\ b^3 \end{bmatrix},$$

hence

$$v \times w = \det\left(\begin{bmatrix} e_1 & a^1 & b^1 \\ e_2 & a^2 & b^2 \\ e_3 & a^3 & b^3 \end{bmatrix}\right)$$

$$= \left(\det\left(\begin{bmatrix} a^2 & b^2 \\ a^3 & b^3 \end{bmatrix}\right), -\det\left(\begin{bmatrix} a^1 & b^1 \\ a^3 & b^3 \end{bmatrix}\right), \det\left(\begin{bmatrix} a^1 & b^1 \\ a^2 & b^2 \end{bmatrix}\right)\right)$$

$$= (a^2 b^3 - a^3 b^2, a^3 b^1 - a^1 b^3, a^1 b^2 - a^2 b^1).$$

In particular,

$$e_1 \times e_2 = e_3, \qquad e_1 \times e_3 = -e_2, \qquad \text{and} \qquad e_2 \times e_3 = e_1. \qquad \Diamond$$

Example 8.4.2 (Vector Product on $\mathbb{R}_1^2$ and $\mathbb{R}_1^3$). Let (e_1, e_2) be the standard basis for $\mathbb{R}_1^2$, and let (a^1, a^2) be a vector in $\mathbb{R}_1^2$. The "vector product" of (a^1, a^2) is (a^2, a^1).

Now let (e_1, e_2, e_3) be the standard basis for $\mathbb{R}_1^3$, and let $v = (a^1, a^2, a^3)$ and $w = (b^1, b^2, b^3)$ be vectors in $\mathbb{R}_1^3$. Then

$$[v]_{\mathcal{E}} = \begin{bmatrix} a^1 \\ a^2 \\ a^3 \end{bmatrix} \qquad \text{and} \qquad [w]_{\mathcal{E}} = \begin{bmatrix} b^1 \\ b^2 \\ b^3 \end{bmatrix},$$

hence

$$v \times w = \det\left(\begin{bmatrix} e_1 & a^1 & b^1 \\ e_2 & a^2 & b^2 \\ -e_3 & a^3 & b^3 \end{bmatrix}\right)$$

$$= \left(\det\left(\begin{bmatrix} a^2 & b^2 \\ a^3 & b^3 \end{bmatrix}\right), -\det\left(\begin{bmatrix} a^1 & b^1 \\ a^3 & b^3 \end{bmatrix}\right), -\det\left(\begin{bmatrix} a^1 & b^1 \\ a^2 & b^2 \end{bmatrix}\right)\right)$$

$$= (a^2 b^3 - a^3 b^2, a^3 b^1 - a^1 b^3, a^2 b^1 - a^1 b^2).$$

In particular,

$$e_1 \times e_2 = -e_3, \qquad e_1 \times e_3 = -e_2, \qquad \text{and} \qquad e_2 \times e_3 = e_1. \qquad \Diamond$$

The vector product has a number of interesting algebraic properties, several of which, not surprisingly, are expressed in terms of determinants and wedge products.

Theorem 8.4.3. *Let $(V, \mathfrak{g})$ be a scalar product space of dimension $m \geq 2$, let $\mathcal{E}$ be an orthonormal basis for V, let $v_1, \ldots, v_{m-1}, w$ be vectors in V, and let c be a real number. Then, for $i, j = 1, \ldots, m-1$:*
(a)

$$\langle v_i, v_1 \times \cdots \times v_{m-1} \rangle = 0.$$

(b)

$$v_1 \times \cdots \times c v_i \times \cdots \times v_{m-1} = c(v_1 \times \cdots \times v_i \times \cdots \times v_{m-1}).$$

(c)

$$v_1 \times \cdots \times (c v_i + w) \times \cdots \times v_{m-1} = c(v_1 \times \cdots \times v_i \times \cdots \times v_{m-1})$$
$$+ v_1 \times \cdots \times w \times \cdots \times v_{m-1}.$$

(d)

$$v_1 \times \cdots \times v_i \times \cdots \times v_j \times \cdots \times v_{m-1}$$
$$= -(v_1 \times \cdots \times v_j \times \cdots \times v_i \times \cdots \times v_{m-1}).$$

(e) *If $v_i = v_j$, then*

$$v_1 \times \cdots \times v_i \times \cdots \times v_j \times \cdots \times v_{m-1} = 0.$$

Proof. (a): From (8.4.1) and (8.4.2),

$$\langle v_i, v_1 \times \cdots \times v_{m-1} \rangle$$

$$= \left\langle \sum_j a_i^j e_j, \sum_k \varepsilon_k (-1)^{k-1} \det\left(\begin{bmatrix} [v_1]_{\mathcal{E}} & \cdots & [v_i]_{\mathcal{E}} & \cdots & [v_{m-1}]_{\mathcal{E}} \end{bmatrix}^{(k)^c}\right) e_k \right\rangle$$

$$= \sum_{jk} a_i^j \langle e_j, e_k \rangle \varepsilon_k (-1)^{k-1} \det\left(\begin{bmatrix} [v_1]_{\mathcal{E}} & \cdots & [v_i]_{\mathcal{E}} & \cdots & [v_{m-1}]_{\mathcal{E}} \end{bmatrix}^{(k)^c}\right)$$

$$= \sum_j (-1)^{j-1} a_i^j \det\left(\begin{bmatrix} [v_1]_{\mathcal{E}} & \cdots & [v_i]_{\mathcal{E}} & \cdots & [v_{m-1}]_{\mathcal{E}} \end{bmatrix}^{(j)^c}\right)$$

$$= \det\left(\begin{bmatrix} [v_i]_{\mathcal{E}} & [v_1]_{\mathcal{E}} & \cdots & [v_i]_{\mathcal{E}} & \cdots & [v_{m-1}]_{\mathcal{E}} \end{bmatrix}\right).$$

Since (at least) two columns of $\left[\begin{bmatrix} v_i \end{bmatrix}_{\mathcal{E}} \quad \begin{bmatrix} v_1 \end{bmatrix}_{\mathcal{E}} \quad \cdots \quad \begin{bmatrix} v_i \end{bmatrix}_{\mathcal{E}} \quad \cdots \quad \begin{bmatrix} v_{m-1} \end{bmatrix}_{\mathcal{E}} \right]$ are equal, it follows from Theorem 2.4.2 that its determinant equals 0.

(b)–(d): Straightforward.   □

In light of Theorem 8.4.3(b), we drop parentheses and denote

$$c(v_1 \times \cdots \times v_i \times \cdots \times v_{m-1}) \qquad \text{by} \qquad c\, v_1 \times \cdots \times v_i \times \cdots \times v_{m-1}.$$

The next result is the main reason for considering vector products.

Theorem 8.4.4. *Let* $(V, \mathfrak{g})$ *be a scalar product space of dimension* $m \geq 2$, *let* $\mathcal{E}$ *be an orthonormal basis for* V, *let* $v_1, \ldots, v_{m-1}$ *be vectors in* V, *and let* U *be the subspace of* V *spanned by* $v_1, \ldots, v_{m-1}$. *Then* $v_1 \times \cdots \times v_{m-1}$ *is a vector in* $U^{\perp}$.

Proof. This follows from Theorem 8.4.3(a).   □

At first glance, Theorem 8.4.4 appears to promise more than it actually delivers. To make the result truly informative, we need $v_1 \times \cdots \times v_{m-1}$ to be a nonzero vector and $U^{\perp}$ to be a 1-dimensional subspace of V. These conditions can be met with further assumptions: see Theorem 8.4.8 and Theorem 8.4.10(a).

It was remarked above that the vector product depends on the order in which vectors are taken and the choice of orthonormal basis. According to Theorem 8.4.3(d), changing the order of vectors at most affects the sign of the resulting vector product. In view of (8.4.1), it might be expected that computing with respect to a different orthonormal basis would have a significant impact on results. As Theorem 8.4.4 shows, this is not necessarily the case: regardless of the choice of orthonormal basis, the vector product is always in the perp of the subspace spanned by the constituent vectors. In fact, several of the results to follow exhibit this same feature: see Theorem 8.4.5, Theorem 8.4.6, Theorem 8.4.8, and Theorem 8.4.9.

Theorem 8.4.5. *Let* $(V, \mathfrak{g})$ *be a scalar product space of dimension* $m \geq 2$ *and index* ν. *Let* $\mathcal{E}$ *be an orthonormal basis for* V, *let* $(\xi^1, \ldots, \xi^m)$ *be its dual basis, and let* $u, v_1, \ldots, v_{m-1}$ *be vectors in* V. *Then:*
(a)

$$\langle u, v_1 \times \cdots \times v_{m-1} \rangle = \det\left(\begin{bmatrix} [u]_{\mathcal{E}} & [v_1]_{\mathcal{E}} & \cdots & [v_{m-1}]_{\mathcal{E}} \end{bmatrix} \right)$$
$$= \xi^1 \wedge \cdots \wedge \xi^m (u, v_1, \ldots, v_{m-1}).$$

(b)

$$\langle u, v_1 \times \cdots \times v_{m-1} \rangle^2$$
$$= (-1)^{\nu} \det\left(\begin{bmatrix} \langle u, u \rangle & \langle u, v_1 \rangle & \cdots & \langle u, v_{m-1} \rangle \\ \langle v_1, u \rangle & \langle v_1, v_1 \rangle & \cdots & \langle v_1, v_{m-1} \rangle \\ \vdots & \vdots & \ddots & \vdots \\ \langle v_{m-1}, u \rangle & \langle v_{m-1}, v_1 \rangle & \cdots & \langle v_{m-1}, v_{m-1} \rangle \end{bmatrix} \right).$$

Proof. Let $\mathcal{E} = (e_1, \ldots, e_m)$, and let V have signature $(\varepsilon_1, \ldots, \varepsilon_m)$.

(a): Let $u = \sum_i b^i e_i$, so that $[u]_{\mathcal{E}} = \begin{bmatrix} b^1 & \cdots & b^m \end{bmatrix}^{\mathrm{T}}$. We have from Theorem 2.4.6 and (8.4.1) that

$$\langle u, v_1 \times \cdots \times v_{m-1} \rangle$$

$$= \left\langle \sum_i b^i e_i, \sum_j \varepsilon_j (-1)^{j-1} \det\left(\begin{bmatrix} [v_1]_{\mathcal{E}} & \cdots & [v_{m-1}]_{\mathcal{E}} \end{bmatrix}^{(j)^c} \right) e_j \right\rangle$$

$$= \sum_{ij} b^i \langle e_i, e_j \rangle \varepsilon_j (-1)^{j-1} \det\left(\begin{bmatrix} [v_1]_{\mathcal{E}} & \cdots & [v_{m-1}]_{\mathcal{E}} \end{bmatrix}^{(j)^c} \right)$$

$$= \sum_i (-1)^{i-1} b^i \det\left(\begin{bmatrix} [v_1]_{\mathcal{E}} & \cdots & [v_{m-1}]_{\mathcal{E}} \end{bmatrix}^{(i)^c} \right)$$

$$= \det\left(\begin{bmatrix} [u]_{\mathcal{E}} & [v_1]_{\mathcal{E}} & \cdots & [v_{m-1}]_{\mathcal{E}} \end{bmatrix} \right),$$

which proves the first equality. By Theorem 1.2.1(d), $u = \sum_i \xi^i(u) e_i$, hence

$$[u]_{\mathcal{E}} = \begin{bmatrix} \xi^1(u) & \cdots & \xi^m(u) \end{bmatrix}^{\mathrm{T}},$$

with corresponding expressions for $[v_1]_{\mathcal{E}}, \ldots, [v_{m-1}]_{\mathcal{E}}$. Then Theorem 7.2.8 gives

$$\xi^1 \wedge \cdots \wedge \xi^m (u, v_1, \ldots, v_{m-1}) = \det\left(\begin{bmatrix} \xi^1(u) & \xi^1(v_1) & \cdots & \xi^1(v_{m-1}) \\ \vdots & \vdots & \ddots & \vdots \\ \xi^m(u) & \xi^m(v_1) & \cdots & \xi^m(v_{m-1}) \end{bmatrix} \right)$$

$$= \det\left(\begin{bmatrix} [u]_{\mathcal{E}} & [v_1]_{\mathcal{E}} & \cdots & [v_{m-1}]_{\mathcal{E}} \end{bmatrix} \right),$$

which proves the second equality.

(b): Let

$$P = \begin{bmatrix} [u]_{\mathcal{E}} & [v_1]_{\mathcal{E}} & \cdots & [v_{m-1}]_{\mathcal{E}} \end{bmatrix},$$

and note that $\mathfrak{g}_{\mathcal{E}} = \mathrm{diag}(\varepsilon_1, \ldots, \varepsilon_m)$. Then

$$\det\left(\begin{bmatrix} \langle u, u \rangle & \langle u, v_1 \rangle & \cdots & \langle u, v_{m-1} \rangle \\ \langle v_1, u \rangle & \langle v_1, v_1 \rangle & \cdots & \langle v_1, v_{m-1} \rangle \\ \vdots & \vdots & \ddots & \vdots \\ \langle v_{m-1}, u \rangle & \langle v_{m-1}, v_1 \rangle & \cdots & \langle v_{m-1}, v_{m-1} \rangle \end{bmatrix} \right)$$

$$= \det(P^{\mathrm{T}} \mathfrak{g}_{\mathcal{E}} P) \qquad\qquad\qquad\qquad \text{[Th 3.1.2(b)]}$$

$$= \det(\mathfrak{g}_{\mathcal{E}}) \det(P)^2$$

$$= (-1)^{\nu} \det(P)^2 \qquad\qquad\qquad\qquad\quad\; \text{[(4.2.3)]}$$

$$= (-1)^{\nu} \langle u, v_1 \times \cdots \times v_{m-1} \rangle^2. \qquad\quad \text{[part (a)]} \qquad \square$$

Theorem 8.4.6. *Let $(V, \mathfrak{g})$ be a scalar product space of dimension $m \geq 2$ and index ν, let $\mathcal{E}$ be an orthonormal basis for V, and let $v_1, \ldots, v_{m-1}, w_1, \ldots, w_{m-1}$ be vectors in V. Then:*

(a)

$$\langle v_1 \times \cdots \times v_{m-1}, w_1 \times \cdots \times w_{m-1} \rangle$$

$$= (-1)^\nu \det\left(\begin{bmatrix} \langle v_1, w_1 \rangle & \cdots & \langle v_1, w_{m-1} \rangle \\ \vdots & \ddots & \vdots \\ \langle v_{m-1}, w_1 \rangle & \cdots & \langle v_{m-1}, w_{m-1} \rangle \end{bmatrix}\right).$$

(b)

$$\|v_1 \times \cdots \times v_{m-1}\| = \sqrt{\left| \det\left(\begin{bmatrix} \langle v_1, v_1 \rangle & \cdots & \langle v_1, v_{m-1} \rangle \\ \vdots & \ddots & \vdots \\ \langle v_{m-1}, v_1 \rangle & \cdots & \langle v_{m-1}, v_{m-1} \rangle \end{bmatrix}\right) \right|}.$$

Proof. Let $\mathcal{E} = (e_1, \ldots, e_m)$, and let V have signature $(\varepsilon_1, \ldots, \varepsilon_m)$.
(a): Let

$$P = \begin{bmatrix} [v_1]_\mathcal{E} & \cdots & [v_{m-1}]_\mathcal{E} \end{bmatrix} \qquad \text{and} \qquad Q = \begin{bmatrix} [w_1]_\mathcal{E} & \cdots & [w_{m-1}]_\mathcal{E} \end{bmatrix},$$

and note that $\mathfrak{g}_\mathcal{E} = \mathrm{diag}(\varepsilon_1, \ldots, \varepsilon_m)$. It is easily shown that

$$\det\big((\mathfrak{g}_\mathcal{E} Q)^{(i)^c}\big) = (-1)^\nu \varepsilon_i \det\big(Q^{(i)^c}\big). \tag{8.4.4}$$

Then

$$\langle v_1 \times \cdots \times v_{m-1}, w_1 \times \cdots \times w_{m-1} \rangle$$

$$= \left\langle \sum_i \varepsilon_i (-1)^{i-1} \det\big(P^{(i)^c}\big) e_i, \sum_j \varepsilon_j (-1)^{j-1} \det\big(Q^{(j)^c}\big) e_j \right\rangle \qquad [(8.4.1)]$$

$$= \sum_{ij} \varepsilon_i \varepsilon_j (-1)^{i+j} \langle e_i, e_j \rangle \det\big(P^{(i)^c}\big) \det\big(Q^{(j)^c}\big)$$

$$= \sum_i \varepsilon_i \det\big(P^{(i)^c}\big) \det\big(Q^{(i)^c}\big)$$

$$= (-1)^\nu \sum_i \det\big(P^{(i)^c}\big) \det\big((\mathfrak{g}_\mathcal{E} Q)^{(i)^c}\big) \qquad [(8.4.4)]$$

$$= (-1)^\nu \sum_{I \in \mathcal{I}_{m-1,m}} \det(P^I) \det\big((\mathfrak{g}_\mathcal{E} Q)^I\big)$$

$$= (-1)^\nu \det(P^\mathrm{T} \mathfrak{g}_\mathcal{E} Q) \qquad [\text{Th } 2.4.10(\text{a})]$$

$$= (-1)^\nu \det\left(\begin{bmatrix} \langle v_1, w_1 \rangle & \cdots & \langle v_1, w_{m-1} \rangle \\ \vdots & \ddots & \vdots \\ \langle v_{m-1}, w_1 \rangle & \cdots & \langle v_{m-1}, w_{m-1} \rangle \end{bmatrix}\right). \qquad [\text{Th } 3.1.2(\text{b})]$$

(b): This follows from (4.1.1) and part (a). $\qquad\qquad\square$

Theorem 8.4.7. *Let $(V, \mathfrak{g})$ be a scalar product space of dimension $m \geq 2$, and let $\mathcal{E} = (e_1, \ldots, e_m)$ be an orthonormal basis for V. Let U be a subspace of V*

of dimension $m - 1$, *let* $u_1, \ldots, u_{m-1}$ *be vectors in* U, *and let* $A : U \longrightarrow U$ *be a linear map. Then*

$$A(u_1) \times \cdots \times A(u_{m-1}) = \det(A)\, u_1 \times \cdots \times u_{m-1}.$$

Proof. Let V have signature $(\varepsilon_1, \ldots, \varepsilon_m)$, and consider the function $\eta^i : U^{m-1} \longrightarrow \mathbb{R}$ defined by

$$\eta^i(v_1, \ldots, v_{m-1}) = \langle v_1 \times \cdots \times v_{m-1}, e_i \rangle$$

for all vectors $v_1, \ldots, v_{m-1}$ in U for $i = 1, \ldots, m$. It follows from parts (b) and (c) of Theorem 8.4.3 that η^i is a multicovector in $\Lambda^{m-1}(U)$. Then

$$
\begin{aligned}
\langle A(u_1) \times \cdots \times A(u_{m-1}), e_i \rangle &= \eta^i\big(A(u_1), \ldots, A(u_{m-1})\big) \\
&= A^*(\eta^i)(u_1, \ldots, u_{m-1}) \\
&= \det(A)\, \eta^i(u_1, \ldots, u_{m-1}) \qquad \text{[Th 7.3.4(a)]} \\
&= \det(A)\, \langle u_1 \times \cdots \times u_{m-1}, e_i \rangle,
\end{aligned}
$$

hence

$$
\begin{aligned}
A(u_1) &\times \cdots \times A(u_{m-1}) \\
&= \sum_i \varepsilon_i \langle A(u_1) \times \cdots \times A(u_{m-1}), e_i \rangle e_i \qquad \text{[Th 4.2.7]} \\
&= \det(A) \sum_i \varepsilon_i \langle u_1 \times \cdots \times u_{m-1}, e_i \rangle e_i \\
&= \det(A)\, u_1 \times \cdots \times u_{m-1}. \qquad \text{[Th 4.2.7]} \qquad \square
\end{aligned}
$$

Theorems 8.4.3–8.4.7 are of little interest when the vector product equals the zero vector. The next result gives a straightforward condition that avoids this situation.

Theorem 8.4.8. *Let* $(V, \mathfrak{g})$ *be a scalar product space of dimension* $m \geq 2$, *let* $\mathcal{E}$ *be an orthonormal basis for* V, *and let* $v_1, \ldots, v_{m-1}$ *be vectors in* V. *Then* $v_1, \ldots, v_{m-1}$ *are linearly independent if and only if* $v_1 \times \cdots \times v_{m-1} \neq 0$.

Proof. We prove the logically equivalent assertion: $v_1, \ldots, v_{m-1}$ are linearly dependent if and only if $v_1 \times \cdots \times v_{m-1} = 0$.

($\Rightarrow$): Since $v_1, \ldots, v_{m-1}$ are linearly dependent, one of them can be expressed as a linear combination of the others, say, $v_1 = \sum_{j=2}^{m-1} a_j v_j$. Then

$$
\begin{aligned}
\det&\left(\begin{bmatrix} [v_1]_{\mathcal{E}} & [v_2]_{\mathcal{E}} & \cdots & [v_{m-1}]_{\mathcal{E}} \end{bmatrix}^{(i)^c} \right) \\
&= \det\left(\begin{bmatrix} [\sum_{j=2}^{m-1} a_j v_j]_{\mathcal{E}} & [v_2]_{\mathcal{E}} & \cdots & [v_{m-1}]_{\mathcal{E}} \end{bmatrix}^{(i)^c} \right) \\
&= \sum_{j=2}^{m-1} a_j \det\left(\begin{bmatrix} [v_j]_{\mathcal{E}} & [v_2]_{\mathcal{E}} & \cdots & [v_{m-1}]_{\mathcal{E}} \end{bmatrix}^{(i)^c} \right).
\end{aligned}
$$

Since $\left[\begin{array}{cccc}[v_j]_\mathcal{E} & [v_2]_\mathcal{E} & \cdots & [v_{m-1}]_\mathcal{E}\end{array}\right]$ has $[v_j]_\mathcal{E}$ in (at least) two positions, it follows from Theorem 2.4.2 that $\det\left(\left[\begin{array}{cccc}[v_j]_\mathcal{E} & [v_2]_\mathcal{E} & \cdots & [v_{m-1}]_\mathcal{E}\end{array}\right]^{(i)^c}\right) = 0$ for $j = 2, \ldots, m-1$. Thus,

$$\det\left(\left[\begin{array}{cccc}[v_1]_\mathcal{E} & [v_2]_\mathcal{E} & \cdots & [v_{m-1}]_\mathcal{E}\end{array}\right]^{(i)^c}\right) = 0$$

for $i = 1, \ldots, m$. It follows from (8.4.1) that $v_1 \times \cdots \times v_{m-1} = 0$.

($\Leftarrow$): Since the dimension of $\mathrm{span}(\{v_1, \ldots, v_{m-1}\})$ is at most $m-1$, there is a vector v_0 in $V \backslash \mathrm{span}(\{v_1, \ldots, v_{m-1}\})$. Then

$$\langle v_0, v_1 \times \cdots \times v_{m-1} \rangle = 0$$

$\Leftrightarrow \quad \det\left(\left[\begin{array}{cccc}[v_0]_\mathcal{E} & [v_1]_\mathcal{E} & \cdots & [v_{m-1}]_\mathcal{E}\end{array}\right]\right) = 0$ [Th 8.4.5(a)]

$\Leftrightarrow \quad [v_0]_\mathcal{E}, [v_1]_\mathcal{E} \ldots, [v_0]_\mathcal{E}$ are linearly dependent [Th 2.4.9]

$\Leftrightarrow \quad v_0, v_1, \ldots, v_{m-1}$ are linearly dependent [Th 2.2.1]

$\Leftrightarrow \quad v_1, \ldots, v_{m-1}$ are linearly dependent. [Th 1.1.2]

Thus, if $v_1 \times \cdots \times v_{m-1} = 0$, then $v_1, \ldots, v_{m-1}$ are linearly dependent. $\square$

Whether vectors in a vector space are linearly independent or linearly dependent is unrelated to the presence or absence of a scalar product. This means that in Theorem 8.4.8, linear independence can be checked using any convenient choice of scalar product and orthonormal basis.

Theorem 8.4.9. *Let $(V, \mathfrak{g})$ be a scalar product space of dimension $m \geq 2$, let $\mathcal{E}$ be an orthonormal basis for V, let U be a subspace of V of dimension $m-1$, and let $(u_1, \ldots, u_{m-1})$ be a basis for U. Then $\mathfrak{g}$ is nondegenerate on U if and only if $\|u_1 \times \cdots \times u_{m-1}\| \neq 0$.*

Proof. The matrix of $\mathfrak{g}|_U$ with respect to $(u_1, \ldots, u_{m-1})$ is $[\langle u_i, u_j \rangle]$, so

$\mathfrak{g}$ is nondegenerate on U

$\Leftrightarrow \quad \mathfrak{g}|_U$ is nondegenerate on U [by definition]

$\Leftrightarrow \quad \det\left([\langle u_i, u_j \rangle]\right) \neq 0$ [Th 3.3.3]

$\Leftrightarrow \quad \langle u_1 \times \cdots \times u_{m-1}, u_1 \times \cdots \times u_{m-1} \rangle \neq 0$ [Th 8.4.6(a)]

$\Leftrightarrow \quad \|u_1 \times \cdots \times u_{m-1}\| \neq 0.$ $\square$

Theorem 8.4.10. *Let $(V, \mathfrak{g})$ be a scalar product space of dimension $m \geq 2$, and let $\mathcal{E}$ be an orthonormal basis for V. Let U be a subspace of V of dimension $m-1$ on which $\mathfrak{g}$ is nondegenerate, let $(u_1, \ldots, u_{m-1})$ be a basis for U, and let ϵ be the sign of $\langle u_1 \times \cdots \times u_{m-1}, u_1 \times \cdots \times u_{m-1} \rangle$, so that*

$$\epsilon = \left\langle \frac{u_1 \times \cdots \times u_{m-1}}{\|u_1 \times \cdots \times u_{m-1}\|}, \frac{u_1 \times \cdots \times u_{m-1}}{\|u_1 \times \cdots \times u_{m-1}\|} \right\rangle. \qquad (8.4.5)$$

Then:

(a) $U^\perp = \mathbb{R}(u_1 \times \cdots \times u_{m-1})$ is a 1-dimensional subspace of V.

(b) $(\epsilon u_1 \times \cdots \times u_{m-1}, u_1, \ldots, u_{m-1})$ is a basis for V that is consistent with $\mathcal{E}$.

(c) If $(u_1, \ldots, u_{m-1})$ is an orthonormal basis for U, then $(\epsilon u_1 \times \cdots \times u_{m-1}, u_1, \ldots, u_{m-1})$ is an orthonormal basis for V.

Remark. By Theorem 8.4.9, $\langle u_1 \times \cdots \times u_{m-1}, u_1 \times \cdots \times u_{m-1} \rangle \neq 0$, so ϵ is defined and the denominators in (8.4.5) are nonzero.

Proof. (a): We have from Theorem 4.1.2(b) that $U^\perp$ is a 1-dimensional subspace of V, from Theorem 8.4.3(a) that $u_1 \times \cdots \times u_{m-1}$ is in $U^\perp$, and from Theorem 8.4.8 that $u_1 \times \cdots \times u_{m-1} \neq 0$. The result follows.

(b): By Theorem 4.1.3, $V = U \oplus U^\perp$. Since $(u_1, \ldots, u_{m-1})$ is a basis for U, it follows from Theorem 1.1.2 and part (a) that $\mathcal{U} = (\epsilon u_1 \times \cdots \times u_{m-1}, u_1, \ldots, u_{m-1})$ is a basis for V. We have from (2.2.7) and Theorem 8.4.5(a) that

$$\det\left(\left[\mathrm{id}_V\right]_{\mathcal{U}}^{\mathcal{E}}\right) = \det\left(\left[\epsilon u_1 \times \cdots \times u_{m-1}\right]_{\mathcal{E}} \quad \left[u_1\right]_{\mathcal{E}} \quad \cdots \quad \left[u_{m-1}\right]_{\mathcal{E}}\right)$$

$$= \epsilon \langle u_1 \times \cdots \times u_{m-1}, u_1 \times \cdots \times u_{m-1} \rangle$$

$$= |\langle u_1 \times \cdots \times u_{m-1}, u_1 \times \cdots \times u_{m-1} \rangle| > 0,$$

so $\mathcal{U}$ is consistent with $\mathcal{E}$.

(c): Since $(u_1, \ldots, u_{m-1})$ is orthonormal, we have from Theorem 8.4.6(b) that $\|\epsilon u_1 \times \cdots \times u_{m-1}\| = 1$, so $\epsilon u_1 \times \cdots \times u_{m-1}$ is a unit vector. By Theorem 8.4.3(a), $\langle \epsilon u_1 \times \cdots \times u_{m-1}, u_i \rangle = 0$ for $j = 1, \ldots, m-1$. The result follows. $\square$

Theorem 8.4.11. *Let $(V, \mathfrak{g})$ be a scalar product space of dimension $m \geq 2$, let $\mathcal{E}$ be an orthonormal basis for V, let U be a subspace of V of dimension $m-1$ on which $\mathfrak{g}$ is nondegenerate, and let $\mathcal{U} = (u_1, \ldots, u_{m-1})$ and $\mathcal{V} = (v_1, \ldots, v_{m-1})$ be bases for U. Then:*

(a)

$$u_1 \times \cdots \times u_{m-1} = \det\left(\left[\mathrm{id}_U\right]_{\mathcal{U}}^{\mathcal{V}}\right) v_1 \times \cdots \times v_{m-1}.$$

(b)

$$\frac{u_1 \times \cdots \times u_{m-1}}{\|u_1 \times \cdots \times u_{m-1}\|} = \pm \frac{v_1 \times \cdots \times v_{m-1}}{\|v_1 \times \cdots \times v_{m-1}\|},$$

where the positive (negative) sign is chosen if $\mathcal{U}$ and $\mathcal{V}$ are consistent (not consistent).

Remark. By Theorem 8.4.9, the denominators in part (b) are nonzero.

Proof. (a): Let $P = \left[\mathrm{id}_U\right]_{\mathcal{U}}^{\mathcal{V}}$, and let $A : U \longrightarrow U$ be the linear isomorphism defined by $A(\mathcal{U}) = \mathcal{V}$. By Theorem 8.4.7,

$$v_1 \times \cdots \times v_{m-1} = \det(A)\, u_1 \times \cdots \times u_{m-1},$$

and by Theorem 2.2.5,

$$P = \left[\mathrm{id}_U\right]_{A^{-1}(\mathcal{V})}^{\mathcal{V}} = \left[A^{-1}\right]_{\mathcal{V}}^{\mathcal{V}} = \left(\left[A\right]_{\mathcal{V}}^{\mathcal{V}}\right)^{-1},$$

hence $\det(A) = \det(P)^{-1}$. The result follows.

(b):   We have from Theorem 2.4.12 and Theorem 8.4.9 that $\det(P)$, $\|u_1 \times \cdots \times u_{m-1}\|$, and $\|v_1 \times \cdots \times v_{m-1}\|$ are each nonzero. Then part (a) gives

$$\frac{u_1 \times \cdots \times u_{m-1}}{\|u_1 \times \cdots \times u_{m-1}\|} = \frac{\det(P)}{|\det(P)|} \frac{v_1 \times \cdots \times v_{m-1}}{\|v_1 \times \cdots \times v_{m-1}\|}$$

$$= \mathrm{sgn}(\det(P)) \frac{v_1 \times \cdots \times v_{m-1}}{\|v_1 \times \cdots \times v_{m-1}\|}. \qquad \square$$

Theorem 8.4.12. *Let $(V, \mathfrak{g})$ be a scalar product space of dimension $m \geq 2$, let $\mathcal{E} = (e_1, \ldots, e_m)$ and $\mathcal{F}$ be orthonormal bases for V, and let $\boxtimes$ denote the vector product operation corresponding to $\mathcal{F}$. Then*

$$e_1 \boxtimes \cdots \boxtimes \widehat{e_i} \boxtimes \cdots \boxtimes e_m = \varepsilon_i (-1)^{i-1} \det\left([\mathrm{id}_V]_{\mathcal{E}}^{\mathcal{F}} \right) e_i$$

for $i = 1, \ldots, m$, where $\,\widehat{}\,$ indicates that an expression is omitted.

Proof. Let U_i be the subspace of V spanned by $e_1, \ldots, \widehat{e_i}, \ldots, e_m$. Since $\mathfrak{g}$ is nondegenerate on $\mathbb{R}e_i$, we have from Theorem 4.1.3 that $V = \mathbb{R}e_i \oplus (\mathbb{R}e_i)^{\perp}$ and $\mathfrak{g}$ is nondegenerate on $(\mathbb{R}e_i)^{\perp}$. By definition, each of $e_1, \ldots, \widehat{e_i}, \ldots, e_m$ is orthogonal to e_i, so $U_i \subseteq (\mathbb{R}e_i)^{\perp}$. From Theorem 1.1.18,

$$m = \dim(V) = \dim(\mathbb{R}e_i) + \dim((\mathbb{R}e_i)^{\perp}) = 1 + \dim((\mathbb{R}e_i)^{\perp}),$$

hence

$$\dim((\mathbb{R}e_i)^{\perp}) = m - 1 = \dim(U_i).$$

By Theorem 1.1.7(b), $U_i = (\mathbb{R}e_i)^{\perp}$, so $\mathfrak{g}$ is nondegenerate on U_i. It follows from Theorem 4.1.2(c) and Theorem 8.4.10(a) that

$$\mathbb{R}e_i = U_i^{\perp} = \mathbb{R}(e_1 \boxtimes \cdots \boxtimes \widehat{e_i} \boxtimes \cdots \boxtimes e_m),$$

so

$$e_1 \boxtimes \cdots \boxtimes \widehat{e_i} \boxtimes \cdots \boxtimes e_m = c_i e_i$$

for some real number c_i. We have

$$c_i \varepsilon_i = \langle e_i, c_i e_i \rangle = \langle e_i, e_1 \boxtimes \cdots \boxtimes \widehat{e_i} \boxtimes \cdots \boxtimes e_m \rangle$$

$$= \det\left(\left[[e_i]_{\mathcal{F}} \quad [e_1]_{\mathcal{F}} \quad \cdots \quad \widehat{[e_i]}_{\mathcal{F}} \quad \cdots \quad [e_m]_{\mathcal{F}} \right] \right) \qquad [\text{Th 8.4.5(a)}]$$

$$= (-1)^{i-1} \det\left(\left[[e_1]_{\mathcal{F}} \quad \cdots \quad [e_i]_{\mathcal{F}} \quad \cdots \quad [e_m]_{\mathcal{F}} \right] \right) \qquad [\text{Th 8.4.3(d)}]$$

$$= (-1)^{i-1} \det\left([\mathrm{id}_V]_{\mathcal{E}}^{\mathcal{F}} \right), \qquad [(2.2.7)]$$

hence

$$c_i = \varepsilon_i (-1)^{i-1} \det\left([\mathrm{id}_V]_{\mathcal{E}}^{\mathcal{F}} \right).$$

The result follows. $\qquad \square$

Theorem 8.4.13. *Let* $(V, \mathfrak{g})$ *be a scalar product space of dimension* $m \geq 2$, *let* $\mathcal{E}$ *and* $\mathcal{F}$ *be orthonormal bases for* V, *and let* $\times$ *and* $\boxtimes$ *denote the corresponding vector product operations. Let* U *be a subspace of* V *of dimension* $m-1$ *on which* $\mathfrak{g}$ *is nondegenerate, and let* $(u_1, \ldots, u_{m-1})$ *be a basis for* U. *Then:*
(a)

$$u_1 \boxtimes \cdots \boxtimes u_{m-1} = \det\left([\mathrm{id}_V]_{\mathcal{E}}^{\mathcal{F}}\right) u_1 \times \cdots \times u_{m-1}.$$

(b)

$$\frac{u_1 \boxtimes \cdots \boxtimes u_{m-1}}{\|u_1 \boxtimes \cdots \boxtimes u_{m-1}\|} = \pm \frac{u_1 \times \cdots \times u_{m-1}}{\|u_1 \times \cdots \times u_{m-1}\|},$$

where the positive (negative) sign is chosen if $\mathcal{E}$ *and* $\mathcal{F}$ *are consistent (not consistent).*

Remark. By Theorem 8.4.9, the denominators in part (b) are nonzero.

Proof. (a): Let $\mathcal{E} = (e_1, \ldots, e_m)$, and let

$$P = [\mathrm{id}_V]_{\mathcal{E}}^{\mathcal{F}} = \left[[e_1]_{\mathcal{F}} \quad \cdots \quad [e_m]_{\mathcal{F}} \right],$$

where the second equality comes from (2.2.7). Let

$$Q = \left[[u_1]_{\mathcal{E}} \quad \cdots \quad [u_{m-1}]_{\mathcal{E}} \right],$$

and let

$$R_i = \left(P^{(i)^c}\right)^{\mathrm{T}}$$

for $i = 1, \ldots, m$. By Theorem 2.1.6(b),

$$R_i^{(j)^c} = \left(P_{(j)^c}^{(i)^c}\right)^{\mathrm{T}}$$

for $j = 1, \ldots, m$. It follows from (2.2.5) that

$$\left[[u_1]_{\mathcal{F}} \quad \cdots \quad [u_{m-1}]_{\mathcal{F}} \right] = [\mathrm{id}_V]_{\mathcal{E}}^{\mathcal{F}} \left[[u_1]_{\mathcal{E}} \quad \cdots \quad [u_{m-1}]_{\mathcal{E}} \right] = PQ,$$

and then from Theorem 2.1.6(c) that

$$\left[[u_1]_{\mathcal{F}} \quad \cdots \quad [u_{m-1}]_{\mathcal{F}} \right]^{(i)^c} = P^{(i)^c} Q.$$

Then (8.4.1) gives

$$u_1 \boxtimes \cdots \boxtimes u_{m-1} = \sum_i \varepsilon_i (-1)^{i-1} \det\left(\left[[u_1]_{\mathcal{F}} \quad \cdots \quad [u_{m-1}]_{\mathcal{F}} \right]^{(i)^c} \right) e_i$$

$$= \sum_i \varepsilon_i (-1)^{i-1} \det\left(P^{(i)^c} Q \right) e_i.$$

By Theorem 2.4.10(a),

$$\det\left(P^{(i)^c} Q\right) = \det\left(R_i^T Q\right) = \sum_j \det\left(R_i^{(j)^c}\right) \det\left(Q^{(j)^c}\right)$$

$$= \sum_j \det\left(P_{(j)^c}^{(i)^c}\right) \det\left(Q^{(j)^c}\right).$$

We have

$$P_{(j)^c}^{(i)^c} = \left(\left[\,[e_1]_{\mathcal{F}} \quad \cdots \quad [e_m]_{\mathcal{F}}\,\right]_{(j)^c}\right)^{(i)^c}$$

$$= \left[\,[e_1]_{\mathcal{F}} \quad \cdots \quad \widehat{[e_j]}_{\mathcal{F}} \quad \cdots \quad [e_m]_{\mathcal{F}}\,\right]^{(i)^c},$$

where $\widehat{}$ indicates that an expression is omitted. The preceding three identities combine to give

$$u_1 \boxtimes \cdots \boxtimes u_{m-1}$$

$$= \sum_i \varepsilon_i(-1)^{i-1}\left(\sum_j \det\left(P_{(j)^c}^{(i)^c}\right) \det\left(Q^{(j)^c}\right)\right) e_i$$

$$= \sum_j \det\left(Q^{(j)^c}\right)\left(\sum_i \varepsilon_i(-1)^{i-1}\det\left(\left[\,[e_1]_{\mathcal{F}} \quad \cdots \quad \widehat{[e_j]}_{\mathcal{F}} \quad \cdots \quad [e_m]_{\mathcal{F}}\,\right]^{(i)^c}\right) e_i\right)$$

$$= \sum_j \det\left(Q^{(j)^c}\right) e_1 \boxtimes \cdots \boxtimes \widehat{e_j} \boxtimes \cdots \boxtimes e_m,$$

where the last equality follows from (8.4.1). By Theorem 8.4.12,

$$e_1 \boxtimes \cdots \boxtimes \widehat{e_j} \boxtimes \cdots \boxtimes e_m = \varepsilon_j(-1)^{j-1}\det(P)\, e_j$$

for $j = 1, \ldots, m$. Thus,

$$u_1 \boxtimes \cdots \boxtimes u_{m-1} = \det(P) \sum_j \varepsilon_j(-1)^{j-1}\det\left(Q^{(j)^c}\right) e_j$$

$$= \det(P) \sum_j \varepsilon_j(-1)^{j-1}\det\left(\left[\,[u_1]_{\mathcal{E}} \quad \cdots \quad [u_{m-1}]_{\mathcal{E}}\,\right]^{(j)^c}\right) e_j$$

$$= \det(P)\, u_1 \times \cdots \times u_{m-1},$$

where the last equality follows from (8.4.1).

(b):   We have from Theorem 2.4.12 and Theorem 8.4.9 that $\det(P)$, $\|u_1 \times \cdots \times u_{m-1}\|$, and $\|u_1 \boxtimes \cdots \boxtimes u_{m-1}\|$ are each nonzero. Then Theorem 8.3.3 and part (a) give

$$\frac{u_1 \boxtimes \cdots \boxtimes u_{m-1}}{\|u_1 \boxtimes \cdots \boxtimes u_{m-1}\|} = \frac{\det(P)}{|\det(P)|}\frac{u_1 \times \cdots \times u_{m-1}}{\|u_1 \times \cdots \times u_{m-1}\|}$$

$$= \operatorname{sgn}\left(\det(P)\right)\frac{u_1 \times \cdots \times u_{m-1}}{\|u_1 \times \cdots \times u_{m-1}\|}. \qquad \square$$

8.5   Hodge Star

In this section, we define a map that assigns to a given multicovector another multicovector that "complements" the first. The methods that result add to our growing armamentarium of techniques for computing with multicovectors.

Theorem 8.5.1. *Let $(V, \mathfrak{g}, \mathfrak{O})$ be an oriented scalar product space of dimension m, let Ω_V be its volume multicovector, and let η be a multicovector in $\Lambda^s(V)$, where $s \leq m$. Then there is a unique multicovector $\star(\eta)$ in $\Lambda^{m-s}(V)$ such that*

$$\eta \wedge \zeta = \langle \star(\eta), \zeta \rangle^\Lambda \, \Omega_V$$

for all multicovectors ζ in $\Lambda^{m-s}(V)$, where $\langle \cdot, \cdot \rangle^\Lambda$ is the scalar product defined in Section 7.5.

Proof. Existence. By Theorem 7.2.12(d), (Ω_V) is a basis for $\Lambda^m(V)$, so

$$\eta \wedge \zeta = f_\eta(\zeta) \, \Omega_V$$

for some real number $f_\eta(\zeta)$. The assignment $\zeta \longmapsto f_\eta(\zeta)$ defines a function $f_\eta : \Lambda^{m-s}(V) \longrightarrow \mathbb{R}$. Evidently, f_η is linear, so f_η is a multicovector in $\Lambda^{m-s}(V)^*$. Recall the flat map and sharp map introduced at the end of Section 7.5:

$$\mathsf{F}^\Lambda : \Lambda^{m-s}(V) \longrightarrow \Lambda^{m-s}(V)^* \qquad \text{and} \qquad \mathsf{S}^\Lambda : \Lambda^{m-s}(V)^* \longrightarrow \Lambda^{m-s}(V).$$

Let us define $\star(\eta) = (f_\eta)^{\mathsf{S}^\Lambda}$. Then $\star(\eta)$ is a multicovector in $\Lambda^{m-s}(V)$ and $f_\eta = \big(\star(\eta)\big)^{\mathsf{F}^\Lambda}$. According to (7.5.12), $f_\eta(\zeta) = \langle \star(\eta), \zeta \rangle^\Lambda$, so

$$\eta \wedge \zeta = \langle \star(\eta), \zeta \rangle^\Lambda \, \Omega_V \qquad\qquad (8.5.1)$$

for all multicovectors ζ in $\Lambda^{m-s}(V)$. Thus, $\star(\eta)$ satisfies the desired property.

   Uniqueness. Suppose ξ is a multicovector in $\Lambda^{m-s}(V)$ satisfying the specified property; that is,

$$\eta \wedge \zeta = \langle \xi, \zeta \rangle^\Lambda \, \Omega_V \qquad\qquad (8.5.2)$$

for all multicovectors ζ in $\Lambda^{m-s}(V)$. Since (Ω_V) is a basis for $\Lambda^m(V)$, we have from (8.5.1) and (8.5.2) that $\langle \xi, \zeta \rangle^\Lambda = \langle \star(\eta), \zeta \rangle^\Lambda$, hence $\langle \xi - \star(\eta), \zeta \rangle^\Lambda = 0$. Because ζ was arbitrary and, by Theorem 7.5.3(a), $\langle \cdot, \cdot \rangle^\Lambda$ is nondegenerate, $\xi - \star(\eta) = 0$.   □

   Hodge star is the family of linear maps

$$\star : \Lambda^s(V) \longrightarrow \Lambda^{m-s}(V)$$

defined for $s \leq m$ by the assignment

$$\eta \longmapsto \star(\eta)$$

for all multicovectors η in $\Lambda^s(V)$. Let us denote

$$\star \circ \star \qquad \text{by} \qquad \star^2 .$$

Theorem 8.5.2. *Let $(V, \mathfrak{g}, \mathfrak{O})$ be an oriented scalar product space, let $(e_1, \ldots, e_m)$ be an orthonormal basis for V that is positively oriented with respect to $\mathfrak{O}$, let $(\xi^1, \ldots, \xi^m)$ be its dual basis, and let I be a multi-index in $\mathcal{I}_{s,m}$. Then*

$$\star(\xi^I) = \operatorname{sgn}(\sigma_{(I,I^c)}) \left\langle \xi^{I^c}, \xi^{I^c} \right\rangle^\Lambda \xi^{I^c},$$

where $\sigma_{(I,I^c)}$ is the permutation defined in Section B.2.

Proof. We have from Theorem 7.5.3(b) that $(\xi^I : I \in \mathcal{I}_{s,m})$ is an orthonormal basis for $(\Lambda^s(V), \mathfrak{g}^\Lambda)$, and then from Theorem 4.2.7 that

$$\star(\xi^I) = \sum_{J \in \mathcal{I}_{m-s,m}} \left\langle \xi^J, \xi^J \right\rangle^\Lambda \left\langle \star(\xi^I), \xi^J \right\rangle^\Lambda \xi^J. \tag{8.5.3}$$

According to Theorem 8.5.1,

$$\left\langle \star(\xi^I), \xi^J \right\rangle^\Lambda \Omega_V = \xi^I \wedge \xi^J.$$

For $J = I^c$, we have

$$\begin{aligned}
\xi^I \wedge \xi^{I^c} &= \operatorname{sgn}(\sigma_{(I,I^c)}) \, \xi^1 \wedge \cdots \wedge \xi^m && \text{[Th 7.2.10]} \\
&= \operatorname{sgn}(\sigma_{(I,I^c)}) \, \Omega_V, && \text{[Th 8.3.4]}
\end{aligned}$$

hence

$$\left\langle \star(\xi^I), \xi^{I^c} \right\rangle^\Lambda \Omega_V = \operatorname{sgn}(\sigma_{(I,I^c)}) \, \Omega_V.$$

Since (Ω_V) is a basis for $\Lambda^m(V)$,

$$\left\langle \star(\xi^I), \xi^{I^c} \right\rangle^\Lambda = \operatorname{sgn}(\sigma_{(I,I^c)}).$$

On the other hand, for $J \neq I^c$, we have from Theorem 7.2.6 that $\xi^I \wedge \xi^J = 0$. In summary,

$$\left\langle \star(\xi^I), \xi^J \right\rangle^\Lambda = \begin{cases} \operatorname{sgn}(\sigma_{(I,I^c)}) & \text{if } J = I^c \\ 0 & \text{if } J \neq I^c. \end{cases}$$

Substituting into (8.5.3) gives the result. $\qquad\square$

Theorem 8.5.2 provides a way to compute with $\star$ on an orthonormal basis for $\Lambda^s(V)$. Computations are then extended to all of $\Lambda^s(V)$ by the linearity of $\star$.

Theorem 8.5.3. *Let $(V, \mathfrak{g}, \mathfrak{O})$ be an oriented scalar product space of dimension m and index ν, let Ω_V be its volume multicovector, and let η, ζ be multicovectors in $\Lambda^s(V)$. Then:*
(a) $\star(1) = (-1)^\nu \Omega_V$.
(b) $\star(\Omega_V) = 1$.
(c) $\star^2(\eta) = (-1)^{s(m-s)+\nu} \eta$.
(d) $\eta \wedge \star(\zeta) = (-1)^\nu \left\langle \eta, \zeta \right\rangle^\Lambda \Omega_V$.

Proof. (b): Recall from (7.1.3) that $\Lambda^0(V) = \mathbb{R}$. Since 1 and $\star(\Omega_V)$ are in $\Lambda^0(V)$, by Theorem 8.5.1,

$$\Omega_V = \Omega_V \wedge 1 = \langle \star(\Omega_V), 1 \rangle^\Lambda \Omega_V = \star(\Omega_V) \Omega_V,$$

hence $\star(\Omega_V) = 1$.

(c): We have

$$
\begin{aligned}
\star^2(\xi^I) &= \star(\star(\xi^I)) \\
&= \star\left(\mathrm{sgn}(\sigma_{(I,I^c)}) \langle \xi^{I^c}, \xi^{I^c} \rangle^\Lambda \xi^{I^c}\right) && \text{[Th 8.5.2]} \\
&= \mathrm{sgn}(\sigma_{(I,I^c)}) \langle \xi^{I^c}, \xi^{I^c} \rangle^\Lambda \star(\xi^{I^c}) \\
&= \mathrm{sgn}(\sigma_{(I,I^c)}) \langle \xi^{I^c}, \xi^{I^c} \rangle^\Lambda \mathrm{sgn}(\sigma_{(I^c,I)}) \langle \xi^I, \xi^I \rangle^\Lambda \xi^I && \text{[Th 8.5.2]} \\
&= (-1)^{s(m-s)+\nu} \xi^I. && \text{[Th 7.5.2, Th B.2.5]}
\end{aligned}
$$

Since the ξ^I span $\Lambda^s(V)$, the result for η follows from the linearity of $\star$.

(a): We have

$$
\begin{aligned}
\star(1) &= \star^2(\Omega_V) && \text{[part (b)]} \\
&= (-1)^\nu \Omega_V. && \text{[part (c), } s = m\text{]}
\end{aligned}
$$

(d): We have

$$
\begin{aligned}
\eta \wedge \star(\zeta) &= (-1)^{s(m-s)} \star(\zeta) \wedge \eta && \text{[Th 7.2.3(a)]} \\
&= (-1)^{s(m-s)} \langle \star^2(\zeta), \eta \rangle^\Lambda \Omega_V && \text{[Th 8.5.1]} \\
&= (-1)^{s(m-s)}(-1)^{s(m-s)+\nu} \langle \zeta, \eta \rangle^\Lambda \Omega_V && \text{[part (c)]} \\
&= (-1)^\nu \langle \eta, \zeta \rangle^\Lambda \Omega_V. && \qquad\square
\end{aligned}
$$

Example 8.5.4 (Hodge Star on $\mathbb{R}_0^3$). Let $\mathcal{E}$ be the standard basis for $\mathbb{R}_0^3$, and let (ξ^1, ξ^2, ξ^3) be its dual basis. Using Theorem 7.5.2 and Theorem 8.5.2, we obtain the following tables:

I	ξ^I	$\star(\xi^I)$	I^c	$\mathrm{sgn}(\sigma_{(I,I^c)})$	$\langle \xi^{I^c}, \xi^{I^c} \rangle^\Lambda$
(1)	ξ^1	$\xi^2 \wedge \xi^3$	(2,3)	1	1
(2)	ξ^2	$-\xi^1 \wedge \xi^3$	(1,3)	-1	1
(3)	ξ^3	$\xi^1 \wedge \xi^2$	(1,2)	1	1

I	ξ^I	$\star(\xi^I)$	I^c	$\mathrm{sgn}(\sigma_{(I,I^c)})$	$\langle \xi^{I^c}, \xi^{I^c} \rangle^\Lambda$
(1,2)	$\xi^1 \wedge \xi^2$	ξ^3	(3)	1	1
(1,3)	$\xi^1 \wedge \xi^3$	$-\xi^2$	(2)	-1	1
(2,3)	$\xi^2 \wedge \xi^3$	ξ^1	(1)	1	1

$\Diamond$

Example 8.5.5 (Hodge Star on $\mathbb{R}^3_1$). Let $\mathcal{E}$ be the standard basis for $\mathbb{R}^3_1$, and let (ξ^1, ξ^2, ξ^3) be its dual basis. Using Theorem 7.5.2 and Theorem 8.5.2, we obtain the following tables:

I	ξ^I	$\star(\xi^I)$	I^{c}	$\text{sgn}(\sigma_{(I,I^{\text{c}})})$	$\langle \xi^{I^{\text{c}}}, \xi^{I^{\text{c}}} \rangle^\Lambda$
(1)	ξ^1	$-\xi^2 \wedge \xi^3$	(2,3)	1	-1
(2)	ξ^2	$\xi^1 \wedge \xi^3$	(1,3)	-1	-1
(3)	ξ^3	$\xi^1 \wedge \xi^2$	(1,2)	1	1

I	ξ^I	$\star(\xi^I)$	I^{c}	$\text{sgn}(\sigma_{(I,I^{\text{c}})})$	$\langle \xi^{I^{\text{c}}}, \xi^{I^{\text{c}}} \rangle^\Lambda$
(1,2)	$\xi^1 \wedge \xi^2$	$-\xi^3$	(3)	1	-1
(1,3)	$\xi^1 \wedge \xi^3$	$-\xi^2$	(2)	-1	1
(2,3)	$\xi^2 \wedge \xi^3$	ξ^1	(1)	1	1

$\Diamond$

Example 8.5.6 (Hodge Star on $\mathbb{R}^4_1$). Let $\mathcal{E}$ be the standard basis for $\mathbb{R}^4_1$, and let $(\xi^1, \xi^2, \xi^3, \xi^4)$ be its dual basis. Using Theorem 7.5.2 and Theorem 8.5.2, we obtain the following tables:

I	ξ^I	$\star(\xi^I)$	I^{c}	$\text{sgn}(\sigma_{(I,I^{\text{c}})})$	$\langle \xi^{I^{\text{c}}}, \xi^{I^{\text{c}}} \rangle^\Lambda$
(1)	ξ^1	$-\xi^2 \wedge \xi^3 \wedge \xi^4$	(2,3,4)	1	-1
(2)	ξ^2	$\xi^1 \wedge \xi^3 \wedge \xi^4$	(1,3,4)	-1	-1
(3)	ξ^3	$-\xi^1 \wedge \xi^2 \wedge \xi^4$	(1,2,4)	1	-1
(4)	ξ^4	$-\xi^1 \wedge \xi^2 \wedge \xi^3$	(1,2,3)	-1	1

I	ξ^I	$\star(\xi^I)$	I^{c}	$\text{sgn}(\sigma_{(I,I^{\text{c}})})$	$\langle \xi^{I^{\text{c}}}, \xi^{I^{\text{c}}} \rangle^\Lambda$
(1,2)	$\xi^1 \wedge \xi^2$	$-\xi^3 \wedge \xi^4$	(3,4)	1	-1
(1,3)	$\xi^1 \wedge \xi^3$	$\xi^2 \wedge \xi^4$	(2,4)	-1	-1
(1,4)	$\xi^1 \wedge \xi^4$	$\xi^2 \wedge \xi^3$	(2,3)	1	1
(2,3)	$\xi^2 \wedge \xi^3$	$-\xi^1 \wedge \xi^4$	(1,4)	1	-1
(2,4)	$\xi^2 \wedge \xi^4$	$-\xi^1 \wedge \xi^3$	(1,3)	-1	1
(3,4)	$\xi^3 \wedge \xi^4$	$\xi^1 \wedge \xi^2$	(1,2)	1	1

I	ξ^I	$\star(\xi^I)$	I^c	$\mathrm{sgn}(\sigma_{(I,I^c)})$	$\langle \xi^{I^c}, \xi^{I^c} \rangle^\Lambda$
$(1,2,3)$	$\xi^1 \wedge \xi^2 \wedge \xi^3$	$-\xi^4$	(4)	1	-1
$(1,2,4)$	$\xi^1 \wedge \xi^2 \wedge \xi^4$	$-\xi^3$	(3)	-1	1
$(1,3,4)$	$\xi^1 \wedge \xi^3 \wedge \xi^4$	ξ^2	(2)	1	1
$(2,3,4)$	$\xi^2 \wedge \xi^3 \wedge \xi^4$	$-\xi^1$	(1)	-1	1

$\Diamond$

Theorem 8.5.7. *Let* $(V, \mathfrak{g}, \mathfrak{O})$ *be an oriented scalar product space of index* ν, *let* Ω_V *be its volume multicovector, and let* w *be a vector in* V. *Then*

$$i_w(\Omega_V) = (-1)^\nu \star(w^\mathsf{F}).$$

Proof. Let V have signature $(\varepsilon_1, \ldots, \varepsilon_m)$, let $(e_1, \ldots, e_m)$ be an orthonormal basis for V that is positively oriented with respect to $\mathfrak{O}$, and let $(\xi^1, \ldots, \xi^m)$ be its dual basis. Since interior multiplication, Hodge star, and the flat map are all linear, it suffices to show that

$$i_{e_j}(\Omega_V) = (-1)^\nu \star(e_j^\mathsf{F})$$

for $j = 1, \ldots, m$. By Theorem 8.3.4, $\Omega_V = \xi^1 \wedge \cdots \wedge \xi^m$, so Theorem 7.4.2 gives

$$i_{e_j}(\Omega_V) = \sum_k (-1)^{k-1} \xi^k(e_j) \xi^1 \wedge \cdots \wedge \widehat{\xi^k} \wedge \cdots \wedge \xi^m \tag{8.5.4}$$
$$= (-1)^{j-1} \xi^1 \wedge \cdots \wedge \widehat{\xi^j} \wedge \cdots \wedge \xi^m,$$

where $\widehat{\ }$ indicates that an expression is omitted. Since $\xi^{(j)^c} = \xi^1 \wedge \cdots \wedge \widehat{\xi^j} \wedge \cdots \wedge \xi^m$ and $((j),(j)^c) = (j, 1, \ldots, j-1, j+1, \ldots, m)$, it is easily shown that

$$\varepsilon_j \langle \xi^{(j)^c}, \xi^{(j)^c} \rangle^\Lambda = (-1)^\nu \quad \text{and} \quad \mathrm{sgn}(\sigma_{((j),(j)^c)}) = (-1)^{j-1}.$$

Then

$$\star(e_j^\mathsf{F}) = \varepsilon_j \star(\xi^j) \qquad\qquad\qquad\qquad \text{[Th 4.5.5(a)]}$$
$$= \varepsilon_j \, \mathrm{sgn}(\sigma_{((j),(j)^c)}) \langle \xi^{(j)^c}, \xi^{(j)^c} \rangle^\Lambda \xi^{(j)^c} \quad \text{[Th 8.5.2]} \tag{8.5.5}$$
$$= (-1)^{j-1+\nu} \xi^1 \wedge \cdots \wedge \widehat{\xi^j} \wedge \cdots \wedge \xi^m.$$

The result now follows from (8.5.4) and (8.5.5). $\qquad\qquad\qquad\qquad\qquad \square$

Chapter 9

Topology

9.1 Topology

Having completed an overview of linear and multilinear algebra, we now turn our attention to topology. This is an extensive area of mathematics, and of necessity the coverage presented here is highly selective. To the uninitiated, topology can be dauntingly abstract. The best example of a topology (and the motivation for much of what follows) is the Euclidean topology on $\mathbb{R}^m$, covered briefly in Section 9.4 and preceded by preparatory material in Section 9.2 and Section 9.3. Readers new to topology might find it helpful to peruse these sections early on to get a glimpse of where the discussion below is heading.

A **topology** on a set X is a collection $\mathcal{T}$ of subsets of X such that:

[**T1**] $\varnothing$ and X are in $\mathcal{T}$.
[**T2**] The union of any subcollection of elements of $\mathcal{T}$ is in $\mathcal{T}$.
[**T3**] The intersection of any *finite* subcollection of elements of $\mathcal{T}$ is in $\mathcal{T}$.

The pair $(X, \mathcal{T})$ is referred as a **topological space** and each element of $\mathcal{T}$ is said to be an **open set in** $\mathcal{T}$ or simply **open in** $\mathcal{T}$. Each element of X is called a **point in** X. Any open set in $\mathcal{T}$ containing a given point x in X is said to be a **neighborhood of x in** X. We say that a subset K of X is a **closed set in** $\mathcal{T}$ or simply **closed in** $\mathcal{T}$ if $X \backslash K$ is an open set in X. It is often convenient to adopt the shorthand of referring to X as a topological space, with $\mathcal{T}$ understood from the context. Accordingly, if U is an open set in $\mathcal{T}$ and K is a closed set in $\mathcal{T}$, we say that U is an **open set in** X or simply **open in** X, and that K is a **closed set in** X or simply **closed in** X.

Theorem 9.1.1. *If X is a topological space, then the following are closed sets in X:*

(a) $\varnothing$ *and* $\mathbb{R}^m$.

Semi-Riemannian Geometry, First Edition. Stephen C. Newman.
© 2019 John Wiley & Sons, Inc. Published 2019 by John Wiley & Sons, Inc.

(b) *The intersection of any collection of closed sets in X.*
(c) *The union of any finite collection of closed sets in X.*

Proof. (a): Straightforward.

(b): Let $\{K_\alpha : \alpha \in A\}$ be a collection of closed sets in X, and let $K = \bigcap_{\alpha \in A} K_\alpha$. Since each $X \backslash K_\alpha$ is open in X and, by Theorem A.1(d),

$$X \backslash K = \bigcup_{\alpha \in A} (X \backslash K_\alpha),$$

it follows that $X \backslash K$ is open in X. Thus, $K = X \backslash (X \backslash K)$ is closed in X.

(c): Let $\{K_i : i = 1, \ldots, m\}$ be a finite collection of closed sets in X, and let $K = \bigcup_{i=1}^{n} K_i$. Since each $X \backslash K_i$ is open in X and, by Theorem A.1(c),

$$X \backslash K = \bigcap_{i=1}^{m} (X \backslash K_i),$$

it follows that $X \backslash K$ is open in X. Thus, $K = X \backslash (X \backslash K)$ is closed in X. $\square$

Let X be a topological space, and let S be a subset of X. The **interior of S in X** is denoted by $\text{int}_X(S)$ and defined to be the union of all open sets in X contained in S. The **exterior of S in X** is denoted by $\text{ext}_X(S)$ and defined to be the union of all open sets in X contained in $X \backslash S$. The **boundary of S in X** is denoted by $\text{bd}_X(S)$ and defined to be the set of all points in X that are neither in $\text{int}_X(S)$ nor in $\text{ext}_X(S)$. Thus, X is the disjoint union

$$X = \text{int}_X(S) \cup \text{ext}_X(S) \cup \text{bd}_X(S).$$

The **closure of S in X** is denoted by $\text{cl}_X(S)$ and defined to be the intersection of all closed sets in X containing S.

Here are some of the basic facts about interiors, exteriors, boundaries, and closures.

Theorem 9.1.2. *If X is a topological space and S is a subset of X, then:*
(a) $\text{int}_X(S)$ *is the largest open set in X contained in S.*
(b) $\text{ext}_X(S)$ *is the largest open set in X contained in $X \backslash S$.*
(c) $\text{cl}_X(S)$ *is the smallest closed set in X containing S.*
(d) $\text{bd}_X(S)$ *is a closed set in X.*
(e) *S is closed in X if and only if $S = \text{cl}_X(S)$.*
(f) $\text{bd}_X(S)$ *is the set of points x in X such that every neighborhood of x intersects S and $X \backslash S$.*
(g) $\text{cl}_X(S)$ *is the set of points x in X such that every neighborhood of x intersects S.*

Proof. (a)–(c): Straightforward.

(d): Since $\text{int}_X(S)$ and $\text{ext}_X(S)$ are open in X,

$$\text{bd}_X(S) = X \backslash [\text{int}_X(S) \cup \text{ext}_X(S)]$$

is closed in X.

(e)($\Rightarrow$): Since S is closed in X, the intersection of all closed sets in X containing S is simply S.

(e)($\Leftarrow$): This follows from part (c).

(f): We prove the logically equivalent assertion:

(f') A point x in X is not in $\text{bd}_X(S)$ if and only if it has a neighborhood in M that does not intersect either S or $X\backslash S$.

(f')($\Rightarrow$): Since x is not in $\text{bd}_X(S)$, it is in either $\text{int}_X(S)$ or $\text{ext}_X(S)$. If x is in $\text{int}_X(S)$, then it is contained in an open set U in M that is a subset of S. Thus, U is a neighborhood of x that does not intersect $X\backslash S$. The argument when x is in $\text{ext}_X(S)$ is similar.

(f')($\Leftarrow$): Let U be a neighborhood of x in X. If U does not intersect S, then U is an open set in $X\backslash S$, hence $U \subseteq \text{ext}_X(S)$. Thus, x is in $\text{ext}_X(S)$, which is disjoint from $\text{bd}_X(S)$. The argument when U does not intersect $X\backslash S$ is similar.

(g): The proof is similar to that of part (f). $\qquad\qquad\qquad\qquad\square$

Rather than having to deal with all open sets in X, it is often convenient to work with a smaller collection that contains the essential information on "openness". A **basis for X** is a collection $\mathcal{B}$ of subsets of X, each of which is called a **basis element**, such that:

[B1] $X = \bigcup_{B \in \mathcal{B}} B$.

[B2] If B_1, B_2 are basis elements in $\mathcal{B}$ and x is a point in $B_1 \cap B_2$, then there is a basis element B in $\mathcal{B}$ containing x such that $B \subseteq B_1 \cap B_2$.

Theorem 9.1.3. *If X is a set and $\mathcal{B}$ is a basis for X, then the collection $\mathcal{T}_{\mathcal{B}}$ consisting of all unions of basis elements in $\mathcal{B}$ is a topology on X, called the* **topology generated by $\mathcal{B}$**.

Proof. We need to show that [T1]–[T3] are satisfied. It is vacuously true that $\varnothing$ is in $\mathcal{T}_{\mathcal{B}}$ and, by [B1], so is X. Thus, [T1] is satisfied. Evidently, [T2] is satisfied. For [T3], let B_1, B_2 be basis elements in $\mathcal{B}$. We claim that $B_1 \cap B_2$ is in $\mathcal{T}_{\mathcal{B}}$. This is clearly the case if $B_1 \cap B_2 = \varnothing$, so suppose otherwise. By [B2], for each x in $B_1 \cap B_2$, there is a basis element B_x in $\mathcal{B}$ such that x is in B_x and $B_x \subseteq B_1 \cap B_2$. It follows that $B_1 \cap B_2 = \bigcup_{x \in B_1 \cap B_2} B_x$, hence $B_1 \cap B_2$ is in $\mathcal{T}_{\mathcal{B}}$. This proves the claim. Let U_1 and U_2 be elements of $\mathcal{T}_{\mathcal{B}}$, with $U_1 = \bigcup_{\alpha_1 \in A_1} B_{\alpha_1}$ and $U_2 = \bigcup_{\alpha_2 \in A_2} B_{\alpha_2}$. By Theorem A.1(a),

$$U_1 \cap U_2 = \left(\bigcup_{\alpha_1 \in A_1} B_{\alpha_1} \right) \cap \left(\bigcup_{\alpha_2 \in A_2} B_{\alpha_2} \right) = \bigcup_{\alpha_1 \in A_1,\, \alpha_2 \in A_2} (B_{\alpha_1} \cap B_{\alpha_2}).$$

Since each $B_{\alpha_1} \cap B_{\alpha_2}$ is in $\mathcal{T}_{\mathcal{B}}$, so is $U_1 \cap U_2$. Using an inductive argument, it is easily shown that the intersection of any finite subcollection of elements of $\mathcal{T}_{\mathcal{B}}$ is in $\mathcal{T}_{\mathcal{B}}$. Thus, [T3] is satisfied. $\qquad\qquad\qquad\qquad\square$

Theorem 9.1.4 (Subspace Topology). *If $(X, \mathcal{T})$ is a topological space and S is a subset of X, then*

$$\mathcal{T}_S = \{U \cap S : U \in \mathcal{T}\}$$

*is a topology on S, called the **subspace topology (induced by $\mathcal{T}$)**.*

Proof. That [T1] is satisfied is obvious, and that [T2] and [T3] are satisfied is easily shown using Theorem A.1.   □

In the notation of Theorem 9.1.4, we say that $(S, \mathcal{T}_S)$ is a **topological subspace** of $(X, \mathcal{T})$ or simply that S is a topological subspace of X.

Throughout, any subset of a topological space is viewed as a topological subspace.

Theorem 9.1.5. *Let X be a topological space, let U be an open set in X, let S be a subset of U, and view U as a topological subspace of X. Then S is open in U if and only if S is open in X.*

Proof. ($\Rightarrow$): Since S is open in U, by definition, there is an open set V in X such that $S = V \cap U$. Because V and U are open in X, so is S.

($\Leftarrow$): Since S is open in X, by definition, $S = S \cap U$ is open in U.   □

Theorem 9.1.6 (Product Topology). *If $(X_1, \mathcal{T}_1), \ldots, (X_m, \mathcal{T}_m)$ are topological spaces, then*

$$\mathcal{B}_\times = \{U_1 \times \cdots \times U_m : U_i \in \mathcal{T}_i \text{ for } i = 1, \ldots, m\}$$

*is a basis for $X_1 \times \cdots \times X_m$, and the topology it generates is called the **product topology**.*

Proof. It is clear that $X_1 \times \cdots \times X_m$ is in $\mathcal{B}_\times$, so [B1] is satisfied. Let $U_1 \times \cdots \times U_m$ and $V_1 \times \cdots \times V_m$ be sets in $\mathcal{B}_\times$. By Theorem A.2,

$$(U_1 \times \cdots \times U_m) \cap (V_1 \times \cdots \times V_m) = (U_1 \cap V_1) \times \cdots \times (U_m \cap V_m).$$

Since $U_i \cap V_i$ is in $\mathcal{T}_i$ for $i = 1, \ldots, m$, it follows that $(U_1 \times \cdots \times U_m) \cap (V_1 \times \cdots \times V_m)$ is in $\mathcal{B}_\times$, so [B2] is satisfied.   □

Let X and Y be topological spaces, let $F : X \longrightarrow Y$ be a map, and let x be a point in X. We say that F is **continuous at x** if for every neighborhood V of $F(x)$ in Y, there is a neighborhood U of x in X (possibly depending on V) such that $F(U) \subseteq V$. Equivalently, F is continuous at x if for every neighborhood V of $F(x)$ in Y, $F^{-1}(V)$ contains a neighborhood of x in X. We say that F is **continuous on X** or simply **continuous** if it is continuous at every x in X.

Theorem 9.1.7. *Let X and Y be topological spaces, and let $F : X \longrightarrow Y$ be a map. Then the following are equivalent:*
(a) *F is continuous.*
(b) *If V is an open set in Y, then $F^{-1}(V)$ is an open set in X.*
(c) *If K is a closed set in Y, then $F^{-1}(K)$ is a closed set in X.*

Proof. (a) ⇒ (b): Let x be a point in $F^{-1}(V)$. Since F is continuous at x and V is a neighborhood of $F(x)$ in Y, $F^{-1}(V)$ contains a neighborhood U_x of x in X. Then $F^{-1}(V) = \bigcup_{x \in U} U_x$ is open in X.

(b) ⇒ (a): Let x be a point in X, and let V be a neighborhood of $F(x)$ in Y. Since V is open in Y, $F^{-1}(V)$ is open in X and is therefore is a neighborhood of x in X.

(b) ⇒ (c): We have

$$K \text{ is closed in } Y$$
$$\Rightarrow \quad Y \backslash K \text{ is open in } Y$$
$$\Rightarrow \quad F^{-1}(Y \backslash K) = X \backslash F^{-1}(K) \text{ is open in } X$$
$$\Rightarrow \quad X \backslash [X \backslash F^{-1}(K)] = F^{-1}(K) \text{ is closed in } X,$$

where the second implication follows from Theorem A.3(e).

(c) ⇒ (b): We have

$$V \text{ is open in } Y$$
$$\Rightarrow \quad Y \backslash V \text{ is closed in } Y$$
$$\Rightarrow \quad F^{-1}(Y \backslash V) = X \backslash F^{-1}(V) \text{ is closed in } X$$
$$\Rightarrow \quad X \backslash [X \backslash F^{-1}(V)] = F^{-1}(V) \text{ is open in } X,$$

where the second implication follows from Theorem A.3(e). $\square$

Theorem 9.1.8. *Let X, Y, and Z be topological spaces, and let $F : X \longrightarrow Y$ and $G : Y \longrightarrow Z$ be maps. If F and G are continuous, then so is $G \circ F$.*

Proof. Let V be an open set in Z. Since G is continuous, by Theorem 9.1.7, $G^{-1}(V)$ is open in Y, and because F is continuous, again by Theorem 9.1.7, $F^{-1}\big(G^{-1}(V)\big) = (G \circ F)^{-1}(V)$ is open in X. The result now follows from Theorem 9.1.7. $\square$

Theorem 9.1.9. *Let X and Y be topological spaces, let $F : X \longrightarrow Y$ be a map, and let $\mathcal{B}$ be a basis that generates the topology on Y. If $F^{-1}(B)$ is open in X for every basis element B in $\mathcal{B}$, then F is continuous.*

Proof. Let V be an open set in Y. According to Theorem 9.1.3, V can be expressed as a union $V = \bigcup_{\alpha \in A} B_\alpha$ of basis elements, and then Theorem A.3(b) gives $F^{-1}(V) = \bigcup_{\alpha \in A} F^{-1}(B_\alpha)$. By assumption, each $F^{-1}(B_\alpha)$ is open in X, and therefore, so is $F^{-1}(V)$. The result now follows from Theorem 9.1.7. $\square$

Theorem 9.1.10. *Let $X_1, \ldots, X_m$ be topological spaces, and suppose $X_1 \times \cdots \times X_m$ has the product topology. Define a map $\mathcal{P}_i : X_1 \times \cdots \times X_m \longrightarrow X_i$, called the **ith projection map** on $X_1 \times \cdots \times X_m$, by*

$$\mathcal{P}_i(x_1, \ldots, x_i, \ldots, x_m) = x_i$$

for all $(x_1, \ldots, x_m)$ in $X_1 \times \cdots \times X_m$ for $i = 1, \ldots, m$. Then each $\mathcal{P}_i$ is continuous.

Proof. Let U_i be an open set in X_i. Since

$$\mathcal{P}_i^{-1}(U_i) = X_1 \times \cdots \times U_i \times \cdots \times X_m$$

is in the basis $\mathcal{B}_\times$ for the product topology on $X_1 \times \cdots \times X_m$, it is open in $X_1 \times \cdots \times X_m$. It follows from Theorem 9.1.7 that $\mathcal{P}_i$ is continuous. $\qquad\square$

Theorem 9.1.11. *Let $X, Y_1, \ldots, Y_m$ be topological spaces, let $F = (F_1, \ldots, F_m) : X \longrightarrow Y_1 \times \cdots \times Y_m$ be a map, and suppose $Y_1 \times \cdots \times Y_m$ has the product topology. Then F is continuous if and only if $F_i : X \longrightarrow Y_i$ is continuous for $i = 1, \ldots, m$.*

Proof. ($\Rightarrow$): Let $\mathcal{P}_i$ be the ith projection map on $Y_1 \times \cdots \times Y_m$. We have from Theorem 9.1.10 that $\mathcal{P}_i$ is continuous, and then from Theorem 9.1.8 that so is $F_i = \mathcal{P}_i \circ F$.

($\Leftarrow$): Let $V_1 \times \cdots \times V_m$ be in the basis $\mathcal{B}_\times$ for the product topology on $Y_1 \times \cdots \times Y_m$. Then Theorem A.5 gives

$$F^{-1}(V_1 \times \cdots \times V_m) = \bigcap_{i=1}^{m} F_i^{-1}(V_i).$$

Since V_i is open in Y_i and F_i is continuous, by Theorem 9.1.7, $F_i^{-1}(V_i)$ is open in X for $i = 1, \ldots, m$. It follows that $F^{-1}(V_1 \times \cdots \times V_m)$ is open in X. By Theorem 9.1.9, F is continuous. $\qquad\square$

Theorem 9.1.12. *Let X_i and Y_i be topological spaces, let $F_i : X_i \longrightarrow Y_i$ be a continuous map for $i = 1, \ldots, m$, and suppose $X_1 \times \cdots \times X_m$ and $Y_1 \times \cdots \times Y_m$ have the respective product topologies. Then the map*

$$F_1 \times \cdots \times F_m : X_1 \times \cdots \times X_m \longrightarrow Y_1 \times \cdots \times Y_m$$

defined by

$$F_1 \times \cdots \times F_m(x_1, \ldots, x_m) = \big(F_1(x_1), \ldots, F_m(x_m)\big)$$

for all $(x_1, \ldots, x_m)$ in $X_1 \times \cdots \times X_m$ is continuous.

Proof. Let $\mathcal{P}_i$ be the ith projection map on $X_1 \times \cdots \times X_m$, and consider the map

$$F_i \circ \mathcal{P}_i : X_1 \times \cdots \times X_m \longrightarrow Y_i$$

for $i = 1, \ldots, m$. Then

$$F_1 \times \cdots \times F_m = (F_1 \circ \mathcal{P}_1, \ldots, F_m \circ \mathcal{P}_m).$$

We have from Theorem 9.1.10 that $\mathcal{P}_i$ is continuous, and then from Theorem 9.1.8 that $F_i \circ \mathcal{P}_i$ is continuous for $i = 1, \ldots, m$. The result now follows from Theorem 9.1.11. $\qquad\square$

Let X and Y be topological spaces, let $F : X \longrightarrow Y$ be a map, and let S be a subset (topological subspace) of X. We say that F is **continuous on S** if the restriction map $F|_S : S \longrightarrow Y$ is continuous on S.

Theorem 9.1.13. *With the above setup, if F is continuous, then so is $F|_S$.*

Proof. Let V be an open set in Y. Since F is continuous, by Theorem 9.1.7, $F^{-1}(V)$ is open in X, hence $F^{-1}(V) \cap S = (F|_S)^{-1}(V)$ is open in S. The result now follows from Theorem 9.1.7. $\qquad\square$

Let X and Y be topological spaces, and let $F : X \longrightarrow Y$ be a continuous map. We say that F is a **homeomorphism**, and that X and Y are **homeomorphic**, if F is bijective and F^{-1} is continuous. The next result shows that a homeomorphism is a map that preserves topological structure (in much the same way that a linear isomorphism preserves linear structure).

Theorem 9.1.14. *Let X and Y be topological spaces, and let $F : X \longrightarrow Y$ be a bijective map. Then the following are equivalent:*
(a) *F is homeomorphism.*
(b) *U is an open set in X if and only if $F(U)$ is an open set in Y.*
(c) *K is a closed set in X if and only if $F(K)$ is a closed set in Y.*

Proof. This follows from Theorem 9.1.7. $\qquad\square$

We say that a topological space X is **disconnected** if there are disjoint nonempty open sets U and V in X such that $X = U \cup V$. In that case, U and V are said to **disconnect** X. If X is not disconnected, we say it is **connected**. A subset S of X is said to be **disconnected (connected) in X** or simply **disconnected (connected)** if it is disconnected (connected) as a topological subspace of X. We say that a subset C of X is a **connected component** of X if it is a connected set in X that is maximal, in the sense that C is not properly contained in any other connected set in X.

The next result gives equivalent conditions for a subset of a topological space to be disconnected in the topological space.

Theorem 9.1.15. *Let X be a topological space, and let S be a subset of X. Then:*
(a) *S is disconnected in X if and only if there are nonempty open sets U and V in X such that (i) $(U \cap V) \cap S = \varnothing$, (ii) $S \subseteq (U \cup V)$, and (iii) $U \cap S \neq \varnothing$ and $V \cap S \neq \varnothing$.*
(b) *S is connected in X if and only if for all nonempty open sets U and V in X such that $(U \cap V) \cap S = \varnothing$ and $S \subseteq (U \cup V)$, either $S \subseteq U$ or $S \subseteq V$.*

Proof. Let U and V be open sets in X. Then

$$(U \cap V) \cap S = \varnothing \quad \Leftrightarrow \quad (U \cap S) \cap (V \cap S) = \varnothing, \qquad (9.1.1)$$

and by Theorem A.1(a),

$$S \subseteq (U \cup V) \quad \Leftrightarrow \quad S = (U \cup V) \cap S = (U \cap S) \cup (V \cap S). \qquad (9.1.2)$$

(a): By definition, S is disconnected if and only if there are open sets U' and V' in S such that (i) $U' \cap V' = \varnothing$, (ii) $S = U' \cup V'$, and (iii) $U' \neq \varnothing$ and $V' \neq \varnothing$. Since any such U' and V' are of the form $U' = U \cap S$ and $V' = V \cap S$ for some open sets U and V in X, the result follows from (9.1.1) and (9.1.2).

(b): We have from part (a) that S is connected in X if and only if for all nonempty open sets U and V in X such that $(U \cap V) \cap S = \varnothing$ and $S \subseteq (U \cup V)$, either $U \cap S = \varnothing$ or $V \cap S = \varnothing$, which in turn, from (9.1.1) and (9.1.2), is equivalent to the statement of part (b). $\qquad\square$

Theorem 9.1.16. *Let X be a topological space, and let $\{S_\alpha : \alpha \in A\}$ be a collection of subsets of X that are connected in X. If $\bigcap_{\alpha \in A} S_\alpha$ is nonempty, then $\bigcup_{\alpha \in A} S_\alpha$ is connected in X.*

Proof. Let $S = \bigcup_{\alpha \in A} S_\alpha$, and let U and V be nonempty open sets in X such that $(U \cap V) \cap S = \varnothing$ and $S \subseteq (U \cup V)$. Then $(U \cap V) \cap S_\alpha = \varnothing$ and $S_\alpha \subseteq (U \cup V)$ for all α in A. Since each S_α is connected in X, we have from Theorem 9.1.15(b) that either $S_\alpha \subseteq U$ or $S_\alpha \subseteq V$ for all α in A. By assumption, there is a point x in $\bigcap_{\alpha \in A} S_\alpha \subseteq (U \cup V)$. Suppose without loss of generality that x is in U, hence $U \cap S_\alpha \neq \varnothing$ for all α in A. If $S_\alpha \subseteq V$ for some α, then $(U \cap V) \cap S_\alpha = \varnothing$ simplifies to $U \cap S_\alpha = \varnothing$, which is a contradiction. It follows that $S_\alpha \subseteq U$ for all α in A, hence $S \subseteq U$. Again by Theorem 9.1.15(b), S is connected in X. $\quad\square$

Theorem 9.1.17. *The distinct connected components of a topological space form a partition of the topological space.*

Proof. Let X be a topological space, and let x be a point in X. Clearly, $\{x\}$ is connected in X. Let C be the union of all connected subsets of X containing x. By Theorem 9.1.16, C is connected in X, and is evidently a connected component of X. Thus, X has one or more connected components, the union of which equals X. It remains to show that distinct connected components are disjoint. We prove the logically equivalent assertion: if connected components are not disjoint, then they are equal. Suppose C and D are connected components that are not disjoint. By Theorem 9.1.16, $C \cup D$ is connected in X. It follows from the maximality property that $C = C \cup D = D$. $\qquad\square$

Theorem 9.1.18. *Let X and Y be topological spaces, and let $F : X \longrightarrow Y$ be a continuous map. If X is connected, then $F(X)$ is connected in Y.*

Proof. We prove the logically equivalent assertion: if $F(X)$ is disconnected in Y, then X is disconnected. By Theorem 9.1.15(a), there are nonempty open sets U and V in Y such that (i) $(U \cap V) \cap F(X) = \varnothing$, (ii) $F(X) \subseteq (U \cup V)$, and (iii) $U \cap F(X) \neq \varnothing$ and $V \cap F(X) \neq \varnothing$. It follows from Theorem 9.1.7 and parts (b) and (c) of Theorem A.3 that $F^{-1}(U)$ and $F^{-1}(V)$ are disjoint nonempty open sets in X such that $X = F^{-1}(U) \cup F^{-1}(V)$. Thus, X is disconnected. $\quad\square$

Theorem 9.1.19 (Intermediate Value Theorem). *Let X be a connected topological space, and let $f : X \longrightarrow \mathbb{R}$ be a continuous function. If x_1, x_2 are points in X and c is a real number such that $f(x_1) < c < f(x_2)$, then there is a point x_0 in X such that $f(x_0) = c$.*

Proof. Consider the sets $S_1 = f(X) \cap (-\infty, c)$ and $S_2 = f(X) \cap (c, +\infty)$. It is clear that S_1 and S_2 are disjoint and nonempty. Since $(-\infty, c)$ and $(c, +\infty)$ are open in $\mathbb{R}$, S_1 and S_2 are open in $f(X)$. Suppose there is no point x_0 in X such that $f(x_0) = c$. Then $f(X) = S_1 \cup S_2$. Thus, $f(X)$ is disconnected in $\mathbb{R}$, which contradicts Theorem 9.1.18. $\square$

Theorem 9.1.20. *If X is a connected topological space and $f : X \longrightarrow \mathbb{R}$ is a nowhere-vanishing continuous function, then f is either strictly positive or strictly negative.*

Proof. Let x_1 and x_2 be distinct points in X. Since f is nowhere-vanishing, $f(x_1), f(x_2) \neq 0$. If $f(x_1) > 0$, then $f(x_2) > 0$; for if not, by Theorem 9.1.19, there is a point x_0 in X such that $f(x_0) = 0$, which contradicts the nowhere-vanishing assumption on f. Similarly, if $f(x_1) < 0$, then $f(x_2) < 0$. Since x_1 and x_2 were arbitrary, the result follows. $\square$

Let X be a topological space, and let $\mathcal{U} = \{U_\alpha : \alpha \in A\}$ be a collection of open sets in X. We say that $\mathcal{U}$ is an **open cover of X** if $X = \bigcup_{\alpha \in A} U_\alpha$. A subcollection of $\mathcal{U}$ that is also an open cover of X is said to be a **subcover of $\mathcal{U}$**. A subcover that is a finite set is called a **finite subcover**. We say that X is a **compact** topological space if every open cover of X has a finite subcover. A subset S of X is said to be a **compact set in X** or simply **compact in X** if it is compact as a topological subspace.

Theorem 9.1.21. *Let X and Y be topological spaces, and let $F : X \longrightarrow Y$ be a continuous map. If X is compact, then $F(X)$ is compact in Y.*

Proof. Let $\mathcal{V}$ be an open cover of $F(X)$. Then $\mathcal{V} = \{V_\alpha \cap F(X) : \alpha \in A\}$, where each V_α is an open set in Y. By Theorem A.1(a),

$$F(X) = \left(\bigcup_{\alpha \in A} V_\alpha \right) \cap F(X) = \bigcup_{\alpha \in A} [V_\alpha \cap F(X)],$$

and by Theorem A.3(b),

$$X = F^{-1}(F(X)) = \bigcup_{\alpha \in A} F^{-1}(V_\alpha \cap F(X)) = \bigcup_{\alpha \in A} F^{-1}(V_\alpha).$$

It follows from Theorem 9.1.7 that $\{F^{-1}(V_\alpha) : \alpha \in A\}$ is an open cover of X. Since X is compact, there is a finite subcover $\{F^{-1}(V_i) : i = 1, \ldots, k\}$, so

$$X = \bigcup_i F^{-1}(V_i).$$

By parts (a) and (d) of Theorem A.3,

$$F(X) = \bigcup_i F(F^{-1}(V_i)) = \bigcup_i [V_i \cap F(X)].$$

Thus, $\{V_i \cap F(X) : i = 1, \ldots, k\}$ is a finite subcover of $\mathcal{V}$. $\square$

Let X be a topological space, and let $f : X \longrightarrow \mathbb{R}$ be a function. We say that f is **bounded on X** or simply **bounded** if there is a real number $c > 0$ such that $|f(x)| < c$ for all points x in X. Let S be a nonempty subset of $\mathbb{R}$. The element s_1 in S is said to be the **smallest element** in S if $s_1 \leq s$ for all s in S. Similarly, the element s_2 in S is said to be the **largest element** in S if $s \leq s_2$ for all s in S. If S has a smallest element s_1 and a largest element s_2, then $s_1 \leq s \leq s_2$ for all s in S. Consider the collection $\mathcal{U} = \{(-\infty, s) \cap S : s \in S\}$ of open sets in S. If S has a largest element s_2, then s_2 is not in $\bigcup_{s \in S}(-\infty, s)$, so $\mathcal{U}$ is not an open cover of S in $\mathbb{R}$. On the other hand, if S does not have a largest element, then $\mathcal{U}$ is an open cover of S in $\mathbb{R}$.

Theorem 9.1.22 (Extreme Value Theorem). *If X is a compact topological space and $f : X \longrightarrow \mathbb{R}$ is a continuous function, then:*
(a) *There are points x_1 and x_2 in X such that $f(x_1) \leq f(x) \leq f(x_2)$ for all x in X.*
(b) *f is bounded.*

Proof. (a): The assertion is equivalent to: There are points x_1 and x_2 in X such that $f(x_1)$ is the smallest element in $f(X)$, and $f(x_2)$ is the largest element in $f(X)$. Suppose, for a contradiction, that $f(X)$ does not have a largest element. In light of above remarks, $\{(-\infty, f(x)) \cap f(X) : x \in X\}$ is an open cover of $f(X)$. Since X is compact and f is continuous, by Theorem 9.1.21, $f(X)$ is compact in $\mathbb{R}$. It follows that there is a finite subcover

$$\{(-\infty, f(x_i)) \cap f(X) : i = 1, \ldots, m\},$$

where we assume without loss of generality that $f(x_1) < \cdots < f(x_m)$. By Theorem A.1(a),

$$f(X) = \bigcup_{i=1}^{m} [(-\infty, f(x_i)) \cap f(X)] = \left[\bigcup_{i=1}^{m}(-\infty, f(x_i))\right] \cap f(X)$$
$$= (-\infty, f(x_m)) \cap f(X),$$

hence $f(X) \subseteq (-\infty, f(x_m))$. Since $f(x_m)$ is in $f(X)$ but not in $(-\infty, f(x_m))$, we have a contradiction. Thus, $f(X)$ has a largest element. The proof that $f(X)$ has a smallest element is similar.
  (b): This follows from part (a). $\square$

Let X be a topological space, and let $f : X \longrightarrow \mathbb{R}$ be a function. The **support of f** is denoted by $\mathrm{supp}(f)$ and defined to be the closure in X of the set of points at which f is nonvanishing:

$$\mathrm{supp}(f) = \mathrm{cl}_X(\{x \in X : f(x) \neq 0\}).$$

Thus, $\mathrm{supp}(f)$ is the smallest closed set in X containing those points at which f is nonvanishing. We say that f has **compact support** if $\mathrm{supp}(f)$ is compact in X. Given a subset S of X, we say that f has **support in S** if $\mathrm{supp}(f) \subseteq S$.

9.2   Metric Spaces

Let X be a nonempty set X. A function

$$\mathsf{d} : X \times X \longrightarrow \mathbb{R}$$

is said to be a **distance function** on X if for all x, y, z in X:

[D1] $\mathsf{d}(x, y) \geq 0$, with $\mathsf{d}(x, y) = 0$ if and only if $x = y$.
[D2] $\mathsf{d}(x, y) = \mathsf{d}(y, x)$.
[D3] $\mathsf{d}(x, y) \leq \mathsf{d}(x, z) + \mathsf{d}(z, y)$.   (triangle inequality)

A **metric space** is a pair (X, d) consisting of a nonempty set X and a distance function d on X. For a given point x in X and real number $r > 0$, the **open ball** of radius r centered at x is defined by

$$B_r(x) = \{y \in X : \mathsf{d}(x, y) < r\},$$

and the **closed ball** of radius r centered at x by

$$\overline{B}_r(x) = \{y \in X : \mathsf{d}(x, y) \leq r\}.$$

Theorem 9.2.1. *With the above setup, if y is a point in $B_r(x)$, then there is a real number s such that $B_s(y) \subseteq B_r(x)$.*

Proof. Let $s = r - \mathsf{d}(x, y)$. For a point z in $B_s(y)$, we have $\mathsf{d}(y, z) < s$. Then [D3] gives

$$\mathsf{d}(x, z) \leq \mathsf{d}(x, y) + \mathsf{d}(y, z) < r,$$

hence z is in $B_r(x)$. Thus, $B_s(y) \subseteq B_r(x)$.   □

Theorem 9.2.2. *If (X, d) is a metric space, then*

$$\mathcal{B}_\mathsf{d} = \{B_r(x) : x \in X, r > 0\}$$

*is a basis for X, and the topology it generates is called the **metric topology induced by d**.*

Proof. We need to show that conditions [B1] and [B2] of Section 9.1 are satisfied. [B1] is trivial. For [B2], let $B_{r_1}(x_1), B_{r_2}(x_2)$ be elements of $\mathcal{B}_\mathsf{d}$, and let x be a point in $B_{r_1}(x_1) \cap B_{r_2}(x_2)$. By Theorem 9.2.1, there are real numbers $s_1, s_2 > 0$ such that $B_{s_1}(x) \subseteq B_{r_1}(x_1)$ and $B_{s_2}(x) \subseteq B_{r_2}(x_2)$. Setting $s = \min(s_1, s_2)$, we have $B_s(x) \subseteq B_{r_1}(x_1) \cap B_{r_2}(x_2)$.   □

The next result justifies our use of the terms "open" and "closed" to describe balls in a metric space.

Theorem 9.2.3. *If (X, d) is a metric space, then open (closed) balls in X are open (closed) with respect to the metric topology induced by d.*

Proof. For open balls, this is true by definition. Let $\overline{B}_r(x)$ be a closed ball in X, and let y be a point in $X\backslash\overline{B}_r(x)$, so that $\mathsf{d}(x,y) > r$. Let $s = \mathsf{d}(x,y) - r$, and let z be a point in $B_s(y)$, which means $\mathsf{d}(y,z) < s$. See Figure 9.2.1. From [D2] and [D3],

$$r + s = \mathsf{d}(x,y) \le \mathsf{d}(x,z) + \mathsf{d}(z,y) < \mathsf{d}(x,z) + s,$$

hence $\mathsf{d}(x,z) > r$, so z is in $X\backslash\overline{B}_r(x)$. Thus, $B_s(y) \subseteq X\backslash\overline{B}_r(x)$. It follows that $X\backslash\overline{B}_r(x)$ is the union of a collection of basis elements of $\mathcal{B}_\mathsf{d}$. By Theorem 9.1.3 and Theorem 9.2.2, $X\backslash\overline{B}_r(x)$ is open with respect to the metric topology induced by d, hence $\overline{B}_r(x) = X\backslash[X\backslash\overline{B}_r(x)]$ is closed with respect to that topology. $\qquad\square$

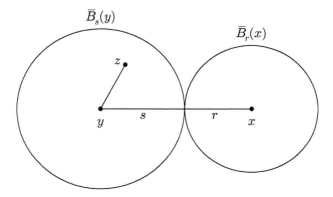

Figure 9.2.1. Diagram for Theorem 9.2.3

The next result expresses continuity in metric spaces in terms familiar from the differential calculus of one real variable.

Theorem 9.2.4 (ε-δ Criterion). *Let (X,d) and (Y,e) be metric spaces, each with its induced metric topology, let $F : X \longrightarrow Y$ be a map, and let x be a point in X. Then F is continuous at x if and only if for every real number $\varepsilon > 0$, there is a real number $\delta > 0$ (possibly depending on ε) such that:*

$$\mathsf{d}(x,y) < \delta \qquad \Rightarrow \qquad \mathsf{e}\big(F(x),F(y)\big) < \varepsilon,$$

or equivalently,

$$y \text{ is in } B_\delta(x) \qquad \Rightarrow \qquad F(y) \text{ is in } B_\varepsilon\big(F(x)\big),$$

or equivalently,

$$B_\delta(x) \subseteq F^{-1}\big(B_\varepsilon\big(F(x)\big)\big).$$

Proof. The equivalence of the three conditions is clear. Let us work with the last one.

($\Rightarrow$): Since F is continuous at x and $B_\varepsilon\big(F(x)\big)$ is a neighborhood of $F(x)$ in Y, by definition, there is a neighborhood U of x in X such that $F(U) \subseteq B_\varepsilon\big(F(x)\big)$. By Theorem 9.1.3 and Theorem 9.2.2, there is an open ball $B_\delta(x)$ such that

$$B_\delta(x) \subseteq U \subseteq F^{-1}\big(B_\varepsilon\big(F(x)\big)\big).$$

($\Leftarrow$): Let V be a neighborhood of $F(x)$ in Y. We have from Theorem 9.1.3 and Theorem 9.2.2 that there is an open ball $B_\varepsilon\big(F(x)\big)$ such that $B_\varepsilon\big(F(x)\big) \subseteq V$. By assumption, there is an open ball $B_\delta(x)$ with $B_\delta(x) \subseteq F^{-1}\big(B_\varepsilon\big(F(x)\big)\big)$. Thus, $B_\delta(x)$ is a neighborhood of x in X such that $F\big(B_\delta(x)\big) \subseteq V$. □

9.3 Normed Vector Spaces

Let V be a vector space. A function

$$\|\cdot\| : V \longrightarrow \mathbb{R}$$

is said to be a **norm** on V if for all vectors v, w in V and all real numbers c:
[**N1**] $\|v\| \geq 0$, with $\|v\| = 0$ if and only if $v = 0$.
[**N2**] $\|cv\| = |c|\,\|v\|$.
[**N3**] $\|v + w\| \leq \|v\| + \|w\|$.   (triangle equality)

A **normed vector space** is a pair $(V, \|\cdot\|)$ consisting of a vector space V and a norm $\|\cdot\|$ on V.

Theorem 9.3.1. *If $(V, \mathfrak{g})$ is an inner product space and $\|\cdot\|$ is the norm defined by $\|v\| = \sqrt{\langle v, v\rangle}$ for all v in V, then $(V, \|\cdot\|)$ is a normed vector space.*

Proof. Using Theorem 4.6.5 and the basic properties of inner products, it is easily shown that [N1]–[N3] are satisfied. □

Theorem 9.3.1 justifies our use of the term "norm" in Section 4.6 (and to a lesser extent its use in Section 4.1).

Theorem 9.3.2. *If $(V, \|\cdot\|)$ is a normed vector space and $\mathrm{d} : V \times V \longrightarrow \mathbb{R}$ is the function defined by $\mathrm{d}(v, w) = \|v - w\|$ for all vectors v, w in V, then (V, d) is a metric space.*

Proof. It is easily shown that conditions [D1]–[D3] of Section 9.2 are satisfied. □

9.4 Euclidean Topology on $\mathbb{R}^m$

Let us put the results of Sections 9.1–9.3 to work constructing a series of "spaces" starting with the inner product space $\mathbb{R}_0^m = (\mathbb{R}^m, \mathfrak{e})$, as defined in Example 4.6.1. We first apply Theorem 9.3.1 to $(\mathbb{R}^m, \mathfrak{e})$ and obtain a normed vector space $(\mathbb{R}^m, \|\cdot\|)$, where

$$\|\cdot\| : \mathbb{R}^m \times \mathbb{R}^m \longrightarrow \mathbb{R}$$

is the norm introduced in Example 4.6.1 and defined by

$$\|(x^1, \ldots, x^m)\| = \sqrt{(x^1)^2 + \cdots + (x^m)^2}.$$

We next apply Theorem 9.3.2 to $(\mathbb{R}^m, \|\cdot\|)$ and obtain a metric space $(\mathbb{R}^m, \mathfrak{d})$, where

$$\mathfrak{d} : \mathbb{R}^m \times \mathbb{R}^m \longrightarrow \mathbb{R}$$

is the distance function defined by

$$\mathfrak{d}\big((x^1, \ldots, x^m), (y^1, \ldots, y^m)\big) = \|(x^1, \ldots, x^m) - (y^1, \ldots, y^m)\|$$
$$= \sqrt{(x^1 - y^1)^2 + \cdots + (x^m - y^m)^2}.$$

Lastly, we apply Theorem 9.2.2 to $(\mathbb{R}^m, \mathfrak{d})$ and obtain a topological space $(\mathbb{R}^m, \mathfrak{T})$.

Recall from Example 4.6.1 that $\mathfrak{e}$ is called the Euclidean inner product. In a corresponding fashion, we refer to $\|\cdot\|$ as the **Euclidean norm** on $\mathbb{R}^m$, to $\mathfrak{d}$ as the **Euclidean distance function** on $\mathbb{R}^m$, and to $\mathfrak{T}$ as the **Euclidean topology** on $\mathbb{R}^m$. With the above constructions, there is now a direct path from $(\mathbb{R}^m, \mathfrak{e})$ to $(\mathbb{R}^m, \|\cdot\|)$ to $(\mathbb{R}^m, \mathfrak{d})$ to $(\mathbb{R}^m, \mathfrak{T})$. We generally simplify notation by using $\mathbb{R}^m$ to denoted any of the latter three spaces, allowing the context to make it clear whether $\mathbb{R}^m$ is being thought of as a normed vector space, a metric space, or a topological space. On occasion, we also adopt $\mathbb{R}^m$ as notation for $(\mathbb{R}^m, \mathfrak{e})$.

Let us discuss a few examples based on the Euclidean topologies on $\mathbb{R}$ and $\mathbb{R}^2$ to illustrate some of the material in Sections 9.1–9.3.

We first consider $\mathbb{R}$. An open (closed) "ball" in $\mathbb{R}$ is nothing other than a finite open (closed) interval. By Theorem 9.1.3 and Theorem 9.2.2, the collection of open intervals in $\mathbb{R}$ is a basis for the Euclidean topology of $\mathbb{R}$. An open set in $\mathbb{R}$ is obtained by forming an arbitrary union of open intervals, and a closed set in $\mathbb{R}$ results from taking the complement in $\mathbb{R}$ of such an open set. For example, $(0, 1)$ and $(0, +\infty)$ are open in $\mathbb{R}$, whereas $[0, 1]$ and $\{0\}$ are closed in $\mathbb{R}$. Working through the definitions, we find that $(0, 1)$ is the interior in $\mathbb{R}$ of both $(0, 1)$ and $[0, 1]$, and that $\{0, 1\}$ is the boundary in $\mathbb{R}$ of both $(0, 1)$ and $[0, 1]$. Consider the functions $f_1, f_2, f_3 : \mathbb{R} \longrightarrow \mathbb{R}$ given by $f_1(x) = x^3$, $f_2(x) = |x|$, and

$$f_3(x) = \begin{cases} 1 & \text{if } x \neq 0 \\ 0 & \text{if } x = 0. \end{cases}$$

Then f_1 and f_2 are continuous, but f_3 is not. In fact, f_1 is a homeomorphism.

We now turn our attention to $\mathbb{R}^2$. The open "ball" (actually "disk") in $\mathbb{R}^2$ of radius R centered at (x_0, y_0) is

$$B_R\big((x_0, y_0)\big) = \{(x, y) \in \mathbb{R}^2 : (x - x_0)^2 + (y - y_0)^2 < R^2\}.$$

For example,

$$D = \{(x, y) \in \mathbb{R}^2 : x^2 + y^2 < 1\},$$

is the **unit open disk** in $\mathbb{R}^2$ (centered at the origin). By Theorem 9.1.3 and Theorem 9.2.2, the collection of open disks in $\mathbb{R}^2$ is a basis for the Euclidean topology of $\mathbb{R}^2$. An open set in $\mathbb{R}^2$ is obtained by forming an arbitrary union of open disks, and a closed set in $\mathbb{R}^2$ results from taking the complement in $\mathbb{R}^2$ of such an open set. For example,

$$(0,1) \times (0,1) = \{(x,y) \in \mathbb{R}^2 : 0 < x, y < 1\}$$

is open in $\mathbb{R}^2$, whereas $[0,1]$ and

$$[0,1] \times [0,1] = \{(x,y) \in \mathbb{R}^2 : 0 \le x, y \le 1\}$$

are closed in $\mathbb{R}^2$.

Suppose D, as given above, has the subspace topology induced by $\mathbb{R}^2$. Consider the map $F : D \longrightarrow \mathbb{R}^2$ (between topological spaces) defined by

$$F(x,y) = \frac{1}{\sqrt{1 - x^2 - y^2}} (x,y).$$

It can be shown that F is a homeomorphism, with inverse $F^{-1} : \mathbb{R}^2 \longrightarrow D$ given by

$$F^{-1}(x,y) = \frac{1}{\sqrt{1 + x^2 + y^2}} (x,y).$$

Thus, D is homeomorphic to all of $\mathbb{R}^2$. Now consider the subsets $\{(x,0) : x \in \mathbb{R}\}$ and $\{(x,|x|) : x \in \mathbb{R}\}$ of $\mathbb{R}^2$, and suppose each has the subspace topology induced by $\mathbb{R}^2$. It can be shown that the map

$$G : \{(x,0) : x \in \mathbb{R}\} \longrightarrow \{(x,|x|) \in \mathbb{R}^2 : x \in \mathbb{R}\}$$

(between topological spaces) given by $G(x,0) = (x,|x|)$ is a homeomorphism. Thus, a "straight" line is homeomorphic to a line with a "corner".

The next result shows that in the Euclidean setting a homeomorphism has something to say about "dimension".

Theorem 9.4.1 (Homeomorphism Invariance of Dimension). *For an open set in* $\mathbb{R}^m$ *to be homeomorphic to an open set in* $\mathbb{R}^n$, *it is necessary that* $m = n$.   □

We close this section with a few remarks on compactness, perhaps the least intuitive of the concepts introduced in Section 9.1. As it stands, determining whether a subset of a topological space is compact is a seemingly complicated task. However, in the Euclidean case, matters are more straightforward. A subset S of $\mathbb{R}^m$ is said to be **bounded** if it is contained in some open ball of finite radius.

Theorem 9.4.2 (Compactness in $\mathbb{R}^m$**).** *A subset of* $\mathbb{R}^m$ *is compact in* $\mathbb{R}^m$ *if and only if it is closed in* $\mathbb{R}^m$ *and bounded.*   □

Thus, $[0,1] \times [0,1]$ is compact in $\mathbb{R}^2$, but $(0,1) \times (0,1)$ and $\mathbb{R} \times \{0\}$ are not. It should be emphasized that the preceding theorem rests on unique features of $\mathbb{R}^m$ and does not extend to an arbitrary metric space.

Chapter 10

Analysis in $\mathbb{R}^m$

10.1 Derivatives

In this section, we review some of the key results in the differential calculus of one or more real variables. For the most part, proofs are not provided.

Let U be an open set in $\mathbb{R}^m$, let $F : U \longrightarrow \mathbb{R}^n$ be a map, and let p be a point in U. We say that F is **differentiable at p** if there is a linear map $L_p : \mathbb{R}^m \longrightarrow \mathbb{R}^n$ such that

$$\lim_{\|v\| \to 0} \frac{\|F(p+v) - F(p) - L_p(v)\|}{\|v\|} = 0.$$

It can be shown that if such a map exists, it is unique. We call this map the **differential of F at p** and henceforth denote it by

$$d_p(F) : \mathbb{R}^m \longrightarrow \mathbb{R}^n. \tag{10.1.1}$$

In the literature, the differential of F at p is also called the derivative of F at p or the total derivative of F at p, and is denoted variously by dF_p, $\mathrm{d}F_p$, $DF(p)$, $D_p(F)$, or $F'(p)$. We have chosen to include parentheses in the notation $d_p(F)$ to set the stage for viewing d_p as a type of map.

It is usual to characterize $d_p(F)$ as being a "linear approximation" to F in the vicinity of p. For a more geometric interpretation, let us define the **graph of F** by

$$\mathrm{graph}(F) = \big\{ (p, F(p)) \in \mathbb{R}^{m+n} : p \in U \big\}.$$

We can think of $d_p(F)(\mathbb{R}^m)$, which is a vector space, as being the "tangent space" to $\mathrm{graph}(F)$ at $F(p)$. This is a generalization of the tangent line and tangent plane familiar from the differential calculus of one or two real variables.

We say that F is **differentiable (on U)** if it is differentiable at every p in U. To illustrate the difference between continuity and differentiability, consider

Semi-Riemannian Geometry, First Edition. Stephen C. Newman.
© 2019 John Wiley & Sons, Inc. Published 2019 by John Wiley & Sons, Inc.

the function $f : \mathbb{R} \longrightarrow \mathbb{R}$ given by $f(x) = |x|$. As was observed in Section 9.4, f is continuous; however, it is not differentiable because there are two candidates for "tangent line" at $x = 0$.

We now specialize to the case $n = 1$ and consider differentiable functions, turning our attention to differentiable maps below. Let $f : U \longrightarrow \mathbb{R}$ be a function, and let p be a point in U. Consistent with (10.1.1), the differential of f at p is denoted by

$$d_p(f) : \mathbb{R}^m \longrightarrow \mathbb{R}.$$

Theorem 10.1.1. *Let U be an open set in $\mathbb{R}^m$, let $f : U \longrightarrow \mathbb{R}$ be a function, and let p be a point in U. If f is differentiable at p, then it is continuous at p.*
□

We need to establish notation for coordinates on $\mathbb{R}^m$. This notation will serve for the present chapter, but will need to be revised later on.

In this chapter, coordinates on $\mathbb{R}^m$ are denoted
by $(x^1, \ldots, x^m)$ or $(y^1, \ldots, y^m)$.

Let U be an open set in $\mathbb{R}^m$, let $f : U \longrightarrow \mathbb{R}$ be a function, and let $p = (p^1, \ldots, p^m)$ be a point in U. The **partial derivative of f with respect to x^i at p** is defined by

$$\frac{\partial f}{\partial x^i}(p) = \lim_{h \to 0} \frac{f(p^1, \ldots, p^i + h, \ldots, p^m) - f(p^1, \ldots, p^m)}{h}$$

for $i = 1, \ldots, m$, provided the limit exists. It is sometimes convenient to denote

$$\frac{\partial f}{\partial x^i}(p) \qquad \text{by} \qquad \frac{\partial}{\partial x^i}(f)(p).$$

When $m = 1$, we denote

$$\frac{\partial f}{\partial x}(p) \qquad \text{by} \qquad \frac{df}{dx}(p).$$

In this notation, the linear function $d_p(f) : \mathbb{R} \longrightarrow \mathbb{R}$ is given by

$$d_p(f)(x) = x \frac{df}{dx}(p)$$

for all x in $\mathbb{R}$. More generally, we have the following result.

Theorem 10.1.2. *Let U be an open set in $\mathbb{R}^m$, let $f : U \longrightarrow \mathbb{R}$ be a function, let p be a point in U, and let $v = (a^1, \ldots, a^m)$ be a vector in $\mathbb{R}^m$. If f is differentiable at p, then $(\partial f / \partial x^i)(p)$ exists for $i = 1, \ldots, m$ and*

$$d_p(f)(v) = \sum_i a^i \frac{\partial f}{\partial x^i}(p).$$
□

Thus, $d_p(f)(v)$ is nothing other than the directional derivative of f at p in the direction v, familiar from the differential calculus of two or more real variables.

Theorem 10.1.3. *Let U be an open set in $\mathbb{R}^m$, let p be a point in U, and let c be a real number. If the functions $f, g : U \longrightarrow \mathbb{R}$ are differentiable at p, then:*
(a) $d_p(cf + g) = c\,d_p(f) + d_p(g)$.
(b) $d_p(fg) = f(p)\,d_p(g) + g(p)\,d_p(f)$.

Proof. For a vector $v = (a^1, \ldots, a^m)$ in $\mathbb{R}^m$, we have from Theorem 10.1.2 and the properties of partial derivatives that

$$d_p(cf + g)(v) = \sum_i a^i \frac{\partial(cf + g)}{\partial x^i}(p) = \sum_i a^i \left[c\frac{\partial f}{\partial x^i}(p) + \frac{\partial g}{\partial x^i}(p) \right]$$

$$= c\sum_i a^i \frac{\partial f}{\partial x^i}(p) + \sum_i a^i \frac{\partial g}{\partial x^i}(p) = c\,d_p(f)(v) + d_p(g)(v)$$

$$= \big(c\,d_p(f) + d_p(g)\big)(v)$$

and

$$d_p(fg)(v) = \sum_i a^i \frac{\partial(fg)}{\partial x^i}(p) = \sum_i a^i \left[f(p)\frac{\partial g}{\partial x^i}(p) + g(p)\frac{\partial f}{\partial x^i}(p) \right]$$

$$= f(p)\sum_i a^i \frac{\partial g}{\partial x^i}(p) + g(p)\sum_i a^i \frac{\partial f}{\partial x^i}(p)$$

$$= f(p)\,d_p(f)(v) + g(p)\,d_p(f)(v)$$

$$= \big(f(p)\,d_p(f) + g(p)\,d_p(f)\big)(v).$$

Since v was arbitrary, the result follows. $\square$

Let U be an open set in $\mathbb{R}^m$, let $f : U \longrightarrow \mathbb{R}$ be a function, and suppose $(\partial f/\partial x^i)(p)$ exists for all p in U for $i = 1, \ldots, m$. The **partial derivative of f with respect to x^i** is the function

$$\frac{\partial f}{\partial x^i} : U \longrightarrow \mathbb{R}$$

defined by the assignment

$$p \longmapsto \frac{\partial f}{\partial x^i}(p)$$

for all p in U. Now suppose $(\partial/\partial x^i)(\partial f/\partial x^j)(p)$ exists for all p in U for $i, j = 1, \ldots, m$. The **second order partial derivative of f with respect to x^i and x^j** is the function

$$\frac{\partial^2 f}{\partial x^i \partial x^j} : U \longrightarrow \mathbb{R}$$

defined by the assignment

$$p \longmapsto \frac{\partial^2 f}{\partial x^i \partial x^j}(p)$$

for all p in U for $i, j = 1, \ldots, m$, where we denote

$$\frac{\partial}{\partial x^i} \left(\frac{\partial f}{\partial x^j} \right)(p) \qquad \text{by} \qquad \frac{\partial^2 f}{\partial x^i \partial x^j}(p).$$

Iterating in an obvious way, higher-order partial derivatives of f are obtained. For $k \geq 2$, a kth-order partial derivative of f is denoted by

$$\frac{\partial^k f}{\partial x^{i_1} \partial x^{i_2} \cdots \partial x^{i_k}},$$

where the integers $1 \leq i_1, i_2, \ldots, i_k \leq m$ are not necessarily distinct.

Let us denote by $C^0(U)$ the set of functions $f : U \longrightarrow \mathbb{R}$ that are continuous on U. For $k \geq 1$, we define $C^k(U)$ to be the set of functions $f : U \longrightarrow \mathbb{R}$ such that all partial derivatives of f of order $\leq k$ exist and are continuous on U.

Theorem 10.1.4 (Criterion for Differentiability of Functions). *If U is an open set in $\mathbb{R}^m$ and f is a function in $C^1(U)$, that is, if $\partial f / \partial x^i$ exists and is continuous on U for $i = 1, \ldots, m$, then f is differentiable on U.* $\square$

From Theorem 10.1.1, Theorem 10.1.2, and Theorem 10.1.4, we obtain the following result, which summarizes the relationship of continuity to differentiability for a function on an open set in $\mathbb{R}^m$.

Theorem 10.1.5. *Let U be an open set in $\mathbb{R}^m$, and let $f : U \longrightarrow \mathbb{R}$ be a function. Then:*
(a) *If f is differentiable on U, then f is continuous on U, and $\partial f / \partial x^i$ exists on U for $i = 1, \ldots, m$.*
(b) *If $\partial f / \partial x^i$ exists and is continuous on U for $i = 1, \ldots, m$, then f is differentiable on U.* $\square$

Theorem 10.1.6 (Equality of Mixed Partial Derivatives). *Let U be an open set in $\mathbb{R}^m$, let f be a function in $C^r(U)$, where $r \geq 1$, and let $1 \leq k \leq r$ be an integer. Then the mixed partial derivatives of f of order k are independent of the order of differentiation. That is, if $1 \leq i_1, i_2, \ldots, i_k \leq m$ are (not necessarily distinct) integers and $j_1, j_2, \ldots, j_k$ are the same integers in some order, then*

$$\frac{\partial^k f}{\partial x^{i_1} \partial x^{i_2} \cdots \partial x^{i_k}} = \frac{\partial^k f}{\partial x^{j_1} \partial x^{j_2} \cdots \partial x^{j_k}}.$$ $\square$

Let U be an open set in $\mathbb{R}^m$, and let $f : U \longrightarrow \mathbb{R}$ be a function. We say that f is **(Euclidean) smooth (on U)** if f is in $C^k(U)$ for all $k \geq 0$. The set of smooth functions on U is denoted by $C^\infty(U)$. In view of Theorem 10.1.5, $C^\infty(U)$ is the set of functions f on U with partial derivatives of all orders on U. It is sometimes said that smooth functions are "infinitely differentiable".

We make $C^\infty(U)$ into both a vector space and a ring by defining operations as follows: for all functions f, g in $C^\infty(U)$ and all real numbers c, let

$$(f + g)(p) = f(p) + g(p),$$

$$(fg)(p) = f(p)g(p),$$

and

$$(cf)(p) = cf(p)$$

for all p in U. The identity element of the ring is the constant function 1_U that sends all points in U to the real number 1.

We now turn our attention to differentiable maps.

Theorem 10.1.7. *Let U be an open set in $\mathbb{R}^m$, let $F : U \longrightarrow \mathbb{R}^n$ be a map, and let p be a point in U. If F is differentiable at p, then it is continuous at p.* □

Theorem 10.1.8. *Let U be an open set in $\mathbb{R}^m$, let $F = (F^1, \ldots, F^n) : U \longrightarrow \mathbb{R}^n$ be a map, and let p be a point in U. Then:*
(a) *F is differentiable at p if and only if F^i is differentiable at p for $i = 1, \ldots, m$.*
(b) *If F is differentiable at p and v is a vector in $\mathbb{R}^m$, then*

$$d_p(F)(v) = \big(d_p(F^1)(v), \ldots, d_p(F^n)(v)\big).$$ □

The next result is one of the workhorses of analysis and will be called upon frequently.

Theorem 10.1.9 (Chain Rule). *Let U and V be open sets in $\mathbb{R}^m$ and $\mathbb{R}^n$, respectively, let $F : U \longrightarrow \mathbb{R}^n$ and $G : V \longrightarrow \mathbb{R}^k$ be maps such that $F(U) \subseteq V$, and let p be a point in U. If F is differentiable at p and G is differentiable at $F(p)$, then $G \circ F$ is differentiable at p and*

$$d_p(G \circ F) = d_{F(p)}(G) \circ d_p(F).$$ □

A remark on the above notation is in order. Instead of $G \circ F$, it would be more precise, although somewhat cluttered, to write $G|_{F(U)} \circ F$. When there is a possibility of confusion or if it improves exposition, notation will be modified in this way.

Let U be an open set in $\mathbb{R}^m$, let $F = (F^1, \ldots, F^n) : U \longrightarrow \mathbb{R}^n$ be a differentiable map, and let p be a point in U. The **Jacobian matrix of F at p** is the $n \times m$ matrix defined by

$$J_F(p) = \begin{bmatrix} \dfrac{\partial F^1}{\partial x^1}(p) & \cdots & \dfrac{\partial F^1}{\partial x^m}(p) \\ \vdots & \ddots & \vdots \\ \dfrac{\partial F^n}{\partial x^1}(p) & \cdots & \dfrac{\partial F^n}{\partial x^m}(p) \end{bmatrix}. \tag{10.1.2}$$

Denoting by $\mathcal{E}$ and $\mathcal{F}$ the standard bases for $\mathbb{R}_m$ and $\mathbb{R}_n$, respectively, we have from (2.2.3), Theorem 10.1.2, and Theorem 10.1.8(b) that

$$J_F(p) = [d_p(F)]_{\mathcal{E}}^{\mathcal{F}}. \tag{10.1.3}$$

When $m = n$, the determinant $\det\big(J_F(p)\big)$ is called the **Jacobian determinant of F at p.** Let $1 \leq i_1 < \cdots < i_s \leq m$ and $1 \leq j_1 < \cdots < j_s \leq n$ be integers, where $1 \leq s \leq \min(m, n)$. Using multi-index notation, the submatrix of $J_F(p)$ consisting of the intersection of rows $i_1, \ldots, i_s$ and columns $j_1, \ldots, j_s$ (in that order) is

$$J_F(p)^{(i_1,\ldots,i_s)}_{(j_1,\ldots,j_s)} = \begin{bmatrix} \dfrac{\partial F^{i_1}}{\partial x^{j_1}}(p) & \cdots & \dfrac{\partial F^{i_1}}{\partial x^{j_s}}(p) \\ \vdots & \ddots & \vdots \\ \dfrac{\partial F^{i_s}}{\partial x^{j_1}}(p) & \cdots & \dfrac{\partial F^{i_s}}{\partial x^{j_s}}(p) \end{bmatrix}.$$

In the literature, the above matrix is commonly denoted by

$$\frac{\partial(F^{i_1}, \ldots, F^{i_s})}{\partial(x^{j_1}, \ldots, x^{j_s})}(p).$$

Theorem 10.1.10 (Classical Chain Rule). *Let U and V be open sets in $\mathbb{R}^m$ and $\mathbb{R}^n$, respectively, let $F : U \longrightarrow \mathbb{R}^n$ and $G : V \longrightarrow \mathbb{R}^k$ be maps such that $F(U) \subseteq V$, and let p be a point in U. If F is differentiable at p and G is differentiable at $F(p)$, then $G \circ F$ is differentiable at p and*

$$J_{G \circ F}(p) = J_G\big(F(p)\big)\, J_F(p).$$

Equivalently, let $F = (F^1, \ldots, F^n)$ and $G = (G^1, \ldots, G^k)$, and let $(x^1, \ldots, x^m)$ and $(y^1, \ldots, y^n)$ be coordinates on $\mathbb{R}^m$ and $\mathbb{R}^n$, respectively. Then

$$\frac{\partial(G^i \circ F)}{\partial x^j}(p) = \sum_{l=1}^{n} \frac{\partial G^i}{\partial y^l}\big(F(p)\big)\, \frac{\partial F^l}{\partial x^j}(p) \tag{10.1.4}$$

for $i = 1, \ldots, k$ and $j = 1, \ldots, m$.

Proof. Let $\mathcal{E}$, $\mathcal{F}$, and $\mathcal{G}$ be the standard bases for $\mathbb{R}^m$, $\mathbb{R}^n$, and $\mathbb{R}^k$, respectively. The Jacobian matrix of G at $F(p)$ is the $k \times n$ matrix

$$J_G\big(F(p)\big) = \begin{bmatrix} \dfrac{\partial G^1}{\partial y^1}\big(F(p)\big) & \cdots & \dfrac{\partial G^1}{\partial y^n}\big(F(p)\big) \\ \vdots & \ddots & \vdots \\ \dfrac{\partial G^k}{\partial y^1}\big(F(p)\big) & \cdots & \dfrac{\partial G^k}{\partial y^n}\big(F(p)\big) \end{bmatrix}. \tag{10.1.5}$$

We have

$$\begin{aligned}
\big[J_{G \circ F}(p)\big]^{\mathcal{G}}_{\mathcal{E}} &= \big[d_p(G \circ F)\big]^{\mathcal{G}}_{\mathcal{E}} && [(10.1.3)] \\
&= \big[d_{F(p)}(G) \circ d_p(F)\big]^{\mathcal{G}}_{\mathcal{E}} && [\text{Th } 10.1.9] \\
&= \big[d_{F(p)}(G)\big]^{\mathcal{G}}_{\mathcal{F}}\, \big[d_p(F)\big]^{\mathcal{F}}_{\mathcal{E}} && [\text{Th } 2.2.2(\text{c})] \\
&= J_G\big(F(p)\big)\, J_F(p) && [(10.1.3)] \\
&= \Big[\sum_l \frac{\partial G^i}{\partial y^l}\big(F(p)\big)\, \frac{\partial F^l}{\partial x^j}(p)\Big]. && [(10.1.2),\ (10.1.5)] \qquad \square
\end{aligned}$$

To give (10.1.4) a more traditional appearance, we continue with the above notation, let $(z^1, \ldots, z^k)$ be standard coordinates on $\mathbb{R}^k$, and replace

$$G^i \circ F(p) = G^i(F^1(p), \ldots, F^n(p))$$

with

$$z^i = z^i(y^1(x^1, \ldots, x^m), \ldots, y^n(x^1, \ldots, x^m)).$$

Then (10.1.4) can be expressed as

$$\frac{\partial z^i}{\partial x^j} = \sum_{l=1}^{n} \frac{\partial z^i}{\partial y^l} \frac{\partial y^l}{\partial x^j}.$$

Let U be an open set in $\mathbb{R}^m$, and let $F = (F^1, \ldots, F^n) : U \longrightarrow \mathbb{R}^n$ be a map. We say that F is **smooth (on U)** if the function F^i is smooth for $i = 1, \ldots, m$. Smooth maps will be our focus for the rest of the book.

The next result, which says that smoothness is ultimately a local phenomenon, will be used frequently, but usually without attribution.

Theorem 10.1.11. *Let U be an open set in $\mathbb{R}^m$, and let $F : U \longrightarrow \mathbb{R}^m$ be a map. Then:*
(a) *If F is smooth and U' is an open set in U, then $F|_{U'}$ is smooth.*
(b) *Conversely, if every point p in U has a neighborhood U' in U such that $F|_{U'}$ is smooth, then F is smooth.* □

Theorem 10.1.12. *Let U and V be open sets in $\mathbb{R}^m$ and $\mathbb{R}^n$, respectively, and let $F : U \longrightarrow \mathbb{R}^n$ and $G : V \longrightarrow \mathbb{R}^k$ be maps such that $F(U) \subseteq V$. If F and G are smooth, then so is $G \circ F$.* □

A **(parametrized) curve** in $\mathbb{R}^m$ is a map $\lambda : I \longrightarrow \mathbb{R}^m$, where I is an interval in $\mathbb{R}$ that is either open, closed, half-open, or half-closed, and where the possibility that I is infinite is not excluded. Our focus will be on the case where I is a finite open interval, usually denoted by (a, b). Rather than provide a separate statement identifying the independent variable for the curve, most often denoted by t, and sometimes by u, it is convenient to incorporate this into the notation for λ, as in $\lambda(t) : (a, b) \longrightarrow \mathbb{R}^3$. Let $\lambda = (\lambda^1, \ldots, \lambda^m)$. By definition, λ is **smooth [on (a, b)]** if and only if λ^i is smooth for $i = 1, \ldots, m$.

Suppose λ is in fact smooth. The **(Euclidean) velocity of λ** and the **(Euclidean) acceleration of λ** are the smooth curves

$$\frac{d\lambda}{dt}(t) : (a, b) \longrightarrow \mathbb{R}^m \quad \text{and} \quad \frac{d^2\lambda}{dt^2}(t) : (a, b) \longrightarrow \mathbb{R}^m,$$

respectively; that is,

$$\frac{d\lambda}{dt}(t) = \left(\frac{d\lambda^1}{dt}(t), \ldots, \frac{d\lambda^m}{dt}(t)\right)$$

and

$$\frac{d^2\lambda}{dt^2}(t) = \left(\frac{d^2\lambda^1}{dt^2}(t), \ldots, \frac{d^2\lambda^m}{dt^2}(t)\right)$$

for all t in (a, b).

Theorem 10.1.13. *Let U be an open set in $\mathbb{R}^m$, let p be a point in U, and let v be a vector in $\mathbb{R}^m$. Then there is a real number $\varepsilon > 0$ and a smooth curve $\lambda(t) : (-\varepsilon, \varepsilon) \longrightarrow U$ such that $\lambda(0) = p$ and $(d\lambda/dt)(0) = v$.*

Proof. Define a smooth curve $\lambda(t) : (-\varepsilon, \varepsilon) \longrightarrow U$ by $\lambda(t) = p + tv$, where ε is chosen small enough that $\lambda((-\varepsilon, \varepsilon)) \subset U$. Clearly, λ has the desired properties. $\square$

The next result is the key to later discussions about "vectors" and "tangent spaces".

Theorem 10.1.14. *Let U be an open set in $\mathbb{R}^m$, let $F : U \longrightarrow \mathbb{R}^n$ be a map, let p be a point in U, and let v be a vector in $\mathbb{R}^m$. Then:*
(a) *If F is differentiable at p, then*

$$d_p(F)(v) = \frac{d(F \circ \lambda)}{dt}(t_0),$$

where $\lambda(t) : (a, b) \longrightarrow U$ is any smooth curve such that $\lambda(t_0) = p$ and $(d\lambda/dt)(t_0) = v$ for some t_0 in (a, b) (as given by Theorem 10.1.13).
(b) *If $\psi(t) : (a, b) \longrightarrow U$ is a smooth curve, then*

$$d_{\psi(t)}(F)\left(\frac{d\psi}{dt}(t)\right) = \frac{d(F \circ \psi)}{dt}(t)$$

for all t in (a, b).

Proof. (a): Let $v = (a^1, \ldots, a^m)$, $\lambda = (\lambda^1, \ldots, \lambda^m)$, and $F = (F^1, \ldots, F^n)$. Then $F \circ \lambda = (F^1 \circ \lambda, \ldots, F^n \circ \lambda)$, so

$$\frac{d(F \circ \lambda)}{dt}(t_0) = \left(\frac{d(F^1 \circ \lambda)}{dt}(t_0), \ldots, \frac{d(F^n \circ \lambda)}{dt}(t_0)\right).$$

We have

$$\frac{d(F^j \circ \lambda)}{dt}(t_0) = \sum_i \frac{\partial F^j}{\partial x^i}\big(\lambda(t_0)\big) \frac{d\lambda^i}{dt}(t_0) \qquad \text{[Th 10.1.10]}$$

$$= \sum_i a^i \frac{\partial F^j}{\partial x^i}(p)$$

$$= d_p(F^j)(v) \qquad\qquad\qquad \text{[Th 10.1.2]}$$

for $j = 1, \ldots, n$. Thus,

$$\frac{d(F \circ \lambda)}{dt}(t_0) = \big(d_p(F^1)(v), \ldots, d_p(F^n)(v)\big) = d_p(F)(v),$$

where the last equality follows from Theorem 10.1.8(b).
  (b): This follows from part (a). $\square$

Not all maps of interest have domains that are open sets. For this reason we need an "extended" definition of smoothness to handle maps defined on sets that are not necessarily open. Let S be an arbitrary subset of $\mathbb{R}^m$, and let $F : S \longrightarrow \mathbb{R}^n$ be a map. We say that F is **(extended) smooth (on S)** if for every point p in S, there is a neighborhood $\mathcal{U}$ of p in $\mathbb{R}^m$ and a (Euclidean) smooth map $\widetilde{F} : \mathcal{U} \longrightarrow \mathbb{R}^n$ such that F and $\widetilde{F}$ agree on $S \cap \mathcal{U}$; that is, $F|_{S \cap \mathcal{U}} = \widetilde{F}|_{S \cap \mathcal{U}}$. Although in the preceding definition both $\mathcal{U}$ and $\widetilde{F}$ might very well depend on p, the next result shows that this dependence can be avoided.

Theorem 10.1.15. *Let S be a subset of $\mathbb{R}^m$, and let $F : S \longrightarrow \mathbb{R}^n$ be a map. Then F is (extended) smooth if and only if there is an open set $\mathcal{U}$ in $\mathbb{R}^m$ containing S and a (Euclidean) smooth map $\widetilde{F} : \mathcal{U} \longrightarrow \mathbb{R}^n$ such that F and $\widetilde{F}$ agree on S; that is, $F = \widetilde{F}|_S$.* $\square$

For example, a curve $\lambda(t) : [a, b] \longrightarrow \mathbb{R}^m$ is smooth if and only if there is an interval $(\widetilde{a}, \widetilde{b})$ containing $[a, b]$ and a smooth curve $\widetilde{\lambda}(t) : (\widetilde{a}, \widetilde{b}) \longrightarrow \mathbb{R}^m$ such that $\lambda = \widetilde{\lambda}|_{[a,b]}$.

Theorem 10.1.16. *Let S be a subset of $\mathbb{R}^m$, and let $F : S \longrightarrow \mathbb{R}^n$ be a map. If F is (extended) smooth, then it is continuous.* $\square$

Theorem 10.1.17. *Let S and T be subsets of $\mathbb{R}^m$ and $\mathbb{R}^n$, respectively, and let $F : S \longrightarrow \mathbb{R}^n$ and $G : T \longrightarrow \mathbb{R}^k$ be maps such that $F(S) \subseteq T$. If F and G are (extended) smooth, then so is $G \circ F$.* $\square$

10.2   Immersions and Diffeomorphisms

Let U be an open set in $\mathbb{R}^m$, let $F : U \longrightarrow \mathbb{R}^n$ be a smooth map, where $m \leq n$, and let p be a point in U. Since F is smooth on U, hence differentiable at p, we have from remarks in Section 10.1 that $d_p(F)(\mathbb{R}^m)$ can be viewed as the "tangent space" to the graph of F at $F(p)$. We say that F is an **immersion at p** if the differential map $d_p(F) : \mathbb{R}^m \longrightarrow \mathbb{R}^n$ is injective, and that F is an **immersion (on U)** if it is an immersion at every p in U.

The next result gives alternative ways of characterizing an immersion.

Theorem 10.2.1. *Let U be an open set in $\mathbb{R}^m$, let $F : U \longrightarrow \mathbb{R}^n$ be a smooth map, where $m \leq n$, and let p be a point in U. Then the following are equivalent:*
(a) *F is an immersion at p.*
(b) *$d_p(F)(\mathbb{R}^m)$ is m-dimensional.*
(c) *$J_F(p)$ has rank m.*
If $m = n$, then each of the following is equivalent to each of the above:
(d) *$d_p(F)$ is a linear isomorphism.*
(e) *$\det\big(J_F(p)\big) \neq 0$.*

Proof. The equivalence of (a) and (b) follows from Theorem 1.1.12, and the equivalence of (b) and (c) from (10.1.3). The equivalence of (d) and (e) follows from Theorem 2.5.3. When $m = n$, the equivalence of (c) and (e) follows from Theorem 2.4.8. □

Let U be an open set in $\mathbb{R}^m$, let $F, G : U \longrightarrow \mathbb{R}^m$ be smooth maps, and let p be a point in U, where we note that both the domain and codomain of F and G are subsets of $\mathbb{R}^m$. We say that $F : U \longrightarrow F(U)$ is a **diffeomorphism**, and that U and $F(U)$ are **diffeomorphic**, if $F(U)$ is open in $\mathbb{R}^m$, $F : U \longrightarrow F(U)$ is bijective, and $F^{-1} : F(U) \longrightarrow \mathbb{R}^m$ is smooth. We say that G is a **local diffeomorphism at** p if there is a neighborhood $U' \subseteq U$ of p in $\mathbb{R}^m$ and a neighborhood V of $G(p)$ in $\mathbb{R}^m$ such that $G|_{U'} : U' \longrightarrow V$ is a diffeomorphism. Then G is said to be a **local diffeomorphism (on** U**)** if it is a local diffeomorphism at every p in U. Evidently, every diffeomorphism is a local diffeomorphism.

It is straightforward to give "extended" versions of the preceding definitions using the extended definition of smoothness presented in Section 10.1.

Example 10.2.2. Let $D = \{(x, y) \in \mathbb{R}^2 : \sqrt{x^2 + y^2} < 1\}$ be the unit open disk in $\mathbb{R}^2$, and recall the map $F : D \longrightarrow \mathbb{R}^2$ from Section 9.4 defined by

$$F(x, y) = \frac{1}{\sqrt{1 - x^2 - y^2}} (x, y).$$

It can be shown that F is a diffeomorphism. Thus, D is diffeomorphic to all of $\mathbb{R}^2$. ◇

Theorem 10.2.3 (Inverse Map Theorem). *Let U be an open set in $\mathbb{R}^m$, let $F : U \longrightarrow \mathbb{R}^m$ be a smooth map, and let p be a point in U. Then F is a local diffeomorphism at p if and only if it is an immersion at p. Thus, F is a local diffeomorphism if and only if it is an immersion.* □

The inverse map theorem (also called the inverse function theorem) underscores the close relationship between "smoothness" and "tangent space". It is one of the most important results in analysis and will be called upon repeatedly.

Theorem 10.2.4. *Let $f(t) : (a, b) \longrightarrow (c, d)$ be a smooth function. Then:*
(a) *If f is a diffeomorphism, then df/dt is nowhere-vanishing.*
(b) *If df/dt is nowhere-vanishing, then f is either strictly increasing or strictly decreasing.*
(c) *If f is a diffeomorphism, then it is either strictly increasing or strictly decreasing.*

Proof. (a): Since $f^{-1} \circ f(t) = t$, by Theorem 10.1.10,

$$\frac{df^{-1}}{dt} (f(t)) \frac{df}{dt} (t) = 1$$

for all t in (a, b). The result follows.

(b): For a point t_2 in (a, b), we have $(df/dt)(t_2) \neq 0$. Suppose $(df/dt)(t_2) > 0$. Since f is smooth, so is df/dt, and therefore, by Theorem 10.1.1, df/dt is continuous. If there is a point t_1 in (a, b) such that $(df/dt)(t_1) < 0$, then by Theorem 9.1.19, there is a point t_0 in (c, d) such that $(df/dt)(t_0) = 0$, which contradicts the nowhere-vanishing assumption on df/dt. Thus, $(df/dt)(t) > 0$ for all t in (a, b); that is, f is strictly increasing. Similarly, if $(df/dt)(t_2) < 0$, then f is strictly decreasing.

(c): This follows from parts (a) and (b). $\square$

10.3   Euclidean Derivative and Vector Fields

Let U be an open set in $\mathbb{R}^m$, and let $X : U \longrightarrow \mathbb{R}^m$ be a map, where we note that both the domain and codomain of X are subsets of $\mathbb{R}^m$. In the present context, we refer to X as a **vector field (on U)**. To highlight the appearance of vector fields and to distinguish them from other types of maps, let us denote

$$X(p) \qquad \text{by} \qquad X_p$$

for all p in U. We say that X **vanishes at p** if $X_p = (0, \dots, 0)$, is **nonvanishing at p** if $X_p \neq (0, \dots, 0)$, and is **nowhere-vanishing (on U)** if it is nonvanishing at every p in U.

Let us denote the set of smooth vector fields on U by $\mathfrak{X}(U)$. We make $\mathfrak{X}(U)$ into both a vector space over $\mathbb{R}$ and a module over $C^\infty(U)$ by defining operations as follows: for all vector fields X, Y in $\mathfrak{X}(U)$, all functions f in $C^\infty(U)$, and all real numbers c, let

$$(X + Y)_p = X_p + Y_p,$$
$$cX_p = cX_p,$$

and

$$(fX)_p = f(p)X_p$$

for all p in U. Let

$$X = (\alpha^1, \dots, \alpha^m) \qquad \text{and} \qquad Y = (\beta^1, \dots, \beta^m), \qquad (10.3.1)$$

where α^i, β^j are functions in $C^\infty(U)$ for $i, j = 1, \dots, m$. Then

$$(X + Y)_p = \big(\alpha^1(p) + \beta^1(p), \dots, \alpha^m(p) + \beta^m(p)\big)$$

and

$$(fX)_p = \big(f(p)\,\alpha^1(p), \dots, f(p)\,\alpha^m(p)\big).$$

The α^i are called the **components of X**. Let E_i be the (constant) vector field in $\mathfrak{X}(U)$ defined by

$$E_i = (0, \dots, 1, \dots, 0),$$

where 1 is in the ith position and 0s are elsewhere for $i = 1, \dots, m$.

The **Euclidean derivative with respect to X** consists of two maps, both denoted by D_X. The first is

$$D_X : C^\infty(U) \longrightarrow C^\infty(U)$$

defined by

$$D_X(f)(p) = d_p(f)(X_p) \tag{10.3.2}$$

for all functions f in $C^\infty(U)$ and all p in U. The second is

$$D_X : \mathfrak{X}(U) \longrightarrow \mathfrak{X}(U)$$

defined by

$$D_X(Y)_p = d_p(Y)(X_p) \tag{10.3.3}$$

for all vector fields Y in $\mathfrak{X}(U)$ and all p in U. It follows from Theorem 10.1.14(a) that (10.3.2) and (10.3.3) can be expressed as

$$D_X(f)(p) = \frac{d(f \circ \lambda)}{dt}(t_0)$$

and

$$D_X(Y)_p = \frac{d(Y \circ \lambda)}{dt}(t_0),$$

respectively, where $\lambda(t) : (a, b) \longrightarrow U$ is any smooth curve such $\lambda(t_0) = p$ and $(d\lambda/dt)(t_0) = X_p$ for some t_0 in (a, b). The existence of such a smooth curve is guaranteed by Theorem 10.1.13.

Evidently, the Euclidean derivatives of f and Y with respect to X evaluated at p have the same mathematical content as the differentials at p of f and Y evaluated at X_p. Their difference, such as it is, amounts to a change of notation that emphasizes the role of X in the Euclidean derivative.

For vector fields X, Y in $\mathfrak{X}(U)$, we define a function

$$\langle X, Y \rangle : U \longrightarrow \mathbb{R}$$

in $C^\infty(U)$ by the assignment

$$p \longmapsto \langle X_p, Y_p \rangle$$

for all p in U.

The Euclidean derivative with respect to X satisfies fundamental algebraic properties, versions of which will reappear later in several settings.

Theorem 10.3.1. *Let U be an open set in $\mathbb{R}^m$, let X, Y, Z be vector fields in $\mathfrak{X}(U)$, and let f be a function in $C^\infty(U)$. Then:*
(a) $D_{X+Y}(Z) = D_X(Z) + D_Y(Z)$.
(b) $D_{fX}(Y) = f D_X(Y)$.
(c) $D_X(Y + Z) = D_X(Y) + D_X(Z)$.
(d) $D_X(fY) = D_X(f) Y + f D_X(Y)$.
(e) $D_X(\langle Y, Z \rangle) = \langle D_X(Y), Z \rangle + \langle Y, D_X(Z) \rangle$.

Proof. Using Theorem 10.1.3 and Theorem 10.1.8(b) gives the result.   □

Theorem 10.3.2. *Let U be an open set in $\mathbb{R}^m$, let f be a function in $C^\infty(U)$, and let X, Y be vector fields in $\mathfrak{X}(U)$, with*

$$X = (\alpha^1, \ldots, \alpha^m) \qquad and \qquad Y = (\beta^1, \ldots, \beta^m).$$

Then:

(a)

$$D_X(f) = \sum_i \alpha^i \frac{\partial f}{\partial x^i}.$$

(b)

$$D_X(D_Y(f)) = \sum_{ij} \left(\alpha^i \frac{\partial \beta^j}{\partial x^i} \frac{\partial f}{\partial x^j} + \alpha^i \beta^j \frac{\partial^2 f}{\partial x^i \partial x^j} \right).$$

Proof. (a): For a point p in U, we have

$$
\begin{aligned}
D_X(f)(p) &= d_p(f)(X_p) && [(10.3.2)] \\
&= d_p(f)(\alpha^1(p), \ldots, \alpha^m(p)) \\
&= \sum_i \alpha^i(p) \frac{\partial f}{\partial x^i}(p). && [\text{Th } 10.1.2]
\end{aligned}
$$

Since p was arbitrary, the result follows.

(b): By part (a),

$$
D_X(D_Y(f)) = D_X\left(\sum_j \beta^j \frac{\partial f}{\partial x^j} \right) = \sum_i \alpha^i \frac{\partial}{\partial x^i}\left(\sum_j \beta^j \frac{\partial f}{\partial x^j} \right)
$$

$$
= \sum_{ij} \alpha^i \frac{\partial}{\partial x^i}\left(\beta^j \frac{\partial f}{\partial x^j} \right) = \sum_{ij}\left(\alpha^i \frac{\partial \beta^j}{\partial x^i} \frac{\partial f}{\partial x^j} + \alpha^i \beta^j \frac{\partial^2 f}{\partial x^i \partial x^j} \right). \quad \square
$$

Theorem 10.3.3. *Let U be an open set in $\mathbb{R}^m$, and let X, Y, Z be vector fields in $\mathfrak{X}(U)$, with*

$$X = (\alpha^1, \ldots, \alpha^m), \qquad Y = (\beta^1, \ldots, \beta^m), \qquad and \qquad Z = (\gamma^1, \ldots, \gamma^m).$$

Then:

(a)

$$D_X(Y) = \sum_j \left(\sum_i \alpha^i \frac{\partial \beta^j}{\partial x^i} \right) E_j.$$

(b)

$$D_X(D_Y(Z)) = \sum_k \left[\sum_{ij} \left(\alpha^i \frac{\partial \beta^j}{\partial x^i} \frac{\partial \gamma^k}{\partial x^j} + \alpha^i \beta^j \frac{\partial^2 \gamma^k}{\partial x^i \partial x^j} \right) \right] E_k.$$

(c)

$$D_{D_X(Y)}(Z) = \sum_k \left(\sum_{ij} \alpha^i \frac{\partial \beta^j}{\partial x^i} \frac{\partial \gamma^k}{\partial x^j} \right) E_k.$$

Proof. We make repeated use of Theorem 10.3.1 and Theorem 10.3.2. Let f be a function in $C^\infty(U)$, and observe that

$$D_X(E_i) = 0 \qquad \text{and} \qquad D_{E_i}(f) = \frac{\partial f}{\partial x^i}$$

for $i = 1, \ldots, m$.

(a): We have

$$\begin{aligned}
D_X(Y) &= D_{\sum_i \alpha^i E_i} \left(\sum_j \beta^j E_j \right) = \sum_{ij} \alpha^i D_{E_i}(\beta^j E_j) \\
&= \sum_{ij} \alpha^i [D_{E_i}(\beta^j) E_j + \beta^j D_{E_i}(E_j)] \\
&= \sum_{ij} \alpha^i \frac{\partial \beta^j}{\partial x^i} E_j = \sum_j \left(\sum_i \alpha^i \frac{\partial \beta^j}{\partial x^i} \right) E_j.
\end{aligned}$$

(b): By part (a),

$$D_Y(Z) = \sum_k \left(\sum_j \beta^j \frac{\partial \gamma^k}{\partial x^j} \right) E_k,$$

hence

$$\begin{aligned}
D_X(D_Y(Z)) &= D_{\sum_i \alpha^i E_i} \left(\sum_{jk} \beta^j \frac{\partial \gamma^k}{\partial x^j} E_k \right) = \sum_{ijk} \alpha^i D_{E_i} \left(\beta^j \frac{\partial \gamma^k}{\partial x^j} E_k \right) \\
&= \sum_{ijk} \alpha^i \left[D_{E_i}(\beta^j) \frac{\partial \gamma^k}{\partial x^j} E_k + \beta^j D_{E_i} \left(\frac{\partial \gamma^k}{\partial x^j} \right) E_k + \beta^j \frac{\partial \gamma^k}{\partial x^j} D_{E_i}(E_k) \right] \\
&= \sum_{ijk} \alpha^i \left(\frac{\partial \beta^j}{\partial x^i} \frac{\partial \gamma^k}{\partial x^j} E_k + \beta^j \frac{\partial^2 \gamma^k}{\partial x^i \partial x^j} E_k \right) \\
&= \sum_k \left[\left(\sum_{ij} \alpha^i \frac{\partial \beta^j}{\partial x^i} \frac{\partial \gamma^k}{\partial x^j} \right) + \sum_{ij} \alpha^i \beta^j \frac{\partial^2 \gamma^k}{\partial x^i \partial x^j} \right] E_k.
\end{aligned}$$

(c): By part (a),

$$\begin{aligned}
D_{D_X(Y)}(Z) &= D_{\sum_{ij} \alpha^i (\partial \beta^j / \partial x^i) E_j} \left(\sum_k \gamma^k E_k \right) \\
&= \sum_{ijk} \left(\alpha^i \frac{\partial \beta^j}{\partial x^i} \right) D_{E_j}(\gamma^k E_k) \\
&= \sum_{ijk} \left(\alpha^i \frac{\partial \beta^j}{\partial x^i} \right) [D_{E_j}(\gamma^k) E_k + \gamma^k D_{E_i}(E_k)] \\
&= \sum_{ijk} \alpha^i \frac{\partial \beta^j}{\partial x^i} \frac{\partial \gamma^k}{\partial x^j} E_k = \sum_k \left(\sum_{ij} \alpha^i \frac{\partial \beta^j}{\partial x^i} \frac{\partial \gamma^k}{\partial x^j} \right) E_k. \qquad \square
\end{aligned}$$

Let U be an open set in $\mathbb{R}^m$, and let X, Y be vector fields in $\mathfrak{X}(U)$. The **second order Euclidean derivative with respect to X and Y** consists of two maps, both denoted by $D^2_{X,Y}$. The first is

$$D^2_{X,Y} : C^\infty(U) \longrightarrow C^\infty(U)$$

defined by

$$D^2_{X,Y}(f) = D_X\big(D_Y(f)\big) - D_{D_X(Y)}(f)$$

for all functions f in $C^\infty(U)$. The second is

$$D^2_{X,Y}(Z) : \mathfrak{X}(U) \longrightarrow \mathfrak{X}(U)$$

defined by

$$D^2_{X,Y}(Z) = D_X\big(D_Y(Z)\big) - D_{D_X(Y)}(Z)$$

for all vector fields Z in $\mathfrak{X}(U)$.

Theorem 10.3.4. *Let U be an open set in $\mathbb{R}^m$, and let X, Y, Z be vector fields in $\mathfrak{X}(U)$, with*

$$X = (\alpha^1, \ldots, \alpha^m), \qquad Y = (\beta^1, \ldots, \beta^m), \qquad and \qquad Z = (\gamma^1, \ldots, \gamma^m).$$

Then
(a)

$$D^2_{X,Y}(Z) = \sum_k \left(\sum_{ij} \alpha^i \beta^j \frac{\partial^2 \gamma^k}{\partial x^i \partial x^j} \right) E_k.$$

(b)

$$D^2_{Y,X}(Z) = D^2_{X,Y}(Z).$$

Proof. (a): This follows from Theorem 10.3.3(b) and Theorem 10.3.3(c).
(b): This follows from Theorem 10.1.6 and part (a).   □

10.4   Lie Bracket

Let U be an open set in $\mathbb{R}^m$. **Lie bracket** is the map

$$[\cdot,\cdot] : \mathfrak{X}(U) \times \mathfrak{X}(U) \longrightarrow \mathfrak{X}(U)$$

defined by

$$[X,Y] = D_X(Y) - D_Y(X)$$

for all vector fields X, Y in $\mathfrak{X}(U)$. We refer to $[X,Y]$ as the **Lie bracket of X and Y**.

Theorem 10.4.1. *Let U be an open set in $\mathbb{R}^m$, and let X, Y be vector fields in $\mathfrak{X}(U)$, with*

$$X = (\alpha^1, \ldots, \alpha^m) \qquad and \qquad Y = (\beta^1, \ldots, \beta^m).$$

Then

$$[X,Y] = \sum_j \left\{ \sum_i \left(\alpha^i \frac{\partial \beta^j}{\partial x^i} - \beta^i \frac{\partial \alpha^j}{\partial x^i} \right) \right\} E_j.$$

Proof. By Theorem 10.3.3(a),

$$D_X(Y) = \sum_j \left(\sum_i \alpha^i \frac{\partial \beta^j}{\partial x^i} \right) E_j \qquad \text{and} \qquad D_Y(X) = \sum_j \left(\sum_i \beta^i \frac{\partial \alpha^j}{\partial x^i} \right) E_j,$$

from which the result follows.   □

Theorem 10.4.2. *Let U be an open set in $\mathbb{R}^m$, let X, Y, Z be vector fields in $\mathfrak{X}(U)$, and let f, g be functions in $C^\infty(U)$. Then:*
(a) $[Y, X] = -[X, Y]$.
(b) $[X + Y, Z] = [X, Z] + [Y, Z]$.
(c) $[X, Y + Z] = [X, Y] + [X, Z]$.
(d) $[fX, gY] = fg[X, Y] + fD_X(g)Y - gD_Y(f)X$.
(e) $\big[X, [Y, Z]\big] + \big[Y, [Z, X]\big] + \big[Z, [X, Y]\big] = (0, \ldots, 0)$.   (Jacobi's identity)
(f) $[E_i, E_j] = (0, \ldots, 0)$ *for $i, j = 1, \ldots, m$.*

Proof. (a): Straightforward.
    (b), (c): This follows from parts (a) and (c) of Theorem 10.3.1.
    (d): Let

$$X = (\alpha^1, \ldots, \alpha^m), \qquad Y = (\beta^1, \ldots, \beta^m), \qquad \text{and} \qquad Z = (\gamma^1, \ldots, \gamma^m),$$

so that $fX = (f\alpha^1, \ldots, f\alpha^m)$ and $gY = (g\beta^1, \ldots, g\beta^m)$. By Theorem 10.3.2 and Theorem 10.4.1,

$$
\begin{aligned}
[fX, gY] &= \sum_j \left\{ \sum_i \left(f\alpha^i \frac{\partial(g\beta^j)}{\partial x^i} - g\beta^i \frac{\partial(f\alpha^j)}{\partial x^i} \right) \right\} E_j \\
&= \sum_j \left\{ \sum_i \left(f\alpha^i \frac{\partial g}{\partial x^i} \beta^j + f\alpha^i g \frac{\partial \beta^j}{\partial x^i} - g\beta^i \frac{\partial f}{\partial x^i} \alpha^j - g\beta^i f \frac{\partial \alpha^j}{\partial x^i} \right) \right\} E_j \\
&= fg \sum_j \left\{ \sum_i \left(\alpha^i \frac{\partial \beta^j}{\partial x^i} - \beta^i \frac{\partial \alpha^j}{\partial x^i} \right) \right\} E_j + f \left(\sum_i \alpha^i \frac{\partial g}{\partial x^i} \right) \left(\sum_j \beta^j E_j \right) \\
&\quad - g \left(\sum_i \beta^i \frac{\partial f}{\partial x^i} \right) \left(\sum_j \alpha^j E_j \right) \\
&= fg[X, Y] + fD_X(g)Y - gD_Y(f)X.
\end{aligned}
$$

    (f): This follows from Theorem 10.4.1.
    (e): It follows from parts (d) and (f) that

$$
\begin{aligned}
[\beta^j E_j, \gamma^k E_k] &= \beta^j \gamma^k [E_j, E_k] + \beta^j D_{E_j}(\gamma^k) E_k - \gamma^k D_{E_k}(\beta^j) E_j \\
&= \beta^j \frac{\partial \gamma^k}{\partial x^j} E_k - \gamma^k \frac{\partial \beta^j}{\partial x^k} E_j
\end{aligned}
$$

for $j, k = 1, \ldots, m$. We have

$$
\begin{aligned}
[X, [Y, Z]] &= \left[\sum_i \alpha^i E_i, \left[\sum_j \beta^j E_j, \sum_k \gamma^k E_k \right] \right] \\
&= \sum_{ijk} [\alpha^i E_i, [\beta^j E_j, \gamma^k E_k]] \\
&= \sum_{ijk} \left[\alpha^i E_i, \beta^j \frac{\partial \gamma^k}{\partial x^j} E_k - \gamma^k \frac{\partial \beta^j}{\partial x^k} E_j \right] \\
&= \overset{(1)}{\sum_{ijk}} \left[\alpha^i E_i, \beta^j \frac{\partial \gamma^k}{\partial x^j} E_k \right] - \overset{(2)}{\sum_{ijk}} \left[\alpha^i E_i, \gamma^k \frac{\partial \beta^j}{\partial x^k} E_j \right],
\end{aligned}
\tag{10.4.1}
$$

where summations are numbered for later reference. We also have

$$
\left[\alpha^i E_i, \beta^j \frac{\partial \gamma^k}{\partial x^j} E_k \right]
$$
$$
= \alpha^i \beta^j [E_i, E_k] + \alpha^i D_{E_i} \left(\beta^j \frac{\partial \gamma^k}{\partial x^j} \right) E_k - \beta^j \frac{\partial \gamma^k}{\partial x^j} D_{E_k}(\alpha^i) E_i
$$
$$
= \alpha^i \frac{\partial}{\partial x^i} \left(\beta^j \frac{\partial \gamma^k}{\partial x^j} \right) E_k - \beta^j \frac{\partial \gamma^k}{\partial x^j} \frac{\partial \alpha^i}{\partial x^k} E_i
$$
$$
= \alpha^i \frac{\partial \beta^j}{\partial x^i} \frac{\partial \gamma^k}{\partial x^j} E_k + \alpha^i \beta^j \frac{\partial^2 \gamma^k}{\partial x^i \partial x^j} E_k - \beta^j \frac{\partial \gamma^k}{\partial x^j} \frac{\partial \alpha^i}{\partial x^k} E_i,
$$

hence

$$
\overset{(1)}{\sum_{ijk}} \left[\alpha^i E_i, \beta^j \frac{\partial \gamma^k}{\partial x^j} E_k \right]
$$
$$
= \sum_{ijk} \alpha^i \beta^j \frac{\partial^2 \gamma^k}{\partial x^i \partial x^j} E_k + \sum_{ijk} \alpha^i \frac{\partial \beta^j}{\partial x^i} \frac{\partial \gamma^k}{\partial x^j} E_k - \sum_{ijk} \beta^j \frac{\partial \gamma^k}{\partial x^j} \frac{\partial \alpha^i}{\partial x^k} E_i
$$
$$
= \sum_{ijk} \alpha^i \beta^j \frac{\partial^2 \gamma^k}{\partial x^i \partial x^j} E_k + \sum_{ijk} \alpha^i \frac{\partial \beta^j}{\partial x^i} \frac{\partial \gamma^k}{\partial x^j} E_k - \sum_{ijk} \beta^i \frac{\partial \gamma^j}{\partial x^i} \frac{\partial \alpha^k}{\partial x^j} E_k
$$
$$
= \sum_k \left\{ \sum_{ij} \left(\alpha^i \beta^j \frac{\partial^2 \gamma^k}{\partial x^i \partial x^j} + \alpha^i \frac{\partial \beta^j}{\partial x^i} \frac{\partial \gamma^k}{\partial x^j} - \frac{\partial \alpha^k}{\partial x^j} \beta^i \frac{\partial \gamma^j}{\partial x^i} \right) \right\} E_k.
\tag{10.4.2}
$$

Replacing (α, β, γ) with (α, γ, β) in $\sum^{(1)}$ gives

$$
\overset{(2)}{\sum_{ijk}} \left[\alpha^i E_i, \gamma^j \frac{\partial \beta^k}{\partial x^j} E_k \right] = \sum_k \left\{ \sum_{ij} \left(\alpha^i \gamma^j \frac{\partial^2 \beta^k}{\partial x^i \partial x^j} + \alpha^i \frac{\partial \gamma^j}{\partial x^i} \frac{\partial \beta^k}{\partial x^j} - \frac{\partial \alpha^k}{\partial x^j} \gamma^i \frac{\partial \beta^j}{\partial x^i} \right) \right\} E_k,
$$

so

$$\sum_{ijk}^{(2)}\left[\alpha^i E_i,\gamma^k\frac{\partial\beta^j}{\partial x^k}E_j\right]$$

$$=\sum_k\left\{\sum_{ij}\left(\alpha^i\gamma^j\frac{\partial^2\beta^k}{\partial x^i\partial x^j}+\alpha^i\frac{\partial\gamma^j}{\partial x^i}\frac{\partial\beta^k}{\partial x^j}-\frac{\partial\alpha^k}{\partial x^j}\gamma^i\frac{\partial\beta^j}{\partial x^i}\right)\right\}E_k \qquad (10.4.3)$$

$$=\sum_k\left\{\sum_{ij}\left(\alpha^i\frac{\partial^2\beta^k}{\partial x^i\partial x^j}\gamma^j+\alpha^i\frac{\partial\beta^k}{\partial x^j}\frac{\partial\gamma^j}{\partial x^i}-\frac{\partial\alpha^k}{\partial x^j}\frac{\partial\beta^j}{\partial x^i}\gamma^i\right)\right\}E_k.$$

It follows from (10.4.1)–(10.4.3) that the kth component of $\big[X,[Y,Z]\big]$ is

$$[X,[Y,Z]]^k=\sum_{ij}^{(3)}\left(\alpha^i\beta^j\frac{\partial^2\gamma^k}{\partial x^i\partial x^j}+\alpha^i\frac{\partial\beta^j}{\partial x^i}\frac{\partial\gamma^k}{\partial x^j}-\frac{\partial\alpha^k}{\partial x^j}\beta^i\frac{\partial\gamma^j}{\partial x^i}\right.$$
$$\left.-\alpha^i\frac{\partial^2\beta^k}{\partial x^i\partial x^j}\gamma^j-\alpha^i\frac{\partial\beta^k}{\partial x^j}\frac{\partial\gamma^j}{\partial x^i}+\frac{\partial\alpha^k}{\partial x^j}\frac{\partial\beta^j}{\partial x^i}\gamma^i\right). \qquad (10.4.4)$$

Replacing (α,β,γ) with (β,γ,α) in $\sum^{(3)}$ gives

$$[Y,[Z,X]]^k=\sum_{ij}\left(\beta^i\gamma^j\frac{\partial^2\alpha^k}{\partial x^i\partial x^j}+\beta^i\frac{\partial\gamma^j}{\partial x^i}\frac{\partial\alpha^k}{\partial x^j}-\frac{\partial\beta^k}{\partial x^j}\gamma^i\frac{\partial\alpha^j}{\partial x^i}\right.$$
$$\left.-\beta^i\frac{\partial^2\gamma^k}{\partial x^i\partial x^j}\alpha^j-\beta^i\frac{\partial\gamma^k}{\partial x^j}\frac{\partial\alpha^j}{\partial x^i}+\frac{\partial\beta^k}{\partial x^j}\frac{\partial\gamma^j}{\partial x^i}\alpha^i\right)$$
$$=\sum_{ij}\left(\frac{\partial^2\alpha^k}{\partial x^i\partial x^j}\beta^i\gamma^j+\frac{\partial\alpha^k}{\partial x^j}\beta^i\frac{\partial\gamma^j}{\partial x^i}-\frac{\partial\alpha^j}{\partial x^i}\frac{\partial\beta^k}{\partial x^j}\gamma^i\right. \qquad (10.4.5)$$
$$\left.-\alpha^j\beta^i\frac{\partial^2\gamma^k}{\partial x^i\partial x^j}-\frac{\partial\alpha^j}{\partial x^i}\beta^i\frac{\partial\gamma^k}{\partial x^j}+\alpha^i\frac{\partial\beta^k}{\partial x^j}\frac{\partial\gamma^j}{\partial x^i}\right).$$

Replacing (α,β,γ) with (γ,α,β) in $\sum^{(3)}$ yields

$$[Z,[X,Y]]^k=\sum_{ij}\left(\gamma^i\alpha^j\frac{\partial^2\beta^k}{\partial x^i\partial x^j}+\gamma^i\frac{\partial\alpha^j}{\partial x^i}\frac{\partial\beta^k}{\partial x^j}-\frac{\partial\gamma^k}{\partial x^j}\alpha^i\frac{\partial\beta^j}{\partial x^i}\right.$$
$$\left.-\gamma^i\frac{\partial^2\alpha^k}{\partial x^i\partial x^j}\beta^j-\gamma^i\frac{\partial\alpha^k}{\partial x^j}\frac{\partial\beta^j}{\partial x^i}+\frac{\partial\gamma^k}{\partial x^j}\frac{\partial\alpha^j}{\partial x^i}\beta^i\right)$$
$$=\sum_{ij}\left(\alpha^j\frac{\partial^2\beta^k}{\partial x^i\partial x^j}\gamma^i+\frac{\partial\alpha^j}{\partial x^i}\frac{\partial\beta^k}{\partial x^j}\gamma^i-\alpha^i\frac{\partial\beta^j}{\partial x^i}\frac{\partial\gamma^k}{\partial x^j}\right. \qquad (10.4.6)$$
$$\left.-\frac{\partial^2\alpha^k}{\partial x^i\partial x^j}\beta^j\gamma^i-\frac{\partial\alpha^k}{\partial x^j}\frac{\partial\beta^j}{\partial x^i}\gamma^i+\frac{\partial\alpha^j}{\partial x^i}\beta^i\frac{\partial\gamma^k}{\partial x^j}\right).$$

Combining (10.4.4)–(10.4.6) and using of Theorem 10.1.6 gives the result. $\square$

Theorem 10.4.3. *If U is an open set in $\mathbb{R}^m$ and X, Y, Z are vector fields in $\mathfrak{X}(U)$, then:*

$$D_{D_X(Y)}(Z) - D_{D_Y(X)}(Z) = D_{[X,Y]}(Z)$$
$$= D_X\big(D_Y(Z)\big) - D_Y\big(D_X(Z)\big).$$

Proof. Let

$$X = (\alpha^1, \ldots, \alpha^m), \qquad Y = (\beta^1, \ldots, \beta^m), \qquad \text{and} \qquad Z = (\gamma^1, \ldots, \gamma^m).$$

(a): By Theorem 10.3.3(c),

$$D_{D_X(Y)}(Z) = \sum_k \left(\sum_{ij} \alpha^i \frac{\partial \beta^j}{\partial x^i} \right) \frac{\partial \gamma^k}{\partial x^j} E_k$$

and

$$D_{D_Y(X)}(Z) = \sum_k \left(\sum_{ij} \beta^i \frac{\partial \alpha^j}{\partial x^i} \right) \frac{\partial \gamma^k}{\partial x^j} E_k,$$

hence

$$D_{D_X(Y)}(Z) - D_{D_Y(X)}(Z)$$
$$= \sum_{kj} \left\{ \sum_i \left(\alpha^i \frac{\partial \beta^j}{\partial x^i} - \beta^i \frac{\partial \alpha^j}{\partial x^i} \right) \frac{\partial \gamma^k}{\partial x^j} \right\} E_k$$
$$= \sum_k \left(\sum_j [X,Y]^j \frac{\partial \gamma^k}{\partial x^j} \right) E_k \qquad \text{[Th 10.4.1]}$$
$$= D_{[X,Y]}(Z), \qquad\qquad\qquad\qquad \text{[Th 10.3.3(a)]}$$

which proves the first equality.

It follows from Theorem 10.3.3(b) that the kth component of $D_X\big(D_Y(Z)\big)$ is

$$\big(D_X(D_Y(Z))\big)^k = \sum_{ij} \left(\alpha^i \frac{\partial \beta^j}{\partial x^i} \frac{\partial \gamma^k}{\partial x^j} + \alpha^i \beta^j \frac{\partial^2 \gamma^k}{\partial x^i \partial x^j} \right). \qquad (10.4.7)$$

Replacing (α, β, γ) with (β, α, γ) in the preceding identity gives

$$\big(D_Y(D_X(Z))\big)^k = \sum_{ij} \left(\beta^i \frac{\partial \alpha^j}{\partial x^i} \frac{\partial \gamma^k}{\partial x^j} + \beta^i \alpha^j \frac{\partial^2 \gamma^k}{\partial x^i \partial x^j} \right). \qquad (10.4.8)$$

The kth components of $D_X(D_Y(Z)) - D_Y(D_X(Z))$ and $D_{[X,Y]}(Z)$ are equal:

$$
\begin{aligned}
\left(D_X(D_Y(Z)) - D_Y(D_X(Z))\right)^k & \\
= \sum_{ij}\left(\alpha^i\frac{\partial\beta^j}{\partial x^i}\frac{\partial\gamma^k}{\partial x^j} - \beta^i\frac{\partial\alpha^j}{\partial x^i}\frac{\partial\gamma^k}{\partial x^j}\right) & \qquad [(10.4.7),\ (10.4.8)] \\
= \sum_{j}\left\{\sum_i\left(\alpha^i\frac{\partial\beta^j}{\partial x^i} - \beta^i\frac{\partial\alpha^j}{\partial x^i}\right)\right\}\frac{\partial\gamma^k}{\partial x^j} & \\
= \sum_{j}[X,Y]^j\frac{\partial\gamma^k}{\partial x^j} & \qquad [\text{Th } 10.4.1] \\
= \left(D_{[X,Y]}(Z)\right)^k, & \qquad [\text{Th } 10.3.3(\text{a})]
\end{aligned}
$$

which proves the second equality.   □

10.5   Integrals

Having discussed derivatives in Sections 10.1–10.4, we now briefly turn our attention to integrals.

Let $[a_i, b_i]$ be a closed interval in $\mathbb{R}$, where $a_i \le b_i$ for $i = 1, \ldots, m$. The set

$$C = [a_1, b_1] \times \cdots \times [a_m, b_m]$$

is called a **closed cell** in $\mathbb{R}^m$. The **content of C** is defined by

$$\text{cont}(C) = (b_1 - a_1)\cdots(b_m - a_m). \qquad (10.5.1)$$

We observe that $\text{cont}(C) = 0$ if and only if $a_i = b_i$ for some $1 \le i \le m$. When $m = 1, 2, 3$, "content" is referred to as "length","area", "volume", respectively. To illustrate, consider the subset $[0, 1]$ of $\mathbb{R}$ and the "geometrically equivalent" subset $[0, 1] \times [0, 0]$ of $\mathbb{R}^2$. Then $[0, 1]$ has a length of 1, while $[0, 1] \times [0, 0]$ has an area of 0. A subset S of $\mathbb{R}^m$ is said to have **content zero** if for every real number $\varepsilon > 0$, there is a *finite* collection $\{C_i : 1 = 1, \ldots, k\}$ of closed cells such that $S \subseteq \bigcup_{i=1}^k C_i$ and $\sum_{i=1}^k \text{cont}(C_i) < \varepsilon$. For example, $[0, 1] \times [0, 0]$ has content zero.

Let C be a closed cell in $\mathbb{R}^m$, and let $f : C \longrightarrow \mathbb{R}$ be a bounded function. We say that a finite collection $\{C_i : 1 = 1, \ldots, k\}$ of closed cells in $\mathbb{R}^m$ is a **partition of C** if each C_i is a subset of C, the C_i intersect only along their boundaries, and $C = \bigcup_{i=1}^k C_i$. For example, $\{[0, 1] \times [0, 1], [1, 2] \times [0, 1]\}$ is a partition of $[0, 2] \times [0, 1]$. Given a point p_i in C_i for $i = 1, \ldots, k$, the sum $\sum_i f(p_i)\,\text{cont}(C_i)$ is an approximation to what we intuitively think of as the "content under the graph of f". For each choice of partition of C and each choice of the p_i, we get a corresponding sum. Using an approach analogous to that adopted in the integral calculus of one real variable, where inscribed and superscribed rectangles are used to approximate the area under the graph of a real-valued function of one real variable, we define lower and upper integrals of f over C, denoted by $\underline{\int}_C f$ and $\overline{\int}_C f$, respectively, to be limits of the preceding approximating sums. It can be shown that since f is bounded, both $\underline{\int}_C f$ and $\overline{\int}_C f$ are finite. We say that f is **(Riemann) integrable** if $\underline{\int}_C f$ and $\overline{\int}_C f$ are equal. In that case, their common value, called the **(Riemann) integral of f over C**, is denoted by

$$\int_C f \, dx^1 \cdots dx^m \qquad \text{or} \qquad \int_C f(x^1, \ldots, x^m) \, dx^1 \cdots dx^m.$$

We need to be able to integrate functions over sets less restrictive than closed cells. Let S be a bounded set in $\mathbb{R}^m$, and let $f : S \longrightarrow \mathbb{R}$ be a bounded function. Since S is bounded, there is a closed cell C containing S. Let us define a function $f_C : C \longrightarrow \mathbb{R}$ by

$$f_C(p) = \begin{cases} f(p) & \text{if } p \in S \\ 0 & \text{if } p \in C \backslash S. \end{cases}$$

We say that f is **(Riemann) integrable** if f_C is integrable. In that case, the **(Riemann) integral of f over S** is denoted by $\int_S f \, dx^1 \cdots dx^m$ and defined by

$$\int_S f \, dx^1 \cdots dx^m = \int_C f_C \, dx^1 \cdots dx^m. \tag{10.5.2}$$

It can be shown that the integrability of f and the value of $\int_S f \, dx^1 \cdots dx^m$ are independent of the choice of closed cell containing S.

We now introduce a type of bounded set in $\mathbb{R}^m$ over which a certain type of function is always integrable. A subset D of $\mathbb{R}^m$ is called a **domain of integration** if it is bounded and its boundary in $\mathbb{R}^m$ has content zero. For example, a closed cell in $\mathbb{R}^m$ is a domain of integration.

Theorem 10.5.1 (Criterion for Integrability). *If D is a domain of integration in $\mathbb{R}^m$ and $f : D \longrightarrow \mathbb{R}$ is a bounded continuous function, then f is integrable.* $\qquad\square$

Theorem 10.5.2. *Let D be a domain of integration in $\mathbb{R}^m$, let $f, g : D \longrightarrow \mathbb{R}$ be bounded continuous functions, and let c be a real number. Then $cf + g$ is integrable and*

$$\int_D (cf + g) \, dx^1 \cdots dx^m = c \int_D f \, dx^1 \cdots dx^m + \int_D g \, dx^1 \cdots dx^m. \qquad\square$$

Let U be an open set in $\mathbb{R}^m$, and let $f : U \longrightarrow \mathbb{R}$ be a continuous function that has compact support; that is, $\text{supp}(f)$ is compact in U. By Theorem 9.2.2, U is the union of a collection of open balls in $\mathbb{R}^m$. Using Theorem 9.1.5, it is easily shown that since $\text{supp}(f)$ is compact in U, it is contained in the union D of a finite subcollection of the open balls. Thus, $\text{supp}(f) \subseteq D \subseteq U$. Furthermore, it can also be shown that D is a domain of integration. The

(Riemann) integral of f over U is denoted by $\int_U f \, dx^1 \cdots dx^m$ and defined by

$$\int_U f \, dx^1 \cdots dx^m = \int_D f \, dx^1 \cdots dx^m, \tag{10.5.3}$$

where the right-hand side is given by (10.5.2). By Theorem 9.1.13, f is continuous on D, and by Theorem 9.1.22(b), f is bounded on $\mathrm{supp}(f)$, hence bounded on D. We have from Theorem 10.5.1 that the integral exists. It can be shown that the value of the integral is independent of the choice of domain of integration containing $\mathrm{supp}(f)$.

Theorem 10.5.3 (Change of Variables). *Let U and V be open sets in $\mathbb{R}^m$, let $F : U \longrightarrow V$ be a diffeomorphism, and let $f : V \longrightarrow \mathbb{R}$ be a continuous function that has compact support. Then*

$$\int_V f \, dx^1 \cdots dx^m = \int_U (f \circ F) \, |\det(J_F)| \, dx^1 \cdots dx^m,$$

where J_F is the Jacobian matrix of F. $\qquad\square$

Theorem 10.5.4 (Iterated Integral). *If $C = [a^1, b^1] \times \cdots \times [a^m, b^m]$ is a closed cell in $\mathbb{R}^m$ and $f : C \longrightarrow \mathbb{R}$ is a continuous function, then*

$$\int_C f = \int_{a^m}^{b^m} \left\{ \cdots \left[\int_{a^2}^{b^2} \left(\int_{a^1}^{b^1} f(x^1, \ldots, x^m) \, dx^1 \right) dx^2 \right] \cdots \right\} dx^m.$$

Furthermore, the value of $\int_C f$ is independent of the order in which the variables are integrated. $\qquad\square$

Theorem 10.5.5 (Differentiating Under Integral Sign). *Let U be an open set in $\mathbb{R}^m$, and let*

$$f(x^1, \ldots, x^m, t) : U \times [a, b] \longrightarrow \mathbb{R}$$

be a continuous function such that $\partial f / \partial x^i$ exists and is continuous on $U \times [a, b]$ for $i = 1, \ldots, m$. Define a function $F : U \longrightarrow \mathbb{R}$ by

$$F(x^1, \ldots, x^m) = \int_a^b f(x^1, \ldots, x^m, t) \, dt$$

for all $(x^1, \ldots, x^m)$ in U. Then F is in $C^1(U)$ and

$$\frac{\partial F}{\partial x^i}(x^1, \ldots, x^m) = \int_a^b \frac{\partial f}{\partial x^i}(x^1, \ldots, x^m, t) \, dt$$

for $i = 1, \ldots, m$. $\qquad\square$

Let D be a domain of integration in $\mathbb{R}^m$. It is clear that the function $1_D : D \longrightarrow \mathbb{R}$ with constant value 1 is continuous and bounded, so Theorem 10.5.1 applies. The **content of D** is defined by

$$\mathrm{cont}(D) = \int_D 1_D \, dx^1 \cdots dx^m. \tag{10.5.4}$$

It can be shown that when D is a closed cell in $\mathbb{R}^m$, the definitions of $\mathrm{cont}(D)$ given by (10.5.1) and (10.5.4) agree.

Theorem 10.5.6 (Mean Value Theorem for Integrals). *If D is a connected domain of integration in $\mathbb{R}^m$ and $f : D \longrightarrow \mathbb{R}$ is a bounded continuous function, then there is a point p_0 in D such that*

$$\int_D f \, dx^1 \cdots dx^m = f(p_0) \operatorname{cont}(D). \qquad \square$$

10.6   Vector Calculus

In this section, we present a brief overview of the basic definitions and results of vector calculus. Much of what follows will be replaced later on with more modern counterparts expressed in the language of differential geometry.

Let U be an open set in $\mathbb{R}^3$, let f, g be functions in $C^\infty(U)$, and let $F = (F^1, F^2, F^3) : U \longrightarrow \mathbb{R}^3$ be a smooth map. The definitions of the classical differential operators in $\mathbb{R}^3$ are given in Table 10.6.1, with the classical vector calculus notation shown alongside the notation to be used in this book. The use of "lap" for the Laplacian is not conventional, but it fits well with notation to be introduced later.

Operator	Definition
Gradient	$\nabla f = \left(\dfrac{\partial f}{\partial x}, \dfrac{\partial f}{\partial y}, \dfrac{\partial f}{\partial z} \right) = \operatorname{grad}(f)$
Divergence	$\nabla \cdot F = \dfrac{\partial F^1}{\partial x} + \dfrac{\partial F^2}{\partial y} + \dfrac{\partial F^3}{\partial z} = \operatorname{div}(F)$
Curl	$\nabla \times F = \left(\dfrac{\partial F^3}{\partial y} - \dfrac{\partial F^2}{\partial z}, \dfrac{\partial F^1}{\partial z} - \dfrac{\partial F^3}{\partial x}, \dfrac{\partial F^2}{\partial x} - \dfrac{\partial F^1}{\partial y} \right) = \operatorname{curl}(F)$
Laplacian	$\nabla^2 f = \dfrac{\partial^2 f}{\partial x^2} + \dfrac{\partial^2 f}{\partial y^2} + \dfrac{\partial^2 f}{\partial z^2} = \operatorname{lap}(f)$
Laplacian	$\nabla^2 F = \left(\operatorname{lap}(F^1), \operatorname{lap}(F^2), \operatorname{lap}(F^3) \right) = \operatorname{lap}(F)$

Table 10.6.1. Classical differential operators

Theorem 10.6.1. *With the above setup:*
(a) $\operatorname{grad}(fg) = f \operatorname{grad}(g) + g \operatorname{grad}(f)$.
(b) $\operatorname{div}(fF) = f \operatorname{div}(F) + \langle \operatorname{grad}(f), F \rangle$.
(c) $\operatorname{curl}\big(\operatorname{grad}(f)\big) = 0$.
(d) $\operatorname{div}\big(\operatorname{curl}(F)\big) = 0$.
(e) $\operatorname{curl}\big(\operatorname{curl}(F)\big) = \operatorname{grad}\big(\operatorname{div}(F)\big) - \operatorname{lap}(F)$. $\qquad \square$

Theorem 10.6.2. *With the above setup, the following are equivalent:*
(a) $\mathrm{curl}(F) = 0$.
(b) $F = \mathrm{grad}(f)$ *for some function* f *in* $C^\infty(U)$.
(c) *For any closed path* $\mathcal{P}$ *in* U, *the line integral* $\int_{\mathcal{P}} F \cdot ds$ *equals* 0.   □

Theorem 10.6.3. *With the above setup, the following are equivalent:*
(a) $\mathrm{div}(F) = 0$.
(b) $F = \mathrm{curl}(G)$ *for some smooth map* $G : U \longrightarrow \mathbb{R}^3$.   □

Part II

Curves and Regular Surfaces

Chapter 11

Curves and Regular Surfaces in $\mathbb{R}^3$

In earlier discussions, the set $\mathbb{R}^m$ appeared in a variety of contexts: as a vector space (also denoted by $\mathbb{R}^m$), an inner product space $(\mathbb{R}^m, \mathfrak{e})$, a normed vector space $(\mathbb{R}^m, \|\cdot\|)$, a metric space $(\mathbb{R}^m, \mathfrak{d})$, and a topological space $(\mathbb{R}^m, \mathfrak{T})$. Section 9.4 outlines the logical connections between these spaces. Looking back at Chapter 10, it would have been more precise, although cumbersome, to use the notation $(\mathbb{R}^m, \mathfrak{e}, \|\cdot\|, \mathfrak{d}, \mathfrak{T})$, or at least $(\mathbb{R}^m, \mathfrak{e})$, instead of simply $\mathbb{R}^m$ when discussing Euclidean derivatives and integrals. In this chapter, we are concerned with many of the same concepts considered in Chapter 10, but this time exclusively for $m = 3$. We use the notation $\mathbb{R}^3$ in the preceding generic manner, allowing the structures relevant to a particular discussion to be left implicit. Aside from notational convenience, this has the added virtue of reserving the notation $\mathbb{R}_0^3 = (\mathbb{R}^m, \mathfrak{e})$, and later $\mathbb{R}_1^3 = (\mathbb{R}^m, \mathfrak{m})$, for more specific purposes in Chapter 12.

11.1 Curves in $\mathbb{R}^3$

Recall from Section 10.1 the definition of a (parametrized) curve in $\mathbb{R}^3$ and what it means for such a curve to be smooth. A smooth curve $\lambda(t) : (a, b) \longrightarrow \mathbb{R}^3$ is said to be **regular** if its velocity $(d\lambda/dt)(t) : (a, b) \longrightarrow \mathbb{R}^3$ is nowhere-vanishing. Let $g(u) : (c, d) \longrightarrow (a, b)$ be a diffeomorphism. Since λ and g are smooth, by Theorem 10.1.12, so is $\lambda \circ g$. We say that the curve $\lambda \circ g(u) : (c, d) \longrightarrow \mathbb{R}^3$ is a **smooth reparametrization of λ**.

Semi-Riemannian Geometry, First Edition. Stephen C. Newman.
© 2019 John Wiley & Sons, Inc. Published 2019 by John Wiley & Sons, Inc.

Theorem 11.1.1. *Let $\lambda(t) : (a, b) \longrightarrow \mathbb{R}^3$ be a smooth curve, and let $g(u) :$ $(c, d) \longrightarrow (a, b)$ be a diffeomorphism. Then λ is regular if and only if $\lambda \circ g$ is regular.*

Proof. By Theorem 10.1.10,

$$\frac{d(\lambda \circ g)}{du}(u) = \frac{dg}{du}(u) \, \frac{d\lambda}{dt}(g(u)),$$

and by Theorem 10.2.4(a), dg/du is nowhere-vanishing. Since g is bijective, as u varies over (c, d), $g(u)$ varies over (a, b). It follows that $d\lambda/dt$ is nowhere-vanishing if and only if $d(\lambda \circ g)/du$ is nowhere-vanishing. $\qquad\square$

Theorem 11.1.1 can be used to define an equivalence relation on the collection of smooth curves as follows: for two such curves λ and ψ, we write $\lambda \sim \psi$ if ψ is a smooth reparametrization of λ. This idea will not be pursued further, but it makes the point that our focus should be on the intrinsic properties of a "curve", for example, whether it is regular, and not the specifics of a particular parametrization.

11.2   Regular Surfaces in $\mathbb{R}^3$

Our immediate goal in this section is to define what we temporarily refer to as a "smooth surface". We all have an intuitive idea of what it means for a geometric object to be "smooth". For example, the sphere definitely has this property, but not the cube. The challenge is to translate such intuition into rigorous mathematical language. A feature of the sphere that gives it "smoothness" is our ability to attach to each of its points a unique "tangent plane", something that is not possible for the cube.

Let U be an open set in $\mathbb{R}^2$, and let $\varphi : U \longrightarrow \mathbb{R}^3$ be a smooth map. In the present context, we refer to φ as a **parametrized surface**. The differential map at a point q in U is $d_q(\varphi) : \mathbb{R}^2 \longrightarrow \mathbb{R}^3$. It follows from Theorem 10.2.1 that if φ is an immersion at q, then $d_q(\varphi)(\mathbb{R}^2)$ is a 2-dimensional vector space, which we can view as a "tangent plane" to the graph of φ at $\varphi(q)$. This suggests that a "smooth surface" might reasonably be defined to be the image of a parametrized surface when the latter has the added feature of being an immersion. Before exploring this concept, we need to establish the notation for coordinates in $\mathbb{R}^2$ and $\mathbb{R}^3$.

In this chapter and the next, coordinates on $\mathbb{R}^2$ are denoted by (r^1, r^2) or (r, s), and those on $\mathbb{R}^3$ by (x^1, x^2, x^3) or (x, y, z).

We sometimes, especially in the examples, identify $\mathbb{R}^2$ with the xy-plane in $\mathbb{R}^3$. In that setting, coordinates on $\mathbb{R}^2$ are denoted by (x, y).

Let U be an open set in $\mathbb{R}^2$, and let $\varphi = (\varphi^1, \varphi^2, \varphi^3) : U \longrightarrow \mathbb{R}^3$ be a parametrized surface. Since φ is smooth, by definition, so are φ^1, φ^2, and φ^3. For each q in U, we have

$$\varphi(q) = \big(\varphi^1(q), \varphi^2(q), \varphi^3(q)\big),$$

hence

$$\frac{\partial \varphi}{\partial r^i}(q) = \left(\frac{\partial \varphi^1}{\partial r^i}(q), \frac{\partial \varphi^2}{\partial r^i}(q), \frac{\partial \varphi^3}{\partial r^i}(q) \right) \tag{11.2.1}$$

for $i = 1, 2$. For brevity, let us denote

$$\frac{\partial \varphi}{\partial r^i}(q) \qquad \text{by} \qquad H_i|_q.$$

Theorem 11.2.1. *Let* $\varphi : U \longrightarrow \mathbb{R}^3$ *be a parametrized surface, let* q *be a point in* U, *and let* (e_1, e_2) *be the standard basis for* $\mathbb{R}^2$. *Then:*
(a) $d_q(\varphi)(e_i) = H_i|_q$ *for* $i = 1, 2$.
(b) $d_q(\varphi)(\mathbb{R}^2) = \text{span}(\{H_1|_q, H_2|_q\})$.

Proof. For a vector $v = (a^1, a^2)$ in $\mathbb{R}^2$, we have

$$\begin{aligned}
d_q(\varphi)(v) &= \left(d_q(\varphi^1)(v), d_q(\varphi^2)(v), d_q(\varphi^3)(v) \right) && \text{[Th 10.1.8(b)]} \\
&= \left(\sum_i a^i \frac{\partial \varphi^1}{\partial r^i}(q), \sum_i a^i \frac{\partial \varphi^2}{\partial r^i}(q), \sum_i a^i \frac{\partial \varphi^3}{\partial r^i}(q) \right) && \text{[Th 10.1.2]} \\
&= \sum_i a^i \left(\frac{\partial \varphi^1}{\partial r^i}(q), \frac{\partial \varphi^2}{\partial r^i}(q), \frac{\partial \varphi^3}{\partial r^i}(q) \right) \\
&= \sum_i a^i H_i|_q,
\end{aligned}$$

from which the result follows.   □

Theorem 11.2.2. *If* $\varphi : U \longrightarrow \mathbb{R}^3$ *is a parametrized surface and* q *is a point in* U, *then the following are equivalent:*
(a) φ *is an immersion at* q.
(b) $d_q(\varphi)(\mathbb{R}^2)$ *is 2-dimensional.*
(c) $J_\varphi(q)$ *has rank 2.*
(d) $H_1|_q$ *and* $H_2|_q$ *are linearly independent.*
(e) $H_1|_q \times H_2|_q \neq (0, 0, 0)$.

Remark. As noted in connection with Theorem 8.4.8, we are free to compute the vector product in part (e) using any choice of scalar product and orthonormal basis. It is convenient in the present context to work with the Euclidean inner product and the standard basis. In other words, computations will be performed in $\mathbb{R}_0^3$.

Proof. (a) ⇔ (b) ⇔ (c): This follows from Theorem 10.2.1.
    (c) ⇔ (d): We have from (10.1.2) that

$$J_\varphi(q) = \begin{bmatrix} \dfrac{\partial \varphi^1}{\partial r^1}(q) & \dfrac{\partial \varphi^1}{\partial r^2}(q) \\[2ex] \dfrac{\partial \varphi^2}{\partial r^1}(q) & \dfrac{\partial \varphi^2}{\partial r^2}(q) \\[2ex] \dfrac{\partial \varphi^3}{\partial r^1}(q) & \dfrac{\partial \varphi^3}{\partial r^2}(q) \end{bmatrix},$$

from which the result follows.

(d) $\Leftrightarrow$ (e): This follows from Theorem 8.4.8. $\square$

The vector product approach in part (e) of Theorem 11.2.2 is a computationally convenient way of determining whether a parametrized surface is an immersion, and we will use it often.

For simplicity, the figures for the next two examples have been drawn in the xy-plane of $\mathbb{R}^3$, leaving it to the reader to imagine the suppressed z-axis.

Example 11.2.3. Consider the parametrized surface $\varphi : \mathbb{R}^2 \longrightarrow \mathbb{R}^3$ given by

$$\varphi(r, s) = (r^3, r^2, s).$$

In the 3-dimensional version of Figure 11.2.1, $\varphi(U)$ is not "smooth" along the z-axis because of a cusp. The Jacobian matrix is

$$J_\varphi(r, s) = \begin{bmatrix} 3r^2 & 0 \\ 2r & 0 \\ 0 & 1 \end{bmatrix},$$

and the corresponding vector product is

$$(3r^2, 2r, 0) \times (0, 0, 1) = (2r, -3r^2, 0),$$

which equals $(0, 0, 0)$ when $r = 0$. It follows from Theorem 11.2.2 that φ is not an immersion. $\Diamond$

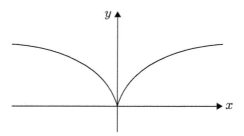

Figure 11.2.1. Diagram for Example 11.2.3

Example 11.2.4. Consider the parametrized surface $\varphi : \mathbb{R}^2 \longrightarrow \mathbb{R}^3$ given by

$$\varphi(r, s) = (r^3 - r, r^2 - 1, s).$$

In the 3-dimensional version of Figure 11.2.2, $\varphi(U)$ is not "smooth" along the z-axis because of self-intersection. The Jacobian matrix is

$$J_\varphi(r, s) = \begin{bmatrix} 3r^2 - 1 & 0 \\ 2r & 0 \\ 0 & 1 \end{bmatrix},$$

and the corresponding vector product is

$$(3r^2 - 1, 2r, 0) \times (0, 0, 1) = (2r, -3r^2 + 1, 0),$$

which never equals $(0, 0, 0)$. By Theorem 11.2.2, φ is an immersion. The self-intersection is parametrized by

$$\varphi(1, s) = (0, 0, s) = \varphi(-1, s)$$

for all real numbers s. We have from Theorem 11.2.1(b) that

$$d_{(1,s)}(\varphi)(\mathbb{R}^2) = \mathrm{span}\{(2, 2, 0), (0, 0, 1)\}$$

and

$$d_{(-1,s)}(\varphi)(\mathbb{R}^2) = \mathrm{span}\{(2, -2, 0), (0, 0, 1)\}.$$

This shows that along the line of self-intersection, there are two candidates for "tangent plane" at each point.   ◇

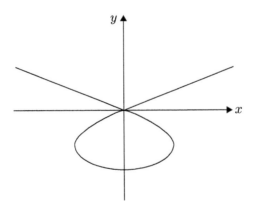

Figure 11.2.2. Diagram for Example 11.2.4

The upshot of the preceding examples is that φ being an immersion is necessary for $\varphi(U)$ to be "smooth", but not sufficient. At a minimum, we need to add the requirement that $\varphi(U)$ does not self-intersect, or equivalently, that φ is injective. Further examples (that will not be presented) reveal additional deficiencies inherent in defining a "smooth surface" to be the image of some type of parametrized surface.

We now take a different approach to the problem that can be loosely described as follows: a "smooth surface" is defined to be a topological subspace of $\mathbb{R}^3$ that can be covered in a piecewise fashion by a collection of parametrized surfaces in such a way that the pieces "fit together nicely". We need to make all this precise.

Let M be a topological subspace of $\mathbb{R}^3$. A **chart (on M)** is a pair (U, φ), where U is an open set in $\mathbb{R}^2$ and $\varphi : U \longrightarrow \mathbb{R}^3$ is a parametrized surface such that:

[C1] $\varphi : U \longrightarrow \mathbb{R}^3$ is an immersion.
[C2] $\varphi(U)$ is an open set in M.
[C3] $\varphi : U \longrightarrow \varphi(U)$ is a homeomorphism.

Condition [C1] has been discussed in detail. Conditions [C2] and [C3] are far from intuitive, but we can at least say about [C3] that it ensures φ is injective, thereby avoiding the problem of self-intersection discussed above.

When it is necessary to make the components of φ explicit in (U, φ), we use the notation $(U, \varphi = (\varphi^i))$ or $(U, \varphi = (\varphi^1, \varphi^2, \varphi^3))$. We refer to U as the **coordinate domain** of the chart, and to φ as its **coordinate map**. For each point p in $\varphi(U)$, (U, φ) is said to be a **chart at p**. When $\varphi(U) = M$, we say that (U, φ) is a **covering chart** on M, and that M is **covered** by (U, φ). Two charts, (U, φ) and $(\widetilde{U}, \widetilde{\varphi})$, on M are said to be **overlapping** if $V = \varphi(U) \cap \widetilde{\varphi}(\widetilde{U})$ is nonempty. In that case, the map

$$\widetilde{\varphi}^{-1} \circ \varphi|_{\varphi^{-1}(V)} : \varphi^{-1}(V) \longrightarrow \widetilde{\varphi}^{-1}(V)$$

is called a **transition map**. For brevity, we usually denote

$$\widetilde{\varphi}^{-1} \circ \varphi|_{\varphi^{-1}(V)} \qquad \text{by} \qquad \widetilde{\varphi}^{-1} \circ \varphi.$$

An **atlas** for M is a collection $\mathfrak{A} = \{(U_\alpha, \varphi_\alpha) : \alpha \in A\}$ of charts on M such that the $\varphi_\alpha(U_\alpha)$ form an open cover of M; that is,

$$M = \bigcup_{\alpha \in A} \varphi_\alpha(U_\alpha).$$

We are now in a position to replace our preliminary attempt at describing a "smooth surface" with something definitive. A **regular surface (in $\mathbb{R}^3$)** is a pair $(M, \mathfrak{A})$, where M is a topological subspace of $\mathbb{R}^3$ and $\mathfrak{A}$ is an atlas for M. A noteworthy feature of this definition is that it places no requirements on the choice of charts making up the atlas other than that their coordinate domains cover M. We usually adopt the shorthand of referring to M as a regular surface, with $\mathfrak{A}$ understood from the context.

Example 11.2.5 (Chart). Let M be a regular surface, and let (U, φ) be a chart on M. Then $\varphi(U)$ is a regular surface and (U, φ) is a covering chart. ◊

Throughout, any chart on a regular surface is viewed as a regular surface.

Example 11.2.6 ($\mathcal{S}^2$). The **unit sphere** (centered at the origin) is

$$\mathcal{S}^2 = \{(x, y, z) \in \mathbb{R}^3 : x^2 + y^2 + z^2 = 1\}.$$

Recall that

$$D = \{(r, s) \in \mathbb{R}^2 : r^2 + s^2 < 1\}$$

is the unit open disk. In what follows, we identify $\mathbb{R}^2$ with the xy-plane in $\mathbb{R}^3$. Let us define functions $\varphi_1, \ldots, \varphi_6 : D \longrightarrow \mathbb{R}^3$ by:

$$\varphi_1(x, y) = \left(x, y, \sqrt{1 - x^2 - y^2}\right)$$
$$\varphi_2(x, y) = \left(x, y, -\sqrt{1 - x^2 - y^2}\right)$$
$$\varphi_3(x, y) = \left(x, \sqrt{1 - x^2 - y^2}, y\right)$$
$$\varphi_4(x, y) = \left(x, -\sqrt{1 - x^2 - y^2}, y\right)$$
$$\varphi_5(x, y) = \left(\sqrt{1 - x^2 - y^2}, x, y\right)$$
$$\varphi_6(x, y) = \left(-\sqrt{1 - x^2 - y^2}, x, y\right).$$

It can be shown that $(D, \varphi_1), \ldots, (D, \varphi_6)$ are charts on $\mathcal{S}^2$, with $\mathcal{S}^2 = \varphi^1(D) \cup \cdots \cup \varphi^6(D)$. Thus, $\{(D, \varphi_1), \ldots, (D, \varphi_6)\}$ is an atlas for $\mathcal{S}^2$, hence $\mathcal{S}^2$ is a regular surface. $\diamondsuit$

Theorem 11.2.7. *A regular surface is not an open set in* $\mathbb{R}^3$.

Proof. Let M be a regular surface, and let (U, φ) be a chart on M. According to [C2], $\varphi(U)$ is open in M. By definition, there is an open set $\mathcal{U}$ in $\mathbb{R}^3$ such that $\varphi(U) = M \cap \mathcal{U}$. It follows that M is not open in $\mathbb{R}^3$; for if it were, then $\varphi(U)$ would be open in $\mathbb{R}^3$, and then Theorem 9.4.1 and [C3] would give a contradiction. $\square$

An implication of Theorem 11.2.7 is that in order to investigate whether a function or map that has a regular surface as its domain is smooth, we need to rely on the extended version of smoothness described at the end of Section 10.1. The next result is an important case in point.

Theorem 11.2.8 (Smoothness of Inverse Coordinate Map). *If M is a regular surface and (U, φ) is a chart on M, then $\varphi^{-1} : \varphi(U) \longrightarrow U$ is (extended) smooth.*

Proof. Let $\varphi = (\varphi^1, \varphi^2, \varphi^3)$, and let q be a point in U. According to [C2], there is a neighborhood $\mathcal{V}$ of $\varphi(q)$ in $\mathbb{R}^3$ such that $\varphi(U) = M \cap \mathcal{V}$. It follows from [C1] and Theorem 11.2.2 that $J_\varphi(q)$ has rank 2. Relabeling coordinates in $\mathbb{R}^3$ if necessary, we have from Theorem 2.3.2 and Theorem 2.4.8 that $\det(J_\varphi(q)^{(1,2)}) \neq 0$. Define a smooth map $F : U \times \mathbb{R} \longrightarrow \mathbb{R}^3$ by

$$F(r, s, t) = \varphi(r, s) + (0, 0, t) = \left(\varphi^1(r, s), \varphi^2(r, s), \varphi^3(r, s) + t\right),$$

so that $F(r, s, 0) = \varphi(r, s)$ for all (r, s) in U. In geometric terms, F can be thought of as sending "horizontal slices" of an "infinite cylinder" over U to "horizontal slices" of an "infinite cylinder" over $\varphi(U)$. In particular, F sends

$U \times \{(0,0,0)\}$ to $\varphi(U)$. See Figure 11.2.3. Since φ is smooth, so is F. We have

$$
J_F(q,0) = \begin{bmatrix}
\dfrac{\partial \varphi^1}{\partial r}(q) & \dfrac{\partial \varphi^1}{\partial s}(q) & 0 \\[2ex]
\dfrac{\partial \varphi^2}{\partial r}(q) & \dfrac{\partial \varphi^2}{\partial s}(q) & 0 \\[2ex]
\dfrac{\partial \varphi^3}{\partial r}(q) & \dfrac{\partial \varphi^3}{\partial s}(q) & 1
\end{bmatrix},
$$

hence $\det\big(J_F(q,0)\big) \neq 0$. By Theorem 10.2.1 and Theorem 10.2.3, there is a neighborhood $W \subseteq U \times \mathbb{R}$ of $(q,0)$ in $\mathbb{R}^3$ and a neighborhood V of $\varphi(q)$ in $\mathbb{R}^3$ such that $F|_W : W \longrightarrow V$ is a diffeomorphism. Since V and $\mathcal{V}$ are open sets in $\mathbb{R}^3$, so is $V \cap \mathcal{V}$, hence $V \cap \mathcal{V}$ is a neighborhood of $\varphi(q)$ in $\mathbb{R}^3$. Replacing V with $V \cap \mathcal{V}$, and W with $(F|_W)^{-1}(V \cap \mathcal{V})$ if necessary, we assume without loss of generality that $V \subseteq \mathcal{V}$. In a similar way, we assume without loss of generality that W is a "finite cylinder" over a neighborhood $U' \subseteq U$ of q in the rs-plane. It follows that $V \cap M$ is a neighborhood of $\varphi(q)$ in M and $\varphi(U') = V \cap M$. Let $\mathcal{P} : \mathbb{R}^3 \longrightarrow \mathbb{R}^2$ be the projection map defined by $\mathcal{P}(x,y,z) = (x,y)$. Since $\mathcal{P}$ and $(F|_W)^{-1}$ are smooth, by Theorem 10.1.12, so is $\mathcal{P} \circ (F|_W)^{-1} : V \longrightarrow U'$. We have

$$
\mathcal{P} \circ (F|_W)^{-1}\big(\varphi(r,s)\big) = \mathcal{P} \circ (F|_W)^{-1}\big(F(r,s,0)\big) = \mathcal{P}(r,s,0) = (r,s)
$$

for all (r,s) in U', hence $\mathcal{P} \circ (F|_W)^{-1}|_{\varphi(U')} = \varphi^{-1}|_{\varphi(U')}$. Thus, $\mathcal{P} \circ (F|_W)^{-1}$ is a smooth map on a neighborhood of $\varphi(q)$ in $\mathbb{R}^3$ that restricts to φ^{-1} on a neighborhood of $\varphi(q)$ in M. Since q was arbitrary, the result follows. $\qquad\square$

Theorem 11.2.9. *If M is a regular surface, and (U,φ) and $(\widetilde{U},\widetilde{\varphi})$ are overlapping charts on M, then the transition map*

$$
\widetilde{\varphi}^{-1} \circ \varphi : \varphi^{-1}(V) \longrightarrow \widetilde{\varphi}^{-1}(V)
$$

is a (Euclidean) diffeomorphism, where $V = \varphi(U) \cap \widetilde{\varphi}(\widetilde{U})$.

Proof. We have from [C3] that $\widetilde{\varphi}^{-1} \circ \varphi$ is bijective. By definition, φ is (Euclidean) smooth, and according to Theorem 11.2.8, $\widetilde{\varphi}^{-1}$ is (extended) smooth. It follows from Theorem 10.1.17 that $\widetilde{\varphi}^{-1} \circ \varphi$ is (Euclidean) smooth. Similarly, so is $\varphi^{-1} \circ \widetilde{\varphi} = (\widetilde{\varphi}^{-1} \circ \varphi)^{-1}$. $\qquad\square$

Theorem 11.2.10. *Let M be a regular surface, and let (U,φ) be a chart on M. If U' is an open set in U, then $(U',\varphi|_{U'})$ is a chart on M.*

Proof. Since (U,φ) is a chart on M, we have [C1]–[C3] at our disposal. With U' open in U, and U open in $\mathbb{R}^2$, by Theorem 9.1.5, U' is open in $\mathbb{R}^2$. According to Theorem 10.1.11(a), $\varphi|_{U'}$ is smooth. Thus, $\varphi|_{U'} : U' \longrightarrow \mathbb{R}^3$ is a parametrized surface. To show that $(U',\varphi|_{U'})$ is a chart on M, we need to prove that:

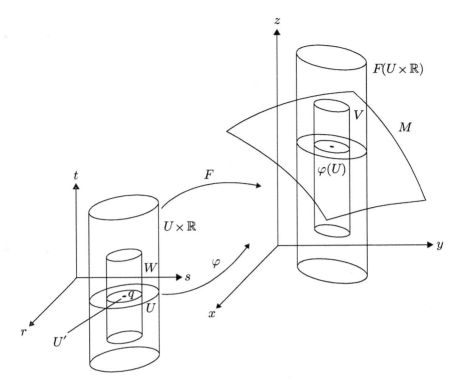

Figure 11.2.3. Diagram for Theorem 11.2.8

[C1′] $\varphi|_{U'} : U' \longrightarrow \mathbb{R}^3$ is an immersion.
[C2′] $\varphi|_{U'}(U')$ is an open set in M.
[C3′] $\varphi|_{U'} : U' \longrightarrow \varphi|_{U'}(U')$ is a homeomorphism.

The proofs are as follows:

[C1′]: This follows from [C1].

[C2′]: Since U' is open in U, it follows from [C3] and Theorem 9.1.14 that $\varphi(U')$ is open in $\varphi(U)$. According to [C2], $\varphi(U)$ is open in M. By Theorem 9.1.5, $\varphi|_{U'}(U') = \varphi(U')$ is open in M.

[C3′]: Let U'' be a subset of U'. By assumption, U' is open in U, and according to [C3], $\varphi : U \longrightarrow \varphi(U)$ is a homeomorphism. We therefore have from Theorem 9.1.14 that U'' is open in U' if and only if $\varphi|_{U'}(U'') = \varphi(U'')$ is open in $\varphi|_{U'}(U') = \varphi(U')$. The result now follows from Theorem 9.1.14. $\square$

By definition, a regular surface is a patchwork of images of parametrized surfaces. The next result shows that instead of images of parametrized surfaces, we can use graphs of smooth functions.

Theorem 11.2.11. *If M is a regular surface and p is a point in M, then there is a chart (V, ψ) at p and a function f in $C^\infty(V)$ such that $\psi(V) = \mathrm{graph}(f)$,* *where*

$$\text{graph}(f) = \{(x,y,f(x,y)) \in \mathbb{R}^3 : (x,y) \in V\}$$

and $\mathbb{R}^2$ *is identified with the xy-plane in* $\mathbb{R}^3$.

Proof. Let $(U, \varphi = (\varphi^1, \varphi^2, \varphi^3))$ be a chart at p, and let $q = \varphi^{-1}(p)$. It follows from [C1] and Theorem 11.2.2 that $J_\varphi(q)$ has rank 2. Relabeling coordinates in $\mathbb{R}^3$ if necessary, we have from Theorem 2.3.2 and Theorem 2.4.8 that $\det(J_\varphi(q)^{(1,2)}) \neq 0$. Consider the projection map $\mathcal{P} : \mathbb{R}^3 \longrightarrow \mathbb{R}^2$ defined by $\mathcal{P}(x,y,z) = (x,y)$. See Figure 11.2.4. Since φ and $\mathcal{P}$ are smooth, by Theorem 10.1.12, so is $\mathcal{P} \circ \varphi : U \longrightarrow \mathbb{R}^2$. The Jacobian matrix is

$$J_{\mathcal{P} \circ \varphi}(q) = J_\varphi(q)^{(1,2)} = \begin{bmatrix} \dfrac{\partial \varphi^1}{\partial x}(q) & \dfrac{\partial \varphi^1}{\partial y}(q) \\[2mm] \dfrac{\partial \varphi^2}{\partial x}(q) & \dfrac{\partial \varphi^2}{\partial y}(q) \end{bmatrix},$$

hence $\det(J_{\mathcal{P} \circ \varphi}(q)) \neq 0$. By Theorem 10.2.1 and Theorem 10.2.3, there is a neighborhood $U' \subseteq U$ of q in $\mathbb{R}^2$ and a neighborhood V of $\mathcal{P} \circ \varphi(q)$ in $\mathbb{R}^2$ such that $\mathcal{P} \circ \varphi|_{U'} : U' \longrightarrow V$ is a diffeomorphism. Since $(\mathcal{P} \circ \varphi|_{U'})^{-1} : V \longrightarrow U'$ is a diffeomorphism, hence smooth, it follows from Theorem 10.1.12 that $\psi = \varphi \circ (\mathcal{P} \circ \varphi|_{U'})^{-1} : V \longrightarrow M$ is smooth. Thus, (V, ψ) is a chart on M. Since φ is smooth, by definition, so is φ^3, and therefore, by Theorem 10.1.12, so is

$$f = \varphi^3 \circ (\mathcal{P} \circ \varphi|_{U'})^{-1} : V \longrightarrow \mathbb{R}.$$

Then

$$\begin{aligned}
\psi(V) &= \varphi \circ (\mathcal{P} \circ \varphi|_{U'})^{-1}(V) \\
&= \{(\varphi^1 \circ (\mathcal{P} \circ \varphi|_{U'})^{-1}(x,y), \varphi^2 \circ (\mathcal{P} \circ \varphi|_{U'})^{-1}(x,y), \\
&\qquad \varphi^3 \circ (\mathcal{P} \circ \varphi|_{U'})^{-1}(x,y)) : (x,y) \in V\} \\
&= \{(x,y,f(x,y)) : (x,y) \in V\} = \text{graph}(f). \qquad \square
\end{aligned}$$

Theorem 11.2.12. *If M is a regular surface and $F : M \longrightarrow \mathbb{R}^n$ is a map, where $n \geq 1$, then the following are equivalent:*
(a) *F is (extended) smooth.*
(b) *For every point p in M, there is a chart (U, φ) on M at p such that the map $F \circ \varphi : U \longrightarrow \mathbb{R}^n$ is (Euclidean) smooth.*
(c) *For every chart (U, φ) on M, the map $F \circ \varphi : U \longrightarrow \mathbb{R}^n$ is (Euclidean) smooth.*

Proof. (a)$\Rightarrow$(c): Let q be a point in U. By assumption, there is a neighborhood $\mathcal{U}$ of $\varphi(q)$ in $\mathbb{R}^3$ and a (Euclidean) smooth map $\widetilde{F} : \mathcal{U} \longrightarrow \mathbb{R}^n$ such that F and $\widetilde{F}$ agree on $M \cap \mathcal{U}$. According to [C2], $\varphi(U)$ is open in M. Since $M \cap \mathcal{U}$ is open in M, we have that $\varphi(U) \cap \mathcal{U} = \varphi(U) \cap (M \cap \mathcal{U})$ is open in $\varphi(U)$. Let $U' = \varphi^{-1}(\varphi(U) \cap \mathcal{U})$ and observe that q is in U'. It follows from [C3] and

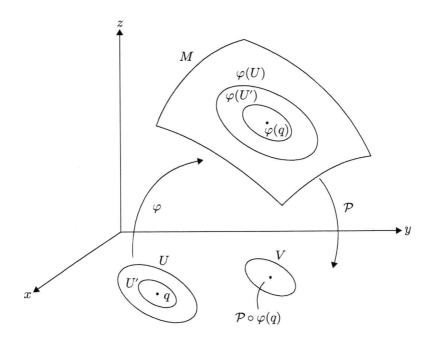

Figure 11.2.4. Diagram for Theorem 11.2.11

Theorem 9.1.7 that U' is open in U. Since φ and $\widetilde{F}$ are (Euclidean) smooth, by Theorem 10.1.11(a) and Theorem 10.1.12, so is $F \circ \varphi|_{U'} = \widetilde{F} \circ \varphi|_{U'} : U' \longrightarrow \mathbb{R}^n$. Because q was arbitrary, the result follows from Theorem 10.1.11(b).

(c)$\Rightarrow$(b): Straightforward.

(b)$\Rightarrow$(a): It follows from Theorem 10.1.17 and Theorem 11.2.8 that

$$F|_{\varphi(U)} = (F \circ \varphi) \circ \varphi^{-1} : \varphi(U) \longrightarrow \mathbb{R}^n$$

is (extended) smooth. By definition, there is a neighborhood $\mathcal{U}$ of p in $\mathbb{R}^3$ and a (Euclidean) smooth map $\widetilde{F|_{\varphi(U)}} : \mathcal{U} \longrightarrow \mathbb{R}^n$ such that $F|_{\varphi(U)}$ and $\widetilde{F|_{\varphi(U)}}$ agree on $\varphi(U) \cap \mathcal{U}$. Then F and $F|_{\varphi(U)}$ agree on $\varphi(U) \cap \mathcal{U}$. Since p was arbitrary, the result follows. $\qquad \square$

Theorem 11.2.12 shows that the existence of charts on regular surfaces makes it possible to answer questions about extended smoothness of maps on regular surfaces using methods developed for Euclidean smoothness.

We close this section with an example of a chart on the unit sphere that is strikingly different from the charts constructed in Example 11.2.6.

Example 11.2.13 (Stereographic Projection). Consider the set $\Sigma^2 = \mathcal{S}^2 \backslash \{(0,0,1)\}$, which is the unit sphere with the "north pole" removed. We define a map $\sigma : \Sigma^2 \longrightarrow \mathbb{R}^2$, called stereographic projection, as follows. For each (x, y, z) in Σ^2, let $\sigma(x, y, z) = (r, s)$ be the point where the straight line through $(0, 0, 1)$ and (x, y, z) intersects the xy-plane, where we identify the latter with $\mathbb{R}^2$. See Figure 11.2.5. Then

$$(x, y, z) - (0, 0, 1) = c[(r, s, 0) - (0, 0, 1)]$$

for some real number c, hence $x = cr$, $y = cs$, and $z = 1 - c$. This gives

$$1 = x^2 + y^2 + z^2 = c^2 r^2 + c^2 s^2 + (1 - c)^2,$$

so

$$c = 1 - z = \frac{2}{r^2 + s^2 + 1}.$$

Thus,

$$\sigma(x, y, z) = (r, s) = \frac{1}{1 - z}(x, y) \tag{11.2.2}$$

for all (x, y, z) in Σ^2. It is evident from Figure 11.2.5 that σ is bijective. Let $\varphi = \sigma^{-1}$. Then $\varphi(\mathbb{R}^2) = \Sigma^2$ and

$$\varphi(r, s) = (x, y, z) = \frac{1}{r^2 + s^2 + 1}(2r, 2s, r^2 + s^2 - 1) \tag{11.2.3}$$

for all (r, s) in $\mathbb{R}^2$.

We claim that $(\mathbb{R}^2, \varphi)$ is a chart on $\mathcal{S}^2$. It is easily shown that $\varphi : \mathbb{R}^2 \longrightarrow \mathbb{R}^3$ is a parametrized surface. We need to prove that:
[C1] $\varphi : \mathbb{R}^2 \longrightarrow \mathbb{R}^3$ is an immersion.
[C2] Σ^2 is an open set in $\mathcal{S}^2$.
[C3] $\varphi : \mathbb{R}^2 \longrightarrow \Sigma^2$ is a homeomorphism.
The proofs are as follows:
[C1]: The Jacobian matrix is

$$J_\varphi(r, s) = \frac{2}{(r^2 + s^2 + 1)^2}\begin{bmatrix} -r^2 + s^2 + 1 & -2rs \\ -2rs & r^2 - s^2 + 1 \\ 2r & 2s \end{bmatrix},$$

and the corresponding vector product is

$$\frac{4}{(r^2 + s^2 + 1)^4}(-r^2 + s^2 + 1, -2rs, 2r) \times (-2rs, r^2 - s^2 + 1, 2s)$$

$$= \frac{4}{(r^2 + s^2 + 1)^4}(-2r(r^2 + s^2 + 1), -2s(r^2 + s^2 + 1), -(r^2 + s^2)^2 + 1),$$

which never equals $(0, 0, 0)$. By Theorem 11.2.2, φ is an immersion.
[C2]:   Since $\mathbb{R}^3 \backslash \{(0, 0, 1)\}$ is open in $\mathbb{R}^3$, it follows that $\Sigma^2 = \mathcal{S}^2 \cap [\mathbb{R}^3 \backslash \{(0, 0, 1)\}]$ is open in $\mathcal{S}^2$.
[C3]: We observed above that φ is bijective, and it is clear from the form of (11.2.2) and (11.2.3) that φ and φ^{-1} are continuous.
This proves the claim.   ◊

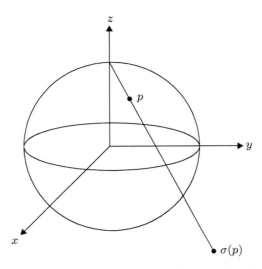

Figure 11.2.5. Stereographic projection: Diagram for Example 11.2.13

11.3 Tangent Planes in $\mathbb{R}^3$

Having defined a regular surface and established some of its basic properties, we are now in a position to present a rigorous definition of "tangent plane".

Let M be a regular surface. A **curve (on M)** is a curve $\lambda(t) : I \longrightarrow M$ as defined in Section 11.1, with the additional feature that it takes values in M. Let p be a point in M. We say that a vector v in $\mathbb{R}^3$ is a **tangent vector to M at p** if there is a smooth curve $\lambda(t) : (a,b) \longrightarrow M$ such that $\lambda(t_0) = p$ and $(d\lambda/dt)(t_0) = v$ for some t_0 in (a,b). The **tangent plane of M at p** is denoted by $T_p(M)$ and defined to be the set of all such tangent vectors:

$$T_p(M) = \left\{ \frac{d\lambda}{dt}(t_0) : \lambda(t) : (a,b) \longrightarrow M \text{ is smooth}, \ \lambda(t_0) = p, \ t_0 \in (a,b) \right\}.$$

Theorem 11.3.1. *Let M be a regular surface, let p be a point in M, let (U, φ) be a chart at p, and let $q = \varphi^{-1}(p)$. Then:*

(a) *$T_p(M)$ is a 2-dimensional subspace of $\mathbb{R}^3$.*

(b) *$(H_1|_q, H_2|_q)$ is a basis for $T_p(M)$, called the **coordinate basis at p** corresponding to (U, φ).*

(c) *$T_p(M) = d_q(\varphi)(\mathbb{R}^2)$.*

(d) *$d_q(\varphi) : \mathbb{R}^2 \longrightarrow T_p(M)$ is a linear isomorphism.*

Proof. (a), (b): We claim that $T_p(M) = \text{span}(\{H_1|_q, H_2|_q\})$.

($\subseteq$): Let v be a vector in $T_p(M)$. By definition, there is a smooth curve $\lambda(t) : (a,b) \longrightarrow M$ such that $\lambda(t_0) = p$ and $(d\lambda/dt)(t_0) = v$ for some t_0 in (a,b). Suppose without loss of generality that $\lambda\big((a,b)\big) \subset \varphi(U)$. It follows from

Theorem 10.1.17 and Theorem 11.2.8 that the map

$$\mu = (\mu^1, \mu^2) = \varphi^{-1} \circ \lambda : (a, b) \longrightarrow U$$

is smooth. By Theorem 10.1.10,

$$v = \frac{d\lambda}{dt}(t_0) = \frac{d(\varphi \circ \mu)}{dt}(t_0) = \sum_i \frac{d\mu^i}{dt}(t_0) H_i|_q,$$

so v is in the span of $H_1|_q$ and $H_2|_q$. Thus, $T_p(M) \subseteq \text{span}(\{H_1|_q, H_2|_q\})$.

($\supseteq$): Let (e_1, e_2) be the standard basis for $\mathbb{R}^2$. For given $1 \le i \le 2$, we define a smooth map

$$\zeta = (\zeta^1, \zeta^2) : (-\varepsilon, \varepsilon) \longrightarrow U$$

by $\zeta(t) = q + te_i$, where $\varepsilon > 0$ is chosen small enough that $(q - \varepsilon e_i, q + \varepsilon e_i) \subset U$. Consider the smooth curve $\lambda(t) : (-\varepsilon, \varepsilon) \longrightarrow M$ defined by $\lambda = \varphi \circ \zeta$, and observe that $\lambda(0) = \varphi(q)$. By Theorem 10.1.10,

$$\frac{d\lambda}{dt}(0) = \frac{d(\varphi \circ \zeta)}{dt}(0) = \sum_j \frac{d\zeta^j}{dt}(0) H_j|_q = H_i|_q, \qquad (11.3.1)$$

so $H_i|_q$ is in $T_p(M)$ for $i = 1, 2$. Thus, $\text{span}(\{H_1|_q, H_2|_q\}) \subseteq T_p(M)$.

This proves the claim. We have from [C1] of Section 11.2 and Theorem 11.2.2 that $H_1|_q$ and $H_2|_q$ are linearly independent. The result follows.

(c): This follows from Theorem 11.2.1(b) and part (b).

(d): We have from part (c) that $d_q(\varphi) : \mathbb{R}^2 \longrightarrow T_p(M)$ is well-defined, and from [C1] of Section 11.2 that it is injective. The result now follows from part (a) and Theorem 1.1.14.   □

In the notation of Theorem 11.3.1, let us denote

$$(H_1, H_2) \qquad \text{by} \qquad \mathcal{H},$$

and

$$(H_1|_q, H_2|_q) \qquad \text{by} \qquad \mathcal{H}_q.$$

We refer to $\mathcal{H}$ as the **coordinate frame** corresponding to (U, φ). Although there is a tendency to think of $T_p(M)$ as literally "tangent" to M at the point p, by Theorem 11.3.1(a), $T_p(M)$ is a subspace of $\mathbb{R}^3$. As such, $T_p(M)$ passes through the origin $(0, 0, 0)$ of $\mathbb{R}^3$. In geometric terms, it is $T_p(M) + p$, the translation of $T_p(M)$ by p, that is tangent to M at p. That said, it is convenient in the figures to label tangent planes as $T_p(M)$ rather than $T_p(M) + p$.

Example 11.3.2 (S^2). Continuing with Example 11.2.6, consider the chart (D, φ_1), where $\varphi_1(x, y) = (x, y, \sqrt{1 - x^2 - y^2})$. The corresponding coordinate frame is given by

$$\left(\frac{\partial \varphi_1}{\partial x}(x, y), \frac{\partial \varphi_1}{\partial y}(x, y) \right) = \left(\left(1, 0, -\frac{x}{\sqrt{1 - x^2 - y^2}} \right), \left(0, 1, -\frac{y}{\sqrt{1 - x^2 - y^2}} \right) \right).$$

For example, at $\varphi_1(0, 0) = (0, 0, 1)$, the coordinate basis is $((1, 0, 0), (0, 1, 0))$, which spans the xy-plane in $\mathbb{R}^3$.   ◇

Theorem 11.3.3 (Change of Coordinate Basis). *Let M be a regular surface, let p be a point in M, let (U, φ) and $(\widetilde{U}, \widetilde{\varphi})$ be charts at p, and let $\mathcal{H}$ and $\widetilde{\mathcal{H}}$ be the corresponding coordinate frames. Then*

$$\left[\mathrm{id}_{T_p(M)}\right]_{\mathcal{H}_q}^{\widetilde{\mathcal{H}}_{\tilde{q}}} = J_{\widetilde{\varphi}^{-1} \circ \varphi}(q),$$

where $q = \varphi^{-1}(p)$ and $\tilde{q} = \widetilde{\varphi}^{-1}(p)$.

Proof. Let $\mathcal{H} = (H_1, H_2)$ and $\widetilde{\mathcal{H}} = (\widetilde{H}_1, \widetilde{H}_2)$, and let (r^1, r^2) and $(\tilde{r}^1, \tilde{r}^2)$ be coordinates on U and $\widetilde{U}$, respectively. By Theorem 11.2.9, the transition map $F = (F^1, F^2) = \widetilde{\varphi}^{-1} \circ \varphi$ is a diffeomorphism, hence smooth. It follows from Theorem 10.1.10 that

$$H_j|_q = \frac{\partial \varphi}{\partial r^j}(q) = \frac{\partial(\widetilde{\varphi} \circ F)}{\partial r^j}(q) = \sum_i \frac{\partial F^i}{\partial r^j}(q)\widetilde{H}_i|_{\tilde{q}},$$

and then from (2.2.6), (2.2.7), and (10.1.2) that

$$\left[\mathrm{id}_{T_p(M)}\right]_{\mathcal{H}_q}^{\widetilde{\mathcal{H}}_{\tilde{q}}} = \left[\frac{\partial F^i}{\partial r^j}(q)\right] = J_{\widetilde{\varphi}^{-1} \circ \varphi}(q). \qquad \square$$

Example 11.3.4 (Polar Coordinates). The xy-plane in $\mathbb{R}^3$, denoted here by Pln, is clearly a regular surface. Consider the open set

$$U = \{(\rho, \phi) \in \mathbb{R}^2 : \rho > 0, 0 < \phi < 2\pi\}$$

in $\mathbb{R}^2$, and define a map $\varphi : U \longrightarrow \mathbb{R}^3$ by

$$\varphi(\rho, \phi) = \big(\rho\cos(\phi), \rho\sin(\phi), 0\big)$$

for all (ρ, ϕ) in U. The image of φ is Pln with the nonnegative x-axis removed. It is easily shown that (U, φ) is a chart on Pln, which we call **polar coordinates**. A covering chart on Pln is given by $(\widetilde{U} = \mathbb{R}^2, \widetilde{\varphi})$, where $\widetilde{\varphi} : \mathbb{R}^2 \longrightarrow$ Pln is defined by $\widetilde{\varphi}(x, y) = (x, y, 0)$. The coordinate frames corresponding to (U, φ) and $(\widetilde{U}, \widetilde{\varphi})$ are given by

$$\mathcal{H}_{(\rho,\phi)} = \left(\frac{\partial \varphi}{\partial \rho}(\rho, \phi), \frac{\partial \varphi}{\partial \phi}(\rho, \phi)\right) = \Big(\big(\cos(\phi), \sin(\phi), 0\big), \big(-\rho\sin(\phi), \rho\cos(\phi), 0\big)\Big)$$

and

$$\widetilde{\mathcal{H}}_{(x,y)} = \left(\frac{\partial \widetilde{\varphi}}{\partial x}(x, y), \frac{\partial \widetilde{\varphi}}{\partial y}(x, y)\right) = \big((1, 0, 0), (0, 1, 0)\big),$$

respectively. The transition map $\widetilde{\varphi}^{-1} \circ \varphi : U \longrightarrow \widetilde{U}$ is defined by

$$\widetilde{\varphi}^{-1} \circ \varphi(\rho, \phi) = \big(\rho\cos(\phi), \rho\sin(\phi)\big).$$

Classically, x and y are viewed as functions of ρ and ϕ, and expressed as

$$x(\rho, \phi) = \rho\cos(\phi) \qquad \text{and} \qquad y(\rho, \phi) = \rho\sin(\phi).$$

By Theorem 11.3.3, the change of coordinates matrix is

$$J_{\tilde{\varphi}^{-1} \circ \varphi}(\rho, \phi) = \begin{bmatrix} \dfrac{\partial x}{\partial \rho}(\rho, \phi) & \dfrac{\partial x}{\partial \phi}(\rho, \phi) \\[2mm] \dfrac{\partial y}{\partial \rho}(\rho, \phi) & \dfrac{\partial y}{\partial \phi}(\rho, \phi) \end{bmatrix} = \begin{bmatrix} \cos(\phi) & -\rho \sin(\phi) \\ \sin(\phi) & \rho \cos(\phi) \end{bmatrix}. \qquad \diamond$$

The next result is reminiscent of Theorem 10.1.13.

Theorem 11.3.5. *Let M be a regular surface, let p be a point in M, and let v be a vector in $T_p(M)$. Then there is a real number $\varepsilon > 0$ and a smooth curve $\lambda(t) : (-\varepsilon, \varepsilon) \longrightarrow M$ such that $\lambda(0) = p$ and $(d\lambda/dt)(0) = v$.*

Proof. By definition, there is a smooth curve $\psi(u) : (a, b) \longrightarrow M$ such that $\psi(u_0) = p$ and $(d\psi/du)(u_0) = v$ for some u_0 in (a, b). Let $g(t) : \mathbb{R}^2 \longrightarrow \mathbb{R}^2$ be the function defined by $g(t) = t + u_0$. Then for sufficiently small $\varepsilon > 0$, by Theorem 10.1.12, $\lambda(t) = \psi \circ g(t) : (-\varepsilon, \varepsilon) \longrightarrow M$ is the desired smooth curve.

Here is an alternative argument that anticipates the proof of Theorem 14.7.2. Let (U, φ) be a chart at p such that $\varphi(0, 0) = p$, let $(H_1|_{(0,0)}, H_2|_{(0,0)})$ be the corresponding coordinate basis at p, and let $v = \sum_i a^i H_i|_{(0,0)}$. Define a smooth curve $\lambda(t) : (-\varepsilon, \varepsilon) \longrightarrow M$ by $\lambda(t) = \varphi(ta^1, ta^2)$, where ε is chosen small enough that $\lambda\big((-\varepsilon, \varepsilon)\big) \subset \varphi(U)$. Clearly, $\lambda(0) = p$. It follows from Theorem 10.1.10 that

$$\frac{d\lambda}{dt}(0) = \sum_i a^i \frac{\partial \varphi}{\partial r^i}(0, 0) = \sum_i a^i H_i|_{(0,0)} = v. \qquad \square$$

11.4   Types of Regular Surfaces in $\mathbb{R}^3$

In this section, we define four types of regular surfaces: open sets in regular surfaces, graphs of functions, surfaces of revolution, and level sets of functions. The table below provides a list of the worked examples of graphs of functions and surfaces of revolution presented in Chapter 13.

Section	Geometric object	Parametrization
13.1	plane	graph of function
13.2	cylinder	surface of revolution
13.3	cone	surface of revolution
13.4	sphere	surface of revolution
13.5	tractoid	surface of revolution
13.6	hyperboloid of one sheet	graph of function
13.7	hyperboloid of two sheets	graph of function
13.8	torus	surface of revolution

Open set in a regular surface. As we now show, an open set in a regular surface is itself a regular surface.

Theorem 11.4.1 (Open Set). *Let M be a regular surface, let V be an open set in M, and view V as a topological subspace of $\mathbb{R}^3$. Then:*
(a) *V is a regular surface.*
(b) *For all points p in V,*
$$T_p(V) = T_p(M).$$

Proof. (a): Let p be a point in V, and let (U, φ) be a chart on M at p. We have properties [C1]–[C3] of Section 11.2 at our disposal. According to [C2], $\varphi(U)$ is open in M, and therefore, so is $\varphi(U) \cap V$. It follows from [C3] and Theorem 9.1.14 that $U' = \varphi^{-1}\big(V \cap \varphi(U)\big)$ is open in U. By Theorem 11.2.10, $(U', \varphi|_{U'})$ is a chart at p such that $\varphi|_{U'}(U') \subseteq V$. Since p was arbitrary, the result follows.

(b): Let v be a vector in $T_p(V)$, and let $\lambda(t) : (a, b) \longrightarrow V$ be a smooth curve such that $\lambda(t_0) = p$ and $(d\lambda/dt)(t_0) = v$ for some t_0 in (a, b). Since $V \subseteq M$, the same smooth curve shows that v is also a vector in $T_p(M)$. Thus, $T_p(V) \subseteq T_p(M)$. By Theorem 11.3.1(a), both $T_p(V)$ and $T_p(M)$ are 2-dimensional, and then Theorem 1.1.7(b) gives $T_p(V) = T_p(M)$. $\qquad\square$

**Throughout, any open set in a regular surface is
viewed as a regular surface.**

Graph of a function. According to Theorem 11.2.11, a regular surface is covered by graphs of functions. We now consider the graph of a function in isolation. Let U be an open set in $\mathbb{R}^2$, and let f be a function in $C^\infty(U)$. Recall from Theorem 11.2.11 that

$$\mathrm{graph}(f) = \big\{ (x, y, f(x, y)) \in \mathbb{R}^3 : (x, y) \in U \big\},$$

where we identify $\mathbb{R}^2$ with the xy-plane in $\mathbb{R}^3$. Defining a map $\varphi : U \longrightarrow \mathbb{R}^3$ by

$$\varphi(x, y) = (x, y, f(x, y))$$

for all (x, y) in U, we see that $\mathrm{graph}(f)$ is the image of φ; that is,

$$\mathrm{graph}(f) = \varphi(U).$$

Theorem 11.4.2 (Graph of Function). *With the above setup, $\mathrm{graph}(f)$, viewed as a topological subspace of $\mathbb{R}^3$, is a regular surface, and (U, φ) is a covering chart on $\mathrm{graph}(f)$.*

Proof. Clearly, $\varphi : U \longrightarrow \mathbb{R}^3$ is a parametrized surface. We need to prove that:
[C1] $\varphi : U \longrightarrow \mathbb{R}^3$ is an immersion.
[C2] $\varphi(U)$ is an open set in $\mathrm{graph}(f)$.
[C3] $\varphi : U \longrightarrow \varphi(U)$ is a homeomorphism.
   The proofs are as follows:

[C1]: The Jacobian matrix is

$$J_\varphi(x,y) = \begin{bmatrix} 1 & 0 \\ 0 & 1 \\ \dfrac{\partial f}{\partial x}(x,y) & \dfrac{\partial f}{\partial y}(x,y) \end{bmatrix},$$

and the corresponding vector product is

$$\left(1,0,\frac{\partial f}{\partial x}(x,y)\right) \times \left(1,0,\frac{\partial f}{\partial y}(x,y)\right) = \left(-\frac{\partial f}{\partial x}(x,y), -\frac{\partial f}{\partial y}(x,y), 1\right),$$

which never equals $(0,0,0)$. By Theorem 11.2.2, φ is an immersion.

[C2]: This follows from $\varphi(U) = \text{graph}(f)$.

[C3]: Clearly, φ is bijective. Since φ is smooth, by Theorem 10.1.7 it is continuous. Let $\mathcal{P} : \mathbb{R}^3 \longrightarrow \mathbb{R}^2$ be the projection map defined by $\mathcal{P}(x,y,z) = (x,y)$, where we identify $\mathbb{R}^2$ with the xy-plane in $\mathbb{R}^3$. Since $\mathcal{P}$ is continuous and $\varphi^{-1}|_{\text{graph}(f)} = \mathcal{P}|_{\text{graph}(f)}$, the result follows.

Thus, (U,φ) is a chart on $\text{graph}(f)$, and since $\varphi(U) = \text{graph}(f)$, it is a covering chart. $\qquad\square$

Surface of revolution. Let $\rho(t), h(t) : (a,b) \longrightarrow \mathbb{R}$ be smooth functions such that:

[R1] ρ is strictly positive on (a,b).

[R2] h is strictly increasing or strictly decreasing on (a,b).

We refer to ρ and h as the **radius function** and **height function**, respectively. Throughout, it is convenient to denote the derivatives of ρ and h with respect to t by an overdot. Consider the smooth curve $\sigma(t) : (a,b) \longrightarrow \mathbb{R}^3$ defined by

$$\sigma(t) = \big(\rho(t), 0, h(t)\big)$$

for all t in (a,b). We observe that [R2] is equivalent to $\dot{h}$ being strictly positive or strictly negative on (a,b), from which it follows that σ is a regular curve. Let

$$U = (a,b) \times (-\pi, \pi),$$

and consider the smooth map $\varphi : U \longrightarrow \mathbb{R}^3$ defined by

$$\varphi(t,\phi) = \big(\rho(t)\cos(\phi), \rho(t)\sin(\phi), h(t)\big).$$

The **surface of revolution** corresponding to σ is denoted by $\text{rev}(\sigma)$ and defined to be the image of φ:

$$\text{rev}(\sigma) = \varphi(U).$$

Thus, $\text{rev}(\sigma)$ is obtained by revolving the image of σ around the z-axis. A remark is that $(-\pi,\pi)$ was chosen when defining U rather than, for example, $[-\pi,\pi)$ or $[0,2\pi)$, to ensure that U is an open set in $\mathbb{R}^2$. As a result, a surface of revolution does not quite make a complete circuit around the z-axis.

For a given point t in (a, b), we define a smooth curve

$$\varphi_t(\phi) : (-\pi, \pi) \longrightarrow \mathbb{R},$$

called the **latitude curve** corresponding to t, by

$$\varphi_t(\phi) = \varphi(t, \phi).$$

Similarly, for a given point ϕ in $(-\pi, \pi)$, we define a smooth curve

$$\varphi_\phi(t) : (a, b) \longrightarrow \mathbb{R},$$

called the **longitude curve** corresponding to ϕ, by

$$\varphi_\phi(t) = \varphi(t, \phi).$$

From

$$\varphi_t(\phi) = \rho(t)\big(\cos(\phi), \sin(\phi), 0\big) + \big(0, 0, h(t)\big),$$

we see that the image of φ_t is, except for a single missing point, a circle of radius $\rho(t)$ centered on the z-axis and lying in the plane parallel to the xy-plane at a height $h(t)$.

Theorem 11.4.3 (Surface of Revolution). *With the above setup, $\mathrm{rev}(\sigma)$, viewed as a topological subspace of $\mathbb{R}^3$, is a regular surface and (U, φ) is a covering chart on $\mathrm{rev}(\sigma)$.*

Proof. Since ρ and h are smooth, by definition, so is φ. Thus, $\varphi : U \longrightarrow \mathbb{R}^3$ is a parametrized surface. We need to prove that:
[C1] $\varphi : U \longrightarrow \mathbb{R}^3$ is an immersion.
[C2] $\varphi(U)$ is an open set in $\mathrm{rev}(\sigma)$.
[C3] $\varphi : U \longrightarrow \varphi(U)$ is a homeomorphism.
     The proofs are as follows:
     [C1]: The Jacobian matrix is

$$J_\varphi(t, \phi) = \begin{bmatrix} \dot{\rho}(t)\cos(\phi) & -\rho(t)\sin(\phi) \\ \dot{\rho}(t)\sin(\phi) & \rho(t)\cos(\phi) \\ \dot{h}(t) & 0 \end{bmatrix},$$

and the corresponding vector product is

$$\big(\dot{\rho}(t)\cos(\phi), \dot{\rho}(t)\sin(\phi), \dot{h}(t)\big) \times \big(-\rho(t)\sin(\phi), \rho(t)\cos(\phi), 0\big)$$
$$= \rho(t)\big(-\dot{h}(t)\cos(\phi), -\dot{h}(t)\sin(\phi), \dot{\rho}(t)\big),$$

which never equals $(0, 0, 0)$; for if it did, then taking the Euclidean inner product of the preceding vector with itself gives

$$0 = \rho(t)^2\big([-\dot{h}(t)\cos(\phi)]^2 + [-\dot{h}(t)\sin(\phi)]^2 + \dot{\rho}(t)^2\big) = \rho(t)^2[\dot{h}(t)^2 + \dot{\rho}(t)^2]$$

for some t in (a, b), which contradicts either [R1] or [R2]. By Theorem 11.2.2, φ is an immersion.

[C2]: This follows from $\varphi(U) = \text{rev}(\sigma)$.

[C3]: We provide only a sketch of the proof. Let

$$(x, y, z) = \big(\rho(t)\cos(\phi), \rho(t)\sin(\phi), h(t)\big),$$

so that

$$\rho(t) = \sqrt{x^2 + y^2}. \tag{11.4.1}$$

Substituting into the trigonometric identity

$$\tan\left(\frac{\phi}{2}\right) = \frac{\sin(\phi)}{1 + \cos(\phi)}$$

yields

$$\tan\left(\frac{\phi}{2}\right) = \frac{y}{x + \sqrt{x^2 + y^2}}. \tag{11.4.2}$$

Using (11.4.1) and (11.4.2), it can be shown that $\varphi^{-1} : \text{rev}(\sigma) \longrightarrow U$ is continuous.

Thus, (U, φ) is a chart on $\text{rev}(\sigma)$, and since $\varphi(U) = \text{rev}(\sigma)$, it is a covering chart. $\qquad\square$

Level set of a function. Let $\mathcal{U}$ be an open set in $\mathbb{R}^3$, and let f be a function in $C^\infty(\mathcal{U})$. The **gradient of f (in $\mathbb{R}^3$)** is the map

$$\text{grad}(f) : \mathcal{U} \longrightarrow \mathbb{R}^3$$

defined by

$$\text{grad}(f)_p = \left(\frac{\partial f}{\partial x}(p), \frac{\partial f}{\partial y}(p), \frac{\partial f}{\partial z}(p)\right) \tag{11.4.3}$$

for all p in $\mathcal{U}$. Given a real number c in $f(\mathcal{U})$, the corresponding **level set of f** is

$$f^{-1}(c) = \{p \in \mathcal{U} : f(p) = c\}.$$

Theorem 11.4.4 (Level Set of Function). *With the above setup, if* $\text{grad}(f)_p \neq (0, 0, 0)$ *for all p in $f^{-1}(c)$, then $f^{-1}(c)$, viewed as a topological subspace of $\mathbb{R}^3$, is a regular surface.*

Proof. Since $\text{grad}(f)_p \neq (0, 0, 0)$, relabeling coordinates in $\mathbb{R}^3$ if necessary, we have $(\partial f/\partial z)(p) \neq 0$. Let us define a map $F : \mathcal{U} \longrightarrow \mathbb{R}^3$ by $F(x, y, z) = \big(x, y, f(x, y, z)\big)$ for all (x, y, z) in $\mathcal{U}$. The Jacobian matrix is

$$J_F(p) = \begin{bmatrix} 1 & 0 & 0 \\ 0 & 1 & 0 \\ \dfrac{\partial f}{\partial x}(p) & \dfrac{\partial f}{\partial y}(p) & \dfrac{\partial f}{\partial z}(p) \end{bmatrix},$$

so $\det\big(J_F(p)\big) \neq 0$. By Theorem 10.2.1 and Theorem 10.2.3, there is a neighborhood $\mathcal{V}$ of p in $\mathbb{R}^3$ and a neighborhood $\mathcal{W}$ of $q = F(p)$ in $\mathbb{R}^3$ such that $F|_{\mathcal{V}} : \mathcal{V} \longrightarrow \mathcal{W}$ is a diffeomorphism. Then $G = (F|_{\mathcal{V}})^{-1} : \mathcal{W} \longrightarrow \mathcal{V}$ is a diffeomorphism, hence smooth. See Figure 11.4.1, where we note that G takes points in $\{(x, y, z) \in \mathbb{R}^3 : z = c\} \cap \mathcal{W}$ to $f^{-1}(c) \cap \mathcal{V}$. Let $\mathcal{P} : \mathbb{R}^3 \longrightarrow \mathbb{R}^2$ be the projection map defined by $\mathcal{P}(x, y, z) = (x, y)$, where we identify $\mathbb{R}^2$ with the xy-plane in $\mathbb{R}^3$. Since $\mathcal{W}$ is open in $\mathbb{R}^3$, there is an open ball $B_\varepsilon(q) \subseteq \mathcal{W}$ for sufficiently small $\varepsilon > 0$. Then $\mathcal{P}\big(B_\varepsilon(q)\big) = B_\varepsilon\big(\mathcal{P}(q)\big)$. Let $U = B_\varepsilon\big(\mathcal{P}(q)\big)$, and define a parametrized surface $\varphi : U \longrightarrow \mathbb{R}^3$ by $\varphi(x, y) = G(x, y, c)$. In light of the preceding remarks, $G(x, y, c)$ is in $f^{-1}(c) \cap \mathcal{V}$ for all (x, y) in U, hence $\varphi(U) \subseteq f^{-1}(c) \cap \mathcal{V}$. Let $G = (G^1, G^2, G^3)$, and consider the function $g : U \longrightarrow \mathbb{R}$ given by $g(x, y) = G^3(x, y, c)$. Since G is smooth, by definition, so is G^3. Thus, g is in $C^\infty(U)$. We have

$$G(x, y, c) = \big(G^1(x, y, c), G^2(x, y, c), G^3(x, y, c)\big) = \big(x, y, g(x, y)\big)$$

for all (x, y) in U, hence $\varphi(U) = \mathrm{graph}(g)$. By Theorem 11.4.2, (U, φ) is a chart at p. Since p was arbitrary, $f^{-1}(c)$ is a regular surface. □

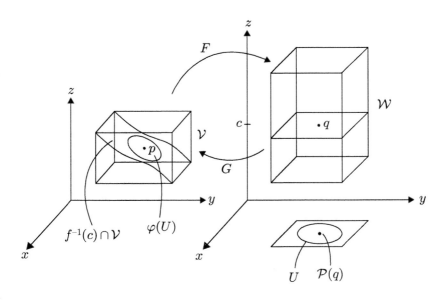

Figure 11.4.1. Diagram for Theorem 11.4.4

Example 11.4.5 ($\mathcal{S}^2$). Continuing with Example 11.2.6, let f be the function in $C^\infty(\mathbb{R}^3)$ given by $f(x, y, z) = x^2 + y^2 + z^2$. Then $\mathcal{S}^2 = f^{-1}(1)$ and

$$\mathrm{grad}(f)_{(x,y,z)} = (2x, 2y, 2z),$$

so that $\mathrm{grad}(f)_{(x,y,z)} \neq (0, 0, 0)$ for all (x, y, z) in $\mathcal{S}^2$. By Theorem 11.4.4, $\mathcal{S}^2$ is a regular surface, something previously established in Example 11.2.6. ◇

11.5   Functions on Regular Surfaces in $\mathbb{R}^3$

Theorem 11.5.1 (Smoothness Criteria for Functions). *If M is a regular surface and $f : M \longrightarrow \mathbb{R}$ is a function, then the following are equivalent:*
(a) *f is (extended) smooth.*
(b) *For every point p in M, there is a chart (U, φ) on M at p such that the function $f \circ \varphi : U \longrightarrow \mathbb{R}$ is (Euclidean) smooth.*
(c) *For every chart (U, φ) on M, the function $f \circ \varphi : U \longrightarrow \mathbb{R}$ is (Euclidean) smooth.*

Proof. Setting $n = 1$ and $F = f$ in Theorem 11.2.12 gives the result.   □

Let M be a regular surface. The set of smooth functions on M is denoted by $C^\infty(M)$. We make $C^\infty(M)$ into both a vector space over $\mathbb{R}$ and a ring by defining operations as follows: for all functions f, g in $C^\infty(M)$ and all real numbers c, let

$$(f + g)(p) = f(p) + g(p),$$
$$(fg)(p) = f(p)g(p),$$

and

$$(cf)(p) = cf(p)$$

for all p in M. The identity element of the ring is the constant function 1_M that sends all points in M to the real number 1.

Let f be a function in $C^\infty(M)$, and let p be a point in M. The **differential of f at p** is the map

$$d_p(f) : T_p(M) \longrightarrow \mathbb{R}$$

defined by

$$d_p(f)(v) = \frac{d(f \circ \lambda)}{dt}(t_0) \tag{11.5.1}$$

for all vectors v in $T_p(M)$, where $\lambda(t) : (a, b) \longrightarrow M$ is any smooth curve such that $\lambda(t_0) = p$ and $(d\lambda/dt)(t_0) = v$ for some t_0 in (a, b).

Theorem 11.5.2. *Let M be a regular surface, let f be a function in $C^\infty(M)$, let p be a point in M, and let v be a vector in $T_p(M)$. Then:*
(a) *$d_p(f)(v)$ is independent of the choice of smooth curve used to express v.*
(b) *$d_p(f)$ is linear.*

Proof. Let (U, φ) be a chart at p, and let $\lambda(t) : (a, b) \longrightarrow M$ be a smooth curve such that $\lambda(t_0) = p$ and $(d\lambda/dt)(t_0) = v$ for some t_0 in (a, b). Suppose without loss of generality that $\lambda\big((a, b)\big) \subset \varphi(U)$, and let $q = \varphi^{-1}(p)$. It follows from Theorem 10.1.17 and Theorem 11.2.8 that the map

$$\mu = (\mu^1, \mu^2) = \varphi^{-1} \circ \lambda : (a, b) \longrightarrow U$$

is (Euclidean) smooth, and from Theorem 10.1.9 that

$$v = \frac{d\lambda}{dt}(t_0) = \frac{d(\varphi \circ \mu)}{dt}(t_0) = d_q(\varphi)\left(\frac{d\mu}{dt}(t_0)\right).$$

By Theorem 11.3.1(d), $d_q(\varphi)$ is invertible, so

$$\frac{d\mu}{dt}(t_0) = d_q(\varphi)^{-1}(v). \tag{11.5.2}$$

Then

$$d_p(f)(v) = \frac{d(f \circ \lambda)}{dt}(t_0) \qquad [(11.5.1)]$$

$$= \frac{d(f \circ \varphi \circ \mu)}{dt}(t_0) \tag{11.5.3}$$

$$= d_{\mu(t_0)}(f \circ \varphi)\left(\frac{d\mu}{dt}(t_0)\right) \qquad [\text{Th } 10.1.9]$$

$$= d_q(f \circ \varphi) \circ d_q(\varphi)^{-1}(v). \qquad [(11.5.2)]$$

The preceding identity makes sense because we have the maps $d_q(\varphi)^{-1} : T_p(M) \longrightarrow \mathbb{R}^2$ and $d_q(f \circ \varphi) : \mathbb{R}^2 \longrightarrow \mathbb{R}$, hence $d_p(f) : T_p(M) \longrightarrow \mathbb{R}$.

(a): This follows from (11.5.3).

(b): Since v was arbitrary, we have from (11.5.3) that

$$d_p(f) = d_q(f \circ \varphi) \circ d_q(\varphi)^{-1}.$$

Thus, $d_p(f)$ is the composition of linear maps, so it too is linear. $\square$

The next result is a counterpart of Theorem 10.1.3.

Theorem 11.5.3. *Let M be a regular surface, let f, g be functions in $C^\infty(M)$, let p be a point in M, and let c be a real number. Then:*

(a) $d_p(cf + g) = c\,d_p(f) + d_p(g)$.

(b) $d_p(fg) = f(p)\,d_p(g) + g(p)\,d_p(f)$.

Proof. Let v be a vector in $T_p(M)$, and let $\lambda(t) : (a, b) \longrightarrow M$ be a smooth curve such that $\lambda(t_0) = p$ and $(d\lambda/dt)(t_0) = v$ for some t_0 in (a, b). Then

$$d_p(cf + g)(v) = \frac{d\big((cf + g) \circ \lambda\big)}{dt}(t_0) \qquad [(11.5.1)]$$

$$= \frac{d\big((cf \circ \lambda) + (g \circ \lambda)\big)}{dt}(t_0)$$

$$= c\frac{d(f \circ \lambda)}{dt}(t_0) + \frac{d(g \circ \lambda)}{dt}(t_0)$$

$$= c\,d_p(f)(v) + d_p(g)(v) \qquad [(11.5.1)$$

$$= \big(c\,d_p(f) + d_p(g)\big)(v)$$

and

$$d_p(fg)(v) = \frac{d(fg \circ \lambda)}{dt}(t_0) \qquad\qquad [(11.5.1)]$$

$$= \frac{d((f \circ \lambda)(g \circ \lambda))}{dt}(t_0)$$

$$= [f \circ \lambda(t_0)] \frac{d(g \circ \lambda)}{dt}(t_0) + [g \circ \lambda(t_0)] \frac{d(f \circ \lambda)}{dt}(t_0)$$

$$= f(p)\, d_p(g)(v) + g(p)\, d_p(f)(v) \qquad\qquad [(11.5.1)]$$

$$= (f(p)\, d_p(g) + g(p)\, d_p(f))(v).$$

Since v was arbitrary, the result follows. $\square$

11.6   Maps on Regular Surfaces in $\mathbb{R}^3$

Theorem 11.6.1 (Smoothness Criteria for Maps). *If M and N are regular surfaces and $F : M \longrightarrow N$ is a map, then the following are equivalent:*
(a) *F is (extended) smooth.*
(b) *For every point p in M, there is a chart (U, φ) on M at p and a chart (V, ψ) on N at $F(p)$ such that the map $\psi^{-1} \circ F \circ \varphi : \varphi^{-1}(W) \longrightarrow \mathbb{R}^2$ is (Euclidean) smooth, where $W = \varphi(U) \cap F^{-1}(\psi(V))$.*
(c) *F is continuous, and for every chart (U, φ) on M and every chart (V, ψ) on N such that $F(\varphi(U)) \subseteq \psi(V)$, the map $\psi^{-1} \circ F \circ \varphi : \varphi(U) \longrightarrow \mathbb{R}^2$ is (Euclidean) smooth.*

Proof. (a) $\Rightarrow$ (c): By Theorem 10.1.16, F is continuous. It follows from Theorem 11.2.12 that $F \circ \varphi$ is (Euclidean) smooth and from Theorem 11.2.8 that ψ^{-1} is (extended) smooth. By Theorem 10.1.17, $\psi^{-1} \circ F \circ \varphi$ is (Euclidean) smooth.

(c) $\Rightarrow$ (b): Let (U', φ') be a chart on M at p, and let (V, ψ) be a chart on N at $F(p)$. According to [C2] of Section 11.2, $\varphi'(U')$ is open in M, and $\psi(V)$ is open in N. Since F is continuous, by Theorem 9.1.7, $F^{-1}(\psi(V))$ is open in M, and therefore, so is $W = \varphi'(U') \cap F^{-1}(\psi(V))$. Both $\varphi'(U')$ and $F^{-1}(\psi(V))$ contain p, so W is nonempty. It follows from [C3] of Section 11.2 and Theorem 9.1.14 that $U = (\varphi')^{-1}(W)$ is open in U'. Let $\varphi = \varphi'|_U$. We have from Theorem 11.2.10 that (φ, U) is a chart at p such that $F(\varphi(U)) \subseteq \psi(V)$. The result now follows from part (c).

(b) $\Rightarrow$ (a): By Theorem 11.2.8, φ^{-1} is (extended) smooth, and by assumption, ψ and $\psi^{-1} \circ F \circ \varphi$ are (Euclidean) smooth. It follows from Theorem 10.1.17 that

$$F|_{\varphi(U)} = \psi \circ (\psi^{-1} \circ F \circ \varphi) \circ \varphi^{-1} : \varphi(U) \longrightarrow \mathbb{R}^3$$

is (extended) smooth. By definition, there is a neighborhood $\mathcal{U}$ of p in $\mathbb{R}^3$ and a (Euclidean) smooth map $\widetilde{F|_{\varphi(U)}} : \mathcal{U} \longrightarrow \mathbb{R}^3$ such that $F|_{\varphi(U)}$ and $\widetilde{F|_{\varphi(U)}}$ agree on $\varphi(U) \cap \mathcal{U}$. Then F and $\widetilde{F|_{\varphi(U)}}$ agree on $\varphi(U) \cap \mathcal{U}$. According to [C2] of

Section 11.2, $\varphi(U)$ is open in M, so there is an open set $\mathcal{V}$ in $\mathbb{R}^3$ such that $\varphi(U) = M \cap \mathcal{V}$. Then

$$M \cap (\mathcal{U} \cap \mathcal{V}) = (M \cap \mathcal{V}) \cap \mathcal{U} = \varphi(U) \cap \mathcal{U},$$

so F and $\widetilde{F|_{\varphi(U)}}$ agree on $M \cap (\mathcal{U} \cap \mathcal{V})$, where we observe that $\mathcal{U} \cap \mathcal{V}$ is open in $\mathbb{R}^3$. Since p was arbitrary, the result follows. $\square$

Theorem 11.6.2. *Let M, N, and P be regular surfaces, and let $F : M \longrightarrow N$ and $G : N \longrightarrow P$ be maps. If F and G are (extended) smooth, then so is $G \circ F$.*

Proof. Let p be a point in M. By Theorem 11.6.1, there is a chart (U, φ) on M at p and a chart (V_1, ψ_1) on N at $F(p)$ such that $\psi_1^{-1} \circ F \circ \varphi$ is (Euclidean) smooth. For the same reason, there is a chart (V_2, ψ_2) on N at $F(p)$ and a chart (W, μ) at $G\big(F(p)\big)$ such that $\mu^{-1} \circ G \circ \psi_2$ is (Euclidean) smooth. By Theorem 11.2.9, $\psi_2^{-1} \circ \psi_1$ is (Euclidean) smooth. It follows from Theorem 10.1.12 that

$$\mu^{-1} \circ (G \circ F) \circ \varphi = (\mu^{-1} \circ G \circ \psi_2) \circ (\psi_2^{-1} \circ \psi_1) \circ (\psi_1^{-1} \circ F \circ \varphi)$$

is (Euclidean) smooth. Since p was arbitrary, the result follows from Theorem 11.6.1. $\square$

Let M and N be regular surfaces, let $F : M \longrightarrow N$ be a smooth map, and let p be a point in M. The **differential of F at p** is the map

$$d_p(F) : T_p(M) \longrightarrow T_{F(p)}(N)$$

defined by

$$d_p(F)(v) = \frac{d(F \circ \lambda)}{dt}(t_0) \tag{11.6.1}$$

for all vectors v in $T_p(M)$, where $\lambda(t) : (a, b) \longrightarrow M$ is any smooth curve such that $\lambda(t_0) = p$ and $(d\lambda/dt)(t_0) = v$ for some t_0 in (a, b). See Figure 11.6.1.

Theorem 11.6.3. *With the above setup:*
(a) $d_p(F)(v)$ is independent of the choice of smooth curve used to express v.
(b) $d_p(F)$ is linear.

Proof. The proof is similar to that given for Theorem 11.5.2. Let (U, φ) be a chart on M at p, let (V, ψ) be a chart on N at $F(p)$, and let $\lambda(t) : (a, b) \longrightarrow M$ be a smooth curve such that $\lambda(t_0) = p$ and $(d\lambda/dt)(t_0) = v$ for some t_0 in (a, b). Suppose without loss of generality that $\lambda\big((a, b)\big) \subset \varphi(U)$, and let $q = \varphi(p)$. It follows from Theorem 10.1.17 and Theorem 11.2.8 that the map

$$\mu = (\mu^1, \mu^2) = \varphi^{-1} \circ \lambda : (a, b) \longrightarrow U$$

is (Euclidean) smooth, and from Theorem 10.1.9 that

$$v = \frac{d\lambda}{dt}(t_0) = \frac{d(\varphi \circ \mu)}{dt}(t_0) = d_q(\varphi)\left(\frac{d\mu}{dt}(t_0)\right).$$

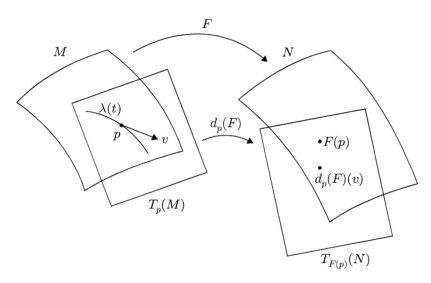

Figure 11.6.1. Differential map

By Theorem 11.3.1(d), $d_q(\varphi)$ is invertible, so

$$\frac{d\mu}{dt}(t_0) = d_q(\varphi)^{-1}(v). \tag{11.6.2}$$

We have from Theorem 11.6.1 that $G = \psi^{-1} \circ F \circ \varphi$ is (Euclidean) smooth, and then from Theorem 10.1.12 that so is

$$F \circ \lambda = F \circ \varphi \circ \mu = \psi \circ G \circ \mu.$$

Then

$$
\begin{aligned}
d_p(F)(v) &= \frac{d(F \circ \lambda)}{dt}(t_0) && [(11.6.1)] \\
&= \frac{d(\psi \circ G \circ \mu)}{dt}(t_0) \\
&= d_{\mu(t_0)}(\psi \circ G)\left(\frac{d\mu}{dt}(t_0)\right) && [\text{Th } 10.1.9] \\
&= d_q(\psi \circ G) \circ d_q(\varphi)^{-1}(v). && [(11.6.2)]
\end{aligned}
\tag{11.6.3}
$$

(a): This follows from (11.6.3).

(b): Since v was arbitrary, we have from (11.6.3) that

$$d_p(F) = d_q(\psi \circ G) \circ d_q(\varphi)^{-1}.$$

Thus, $d_p(F)$ is the composition of linear maps, so it too is linear. $\qquad\square$

Theorem 11.6.4 (Chain Rule). *Let M, N, and P be regular surfaces, let $F : M \longrightarrow N$ and $G : N \longrightarrow P$ be smooth maps, and let p be a point in M. Then*

$$d_p(G \circ F) = d_{F(p)}(G) \circ d_p(F).$$

Proof. Let v be a vector in $T_p(M)$, and let $\lambda(t) : (a, b) \longrightarrow M$ be a smooth curve such that $\lambda(t_0) = p$ and $(d\lambda/dt)(t_0) = v$ for some t_0 in (a, b). Then

$$d_p(G \circ F)(v) = \frac{d(G \circ F \circ \lambda)}{dt}(t_0) \qquad\qquad [(11.6.1)]$$

$$= d_{F \circ \lambda(t_0)}(G)\left(\frac{d(F \circ \lambda)}{dt}(t_0)\right) \qquad\qquad [(11.6.1)]$$

$$= d_{F(\lambda(t_0))}(G)\left(d_{\lambda(t_0)}(F)\left(\frac{d\lambda}{dt}(t_0)\right)\right) \qquad\qquad [(11.6.1)]$$

$$= d_{F(p)}(G) \circ d_p(F)(v).$$

Since v was arbitrary, the result follows. $\qquad\qquad\qquad\qquad\qquad\qquad\square$

The next result is a generalization of Theorem 11.3.3.

Theorem 11.6.5. *Let M and $\widetilde{M}$ be regular surfaces, let $F : M \longrightarrow \widetilde{M}$ be a smooth map, and let p be a point in M. Let (U, φ) be a chart on M at p, let $(\widetilde{U}, \widetilde{\varphi})$ be a chart on $\widetilde{M}$ at $F(p)$, and let $\mathcal{H}$ and $\widetilde{\mathcal{H}}$ be the corresponding coordinate frames. Then*

$$[d_p(F)]_{\mathcal{H}_q}^{\widetilde{\mathcal{H}}_{\widetilde{q}}} = J_{\widetilde{\varphi}^{-1} \circ F \circ \varphi}(q),$$

where $q = \varphi^{-1}(p)$ and $\widetilde{q} = \widetilde{\varphi}^{-1} \circ F(p)$.

Proof. Let $\mathcal{H} = (H_1, H_2)$ and $\widetilde{\mathcal{H}} = (\widetilde{H}_1, \widetilde{H}_2)$, let (r^1, r^2) and $(\widetilde{r}^1, \widetilde{r}^2)$ be coordinates on U and $\widetilde{U}$, respectively, and let (e_1, e_2) be the standard basis for $\mathbb{R}^2$. For given $1 \leq j \leq 2$, we define a (Euclidean) smooth map

$$\zeta = (\zeta^1, \zeta^2) : (-\varepsilon, \varepsilon) \longrightarrow U$$

by $\zeta(t) = q + t e_j$, where $\varepsilon > 0$ is chosen small enough that $(q - \varepsilon e_j, q + \varepsilon e_j) \subset U$. Consider the smooth curve $\lambda(t) : (-\varepsilon, \varepsilon) \longrightarrow M$ defined by $\lambda = \varphi \circ \zeta$, and observe that $\lambda(0) = \varphi(q)$. By Theorem 10.1.10,

$$\frac{d\lambda}{dt}(0) = \frac{d(\varphi \circ \zeta)}{dt}(0) = \sum_k \frac{d\zeta^k}{dt}(0) H_k|_q = H_j|_q \qquad (11.6.4)$$

and

$$\frac{d(F \circ \lambda)}{dt}(0) = \frac{d(F \circ \varphi \circ \zeta)}{dt}(0) = \sum_k \frac{d\zeta^k}{dt}(0) \frac{\partial(F \circ \varphi)}{\partial r^k}(q)$$

$$= \frac{\partial(F \circ \varphi)}{\partial r^j}(q). \qquad\qquad\qquad (11.6.5)$$

It follows from Theorem 11.6.1 that the map

$$G = (G^1, G^2) = \widetilde{\varphi}^{-1} \circ F \circ \varphi$$

is smooth. By Theorem 10.1.10,

$$\frac{\partial(F \circ \varphi)}{dr^j}(q) = \frac{\partial(\widetilde{\varphi} \circ G)}{\partial r^j}(q) = \sum_i \frac{\partial G^i}{\partial r^j}(q)\widetilde{H}_i|_{\widetilde{q}}. \qquad (11.6.6)$$

We have

$$
\begin{aligned}
d_p(F)(H_j|_q) &= \frac{d(F \circ \lambda)}{dt}(0) && [(11.6.1),\ (11.6.4)] \\
&= \frac{\partial(F \circ \varphi)}{dr^j}(q) && [(11.6.5)] \\
&= \sum_i \frac{\partial G^i}{\partial r^j}(q)\widetilde{H}_i|_{\widetilde{q}}. && [(11.6.6)]
\end{aligned}
$$

Then (2.2.2), (2.2.3), and (10.1.2) give

$$\left[d_p(F)\right]_{\mathcal{H}_q}^{\widetilde{\mathcal{H}}_{\widetilde{q}}} = \left[\frac{\partial G^i}{\partial r^j}(q)\right] = J_{\widetilde{\varphi}^{-1} \circ F \circ \varphi}(q). \qquad \square$$

11.7   Vector Fields Along Regular Surfaces in $\mathbb{R}^3$

Let M be a regular surface, let $V : M \longrightarrow \mathbb{R}^3$ be a map, and let p be a point in M. In the present context, we refer to V as a **vector field** *along* **M**. We say that V **vanishes at** p if $V_p = (0,0,0)$, is **nonvanishing at** p if $V_p \neq (0,0,0)$, and is **nowhere-vanishing (on M)** if it is nonvanishing at every p in M.

Let us denote by $\mathfrak{X}_{\mathbb{R}^3}(M)$ the set of smooth vector fields along M. Then $\mathfrak{X}_{\mathbb{R}^3}(M)$ is nothing more than the set of (extended) smooth maps from M to $\mathbb{R}^3$. With operations on $\mathfrak{X}_{\mathbb{R}^3}(M)$ defined in a manner analogous to those in Section 10.3, $\mathfrak{X}_{\mathbb{R}^3}(M)$ is a vector space over $\mathbb{R}$ and a module over $C^\infty(M)$. We say that a vector field $X : M \longrightarrow \mathbb{R}^3$ along M is a **(tangent) vector field** *on* **M** if X_p is in $T_p(M)$ for all p in M. The set of smooth vector fields on M is denoted by $\mathfrak{X}(M)$. Clearly, $\mathfrak{X}(M) \subset \mathfrak{X}_{\mathbb{R}^3}(M)$. In fact, $\mathfrak{X}(M)$ is a vector subspace and a $C^\infty(M)$-submodule of $\mathfrak{X}_{\mathbb{R}^3}(M)$. As an example, for a regular surface that has a covering chart, each of the components of the corresponding coordinate frame is a tangent vector field.

Theorem 11.7.1 (Smoothness Criteria for Vector Fields). *If M is a regular surface and $V : M \longrightarrow \mathbb{R}^3$ is a vector field along M, then the following are equivalent:*

(a) *V is (extended) smooth.*
(b) *For every point p in M, there is a chart (U, φ) on M at p such that the map $V \circ \varphi : U \longrightarrow \mathbb{R}^3$ is (Euclidean) smooth.*
(c) *For every chart (U, φ) on M, the map $V \circ \varphi : U \longrightarrow \mathbb{R}^3$ is (Euclidean) smooth.*

Proof. Setting $n = 3$ and $F = V$ in Theorem 11.2.12 gives the result. $\square$

Let M be a regular surface, let V be a vector field in $\mathfrak{X}_{\mathbb{R}^3}(M)$, and let p be a point in M. The **differential of** V **at** p is the map

$$d_p(V) : T_p(M) \longrightarrow \mathbb{R}^3$$

defined by

$$d_p(V)(v) = \frac{d(V \circ \lambda)}{dt}(t_0) \tag{11.7.1}$$

for all vectors v in $T_p(M)$, where $\lambda(t) : (a, b) \longrightarrow M$ is any smooth curve such that $\lambda(t_0) = p$ and $(d\lambda/dt)(t_0) = v$ for some t_0 in (a, b). Observe that this is not a special case of (11.6.1) because $\mathbb{R}^3$ is not a regular surface.

Theorem 11.7.2. *With the above setup, if* $V = (\alpha^1, \alpha^2, \alpha^3)$, *then*

$$d_p(V)(v) = \big(d_p(\alpha^1)(v), d_p(\alpha^2)(v), d_p(\alpha^3)(v)\big).$$

Proof. Since $V \circ \lambda = (\alpha^1 \circ \lambda, \alpha^2 \circ \lambda, \alpha^3 \circ \lambda)$, we have

$$
\begin{aligned}
d_p(V)(v) &= \frac{d(V \circ \lambda)}{dt}(t_0) & [(11.7.1)] \\
&= \left(\frac{d(\alpha^1 \circ \lambda)}{dt}(t_0), \frac{d(\alpha^2 \circ \lambda)}{dt}(t_0), \frac{d(\alpha^3 \circ \lambda)}{dt}(t_0) \right) \\
&= \big(d_p(\alpha^1)(v), d_p(\alpha^2)(v), d_p(\alpha^3)(v)\big). & [(11.5.1)] \quad \square
\end{aligned}
$$

Theorem 11.7.3. *With the above setup:*
(a) $d_p(V)(v)$ *is independent of the choice of smooth curve used to express* v.
(b) $d_p(V)$ *is linear.*

Proof. In light of Theorem 11.7.2, the result follows from applying Theorem 11.5.2 to each component of $d_p(V)(v)$. $\quad\square$

Chapter 12

Curves and Regular Surfaces in $\mathbb{R}^3_\nu$

Chapter 11 was devoted to a discussion of curves and regular surfaces in $\mathbb{R}^3$. A regular surface was defined to be a subset of $\mathbb{R}^3$ with certain properties specified in terms of the subspace topology, smooth maps, immersions, homeomorphisms, and so on. The fact that $\mathbb{R}^3$ has an inner product (which gives rise to a norm, which in turn gives rise to a distance function, which in turn gives rise to a topology) was relegated to the background—present but largely unacknowledged. The topological and metric aspects of $\mathbb{R}^3$ were central to our discussion of what it means for a regular surface to be "smooth", and in that way the inner product (through the distance function) was involved.

In this chapter, we continue our discussion of regular surfaces, but this time endow each tangent plane with additional linear structure induced by the linear structure on $\mathbb{R}^3$. Specifically, we view $\mathbb{R}^3$ as either Euclidean 3-space, that is, $\mathbb{R}^3_0 = (\mathbb{R}^3, \mathfrak{e})$, or Minkowski 3-space, that is, $\mathbb{R}^3_1 = (\mathbb{R}^3, \mathfrak{m})$, and give each tangent plane the corresponding inner product or Lorentz scalar product obtained by restriction. It must be stressed that this additional linear structure changes nothing regarding the underlying regular surface. The definitions introduced in Chapter 11 remain in force, but we now express them somewhat differently. To that end, let us denote $\mathbb{R}^3_0$ and $\mathbb{R}^3_1$ collectively by $\mathbb{R}^3_\nu$, with the understanding that $\nu = 0$ or 1 depending on the context. After introducing a series of definitions, we will speak of a regular surface as being a "regular surface in $\mathbb{R}^3_\nu$". Again it must be emphasized that aside from the additional structure given to tangent planes, a regular surface in $\mathbb{R}^3_\nu$ is the same underlying regular surface considered in Chapter 11.

Semi-Riemannian Geometry, First Edition. Stephen C. Newman.

12.1   Curves in $\mathbb{R}^3_\nu$

Let $\lambda = (\lambda^1, \lambda^2, \lambda^3) : (a, b) \longrightarrow \mathbb{R}^3_\nu$ be a smooth curve, and recall from Section 10.1 that the velocity of λ is the smooth curve $d\lambda/dt : (a, b) \longrightarrow \mathbb{R}^3_\nu$. When $\nu = 1$, we say that λ is **spacelike** (resp., **timelike, lightlike**) if $(d\lambda/dt)(t)$ is spacelike (resp., timelike, lightlike) for all t in (a, b). According to (4.1.1), the norm of $(d\lambda/dt)(t)$ is

$$\left\| \frac{d\lambda}{dt}(t) \right\| = \sqrt{\left| \left(\frac{d\lambda^1}{dt}(t) \right)^2 + \left(\frac{d\lambda^2}{dt}(t) \right)^2 + (-1)^\nu \left(\frac{d\lambda^3}{dt}(t) \right)^2 \right|},$$

where we note the presence of $(-1)^\nu$ and the absolute value bars. The function $\|d\lambda/dt\| : (a, b) \longrightarrow \mathbb{R}$ is called the **speed of λ**. Recall that λ is said to be regular if its velocity is nowhere-vanishing. When $\nu = 0$, this is equivalent to its speed being nowhere-vanishing. We say that λ has **constant speed** if there is a real number c such that $\|(d\lambda/dt)(t)\| = c$ for all t in (a, b).

Let $\lambda(t) : [a, b] \longrightarrow \mathbb{R}^3_\nu$ be an (extended) smooth curve. The **length of λ** (more precisely, the length of the image of λ) is defined by

$$L(\lambda) = \int_a^b \left\| \frac{d\lambda}{dt}(t) \right\| dt.$$

Other than their role in defining the above integral, we have little interest in the endpoints of $[a, b]$. In order to avoid having to consider one-sided limits, we continue to frame the discussion in terms of $\lambda(t) : [a, b] \longrightarrow \mathbb{R}^3_\nu$ but compute with $\lambda(t) : (a, b) \longrightarrow \mathbb{R}^3_\nu$. In short, we systematically confuse the distinction between $[a, b]$ and (a, b).

As the next result shows, the length of a smooth curve does not depend on the choice of parametrization.

Theorem 12.1.1 (Diffeomorphism Invariance of Length). *If $\lambda(t) : [a, b] \longrightarrow \mathbb{R}^3_\nu$ is a smooth curve and $g(u) : [c, d] \longrightarrow [a, b]$ is a diffeomorphism, then $L(\lambda) = L(\lambda \circ g)$.*

Proof. It follows from Theorem 10.1.10 that

$$\left\| \frac{d(\lambda \circ g)}{du}(u) \right\| = \left\| \frac{d\lambda}{dt}(g(u)) \right\| \left| \frac{dg}{du}(u) \right|$$

for all u in $[c, d]$. By the change of variables theorem from the differential calculus of one real variable,

$$L(\lambda) = \int_a^b \left\| \frac{d\lambda}{dt}(t) \right\| dt = \int_{g^{-1}(a)}^{g^{-1}(b)} \left\| \frac{d\lambda}{dt}(g(u)) \right\| \frac{dg}{du}(u)\, du.$$

According to Theorem 10.2.4(c), g is either strictly increasing or strictly decreasing. In the former case,

$$g^{-1}(a) = c, \qquad g^{-1}(b) = d, \qquad \text{and} \qquad \frac{dg}{du}(u) = \left| \frac{dg}{du}(u) \right|;$$

and in the latter case,

$$g^{-1}(a) = d, \qquad g^{-1}(b) = c, \qquad \text{and} \qquad \frac{dg}{du}(u) = -\left|\frac{dg}{du}(u)\right|.$$

Either way,

$$\int_{g^{-1}(a)}^{g^{-1}(b)} \left\|\frac{d\lambda}{dt}(g(u))\right\| \frac{dg}{du}(u)\,du = \int_d^c \left\|\frac{d\lambda}{dt}(g(u))\right\| \left|\frac{dg}{du}(u)\right|\,du$$

$$= \int_c^d \left\|\frac{d(\lambda \circ g)}{du}(u)\right\|\,du$$

$$= L(\lambda \circ g).$$

The result follows. $\qquad\qquad\qquad\qquad\qquad\qquad\qquad\qquad\qquad\qquad\qquad\quad\square$

12.2   Regular Surfaces in $\mathbb{R}^3_\nu$

A regular surface is by definition a subset of $\mathbb{R}^3$. We now view a regular surface as a subset of $\mathbb{R}^3_\nu$, where ν is left unspecified. The scalar product on $\mathbb{R}^3_\nu$ is given by

$$\langle\cdot,\cdot\rangle = \begin{cases} \mathfrak{e} & \text{if } \nu = 0 \\ \mathfrak{m} & \text{if } \nu = 1, \end{cases}$$

where $\mathfrak{e}$ and $\mathfrak{m}$ are the Euclidean inner product and Minkowski scalar product, respectively.

Let M be a regular surface, and let p be a point in M. We obtain a symmetric tensor $\mathfrak{g}_p$ in $T^0_2(T_p(M))$ by restricting the scalar product on $\mathbb{R}^3_\nu$ to $T_p(M) \times T_p(M)$:

$$\mathfrak{g}_p = \begin{cases} \mathfrak{e}|_{T_p(M)} & \text{if } \nu = 0 \\ \mathfrak{m}|_{T_p(M)} & \text{if } \nu = 1. \end{cases} \qquad (12.2.1)$$

For brevity, we usually denote

$$\mathfrak{g}_p(\cdot,\cdot) \qquad \text{by} \qquad \langle\cdot,\cdot\rangle.$$

Whether the notation $\langle\cdot,\cdot\rangle$ refers to the scalar product on $\mathbb{R}^3_\nu$ or the tensor $\mathfrak{g}_p$ will be clear from the context.

The **first fundamental form on M** is the map denoted by $\mathfrak{g}$ and defined by the assignment $p \longmapsto \mathfrak{g}_p$ for all p in M. In the literature, $\mathfrak{g}$ is often denoted by I. For vector fields X, Y in $\mathfrak{X}(M)$, we define a function

$$\mathfrak{g}(X,Y) = \langle X, Y \rangle : M \longrightarrow \mathbb{R}$$

in $C^\infty(M)$ by the assignment

$$p \longmapsto \mathfrak{g}_p(X_p, Y_p) = \langle X_p, Y_p \rangle$$

for all p in M.

Since a subspace of an inner product space is itself an inner product space, when $\nu = 0$, $\mathfrak{g}_p$ is an inner product on $T_p(M)$ for all p in M. On the other hand, when $\nu = 1$, $\mathfrak{g}_p$ is bilinear and symmetric on $T_p(M)$, but there is no guarantee that it is nondegenerate. Furthermore, even if $\mathfrak{g}_p$ is nondegenerate on each $T_p(M)$, it might be an inner product for some p and a Lorentz scalar product for others. In other words, $\mathfrak{g}_p$ might not have the same index for all p in M. For these reasons, we make the following definition.

We say that $\mathfrak{g}$ is a **metric (on M)** if:

[G1] $\mathfrak{g}_p$ is nondegenerate on $T_p(M)$ for all p in M.
[G2] $\mathrm{ind}(\mathfrak{g}_p)$ is independent of p in M.

When [G1] is satisfied, $\mathfrak{g}_p$ is a scalar product on $T_p(M)$ for all p in M. [G1] and [G2] are automatically satisfied when $\nu = 0$.

We say that a vector v in $\mathbb{R}^3_\nu$ is **normal at p** if v is in $T_p(M)^\perp$, where $\perp$ is computed using the scalar product in $\mathbb{R}^3_\nu$. If v is also a unit vector, it is said to be **unit normal at p**.

Let V be a vector field along M. Recall that this means nothing more than V is a map from M to $\mathbb{R}^3_\nu$. Looked at another way, V is effectively a collection of vectors in $\mathbb{R}^3_\nu$, one for each p in M. Without further assumptions, there is no reason to expect V to be smooth; that is, V is not necessarily a vector field in $\mathfrak{X}_{\mathbb{R}^3_\nu}(M)$. We say that V is a **unit vector field** if V_p is a unit vector for all p in M, and that V is a **normal vector field** if V_p is normal at p for all p in M. Clearly, a unit vector field is nowhere-vanishing. When V is both a unit vector field and a normal vector field, it is said to be a **unit normal vector field**.

For vector fields V, W along M, let us define the function

$$\langle V, W \rangle : M \longrightarrow \mathbb{R}$$

by the assignment

$$p \longmapsto \mathfrak{g}_p(V_p, W_p) = \langle V_p, W_p \rangle \tag{12.2.2}$$

for all p in M. Let us also define the function

$$\|V\| : M \longrightarrow \mathbb{R}$$

by the assignment

$$p \longmapsto \|V_p\| = \sqrt{|\langle V_p, W_p \rangle|}$$

for all p in M. When $\|V\|$ is nowhere-vanishing, we define $V/\|V\|$ to be the vector field along M given by the assignment $p \longmapsto V_p/\|V_p\|$ for all p in M.

Here are two properties that a vector field V along M might satisfy:

[V1] $T_p(M)^\perp = \mathbb{R}V_p$ for all p in M.
[V2] $\langle V_p, V_p \rangle$ is positive for all p in M, or negative for all p in M.

We observe that [V2] is equivalent to V_p being either nonzero spacelike for all p in M, or timelike for all p in M.

Theorem 12.2.1. *With the above setup, if V satisfies [V2], then:*
(a) *V is nowhere-vanishing on M; that is, $V_p \neq (0,0,0)$ for all p in M.*
(b) *$\|V\|$ is nowhere-vanishing on M; that is, $\|V_p\| \neq 0$ for all p in M.*
(c) *$V/\|V\|$ is a unit vector field along M, and if V is smooth on M, then so is $V/\|V\|$.*
(d) *If $\nu = 0$, then the converse of part (a) holds: V satisfies [V2] if and only if V is nowhere-vanishing.*

Proof. (a), (b), (d): Straightforward.
   (c): The first assertion follows from part (b). For the second assertion, since V satisfies [V2], either $|\langle V, V \rangle| = \langle V, V \rangle$ on M, or $|\langle V, V \rangle| = -\langle V, V \rangle$ on M. Since V is smooth on M, so is $|\langle V, V \rangle|$, and therefore, so is $\|V\| = \sqrt{|\langle V, V \rangle|}$. It follows that $V/\|V\|$ is smooth. $\qquad\square$

We now show that properties [G1]–[G2] and [V1]–[V2] are closely related. For convenience of exposition, most of the results to follow are presented for arbitrary ν. However, the findings for $\nu = 0$ are essentially trivial; it is the case $\nu = 1$ that is of primary interest.

Theorem 12.2.2. *Let M be a regular surface, let $\mathfrak{g}$ be the first fundamental form on M, and let p be a point in M such that $\mathfrak{g}_p$ is nondegenerate. Then:*
(a) *There are precisely two unit normal vectors at p.*
(b) *If v is a unit normal vector at p, then*

$$T_p(M)^\perp = \mathbb{R}v \qquad \text{and} \qquad \langle v, v \rangle = (-1)^{\nu - \text{ind}(\mathfrak{g}_p)}.$$

Proof. It follows from Theorem 4.1.3 that $\mathbb{R}^3_\nu = T_p(M) \oplus T_p(M)^\perp$, and then from Theorem 1.1.18 that $T_p(M)^\perp$ is 1-dimensional. Thus, $T_p(M)^\perp = \mathbb{R}v$, where v is one of the two unit vectors in $T_p(M)^\perp$, the other being $-v$. Let (e_1, e_2) be an orthonormal basis for $T_p(M)$. Then (e_1, e_2, v) is an orthonormal basis for $\mathbb{R}^3_\nu$. Using (4.2.3) twice yields

$$(-1)^\nu = \langle e_1, e_1 \rangle \langle e_2, e_2 \rangle \langle v, v \rangle = (-1)^{\text{ind}(\mathfrak{g}_p)} \langle v, v \rangle. \qquad\square$$

Theorem 12.2.3. *Let M be a regular surface, and let $\mathfrak{g}$ be the first fundamental form on M. Then:*
(a) *$\mathfrak{g}$ satisfies [G1] if and only if there is a unit normal vector field V along M satisfying [V1].*
(b) *If V and $\widetilde{V}$ are unit normal vector fields along M satisfying [V1], then $\langle V_p, V_p \rangle = \langle \widetilde{V}_p, \widetilde{V}_p \rangle$ for all p in M.*

Proof. (a)($\Rightarrow$): This follows from parts (a) and (b) of Theorem 12.2.2.
   (a)($\Leftarrow$): Let p be a point in M. If v is a vector in $T_p(M) \cap T_p(M)^\perp$, then $\langle v, v \rangle = 0$. Since $T_p(M)^\perp = \mathbb{R}V_p$, we have $v = cV_p$ for some real number c. Then

$$0 = \langle cV_p, cV_p \rangle = c^2 \langle V_p, V_p \rangle = \pm c^2,$$

hence $c = 0$. Thus, $T_p(M) \cap T_p(M)^\perp = \{0\}$. By Theorem 4.1.3, $\mathfrak{g}_p$ is nondegenerate on $T_p(M)$. Since p was arbitrary, the result follows.

(b): Let p be a point in M. Since $\mathbb{R}V_p = T_p(M)^\perp = \mathbb{R}\widetilde{V}_p$, we have $\widetilde{V}_p = cV_p$ for some real number c. Then

$$\pm 1 = \langle \widetilde{V}_p, \widetilde{V}_p \rangle = c^2 \langle V_p, V_p \rangle = \pm c^2,$$

hence $c = \pm 1$. Thus, $\langle \widetilde{V}_p, \widetilde{V}_p \rangle = \langle \pm V_p, \pm V_p \rangle = \langle V_p, V_p \rangle$. Since p was arbitrary, the result follows. $\qquad\square$

Theorem 12.2.4. *Let M be a regular surface, let V be a unit normal vector field along M, and suppose $\mathfrak{g}$, the first fundamental form on M, satisfies [G1]. Then $\mathfrak{g}$ satisfies [G2] if and only if V satisfies [V2].*

Proof. Since V is a unit vector field along M, [V2] is equivalent to: $\langle V_p, V_p \rangle = 1$ for all p in M, or $\langle V_p, V_p \rangle = -1$ for all p in M. By Theorem 12.2.2(b),

$$(-1)^{\nu - \mathrm{ind}(\mathfrak{g}_p)} = \langle V_p, V_p \rangle.$$

The result follows. $\qquad\square$

Let M be a regular surface, and let $\mathfrak{g}$ be the first fundamental form on M. When $\mathfrak{g}$ is a metric, the pair $(M, \mathfrak{g})$ is called a **regular surface in** $\mathbb{R}^3_\nu$. In that case, we ascribe to $\mathfrak{g}$ those properties of $\mathfrak{g}_p$ that are independent of p. Accordingly, $\mathfrak{g}$ is said to be bilinear, symmetric, nondegenerate, and so on. The common value of the $\mathrm{ind}(\mathfrak{g}_p)$ is denoted by $\mathrm{ind}(\mathfrak{g})$ and called the **index of $\mathfrak{g}$** or the **index of M**.

The next result shows that we could have defined a regular surface in $\mathbb{R}^3_\nu$ using properties [V1] and [V2] instead of [G1] and [G2].

Theorem 12.2.5. *Let M be a regular surface. Then $(M, \mathfrak{g})$ is a regular surface in $\mathbb{R}^3_\nu$ if and only if there is a unit normal vector field along M satisfying [V1] and [V2].*

Proof. This follows from Theorem 12.2.3(a) and Theorem 12.2.4. $\qquad\square$

Theorem 12.2.6. *Let M be a regular surface, and let $\mathfrak{g}$ be the first fundamental form on M. Let (U, φ) be a chart on M, let (H_1, H_2) be the corresponding coordinate frame, and, using Example 11.2.5, view $\varphi(U)$ as a regular surface. Then $\big(\varphi(U), \mathfrak{g}|_{\varphi(U)}\big)$ is a regular surface in $\mathbb{R}^3_\nu$ if and only if $\langle H_1|_q \times H_2|_q, H_1|_q \times H_2|_q \rangle$ is positive for all q in U, or negative for all q in U, where the vector product is computed with respect to the standard basis for $\mathbb{R}^3_\nu$.*

Remark. Since $\mathfrak{g}|_{\varphi(U)}$ is the first fundamental form on $\varphi(U)$, the assertion makes sense. In light of earlier remarks, if $\nu = 0$, then $\big(\varphi(U), \mathfrak{g}|_{\varphi(U)}\big)$ is a regular surface in $\mathbb{R}^3_0$ without any assumptions on the behavior of H_1 and H_2.

Proof. For brevity, we refer to the statement "$\langle H_1|_q \times H_2|_q, H_1|_q \times H_2|_q \rangle$ is positive for all q in U, or negative for all q in U" as property $(*)$.

($\Rightarrow$): By Theorem 12.2.5, there is a unit normal vector field W along $\varphi(U)$ satisfying [V1] and [V2]. Since W satisfies [V1], we have from Theorem 8.4.4,

Theorem 8.4.8, and Theorem 11.4.1(b) that $H_1|_{\varphi^{-1}(p)} \times H_2|_{\varphi^{-1}(p)} = c_p W_p$ for some nonzero real number c_p, hence

$$\langle H_1|_{\varphi^{-1}(p)} \times H_2|_{\varphi^{-1}(p)}, H_1|_{\varphi^{-1}(p)} \times H_2|_{\varphi^{-1}(p)} \rangle = c_p^2 \langle W_p, W_p \rangle$$

for all p in $\varphi(U)$. Since W satisfies [V2], property $(*)$ is satisfied.

($\Leftarrow$): We have from property $(*)$ that $\|H_1|_{\varphi^{-1}(p)} \times H_2|_{\varphi^{-1}(p)}\| \neq 0$ for all p in $\varphi(U)$. Setting $(V, \mathfrak{g}) = \mathbb{R}^3_\nu$, $U = T_p(M)$, and $(u_1, u_2) = (H_1|_{\varphi^{-1}(p)}, H_2|_{\varphi^{-1}(p)})$ in Theorem 8.4.9, and using Theorem 11.4.1(b), shows that for all p in $\varphi(U)$, $(\mathfrak{g}|_{\varphi(U)})|_p$ is nondegenerate on $T_p(\varphi(U))$ if and only if $\|H_1|_{\varphi^{-1}(p)} \times H_2|_{\varphi^{-1}(p)}\| \neq 0$. Thus, $\mathfrak{g}|_{\varphi(U)}$ satisfies [G1]. For a given point p in $\varphi(U)$, consider the vector

$$V_p = \frac{H_1|_{\varphi^{-1}(p)} \times H_2|_{\varphi^{-1}(p)}}{\|H_1|_{\varphi^{-1}(p)} \times H_2|_{\varphi^{-1}(p)}\|}.$$

We have from Theorem 8.4.8 that $V_p \neq (0, 0, 0)$, and from Theorem 8.4.4 and Theorem 11.4.1(b) that V_p is in $T_p(\varphi(U))^\perp$. Thus, the assignment $p \longmapsto V_p$ defines a unit normal vector field V along $\varphi(U)$. It follows from property $(*)$ that V satisfies [V2]. By Theorem 12.2.4, $\mathfrak{g}|_{\varphi(U)}$ satisfies [G2]. $\square$

Theorem 12.2.7 (Open Set). *If $(M, \mathfrak{g})$ is a regular surface in $\mathbb{R}^3_\nu$ and W is an open set in M, then $(W, \mathfrak{g}|_W)$ is a regular surface in $\mathbb{R}^3_\nu$. In particular, if (U, φ) is a chart on M, then $(\varphi(U), \mathfrak{g}|_{\varphi(U)})$ is a regular surface in $\mathbb{R}^3_\nu$.*

Proof. By Theorem 11.4.1, W is a regular surface and $T_p(W) = T_p(M)$ for all p in W. It follows that $(\mathfrak{g}|_W)|_p$ is nondegenerate on $T_p(W)$ and $\mathrm{ind}(\mathfrak{g}_p) = \mathrm{ind}((\mathfrak{g}|_W)_p)$ for all p in W. Thus, $\mathfrak{g}|_W$ satisfies [G1] and [G2], which proves the first assertion. According to [C2] of Section 11.2, $\varphi(U)$ is an open set in M, so the second assertion follows from the first. $\square$

Theorem 12.2.8 (Graph of Function). *In the notation of Theorem 11.4.2, graph(f) is a regular surface in $\mathbb{R}^3_\nu$ if and only if*

$$\left(\frac{\partial f}{\partial x}(x, y) \right)^2 + \left(\frac{\partial f}{\partial y}(x, y) \right)^2 - 1$$

is positive for all (x, y) in U, or negative for all (x, y) in U.

Proof. As remarked in connection with Theorem 12.2.6, the case $\nu = 0$ is straightforward, so assume $\nu = 1$. We have from Example 11.2.5 that graph(f) is a regular surface, from Theorem 11.4.2 that (U, φ) is a covering chart, and from the proof of part (b) of Theorem 12.10.2 that

$$\langle H_1|_{(x,y)} \times H_2|_{(x,y)}, H_1|_{(x,y)} \times H_2|_{(x,y)} \rangle = \left(\frac{\partial f}{\partial x}(x, y) \right)^2 + \left(\frac{\partial f}{\partial y}(x, y) \right)^2 - 1.$$

The result now follows from Theorem 12.2.6. $\square$

Theorem 12.2.9 (Surface of Revolution). *In the notation of Theorem 11.4.3,* $\mathrm{rev}(\sigma)$ *is a regular surface in* $\mathbb{R}^3_\nu$ *if and only if* $\dot{h}(t)^2 - \dot{\rho}(t)^2$ *is positive for all* t *in* (a, b), *or negative for all* t *in* (a, b).

Proof. As remarked in connection with Theorem 12.2.6, the case $\nu = 0$ is straightforward, so assume $\nu = 1$. We have from Example 11.2.5 that $\mathrm{rev}(\sigma)$ is a regular surface, from Theorem 11.4.3 that (U, φ) is a covering chart, and from the proof of part (b) of Theorem 12.10.4 that

$$\langle H_1|_{(t,\phi)} \times H_2|_{(t,\phi)}, H_1|_{(t,\phi)} \times H_2|_{(t,\phi)} \rangle = \rho(t)^2[\dot{h}(t)^2 - \dot{\rho}(t)^2].$$

By definition, $\rho(t)$ is positive, hence nonzero, for all t in (a, b). The result now follows from Theorem 12.2.6. □

Let $\mathcal{U}$ be an open set in $\mathbb{R}^3_\nu$, and let f be a function in $C^\infty(\mathcal{U})$. The **gradient of** f (**in** $\mathbb{R}^3_\nu$) is the map

$$\mathrm{Grad}(f) : \mathcal{U} \longrightarrow \mathbb{R}^3$$

defined by

$$\mathrm{Grad}(f)_p = \left(\frac{\partial f}{\partial x}(p), \frac{\partial f}{\partial y}(p), (-1)^\nu \frac{\partial f}{\partial z}(p) \right)$$

for all p in $\mathcal{U}$, where we note the presence of $(-1)^\nu$. When $\nu = 0$, the above identity simplifies to (11.4.3), in which case, $\mathrm{Grad}(f)_p = \mathrm{grad}(f)_p$.

Theorem 12.2.10 (Level Set of Function). *With the above setup, let c be a real number in $f(\mathcal{U})$, and let $M = f^{-1}(c)$. Then:*
(a) *If* $\mathrm{Grad}(f)$ *is nowhere-vanishing on M, then M is a regular surface.*
(b) *If* $\mathrm{Grad}(f)$ *satisfies* [V2], *then* $(M, \mathfrak{g})$ *is a regular surface in* $\mathbb{R}^3_\nu$ *and* $\mathrm{Grad}(f)$ $/\|\mathrm{Grad}(f)\|$ *is a unit normal vector field in* $\mathfrak{X}_{\mathbb{R}^3_\nu}(M)$.

Remark. We have from Theorem 12.2.1(d) that when $\nu = 0$, the assumptions in parts (a) and (b) are equivalent.

Proof. For brevity, let $V = \mathrm{Grad}(f)$.
  (a): Since V is nowhere-vanishing on M, so is $\mathrm{grad}(f)$. By Theorem 11.4.4, M is a regular surface.
  (b): Let p be a point in M, and let v be a vector in $T_p(M)$. By definition, there is a smooth curve $\lambda = (\lambda^1, \lambda^2, \lambda^3) : (a, b) \longrightarrow M$ such that $\lambda(t_0) = p$ and $(d\lambda/dt)(t_0) = v$ for some t_0 in (a, b). Since $f \circ \lambda(t) = c$ for all t in (a, b), by Theorem 10.1.10,

$$
\begin{aligned}
0 &= \frac{d(f \circ \lambda)}{dt}(t_0) = \sum_i \frac{\partial f}{\partial x^i}(p) \frac{d\lambda^i}{dt}(t_0) \\
&= \left\langle \left(\frac{\partial f}{\partial x^1}(p), \frac{\partial f}{\partial x^2}(p), (-1)^\nu \frac{\partial f}{\partial x^3}(p) \right), \left(\frac{d\lambda^1}{dt}(t_0), \frac{d\lambda^2}{dt}(t_0), \frac{d\lambda^3}{dt}(t_0) \right) \right\rangle \\
&= \langle V_p, v \rangle.
\end{aligned}
$$

Because p and v were arbitrary, it follows that V is a normal vector field along M and $T_p(M) \subseteq (\mathbb{R}V_p)^\perp$ for all p in M. We have from parts (a) and (c) of Theorem 12.2.1 that V is nowhere-vanishing on M and $W = V/\|V\|$ is a unit normal vector field along M. Since f is smooth, so is V, and therefore, by Theorem 12.2.1(c), so is W; that is, W is a vector field in $\mathfrak{X}_{\mathbb{R}^3_\nu}(M)$. Clearly, W satisfies [V2]. Since $T_p(M) \subseteq (\mathbb{R}V_p)^\perp = (\mathbb{R}W_p)^\perp$, and by Theorem 11.3.1(a), $T_p(M)$ is 2-dimensional, it follows that $(\mathbb{R}W_p)^\perp$ is either 2-dimensional or 3-dimensional. The latter possibility is excluded because we would then have $(\mathbb{R}W_p)^\perp = \mathbb{R}^3_\nu$, hence $\langle W_p, W_p \rangle = 0$, which contradicts the fact that W satisfies [V2]. Thus, $(\mathbb{R}W_p)^\perp$ is 2-dimensional, so $T_p(M) = (\mathbb{R}W_p)^\perp$. By Theorem 4.1.2(c), $T_p(M)^\perp = (\mathbb{R}W_p)$ for all p in M; that is, W satisfies [V1]. It follows from part (a) and Theorem 12.2.5 that $(M, \mathfrak{g})$ is a regular surface in $\mathbb{R}^3_\nu$. $\qquad\square$

Let $(M, \mathfrak{g})$ be a regular surface in $\mathbb{R}^3_\nu$. We have from Theorem 12.2.5 that there is a (not necessarily smooth) unit normal vector field V along M satisfying [V1] and [V2]. As pointed out in the proof of Theorem 12.2.4, since V is a unit vector field, [V2] is equivalent to: $\langle V_p, V_p \rangle = 1$ for all p in M, or $\langle V_p, V_p \rangle = -1$ for all p in M. The common value of the $\langle V_p, V_p \rangle$ is denoted by ϵ_M and called the **sign of M**. Thus,

$$\epsilon_M = \langle V_p, V_p \rangle \qquad (12.2.3)$$

for all p in M. By Theorem 12.2.3(b), ϵ_M is independent of the choice of unit normal vector field along M satisfying [V1] and [V2]. We have from Theorem 12.2.2(b) that

$$\epsilon_M = (-1)^{\nu - \mathrm{ind}(\mathfrak{g})}. \qquad (12.2.4)$$

A convenient way to determine $\mathrm{ind}(\mathfrak{g})$ that avoids having to construct an orthonormal basis is to find ϵ_M using (12.2.3) and then compute $\mathrm{ind}(\mathfrak{g})$ using (12.2.4). The values of ν, $\mathrm{ind}(\mathfrak{g})$, and ϵ_M are related to each other as follows:

ν	$\mathrm{ind}(\mathfrak{g})$	ϵ_M
0	0	1
1	1	1
1	0	-1

$\qquad (12.2.5)$

Theorem 12.2.11. *Let $(M, \mathfrak{g})$ be a regular surface in $\mathbb{R}^3_\nu$, let (U, φ) be a chart on M, let (H_1, H_2) be the corresponding coordinate frame, and, using Example 11.2.5, view $\varphi(U)$ as a regular surface. Define a map $G : U \longrightarrow \mathbb{R}^3_\nu$, called the* **coordinate unit normal vector field** *corresponding to (U, φ), by*

$$G_q = \epsilon_M \frac{H_1|_q \times H_2|_q}{\|H_1|_q \times H_2|_q\|}$$

for all q in U, where the vector product is computed with respect to the standard basis for $\mathbb{R}^3_\nu$. Then:

(a) G *is smooth.*
(b) $\epsilon_M = \langle G_q, G_q \rangle$ *for all* q *in* U.
(c) $G \circ \varphi^{-1}$ *is a unit normal vector field in* $\mathfrak{X}_{\mathbb{R}^3_\nu}(\varphi(U))$.

Proof. (a), (c): By Theorem 12.2.7, $(\varphi(U), \mathfrak{g}|_{\varphi(U)})$ is a regular surface in $\mathbb{R}^3_\nu$, so $\mathfrak{g}|_{\varphi(U)}$ satisfies [G1] and [G2]. Setting $(V, \mathfrak{g}) = \mathbb{R}^3_\nu$, $U = T_p(M)$, and $(u_1, u_2) = (H_1|_{\varphi^{-1}(p)}, H_2|_{\varphi^{-1}(p)})$ in Theorem 8.4.9, and using Theorem 11.4.1(b), it follows from [G1] that $\|H_1|_{\varphi^{-1}(p)} \times H_2|_{\varphi^{-1}(p)}\| \neq 0$ for all p in $\varphi(U)$. Using similar reasoning, by Theorem 8.4.8, $H_1|_{\varphi^{-1}(p)} \times H_2|_{\varphi^{-1}(p)} \neq (0, 0, 0)$, and by Theorem 8.4.4 and Theorem 11.4.1(b), $H_1|_{\varphi^{-1}(p)} \times H_2|_{\varphi^{-1}(p)}$ is in $T_p(\varphi(U))^\perp$ for all p in $\varphi(U)$. It follows that $G \circ \varphi^{-1}$ is a unit normal vector field along $\varphi(U)$. We have from [G2] and Theorem 12.2.4 that $G \circ \varphi^{-1}$ satisfies [V2]. Since H_1 and H_2 are smooth on U, by Theorem 10.1.17 and Theorem 11.2.8, $H_1 \circ \varphi^{-1} \times H_2|\circ\varphi^{-1}$ is smooth on $\varphi(U)$, and therefore, by Theorem 12.2.1(c), so is $G \circ \varphi^{-1}$. We then have from Theorem 10.1.17 that $G = (G \circ \varphi^{-1}) \circ \varphi$ is smooth on U.
    (b): This follows from (12.2.3) and part (c). □

Continuing with the setup of Theorem 12.2.11, we note that the existence of a unit normal smooth vector field corresponding to each chart on M does not guarantee the existence of a unit normal *smooth* vector field along M. The reason is that the unit normal vector fields corresponding to different charts may not agree on the overlaps of images of their coordinate domains. In Section 12.7, we place additional structure on M that resolves this problem.

Let us now turn our attention to a special class of regular surfaces in $\mathbb{R}^3_\nu$. Recall from Section 3.1 that the quadratic function $\mathfrak{q}$ corresponding to $\mathbb{R}^3_\nu$ is given by $\mathfrak{q}(\cdot) = \langle \cdot, \cdot \rangle$. We consider three level sets of $\mathfrak{q}$, the first of which we have seen previously. For $\nu = 0$, the **unit sphere** is

$$\mathcal{S}^2 = \{p \in \mathbb{R}^3_0 : \mathfrak{q}(p) = 1\} = \mathfrak{q}^{-1}(1). \tag{12.2.6}$$

For $\nu = 1$, we define the **pseudosphere** by

$$\mathcal{P}^2 = \{p \in \mathbb{R}^3_1 : \mathfrak{q}(p) = 1\} = \mathfrak{q}^{-1}(1), \tag{12.2.7}$$

and **hyperbolic space** by

$$\mathcal{H}^2 = \{p \in \mathbb{R}^3_1 : \mathfrak{q}(p) = -1\} = \mathfrak{q}^{-1}(-1). \tag{12.2.8}$$

Thus, $\mathcal{S}^2$ is the set of (spacelike) unit vectors in $\mathbb{R}^3_0$, $\mathcal{P}^2$ is the set of spacelike unit vectors in $\mathbb{R}^3_1$, and $\mathcal{H}^2$ is the set of timelike unit vectors in $\mathbb{R}^3_1$. Taken together, $\mathcal{S}^2$, $\mathcal{P}^2$, and $\mathcal{H}^2$ are called the **hyperquadrics** in $\mathbb{R}^3_\nu$ and are denoted collectively by $\mathcal{Q}^2$. We have the following table:

$\mathcal{Q}^2$	ν	Type of vectors
$\mathcal{S}^2$	0	spacelike
$\mathcal{P}^2$	1	spacelike
$\mathcal{H}^2$	1	timelike

(12.2.9)

Theorem 12.2.12 (Hyperquadrics). *Let* $\mathfrak{q}$ *be the quadratic function corresponding to* $\mathbb{R}^3_\nu$, *let* $\mathcal{Q}^2$ *be the hyperquadrics in* $\mathbb{R}^3_\nu$, *and let* p *be a point in* $\mathcal{Q}^2$. *Then:*

(a) $\mathcal{Q}^2$ *is a regular surface in* $\mathbb{R}^3_\nu$; *that is,* $\mathcal{S}^2$ *is a regular surface in* $\mathbb{R}^3_0$, *and* $\mathcal{P}^2$ *and* $\mathcal{H}^2$ *are regular surfaces in* $\mathbb{R}^3_1$.

(b)
$$\frac{\mathrm{Grad}(\mathfrak{q})_p}{\|\mathrm{Grad}(\mathfrak{q})_p\|} = p.$$

(c) p *is a unit vector in* $T_p(\mathcal{Q}^2)^\perp$, *where the first* p *is viewed as a vector in* $\mathbb{R}^3_\nu$ *and the second* p *is viewed as a point in* $\mathcal{Q}^2$.

(d) *The hyperquadrics have the following features:*

$\mathcal{Q}^2$	ν	Type of vectors	$\mathrm{ind}(\mathfrak{g})$	$\epsilon_{\mathcal{Q}^2}$
$\mathcal{S}^2$	0	spacelike	0	1
$\mathcal{P}^2$	1	spacelike	1	1
$\mathcal{H}^2$	1	timelike	0	-1

Proof. Let $p = (x, y, z)$, and, for brevity, let $V = \mathrm{Grad}(\mathfrak{q})$. Since $(1, 1, (-1)^\nu)$ is the signature of $\mathbb{R}^3_\nu$, by Theorem 4.2.8, $\mathfrak{q}(x, y, z) = x^2 + y^2 + (-1)^\nu z^2$, hence $V_{(x,y,z)} = 2(x, y, z)$; that is,
$$V_p = 2p, \tag{12.2.10}$$
so
$$\langle V_p, V_p \rangle = 4\langle p, p \rangle = 4\,\mathfrak{q}(p). \tag{12.2.11}$$

(b): As remarked in connection with (12.2.6)–(12.2.8), each p in $\mathcal{Q}^2$ is a unit vector. Then (12.2.11) gives
$$\|V_p\| = \sqrt{|4\langle p, p \rangle|} = 2.$$

The result now follows from (12.2.10).

(a), (c): We have from (12.2.6)–(12.2.8) that $\mathcal{Q}^2$ is a level set of $\mathfrak{q}$, and from (12.2.11) that V satisfies [V2]. It follows from Theorem 12.2.10(b) and part (b) that $\mathcal{Q}^2$ is a regular surface in $\mathbb{R}^3_\nu$ and p is a unit normal vector at p.

(d): The entries in columns two and three come directly from (12.2.9). It follows from (12.2.3) and part (c) that $\epsilon_{\mathcal{Q}^2} = \langle p, p \rangle$, so the entries in column five follow from those in column three. Using (12.2.4), the entries in column four follow from those in columns two and five [see (12.2.5)]. $\square$

It is interesting to observe that according to the table in part (d) of Theorem 12.2.12, the index of $\mathcal{H}^2$ is 0. Thus, the tangent plane $T_p(\mathcal{H}^2)$ for each p in $\mathcal{H}^2$ is an inner product space, despite the fact that $T_p(\mathcal{H}^2)$ is a subspace of the Lorentz vector space $\mathbb{R}^3_1$.

Example 12.2.13 ($\mathcal{S}^2$). It was demonstrated in Example 11.2.6 and again in Example 11.4.5 that $\mathcal{S}^2$ a regular surface. By Theorem 12.2.12(a), $\mathcal{S}^2$ is a regular surface in $\mathbb{R}^3_0$.   ◇

We close this section with some definitions that will be used later on. Let $(M, \mathfrak{g})$ be a regular surface in $\mathbb{R}^3_\nu$, let (U, φ) be a chart on M, and let $\mathcal{H} = (H_1, H_2)$ be the corresponding coordinate frame. We define functions $\mathfrak{g}_{ij}$ in $C^\infty(U)$ by

$$\mathfrak{g}_{ij}(q) = \langle H_i|_q, H_j|_q \rangle$$

for all q in U for $i, j = 1, 2$. The **matrix of $\mathfrak{g}$** with respect to $\mathcal{H}$ is denoted by $\mathfrak{g}_\mathcal{H}$ and defined by

$$\mathfrak{g}_\mathcal{H}(q) = \begin{bmatrix} \mathfrak{g}_{11}(q) & \mathfrak{g}_{12}(q) \\ \mathfrak{g}_{21}(q) & \mathfrak{g}_{22}(q) \end{bmatrix}$$

for all q in U. Setting $p = \varphi(q)$, we recall from Section 3.1 that the matrix of $\mathfrak{g}_p$ with respect to $\mathcal{H}_q$ is $(\mathfrak{g}_p)_{\mathcal{H}_q} = [\mathfrak{g}_{ij}(q)]$. Thus, as a matter of notation,

$$\mathfrak{g}_\mathcal{H}(q) = (\mathfrak{g}_p)_{\mathcal{H}_q}.$$

The **inverse matrix of $\mathfrak{g}$** with respect to $\mathcal{H}$ is denoted by $\mathfrak{g}_\mathcal{H}^{-1}$ and defined by

$$\mathfrak{g}_\mathcal{H}^{-1}(q) = \mathfrak{g}_\mathcal{H}(q)^{-1}$$

for all q in U. It is usual to express the entries of $\mathfrak{g}_\mathcal{H}(q)^{-1}$ with superscripts:

$$\mathfrak{g}_\mathcal{H}^{-1}(q) = \begin{bmatrix} \mathfrak{g}_{11}(q) & \mathfrak{g}_{12}(q) \\ \mathfrak{g}_{21}(q) & \mathfrak{g}_{22}(q) \end{bmatrix}^{-1} = \begin{bmatrix} \mathfrak{g}^{11}(q) & \mathfrak{g}^{12}(q) \\ \mathfrak{g}^{21}(q) & \mathfrak{g}^{22}(q) \end{bmatrix}.$$

The assignment $q \longmapsto \mathfrak{g}^{ij}(q)$ defines functions $\mathfrak{g}^{ij}$ in $C^\infty(U)$ for $i, j = 1, 2$. Since $[\mathfrak{g}_{ij}]$ and $[\mathfrak{g}^{ij}]$ are symmetric matrices, the functions $\mathfrak{g}_{ij}$ and $\mathfrak{g}^{ij}$ are symmetric in i, j.

12.3   Induced Euclidean Derivative in $\mathbb{R}^3_\nu$

Let M be a regular surface, and let X be a vector field in $\mathfrak{X}(M)$. The **induced Euclidean derivative with respect to X** consists of two maps, both denoted by D_X. The first is

$$D_X : C^\infty(M) \longrightarrow C^\infty(M)$$

defined by

$$D_X(f)(p) = d_p(f)(X_p) \tag{12.3.1}$$

for all functions f in $C^\infty(M)$ and all p in M. The second is

$$D_X : \mathfrak{X}_{\mathbb{R}^3}(M) \longrightarrow \mathfrak{X}_{\mathbb{R}^3_\nu}(M)$$

defined by

$$D_X(V)_p = d_p(V)(X_p) \tag{12.3.2}$$

for all vector fields V in $\mathfrak{X}_{\mathbb{R}^3_\nu}(M)$ and all p in M. (It will be clear from the context when the notation D_X denotes the induced Euclidean derivative with respect to X as opposed to the Euclidean derivative with respect to X discussed in Section 10.3.)

We have from (11.5.1) and (11.7.1) that $D_X(f)(p)$ and $D_X(V)_p$ can be expressed as

$$D_X(f)(p) = \frac{d(f \circ \lambda)}{dt}(t_0) \tag{12.3.3}$$

and

$$D_X(V)_p = \frac{d(V \circ \lambda)}{dt}(t_0), \tag{12.3.4}$$

where $\lambda(t) : (a, b) \longrightarrow M$ is any smooth curve such $\lambda(t_0) = p$ and $(d\lambda/dt)(t_0) = X_p$. Let $V = (\alpha^1, \alpha^2, \alpha^3)$. It follows from Theorem 11.7.2 and (12.3.2) that $D_X(V)_p$ can also be expressed as

$$D_X(V)_p = \big(d_p(\alpha^1)(X_p), d_p(\alpha^2)(X_p), d_p(\alpha^3)(X_p)\big). \tag{12.3.5}$$

Following (12.2.2), for vector fields V, W in $\mathfrak{X}_{\mathbb{R}^3_\nu}(M)$, we define a function

$$\langle V, W \rangle : M \longrightarrow \mathbb{R}$$

in $C^\infty(M)$ by the assignment

$$p \longmapsto \langle V_p, W_p \rangle$$

for all p in M.

The next result is a counterpart of Theorem 10.3.1.

Theorem 12.3.1. *Let M be a regular surface, let X, Y and V, W be vector fields in $\mathfrak{X}(M)$ and $\mathfrak{X}_{\mathbb{R}^3_\nu}(M)$, respectively, and let f be a function in $C^\infty(M)$. Then:*
(a) $D_{X+Y}(V) = D_X(V) + D_Y(V)$.
(b) $D_{fX}(V) = f D_X(V)$.
(c) $D_X(V + W) = D_X(V) + D_X(W)$.
(d) $D_X(fV) = D_X(f) V + f D_X(V)$.
(e) $D_X(\langle V, W \rangle) = \langle D_X(V), W \rangle + \langle V, D_X(W) \rangle$.

Proof. (a)–(d): Using Theorem 11.5.3 and Theorem 11.7.2 gives the result.

(e): Let p be a point in M, and let $\lambda(t) : (a, b) \longrightarrow M$ be a smooth curve such that $\lambda(t_0) = p$ and $(d\lambda/dt)(t_0) = X_p$ for some t_0 in (a, b). Also, let $(\varepsilon_1, \varepsilon_2, \varepsilon_3) = (1, 1, (-1)^\nu)$ be the signature of $\mathbb{R}^3_\nu$, and let $V = (\alpha^1, \alpha^2, \alpha^3)$ and $W = (\beta^1, \beta^2, \beta^3)$. By Theorem 4.2.8,

$$\langle V, W \rangle \circ \lambda = \langle V \circ \lambda, W \circ \lambda \rangle = \sum_i \varepsilon_i(\alpha^i \circ \lambda)(\beta^i \circ \lambda),$$

hence

$$D_X(\langle V, W \rangle)_p = \frac{d(\langle V, W \rangle \circ \lambda)}{dt}(t_0)$$

$$= \sum_i \varepsilon_i \left[\frac{d(\alpha^i \circ \lambda)}{dt}(t_0)\,(\beta^i \circ \lambda)(t_0) + (\alpha^i \circ \lambda)(t_0)\,\frac{d(\beta^i \circ \lambda)}{dt}(t_0) \right]$$

$$= \left\langle \frac{d(V \circ \lambda)}{dt}(t_0), W \circ \lambda(t_0) \right\rangle + \left\langle V \circ \lambda(t_0), \frac{d(W \circ \lambda)}{dt}(t_0) \right\rangle$$

$$= \langle D_X(V)_p, W_p \rangle + \langle V_p, D_X(W)_p \rangle,$$

where the first equality follows from (12.3.3), the third equality from Theorem 4.2.8, and the last equality from (12.3.4). Since p was arbitrary, the result follows. □

By definition, if V is a vector field in $\mathfrak{X}_{\mathbb{R}^3_\nu}(M)$, then $D_X(V)$ is a vector field in $\mathfrak{X}_{\mathbb{R}^3_\nu}(M)$. In particular, if Y is a vector field in $\mathfrak{X}(M) \subset \mathfrak{X}_{\mathbb{R}^3_\nu}(M)$, then $D_X(Y)$ is a vector field in $\mathfrak{X}_{\mathbb{R}^3_\nu}(M)$. However, as the following example shows, $D_X(Y)$ might not be a vector field in $\mathfrak{X}(M)$. In other words, even though Y_p is a vector in $T_p(M)$ for all p in M, the same might not be true of $D_X(Y)_p$.

Example 12.3.2 (Hemisphere). Continuing with Example 11.2.6 and Example 11.3.2, it follows from Example 11.2.5 that

$$\varphi_1(D) = \{ (x, y, \sqrt{1 - x^2 - y^2}) : (x, y) \in D \}$$

is a regular surface. Then (D, φ_1) is a covering chart on $\varphi_1(D)$, and

$$X = \left(1, 0, -\frac{x}{\sqrt{1 - x^2 - y^2}} \right),$$

which is the first vector field in the coordinate frame corresponding to (D, φ_1), is a tangent vector field on $\varphi_1(D)$. Thus, $X_{(x,y)}$ is in $T_{p(x,y)}(\varphi_1(D))$, where $p(x, y) = (x, y, \sqrt{1 - x^2 - y^2})$. Consider the smooth curve $\lambda(t) : (-1, 1) \longrightarrow \varphi_1(D)$ given by $\lambda(t) = (t, 0, \sqrt{1 - t^2})$, the image of which is the intersection of $\varphi_1(D)$ with the xz-plane in $\mathbb{R}^3_\nu$. We have

$$X_{\lambda(t)} = \left(1, 0, -\frac{t}{\sqrt{1 - t^2}} \right).$$

Let us examine the behavior of $D_X(X)$ on the image of λ. According to (12.3.5),

$$D_X(X)_{\lambda(t)} = \left(0, 0, -\frac{1}{(1 - t^2)^{3/2}} \right).$$

When $t = 0$, for example,

$$\lambda(0) = (0, 0, 1), \qquad X_{\lambda(0)} = (1, 0, 0), \qquad \text{and} \qquad D_X(X)_{\lambda(0)} = (0, 0, -1).$$

This shows that at $\varphi_1(0, 0) = (0, 0, 1)$, $X_{\lambda(0)}$ is tangent to $\varphi_1(D)$, but $D_X(X)_{\lambda(0)}$ is not.   ◇

Let $(M, \mathfrak{g})$ be a regular surface in $\mathbb{R}^3_\nu$, let (U, φ) be a chart on M, and let $\mathcal{H} = (H_1, H_2)$ and G be the corresponding coordinate frame and coordinate unit normal vector field. In keeping with earlier notation for a vector-valued map, we denote

$$\frac{\partial H_i}{\partial r^j}(q) \qquad \text{by} \qquad \frac{\partial H_i}{\partial r^j}\bigg|_q$$

for all q in U for $i, j = 1, 2$. It follows from Theorem 8.4.10(b) and Theorem 12.2.11 that $(G_q, H_1|_q, H_2|_q)$ is a basis for $\mathbb{R}^3_\nu$. Then $(\partial H_i / \partial r^j)$ can be expressed as

$$\frac{\partial H_i}{\partial r^j} = \sum_k \Gamma^k_{ij} H_k + \vartheta_{ij} G, \tag{12.3.6}$$

where the Γ^k_{ij}, called the **Christoffel symbols**, and the ϑ_{ij} are uniquely determined functions on U for $i, j, k = 1, 2$.

Theorem 12.3.3. *With the above setup, for $i, j, k = 1, 2$:*
(a) Γ^k_{ij} *and* ϑ_{ij} *are functions in* $C^\infty(U)$.
(b)

$$\Gamma^k_{ij} = \frac{1}{2} \sum_l \mathfrak{g}^{kl} \left(\frac{\partial \mathfrak{g}_{jl}}{\partial r^i} + \frac{\partial \mathfrak{g}_{il}}{\partial r^j} - \frac{\partial \mathfrak{g}_{ij}}{\partial r^l} \right).$$

(c)

$$\vartheta_{ij} = \epsilon_M \left\langle \frac{\partial H_i}{\partial r^j}, G \right\rangle.$$

(d) Γ^k_{ij} *and* ϑ_{ij} *are symmetric in i, j; that is,*

$$\Gamma^k_{ij} = \Gamma^k_{ji} \qquad \text{and} \qquad \vartheta_{ij} = \vartheta_{ji}.$$

Proof. (b): It follows from (12.3.6) that

$$\left\langle \frac{\partial H_i}{\partial r^j}, H_k \right\rangle = \left\langle \sum_n \Gamma^n_{ij} H_n + \vartheta_{ij} G, H_k \right\rangle = \sum_n \Gamma^n_{ij} \langle H_k, H_n \rangle = \sum_n \mathfrak{g}_{kn} \Gamma^n_{ij},$$

and then from $\mathfrak{g}_{ij} = \langle H_i, H_j \rangle$ that

$$\frac{\partial \mathfrak{g}_{ij}}{\partial r^k} = \left\langle \frac{\partial H_i}{\partial r^k}, H_j \right\rangle + \left\langle H_i, \frac{\partial H_j}{\partial r^k} \right\rangle = \sum_n (\mathfrak{g}_{jn} \Gamma^n_{ik} + \mathfrak{g}_{in} \Gamma^n_{jk}). \tag{12.3.7}$$

Thus,

$$\frac{\partial \mathfrak{g}_{jl}}{\partial r^i} = \sum_n (\mathfrak{g}_{ln} \Gamma^n_{ij} + \mathfrak{g}_{jn} \Gamma^n_{il})$$

$$\frac{\partial \mathfrak{g}_{il}}{\partial r^j} = \sum_n (\mathfrak{g}_{ln} \Gamma^n_{ij} + \mathfrak{g}_{in} \Gamma^n_{jl})$$

$$\frac{\partial \mathfrak{g}_{ij}}{\partial r^l} = \sum_n (\mathfrak{g}_{jn}\Gamma^n_{il} + \mathfrak{g}_{in}\Gamma^n_{jl}),$$

hence

$$\frac{1}{2}\left(\frac{\partial \mathfrak{g}_{jl}}{\partial r^i} + \frac{\partial \mathfrak{g}_{il}}{\partial r^j} - \frac{\partial \mathfrak{g}_{ij}}{\partial r^l}\right) = \sum_n \mathfrak{g}_{ln}\Gamma^n_{ij}.$$

Multiplying both sides of the preceding identity by $\mathfrak{g}^{kl}$ and summing over l gives

$$\frac{1}{2}\sum_l \mathfrak{g}^{kl}\left(\frac{\partial \mathfrak{g}_{jl}}{\partial r^i} + \frac{\partial \mathfrak{g}_{il}}{\partial r^j} - \frac{\partial \mathfrak{g}_{ij}}{\partial r^l}\right) = \sum_l \mathfrak{g}^{kl}\left(\sum_n \mathfrak{g}_{ln}\Gamma^n_{ij}\right) = \sum_n \left(\sum_l \mathfrak{g}^{kl}\mathfrak{g}_{ln}\right)\Gamma^n_{ij}$$

$$= \sum_n \delta^k_n \Gamma^n_{ij} = \Gamma^k_{ij}.$$

(c): We have from Theorem 12.2.11(b) and (12.3.6) that

$$\left\langle \frac{\partial H_i}{\partial r^j}, G \right\rangle = \left\langle \sum_k \Gamma^k_{ij}H_k + \vartheta_{ij}G, G \right\rangle = \sum_k \Gamma^k_{ij}\langle H_k, G\rangle + \vartheta_{ij}\langle G, G\rangle$$

$$= \epsilon_M \vartheta_{ij}.$$

(a): This follows from parts (b) and (c).

(d): The symmetry of the Γ^k_{ij} follows from part (b) and the symmetry of the $\mathfrak{g}_{ij}$. We have from Theorem 10.1.6 and (11.2.1) that

$$\frac{\partial H_i}{\partial r^j} = \frac{\partial H_j}{\partial r^i}.$$

The symmetry of the ϑ_{ij} now follows from part (c).   □

We will make frequent use of the symmetry of the Christoffel symbols given by Theorem 12.3.3(d), usually without attribution. A quantity is said to be **intrinsic** to the geometry of a regular surface in $\mathbb{R}^3_\nu$ if its definition depends only on the metric. Accordingly, Theorem 12.3.3(b) demonstrates that the Christoffel symbols are intrinsic.

We will see later that the Christoffel symbols are closely related to the "curvature" of a regular surface in $\mathbb{R}^3_\nu$. In particular, when all Christoffel symbols have constant value 0, the surface is "flat". For example, consider Pln, the xy-plane in $\mathbb{R}^3_0$ discussed in Section 13.1. Since $\mathfrak{g}_\mathcal{H} = I_2$, it follows from Theorem 12.3.3(b) that each $\Gamma^k_{ij} = 0$. Thus, not surprisingly, Pln is "flat".

Let $(M, \mathfrak{g})$ be a regular surface in $\mathbb{R}^3_\nu$, and let X be a (not necessarily smooth) vector field on M. Let (U, φ) be a chart on M, and let (H_1, H_2) be the corresponding coordinate frame. Then $X \circ \varphi$ can be expressed as

$$X \circ \varphi = \sum_i \alpha^i H_i, \tag{12.3.8}$$

where the α^i are uniquely determined functions on U, called the **components of X** with respect to (U, φ). The right-hand side of (12.3.8) is said to express

X in **local coordinates** with respect to (U, φ). Let us introduce the notation

$$\alpha^i_{;j} = \frac{\partial \alpha^i}{\partial r^j} + \sum_k \alpha^k \Gamma^i_{jk} \tag{12.3.9}$$

for $i, j = 1, 2$.

Theorem 12.3.4. *With the above setup, if X is in $\mathfrak{X}(M)$, then α^i is a function in $C^\infty(U)$ for $i = 1, 2$.*

Proof. We have

$$\langle X \circ \varphi, H_k \rangle = \left\langle \sum_j \alpha^j H_j, H_k \right\rangle = \sum_j \alpha^j \langle H_j, H_k \rangle = \sum_j \alpha^j \mathfrak{g}_{jk},$$

hence

$$\sum_k \mathfrak{g}^{ik} \langle X \circ \varphi, H_k \rangle = \sum_k \mathfrak{g}^{ik} \left(\sum_j \alpha^j \mathfrak{g}_{jk} \right) = \sum_j \alpha^j \left(\sum_k \mathfrak{g}^{ik} \mathfrak{g}_{kj} \right)$$
$$= \sum_j \alpha^j \delta^i_j = \alpha^i.$$

By assumption, X is smooth. It follows from Theorem 11.7.1 that $X \circ \varphi$ is smooth, and therefore, so is α^i. □

Theorem 12.3.5. *Let $(M, \mathfrak{g})$ be a regular surface in $\mathbb{R}^3_\nu$, let (U, φ) be a chart on M, and let (H_1, H_2) be the corresponding coordinate frame. Let X, Y be vector fields in $\mathfrak{X}(M)$, and, in local coordinates, let*

$$X \circ \varphi = \sum_i \alpha^i H_i \qquad and \qquad Y \circ \varphi = \sum_j \beta^j H_j.$$

Then, for $i, j = 1, 2$:
(a)
$$D_{H_i \circ \varphi^{-1}}(H_j \circ \varphi^{-1}) \circ \varphi = \sum_k \Gamma^k_{ij} H_k + \vartheta_{ij} G.$$

(b)
$$D_{H_i \circ \varphi^{-1}}(Y) \circ \varphi = \sum_j \beta^j_{;i} H_j + \left(\sum_j \beta^j \vartheta_{ij} \right) G.$$

(c)
$$D_X(Y) \circ \varphi = \sum_j \left(\sum_i \alpha^i \beta^j_{;i} \right) H_j + \left(\sum_{ij} \alpha^i \beta^j \vartheta_{ij} \right) G.$$

Proof. Theorem 12.3.1 is used repeatedly in what follows.
    (a): Let p be a point in $\varphi(U)$, let $q = \varphi^{-1}(p)$, and let (e_1, e_2) be the standard basis for $\mathbb{R}^2$. For given $1 \le i \le 2$, we define a smooth map

$$\zeta = (\zeta^1, \zeta^2) : (-\varepsilon, \varepsilon) \longrightarrow U$$

by $\zeta(t) = q + te_i$, where $\varepsilon > 0$ is chosen small enough that $(q - \varepsilon e_i, q + \varepsilon e_i) \subset U$. Consider the smooth curve $\lambda(t) : (-\varepsilon, \varepsilon) \longrightarrow M$ defined by $\lambda = \varphi \circ \zeta$, and observe that $\lambda(0) = \varphi(q) = p$. By Theorem 10.1.10,

$$\frac{d\lambda}{dt}(0) = \frac{d(\varphi \circ \zeta)}{dt}(0) = \sum_k \frac{d\zeta^k}{dt}(0) H_k|_q = H_i|_q \qquad (12.3.10)$$

and

$$\frac{d(H_j \circ \zeta)}{dt}(0) = \sum_k \frac{d\zeta^k}{dt}(0) \left.\frac{\partial H_j}{\partial r^k}\right|_q = \left.\frac{\partial H_j}{\partial r^i}\right|_q. \qquad (12.3.11)$$

It follows from Theorem 10.1.17 and Theorem 11.2.8 that $H_j \circ \varphi^{-1}$ is smooth. We have

$$
\begin{aligned}
D_{H_i \circ \varphi^{-1}}&(H_j \circ \varphi^{-1})|_p \\
&= \frac{d(H_j \circ \varphi^{-1} \circ \lambda)}{dt}(0) && [(12.3.4),\ (12.3.10)] \\
&= \frac{d(H_j \circ \zeta)}{dt}(0) \\
&= \left.\frac{\partial H_j}{\partial r^i}\right|_q && [(12.3.11)] \\
&= \sum_k \Gamma^k_{ij}(q) H_k|_q + \vartheta_{ij}(q) G_q && [(12.3.6),\ \text{Th } 12.3.3(d)] \\
&= \left.\left(\sum_k \Gamma^k_{ij} H_k + \vartheta_{ij} G\right)\right|_{\varphi^{-1}(p)} \\
&= \left.\left(\left[\sum_k \Gamma^k_{ij} H_k + \vartheta_{ij} G\right] \circ \varphi^{-1}\right)\right|_p.
\end{aligned}
$$

Since p was arbitrary,

$$D_{H_i \circ \varphi^{-1}}(H_j \circ \varphi^{-1}) = \left(\sum_k \Gamma^k_{ij} H_k + \vartheta_{ij} G\right) \circ \varphi^{-1},$$

from which the result follows.

(b): Arguing as in part (a) gives

$$\frac{d(\beta^j \circ \zeta)}{dt}(0) = \frac{\partial \beta^j}{\partial r^i}(q). \qquad (12.3.12)$$

It follows from Theorem 10.1.17 and Theorem 11.2.8 that $\beta^j \circ \varphi^{-1}$ is smooth. We have

$$
\begin{aligned}
D_{H_i \circ \varphi^{-1}}(\beta^j \circ \varphi^{-1})(p) &= \frac{d(\beta^j \circ \varphi^{-1} \circ \lambda)}{dt}(0) && [(12.3.3)] \\
&= \frac{d(\beta^j \circ \zeta)}{dt}(0) \\
&= \frac{\partial \beta^j}{\partial r^i}(q) && [(12.3.12)] \\
&= \frac{\partial \beta^j}{\partial r^i}(\varphi^{-1}(p)) = \left(\frac{\partial \beta^j}{\partial r^i} \circ \varphi^{-1} \right)(p).
\end{aligned}
$$

Since p was arbitrary,

$$
D_{H_i \circ \varphi^{-1}}(\beta^j \circ \varphi^{-1}) = \frac{\partial \beta^j}{\partial r^i} \circ \varphi^{-1}. \tag{12.3.13}
$$

Then

$$
\begin{aligned}
D_{H_i \circ \varphi^{-1}}(Y) &= D_{H_i \circ \varphi^{-1}}\left(\left[\sum_j \beta^j H_j \right] \circ \varphi^{-1} \right) \\
&= D_{H_i \circ \varphi^{-1}}\left(\sum_j (\beta^j \circ \varphi^{-1})(H_j \circ \varphi^{-1}) \right) \\
&= \sum_j D_{H_i \circ \varphi^{-1}}\left((\beta^j \circ \varphi^{-1})(H_j \circ \varphi^{-1}) \right) \\
&= \sum_j [D_{H_i \circ \varphi^{-1}}(\beta^j \circ \varphi^{-1})(H_j \circ \varphi^{-1}) \\
&\quad + (\beta^j \circ \varphi^{-1}) D_{H_i \circ \varphi^{-1}}(H_j \circ \varphi^{-1})] \\
&= \sum_j \left(\frac{\partial \beta^j}{\partial r^i} \circ \varphi^{-1} \right)(H_j \circ \varphi^{-1}) \\
&\quad + \sum_j (\beta^j \circ \varphi^{-1})\left[\left(\sum_k \Gamma_{ij}^k H_k + \vartheta_{ij} G \right) \circ \varphi^{-1} \right] \\
&= \left[\sum_j \frac{\partial \beta^j}{\partial r^i} H_j + \sum_j \beta^j \left(\sum_k \Gamma_{ij}^k H_k + \vartheta_{ij} G \right) \right] \circ \varphi^{-1},
\end{aligned}
$$

where the fifth equality follows from (12.3.13) and part (a). Thus,

$$
\begin{aligned}
D_{H_i \circ \varphi^{-1}}(Y) \circ \varphi &= \sum_j \frac{\partial \beta^j}{\partial r^i} H_j + \sum_j \beta^j \left(\sum_k \Gamma^k_{ij} H_k + \vartheta_{ij} G \right) \\
&= \sum_j \frac{\partial \beta^j}{\partial r^i} H_j + \sum_k \beta^k \left(\sum_j \Gamma^j_{ik} H_j + \vartheta_{ik} G \right) \\
&= \sum_j \left(\frac{\partial \beta^j}{\partial r^i} + \sum_k \beta^k \Gamma^j_{ik} \right) H_j + \left(\sum_k \beta^k \vartheta_{ik} \right) G \\
&= \sum_j \beta^j_{;i} H_j + \left(\sum_j \beta^j \vartheta_{ij} \right) G. \qquad\qquad [(12.3.9)]
\end{aligned}
$$

(c): We have

$$
\begin{aligned}
D_X(Y) = D_{\sum_i (\alpha^i \circ \varphi^{-1})(H_i \circ \varphi^{-1})}(Y) &= \sum_i (\alpha^i \circ \varphi^{-1}) D_{H_i \circ \varphi^{-1}}(Y) \\
&= \left(\sum_i \alpha^i [D_{H_i \circ \varphi^{-1}}(Y) \circ \varphi] \right) \circ \varphi^{-1},
\end{aligned}
$$

hence

$$
\begin{aligned}
D_X(Y) \circ \varphi &= \sum_i \alpha^i [D_{H_i \circ \varphi^{-1}}(Y) \circ \varphi] \\
&= \sum_i \alpha^i \left[\sum_j \beta^j_{;i} H_j + \left(\sum_j \beta^j \vartheta_{ij} \right) G \right] \qquad \text{[part (b)]} \\
&= \sum_j \left(\sum_i \alpha^i \beta^j_{;i} \right) H_j + \left(\sum_{ij} \alpha^i \beta^j \vartheta_{ij} \right) G. \qquad\qquad \square
\end{aligned}
$$

12.4 Covariant Derivative on Regular Surfaces in $\mathbb{R}^3_\nu$

Let $(M, \mathfrak{g})$ be a regular surface in $\mathbb{R}^3_\nu$, and let X, Y be vector fields in $\mathfrak{X}(M)$. As remarked in conjunction with Example 12.3.2, although the vector field $D_X(Y)$ is in $\mathfrak{X}_{\mathbb{R}^3_\nu}(M)$, it may not be in $\mathfrak{X}(M)$. In other words, even though Y_p is a vector in $T_p(M)$ for all p in M, the same might not be true of $D_X(Y)_p$. We need a definition of "derivative" that sends vector fields in $\mathfrak{X}(M)$ to vector fields in $\mathfrak{X}(M)$, thereby avoiding this problem. Our approach is pragmatic: we modify the induced Euclidean derivative, discussed in Section 12.3, by eliminating the part that is not tangential to M.

For each point p in M, we have by definition that $\mathfrak{g}_p$ is nondegenerate on the subspace $T_p(M)$ of $\mathbb{R}^3_\nu$. It follows from Theorem 4.1.3 that $\mathbb{R}^3_\nu$ is the direct sum $\mathbb{R}^3_\nu = T_p(M) \oplus T_p(M)^\perp$. For brevity, let us denote the projection maps $\mathcal{P}_{T_p(M)}$

and $\mathcal{P}_{T_p(M)^\perp}$ by $\tan_p$ and nor_p, respectively, so that

$$\tan_p : \mathbb{R}^3_\nu \longrightarrow T_p(M) \qquad \text{and} \qquad \text{nor}_p : \mathbb{R}^3_\nu \longrightarrow T_p(M)^\perp.$$

The **covariant derivative with respect to X** consists of two maps, both denoted by ∇_X. The first is

$$\nabla_X : C^\infty(M) \longrightarrow C^\infty(M)$$

defined by

$$\nabla_X(f)(p) = d_p(f)(X_p) \tag{12.4.1}$$

for all functions f in $C^\infty(M)$ and all p in M. The second is

$$\nabla_X : \mathfrak{X}(M) \longrightarrow \mathfrak{X}(M)$$

defined by

$$\nabla_X(Y)_p = \tan_p(D_X(Y)_p) \tag{12.4.2}$$

for all vector fields Y in $\mathfrak{X}(M)$ and all p in M, where $D_X(Y)_p$ is given by (12.3.2). Observe that in the definition of the covariant derivative, all vector fields reside in $\mathfrak{X}(M)$. This is in contrast to the definition in Section 12.3 of the induced Euclidean derivative where vector fields in $\mathfrak{X}_{\mathbb{R}^3_\nu}(M)$ also appear.

For vector fields X, Y in $\mathfrak{X}(M)$, we define a function

$$\langle X, Y \rangle : M \longrightarrow \mathbb{R}$$

in $C^\infty(M)$ by the assignment

$$p \longmapsto \langle X_p, Y_p \rangle$$

for all p in M.

Theorem 12.4.1. *Let $(M, \mathfrak{g})$ be a regular surface in $\mathbb{R}^3_\nu$, let X, Y, Z be vector fields in $\mathfrak{X}(M)$, and let f be a function in $C^\infty(M)$. Then:*
(a) $\nabla_{X+Y}(Z) = \nabla_X(Z) + \nabla_Y(Z).$
(b) $\nabla_{fX}(Y) = f\nabla_X(Y).$
(c) $\nabla_X(Y + Z) = \nabla_X(Y) + \nabla_X(Z).$
(d) $\nabla_X(fY) = \nabla_X(f)\,Y + f\nabla_X(Z).$
(e) $\nabla_X(\langle Y, Z \rangle) = \langle \nabla_X(Y), Z \rangle + \langle Y, \nabla_X(Z) \rangle.$

Proof. (a)–(d): This follows from parts (a)–(d) of Theorem 12.3.1.
(e): Let p be a point in M. We have

$$\nabla_X(\langle Y, Z \rangle)(p) = D_X(\langle Y, Z \rangle)(p) \qquad [(12.3.1),\ (12.4.1)]$$
$$= \langle D_X(Y)_p, Z_p \rangle + \langle Y_p, D_X(Z)_p \rangle. \qquad [\text{Th } 12.3.1(e)]$$

We also have

$$D_X(Y)_p = \tan_p(D_X(Y)_p) + \text{nor}_p(D_X(Y)_p) \qquad [\text{Th } 4.1.4]$$
$$= \nabla_X(Y)_p + \text{nor}_p(D_X(Y)_p), \qquad [(12.4.2)]$$

so

$$\langle D_X(Y)_p, Z_p \rangle = \langle \nabla_X(Y)_p, Z_p \rangle + \langle \mathrm{nor}_p(D_X(Y)_p), Z_p \rangle = \langle \nabla_X(Y)_p, Z_p \rangle.$$

Similarly, $\langle Y_p, D_X(Z)_p \rangle = \langle Y_p, \nabla_X(Z)_p \rangle$. Thus,

$$\nabla_X(\langle Y, Z \rangle)(p) = \langle \nabla_X(Y)_p, Z_p \rangle + \langle Y_p, \nabla_X(Z)_p \rangle.$$

Since p was arbitrary, the result follows. $\square$

Here are the basic formulas for computing with covariant derivatives.

Theorem 12.4.2. *Let $(M, \mathfrak{g})$ be a regular surface in $\mathbb{R}^3_\nu$, let (U, φ) be a chart on M, and let (H_1, H_2) be the corresponding coordinate frame. Let X, Y be vector fields in $\mathfrak{X}(M)$, and, in local coordinates, let*

$$X \circ \varphi = \sum_i \alpha^i H_i \qquad and \qquad Y \circ \varphi = \sum_j \beta^j H_j.$$

Then, for $i, j = 1, 2$:
(a)

$$\nabla_{H_i \circ \varphi^{-1}}(H_j \circ \varphi^{-1}) \circ \varphi = \sum_k \Gamma^k_{ij} H_k.$$

(b)

$$\nabla_{H_i \circ \varphi^{-1}}(Y) \circ \varphi = \sum_j \beta^j_{;i} H_j.$$

(c)

$$\nabla_X(Y) \circ \varphi = \sum_j \left(\sum_i \alpha^i \beta^j_{;i} \right) H_j.$$

Proof. This follows from Theorem 12.3.5. $\square$

Let $(M, \mathfrak{g})$ be a regular surface in $\mathbb{R}^3_\nu$, and let X, Y be vector fields in $\mathfrak{X}(M)$. The **second order covariant derivative with respect to X and Y** consists of two maps, both denoted by $\nabla^2_{X,Y}$. The first is

$$\nabla^2_{X,Y} : C^\infty(M) \longrightarrow C^\infty(M)$$

defined by

$$\nabla^2_{X,Y}(f) = \nabla_X(\nabla_Y(f)) - \nabla_{\nabla_X(Y)}(f) \qquad (12.4.3)$$

for all functions f in $C^\infty(M)$. The second is

$$\nabla^2_{X,Y} : \mathfrak{X}(M) \longrightarrow \mathfrak{X}(M)$$

defined by

$$\nabla^2_{X,Y}(Z) = \nabla_X(\nabla_Y(Z)) - \nabla_{\nabla_X(Y)}(Z) \qquad (12.4.4)$$

for all vector fields Z in $\mathfrak{X}(M)$. These definitions are counterparts of the Euclidean versions given in Section 10.3.

Theorem 12.4.3. *Let* $(M, \mathfrak{g})$ *be a regular surface in* $\mathbb{R}^3_\nu$, *let* X, Y, Z *be vector fields in* $\mathfrak{X}(M)$, *and, in local coordinates, let*

$$X \circ \varphi = \sum_i \alpha^i H_i, \qquad Y \circ \varphi = \sum_j \beta^j H_j, \qquad and \qquad Z \circ \varphi = \sum_k \gamma^k H_k.$$

Then:

(a)

$$
\begin{aligned}
\nabla^2_{X,Y}(Z) \circ \varphi \\
= \sum_l \Bigg[\sum_{ij} \alpha^i \beta^j \frac{\partial^2 \gamma^l}{\partial r^i \partial r^j} + \sum_{ijk} (\alpha^i \beta^j + \alpha^j \beta^i) \frac{\partial \gamma^k}{\partial r^i} \Gamma^l_{jk} - \sum_{ijk} \alpha^i \beta^j \frac{\partial \gamma^l}{\partial r^k} \Gamma^k_{ij} \\
+ \sum_{ijk} \alpha^i \beta^j \gamma^k \left(\frac{\partial \Gamma^l_{jk}}{\partial r^i} + \sum_n (\Gamma^l_{in} \Gamma^n_{jk} - \Gamma^l_{kn} \Gamma^n_{ij}) \right) \Bigg] H_l.
\end{aligned}
$$

(b) *If* $\Gamma^i_{jk} = 0$ *for* $i, j, k = 1, \ldots, m$, *then*

$$
\begin{aligned}
\nabla^2_{X,Y}(Z) \circ \varphi &= \sum_l \left(\sum_{ij} \alpha^i \beta^j \frac{\partial^2 \gamma^l}{\partial r^i \partial r^j} \right) H_l \\
&= \nabla^2_{Y,X}(Z) \circ \varphi.
\end{aligned}
$$

Proof. (a): The proof is a lengthy computation that uses Theorem 12.4.1 repeatedly.

   Step 1. Compute $\nabla_X(\nabla_Y(Z))$.
   Let

$$\mu^l = \sum_j \beta^j \gamma^l_{;j}. \tag{12.4.5}$$

Using Theorem 12.4.2(c) twice gives

$$\nabla_Y(Z) \circ \varphi = \sum_l \mu^l H_l$$

and

$$\nabla_X(\nabla_Y(Z)) \circ \varphi = \sum_l \left(\sum_i \alpha^i \mu^l_{;i} \right) H_l. \tag{12.4.6}$$

We have from (12.3.9) and (12.4.5) that

$$
\begin{aligned}
\mu^l_{;i} &= \frac{\partial \mu^l}{\partial r^i} + \sum_k \mu^k \Gamma^l_{ik} = \frac{\partial \mu^l}{\partial r^i} + \sum_k \left(\sum_j \beta^j \gamma^k_{;j} \right) \Gamma^l_{ik} \\
&= \frac{\partial \mu^l}{\partial r^i} + \sum_{jk} \beta^j \gamma^k_{;j} \Gamma^l_{ik},
\end{aligned} \tag{12.4.7}
$$

where

$$\gamma^l_{;j} = \frac{\partial \gamma^l}{\partial r^j} + \sum_n \gamma^n \Gamma^l_{jn}. \tag{12.4.8}$$

We seek alternative expressions for the two terms in the second row of (12.4.7). It follows from (12.4.5) that

$$\frac{\partial \mu^l}{\partial r^i} = \sum_j \frac{\partial \beta^j}{\partial r^i} \gamma^l_{;j} + \sum_j \beta^j \frac{\partial \gamma^l_{;j}}{\partial r^i}, \tag{12.4.9}$$

and from (12.4.8) that

$$\sum_j \frac{\partial \beta^j}{\partial r^i} \gamma^l_{;j} = \sum_j \frac{\partial \beta^j}{\partial r^i} \frac{\partial \gamma^l}{\partial r^j} + \sum_{jn} \frac{\partial \beta^j}{\partial r^i} \gamma^n \Gamma^l_{jn} \tag{12.4.10}$$

and

$$\frac{\partial \gamma^l_{;j}}{\partial r^i} = \frac{\partial^2 \gamma^l}{\partial r^i \partial r^j} + \sum_n \frac{\partial \gamma^n}{\partial r^i} \Gamma^l_{jn} + \sum_n \gamma^n \frac{\partial \Gamma^l_{jn}}{\partial r^i}. \tag{12.4.11}$$

Then (12.4.11) gives

$$\sum_j \beta^j \frac{\partial \gamma^l_{;j}}{\partial r^i} = \sum_j \beta^j \frac{\partial^2 \gamma^l}{\partial r^i \partial r^j} + \sum_{jn} \beta^j \frac{\partial \gamma^n}{\partial r^i} \Gamma^l_{jn} + \sum_{jn} \beta^j \gamma^n \frac{\partial \Gamma^l_{jn}}{\partial r^i}. \tag{12.4.12}$$

Combining (12.4.9), (12.4.10), and (12.4.12) yields

$$\frac{\partial \mu^l}{\partial r^i} = \sum_j \frac{\partial \beta^j}{\partial r^i} \frac{\partial \gamma^l}{\partial r^j} + \sum_{jn} \frac{\partial \beta^j}{\partial r^i} \gamma^n \Gamma^l_{jn} + \sum_j \beta^j \frac{\partial^2 \gamma^l}{\partial r^i \partial r^j}$$
$$+ \sum_{jn} \beta^j \frac{\partial \gamma^n}{\partial r^i} \Gamma^l_{jn} + \sum_{jn} \beta^j \gamma^n \frac{\partial \Gamma^l_{jn}}{\partial r^i}, \tag{12.4.13}$$

which is the desired expression for the first term in the second row of (12.4.7). We have from (12.4.8) that

$$\sum_{jk} \beta^j \gamma^k_{;j} \Gamma^l_{ik} = \sum_{jk} \beta^j \left(\frac{\partial \gamma^k}{\partial r^j} + \sum_n \gamma^n \Gamma^k_{jn} \right) \Gamma^l_{ik}$$
$$= \sum_{jk} \beta^j \frac{\partial \gamma^k}{\partial r^j} \Gamma^l_{ik} + \sum_{jkn} \beta^j \gamma^n \Gamma^k_{jn} \Gamma^l_{ik}, \tag{12.4.14}$$

which is the desired expression for the second term in the second row of (12.4.7). Substituting (12.4.13) and (12.4.14) into (12.4.7) gives

$$
\mu^l_{;i} = \sum_j \frac{\partial \beta^j}{\partial r^i} \frac{\partial \gamma^l}{\partial r^j} + \sum_{jn} \frac{\partial \beta^j}{\partial r^i} \gamma^n \Gamma^l_{jn} + \sum_j \beta^j \frac{\partial^2 \gamma^l}{\partial r^i \partial r^j}
$$
$$
+ \sum_{jn} \beta^j \frac{\partial \gamma^n}{\partial r^i} \Gamma^l_{jn} + \sum_{jn} \beta^j \gamma^n \frac{\partial \Gamma^l_{jn}}{\partial r^i} \tag{12.4.15}
$$
$$
+ \sum_{jk} \beta^j \frac{\partial \gamma^k}{\partial r^j} \Gamma^l_{ik} + \sum_{jkn} \beta^j \gamma^n \Gamma^k_{jn} \Gamma^l_{ik}.
$$

It follows from (12.4.6) and (12.4.15) that the lth component of $\nabla_X (\nabla_Y(Z)) \circ \varphi$ is

$$
\begin{aligned}
&\left(\nabla_X(\nabla_Y(Z)) \circ \varphi \right)^l \\
&= \sum_i \alpha^i \mu^l_{;i} \\
&= \sum_{ij} \alpha^i \frac{\partial \beta^j}{\partial r^i} \frac{\partial \gamma^l}{\partial r^j} + \sum_{ijn} \alpha^i \frac{\partial \beta^j}{\partial r^i} \gamma^n \Gamma^l_{jn} + \sum_{ij} \alpha^i \beta^j \frac{\partial^2 \gamma^l}{\partial r^i \partial r^j} \\
&\quad + \sum_{ijn} \alpha^i \beta^j \frac{\partial \gamma^n}{\partial r^i} \Gamma^l_{jn} + \sum_{ijn} \alpha^i \beta^j \gamma^n \frac{\partial \Gamma^l_{jn}}{\partial r^i} \\
&\quad + \sum_{ijk} \alpha^i \beta^j \frac{\partial \gamma^k}{\partial r^j} \Gamma^l_{ik} + \sum_{ijkn} \alpha^i \beta^j \gamma^n \Gamma^k_{jn} \Gamma^l_{ik} \\
&= \overset{(1)}{\sum_{ij} \alpha^i \frac{\partial \beta^j}{\partial r^i} \frac{\partial \gamma^l}{\partial r^j}} + \overset{(2)}{\sum_{ijk} \alpha^i \frac{\partial \beta^j}{\partial r^i} \gamma^k \Gamma^l_{jk}} + \overset{(3)}{\sum_{ij} \alpha^i \beta^j \frac{\partial^2 \gamma^l}{\partial r^i \partial r^j}} \\
&\quad + \overset{(4)}{\sum_{ijk} \alpha^i \beta^j \frac{\partial \gamma^k}{\partial r^i} \Gamma^l_{jk}} + \overset{(5)}{\sum_{ijk} \alpha^i \beta^j \gamma^k \frac{\partial \Gamma^l_{jk}}{\partial r^i}} \\
&\quad + \overset{(6)}{\sum_{ijk} \alpha^i \beta^j \frac{\partial \gamma^k}{\partial r^j} \Gamma^l_{ik}} + \overset{(7)}{\sum_{ijkn} \alpha^i \beta^j \gamma^k \Gamma^n_{jk} \Gamma^l_{in}},
\end{aligned} \tag{12.4.16}
$$

where the summations have been numbered for easy reference.

Step 2. Compute $\nabla_{\nabla_X(Y)}(Z)$.
By Theorem 12.4.2(c),

$$
\nabla_X(Y) = \sum_j \left(\sum_i (\alpha^i \circ \varphi^{-1})(\beta^j_{;i} \circ \varphi^{-1}) \right) H_j \circ \varphi^{-1},
$$

so

$$
\begin{aligned}
\nabla_{\nabla_X(Y)}(Z) &= \nabla_{\sum_j \left(\sum_i (\alpha^i \circ \varphi^{-1})(\beta^j_{;i} \circ \varphi^{-1}) \right) H_j \circ \varphi^{-1}}(Z) \\
&= \sum_{ij} (\alpha^i \circ \varphi^{-1})(\beta^j_{;i} \circ \varphi^{-1}) \nabla_{H_j \circ \varphi^{-1}}(Z) \\
&= \sum_{ij} (\alpha^i \circ \varphi^{-1})(\beta^j_{;i} \circ \varphi^{-1}) \left(\sum_l (\gamma^l_{;j} \circ \varphi^{-1}) H_l \circ \varphi^{-1} \right) \\
&= \sum_l \left(\sum_{ij} (\alpha^i \circ \varphi^{-1})(\beta^j_{;i} \circ \varphi^{-1})(\gamma^l_{;j} \circ \varphi^{-1}) \right) H_l \circ \varphi^{-1} \\
&= \left(\sum_l \left[\sum_{ij} \alpha^i \beta^j_{;i} \gamma^l_{;j} \right] H_l \right) \circ \varphi^{-1},
\end{aligned}
$$

where the third equality follows from Theorem 12.4.2(b). Then

$$
\nabla_{\nabla_X(Y)}(Z) \circ \varphi = \sum_l \left(\sum_{ij} \alpha^i \beta^j_{;i} \gamma^l_{;j} \right) H_l, \tag{12.4.17}
$$

where

$$
\beta^j_{;i} = \frac{\partial \beta^j}{\partial r^i} + \sum_k \beta^k \Gamma^j_{ik} \qquad \text{and} \qquad \gamma^l_{;j} = \frac{\partial \gamma^l}{\partial r^j} + \sum_n \gamma^n \Gamma^l_{jn}. \tag{12.4.18}
$$

From (12.4.17) and (12.4.18), the lth component of $\nabla_{\nabla_X(Y)}(Z) \circ \varphi$ is

$$
\begin{aligned}
\left(\nabla_{\nabla_X(Y)}(Z) \circ \varphi \right)^l &= \sum_{ij} \alpha^i \beta^j_{;i} \gamma^l_{;j} \\
&= \sum_{ij} \left[\alpha^i \left(\frac{\partial \beta^j}{\partial r^i} + \sum_k \beta^k \Gamma^j_{ik} \right) \left(\frac{\partial \gamma^l}{\partial r^j} + \sum_n \gamma^n \Gamma^l_{jn} \right) \right] \\
&= \sum_{ij} \alpha^i \frac{\partial \beta^j}{\partial r^i} \frac{\partial \gamma^l}{\partial r^j} + \sum_{ijn} \alpha^i \frac{\partial \beta^j}{\partial r^i} \gamma^n \Gamma^l_{jn} \\
&\quad + \sum_{ijk} \alpha^i \beta^k \frac{\partial \gamma^l}{\partial r^j} \Gamma^j_{ik} + \sum_{ijkn} \alpha^i \beta^k \gamma^n \Gamma^j_{ik} \Gamma^l_{jn} \\
&= \overset{(1)}{\sum_{ij} \alpha^i \frac{\partial \beta^j}{\partial r^i} \frac{\partial \gamma^l}{\partial r^j}} + \overset{(2)}{\sum_{ijk} \alpha^i \frac{\partial \beta^j}{\partial r^i} \gamma^k \Gamma^l_{jk}} \\
&\quad + \overset{(8)}{\sum_{ijk} \alpha^i \beta^j \frac{\partial \gamma^l}{\partial r^k} \Gamma^k_{ij}} + \overset{(9)}{\sum_{ijkn} \alpha^i \beta^j \gamma^k \Gamma^n_{ij} \Gamma^l_{kn}}.
\end{aligned} \tag{12.4.19}
$$

Step 3. Compute $\nabla_{X,Y}^2(Z)$.
By definition,

$$\nabla_{X,Y}^2(Z) = \nabla_X\big(\nabla_Y(Z)\big) - \nabla_{\nabla_X(Y)}(Z).$$

We have from (12.4.16) and (12.4.19) that the lth component of $\nabla_{X,Y}^2(Z) \circ \varphi$ is

$$
\begin{aligned}
\big(&\nabla_{X,Y}^2(Z) \circ \varphi\big)^l \\
&= \left(\big[\nabla_X\big(\nabla_Y(Z)\big) - \nabla_{\nabla_X(Y)}(Z)\big] \circ \varphi\right)^l \\
&= \left(\overset{(1)}{\sum}+\overset{(2)}{\sum}+\overset{(3)}{\sum}+\overset{(4)}{\sum}+\overset{(5)}{\sum}+\overset{(6)}{\sum}+\overset{(7)}{\sum}\right) \\
&\quad - \left(\overset{(1)}{\sum}+\overset{(2)}{\sum}+\overset{(8)}{\sum}+\overset{(9)}{\sum}\right) \\
&= \overset{(3)}{\underset{ij}{\sum}}\alpha^i\beta^j\frac{\partial^2\gamma^l}{\partial r^i\partial r^j} + \overset{(4)}{\underset{ijk}{\sum}}\alpha^i\beta^j\frac{\partial\gamma^k}{\partial r^i}\Gamma_{jk}^l + \overset{(5)}{\underset{ijk}{\sum}}\alpha^i\beta^j\gamma^k\frac{\partial\Gamma_{jk}^l}{\partial r^i} \\
&\quad + \overset{(6)}{\underset{ijk}{\sum}}\alpha^i\beta^j\frac{\partial\gamma^k}{\partial r^j}\Gamma_{ik}^l + \overset{(7)}{\underset{ijkn}{\sum}}\alpha^i\beta^j\gamma^k\Gamma_{jk}^n\Gamma_{in}^l \\
&\quad - \overset{(8)}{\underset{ijk}{\sum}}\alpha^i\beta^j\frac{\partial\gamma^l}{\partial r^k}\Gamma_{ij}^k - \overset{(9)}{\underset{ijkn}{\sum}}\alpha^i\beta^j\gamma^k\Gamma_{ij}^n\Gamma_{nk}^l.
\end{aligned}
\tag{12.4.20}
$$

Since

$$
\overset{(4)}{\underset{ijk}{\sum}}\alpha^i\beta^j\frac{\partial\gamma^k}{\partial r^i}\Gamma_{jk}^l + \overset{(6)}{\underset{ijk}{\sum}}\alpha^j\beta^i\frac{\partial\gamma^k}{\partial r^i}\Gamma_{jk}^l = \overset{(4)+(6)}{\underset{ijk}{\sum}}(\alpha^i\beta^j + \alpha^j\beta^i)\frac{\partial\gamma^k}{\partial r^i}\Gamma_{jk}^l \tag{12.4.21}
$$

and

$$
\begin{aligned}
\overset{(5)}{\underset{ijk}{\sum}}&\alpha^i\beta^j\gamma^k\frac{\partial\Gamma_{jk}^l}{\partial r^i} + \overset{(7)}{\underset{ijkn}{\sum}}\alpha^i\beta^j\gamma^k\Gamma_{jk}^n\Gamma_{in}^l - \overset{(9)}{\underset{ijkn}{\sum}}\alpha^i\beta^j\gamma^k\Gamma_{ij}^n\Gamma_{kn}^l \\
&= \overset{(5)}{\underset{ijk}{\sum}}\alpha^i\beta^j\gamma^k\frac{\partial\Gamma_{jk}^l}{\partial r^i} + \overset{(7)}{\underset{ijk}{\sum}}\left(\alpha^i\beta^j\gamma^k\underset{n}{\sum}\Gamma_{jk}^n\Gamma_{in}^l\right) \\
&\quad - \overset{(9)}{\underset{ijk}{\sum}}\left(\alpha^i\beta^j\gamma^k\underset{n}{\sum}\Gamma_{ij}^n\Gamma_{kn}^l\right) \\
&= \overset{(5)+(7)-(9)}{\underset{ijk}{\sum}}\left[\alpha^i\beta^j\gamma^k\left(\frac{\partial\Gamma_{jk}^l}{\partial r^i} + \underset{n}{\sum}(\Gamma_{in}^l\Gamma_{jk}^n - \Gamma_{kn}^l\Gamma_{ij}^n)\right)\right],
\end{aligned}
\tag{12.4.22}
$$

we have from (12.4.20)–(12.4.22) that

$$\left(\nabla^2_{X,Y}(Z) \circ \varphi\right)^l$$

$$= \overset{(3)}{\sum_{ij} \alpha^i \beta^j \frac{\partial^2 \gamma^l}{\partial r^i \partial r^j}} + \overset{(4)+(6)}{\sum_{ijk} (\alpha^i \beta^j + \alpha^j \beta^i) \frac{\partial \gamma^k}{\partial r^i} \Gamma^l_{jk}} - \overset{(8)}{\sum_{ijk} \alpha^i \beta^j \frac{\partial \gamma^l}{\partial r^k} \Gamma^k_{ij}}$$

$$+ \overset{(5)+(7)-(9)}{\sum_{ijk} \alpha^i \beta^j \gamma^k \left(\frac{\partial \Gamma^l_{jk}}{\partial r^i} + \sum_n (\Gamma^l_{in} \Gamma^n_{jk} - \Gamma^l_{kn} \Gamma^n_{ij}) \right)}.$$

(b): This follows from Theorem 10.1.6 and part (a). $\square$

It was remarked following Theorem 12.3.3 that the Christoffel symbols corresponding to Pln have constant value 0 and this is related to Pln being "flat". We see from Theorem 12.4.3(b) that in the context of Pln, the order of vector fields is immaterial when computing the second order covariant derivative. This is reminiscent of the Euclidean situation in $\mathbb{R}^m$ [see (Theorem 10.3.4(b)]. The following example shows that for the sphere $\mathcal{S}^2$, order is important.

Example 12.4.4 ($\mathcal{S}^2$). In what follows, we use the Christoffel symbol results of Section 13.4. Setting $X = H_1 \circ \varphi^{-1}$ and $Y = H_2 \circ \varphi^{-1}$ in Theorem 12.4.3(a) (so that $\alpha^1 = \beta^2 = 1$ and $\alpha^2 = \beta^1 = 0$), and also setting $r^1 = \theta$ and $r^2 = \phi$, a lengthy but straightforward computation yields

$$\left(\nabla^2_{H_1 \circ \varphi^{-1}, H_2 \circ \varphi^{-1}}(Z) - \nabla^2_{H_2 \circ \varphi^{-1}, H_1 \circ \varphi^{-1}}(Z)\right) \circ \varphi = \sin^2(\theta)\gamma^2 H_1 - \gamma^1 H_2.$$

Due to the symmetry of the sphere, it is not surprising that the above expression is independent of ϕ. $\Diamond$

12.5   Covariant Derivative on Curves in $\mathbb{R}^3_\nu$

Let $(M, \mathfrak{g})$ be a regular surface in $\mathbb{R}^3_\nu$, let $\lambda(t) : (a, b) \longrightarrow M$ be a smooth curve, and let $J(t) : (a, b) \longrightarrow \mathbb{R}^3_\nu$ be a map. In the present context, we refer to J as a **vector field** *along* λ. The set of smooth vector fields along λ is denoted by $\mathfrak{X}_{\mathbb{R}^3_\nu}(\lambda)$. As an example, if V is a vector field in $\mathfrak{X}_{\mathbb{R}^3_\nu}(M)$, then $V \circ \lambda$ is a vector field in $\mathfrak{X}_{\mathbb{R}^3_\nu}(\lambda)$. We say that J is a **(tangent) vector field** *on* λ if $J(t)$ is in $T_{\lambda(t)}(M)$ for all t in (a, b). Let us denote the set of smooth vector fields on λ by $\mathfrak{X}_M(\lambda)$. For example, $d\lambda/dt$, the velocity of λ, is in $\mathfrak{X}_M(\lambda)$. As another example, if X is a vector field in $\mathfrak{X}(M)$, then $X \circ \lambda$ is a vector field in $\mathfrak{X}_M(\lambda)$.

For a vector field J in $\mathfrak{X}_M(\lambda)$, we have by definition that $J(t)$ is a vector in $T_{\lambda(t)}(M)$ for all t in (a, b). But this is not necessarily so for $(dJ/dt)(t)$. In particular, although the velocity of λ is in $\mathfrak{X}_M(\lambda)$, its (Euclidean) acceleration may not be. We need a definition of "derivative" that avoids this problem. Our response is similar to the approach taken in Section 12.4.

The **covariant derivative on** λ consists of two maps, both denoted by ∇/dt. The first is

$$\frac{\nabla}{dt} : C^\infty\big((a, b)\big) \longrightarrow C^\infty\big((a, b)\big)$$

defined by

$$\frac{\nabla f}{dt}(t) = \frac{df}{dt}(t)$$

for all functions f in $C^\infty\big((a,b)\big)$ and all t in (a,b). The second is

$$\frac{\nabla}{dt} : \mathfrak{X}_M(\lambda) \longrightarrow \mathfrak{X}_M(\lambda)$$

defined by

$$\frac{\nabla J}{dt}(t) = \tan_{\lambda(t)}\left(\frac{dJ}{dt}(t)\right) \qquad (12.5.1)$$

for all vector fields J in $\mathfrak{X}_M(\lambda)$ and all t in (a,b), where, following Section 12.4, $\tan_{\lambda(t)}$ denotes the projection map $\mathcal{P}_{T_{\lambda(t)}(M)}$.

The **(covariant) acceleration of λ** is defined to be the smooth curve

$$\frac{\nabla}{dt}\left(\frac{d\lambda}{dt}\right)(t) : (a,b) \longrightarrow M.$$

For vector fields J, K in $\mathfrak{X}_M(\lambda)$, we define a function

$$\langle J, K \rangle : (a,b) \longrightarrow \mathbb{R}$$

in $C^\infty\big((a,b)\big)$ by the assignment

$$t \longmapsto \langle J(t), K(t) \rangle$$

and all t in (a,b).

The definition of covariant derivative on a curve has an appealing physical interpretation. Imagine a "bug" that is confined to the 2-dimensional world of a given regular surface in $\mathbb{R}^3_\nu$. For this creature, there is no "up" or "down", only movements "on" the surface. Suppose the bug is scurrying along, tracing a smooth curve as it goes. From our vantage point in $\mathbb{R}^3_\nu$, and knowing something about Newtonian physics, we determine that the bug has a certain velocity and nonzero (Euclidean) acceleration. For both us and the bug, velocity is entirely a tangential phenomenon. On the other hand, we observe the acceleration to have both tangential and normal components. But not so for the bug, which is oblivious to any such normal phenomena. This suggests that in order to quantify what we presume to be the acceleration felt by the bug, we should confine our attention to the tangential component. This is accomplished by taking the projection onto the tangent plane.

Theorem 12.5.1. *Let $(M, \mathfrak{g})$ be a regular surface in $\mathbb{R}^3_\nu$, and let $\lambda(t) : (a,b) \longrightarrow M$ be a smooth curve. Let J, K be vector fields in $\mathfrak{X}_M(\lambda)$, let f be a function in $C^\infty\big((a,b)\big)$, and let $g(u) : (c,d) \longrightarrow (a,b)$ be a diffeomorphism. Then, for all t in (a,b) and all u in (c,d):*
(a)

$$\frac{\nabla(J+K)}{dt}(t) = \frac{\nabla J}{dt}(t) + \frac{\nabla K}{dt}(t).$$

(b)
$$\frac{\nabla(fJ)}{dt}(t) = \frac{df}{dt}(t)\,J(t) + f(t)\,\frac{\nabla J}{dt}(t).$$

(c)
$$\frac{d(\langle J, K\rangle)}{dt}(t) = \left\langle \frac{\nabla J}{dt}(t), K(t)\right\rangle + \left\langle J(t), \frac{\nabla K}{dt}(t)\right\rangle.$$

(d)
$$\frac{\nabla(J \circ g)}{du}(u) = \frac{dg}{du}(u)\,\frac{\nabla J}{dt}(g(u)). \qquad \square$$

Let $(M, \mathfrak{g})$ be a regular surface in $\mathbb{R}^3_\nu$, let (U, φ) be a chart on M, and let (H_1, H_2) and G be the corresponding coordinate frame and coordinate unit normal vector field. Let $\lambda : (a, b) \longrightarrow M$ be a smooth curve such that $\lambda\big((a, b)\big) \subset U$, and let J be a vector field in $\mathfrak{X}_M(\lambda)$. By Theorem 10.1.17 and Theorem 11.2.8, the map

$$\mu = \varphi^{-1} \circ \lambda = (\mu^1, \mu^1) : (a, b) \longrightarrow U \qquad (12.5.2)$$

is smooth. Then $J(t)$ can be expressed as

$$J(t) = \sum_i \alpha^i(t) H_i|_{\mu(t)}, \qquad (12.5.3)$$

where the α^i are uniquely determined functions in $C^\infty(U)$, called the **components of J** with respect to (U, φ). The right-hand side of (12.5.3) is said to express J in **local coordinates** with respect to (U, φ).

Theorem 12.5.2. *With the above setup, for all t in (a, b):*
(a)
$$\frac{dJ}{dt}(t) = \sum_k \left(\frac{d\alpha^k}{dt}(t) + \sum_{ij}\alpha^i(t)\,\frac{d\mu^j}{dt}(t)\,\Gamma^k_{ij}\big(\mu(t)\big)\right) H_k|_{\mu(t)}$$
$$+ \left(\sum_{ij}\alpha^i(t)\,\frac{d\mu^j}{dt}(t)\,\vartheta_{ij}\big(\mu(t)\big)\right) G_{\mu(t)}.$$

(b)
$$\frac{\nabla J}{dt}(t) = \sum_k \left(\frac{d\alpha^k}{dt}(t) + \sum_{ij}\alpha^i(t)\,\frac{d\mu^j}{dt}(t)\,\Gamma^k_{ij}\big(\mu(t)\big)\right) H_k|_{\mu(t)}.$$

(c)
$$\frac{d^2\lambda}{dt^2}(t) = \sum_k \left(\frac{d^2\mu^k}{dt}(t) + \sum_{ij}\frac{d\mu^i}{dt}(t)\,\frac{d\mu^j}{dt}(t)\,\Gamma^k_{ij}\big(\mu(t)\big)\right) H_k|_{\mu(t)}$$
$$+ \left(\sum_{ij}\frac{d\mu^i}{dt}(t)\,\frac{d\mu^j}{dt}(t)\,\vartheta_{ij}\big(\mu(t)\big)\right) G_{\mu(t)}.$$

(d)
$$\left\langle \frac{d\lambda}{dt}(t), \frac{d\lambda}{dt}(t)\right\rangle = \sum_{ij}\frac{d\mu^i}{dt}(t)\,\frac{d\mu^j}{dt}(t)\,\mathfrak{g}_{ij}\big(\mu(t)\big).$$

Proof. (a): We have from (12.5.3) and Theorem 10.1.10 that

$$\frac{dJ}{dt}(t) = \sum_i \left(\frac{d\alpha^i}{dt}(t)\, H_i|_{\mu(t)} + \alpha^i(t)\frac{d(H_i \circ \mu)}{dt}(t) \right)$$

$$= \sum_i \frac{d\alpha^i}{dt}(t)\, H_i|_{\mu(t)} + \sum_i \left[\alpha^i(t) \left(\sum_j \frac{d\mu^j}{dt}(t) \left.\frac{\partial H_i}{\partial r^j}\right|_{\mu(t)} \right) \right] \qquad (12.5.4)$$

$$= \sum_k \frac{d\alpha^k}{dt}(t)\, H_k|_{\mu(t)} + \sum_{ij} \alpha^i(t)\frac{d\mu^j}{dt}(t) \left.\frac{\partial H_i}{\partial r^j}\right|_{\mu(t)}.$$

Using (12.3.6), the second term in the last row of (12.5.4) can be expressed as

$$\sum_{ij} \alpha^i(t) \frac{d\mu^j}{dt}(t) \left.\frac{\partial H_i}{\partial r^j}\right|_{\mu(t)}$$

$$= \sum_{ij} \alpha^i(t) \frac{d\mu^j}{dt}(t) \left(\sum_k \Gamma^k_{ij}(\mu(t))\, H_k|_{\mu(t)} + \vartheta_{ij}(\mu(t))\, G_{\mu(t)} \right)$$

$$= \sum_k \left(\sum_{ij} \alpha^i(t) \frac{d\mu^j}{dt}(t)\, \Gamma^k_{ij}(\mu(t)) \right) H_k|_{\mu(t)} \qquad (12.5.5)$$

$$+ \left(\sum_{ij} \alpha^i(t) \frac{d\mu^j}{dt}(t)\, \vartheta_{ij}(\mu(t)) \right) G_{\mu(t)}.$$

Combining (12.5.4) and (12.5.5) gives the result.

(b): This follows from (12.5.1) and part (a).

(c): Setting $J = d\lambda/dt$ and $\alpha^i = d\mu^i/dt$ in part (a) gives the result.

(d): By Theorem 10.1.10,

$$\frac{d\lambda}{dt}(t) = \frac{d(\varphi \circ \mu)}{dt} = \sum_i \frac{d\mu^i}{dt}(t)\, H_i|_{\mu(t)},$$

from which the result follows. □

12.6   Lie Bracket in $\mathbb{R}^3_\nu$

Let $(M, \mathfrak{g})$ be a regular surface in $\mathbb{R}^3_\nu$. **Lie bracket** is the map

$$[\cdot,\cdot] : \mathfrak{X}(M) \times \mathfrak{X}(M) \longrightarrow \mathfrak{X}(M)$$

defined by

$$[X, Y] = \nabla_X(Y) - \nabla_Y(X) \qquad (12.6.1)$$

for all vector fields X, Y in $\mathfrak{X}(M)$.

Theorem 12.6.1. *Let* $(M, \mathfrak{g})$ *be a regular surface in* $\mathbb{R}^3_\nu$, *let* (U, φ) *be a chart on* M, *let* X, Y *be vector fields in* $\mathfrak{X}(M)$, *and, in local coordinates, let*

$$X \circ \varphi = \sum_i \alpha^i H_i \qquad and \qquad Y \circ \varphi = \sum_j \beta^j H_j.$$

Then

$$[X, Y] \circ \varphi = \sum_j \left\{ \sum_i \left(\alpha^i \frac{\partial \beta^j}{\partial r^i} - \beta^i \frac{\partial \alpha^j}{\partial r^i} \right) \right\} H_j$$

$$= \sum_j \left(\sum_i (\alpha^i \beta^j_{;i} - \beta^i \alpha^j_{;i}) \right) H_j.$$

Proof. By Theorem 12.4.2(c),

$$\nabla_X(Y) \circ \varphi = \sum_j \left(\sum_i \alpha^i \beta^j_{;i} \right) H_j$$

and

$$\nabla_Y(X) \circ \varphi = \sum_j \left(\sum_i \beta^i \alpha^j_{;i} \right) H_j,$$

hence

$$[X, Y] \circ \varphi = \nabla_X(Y) \circ \varphi - \nabla_Y(X) \circ \varphi = \sum_j \left(\sum_i (\alpha^i \beta^j_{;i} - \beta^i \alpha^j_{;i}) \right) H_j.$$

We have from (12.3.9) that

$$\sum_i (\alpha^i \beta^j_{;i} - \beta^i \alpha^j_{;i}) = \sum_i \left\{ \alpha^i \left(\frac{\partial \beta^j}{\partial r^i} + \sum_k \beta^k \Gamma^j_{ik} \right) - \beta^i \left(\frac{\partial \alpha^j}{\partial r^i} + \sum_k \alpha^k \Gamma^j_{ik} \right) \right\}$$

$$= \sum_i \left(\alpha^i \frac{\partial \beta^j}{\partial r^i} - \beta^i \frac{\partial \alpha^j}{\partial r^i} \right) + \sum_{ik} \alpha^i \beta^k \Gamma^j_{ik} - \sum_{ik} \alpha^k \beta^i \Gamma^j_{ik}$$

$$= \sum_i \left(\alpha^i \frac{\partial \beta^j}{\partial r^i} - \beta^i \frac{\partial \alpha^j}{\partial r^i} \right),$$

where the last equality relies on Theorem 12.3.3(d). The result follows. $\qquad \square$

The next result shows that the Lie bracket on a regular surface in $\mathbb{R}^3_\nu$, formulated above in terms of the covariant derivative, can also be expressed in terms of the induced Euclidean derivative.

Theorem 12.6.2. *If* $(M, \mathfrak{g})$ *is a regular surface in* $\mathbb{R}^3_\nu$ *and* X, Y *are vector fields in* $\mathfrak{X}(M)$, *then*

$$\nabla_X(Y) - \nabla_Y(X) = D_X(Y) - D_Y(X).$$

Proof. We have from Theorem 12.3.5(c) that

$$D_X(Y) \circ \varphi = \sum_j \left(\sum_i \alpha^i \beta^j_{;i} \right) H_j + \left(\sum_{ij} \alpha^i \beta^j \vartheta_{ij} \right) G$$

and

$$D_Y(X) \circ \varphi = \sum_j \left(\sum_i \beta^i \alpha^j_{;i} \right) H_j + \left(\sum_{ij} \beta^i \alpha^j \vartheta_{ij} \right) G.$$

Then Theorem 12.3.3(d) gives

$$\left(D_X(Y) - D_Y(X) \right) \circ \varphi = \sum_j \left(\sum_i \alpha^i \beta^j_{;i} - \beta^i \alpha^j_{;i} \right) H_j.$$

The result now follows from Theorem 12.6.1. $\qquad\square$

Here is a counterpart of Theorem 10.4.2.

Theorem 12.6.3. *Let* $(M, \mathfrak{g})$ *be a regular surface in* $\mathbb{R}^3_\nu$, *let* X, Y, Z *be vector fields in* $\mathfrak{X}(M)$, *and let* f, g *be functions in* $C^\infty(M)$. *Let* (U, φ) *be a chart on* M *and let* (H_1, H_2) *be the corresponding coordinate frame. Then:*
(a) $[Y, X] = -[X, Y]$.
(b) $[X + Y, Z] = [X, Z] + [Y, Z]$.
(c) $[X, Y + Z] = [X, Y] + [X, Z]$.
(d) $[fX, gY] = fg[X, Y] + f\nabla_X(g)Y - g\nabla_Y(f)X$.
(e) $[X, [Y, Z]] + [Y, [Z, X]] + [Z, [X, Y]] = (0, 0, 0)$.   (Jacobi's identity)
(f) $[H_i \circ \varphi^{-1}, H_j \circ \varphi^{-1}] = (0, 0, 0)$ *for* $i, j = 1, 2$.

Proof. (a)–(d): Parts (a)–(d) of Theorem 12.4.1 and (12.6.1) give the result.
(e): We have from parts (a) and (c) of Theorem 12.4.1 and (12.6.1) that

$$\begin{aligned}
[X, [Y, Z]] &= \nabla_X([Y, Z]) - \nabla_{[Y,Z]}(X) \\
&= \nabla_X(\nabla_Y(Z) - \nabla_Z(Y)) - \nabla_{\nabla_Y(Z) - \nabla_Z(Y)}(X) \\
&= \nabla_X(\nabla_Y(Z)) - \nabla_X(\nabla_Z(Y)) - \nabla_{\nabla_Y(Z)}(X) + \nabla_{\nabla_Z(Y)}(X).
\end{aligned}$$

Likewise,

$$[Y, [Z, X]] = \nabla_Y(\nabla_Z(X)) - \nabla_Y(\nabla_X(Z)) - \nabla_{\nabla_Z(X)}(Y) + \nabla_{\nabla_X(Z)}(Y)$$
$$[Z, [X, Y]] = \nabla_Z(\nabla_X(Y)) - \nabla_Z(\nabla_Y(X)) - \nabla_{\nabla_X(Y)}(Z) + \nabla_{\nabla_Y(X)}(Z).$$

Summing the preceding identities and using (12.4.4) yields

$$\begin{aligned}
&[X, [Y, Z]] + [Y, [Z, X]] + [Z, [X, Y]] \\
&= [\nabla^2_{X,Y}(Z) - \nabla^2_{Y,X}(Z)] + [\nabla^2_{Y,Z}(X) - \nabla^2_{Z,Y}(X)] + [\nabla^2_{Z,X}(Y) - \nabla^2_{X,Z}(Y)] \\
&= R(X, Y)Z + R(Y, Z)X + R(Z, X)Y = (0, 0, 0),
\end{aligned}$$

where the second identity follows from Theorem 12.9.1, and the last identity from Theorem 12.9.13.
(f): This follows from Theorem 12.6.1. $\qquad\square$

12.7   Orientation in $\mathbb{R}^3_\nu$

In Section 12.2, we defined a regular surface to be a regular surface in $\mathbb{R}^3_\nu$ provided its first fundamental form satisfies certain properties. We then proceeded to demonstrate an equivalent formulation based on the existence of a particular type of unit normal vector field. Aside from an increase in geometric intuition, the latter approach offers computational advantages. For example, as remarked in connection with (12.2.4), it is usually more convenient to compute the index of a regular surface in $\mathbb{R}^3_\nu$ indirectly using its sign. In this section, we explore orientation in the context of regular surfaces in $\mathbb{R}^3_\nu$. The basic definition is given in terms of atlases, but once again unit normal vector fields play a prominent role. In what follows, we rely heavily on the discussion of orientation of vector spaces given in Section 8.2.

Theorem 12.7.1. *Let $(M, \mathfrak{g})$ be a regular surface in $\mathbb{R}^3_\nu$, let (U, φ) be a chart on M, let $\mathcal{H} = (H_1, H_2)$ and G be the corresponding coordinate frame and coordinate unit normal vector field, and let q be a point in U. Then $(G_q, H_1|_q, H_2|_q)$ is a basis for $\mathbb{R}^3_\nu$ that has the standard orientation.*

Proof. This follows from Theorem 8.4.10(b) and Theorem 12.2.11. $\square$

Let $(M, \mathfrak{g})$ be a regular surface in $\mathbb{R}^3_\nu$, and let (U, φ) and $(\widetilde{U}, \widetilde{\varphi})$ be overlapping charts on M. Let $\mathcal{H}$ and $\widetilde{\mathcal{H}}$ be the corresponding coordinate frames, and let G and $\widetilde{G}$ be the corresponding coordinate unit normal vector fields. Let $W = \varphi(U) \cap \widetilde{\varphi}(\widetilde{U})$, and let p be a point in W. Recall from Section 8.2 that the coordinate bases $\mathcal{H}_{\varphi^{-1}(p)}$ and $\widetilde{\mathcal{H}}_{\widetilde{\varphi}^{-1}(p)}$ are said to be **consistent** if

$$\det\left(\left[\mathrm{id}_{T_p(M)}\right]^{\widetilde{\mathcal{H}}_{\widetilde{\varphi}^{-1}(p)}}_{\mathcal{H}_{\varphi^{-1}(p)}}\right) > 0.$$

We say that (U, φ) and $(\widetilde{U}, \widetilde{\varphi})$ are **consistent** if $\mathcal{H}_{\varphi^{-1}(p)}$ and $\widetilde{\mathcal{H}}_{\widetilde{\varphi}^{-1}(p)}$ are consistent for all p in W.

Theorem 12.7.2. *With the above setup, (U, φ) and $(\widetilde{U}, \widetilde{\varphi})$ are consistent if and only if $G_{\varphi^{-1}(p)} = \widetilde{G}_{\widetilde{\varphi}^{-1}(p)}$ for all p in V.*

Proof. This follows from Theorem 8.4.11(b). $\square$

Let $(M, \mathfrak{g})$ be a regular surface in $\mathbb{R}^3_\nu$. An atlas for M is said to be **consistent** if every pair of overlapping charts in the atlas is consistent. We say that M is **orientable** if it has a consistent atlas. Suppose M is in fact orientable, and let $\mathfrak{A}$ be a consistent atlas for M. The triple $(M, \mathfrak{g}, \mathfrak{A})$ is called an **oriented regular surface in $\mathbb{R}^3_\nu$**. Let p be a point in M, let (U, φ) be a chart in $\mathfrak{A}$ at p, and let $\mathcal{H}$ be the corresponding coordinate frame. Let

$$\mathfrak{O}(p) = [\mathcal{H}_{\varphi^{-1}(p)}],$$

where we recall from Section 8.2 that $[\mathcal{H}_{\varphi^{-1}(p)}]$ is the equivalence class of all bases for $T_p(M)$ (not just coordinate bases) that are consistent with $\mathcal{H}_{\varphi^{-1}(p)}$.

Let $(\widetilde{U}, \widetilde{\varphi})$ be another chart in $\mathfrak{A}$ at p, and let $\widetilde{\mathcal{H}}$ be the corresponding coordinate frame. Since $\mathfrak{A}$ is consistent, (U, φ) and $(\widetilde{U}, \widetilde{\varphi})$ are consistent, hence $[\mathcal{H}_{\varphi^{-1}(p)}] = [\widetilde{\mathcal{H}}_{\widetilde{\varphi}^{-1}(p)}]$. This shows that the definition of $\mathfrak{O}(p)$ is independent of the choice of representative chart at p. We call the set of equivalence classes

$$\mathfrak{O} = \{\mathfrak{O}(p) : p \in M\}$$

the **orientation induced by** $\mathfrak{A}$ and say that M is **oriented by** $\mathfrak{A}$. The notation $(M, \mathfrak{g}, \mathfrak{O})$, and sometimes $(M, \mathfrak{g}, \mathfrak{A}, \mathfrak{O})$, is used as an alternative to $(M, \mathfrak{g}, \mathfrak{A})$.

Consider the map $\iota : \mathbb{R}^2 \longrightarrow \mathbb{R}^2$ given by $\iota(r^1, r^2) = (-r^1, r^2)$. Since ι is a diffeomorphism and $\iota^{-1} = \iota$, $(\iota(U), \varphi \circ \iota)$ is a chart on M, where, for brevity, we denote $\iota|_{\iota(U)}$ by ι. Because

$$\varphi \circ \iota(r^1, r^2) = \big(\varphi^1(-r^1, r^2), \varphi^2(-r^1, r^2), \varphi^3(-r^1, r^2)\big),$$

the corresponding coordinate frame and coordinate unit normal vector field are

$$-\mathcal{H} = (-H_1, H_2)$$

and $-G$. It is easily shown using Theorem 11.3.3 that

$$-\mathfrak{A} = \big\{(\iota(U), \varphi \circ \iota) : (U, \varphi) \in \mathfrak{A}\big\}$$

is a consistent atlas for M. The orientation of M induced by $-\mathfrak{A}$ is

$$-\mathfrak{O} = \{-\mathfrak{O}(p) : p \in M\},$$

where

$$-\mathfrak{O}(p) = [-\mathcal{H}_{\varphi^{-1}(p)}] = [(-H_1|_{\varphi^{-1}(p)}, H_2|_{\varphi^{-1}(p)})].$$

We say that the orientation $-\mathfrak{O}$ is the **opposite of** $\mathfrak{O}$.

Theorem 12.7.3. *If* $(M, \mathfrak{g}, \mathfrak{A})$ *is an oriented regular surface in* $\mathbb{R}^3_\nu$, *then:*

(a) *There is a unit normal vector field* $\mathcal{N}$ *in* $\mathfrak{X}_{\mathbb{R}^3_\nu}(M)$, *called the* **Gauss map**, *such that for every chart* (U, φ) *in* $\mathfrak{A}$, $\mathcal{N} \circ \varphi$ *is the corresponding coordinate unit normal vector field.*

(b) $\mathcal{N}$ *satisfies properties* [V1] *and* [V2] *of Section 12.2.*

Remark. Calling $\mathcal{N}$ a "map" is reasonable because $\mathfrak{X}_{\mathbb{R}^3_\nu}(M)$ is, by definition, a set of maps.

Proof. (a): Since $\mathfrak{A}$ is a consistent atlas for M, it follows from Theorem 12.7.2 that whenever two charts in $\mathfrak{A}$ overlap, the corresponding coordinate unit normal vector fields agree on the overlap. We can therefore assign to each point p in M a vector $\mathcal{N}_p$ that is unit normal at p, and do so in such a way that $\mathcal{N}_p$ is independent of the choice of chart at p selected from $\mathfrak{A}$. More specifically, given the chart (U, φ) in $\mathfrak{A}$ and the corresponding coordinate unit normal vector field G, we define $\mathcal{N}_p = G_{\varphi^{-1}(p)}$ for all p in $\varphi(U)$. Since G is smooth on U, it follows from Theorem 10.1.17 and Theorem 11.2.8 that $\mathcal{N}$ is smooth on $\varphi(U)$. Because coordinate unit normal vector fields agree on overlaps, we can combine across charts to obtain a unit normal vector field $\mathcal{N}$ in $\mathfrak{X}_{\mathbb{R}^3_\nu}(M)$.

(b): This follows from Theorem 12.2.2(b) and Theorem 12.2.4. $\qquad\square$

Theorem 12.7.4. *Every regular surface M has an atlas $\mathfrak{B}$ such that for each chart (U, φ) in $\mathfrak{B}$, $\varphi(U)$ is a connected set in M.*

Proof. Let $\mathfrak{A}$ be an atlas for M, let p be a point in M, and let (W, ψ) be chart in $\mathfrak{A}$ at p. By definition, W is an open set in $\mathbb{R}^2$. Let U be an open connected set in W containing $\psi^{-1}(p)$, for example, an open disk of sufficiently small radius centered at $\psi^{-1}(p)$. According to [C3] of Section 11.2, ψ is a homeomorphism. It follows from Theorem 9.1.18 that $\psi(U)$ is a connected set in M. Let $\varphi = \psi|_U$. By Theorem 9.1.5 and Theorem 11.2.10, (U, φ) is a chart at p with the desired property. Since p was arbitrary, the result follows. $\qquad\square$

Theorem 12.7.5. *If M is a regular surface that is connected as a topological space, then $\mathfrak{X}_{\mathbb{R}^3_\nu}(M)$ contains either no unit vector fields or precisely two unit vector fields. In the latter case, if V is one of them, then $-V$ is the other.*

Proof. Let V and W be unit normal vector fields in $\mathfrak{X}_{\mathbb{R}^3_\nu}(M)$, and let p be a point in M. Since V_p and W_p are both unit normal vectors at p, they differ by at most a sign. Define a function $\phi : M \longrightarrow \{1, -1\}$ by $W_p = \phi(p)V_p$, and define nowhere-vanishing functions f, g in $C^\infty(M)$ by $f(p) = \langle V_p, W_p \rangle$ and $g(p) = \langle V_p, V_p \rangle$ for all p in M. Then $f(p) = \phi(p)g(p)$ for all p in M, hence $f = \phi g$. Since f and g are smooth, by Theorem 10.1.1, they are continuous. Thus, $\phi = f/g$ is a nowhere-vanishing continuous function. It follows from Theorem 9.1.20 that either $\phi = 1$ or $\phi = -1$, so W equals V or $-V$. $\qquad\square$

Theorem 12.7.6. *Let $(M, \mathfrak{g})$ be a regular surface in $\mathbb{R}^3_\nu$. Then M is orientable if and only if there is a unit normal vector field in $\mathfrak{X}_{\mathbb{R}^3_\nu}(M)$.*

Proof. ($\Rightarrow$): This follows from Theorem 12.7.3(a).

($\Leftarrow$): We construct a consistent atlas $\mathfrak{A}$ for M as follows. Let V be a unit normal vector field in $\mathfrak{X}_{\mathbb{R}^3_\nu}(M)$, and let $\mathfrak{B}$ be an atlas for M of the type given by Theorem 12.7.4. Let (U, φ) be a chart in $\mathfrak{B}$, and let G be the corresponding coordinate unit normal vector field. Using Example 11.2.5, we view $\varphi(U)$ as a regular surface. It follows from Theorem 12.2.11(c) that $G \circ \varphi^{-1}$ and $V|_{\varphi(U)}$ are unit normal vector fields in $\mathfrak{X}_{\mathbb{R}^3_\nu}(\varphi(U))$. Since $\varphi(U)$ is a connected set in M, we have from Theorem 12.7.5 that either $G \circ \varphi^{-1} = V|_{\varphi(U)}$ or $-G \circ \varphi^{-1} = V|_{\varphi(U)}$. In the former case, we include (U, φ) in $\mathfrak{A}$, and in the latter case, we include $(\iota(U), \varphi \circ \iota)$, where the map ι is defined above and we note that $-G \circ \varphi^{-1}$ is the corresponding coordinate unit normal vector field. Thus, for each chart in $\mathfrak{A}$, the corresponding coordinate unit normal vector field is a restriction of V to the coordinate domain of the chart. The result now follows from Theorem 12.7.2. $\qquad\square$

Example 12.7.7 ($\mathcal{S}^2$). It follows from parts (a)–(c) of Theorem 12.2.12 and Theorem 12.7.6 that $\mathcal{S}^2$ is an orientable regular surface in $\mathbb{R}^3_0$. $\qquad\diamond$

Reviewing the proof of Theorem 12.2.12, we see that the preceding example rests on the gradient in question satisfying property [V2] of Section 12.2. More generally, we have the following extension of Theorem 12.2.10(b).

Theorem 12.7.8 (Level Set of Function). *Let $\mathcal{U}$ be an open set in $\mathbb{R}^3_\nu$, let f be a function in $C^\infty(\mathcal{U})$, let c be a real number in $f(\mathcal{U})$, and let $M = f^{-1}(c)$. If $\mathrm{Grad}(f)$ satisfies property [V2] of Section 12.2, then $(M, \mathfrak{g})$ is an orientable regular surface in $\mathbb{R}^3_\nu$.*

Proof. This follows from Theorem 12.2.10(b) and Theorem 12.7.6. □

Example 12.7.9 (Möbius Band). The Möbius band is the subset of $\mathbb{R}^3$ defined by

$$\mathrm{M\ddot{o}b} = \{\varphi(t, \phi) \in \mathbb{R}^3 : (t, \phi) \in U\},$$

where

$$\varphi(t, \phi) = \big([1 - t\sin(\phi/2)]\cos(\phi), [1 - t\sin(\phi/2)]\sin(\phi), t\cos(\phi/2)\big)$$

and

$$U = (-1/2, 1/2) \times (-\pi, \pi).$$

A model of Möb can be made by giving a strip of paper a half twist and then pasting the ends together. See Figure 12.7.1. It can be shown that Möb is a regular surface in $\mathbb{R}^3_0$, something that is intuitively clear from the diagram. However, $\mathfrak{X}_{\mathbb{R}^3_0}(\mathrm{M\ddot{o}b})$ does not contain a unit normal vector field. This can be seen by choosing a starting point on the curve that bisects Möb longitudinally (dotted line in Figure 12.7.1), and then sliding a given unit normal vector (viewed as an arrow) along the curve on its base while keeping the shaft normal to the surface. After making a complete circuit, the unit normal vector is at the starting point, but now projects from the "opposite" side of the band. This shows that any unit normal vector field along Möb is not continuous, let alone smooth. However, completing a second circuit produces the original vector. In a manner of speaking, Möb has only one "side". ◇

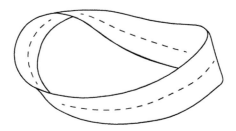

Figure 12.7.1. Möbius band: Diagram for Example 12.7.9

12.8   Gauss Curvature in $\mathbb{R}^3_\nu$

In this section, we describe a way of measuring the "curvature" of a regular surface in $\mathbb{R}^3_\nu$.

As part of the discussion of hyperquadrics $\mathcal{Q}^2$ in $\mathbb{R}^3_\nu$ in Section 12.2, we observed that $\mathcal{S}^2$ is the set of (spacelike) unit vectors in $\mathbb{R}^3_0$, $\mathcal{P}^2$ is the set of spacelike unit vectors in $\mathbb{R}^3_1$, and $\mathcal{H}^2$ is the set of timelike unit vectors in $\mathbb{R}^3_1$. In fact, more than just being sets, according to Theorem 12.2.12(a), $\mathcal{S}^2$ is a regular surface in $\mathbb{R}^3_0$, and $\mathcal{P}^2$ and $\mathcal{H}^2$ are regular surfaces in $\mathbb{R}^3_1$.

Let $(M, \mathfrak{g}, \mathfrak{A}, \mathfrak{O})$ be an oriented regular surface in $\mathbb{R}^3_\nu$, let $\mathcal{N} : M \longrightarrow \mathbb{R}^3_\nu$ be the Gauss map, and let p be a point in M. Since $\mathcal{N}_p$ is a unit normal vector at p, it follows from (12.2.3) that $\mathfrak{q}(\mathcal{N}_p) = \langle \mathcal{N}_p, \mathcal{N}_p \rangle = \epsilon_M$, where $\mathfrak{q}$ is the quadratic function corresponding to $\mathfrak{g}$. Thus, $\mathcal{N}_p$ is in the same hyperquadric for all p in M. Denoting the hyperquadric by $\mathcal{Q}^2$, we can now say that $\mathcal{N}_p$ is in $\mathcal{Q}^2$ for all p in M. Thus, $\mathcal{N}$ can be expressed more precisely as

$$\mathcal{N} : M \longrightarrow \mathcal{Q}^2.$$

The situation for $\mathcal{S}^2$ is depicted in Figure 12.8.1, where $\mathcal{N}_i$ stands for $\mathcal{N}_{p_i}$ for $i = 1, 2, 3$.

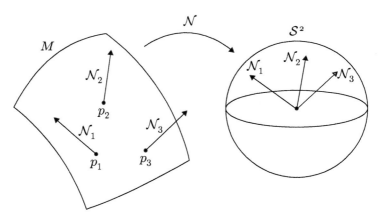

Figure 12.8.1. Gauss map

The differential of $\mathcal{N}$ at p is $d_p(\mathcal{N}) : T_p(M) \longrightarrow T_{\mathcal{N}_p}(\mathcal{Q}^2)$. By definition, $\mathcal{N}_p$ is in $T_p(M)^\perp$. On the other hand, since $\mathcal{N}_p$ is in $\mathcal{Q}^2$, we have from Theorem 12.2.12(c) that $\mathcal{N}_p$ is also in $T_{\mathcal{N}_p}(\mathcal{Q}^2)^\perp$. Since $T_p(M)^\perp$ and $T_{\mathcal{N}_p}(\mathcal{Q}^2)^\perp$ are both 1-dimensional, it follows that $T_p(M)^\perp = T_{\mathcal{N}_p}(\mathcal{Q}^2)^\perp$, and then from Theorem 4.1.2(c) that

$$T_p(M) = T_{\mathcal{N}_p}(\mathcal{Q}^2).$$

We can therefore express the differential of $\mathcal{N}$ at p as

$$d_p(\mathcal{N}) : T_p(M) \longrightarrow T_p(M).$$

Thus, $d_p(\mathcal{N})$ is a linear map from $T_p(M)$ to itself.

For each point p in M, the **Weingarten map at p** is denoted by

$$\mathcal{W}_p : T_p(M) \longrightarrow T_p(M)$$

and defined by

$$\mathcal{W}_p = -d_p(\mathcal{N}). \tag{12.8.1}$$

For all vectors v in $T_p(M)$, we have from (11.6.1) that

$$\mathcal{W}_p(v) = -\frac{d(\mathcal{N} \circ \lambda)}{dt}(t_0), \tag{12.8.2}$$

where $\lambda(t) : (a,b) \longrightarrow M$ is any smooth curve such that $\lambda(t_0) = p$ and $(d\lambda/dt)(t_0) = v$ for some t_0 in (a,b). The **Weingarten map** is the linear map

$$\mathcal{W} : \mathfrak{X}(M) \longrightarrow \mathfrak{X}(M)$$

defined by

$$\mathcal{W}(X)_p = \mathcal{W}_p(X_p) \tag{12.8.3}$$

for all vector fields X in $\mathfrak{X}(M)$ and all p in M.

Let (U, φ) be a chart in $\mathfrak{A}$, let $\mathcal{H} = (H_1, H_2)$ be the corresponding coordinate frame, and let q be a point in U. The vector $\mathcal{W}_p(H_j|_q)$ can be expressed as

$$\mathcal{W}_p(H_j|_q) = \sum_i \mathfrak{w}^i_j(q) H_i|_q, \tag{12.8.4}$$

where the $\mathfrak{w}^i_j$ are uniquely determined functions in $C^\infty(U)$. We then have from (2.2.2) and (2.2.3) that

$$[\mathcal{W}_p]^{\mathcal{H}_q}_{\mathcal{H}_q} = \begin{bmatrix} \mathfrak{w}^1_1(q) & \mathfrak{w}^1_2(q) \\ \mathfrak{w}^2_1(q) & \mathfrak{w}^2_2(q) \end{bmatrix}. \tag{12.8.5}$$

Theorem 12.8.1. *If $(M, \mathfrak{g}, \mathfrak{A}, \mathfrak{O})$ is an oriented regular surface in $\mathbb{R}^3_\nu$ and X is a vector field in $\mathfrak{X}(M)$, then*

$$\mathcal{W}(X) = -D_X(\mathcal{N}).$$

Proof. Let p be a point in M, and let $\lambda(t) : (a,b) \longrightarrow M$ be a smooth curve such that $\lambda(t_0) = p$ and $(d\lambda/dt)(t_0) = X_p$ for some t_0 in (a,b). Then

$$
\begin{aligned}
-D_X(\mathcal{N})_p &= -\frac{d(\mathcal{N} \circ \lambda)}{dt}(t_0) && [(12.3.4)] \\
&= \mathcal{W}_p(X_p) && [(12.8.2)] \\
&= \mathcal{W}(X)_p. && [(12.8.3)]
\end{aligned}
$$

Since p was arbitrary, the result follows. $\square$

Let $(M, \mathfrak{g}, \mathfrak{A}, \mathfrak{D})$ be an oriented regular surface in $\mathbb{R}^3_\nu$, and let p be a point in M. Since $\mathfrak{g}_p$ is bilinear and $\mathcal{W}_p$ is linear, we have the tensor $\mathfrak{h}_p$ in $T_2^0(T_p(M))$ defined by

$$\mathfrak{h}_p(v, w) = \langle \mathcal{W}_p(v), w \rangle \tag{12.8.6}$$

for all vectors v, w in $T_p(M)$. The **second fundamental form on M** is the map denoted by $\mathfrak{h}$ and defined by the assignment $p \longmapsto \mathfrak{h}_p$ for all p in M. In the literature, $\mathfrak{h}$ is often denoted by II.

For vector fields X, Y in $\mathfrak{X}(M)$, we define a function

$$\mathfrak{h}(X, Y) = \langle \mathcal{W}(X), Y \rangle : M \longrightarrow \mathbb{R}$$

in $C^\infty(M)$ by the assignment

$$p \longmapsto \mathfrak{h}_p(X_p, Y_p) = \langle \mathcal{W}(X)_p, Y_p \rangle \tag{12.8.7}$$

for all p in M.

Let (U, φ) be a chart in $\mathfrak{A}$, and let $\mathcal{H} = (H_1, H_2)$ be the corresponding coordinate frame. We define functions $\mathfrak{h}_{ij}$ in $C^\infty(U)$ by

$$\mathfrak{h}_{ij}(q) = \mathfrak{h}_p(H_i|_q, H_j|_q) \tag{12.8.8}$$

for all q in U for $i, j = 1, 2$, where $p = \varphi(q)$. The **matrix of $\mathfrak{h}$** with respect to $\mathcal{H}$ is denoted by $\mathfrak{h}_\mathcal{H}$ and defined by

$$\mathfrak{h}_\mathcal{H}(q) = \begin{bmatrix} \mathfrak{h}_{11}(q) & \mathfrak{h}_{12}(q) \\ \mathfrak{h}_{21}(q) & \mathfrak{h}_{22}(q) \end{bmatrix}$$

for all q in U.

Theorem 12.8.2. *Let $(M, \mathfrak{g}, \mathfrak{A}, \mathfrak{D})$ be an oriented regular surface in $\mathbb{R}^3_\nu$, let $\mathcal{N}$ be the Gauss map, and let p be a point in M. Let (U, φ) be a chart in $\mathfrak{A}$ at p, let (H_1, H_2) be the corresponding coordinate frame, and let $q = \varphi^{-1}(p)$. Then:*
(a)

$$\mathfrak{h}_{ij}(q) = \left\langle \left. \frac{\partial H_i}{\partial r^j} \right|_q, \mathcal{N}_p \right\rangle = \epsilon_M \vartheta_{ij}(q),$$

where ϑ_{ij} is given by (12.3.6).
(b) *$\mathcal{W}_p$ is self-adjoint with respect to $\mathfrak{g}_p$; that is,*

$$\langle \mathcal{W}_p(v), w \rangle = \langle v, \mathcal{W}_p(w) \rangle$$

for all vectors v, w in $T_p(M)$.
(c) *$\mathfrak{h}_p$ is a symmetric bilinear function.*

Proof. (a): Let (e_1, e_2) be the standard basis for $\mathbb{R}^2$. For given $1 \leq i \leq 2$, we define a smooth map

$$\zeta = (\zeta^1, \zeta^2) : (-\varepsilon, \varepsilon) \longrightarrow U$$

by $\zeta(t) = q + te_i$, where $\varepsilon > 0$ is chosen small enough that $(q - \varepsilon e_i, q + \varepsilon e_i) \subset U$. Consider the smooth curve $\lambda(t) : (-\varepsilon, \varepsilon) \longrightarrow M$ defined by $\lambda = \varphi \circ \zeta$, and observe that $\lambda(0) = \varphi(q) = p$. By Theorem 10.1.10,

$$\frac{d\lambda}{dt}(0) = \frac{d(\varphi \circ \zeta)}{dt}\bigg|_0 = \sum_j \frac{d\zeta^j}{dt}(0)H_j|_q = H_i|_q.$$

It also follows from Theorem 10.1.10 that

$$\frac{d(\mathcal{N} \circ \lambda)}{dt}\bigg|_0 = \frac{d(\mathcal{N} \circ \varphi \circ \zeta)}{dt}\bigg|_0 = \sum_j \frac{d\zeta^j}{dt}(0)\frac{\partial(\mathcal{N} \circ \varphi)}{\partial r^j}\bigg|_q$$
$$= \frac{\partial(\mathcal{N} \circ \varphi)}{\partial r^i}\bigg|_q, \tag{12.8.9}$$

and then from (12.8.2) and (12.8.9) that

$$\mathcal{W}_p(H_i|_q) = -\frac{d(\mathcal{N} \circ \lambda)}{dt}\bigg|_0 = -\frac{\partial(\mathcal{N} \circ \varphi)}{\partial r^i}\bigg|_q. \tag{12.8.10}$$

Taking the partial derivative with respect to r^i of both sides of $\langle \mathcal{N}_p, H_j|_q \rangle = 0$ yields

$$0 = \left\langle \frac{\partial(\mathcal{N} \circ \varphi)}{\partial r^i}\bigg|_q, H_j|_q \right\rangle + \left\langle \mathcal{N}_p, \frac{\partial H_j}{\partial r^i}\bigg|_q \right\rangle.$$

We have from Theorem 10.1.6 and (11.2.1) that

$$\frac{\partial H_i}{\partial r^j} = \frac{\partial H_j}{\partial r^i},$$

hence

$$\left\langle -\frac{\partial(\mathcal{N} \circ \varphi)}{\partial r^i}\bigg|_q, H_j|_q \right\rangle = \left\langle \mathcal{N}_p, \frac{\partial H_i}{\partial r^j}\bigg|_q \right\rangle. \tag{12.8.11}$$

Then

$$\mathfrak{h}_{ij}(q) = \langle \mathcal{W}_p(H_i|_q), H_j|_q \rangle \qquad \text{[(12.8.6), (12.8.8)]}$$
$$= \left\langle -\frac{\partial(\mathcal{N} \circ \varphi)}{\partial r^i}\bigg|_q, H_j|_q \right\rangle \qquad \text{[(12.8.10)]}$$
$$= \left\langle \frac{\partial H_i}{\partial r^j}\bigg|_q, \mathcal{N}_p \right\rangle, \qquad \text{[(12.8.11)]}$$

which proves the first equality. We also have

$$\left\langle \frac{\partial H_i}{\partial r^j}\bigg|_q, \mathcal{N}_p \right\rangle = \left\langle \frac{\partial H_i}{\partial r^j}\bigg|_q, G_q \right\rangle \qquad \text{[Th 12.7.3(a)]}$$
$$= \epsilon_M \vartheta_{ij}(q), \qquad \text{[Th 12.3.3(c)]}$$

which proves the second equality.

(b): We have

$$
\begin{aligned}
\langle \mathcal{W}_p(H_i|_q), H_j|_q \rangle &= \mathfrak{h}_p(H_i|_q, H_j|_q) && [(12.8.6)] \\
&= \mathfrak{h}_{ij}(q) && [(12.8.8)] \\
&= \epsilon_M \vartheta_{ij}(q) && [\text{part (a)}] \\
&= \epsilon_M \vartheta_{ji}(q) && [\text{Th 12.3.3(d)}] \\
&= \mathfrak{h}_{ji}(q) && [\text{part (a)}] \\
&= \mathfrak{h}_p(H_j|_q, H_i|_q) && [(12.8.8)] \\
&= \langle \mathcal{W}_p(H_j|_q), H_i|_q \rangle && [(12.8.6)] \\
&= \langle H_i|_q, \mathcal{W}_p(H_j|_q) \rangle.
\end{aligned}
$$

The result now follows from the observations that $(H_1|_q, H_2|_q)$ is a basis for $T_p(M)$, $\mathfrak{g}_p$ is bilinear, and $\mathcal{W}_p$ is linear.

(c): It was established in connection with (12.8.6) that $\mathfrak{h}_p$ is bilinear. For vectors v, w in $T_p(M)$, we have

$$
\begin{aligned}
\mathfrak{h}_p(v, w) &= \langle \mathcal{W}_p(v), w \rangle && [(12.8.6)] \\
&= \langle v, \mathcal{W}_p(w) \rangle && [\text{part (b)}] \\
&= \langle \mathcal{W}_p(w), v \rangle \\
&= \mathfrak{h}_p(w, v). && [(12.8.6)]
\end{aligned}
$$

Since v and w were arbitrary, the result follows. $\square$

Theorem 12.8.3. *Let $(M, \mathfrak{g}, \mathfrak{A}, \mathfrak{O})$ be an oriented regular surface in $\mathbb{R}^3_\nu$, and let p be a point in M. Let (U, φ) be a chart in $\mathfrak{A}$ at p, let $\mathcal{H}$ be the corresponding coordinate frame, and let $q = \varphi^{-1}(p)$. Then:*
(a)

$$
\begin{bmatrix} \mathfrak{h}_{11}(q) & \mathfrak{h}_{12}(q) \\ \mathfrak{h}_{21}(q) & \mathfrak{h}_{22}(q) \end{bmatrix} = \begin{bmatrix} \mathfrak{g}_{11}(q) & \mathfrak{g}_{12}(q) \\ \mathfrak{g}_{21}(q) & \mathfrak{g}_{22}(q) \end{bmatrix} \begin{bmatrix} \mathfrak{w}_1^1(q) & \mathfrak{w}_2^1(q) \\ \mathfrak{w}_1^2(q) & \mathfrak{w}_2^2(q) \end{bmatrix}.
$$

(b)

$$
[\mathcal{W}_p]_{\mathcal{H}_q}^{\mathcal{H}_q} = \begin{bmatrix} \mathfrak{g}^{11}(q) & \mathfrak{g}^{12}(q) \\ \mathfrak{g}^{21}(q) & \mathfrak{g}^{22}(q) \end{bmatrix} \begin{bmatrix} \mathfrak{h}_{11}(q) & \mathfrak{h}_{12}(q) \\ \mathfrak{h}_{21}(q) & \mathfrak{h}_{22}(q) \end{bmatrix}.
$$

Proof. (a): Let $\mathcal{H} = (H_1, H_2)$. We have

$$
\begin{aligned}
\mathfrak{h}_{ij}(q) &= \langle \mathcal{W}_p(H_i|_q), H_j|_q \rangle && [(12.8.6), (12.8.8)] \\
&= \langle H_i|_q, \mathcal{W}_p(H_j|_q) \rangle && [\text{Th 12.8.2(b)}] \\
&= \left\langle H_i|_q, \sum_k \mathfrak{w}_j^k(q) H_k|_q \right\rangle && [(12.8.4)] \\
&= \sum_k \mathfrak{g}_{ik}(q) \mathfrak{w}_j^k(q),
\end{aligned}
$$

from which the result follows.

(b): This follows from (12.8.5) and part (a). $\square$

Let $(M, \mathfrak{g}, \mathfrak{A}, \mathfrak{O})$ be an oriented regular surface in $\mathbb{R}^3_\nu$. The **Gauss curvature** is the smooth function

$$\mathcal{K} : M \longrightarrow \mathbb{R}$$

defined by

$$\mathcal{K}(p) = \epsilon_M \det(\mathcal{W}_p) \tag{12.8.12}$$

for all p in M. An intuitively appealing justification for this definition is provided below. For the moment, we simply observe that from (12.8.1), $\mathcal{W}_p$ is defined in terms of $d_p(\mathcal{N})$, which is related to the "rate of change" of the unit normal vector field $\mathcal{N}$ at p. In geometric terms, the greater the rate of change of $\mathcal{N}$, the greater the "curvature" we expect M to have at p.

It follows from Theorem 4.7.4 and Theorem 12.8.2(b) that $\mathcal{W}_p$ has two (not necessarily distinct) real eigenvalues, which we denote by $\kappa_1(p)$ and $\kappa_2(p)$.

Theorem 12.8.4. *Let* $(M, \mathfrak{g}, \mathfrak{A}, \mathfrak{O})$ *be an oriented regular surface in* $\mathbb{R}^3_\nu$, *and let p be a point in M. Let (U, φ) be a chart in $\mathfrak{A}$ at p, let $\mathcal{H}$ be the corresponding coordinate frame, and let $q = \varphi^{-1}(p)$. Then*

$$\mathcal{K}(p) = \epsilon_M \frac{\det(\mathfrak{h}_\mathcal{H}(q))}{\det(\mathfrak{g}_\mathcal{H}(q))} = \epsilon_M \, \kappa_1(p) \, \kappa_2(p).$$

Proof. By Theorem 4.7.5,

$$\det(\mathcal{W}_p) = \kappa_1(p) \, \kappa_2(p),$$

and by Theorem 12.8.3(b),

$$\det(\mathcal{W}_p) = \frac{\det(\mathfrak{h}_\mathcal{H}(q))}{\det(\mathfrak{g}_\mathcal{H}(q))}.$$

The result now follows from (12.8.12). ☐

The next result uses material on "local diffeomorphisms" from Section 14.6 and "area" from Section 19.10. It is included here because it provides a rationale for the definition of Gauss curvature when $\nu = 0$.

Theorem 12.8.5. *Let* $(M, \mathfrak{g}, \mathfrak{A}, \mathfrak{O})$ *be an oriented regular surface in* $\mathbb{R}^3_0$, *let $\mathcal{N}$ be the Gauss map, and let p be a point in M. Let (U, φ) be a chart in $\mathfrak{A}$ at p, let $q = \varphi^{-1}(p)$, and let $\varepsilon > 0$ be a real number small enough that the open disk $B_\varepsilon = B_\varepsilon(q)$ is contained in U. If $\mathcal{K}(p) \neq 0$, then*

$$|\mathcal{K}(p)| = \lim_{\varepsilon \to 0} \frac{\mathrm{area}(\mathcal{N} \circ \varphi(B_\varepsilon))}{\mathrm{area}(\varphi(B_\varepsilon))}.$$

Proof. By Theorem 11.2.10, $(B_\varepsilon, \varphi|_{B_\varepsilon})$ is a chart on M. Let (H_1, H_2) be the coordinate frame corresponding to (U, φ). Since $\mathcal{K}(p) \neq 0$, we have from (12.8.1) and (12.8.12) that $\det(d_p(\mathcal{N})) \neq 0$, and then from Theorem 2.5.3 that $d_p(\mathcal{N})$ is a linear isomorphism. Thus, $\mathcal{N}$ is an immersion at p. Since M and $\mathcal{S}^2$ are

2-dimensional, by Theorem 14.6.2, $\mathcal{N}$ is a local diffeomorphism at p. Taking ε to be smaller if necessary, it follows that $(B_\varepsilon, \mathcal{N} \circ \varphi|_{B_\varepsilon})$ is a chart on $\mathcal{S}^2$. See Figure 12.8.2. We have from Example 19.10.3 that

$$\text{area}\big(\varphi(B_\varepsilon)\big) = \iint_{B_\varepsilon} \|H_1 \times H_2\| \, dr^1 dr^2$$

and

$$\text{area}\big(\mathcal{N} \circ \varphi(B_\varepsilon)\big) = \iint_{B_\varepsilon} \left\| \frac{\partial(\mathcal{N} \circ \varphi)}{\partial r^1} \times \frac{\partial(\mathcal{N} \circ \varphi)}{\partial r^2} \right\| dr^1 dr^2$$

$$= \iint_{B_\varepsilon} \|\mathcal{W}(H_1) \times \mathcal{W}(H_2)\| \, dr^1 dr^2 \qquad [(12.8.10)]$$

$$= \iint_{B_\varepsilon} |\det(\mathcal{W})| \, \|H_1 \times H_2\| \, dr^1 dr^2. \qquad [\text{Th } 8.4.7]$$

It can be shown using Theorem 10.5.6 that

$$\lim_{\varepsilon \to 0} \frac{\text{area}\big(\mathcal{N} \circ \varphi(B_\varepsilon)\big)}{\text{area}\big(\varphi(B_\varepsilon)\big)} = \lim_{\varepsilon \to 0} \frac{\iint_{B_\varepsilon} |\det(\mathcal{W})| \, \|H_1 \times H_2\| \, dr^1 dr^2}{\iint_{B_\varepsilon} \|H_1 \times H_2\| \, dr^1 dr^2}$$

$$= \big|\det(\mathcal{W}_p)\big| = |\mathcal{K}(p)|. \qquad \square$$

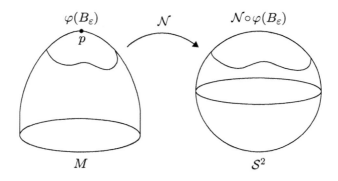

Figure 12.8.2. Diagram for Theorem 12.8.5

Figure 12.8.2 provides the geometric intuition for Theorem 12.8.5. Since M as depicted is highly curved at p, the area of $\mathcal{N} \circ \varphi(B_\varepsilon)$ is correspondingly greater than the area of $\varphi(B_\varepsilon)$, leading to a larger value of $|\mathcal{K}(p)|$.

Example 12.8.6. The tables below present a summary of the Gauss curvatures of the regular surfaces in $\mathbb{R}^3_0$ and $\mathbb{R}^3_1$ detailed in Chapter 13. The notation is explained in the relevant sections of that chapter. A few observations are in order.

In $\mathbb{R}^3_0$, the plane, cylinder, and cone all have a constant Gauss curvature of 0. This is not surprising for the plane, but is perhaps counterintuitive for the cylinder and cone. The explanation is that the cylinder and cone can be obtained from (portions of) the plane by smooth deformations that involve bending but not stretching. This keeps the "intrinsic" geometry of the deformed plane intact, thereby preserving the Gauss curvature at each point. The sphere has constant positive Gauss curvature, while the tractoid, which is shaped like a bugle, has constant negative Gauss curvature. The Gauss curvature of the hyperboloid of one sheet (two sheets) is negative (positive) but nonconstant. The torus has a region where the Gauss curvature is positive, and one where it is negative, with a transition zone in between where the Gauss curvature is 0.

Section	Geometric object in $\mathbb{R}^3_0$	Gauss curvature
13.1	plane	0
13.2	cylinder	0
13.2	cone	0
13.4	sphere	$1/R^2$
13.5	tractoid	-1
13.6	hyperboloid of one sheet	$-1/(2x^2 + 2y^2 - 1)^2$
13.7	hyperboloid of two sheets	$1/(2x^2 + 2y^2 + 1)^2$
13.8	torus	$\cos(\phi)/[\cos(\phi) + R]$

In $\mathbb{R}^3_1$, the pseudosphere has constant positive Gauss curvature, while hyperbolic space has constant negative Gauss curvature. It is interesting to observe that the hyperboloid of one sheet and the pseudosphere are defined in terms of the same underlying surface. The difference in their Gauss curvatures is due entirely to the fact that one resides in the inner product space $\mathbb{R}^3_0$, and the other in the Lorentz vector space $\mathbb{R}^3_1$. A similar remark applies to the hyperboloid of two sheets and hyperbolic space.

Section	Geometric object in $\mathbb{R}^3_1$	Gauss curvature
13.9	pseudosphere	1
13.10	hyperbolic space	-1

12.9 Riemann Curvature Tensor in $\mathbb{R}^3_\nu$

The **Riemann curvature tensor** for a regular surface $(M, \mathfrak{g})$ in $\mathbb{R}^3_\nu$ is the map

$$R : \mathfrak{X}(M)^3 \longrightarrow \mathfrak{X}(M)$$

defined by

$$R(X,Y)Z = \nabla_X\big(\nabla_Y(Z)\big) - \nabla_Y\big(\nabla_X(Z)\big) - \nabla_{[X,Y]}(Z) \qquad (12.9.1)$$

for all vector fields X, Y, Z in $\mathfrak{X}(M)$; that is,

$$\left(R(X,Y)Z\right)_p = \left(\nabla_X(\nabla_Y(Z))\right)_p - \left(\nabla_Y(\nabla_X(Z))\right)_p - \left(\nabla_{[X,Y]}(Z)\right)_p$$

for all p in M. The large parentheses are included to make it clear that each of the four terms in the preceding identity is a vector field in $\mathfrak{X}(M)$ evaluated at the point p, and as such is a vector in $T_p(M)$. Since $(R(X,Y)Z)_p$ is not a real number, using the term "tensor" to describe R is something of a misnomer. This conflict is resolved in Theorem 19.5.5. The expression $R_p(X_p, Y_p)Z_p$ has no meaning—at least not yet.

We presented an instance in Example 12.4.4 where the second order covariant derivatives $\nabla^2_{X,Y}(Z)$ and $\nabla^2_{Y,X}(Z)$ are not equal. As the next result shows, the difference between these two vector fields is precisely $R(X,Y)Z$.

Theorem 12.9.1. *If* $(M, \mathfrak{g})$ *is a regular surface in* $\mathbb{R}^3_\nu$, *then*

$$R(X,Y)Z = \nabla^2_{X,Y}(Z) - \nabla^2_{Y,X}(Z)$$

for all vector fields X, Y, Z *in* $\mathfrak{X}(M)$.

Proof. We have from Theorem 12.4.1(a) and (12.6.1) that

$$\nabla_{[X,Y]}(Z) = \nabla_{\nabla_X(Y)-\nabla_Y(X)}(Z) = \nabla_{\nabla_X(Y)}(Z) - \nabla_{\nabla_Y(X)}(Z).$$

Then

$$\begin{aligned}
R(X,Y)Z &= \nabla_X\left(\nabla_Y(Z)\right) - \nabla_Y\left(\nabla_X(Z)\right) - \nabla_{[X,Y]}(Z) \\
&= \left[\nabla_X\left(\nabla_Y(Z)\right) - \nabla_{\nabla_X(Y)}(Z)\right] - \left[\nabla_Y\left(\nabla_X(Z)\right) - \nabla_{\nabla_Y(X)}(Z)\right] \\
&= \nabla^2_{X,Y}(Z) - \nabla^2_{Y,X}(Z),
\end{aligned}$$

where the last equality follows from (12.4.4). $\qquad\square$

For computational purposes, it is helpful to have a local coordinate expression for R.

Theorem 12.9.2. *Let* $(M, \mathfrak{g})$ *be a regular surface in* $\mathbb{R}^3_\nu$, *let* X, Y, Z *be vector fields in* $\mathfrak{X}(M)$, *and, in local coordinates, let*

$$X \circ \varphi = \sum_i \alpha^i H_i, \qquad Y \circ \varphi = \sum_j \beta^j H_j, \qquad and \qquad Z \circ \varphi = \sum_k \gamma^k H_k.$$

Then

$$R(X,Y)Z \circ \varphi = \sum_l \left(\sum_{ijk} \alpha^i \beta^j \gamma^k R^l_{ijk}\right) H_l, \tag{12.9.2}$$

where

$$R^l_{ijk} = \frac{\partial \Gamma^l_{jk}}{\partial r^i} - \frac{\partial \Gamma^l_{ik}}{\partial r^j} + \sum_n (\Gamma^l_{in}\Gamma^n_{jk} - \Gamma^l_{jn}\Gamma^n_{ik}) \tag{12.9.3}$$

for $i, j, k, l = 1, 2$.

Proof. We have from (12.4.20) that

$$
\left(\nabla^2_{X,Y}(Z) \circ \varphi\right)^l = \overset{(3)}{\sum_{ij}} \alpha^i \beta^j \frac{\partial^2 \gamma^l}{\partial r^i \partial r^j} + \overset{(4)}{\sum_{ijk}} \alpha^i \beta^j \frac{\partial \gamma^k}{\partial r^i} \Gamma^l_{jk} + \overset{(5)}{\sum_{ijk}} \alpha^i \beta^j \gamma^k \frac{\partial \Gamma^l_{jk}}{\partial r^i}
$$

$$
+ \overset{(6)}{\sum_{ijk}} \alpha^i \beta^j \frac{\partial \gamma^k}{\partial r^j} \Gamma^l_{ik} + \overset{(7)}{\sum_{ijkn}} \alpha^i \beta^j \gamma^k \Gamma^n_{jk} \Gamma^l_{in}
$$

$$
- \overset{(8)}{\sum_{ijk}} \alpha^i \beta^j \frac{\partial \gamma^l}{\partial r^k} \Gamma^k_{ij} - \overset{(9)}{\sum_{ijkn}} \alpha^i \beta^j \gamma^k \Gamma^n_{ij} \Gamma^l_{kn}.
$$

Reversing the roles of X and Y yields

$$
\left(\nabla^2_{Y,X}(Z) \circ \varphi\right)^l = \sum_{ij} \alpha^j \beta^i \frac{\partial^2 \gamma^l}{\partial r^i \partial r^j} + \sum_{ijk} \alpha^j \beta^i \frac{\partial \gamma^k}{\partial r^i} \Gamma^l_{jk} + \sum_{ijk} \alpha^j \beta^i \gamma^k \frac{\partial \Gamma^l_{jk}}{\partial r^i}
$$

$$
+ \sum_{ijk} \alpha^j \beta^i \frac{\partial \gamma^k}{\partial r^j} \Gamma^l_{ik} + \sum_{ijkn} \alpha^j \beta^i \gamma^k \Gamma^n_{jk} \Gamma^l_{in}
$$

$$
- \sum_{ijk} \alpha^j \beta^i \frac{\partial \gamma^l}{\partial r^k} \Gamma^k_{ij} - \sum_{ijkn} \alpha^j \beta^i \gamma^k \Gamma^n_{ij} \Gamma^l_{kn}
$$

$$
= \overset{(3)}{\sum_{ij}} \alpha^i \beta^j \frac{\partial^2 \gamma^l}{\partial r^i \partial r^j} + \overset{(6)}{\sum_{ijk}} \alpha^i \beta^j \frac{\partial \gamma^k}{\partial r^j} \Gamma^l_{ik} + \overset{(10)}{\sum_{ijk}} \alpha^i \beta^j \gamma^k \frac{\partial \Gamma^l_{ik}}{\partial r^j}
$$

$$
+ \overset{(4)}{\sum_{ijk}} \alpha^i \beta^j \frac{\partial \gamma^k}{\partial r^i} \Gamma^l_{jk} + \overset{(11)}{\sum_{ijkn}} \alpha^i \beta^j \gamma^k \Gamma^n_{ik} \Gamma^l_{jn}
$$

$$
- \overset{(8)}{\sum_{ijk}} \alpha^i \beta^j \frac{\partial \gamma^l}{\partial r^k} \Gamma^k_{ij} - \overset{(9)}{\sum_{ijkn}} \alpha^i \beta^j \gamma^k \Gamma^n_{ij} \Gamma^l_{kn},
$$

where the numbering of sums that was initiated in the proof of Theorem 12.4.3 is continued. It follows from Theorem 12.9.1 that the lth component of $R(X,Y)Z$ is

$$
\left(\left[\nabla^2_{X,Y}(Z) - \nabla^2_{Y,X}(Z)\right] \circ \varphi\right)^l
$$

$$
= \overset{(5)}{\sum_{ijk}} \alpha^i \beta^j \gamma^k \frac{\partial \Gamma^l_{jk}}{\partial r^i} + \overset{(7)}{\sum_{ijkn}} \alpha^i \beta^j \gamma^k \Gamma^n_{jk} \Gamma^l_{in}
$$

$$
- \overset{(10)}{\sum_{ijk}} \alpha^i \beta^j \gamma^k \frac{\partial \Gamma^l_{ik}}{\partial r^j} - \overset{(11)}{\sum_{ijkn}} \alpha^i \beta^j \gamma^k \Gamma^n_{ik} \Gamma^l_{jn}
$$

$$
= \sum_{ijk} \alpha^i \beta^j \gamma^k \left(\frac{\partial \Gamma^l_{jk}}{\partial r^i} - \frac{\partial \Gamma^l_{ik}}{\partial r^j} + \sum_n (\Gamma^l_{in} \Gamma^n_{jk} - \Gamma^l_{jn} \Gamma^n_{ik}) \right)
$$

$$
= \sum_{ijk} \alpha^i \beta^j \gamma^k R^l_{ijk}. \qquad \square
$$

We observed in Section 12.3 that the Christoffel symbols are intrinsic. It follows from (12.9.2) and (12.9.3) that the same is true of the Riemann curvature tensor.

Example 12.9.3 ($\mathcal{S}^2$). Substituting into (12.9.3) the values of the Christoffel symbols for $\mathcal{S}^2$ given in Section 13.4 yields

$$R^1_{122} = \sin^2(\theta) \qquad R^1_{212} = -\sin^2(\theta)$$
$$R^2_{121} = -1 \qquad R^2_{211} = 1,$$

with the remaining R^l_{ijk} equal to 0. Observe that the R^l_{ijk} are independent of ϕ, as would be expected due to the symmetry of the sphere. $\diamond$

It is a remarkable feature of (12.9.2) that no partial derivatives of the component functions appear in the expression. This crucial observation underlies the next two results.

Theorem 12.9.4. *If $(M, \mathfrak{g})$ is a regular surface in $\mathbb{R}^3_\nu$, then R is **determined pointwise** on M in the following sense: for all points p in M, if X, Y, Z and $\widetilde{X}, \widetilde{Y}, \widetilde{Z}$ are vector fields in $\mathfrak{X}(M)$ such that*

$$(X_p, Y_p, Z_p) = (\widetilde{X}_p, \widetilde{Y}_p, \widetilde{Z}_p),$$

then

$$\big(R(X, Y)Z\big)_p = \big(R(\widetilde{X}, \widetilde{Y})\widetilde{Z}\big)_p.$$

Proof. In local coordinates, let

$$X \circ \varphi = \sum_i \alpha^i H_i \qquad \text{and} \qquad \widetilde{X} \circ \varphi = \sum_i \widetilde{\alpha}^i H_i.$$

Then $X_p = \widetilde{X}_p$ translates into $\alpha^i(p) = \widetilde{\alpha}^i(p)$ for $i = 1, 2$, with corresponding identities for the other two vector fields. The result now follows from (12.9.2). $\square$

Theorem 12.9.5. *If $(M, \mathfrak{g})$ is a regular surface in $\mathbb{R}^3_\nu$, then R is $C^\infty(M)$-**multilinear** in the following sense: for all vectors fields X, Y, Z, W in $\mathfrak{X}(M)$ and all functions f in $C^\infty(M)$,*

$$R(fX + W, Y)Z = fR(X, Y)Z + R(W, Y)Z,$$

with corresponding identities for the other two arguments of R.

Proof. This follows from reasoning similar to that used in the proof of Theorem 12.9.4. $\square$

Theorem 12.9.6. *Let $(M, \mathfrak{g}, \mathfrak{O})$ be an oriented regular surface in $\mathbb{R}^3_\nu$, let $\mathcal{N}$ be the Gauss map, and let X, Y be vector fields in $\mathfrak{X}(M)$. Then:*
(a)

$$\langle D_X(Y), \mathcal{N}\rangle = \langle \mathcal{W}(X), Y\rangle.$$

(b)
$$D_X(Y) = \nabla_X(Y) + \epsilon_M \langle \mathcal{W}(X), Y \rangle \mathcal{N}. \qquad (12.9.4)$$

Proof. (a): Applying D_X to both sides of $\langle Y, \mathcal{N} \rangle = 0$ and using Theorem 12.3.1(e) yields $\langle D_X(Y), \mathcal{N} \rangle = \langle -D_X(\mathcal{N}), Y \rangle$. The result now follows from Theorem 12.8.1.

(b): Let p be a point in M, and recall the definitions of $\tan_p$ and nor_p given in Section 12.4. By Theorem 4.1.4,
$$D_X(Y)_p = \tan_p(D_X(Y)_p) + \text{nor}_p(D_X(Y)_p),$$

and by (12.4.2),
$$\tan_p(D_X(Y)_p) = \nabla_X(Y)_p.$$

It follows from Theorem 12.7.3(b) that $T_p(M)^\perp = \mathbb{R}\mathcal{N}_p$, and then from Theorem 4.1.5 and (12.2.3) that
$$\text{nor}_p(D_X(Y)_p) = \frac{\langle \mathcal{N}_p, D_X(Y)_p \rangle}{\langle \mathcal{N}_p, \mathcal{N}_p \rangle} \mathcal{N}_p = \epsilon_M \langle D_X(Y)_p, \mathcal{N}_p \rangle \mathcal{N}_p.$$

Thus,
$$D_X(Y)_p = \nabla_X(Y)_p + \epsilon_M \langle D_X(Y)_p, \mathcal{N}_p \rangle \mathcal{N}_p.$$

Since p was arbitrary,
$$D_X(Y) = \nabla_X(Y) + \epsilon_M \langle D_X(Y), \mathcal{N} \rangle \mathcal{N}.$$

The result now follows from part (a). $\qquad\qquad\square$

Theorem 12.9.7. *If* $(M, \mathfrak{g}, \mathfrak{D})$ *is an oriented regular surface in* $\mathbb{R}^3_\nu$ *and* X, Y, Z *are vector fields in* $\mathfrak{X}(M)$, *then:*
(a)
$$R(X,Y)Z = \epsilon_M[\mathfrak{h}(Y,Z)\,\mathcal{W}(X) - \mathfrak{h}(X,Z)\,\mathcal{W}(Y)].$$
(b)
$$\nabla_X(\mathcal{W}(Y)) - \nabla_Y(\mathcal{W}(X)) = \mathcal{W}([X,Y]).$$

Proof. By Theorem 12.9.6(b),
$$\begin{aligned} D_X(D_Y(Z)) &= D_X(\nabla_Y(Z) + \epsilon_M \langle \mathcal{W}(Y), Z \rangle \mathcal{N}) \\ &= D_X(\nabla_Y(Z)) + \epsilon_M D_X(\langle \mathcal{W}(Y), Z \rangle \mathcal{N}) \end{aligned} \qquad (12.9.5)$$

and
$$D_X(\nabla_Y(Z)) = \nabla_X(\nabla_Y(Z)) + \epsilon_M \langle \mathcal{W}(X), \nabla_Y(Z) \rangle \mathcal{N}, \qquad (12.9.6)$$

and by parts (d) and (e) of Theorem 12.3.1 and Theorem 12.8.1,
$$\begin{aligned} D_X(\langle \mathcal{W}(Y), Z \rangle \mathcal{N}) &= D_X(\langle \mathcal{W}(Y), Z \rangle)\mathcal{N} + \langle \mathcal{W}(Y), Z \rangle D_X(\mathcal{N}) \\ &= [\langle D_X(\mathcal{W}(Y)), Z \rangle + \langle \mathcal{W}(Y), D_X(Z) \rangle]\mathcal{N} \qquad (12.9.7) \\ &\quad - \langle \mathcal{W}(Y), Z \rangle \mathcal{W}(X). \end{aligned}$$

Substituting (12.9.6) and (12.9.7) into (12.9.5) gives

$$
\begin{aligned}
D_X\big(D_Y(Z)\big) = {} & \nabla_X\big(\nabla_Y(Z)\big) - \epsilon_M\langle \mathcal{W}(Y), Z\rangle \mathcal{W}(X) \\
& + \epsilon_M\big[\langle D_X(\mathcal{W}(Y)), Z\rangle + \langle \mathcal{W}(Y), D_X(Z)\rangle \\
& + \langle \mathcal{W}(X), \nabla_Y(Z)\rangle\big]\mathcal{N}.
\end{aligned}
\tag{12.9.8}
$$

Also by Theorem 12.9.6(b),

$$
\begin{aligned}
\langle D_X(\mathcal{W}(Y)), Z\rangle &= \langle \nabla_X(\mathcal{W}(Y)) + \epsilon_M\langle \mathcal{W}(X), \mathcal{W}(Y)\rangle\mathcal{N}, Z\rangle \\
&= \langle \nabla_X(\mathcal{W}(Y)), Z\rangle + \epsilon_M\langle \mathcal{W}(X), \mathcal{W}(Y)\rangle\langle \mathcal{N}, Z\rangle \\
&= \langle \nabla_X(\mathcal{W}(Y)), Z\rangle
\end{aligned}
\tag{12.9.9}
$$

and

$$
\begin{aligned}
\langle \mathcal{W}(Y), D_X(Z)\rangle &= \langle \mathcal{W}(Y), \nabla_X(Z) + \epsilon_M\langle \mathcal{W}(X), Z\rangle\mathcal{N}\rangle \\
&= \langle \mathcal{W}(Y), \nabla_X(Z)\rangle + \epsilon_M\langle \mathcal{W}(X), Z\rangle\langle \mathcal{W}(Y), \mathcal{N}\rangle \\
&= \langle \mathcal{W}(Y), \nabla_X(Z)\rangle.
\end{aligned}
\tag{12.9.10}
$$

Substituting (12.9.9) and (12.9.10) into (12.9.8) yields

$$
\begin{aligned}
D_X\big(D_Y(Z)\big) = {} & \nabla_X\big(\nabla_Y(Z)\big) - \epsilon_M\langle \mathcal{W}(Y), Z\rangle \mathcal{W}(X) \\
& + \epsilon_M\big[\langle \nabla_X(\mathcal{W}(Y)), Z\rangle + \langle \mathcal{W}(Y), \nabla_X(Z)\rangle \\
& + \langle \mathcal{W}(X), \nabla_Y(Z)\rangle\big]\mathcal{N}.
\end{aligned}
\tag{12.9.11}
$$

Interchanging the roles of X and Y in (12.9.11) gives

$$
\begin{aligned}
D_Y\big(D_X(Z)\big) = {} & \nabla_Y\big(\nabla_X(Z)\big) - \epsilon_M\langle \mathcal{W}(X), Z\rangle \mathcal{W}(Y) \\
& + \epsilon_M\big[\langle \nabla_Y(\mathcal{W}(X)), Z\rangle + \langle \mathcal{W}(X), \nabla_Y(Z)\rangle \\
& + \langle \mathcal{W}(Y), \nabla_X(Z)\rangle\big]\mathcal{N}.
\end{aligned}
\tag{12.9.12}
$$

Yet again by Theorem 12.9.6(b),

$$
D_{[X,Y]}(Z) = \nabla_{[X,Y]}(Z) + \epsilon_M\langle \mathcal{W}([X,Y]), Z\rangle\mathcal{N}.
\tag{12.9.13}
$$

We note that in the present context, $[X,Y]$ denotes $\nabla_X(Y) - \nabla_Y(X)$, not $D_X(Y) - D_Y(X)$. But according to Theorem 12.6.2, the latter two vector fields are equal. By Theorem 10.1.15, there is an open set U in $\mathbb{R}^3$ containing M, and vector fields $\widetilde{X}, \widetilde{Y}, \widetilde{Z}$ in $\mathfrak{X}(U)$ that restrict to X, Y, Z on M, respectively. By Theorem 10.4.3,

$$
\widetilde{D}_{\widetilde{X}}\big(\widetilde{D}_{\widetilde{Y}}(\widetilde{Z})\big) - \widetilde{D}_{\widetilde{Y}}\big(\widetilde{D}_{\widetilde{X}}(\widetilde{Z})\big) - \widetilde{D}_{\widetilde{D}_{\widetilde{X}}(\widetilde{Y}) - \widetilde{D}_{\widetilde{Y}}(\widetilde{X})}(\widetilde{Z}) = (0,0,0)
$$

on U, where $\widetilde{D}$ denotes the Euclidean derivative (Section 10.3). It follows that

$$
\begin{aligned}
(0,0,0) &= D_X\big(D_Y(Z)\big) - D_Y\big(D_X(Z)\big) - D_{D_X(Y)-D_Y(X)}(Z) \\
&= D_X\big(D_Y(Z)\big) - D_Y\big(D_X(Z)\big) - D_{[X,Y]}(Z)
\end{aligned}
\tag{12.9.14}
$$

on M. Substituting (12.9.11)–(12.9.13) into (12.9.14) and using (12.9.1) gives

$$R(X,Y)Z - \epsilon_M \left[\langle \mathcal{W}(Y), Z \rangle \mathcal{W}(X) - \langle \mathcal{W}(X), Z \rangle \mathcal{W}(Y) \right]$$
$$+ \epsilon_M \langle \nabla_X \left(\mathcal{W}(Y) \right) - \nabla_Y \left(\mathcal{W}(X) \right) - \mathcal{W}([X,Y]), Z \rangle \mathcal{N} = (0,0,0).$$

It follows that both the tangential and normal components of the left-hand side of the preceding expression equal $(0,0,0)$, hence

$$R(X,Y)Z = \epsilon_M \left[\langle \mathcal{W}(Y), Z \rangle \mathcal{W}(X) - \langle \mathcal{W}(X), Z \rangle \mathcal{W}(Y) \right] \tag{12.9.15}$$

and

$$\langle \nabla_X \left(\mathcal{W}(Y) \right) - \nabla_Y \left(\mathcal{W}(X) \right) - \mathcal{W}([X,Y]), Z \rangle = 0. \tag{12.9.16}$$

From (12.8.7) and (12.9.15), we obtain

$$R(X,Y)Z = \epsilon_M [\mathfrak{h}(Y,Z)\,\mathcal{W}(X) - \mathfrak{h}(X,Z)\,\mathcal{W}(Y)],$$

which proves part (a). Since $\mathfrak{g}$ is nondegenerate and (12.9.16) holds for all vector fields Z in $\mathfrak{X}(M)$, we have

$$\nabla_X \left(\mathcal{W}(Y) \right) - \nabla_Y \left(\mathcal{W}(X) \right) - \mathcal{W}([X,Y]) = (0,0,0),$$

which proves part (b). $\qquad\qquad\qquad\qquad\qquad\qquad\qquad\qquad\qquad\qquad\square$

Let $(M, \mathfrak{g})$ be a regular surface in $\mathbb{R}^3$, and define a map

$$\mathcal{R} : \mathfrak{X}(M)^4 \longrightarrow C^\infty(M),$$

also called the **Riemann curvature tensor**, by

$$\mathcal{R}(X,Y,Z,W) = \langle R(X,Y)Z, W \rangle$$

for all vector fields X, Y, Z, W in $\mathfrak{X}(M)$; that is,

$$\mathcal{R}(X,Y,Z,W)(p) = \left\langle \left(R(X,Y)Z \right)_p, W_p \right\rangle \tag{12.9.17}$$

for all p in M. By definition, $\mathcal{R}(X,Y,Z,W)$ is a function in $C^\infty(M)$. Since $\mathcal{R}(X,Y,Z,W)(p)$ is a real number, calling $\mathcal{R}$ a "tensor" is perhaps justified. We return to this issue below.

Theorem 12.9.8. *Let $(M, \mathfrak{g})$ be a regular surface in $\mathbb{R}^3_\nu$, let X, Y, Z, W be vector fields in $\mathfrak{X}(M)$, and, in local coordinates, let*

$$X \circ \varphi = \sum_i \alpha^i H_i \qquad Y \circ \varphi = \sum_j \beta^j H_j$$
$$Z \circ \varphi = \sum_k \gamma^k H_k \qquad W \circ \varphi = \sum_l \delta^l H_l.$$

Then

$$\mathcal{R}(X,Y,Z,W) \circ \varphi = \sum_{ijkln} \alpha^i \beta^j \gamma^k \delta^l \, \mathfrak{g}_{ln} R^n_{ijk}. \tag{12.9.18}$$

Proof. We have from (12.9.2) that

$$\mathcal{R}(X,Y,Z,W) \circ \varphi = \langle R(X,Y)Z \circ \varphi, W \circ \varphi \rangle$$

$$= \left\langle \sum_n \left(\sum_{ijk} \alpha^i \beta^j \gamma^k R^n_{ijk} \right) H_n, \sum_l \delta^l H_l \right\rangle$$

$$= \sum_{ijkln} \alpha^i \beta^j \gamma^k \delta^l \mathfrak{g}_{ln} R^n_{ijk}. \qquad \square$$

We noted in conjunction with (12.9.2) and (12.9.3) that the Riemann curvature tensor R is intrinsic. In view of (12.9.18), the same can be said of the Riemann curvature tensor $\mathcal{R}$. Just as was the case for (12.9.2), there are no partial derivatives of the component functions in (12.9.18). This observation underlies the next two results, which are counterparts of Theorem 12.9.4 and Theorem 12.9.5, and are proved similarly.

Theorem 12.9.9. *If* $(M, \mathfrak{g})$ *is a regular surface in* $\mathbb{R}^3_\nu$, *then* $\mathcal{R}$ *is* **determined pointwise** *in the following sense: for all points p in M, if X, Y, Z, W and $\widetilde{X}, \widetilde{Y}, \widetilde{Z}, \widetilde{W}$ are vector fields in $\mathfrak{X}(M)$ such that*

$$(X_p, Y_p, Z_p, W_p) = (\widetilde{X}_p, \widetilde{Y}_p, \widetilde{Z}_p, \widetilde{W}_p),$$

then

$$\mathcal{R}(X,Y,Z,W)(p) = \mathcal{R}(\widetilde{X}, \widetilde{Y}, \widetilde{Z}, \widetilde{W})(p).$$

$\square$

Theorem 12.9.10. *If* $(M, \mathfrak{g})$ *is a regular surface in* $\mathbb{R}^3_\nu$, *then* $\mathcal{R}$ *is* $C^\infty(M)$**-multilinear** *in the following sense: for all vector fields X, Y, Z, V, W in $\mathfrak{X}(M)$ and all functions f in $C^\infty(M)$,*

$$\mathcal{R}(fX + W, Y, Z, V) = f\mathcal{R}(X,Y,Z,V) + \mathcal{R}(W,Y,Z,V),$$

with corresponding identities for the other three arguments of $\mathcal{R}$. $\qquad \square$

Let $(M, \mathfrak{g})$ be a regular surface in $\mathbb{R}^3_\nu$, let p be a point in M, and let v be a vector in $T_p(M)$. According to Theorem 15.1.2, there is a vector field X in $\mathfrak{X}(M)$ such that $X_p = v$. Taken in conjunction with Theorem 12.9.4, Theorem 12.9.5, Theorem 12.9.9, and Theorem 12.9.10, this allows us to give R and $\mathcal{R}$ interesting interpretations. We define a map

$$R_p : T_p(M)^3 \longrightarrow T_p(M)$$

by

$$R_p(v_1, v_2)v_3 = \big(R(X_1, X_2)X_3 \big)_p, \qquad (12.9.19)$$

and a map

$$\mathcal{R}_p : T_p(M)^4 \longrightarrow \mathbb{R}$$

by

$$\mathcal{R}_p(v_1, v_2, v_3, v_4) = \mathcal{R}(X_1, X_2, X_3, X_4)(p)$$

for all vectors v_1, v_2, v_3, v_4 in $T_p(M)$, where X_1, X_2, X_3, X_4 are any vector fields in $\mathfrak{X}(M)$ such that

$$(X_1|_p, X_2|_p, X_3|_p, X_4|_p) = (v_1, v_2, v_3, v_4).$$

By Theorem 12.9.4 and Theorem 12.9.9, respectively, R_p and $\mathcal{R}_p$ are independent of the choice of vector fields, so the definitions makes sense. It follows from (12.9.17) and the above identities that

$$\begin{aligned}
\mathcal{R}_p(v_1, v_2, v_3, v_4) &= \mathcal{R}(X_1, X_2, X_3, X_4)(p) \\
&= \left\langle \left(R(X_1, X_2)X_3\right)_p, X_4|_p \right\rangle \\
&= \langle R_p(v_1, v_2)v_3, v_4 \rangle.
\end{aligned}$$

By Theorem 12.9.10, $\mathcal{R}_p$ is in $\mathcal{T}^0_4(T_p(M))$, and by Theorem 12.9.5, R_p is in $\mathrm{Mult}(T_p(M)^3, T_p(M))$. This provides a justification for calling $\mathcal{R}$ a "tensor", and to a lesser extent a rationale for doing the same with R. Another tensor of interest in $\mathcal{T}^0_4(T_p(M))$ is $\mathcal{D}_p$, as defined by (6.6.7):

$$\mathcal{D}_p(v_1, v_2, v_3, v_4) = \det\left(\begin{bmatrix} \langle v_1, v_3 \rangle & \langle v_1, v_4 \rangle \\ \langle v_2, v_3 \rangle & \langle v_2, v_4 \rangle \end{bmatrix}\right). \tag{12.9.20}$$

Theorem 12.9.11 (Gauss's Equation). *If* $(M, \mathfrak{g}, \mathfrak{O})$ *is an oriented regular surface in* $\mathbb{R}^3_\nu$ *and* X, Y, Z, W *are vector fields in* $\mathfrak{X}(M)$, *then*

$$\mathcal{R}(X, Y, Z, W) = -\epsilon_M \det\left(\begin{bmatrix} \mathfrak{h}(X, Z) & \mathfrak{h}(X, W) \\ \mathfrak{h}(Y, Z) & \mathfrak{h}(Y, W) \end{bmatrix}\right).$$

Proof. It follows from (12.8.7) and Theorem 12.9.7(a) that

$$\begin{aligned}
\mathcal{R}(X, Y, Z, W) &= \epsilon_M \langle \mathfrak{h}(Y, Z)\, \mathcal{W}(X) - \mathfrak{h}(X, Z)\, \mathcal{W}(Y), W \rangle \\
&= \epsilon_M [\mathfrak{h}(Y, Z)\, \langle \mathcal{W}(X), W \rangle - \mathfrak{h}(X, Z)\, \langle \mathcal{W}(Y), W \rangle] \\
&= \epsilon_M [\mathfrak{h}(Y, Z)\, \mathfrak{h}(X, W) - \mathfrak{h}(X, Z)\, \mathfrak{h}(Y, W)] \\
&= -\epsilon_M \det\left(\begin{bmatrix} \mathfrak{h}(X, Z) & \mathfrak{h}(X, W) \\ \mathfrak{h}(Y, Z) & \mathfrak{h}(Y, W) \end{bmatrix}\right). \qquad \square
\end{aligned}$$

Theorem 12.9.12 (Symmetries of Riemann Curvature Tensor). *If* $(M, \mathfrak{g}, \mathfrak{O})$ *is an oriented regular surface in* $\mathbb{R}^3_\nu$ *and* X, Y, Z, W *are vector fields in* $\mathfrak{X}(M)$, *then* $\mathcal{R}$ *satisfies the following symmetries:*
[**S1**] $\mathcal{R}(X, Y, Z, W) = -\mathcal{R}(Y, X, Z, W)$.
[**S2**] $\mathcal{R}(X, Y, Z, W) = -\mathcal{R}(X, Y, W, Z)$.
[**S3**] $\mathcal{R}(X, Y, Z, W) = \mathcal{R}(Z, W, X, Y)$.

Proof. This follows from Theorem 12.9.11. $\square$

Theorem 12.9.13 (First Bianchi Identity). *If* $(M, \mathfrak{g})$ *is a regular surface in* $\mathbb{R}^3_\nu$ *and* X, Y, Z *are vector fields in* $\mathfrak{X}(M)$, *then*

$$R(X,Y)Z + R(Y,Z)X + R(Z,X)Y = (0,0,0).$$

Proof. In local coordinates, let

$$X \circ \varphi = \sum_i \alpha^i H_i, \qquad Y \circ \varphi = \sum_j \beta^j H_j, \qquad \text{and} \qquad Z \circ \varphi = \sum_k \gamma^k H_k.$$

It follows from (12.9.2) that

$$\big(R(X,Y)Z + R(Y,Z)X + R(Z,X)Y\big) \circ \varphi$$
$$= \sum_l \bigg(\sum_{ijk} \alpha^i \beta^j \gamma^k (R^l_{ijk} + R^l_{jki} + R^l_{kij}) \bigg) H_l,$$

and from (12.9.3) that

$$R^l_{ijk} = \frac{\partial \Gamma^l_{jk}}{\partial r^i} - \frac{\partial \Gamma^l_{ik}}{\partial r^j} + \sum_n (\Gamma^l_{in}\Gamma^n_{jk} - \Gamma^l_{jn}\Gamma^n_{ik})$$

$$R^l_{jki} = \frac{\partial \Gamma^l_{ki}}{\partial r^j} - \frac{\partial \Gamma^l_{ji}}{\partial r^k} + \sum_n (\Gamma^l_{jn}\Gamma^n_{ki} - \Gamma^l_{kn}\Gamma^n_{ji})$$

$$R^l_{kij} = \frac{\partial \Gamma^l_{ij}}{\partial r^k} - \frac{\partial \Gamma^l_{kj}}{\partial r^i} + \sum_n (\Gamma^l_{kn}\Gamma^n_{ij} - \Gamma^l_{in}\Gamma^n_{kj}),$$

hence $R^l_{ijk} + R^l_{jki} + R^l_{kij} = 0$ for $i, j, k, l = 1, 2$. The result follows. $\square$

The name traditionally given to the next result is "Theorema Egregium", which is Latin for "remarkable theorem". The rationale for this impressive title is given below.

Theorem 12.9.14 (Theorema Egregium). *Let* $(M, \mathfrak{g}, \mathfrak{O})$ *be an oriented regular surface in* $\mathbb{R}^3_\nu$, *and let* p *be a point in* M.
(a) *If* (h_1, h_2) *is a basis for* $T_p(M)$, *then*

$$\mathcal{K}(p) = -\frac{\mathcal{R}_p(h_1, h_2, h_2, h_1)}{\mathcal{D}_p(h_1, h_2, h_2, h_1)}.$$

(b) *If* (e_1, e_2) *is an orthonormal basis for* $T_p(M)$, *then*

$$\mathcal{K}(p) = (-1)^{\mathrm{ind}(\mathfrak{g})} \mathcal{R}_p(e_1, e_2, e_2, e_1).$$

Proof. Let $q = \varphi^{-1}(p)$ and $(f_1, f_2) = (H_1|_q, H_2|_q)$.
  (a): It follows from Theorem 6.6.4 and Theorem 12.9.12 that

$$\frac{\mathcal{R}_p(h_1, h_2, h_2, h_1)}{\mathcal{D}_p(h_1, h_2, h_2, h_1)} = \frac{\mathcal{R}_p(f_1, f_2, f_2, f_1)}{\mathcal{D}_p(f_1, f_2, f_2, f_1)},$$

and from (12.8.8) and Theorem 12.9.11 that

$$\mathcal{R}_p(f_1, f_2, f_2, f_1) = -\epsilon_M \det\left(\begin{bmatrix} \mathfrak{h}_p(f_1, f_2) & \mathfrak{h}_p(f_1, f_1) \\ \mathfrak{h}_p(f_2, f_2) & \mathfrak{h}_p(f_2, f_1) \end{bmatrix}\right)$$

$$= \epsilon_M \det\left(\begin{bmatrix} \mathfrak{h}_{11}(q) & \mathfrak{h}_{12}(q) \\ \mathfrak{h}_{21}(q) & \mathfrak{h}_{22}(q) \end{bmatrix}\right) = \epsilon_M \det\left(\mathfrak{h}_{\mathcal{H}}(q)\right).$$

We also have

$$\mathcal{D}_p(f_1, f_2, f_2, f_1) = -\det\left(\begin{bmatrix} \mathfrak{g}_{11}(q) & \mathfrak{g}_{12}(q) \\ \mathfrak{g}_{21}(q) & \mathfrak{g}_{22}(q) \end{bmatrix}\right) = -\det\left(\mathfrak{g}_{\mathcal{H}}(q)\right).$$

Combining the above identities yields

$$\frac{\mathcal{R}_p(h_1, h_2, h_2, h_1)}{\mathcal{D}_p(h_1, h_2, h_2, h_1)} = -\epsilon_M \frac{\det\left(\mathfrak{h}_{\mathcal{H}}(q)\right)}{\det\left(\mathfrak{g}_{\mathcal{H}}(q)\right)}.$$

The result now follows from Theorem 12.8.4.

(b): By (4.2.3) and (12.9.20),

$$\mathcal{D}_p(e_1, e_2, e_2, e_1) = -\langle e_1, e_1\rangle\langle e_2, e_2\rangle = (-1)^{\mathrm{ind}(\mathfrak{g})+1}.$$

The result now follows from part (a). $\square$

As remarked earlier, the Riemann curvature is intrinsic, whether we are dealing with R or $\mathcal{R}$. The Gauss curvature is defined using the Gauss map, which in turn is defined using the second fundamental form. For this reason, it would appear that the Gauss curvature depends on factors that are "external". However, part (b) of the Theorema Egregium shows that the Gauss curvature is in fact intrinsic, something that is unexpected and indeed "remarkable". The next result makes the same point using local coordinates.

Theorem 12.9.15. *Let* $(M, \mathfrak{g}, \mathfrak{A}, \mathfrak{D})$ *be an oriented regular surface in* $\mathbb{R}^3_\nu$, *and let* p *be a point in* M. *Then:*
(a)
$$\mathcal{R}_p(v_1, v_2)v_3 = \mathcal{K}(p)\left[\langle v_2, v_3\rangle v_1 - \langle v_1, v_3\rangle v_2\right]$$

for all vectors v_1, v_2, v_3 *in* $T_p(M)$.
(b) *If* (U, φ) *is a chart in* $\mathfrak{A}$ *at* p *and* $q = \varphi^{-1}(p)$, *then, in local coordinates,*

$$R^l_{ijk}(p) = \mathcal{K}(p)\left[\mathfrak{g}_{jk}(q)\delta^l_i - \mathfrak{g}_{ik}(q)\delta^l_j\right]$$

for $i, j, k, l = 1, 2$.

Proof. (a): We recall from Section 6.6 that $\mathcal{S}(T_p(M))$ is the subspace of $T^0_4(T_p(M))$ consisting of tensors satisfying [S1] and [S2]. By Theorem 6.6.3(a), $\mathcal{D}_p$ is such a tensor, and by parts (a) and (b) of Theorem 12.9.12, so is $\mathcal{R}_p$.

It follows from Theorem 6.6.3(b) that $\mathcal{R}_p = c\mathcal{D}_p$ for some real number c. Let (e_1, e_2) be an orthonormal basis for $T_p(M)$. We have

$$
\begin{aligned}
(-1)^{\mathrm{ind}(\mathfrak{g})}\mathcal{K}(p) &= \mathcal{R}_p(e_1, e_2, e_2, e_1) && \text{[Th 12.9.14(b)]} \\
&= c\mathcal{D}_p(e_1, e_2, e_2, e_1) \\
&= -c\langle e_1, e_1\rangle\langle e_2, e_2\rangle && \text{[(12.9.20)]} \\
&= c(-1)^{\mathrm{ind}(\mathfrak{g})+1}, && \text{[(4.2.3)]}
\end{aligned}
$$

hence $c = -\mathcal{K}(p)$, so $\mathcal{R}_p = -\mathcal{K}(p)\mathcal{D}_p$. Then (12.9.20) gives

$$\langle R_p(v_1, v_2)v_3, v_4\rangle = \mathcal{K}(p)\left\langle \langle v_2, v_3\rangle v_1 - \langle v_1, v_3\rangle v_2, v_4\right\rangle$$

for all vectors v_1, v_2, v_3, v_4 in $T_p(M)$. Since $\mathfrak{g}$ is nondegenerate,

$$R_p(v_1, v_2)v_3 = \mathcal{K}(p)\left[\langle v_2, v_3\rangle v_1 - \langle v_1, v_3\rangle v_2\right]$$

for all vectors v_1, v_2, v_3 in $T_p(M)$.

(b): Let (H_1, H_2) be the coordinate frame corresponding to (U, φ). We have from part (a) that

$$
\begin{aligned}
R_p(H_i|_q, H_j|_q)H_k|_q &= \mathcal{K}(p)\left[\langle H_j|_q, H_k|_q\rangle H_i|_q - \langle H_i|_q, H_k|_q\rangle H_j|_q\right] \\
&= \mathcal{K}(p)\left[\mathfrak{g}_{jk}(q)H_i|_q - \mathfrak{g}_{ik}(q)H_j|_q\right].
\end{aligned}
$$

On the other hand, (12.9.2) and (12.9.19) give

$$R_p(H_i|_q, H_j|_q)H_k|_q = \sum_l R^l_{ijk}(p)H_l|_q.$$

It follows that

$$
R^l_{ijk}(p) = \begin{cases}
\mathcal{K}(p)\,\mathfrak{g}_{jk}(q) & \text{if } l = i \\
-\mathcal{K}(p)\,\mathfrak{g}_{ik}(q) & \text{if } l = j \\
0 & \text{otherwise}
\end{cases}
$$

$$= \mathcal{K}(p)\left[\mathfrak{g}_{jk}(q)\delta^l_i - \mathfrak{g}_{ik}(q)\delta^l_j\right]$$

for $i, j, k, l = 1, 2$. $\qquad\qquad\qquad\qquad\qquad\qquad\qquad\qquad\qquad\qquad\square$

12.10   Computations for Regular Surfaces in $\mathbb{R}^3_\nu$

We showed in Theorem 11.4.2 and Theorem 11.4.3 that graphs of surfaces and surfaces of revolution are regular surfaces. In this section, we view them as regular surfaces in $\mathbb{R}^3_\nu$ and develop specific formulas for computing the coordinate frame, Gauss map, first and second fundamental forms, Gauss curvature, and sign. For surfaces of revolution in $\mathbb{R}^3_0$, formulas for the Christoffel symbols and eigenvalues are also provided.

Theorem 12.10.1 (Graph of Function in $\mathbb{R}^3_0$). *Let $\nu = 0$ and continue with the setup of Theorem 11.4.2. Then:*

(a)
$$H_1 = \left(1, 0, \frac{\partial f}{\partial x}\right) \qquad H_2 = \left(0, 1, \frac{\partial f}{\partial y}\right).$$

(b)
$$\mathcal{N} \circ \varphi = \frac{1}{\sqrt{1 + \left(\dfrac{\partial f}{\partial x}\right)^2 + \left(\dfrac{\partial f}{\partial y}\right)^2}} \left(-\frac{\partial f}{\partial x}, -\frac{\partial f}{\partial y}, 1\right).$$

(c)
$$\mathfrak{g}_{\mathcal{H}} = \begin{bmatrix} 1 + \left(\dfrac{\partial f}{\partial x}\right)^2 & \dfrac{\partial f}{\partial x}\dfrac{\partial f}{\partial y} \\[3mm] \dfrac{\partial f}{\partial x}\dfrac{\partial f}{\partial y} & 1 + \left(\dfrac{\partial f}{\partial y}\right)^2 \end{bmatrix}.$$

(d)
$$\mathfrak{h}_{\mathcal{H}} = \frac{1}{\sqrt{1 + \left(\dfrac{\partial f}{\partial x}\right)^2 + \left(\dfrac{\partial f}{\partial y}\right)^2}} \begin{bmatrix} \dfrac{\partial^2 f}{\partial x^2} & \dfrac{\partial^2 f}{\partial x \partial y} \\[3mm] \dfrac{\partial^2 f}{\partial x \partial y} & \dfrac{\partial^2 f}{\partial y^2} \end{bmatrix}.$$

(e)
$$\mathcal{K} \circ \varphi = \frac{\dfrac{\partial^2 f}{\partial x^2}\dfrac{\partial^2 f}{\partial y^2} - \left(\dfrac{\partial^2 f}{\partial x \partial y}\right)^2}{\left[1 + \left(\dfrac{\partial f}{\partial x}\right)^2 + \left(\dfrac{\partial f}{\partial y}\right)^2\right]^2}.$$

(f)
$$\epsilon_{\text{graph}(f)} = 1.$$

Proof. (a): Straightforward.

(b): This follows from Theorem 12.2.11(c), Theorem 12.7.3(a), part (f), and the computations

$$H_1 \times H_2 = \det\left(\begin{bmatrix} e_1 & 1 & 0 \\ e_2 & 0 & 1 \\ e_3 & \dfrac{\partial f}{\partial x} & \dfrac{\partial f}{\partial y} \end{bmatrix}\right) = \left(-\frac{\partial f}{\partial x}, -\frac{\partial f}{\partial y}, 1\right)$$

and

$$\|H_1 \times H_2\| = \sqrt{1 + \left(\frac{\partial f}{\partial x}\right)^2 + \left(\frac{\partial f}{\partial y}\right)^2}.$$

(c): Straightforward.

(d): We have, for example,

$$\frac{\partial H_1}{\partial y} = \left(0, 0, \frac{\partial^2 f}{\partial x \partial y}\right).$$

By Theorem 12.8.2(a) and part (b),

$$\mathfrak{h}_{12} = \left\langle \frac{\partial H_1}{\partial y}, \mathcal{N} \circ \varphi \right\rangle = \frac{\dfrac{\partial^2 f}{\partial x \partial y}}{\sqrt{1 + \left(\dfrac{\partial f}{\partial x}\right)^2 + \left(\dfrac{\partial f}{\partial y}\right)^2}}.$$

(e): By part (c),

$$\det(\mathfrak{g}_\mathcal{H}) = 1 + \left(\frac{\partial f}{\partial x}\right)^2 + \left(\frac{\partial f}{\partial y}\right)^2,$$

and by part (d),

$$\det(\mathfrak{h}_\mathcal{H}) = \frac{\dfrac{\partial^2 f}{\partial x^2} \dfrac{\partial^2 f}{\partial y^2} - \left(\dfrac{\partial^2 f}{\partial x \partial y}\right)^2}{1 + \left(\dfrac{\partial f}{\partial x}\right)^2 + \left(\dfrac{\partial f}{\partial y}\right)^2}.$$

The result now follows from Theorem 12.8.4 and part (f).

(f): This follows from $\nu = 0$ and (12.2.5). $\qquad\qquad\square$

Theorem 12.10.2 (Graph of Function in $\mathbb{R}^3_1$). *Let* $\nu = 1$, *continue with the setup of Theorem 11.4.2, and suppose* graph(f) *is a regular surface in* $\mathbb{R}^3_1$. *Then:*

(a)

$$H_1 = \left(1, 0, \frac{\partial f}{\partial x}\right) \qquad H_2 = \left(0, 1, \frac{\partial f}{\partial y}\right).$$

(b)

$$\mathcal{N} \circ \varphi = -\epsilon_{\text{graph}(f)} \frac{1}{\sqrt{\left|1 - \left(\dfrac{\partial f}{\partial x}\right)^2 - \left(\dfrac{\partial f}{\partial y}\right)^2\right|}} \left(\frac{\partial f}{\partial x}, \frac{\partial f}{\partial y}, 1\right),$$

where $\epsilon_{\text{graph}(f)}$ *is given by part (f).*

(c)

$$\mathfrak{g}_\mathcal{H} = \begin{bmatrix} 1 - \left(\dfrac{\partial f}{\partial x}\right)^2 & -\dfrac{\partial f}{\partial x}\dfrac{\partial f}{\partial y} \\[3mm] -\dfrac{\partial f}{\partial x}\dfrac{\partial f}{\partial y} & 1 - \left(\dfrac{\partial f}{\partial y}\right)^2 \end{bmatrix}.$$

(d)

$$\mathfrak{h}_{\mathcal{H}} = \epsilon_{\mathrm{graph}(f)} \frac{1}{\sqrt{\left|1 - \left(\dfrac{f}{\partial x}\right)^2 - \left(\dfrac{\partial f}{\partial y}\right)^2\right|}} \begin{bmatrix} \dfrac{\partial^2 f}{\partial x^2} & \dfrac{\partial^2 f}{\partial x \partial y} \\[2mm] \dfrac{\partial^2 f}{\partial x \partial y} & \dfrac{\partial^2 f}{\partial y^2} \end{bmatrix}.$$

(e)

$$K \circ \varphi = - \frac{\dfrac{\partial^2 f}{\partial x^2}\dfrac{\partial^2 f}{\partial y^2} - \left(\dfrac{\partial^2 f}{\partial x \partial y}\right)^2}{\left[1 - \left(\dfrac{\partial f}{\partial x}\right)^2 - \left(\dfrac{\partial f}{\partial y}\right)^2\right]^2}.$$

(f)

$$\epsilon_{\mathrm{graph}(f)} = -\mathrm{sgn}\left(1 - \left(\dfrac{\partial f}{\partial x}\right)^2 - \left(\dfrac{\partial f}{\partial y}\right)^2\right).$$

Proof. (a): Straightforward.

(b): By assumption, $\mathrm{graph}(f)$ is a regular surface in $\mathbb{R}^3_1$. The result follows from Theorem 12.2.11(c), Theorem 12.7.3(a), part (f), and the computations

$$H_1 \times H_2 = \det\left(\begin{bmatrix} e_1 & 1 & 1 \\ e_2 & 0 & 1 \\ -e_3 & \dfrac{\partial f}{\partial x} & \dfrac{\partial f}{\partial y} \end{bmatrix}\right) = -\left(\dfrac{\partial f}{\partial x}, \dfrac{\partial f}{\partial y}, 1\right)$$

and

$$\|H_1 \times H_2\| = \sqrt{\left|1 - \left(\dfrac{\partial f}{\partial x}\right)^2 - \left(\dfrac{\partial f}{\partial y}\right)^2\right|}.$$

(c): Straightforward.

(d): We have, for example,

$$\frac{\partial H_1}{\partial y} = \left(0, 0, \dfrac{\partial^2 f}{\partial x \partial y}\right).$$

By Theorem 12.8.2(a) and part (b),

$$\mathfrak{h}_{12} = \left\langle \frac{\partial H_1}{\partial y}, \mathcal{N} \circ \varphi \right\rangle$$

$$= \epsilon_{\text{graph}(f)} \frac{1}{\sqrt{\left| 1 - \left(\frac{\partial f}{\partial x}\right)^2 - \left(\frac{\partial f}{\partial y}\right)^2 \right|}} \left\langle \left(0, 0, \frac{\partial^2 f}{\partial x \partial y}\right), -\left(\frac{\partial f}{\partial x}, \frac{\partial f}{\partial y}, 1\right) \right\rangle$$

$$= \epsilon_{\text{graph}(f)} \frac{\dfrac{\partial^2 f}{\partial x \partial y}}{\sqrt{\left| 1 - \left(\frac{\partial f}{\partial x}\right)^2 - \left(\frac{\partial f}{\partial y}\right)^2 \right|}}.$$

(e): By part (c),

$$\det(\mathfrak{g}_{\mathcal{H}}) = 1 - \left(\frac{\partial f}{\partial x}\right)^2 - \left(\frac{\partial f}{\partial y}\right)^2,$$

and by part (d),

$$\det(\mathfrak{h}_{\mathcal{H}}) = \frac{\dfrac{\partial^2 f}{\partial x^2}\dfrac{\partial^2 f}{\partial y^2} - \left(\dfrac{\partial^2 f}{\partial x \partial y}\right)^2}{\left| 1 - \left(\dfrac{\partial f}{\partial x}\right)^2 - \left(\dfrac{\partial f}{\partial y}\right)^2 \right|}.$$

The result now follows from Theorem 12.8.4 and part (f).

(f): We have

$$\langle H_1 \times H_2, H_1 \times H_2 \rangle = \left(\frac{\partial f}{\partial x}\right)^2 + \left(\frac{\partial f}{\partial y}\right)^2 - 1.$$

Then Theorem 12.2.11(b) gives

$$\epsilon_{\text{graph}(f)} = \text{sgn}\left(\left(\frac{\partial f}{\partial x}\right)^2 + \left(\frac{\partial f}{\partial y}\right)^2 - 1\right) = -\text{sgn}\left(1 - \left(\frac{\partial f}{\partial x}\right)^2 - \left(\frac{\partial f}{\partial y}\right)^2\right).$$

$\square$

Theorem 12.10.3 (Surface of Revolution in $\mathbb{R}^3_0$). *Let $\nu = 0$ and continue with the setup of Theorem 11.4.3. Also, denote derivatives with respect to t by an overdot; let the subscripts 1 and 2 refer to t and ϕ, respectively; and for brevity, drop t and ϕ from the notation (except for ϕ in the case of the trigonometric functions). Then:*

(a)

$$H_1 = \left(\dot{\rho}\cos(\phi), \dot{\rho}\sin(\phi), \dot{h}\right) \qquad H_2 = \left(-\rho\sin(\phi), \rho\cos(\phi), 0\right).$$

(b)
$$\mathcal{N} \circ \varphi = \frac{1}{\sqrt{\dot{\rho}^2 + \dot{h}^2}} \left(-\dot{h} \cos(\phi), -\dot{h} \sin(\phi), \dot{\rho} \right).$$

(c)
$$\mathfrak{g}_\mathcal{H} = \begin{bmatrix} \dot{\rho}^2 + \dot{h}^2 & 0 \\ 0 & \rho^2 \end{bmatrix}.$$

(d)
$$\mathfrak{h}_\mathcal{H} = \frac{1}{\sqrt{\dot{\rho}^2 + \dot{h}^2}} \begin{bmatrix} \dot{\rho}\ddot{h} - \ddot{\rho}\dot{h} & 0 \\ 0 & \rho\dot{h} \end{bmatrix}.$$

(e)
$$K \circ \varphi = \frac{\dot{h}(\dot{\rho}\ddot{h} - \ddot{\rho}\dot{h})}{\rho(\dot{\rho}^2 + \dot{h}^2)^2}.$$

(f)
$$\epsilon_{\mathrm{rev}(\sigma)} = 1.$$

(g)
$$\Gamma^1_{11} = \frac{\dot{\rho}\ddot{\rho} + \dot{h}\ddot{h}}{\dot{\rho}^2 + \dot{h}^2} \qquad \Gamma^1_{12} = \Gamma^1_{21} = 0 \qquad \Gamma^1_{22} = -\frac{\rho\dot{\rho}}{\dot{\rho}^2 + \dot{h}^2}$$

$$\Gamma^2_{11} = 0 \qquad \Gamma^2_{12} = \Gamma^2_{21} = \frac{\dot{\rho}}{\rho} \qquad \Gamma^2_{22} = 0.$$

(h) *The eigenvalues of* $\mathcal{W}$ *are*

$$\kappa_1 = \frac{\dot{\rho}\ddot{h} - \ddot{\rho}\dot{h}}{(\dot{\rho}^2 + \dot{h}^2)^{3/2}} \qquad \kappa_2 = \frac{\dot{h}}{\rho\sqrt{\dot{\rho}^2 + \dot{h}^2}},$$

and H_1 *and* H_2 *are eigenvectors corresponding to* κ_1 *and* κ_2, *respectively.*

If $\dot{\rho}^2 + \dot{h}^2 = 1$ *for all* t *in* (a, b), *then:*

(b')
$$\mathcal{N} \circ \varphi = \left(-\dot{h} \cos(\phi), -\dot{h} \sin(\phi), \dot{\rho} \right).$$

(c')
$$\mathfrak{g}_\mathcal{H} = \begin{bmatrix} 1 & 0 \\ 0 & \rho^2 \end{bmatrix}.$$

(d')
$$\mathfrak{h}_\mathcal{H} = \begin{bmatrix} \dot{\rho}\ddot{h} - \ddot{\rho}\dot{h} & 0 \\ 0 & \rho\dot{h} \end{bmatrix}.$$

(e')
$$K \circ \varphi = -\frac{\ddot{\rho}}{\rho}.$$

(g')

$$\Gamma^1_{11} = \dot\rho\ddot\rho + h\ddot h \qquad \Gamma^1_{12} = \Gamma^1_{21} = 0 \qquad \Gamma^1_{22} = -\rho\dot\rho$$

$$\Gamma^2_{11} = 0 \qquad \Gamma^2_{12} = \Gamma^2_{21} = \frac{\dot\rho}{\rho} \qquad \Gamma^2_{22} = 0.$$

(h') *The eigenvalues of* $\mathcal{W}$ *are*

$$\kappa_1 \circ \varphi = \dot\rho\ddot h - \ddot\rho\dot h \qquad \kappa_2 \circ \varphi = \frac{\dot h}{\rho}.$$

Proof. (a): Straightforward.

(b): This follows from Theorem 12.2.11(c), Theorem 12.7.3(a), part (f), and the computations

$$H_1 \times H_2 = \det\left(\begin{bmatrix} e_1 & \dot\rho\cos(\phi) & -\rho\sin(\phi) \\ e_2 & \dot\rho\sin(\phi) & \rho\cos(\phi) \\ e_3 & \dot h & 0 \end{bmatrix}\right) = \rho\big(-\dot h\cos(\phi), -\dot h\sin(\phi), \dot\rho\big)$$

and

$$\|H_1 \times H_2\| = \rho\sqrt{\dot\rho^2 + \dot h^2}.$$

(c): Straightforward.

(d): We have

$$\frac{\partial H_1}{\partial t} = \big(\ddot\rho\cos(\phi), \ddot\rho\sin(\phi), \ddot h\big)$$

$$\frac{\partial H_1}{\partial \phi} = \big(-\dot\rho\sin(\phi), \dot\rho\cos(\phi), 0\big) = \frac{\partial H_2}{\partial t}$$

$$\frac{\partial H_2}{\partial \phi} = \big(-\rho\cos(\phi), -\rho\sin(\phi), 0\big).$$

By Theorem 12.8.2(a) and part (b),

$$\mathfrak{h}_{11} = \frac{\dot\rho\ddot h - \ddot\rho\dot h}{\sqrt{\dot\rho^2 + \dot h^2}} \qquad \mathfrak{h}_{12} = \mathfrak{h}_{12} = 0 \qquad \mathfrak{h}_{22} = \frac{\rho\dot h}{\sqrt{\dot\rho^2 + \dot h^2}}.$$

(e): By part (c),

$$\det(\mathfrak{g}_{\mathcal{H}}) = \rho^2(\dot\rho^2 + \dot h^2),$$

and by part (d),

$$\det(\mathfrak{h}_{\mathcal{H}}) = \frac{\rho\dot h(\dot\rho\ddot h - \ddot\rho\dot h)}{\dot\rho^2 + \dot h^2}.$$

The result now follows from Theorem 12.8.4 and part (f).

(f): This follows from $\nu = 0$ and (12.2.5).

(g): We make repeated use of part (c). From (12.3.7),

$$\frac{\partial \mathfrak{g}_{ij}}{\partial t} = \mathfrak{g}_{j1}\Gamma^1_{i1} + \mathfrak{g}_{i1}\Gamma^1_{j1} + \mathfrak{g}_{j2}\Gamma^2_{i1} + \mathfrak{g}_{i2}\Gamma^2_{j1}$$

$$\frac{\partial \mathfrak{g}_{ij}}{\partial \phi} = \mathfrak{g}_{j1}\Gamma^1_{i2} + \mathfrak{g}_{i1}\Gamma^1_{j2} + \mathfrak{g}_{j2}\Gamma^2_{i2} + \mathfrak{g}_{i2}\Gamma^2_{j2}$$

for $i, j = 1, 2$. For $(i, j; k) = (1, 1; 1)$,

$$2(\dot{\rho}\ddot{\rho} + \dot{h}\ddot{h}) = \frac{\partial \mathfrak{g}_{11}}{\partial t} = \mathfrak{g}_{11}\Gamma^1_{11} + \mathfrak{g}_{11}\Gamma^1_{11} + \mathfrak{g}_{12}\Gamma^2_{11} + \mathfrak{g}_{12}\Gamma^2_{11}$$
$$= 2(\dot{\rho}^2 + \dot{h}^2)\Gamma^1_{11},$$

hence

$$\Gamma^1_{11} = \frac{\dot{\rho}\ddot{\rho} + \dot{h}\ddot{h}}{\dot{\rho}^2 + \dot{h}^2}.$$

For $(i, j; k) = (1, 2; 1)$,

$$0 = \frac{\partial \mathfrak{g}_{12}}{\partial t} = \mathfrak{g}_{21}\Gamma^1_{11} + \mathfrak{g}_{11}\Gamma^1_{21} + \mathfrak{g}_{22}\Gamma^2_{11} + \mathfrak{g}_{12}\Gamma^2_{21}$$
$$= (\dot{\rho}^2 + \dot{h}^2)\Gamma^1_{21} + \rho^2\Gamma^2_{11}. \tag{12.10.1}$$

For $(i, j; k) = (2, 2; 1)$,

$$2\rho\dot{\rho} = \frac{\partial \mathfrak{g}_{22}}{\partial t} = \mathfrak{g}_{21}\Gamma^1_{21} + \mathfrak{g}_{21}\Gamma^1_{21} + \mathfrak{g}_{22}\Gamma^2_{21} + \mathfrak{g}_{22}\Gamma^2_{21}$$
$$= 2\rho^2\Gamma^2_{21},$$

hence

$$\Gamma^2_{21} = \frac{\dot{\rho}}{\rho} = \Gamma^2_{12}. \tag{12.10.2}$$

For $(i, j; k) = (1, 1; 2)$,

$$0 = \frac{\partial \mathfrak{g}_{11}}{\partial \phi} = \mathfrak{g}_{11}\Gamma^1_{12} + \mathfrak{g}_{11}\Gamma^1_{12} + \mathfrak{g}_{12}\Gamma^2_{12} + \mathfrak{g}_{12}\Gamma^2_{12}$$
$$= 2(\dot{\rho}^2 + \dot{h}^2)\Gamma^1_{12},$$

hence

$$\Gamma^1_{12} = 0 = \Gamma^1_{21}. \tag{12.10.3}$$

For $(i, j; k) = (1, 2; 2)$,

$$0 = \frac{\partial \mathfrak{g}_{12}}{\partial \phi} = \mathfrak{g}_{21}\Gamma^1_{12} + \mathfrak{g}_{11}\Gamma^1_{22} + \mathfrak{g}_{22}\Gamma^2_{12} + \mathfrak{g}_{12}\Gamma^2_{22}$$
$$= (\dot{\rho}^2 + \dot{h}^2)\Gamma^1_{22} + \rho^2\Gamma^2_{12}. \tag{12.10.4}$$

For $(i, j; k) = (2, 2; 2)$,

$$0 = \frac{\partial \mathfrak{g}_{22}}{\partial \phi} = \mathfrak{g}_{21}\Gamma^1_{22} + \mathfrak{g}_{21}\Gamma^1_{22} + \mathfrak{g}_{22}\Gamma^2_{22} + \mathfrak{g}_{22}\Gamma^2_{22}$$
$$= 2\rho^2\Gamma^2_{22},$$

hence

$$\Gamma^2_{22} = 0.$$

It follows from (12.10.1) and (12.10.3) that

$$\Gamma^2_{11} = 0,$$

and from (12.10.2) and (12.10.4) that

$$\Gamma^1_{22} = -\frac{\rho\dot\rho}{\dot\rho^2 + \dot{h}^2}.$$

(h): We have from part (c) that

$$\mathfrak{g}_{\mathcal{H}}^{-1} = \frac{1}{\rho^2(\dot\rho^2 + \dot{h}^2)}\begin{bmatrix} \rho^2 & 0 \\ 0 & \dot\rho^2 + \dot{h}^2 \end{bmatrix},$$

and then from part (d) and Theorem 12.8.3(b) that

$$[\mathcal{W}]^{\mathcal{H}}_{\mathcal{H}} = \mathfrak{g}_{\mathcal{H}}^{-1}\mathfrak{h}_{\mathcal{H}} = \frac{1}{\rho(\dot\rho^2 + \dot{h}^2)^{3/2}}\begin{bmatrix} \rho(\dot\rho\ddot{h} - \ddot\rho\dot{h}) & 0 \\ 0 & \dot{h}(\dot\rho^2 + \dot{h}^2) \end{bmatrix}.$$

Thus,

$$\mathcal{W}(H_1) = \frac{\dot\rho\ddot{h} - \ddot\rho\dot{h}}{(\dot\rho^2 + \dot{h}^2)^{3/2}}\,H_1 \qquad \mathcal{W}(H_2) = \frac{\dot{h}}{\rho\sqrt{\dot\rho^2 + \dot{h}^2}}\,H_2,$$

from which the result follows.

(b′)–(g′): Parts (b′)–(d′), (g′), and (h′) follow from $\dot\rho^2 + \dot{h}^2 = 1$. For part (e′), differentiating both sides of the preceding identity with respect to t gives $\dot\rho\ddot\rho = -\dot{h}\ddot{h}$, hence $\dot{h}(\dot\rho\ddot{h} - \ddot\rho\dot{h}) = -\ddot\rho$.   □

Theorem 12.10.4 (Surface of Revolution in $\mathbb{R}^3_1$). *Let $\nu = 1$, continue with the setup of Theorem 11.4.3, and suppose* rev(σ) *is a regular surface in $\mathbb{R}^3_1$. Also, denote derivatives with respect to t by an overdot; let the subscripts 1 and 2 refer to t and ϕ, respectively; and for brevity, drop t and ϕ from the notation (except for ϕ in the case of trigonometric functions). Then:*
(a)

$$H_1 = \big(\dot\rho\cos(\phi), \dot\rho\sin(\phi), \dot{h}\big) \qquad H_2 = \big(-\rho\sin(\phi), \rho\cos(\phi), 0\big).$$

(b)

$$\mathcal{N} \circ \varphi = -\epsilon_{\text{rev}(\sigma)}\frac{1}{\sqrt{|\dot\rho^2 - \dot{h}^2|}}\big(\dot{h}\cos(\phi), \dot{h}\sin(\phi), \dot\rho\big),$$

where $\epsilon_{\text{rev}(\sigma)}$ is given by part (f).
(c)

$$\mathfrak{g}_{\mathcal{H}} = \begin{bmatrix} \dot\rho^2 - \dot{h}^2 & 0 \\ 0 & \rho^2 \end{bmatrix}.$$

(d)

$$\mathfrak{h}_{\mathcal{H}} = \epsilon_{\text{rev}(\sigma)}\frac{1}{\sqrt{|\dot\rho^2 - \dot{h}^2|}}\begin{bmatrix} \dot\rho\ddot{h} - \ddot\rho\dot{h} & 0 \\ 0 & \rho\dot{h} \end{bmatrix}.$$

(e)
$$\mathcal{K} \circ \varphi = -\frac{\dot{h}(\dot{\rho}\ddot{h} - \ddot{\rho}\dot{h})}{\rho(\dot{\rho}^2 - \dot{h}^2)^2}.$$

(f)
$$\epsilon_{\text{rev}(\sigma)} = -\text{sgn}(\dot{\rho}^2 - \dot{h}^2).$$

If $\dot{\rho}^2 - \dot{h}^2 = 1$ for all t in (a, b), then:

(b')
$$\mathcal{N} \circ \varphi = (\dot{h}\cos(\phi), \dot{h}\sin(\phi), \dot{\rho}).$$

(c')
$$g_{\mathcal{H}} = \begin{bmatrix} 1 & 0 \\ 0 & \rho^2 \end{bmatrix}.$$

(d')
$$\mathfrak{h}_{\mathcal{H}} = -\begin{bmatrix} \dot{\rho}\ddot{h} - \ddot{\rho}\dot{h} & 0 \\ 0 & \rho\dot{h} \end{bmatrix}.$$

(e')
$$\mathcal{K} \circ \varphi = -\frac{\ddot{\rho}}{\rho}.$$

(f')
$$\epsilon_{\text{rev}(\sigma)} = -1.$$

Proof. (a): Straightforward.

(b): By assumption, $\text{rev}(\sigma)$ is a regular surface in $\mathbb{R}^3_1$. The result follows from Theorem 12.2.11(c), Theorem 12.7.3(a), part (f), and the computations

$$H_1 \times H_2 = \det\left(\begin{bmatrix} e_1 & \dot{\rho}\cos(\phi) & -\rho\sin(\phi) \\ e_2 & \dot{\rho}\sin(\phi) & \rho\cos(\phi) \\ -e_3 & \dot{h} & 0 \end{bmatrix}\right) = -\rho(\dot{h}\cos(\phi), \dot{h}\sin(\phi), \dot{\rho})$$

and

$$\|H_1 \times H_2\| = \rho\sqrt{|\dot{\rho}^2 - \dot{h}^2|}.$$

(c): Straightforward.

(d): We have

$$\frac{\partial H_1}{\partial t} = (\ddot{\rho}\cos(\phi), \ddot{\rho}\sin(\phi), \ddot{h})$$

$$\frac{\partial H_1}{\partial \phi} = (-\dot{\rho}\sin(\phi), \dot{\rho}\cos(\phi), 0) = \frac{\partial H_2}{\partial t}$$

$$\frac{\partial H_2}{\partial \phi} = (-\rho\cos(\phi), -\rho\sin(\phi), 0).$$

By Theorem 12.8.2(a) and part (b),

$$\mathfrak{h}_{11} = \epsilon_{\text{rev}(\sigma)} \frac{\dot{\rho}\ddot{h} - \ddot{\rho}\dot{h}}{\sqrt{|\dot{\rho}^2 - \dot{h}^2|}} \qquad \mathfrak{h}_{12} = \mathfrak{h}_{12} = 0 \qquad \mathfrak{h}_{22} = \epsilon_{\text{rev}(\sigma)} \frac{\dot{\rho}\dot{h}}{\sqrt{|\dot{\rho}^2 - \dot{h}^2|}}.$$

(e): By part (c),

$$\det(\mathfrak{g}_{\mathcal{H}}) = \rho^2(\dot{\rho}^2 - \dot{h}^2),$$

and by part (d),

$$\det(\mathfrak{h}_{\mathcal{H}}) = \frac{\dot{\rho}\dot{h}(\dot{\rho}\ddot{h} - \ddot{\rho}\dot{h})}{|\dot{\rho}^2 - \dot{h}^2|}.$$

The result now follows from Theorem 12.8.4 and part (f).

(f): We have

$$\langle H_1 \times H_2, H_1 \times H_2 \rangle = \rho^2(\dot{h}^2 - \dot{\rho}^2).$$

Since ρ has only positive values, Theorem 12.2.11(b) gives

$$\epsilon_{\text{rev}(\sigma)} = \text{sgn}(\dot{h}^2 - \dot{\rho}^2) = -\text{sgn}(\dot{\rho}^2 - \dot{h}^2).$$

(b')–(e'): Parts (b')–(d') and (f') follow from $\dot{\rho}^2 - \dot{h}^2 = 1$. For part (e'), differentiating both sides of the preceding identity with respect to t yields $\dot{\rho}\ddot{\rho} = \dot{h}\ddot{h}$, hence $\dot{h}(\dot{\rho}\ddot{h} - \ddot{\rho}\dot{h}) = \ddot{\rho}$. $\qquad\square$

Chapter 13

Examples of Regular Surfaces

This chapter provides worked examples of graphs of functions in $\mathbb{R}^3_0$ and $\mathbb{R}^3_1$, and surfaces of revolution in $\mathbb{R}^3_1$. The details of computations, which are not included, are based on formulas appearing in Theorems 12.10.1–12.10.4. In this chapter, we identify $\mathbb{R}^2$ with the xy-plane in $\mathbb{R}^3$. See Example 12.8.6 for a summary of Gauss curvatures as well as related comments.

Each of the regular surfaces to be considered can be parametrized as either the graph of a function or a surface of revolution. The former approach has the advantage that the regular surface can be depicted literally as a graph in $\mathbb{R}^3$. On the other hand, when symmetries are present, the surface of revolution parametrization can be quite revealing and computationally convenient. The choice of parametrization made here is somewhat arbitrary. There is a small issue that differentiates the two computational methods. Parameterizing a regular surface as a surface of revolution leaves out certain points compared with the corresponding parametrization as the graph of a function; more specifically, with the former approach, part of a longitude curve is "missing". Since we are interested exclusively in local aspects of regular surfaces, in particular, the Gauss curvature, this is not a concern and will not be discussed further.

13.1   Plane in $\mathbb{R}^3_0$

The set

$$\text{Pln} = \{(x, y, z) \in \mathbb{R}^3 : z = 0\}$$

Semi-Riemannian Geometry, First Edition. Stephen C. Newman.
© 2019 John Wiley & Sons, Inc. Published 2019 by John Wiley & Sons, Inc.

is the xy-plane in $\mathbb{R}^3$. In the notation of Theorem 11.4.2, let

$$f(x, y) = 0$$
$$\varphi(x, y) = (x, y, 0)$$
$$U = \mathbb{R}^2.$$

By Theorem 11.4.2, Pln is a regular surface and (U, φ) is a chart. Viewing Pln as a regular surface in $\mathbb{R}^3_0$, Theorem 12.10.1 gives:

$$H_1 = (1, 0, 0) \qquad H_2 = (0, 1, 0)$$
$$\mathcal{N} \circ \varphi = (0, 0, 1)$$
$$\mathfrak{g}\mathcal{H} = \begin{bmatrix} 1 & 0 \\ 0 & 1 \end{bmatrix}$$
$$\mathfrak{h}\mathcal{H} = \begin{bmatrix} 0 & 0 \\ 0 & 0 \end{bmatrix}$$
$$\mathcal{K} \circ \varphi = 0.$$

13.2   Cylinder in $\mathbb{R}^3_0$

The set

$$\mathrm{Cyl} = \{(x, y, z) \in \mathbb{R}^3 : x^2 + y^2 = 1, z > 0\}$$

is an infinite cylinder standing on the xy-plane in $\mathbb{R}^3$. In the notation of Theorem 11.4.3, let

$$\rho(t) = 1$$
$$h(t) = t$$
$$\varphi(t, \phi) = (\cos(\phi), \sin(\phi), t)$$
$$U = (0, +\infty) \times (-\pi, \pi).$$

By Theorem 11.4.3, Cyl is a regular surface and (U, φ) is a chart. Viewing Cyl as a regular surface in $\mathbb{R}^3_0$, Theorem 12.10.3 gives:

$$H_1 = (0, 0, 1) \qquad H_2 = (-\sin(\phi), \cos(\phi), 0)$$
$$\mathcal{N} \circ \varphi = (-\cos(\phi), -\sin(\phi), 0)$$
$$\mathfrak{g}\mathcal{H} = \begin{bmatrix} 1 & 0 \\ 0 & 1 \end{bmatrix}$$
$$\mathfrak{h}\mathcal{H} = \begin{bmatrix} 0 & 0 \\ 0 & 1 \end{bmatrix}$$
$$\mathcal{K} \circ \varphi = 0$$

$$\Gamma^1_{11} = 0 \qquad \Gamma^1_{12} = \Gamma^1_{21} = 0 \qquad \Gamma^1_{22} = 0$$
$$\Gamma^2_{11} = 0 \qquad \Gamma^2_{12} = \Gamma^2_{21} = 0 \qquad \Gamma^2_{22} = 0$$
$$\kappa_1 \circ \varphi = 0 \qquad\qquad \kappa_2 \circ \varphi = 1.$$

An intuitive explanation for why the Gauss curvature of Cyl equals 0 is given in Example 12.8.6.

13.3   Cone in $\mathbb{R}^3_0$

The set

$$\text{Con} = \{(x, y, z) \in \mathbb{R}^3 : x^2 + y^2 - z^2 = 0, z > 0\}$$

is an inverted infinite cone (minus its vertex) standing on the xy-plane in $\mathbb{R}^3$. See Figure 13.3.1.

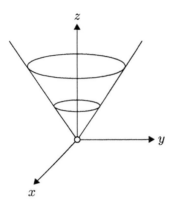

Figure 13.3.1. Con

In the notation of Theorem 11.4.3, let

$$\rho(t) = t$$
$$h(t) = t$$
$$\varphi(t, \phi) = \big(t \cos(\phi), t \sin(\phi), t\big)$$
$$U = (0, +\infty) \times (-\pi, \pi).$$

By Theorem 11.4.3, Con is a regular surface and (U, φ) is a chart. Had the vertex $(0, 0, 0)$ been included as part of Con, the resulting set would not be a regular surface because there is more than one tangent plane at $(0, 0, 0)$. Viewing Con as a regular surface in $\mathbb{R}^3_0$, Theorem 12.10.3 gives:

$$H_1 = \big(\cos(\phi), \sin(\phi), 1\big) \qquad H_2 = \big(-t \sin(\phi), t \cos(\phi), 0\big)$$

$$\mathcal{N} \circ \varphi = \frac{1}{\sqrt{2}} \big(-\cos(\phi), -\sin(\phi), 1\big)$$

$$\mathfrak{g}\mathcal{H} = \begin{bmatrix} 2 & 0 \\ 0 & t^2 \end{bmatrix}$$

$$\mathfrak{h}\mathcal{H} = \frac{1}{\sqrt{2}} \begin{bmatrix} 0 & 0 \\ 0 & t \end{bmatrix}$$

$$\mathcal{K} \circ \varphi = 0$$

$$\Gamma^1_{11} = 0 \qquad \Gamma^1_{12} = \Gamma^1_{21} = 0 \qquad \Gamma^1_{22} = -\frac{t}{2}$$

$$\Gamma^2_{11} = 0 \qquad \Gamma^2_{12} = \Gamma^2_{21} = \frac{1}{t} \qquad \Gamma^2_{22} = 0$$

$$\kappa_1 \circ \varphi = 0 \qquad \kappa_2 \circ \varphi = \frac{1}{\sqrt{2t}}.$$

An intuitive explanation for why the Gauss curvature of Con equals 0 is given in Example 12.8.6.

13.4 Sphere in $\mathbb{R}^3_0$

For a real number $R > 0$, the set

$$S^2_R = \{(x, y, z) \in \mathbb{R}^3 : x^2 + y^2 + z^2 = R^2\}$$

is a sphere of radius R centered at the origin. See Figure 13.4.1.

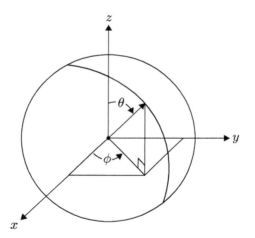

Figure 13.4.1. S^2_R

When $R = 1$, we write S^2 in place of S^2_1. In the notation of Theorem 11.4.3, let

$$\rho(\theta) = R\sin(\theta)$$
$$h(\theta) = R\cos(\theta)$$
$$\varphi(\theta, \phi) = R\big(\sin(\theta)\cos(\phi), \sin(\theta)\sin(\phi), \cos(\theta)\big)$$
$$U = (0, \pi) \times (-\pi, \pi).$$

By Theorem 11.4.3, S^2_R is a regular surface and (U, φ) is a chart. Viewing S^2_R as a regular surface in $\mathbb{R}^3_0$, Theorem 12.10.3 gives:

$$H_1 = R\big(\cos(\theta)\cos(\phi), \cos(\theta)\sin(\phi), -\sin(\theta)\big)$$

$$H_2 = R\big(-\sin(\theta)\sin(\phi), \sin(\theta)\cos(\phi), 0\big)$$

$$\mathcal{N} \circ \varphi = \big(\sin(\theta)\cos(\phi), \sin(\theta)\sin(\phi), \cos(\theta)\big) = \frac{1}{R}\varphi$$

$$g_{\mathcal{H}} = R^2 \begin{bmatrix} 1 & 0 \\ 0 & \sin^2(\theta) \end{bmatrix}$$

$$\flat_{\mathcal{H}} = -\frac{1}{R}\, g_{\mathcal{H}}$$

$$\mathcal{K} \circ \varphi = \frac{1}{R^2}$$

$$\Gamma^1_{11} = 0 \qquad \Gamma^1_{12} = \Gamma^1_{21} = 0 \qquad \Gamma^1_{22} = -\cos(\theta)\sin(\theta)$$
$$\Gamma^2_{11} = 0 \qquad \Gamma^2_{12} = \Gamma^2_{21} = \cot(\theta) \qquad \Gamma^2_{22} = 0$$

$$\kappa_1 \circ \varphi = \kappa_2 \circ \varphi = -\frac{1}{R}.$$

When $R = 1$, we have $\mathcal{N} \circ \varphi = \varphi$; that is, $\mathcal{N}_{\varphi(\theta,\phi)} = \varphi(\theta, \phi)$ for all (θ, ϕ) in U. This is consistent with Theorem 12.2.12(c).

13.5   Tractoid in $\mathbb{R}^3_0$

The set

$$\mathrm{Trc} = \big\{(x, y, z) \in \mathbb{R}^3 : \mathrm{arsech}\big(\sqrt{x^2 + y^2}\big) - \sqrt{1 - x^2 - y^2} - z = 0,$$
$$x^2 + y^2 < 1, z > 0\big\}$$

is the upper portion of the **tractoid**, also known as the **tractricoid**. It is better understood as the surface obtained by revolving around the z-axis the smooth curve $\sigma(t) : (0, +\infty) \longrightarrow \mathbb{R}^3$ given

$$\sigma(t) = \big(\mathrm{sech}(t), 0, t - \tanh(t)\big).$$

See Figure 13.5.1. In the notation of Theorem 11.4.3, let

$$\rho(\phi) = \mathrm{sech}(t)$$
$$h(t) = t - \tanh(t)$$
$$\varphi(t, \phi) = \big(\mathrm{sech}(t)\cos(\phi), \mathrm{sech}(t)\sin(\phi), t - \tanh(t)\big)$$
$$U = (0, +\infty) \times (-\pi, \pi).$$

By Theorem 11.4.3, Trc is a regular surface and (U, φ) is a chart. Viewing Trc as a regular surface in $\mathbb{R}^3_0$, Theorem 12.10.3 gives:

$$H_1 = \big(-\mathrm{sech}(t)\tanh(t)\cos(\phi), -\mathrm{sech}(t)\tanh(t)\sin(\phi), \tanh^2(t)\big)$$

$$H_2 = \big(-\mathrm{sech}(t)\sin(\phi), \mathrm{sech}(t)\cos(\phi), 0\big)$$

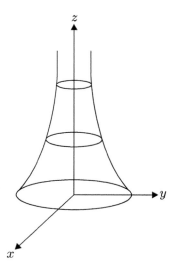

Figure 13.5.1. Trc

$$\mathcal{N} \circ \varphi = -\big(\tanh(t)\cos(\phi), \tanh(t)\sin(\phi), \operatorname{sech}(t)\big)$$

$$\mathfrak{g}_{\mathcal{H}} = \begin{bmatrix} \tanh^2(t) & 0 \\ 0 & \operatorname{sech}^2(t) \end{bmatrix}$$

$$\mathfrak{h}_{\mathcal{H}} = \begin{bmatrix} -\operatorname{sech}(t)\tanh(t) & 0 \\ 0 & \operatorname{sech}(t)\tanh(t) \end{bmatrix}$$

$$\mathcal{K} \circ \varphi = -1$$

$$\Gamma^1_{11} = \frac{\operatorname{sech}^2(t)}{\tanh(t)} \qquad \Gamma^1_{12} = \Gamma^1_{21} = 0 \qquad\qquad \Gamma^1_{22} = \frac{\operatorname{sech}^2(t)}{\tanh(t)}$$

$$\Gamma^2_{11} = 0 \qquad\qquad \Gamma^2_{12} = \Gamma^2_{21} = -\tanh(t) \qquad \Gamma^2_{22} = 0$$

$$\kappa_1 \circ \varphi = -\frac{\operatorname{sech}(t)}{\tanh(t)} \qquad \kappa_2 \circ \varphi = \frac{\tanh(t)}{\operatorname{sech}(t)}.$$

Together with Pln and $\mathcal{S}^2$, Trc makes a trio of regular surfaces in $\mathbb{R}^3_0$ with constant Gauss curvatures of 0, 1, and -1, respectively.

13.6   Hyperboloid of One Sheet in $\mathbb{R}^3_0$

The set

$$\text{One} = \{(x, y, z) \in \mathbb{R}^3 : x^2 + y^2 - z^2 = 1, z > 0\}$$

is the upper half of a hyperboloid of one sheet. See Figure 13.6.1.

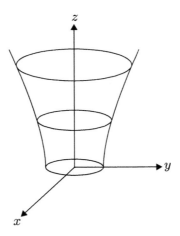

Figure 13.6.1. One, $\mathcal{P}^2$

In the notation of Theorem 11.4.2, let

$$f(x,y) = \sqrt{x^2 + y^2 - 1}$$
$$\varphi(x,y) = (x, y, \sqrt{x^2 + y^2 - 1})$$
$$U = \{(x,y) \in \mathbb{R}^2 : x^2 + y^2 > 1\}.$$

By Theorem 11.4.2, One is a regular surface and (U, φ) is a chart. Viewing One as a regular surface in $\mathbb{R}_0^3$, Theorem 12.10.1 gives:

$$H_1 = \left(1, 0, \frac{x}{\sqrt{x^2 + y^2 - 1}}\right) \qquad H_2 = \left(0, 1, \frac{y}{\sqrt{x^2 + y^2 - 1}}\right)$$

$$\mathcal{N} \circ \varphi = \frac{1}{\sqrt{2x^2 + 2y^2 - 1}} \left(-x, -y, \sqrt{x^2 + y^2 - 1}\right)$$

$$\mathcal{g}_{\mathcal{H}} = \frac{1}{x^2 + y^2 - 1} \begin{bmatrix} 2x^2 + y^2 - 1 & xy \\ xy & x^2 + 2y^2 - 1 \end{bmatrix}$$

$$\mathcal{h}_{\mathcal{H}} = \frac{1}{(x^2 + y^2 - 1)\sqrt{2x^2 + 2y^2 - 1}} \begin{bmatrix} y^2 - 1 & -xy \\ -xy & x^2 - 1 \end{bmatrix}$$

$$\mathcal{K} \circ \varphi = -\frac{1}{(2x^2 + 2y^2 - 1)^2}.$$

We observe that the Gauss curvature is nonconstant and strictly negative.

13.7 Hyperboloid of Two Sheets in $\mathbb{R}_0^3$

The set

$$\text{Two} = \{(x, y, z) \in \mathbb{R}^3 : x^2 + y^2 - z^2 = -1, z > 0\}$$

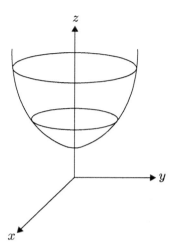

Figure 13.7.1. Two, $\mathcal{H}^2$

is the upper sheet of a hyperboloid of two sheets. See Figure 13.7.1. In the notation of Theorem 11.4.2, let

$$f(x,y) = \sqrt{x^2 + y^2 + 1}$$
$$\varphi(x,y) = \left(x, y, \sqrt{x^2 + y^2 + 1}\right)$$
$$U = \mathbb{R}^2.$$

By Theorem 11.4.2, Two is a regular surface and (U, φ) is a chart. Viewing Two as a regular surface in $\mathbb{R}_0^3$, Theorem 12.10.1 gives:

$$H_1 = \left(1, 0, \frac{x}{\sqrt{x^2 + y^2 + 1}}\right) \qquad H_2 = \left(0, 1, \frac{y}{\sqrt{x^2 + y^2 + 1}}\right)$$

$$\mathcal{N} \circ \varphi = \frac{1}{\sqrt{2x^2 + 2y^2 + 1}} \left(-x, -y, \sqrt{x^2 + y^2 + 1}\right)$$

$$\mathfrak{g}_{\mathcal{H}} = \frac{1}{x^2 + y^2 + 1} \begin{bmatrix} 2x^2 + y^2 + 1 & xy \\ xy & x^2 + 2y^2 + 1 \end{bmatrix}$$

$$\mathfrak{h}_{\mathcal{H}} = \frac{1}{(x^2 + y^2 + 1)\sqrt{2x^2 + 2y^2 + 1}} \begin{bmatrix} y^2 + 1 & -xy \\ -xy & x^2 + 1 \end{bmatrix}$$

$$\mathcal{K} \circ \varphi = \frac{1}{(2x^2 + 2y^2 + 1)^2}.$$

We observe that the Gauss curvature is nonconstant and strictly positive.

13.8   Torus in $\mathbb{R}_0^3$

For a real number $R > 1$, the set

$$\text{Tor} = \{(x, y, z) \in \mathbb{R}^3 : (\sqrt{x^2 + y^2} - R)^2 + z^2 = 1, x^2 + y^2 \geq 0\}$$

is the torus obtained by rotating about the z-axis the unit circle in the xz-plane centered at $(R, 0, 0)$. See Figure 13.8.1.

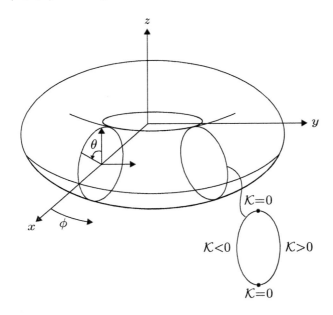

Figure 13.8.1. Tor

In the notation of Theorem 11.4.3, let

$$\rho(\theta) = \cos(\theta) + R$$
$$h(\theta) = \sin(\theta)$$
$$\varphi(\theta, \phi) = ([\cos(\theta) + R]\cos(\phi), [\cos(\theta) + R]\sin(\phi), \sin(\theta))$$
$$U = (-\pi/2, \pi/2) \times (-\pi, \pi).$$

The domain for h was chosen to be $(-\pi/2, \pi/2)$ instead of $(0, 2\pi)$, for example, to ensure that property [R2] of Section 11.4 is satisfied. This parametrizes the "outer" half of the torus; a separate parametrization gives the "inner" half. It follows from Theorem 11.4.3 that Tor is a regular surface and (U, φ) is a chart. Viewing Tor as a regular surface in $\mathbb{R}_0^3$, Theorem 12.10.3 gives:

$$H_1 = (-\sin(\theta)\cos(\phi), -\sin(\theta)\sin(\phi), \cos(\theta))$$

$$H_2 = (-[\cos(\theta) + R]\sin(\phi), [\cos(\theta) + R]\cos(\phi), 0)$$

$$\mathcal{N} \circ \varphi = -\big(\cos(\theta)\cos(\phi), \cos(\theta)\sin(\phi), \sin(\theta)\big)$$

$$g_{\mathcal{H}} = \begin{bmatrix} 1 & 0 \\ 0 & [\cos(\theta) + R]^2 \end{bmatrix}$$

$$\mathfrak{h}_{\mathcal{H}} = \begin{bmatrix} 1 & 0 \\ 0 & [\cos(\theta) + R]\cos(\theta) \end{bmatrix}$$

$$\mathcal{K} \circ \varphi = \frac{\cos(\theta)}{\cos(\theta) + R}$$

$$\Gamma^1_{11} = 0 \qquad \Gamma^1_{12} = \Gamma^1_{21} = 0 \qquad\qquad \Gamma^1_{22} = [\cos(\theta) + R]\sin(\theta)$$

$$\Gamma^2_{11} = 0 \qquad \Gamma^2_{12} = \Gamma^2_{21} = -\frac{\sin(\theta)}{\cos(\theta) + R} \qquad \Gamma^2_{22} = 0$$

$$\kappa_1 \circ \varphi = 1 \qquad \kappa_2 \circ \varphi = \frac{\cos(\theta)}{\cos(\theta) + R}.$$

As depicted in Figure 13.8.1, the Gauss curvature takes positive, negative, and zero values at various points of Tor.

13.9   Pseudosphere in $\mathbb{R}^3_1$

The set

$$\mathcal{P}^2 = \{(x, y, z) \in \mathbb{R}^3 : x^2 + y^2 - z^2 = 1, z > 0\}$$

is the same upper half of the hyperboloid of one sheet described in Section 13.6, and just as before, it is a regular surface. Since

$$\left(\frac{\partial f}{\partial x}(x, y)\right)^2 + \left(\frac{\partial f}{\partial y}(x, y)\right)^2 - 1 = \frac{1}{x^2 + y^2 + 1},$$

the condition of Theorem 12.2.8 is satisfied. Thus, $\mathcal{P}^2$ is a regular surface in $\mathbb{R}^3_1$. Alternatively, this comes directly from Theorem 12.2.12(a). Theorem 12.10.2 gives:

$$H_1 = \left(1, 0, \frac{x}{\sqrt{x^2 + y^2 - 1}}\right) \qquad H_2 = \left(0, 1, \frac{y}{\sqrt{x^2 + y^2 - 1}}\right)$$

$$\mathcal{N} \circ \varphi = \left(-x, -y, -\sqrt{x^2 + y^2 - 1}\right) = -\varphi$$

$$g_{\mathcal{H}} = \mathfrak{h}_{\mathcal{H}} = \frac{1}{x^2 + y^2 - 1}\begin{bmatrix} y^2 - 1 & -xy \\ -xy & x^2 - 1 \end{bmatrix}$$

$$\mathcal{K} \circ \varphi = 1$$

$$\epsilon_{\mathcal{P}^2} = 1.$$

We note that $\epsilon_{\mathcal{P}^2} = 1$ agrees with Theorem 12.2.12(d). In the context of $\mathbb{R}^3_1$, we refer to $\mathcal{P}^2$ as the **pseudosphere**.

13.10   Hyperbolic Space in $\mathbb{R}^3_1$

The set

$$\mathcal{H}^2 = \{(x, y, z) \in \mathbb{R}^3 : x^2 + y^2 - z^2 = -1, z > 0\}$$

is the same upper sheet of the hyperboloid of two sheets described in Section 13.7, and just as before, it is a regular surface. Since

$$\left(\frac{\partial f}{\partial x}(x, y)\right)^2 + \left(\frac{\partial f}{\partial y}(x, y)\right)^2 - 1 = -\frac{1}{x^2 + y^2 - 1}$$

and, by definition, $x^2 + y^2 > 1$, the condition of Theorem 12.2.8 is satisfied. Thus, $\mathcal{H}^2$ is a regular surface in $\mathbb{R}^3_1$. Alternatively, this comes from directly from Theorem 12.2.12(a). Theorem 12.10.2 gives:

$$H_1 = \left(1, 0, \frac{x}{\sqrt{x^2 + y^2 + 1}}\right) \qquad H_2 = \left(0, 1, \frac{y}{\sqrt{x^2 + y^2 + 1}}\right)$$

$$\mathcal{N} \circ \varphi = \left(x, y, \sqrt{x^2 + y^2 + 1}\right) = \varphi$$

$$\mathfrak{g}_{\mathcal{H}} = \frac{1}{x^2 + y^2 + 1} \begin{bmatrix} y^2 + 1 & -xy \\ -xy & x^2 + 1 \end{bmatrix}$$

$$\mathfrak{b}_{\mathcal{H}} = -\mathfrak{g}_{\mathcal{H}}$$

$$\mathcal{K} \circ \varphi = -1$$

$$\epsilon_{\mathcal{H}^2} = -1.$$

We note that $\epsilon_{\mathcal{H}^2} = -1$ agrees with Theorem 12.2.12(d). In the context of $\mathbb{R}^3_1$, we refer to $\mathcal{H}^2$ as **hyperbolic space**.

We can also parametrize $\mathcal{H}^2$ as a surface of revolution. In the notation of Theorem 11.4.3, let

$$\rho(t) = \sinh(t)$$
$$h(t) = \cosh(t)$$
$$\varphi(t, \phi) = \left(\sinh(t) \cos(\phi), \sinh(t) \sin(\phi), \cosh(t)\right)$$
$$U = (0, +\infty) \times (-\pi, \pi).$$

From the well-known identity $\sinh^2(t) - \cosh^2(t) = -1$, we obtain $\dot{\rho}(t)^2 - \dot{h}(t)^2 = 1$. Evidently, the condition of Theorem 12.2.9 is satisfied, so once again we see that $\mathcal{H}^2$ is a regular surface in $\mathbb{R}^3_1$. Theorem 12.10.4 gives:

$$H_1 = \left(\cosh(t) \cos(\phi), \cosh(t) \sin(\phi), \sinh(t)\right)$$

$$H_2 = \left(-\sinh(t) \sin(\phi), \sinh(t) \cos(\phi), 0\right)$$

$$\mathcal{N} \circ \varphi = \left(\sinh(t) \cos(\phi), \sinh(t) \sin(\phi), \cosh(t)\right) = \varphi$$

$$\mathfrak{g}_{\mathcal{H}} = \begin{bmatrix} 1 & 0 \\ 0 & \sinh^2(t) \end{bmatrix}$$

$$\mathfrak{h}_{\mathcal{H}} = -\mathfrak{g}_{\mathcal{H}}$$

$$\mathcal{K} \circ \varphi = -1$$

$$\epsilon_{\mathcal{H}^2} = -1.$$

We observe from $\mathfrak{g}_{\mathcal{H}}$ that the coordinate frame $\mathcal{H}$ is orthogonal, giving this parametrization certain computational advantages.

Part III

Smooth Manifolds and Semi-Riemannian Manifolds

In Part II, we covered a range of topics on curves and surfaces. Despite the sometimes abstract nature of the mathematics, the fact that we were dealing with 1- and 2-dimensional geometric objects residing in 3-dimensional space made the undertaking relatively concrete. The restrictions on dimensions imposed in Part II reflect the classical nature of the study of curves and surfaces. With relatively little effort, it is possible to generalize results to $(m-1)$-dimensional "surfaces" in m-dimensional space. However, this does not resolve an important inherent limitation of this approach—the need for an ambient space. Perhaps the most important application of semi-Riemannian geometry, and the one that historically motivated this area of mathematics, is Einstein's general theory of relativity. According to this cosmological construct, the universe we inhabit has precisely four dimensions; there is no 5-dimensional ambient space in which our universe resides. Of course, it is possible to fashion one mathematically, but that would not reflect physical reality. In order to model the general theory of relativity, we need a mathematical description of "surfaces" that dispenses with ambient space altogether.

Let us consider where $\mathbb{R}^3$ entered (and did not enter) into our discussion of surfaces in an effort to see whether we can circumvent the role of an ambient space. Recall that in the first instance, a surface is defined to be a topological space, something that does not depend on an ambient space. However, after that we quickly run into roadblocks. Any of the concepts that rely on tangent vectors, and especially normal vectors, seem inextricably linked to an ambient space. We came to view Christoffel symbols, covariant derivatives, and even the Gauss curvature as "intrinsic" quantities, but their initial formulations relied on ambient space. It seems that as a first step, we need to find a way to define a "tangent vector" without resorting to this externality. Once that has been accomplished, we can then proceed to "tangent space", "differential map", and other fundamental notions of differential geometry. This is the goal of Chapters 14–17.

One of the "problems" with $\mathbb{R}^3$ is its abundance of mathematical structure. As has been observed previously, $\mathbb{R}^3$ can be viewed as an inner product space, a Lorentz vector space, a topological space, a metric space, and more. In our study of curves and surfaces, this forced us to clarify what assumptions about $\mathbb{R}^3$ were being made at each stage of the discussion. We are about to embark on the development of a mathematical theory that builds on what we know about curves and surfaces, but is in many respects something quite new. This gives us the opportunity to assume only what is essential at each juncture, an approach that is greatly illuminating. Accordingly, we delay introducing the concept of "connection" until Chapter 18, and that of "metric" until Chapters 19 and 20. In Chapter 21, some of the classical results of vector calculus are presented using the new language at our disposal, and in Chapter 22, we conclude with some applications to areas of physics closely related to the special and general theories of relativity.

Chapter 14

Smooth Manifolds

14.1   Smooth Manifolds

Many of the definitions presented in this and subsequent chapters are adaptations of ones we encountered in the study of curves and surfaces.

Let M be a topological space. A **chart on M** is a pair (U, φ), where U is an open set in M and $\varphi : U \longrightarrow \mathbb{R}^m$ is a map such that:

[C1] $\varphi(U)$ is an open set in $\mathbb{R}^m$.
[C2] $\varphi : U \longrightarrow \varphi(U)$ is a homeomorphism.

We refer to U as the **coordinate domain** of the chart, to φ as its **coordinate map**, and to m as the **dimension** of the chart. For each point p in U, we say that (U, φ) is a **chart at p**. (Note that in the definition of regular surfaces, coordinate maps went from open sets in $\mathbb{R}^2$ to open sets in the regular surface, whereas now traffic is in the opposite direction.) When $U = M$, we say that (M, φ) is a **covering chart** on M, and that M is **covered** by (M, φ). In the present context, the component functions of φ are denoted by $\varphi = (x^1, \ldots, x^m)$ rather than $\varphi = (\varphi^1, \ldots, \varphi^m)$, and are said to be **local coordinates** on U. This choice of notation is adopted specifically to encourage the informal identification of the point p in M with its local coordinate counterpart $\varphi(p) = \big(x^1(p), \ldots, x^m(p)\big)$ in $\mathbb{R}^m$. We often denote

$$(U, \varphi) \quad \text{by} \quad (U, (x^i)) \quad \text{or} \quad (U, \varphi = (x^i)),$$

where $(x^i) = (x^1, \ldots, x^m)$. The charts (U, φ) and $(\widetilde{U}, \widetilde{\varphi})$ on M are said to be **overlapping** if $U \cap \widetilde{U}$ is nonempty. In that case, the map

$$\widetilde{\varphi} \circ \varphi^{-1}\big|_{\varphi(U \cap \widetilde{U})} : \varphi(U \cap \widetilde{U}) \longrightarrow \widetilde{\varphi}(U \cap \widetilde{U})$$

Semi-Riemannian Geometry, First Edition. Stephen C. Newman.
© 2019 John Wiley & Sons, Inc. Published 2019 by John Wiley & Sons, Inc.

is called a **transition map**. For brevity, we usually denote

$$\widetilde{\varphi} \circ \varphi^{-1}\big|_{\varphi(U \cap \widetilde{U})} \qquad \text{by} \qquad \widetilde{\varphi} \circ \varphi^{-1},$$

and in a similar manner often (but not always) drop the "restriction" subscript from other notation when the situation is clear from the context. An **atlas** for M is a collection $\mathfrak{A} = \{(U_\alpha, \varphi_\alpha) : \alpha \in A\}$ of charts on M such that the U_α form an open cover of M; that is, $M = \bigcup_{\alpha \in A} U_\alpha$. At this point, there is no requirement that charts have the same dimension.

A **topological m-manifold** is a pair $(M, \mathfrak{A})$, where M is a topological space and $\mathfrak{A}$ is an atlas for M such that:

[**T1**] Each chart in $\mathfrak{A}$ has dimension m.

[**T2**] The topology of M has a countable basis.

[**T3**] For every pair of distinct points p_1, p_2 in M, there are disjoint open sets U_1 and U_2 in M such that p_1 is in U_1 and p_2 is in U_2.

Observe that [T1] refers exclusively to the atlas $\mathfrak{A}$, whereas [T2] and [T3] have to do with the topological structure of M. Properties [T2] and [T3] are technical requirements needed for certain constructions and will not be further elaborated upon.

Example 14.1.1 ($\mathbb{R}^m$). $\mathbb{R}^m$ endowed with the Euclidean topology and paired with the atlas consisting of the single chart $\big(\mathbb{R}^m, \mathrm{id}_{\mathbb{R}^m} = (r^1, \ldots, r^m)\big)$ is a topological m-manifold. $\diamond$

Theorem 14.1.2. *Let M be a topological space. If $\mathfrak{A}$ and $\mathfrak{B}$ are atlases for M such that $(M, \mathfrak{A})$ is a topological m-manifold and $(M, \mathfrak{B})$ is a topological n-manifold, then $m = n$.*

Proof. Let (U, φ) and (V, ψ) be charts in $\mathfrak{A}$ and $\mathfrak{B}$, respectively, such that $U \cap V$ is nonempty. Since $U \cap V$ is open in M, we have from [C2] and Theorem 9.1.14 that $\varphi(U \cap V)$ is open in $\varphi(U)$, and then from [C1] and Theorem 9.1.5 that $\varphi(U \cap V)$ is open in $\mathbb{R}^m$. Similar reasoning shows that $\varphi|_{U \cap V} : U \cap V \longrightarrow \varphi(U \cap V)$ is a homeomorphism. Likewise, $\psi(U \cap V)$ is open in $\mathbb{R}^n$ and $\psi|_{U \cap V} : U \cap V \longrightarrow \psi(U \cap V)$ is a homeomorphism. It follows that $\psi^{-1} \circ \varphi|_{\varphi(U \cap V)} : \varphi(U \cap V) \longrightarrow \psi(U \cap V)$ is a homeomorphism. By Theorem 9.4.1, $m = n$. $\square$

Let $(M, \mathfrak{A})$ be a topological m-manifold. We have from Theorem 14.1.2 that m is an invariant of $(M, \mathfrak{A})$, which we refer to as the **dimension of M** and denote by $\dim(M)$.

It can be shown that the connected components of a topological space are closed sets in the topological space. In the case of a topological manifold, the connected components have additional properties.

Theorem 14.1.3. *If $(M, \mathfrak{A})$ is a topological manifold, then M has countably many connected components, each of which is both an open and a closed set in M.* $\square$

Let $(M, \mathfrak{A})$ be a topological m-manifold, and let (U, φ) and $(\widetilde{U}, \widetilde{\varphi})$ be overlapping charts on M. We say that (U, φ) and $(\widetilde{U}, \widetilde{\varphi})$ are **smoothly compatible** if the transition maps $\widetilde{\varphi} \circ \varphi^{-1}$ and $\varphi \circ \widetilde{\varphi}^{-1}$ are (Euclidean) smooth. Since $(\varphi \circ \widetilde{\varphi}^{-1})^{-1} = \widetilde{\varphi} \circ \varphi^{-1}$, this is equivalent to either $\widetilde{\varphi} \circ \varphi^{-1}$ or $\varphi \circ \widetilde{\varphi}^{-1}$ being a (Euclidean) diffeomorphism. We say that the atlas $\mathfrak{A}$ is **smooth** if any two overlapping charts in $\mathfrak{A}$ are smoothly compatible. A smooth atlas for M is also called a **smooth structure** on M. Recall that in our study of charts on regular surfaces, coordinate maps were assumed to be smooth, where smoothness was defined using relevant properties of the coordinate domain in $\mathbb{R}^2$ and the ambient space $\mathbb{R}^3$. Since M does not have such inherent properties, we have turned Theorem 11.2.9, a result on the smoothness of transition maps for regular surfaces, into a definition of smoothness for topological manifolds. We will see further examples of this approach as we proceed, where a theorem about "surfaces" becomes a definition for "manifolds".

We say that a topological m-manifold $(M, \mathfrak{A})$ is a **smooth m-manifold** if $\mathfrak{A}$ is a smooth atlas. It is often convenient to adopt the shorthand of referring to M as a smooth m-manifold, with $\mathfrak{A}$ understood from the context. Furthermore, when it is not important to specify the dimension of M, we refer to $(M, \mathfrak{A})$ or simply M as a **smooth manifold**.

A smooth atlas $\mathfrak{A}$ for M is said to be **maximal** if it is not properly contained in any larger smooth atlas for M. This means that any chart on M that is smoothly compatible with every chart in $\mathfrak{A}$ is already in $\mathfrak{A}$. It can be shown that every smooth atlas for M is contained in a unique maximal smooth atlas. A given topological space can have distinct smooth atlases that generate the same maximal smooth atlas. On the other hand, it is also possible for a topological space to have distinct smooth atlases that give rise to distinct maximal smooth atlases. Accordingly, we adopt the following convention.

Throughout, any chart on a smooth manifold comes from the underlying smooth atlas or its corresponding maximal atlas.

Example 14.1.4 ($\mathbb{R}^m$). Continuing with Example 14.1.1, it is clear that the atlas consisting of the single chart $\left(\mathbb{R}^m, \mathrm{id}_{\mathbb{R}^m} = (r^1, \ldots, r^m)\right)$ is smooth. Thus, $\mathbb{R}^m$ is not just a topological m-manifold, it is a smooth m-manifold. In this context, the chart $\left(\mathbb{R}^m, \mathrm{id}_{\mathbb{R}^m} = (r^1, \ldots, r^m)\right)$ is referred to as **standard coordinates** on $\mathbb{R}^m$. It is convenient to adopt the shorthand of saying that $(r^1, \ldots, r^m)$ are standard coordinates on $\mathbb{R}^m$. $\Diamond$

Example 14.1.5 (Regular Surface). It can be shown with the help of Theorem 11.2.9 that a regular surface is a smooth 2-manifold. $\Diamond$

We noted in connection with Theorem 1.1.10 that all m-dimensional vector spaces are isomorphic to $\mathbb{R}^m$ (viewed as a vector space). In light of Example 14.1.4, it should come as no surprise that any m-dimensional vector space has a smooth structure induced by $\mathbb{R}^m$ (now viewed as a smooth m-manifold).

Theorem 14.1.6 (Standard Smooth Structure). *If V is an m-dimensional vector space, then there is a unique topology and maximal smooth atlas on V,* called the **standard smooth structure** on V, making V into a smooth m-manifold such that for every linear isomorphism $A : V \longrightarrow \mathbb{R}^m$, the pair (V, A) is a (covering) chart on V.   □

Recall the discussion on product topologies in Theorem 9.1.6 and Theorem 9.1.12.

Theorem 14.1.7 (Product Manifold). Let $(M_i, \mathfrak{A}_i)$ be a smooth m_i-manifold, let (U_i, φ_i) be a chart in $\mathfrak{A}_i$ for $i = 1, \ldots, k$, and suppose $M_1 \times \cdots \times M_k$ has the product topology. Then:

(a) $(U_1 \times \cdots \times U_k, \varphi_1 \times \cdots \times \varphi_k)$ is a chart on $M_1 \times \cdots \times M_k$.
(b) The topological space $M_1 \times \cdots \times M_k$ paired with the smooth atlas consisting of the collection of charts as defined in part (a) is a smooth $(m_1 + \cdots + m_k)$-manifold, called the **product manifold of $M_1, \ldots, M_k$**.   □

For example, it can be shown that the product manifold of m copies of $\mathbb{R}$ is precisely $\mathbb{R}^m$ with the standard smooth structure.

14.2   Functions and Maps

Let M be a smooth manifold, and let $f : M \longrightarrow \mathbb{R}$ be a function. Since M and $\mathbb{R}$ are topological spaces, we know what it means for f to be **continuous (on M)**. Motivated by Theorem 11.5.1, we say that f is **smooth (on M)** if for every point p in M, there is a chart (U, φ) at p such that the function $f \circ \varphi^{-1} : \varphi(U) \longrightarrow \mathbb{R}$ is (Euclidean) smooth. The set of smooth functions on M is denoted by $C^\infty(M)$. We make $C^\infty(M)$ into both a vector space and a ring by defining operations as follows: for all functions f, g in $C^\infty(M)$ and all real numbers c, let

$$(f + g)(p) = f(p) + g(p),$$
$$(fg)(p) = f(p)g(p),$$

and

$$(cf)(p) = cf(p)$$

for all p in M.

The next result is reminiscent of Theorem 11.5.1.

Theorem 14.2.1 (Smoothness Criterion for Functions). Let M be a smooth manifold, and let $f : M \longrightarrow \mathbb{R}$ be a function. Then f is smooth if and only if the function $f \circ \varphi^{-1} : \varphi(U) \longrightarrow \mathbb{R}$ is (Euclidean) smooth for every chart (U, φ) on M.

Proof. ($\Rightarrow$): Let p be a point in M. By assumption, there is a chart $(\widetilde{U}, \widetilde{\varphi})$ at p such that $f \circ \widetilde{\varphi}^{-1} : \varphi(\widetilde{U}) \longrightarrow \mathbb{R}$ is (Euclidean) smooth. Since $\widetilde{\varphi} \circ \varphi^{-1} : \varphi(U \cap \widetilde{U}) \longrightarrow \mathbb{R}^m$ is a transition map, it too is (Euclidean) smooth. It follows from Theorem 10.1.12 that

$$f \circ \varphi^{-1} = (f \circ \widetilde{\varphi}^{-1}) \circ (\widetilde{\varphi} \circ \varphi^{-1}) : \varphi(U \cap \widetilde{U}) \longrightarrow \mathbb{R}$$

is (Euclidean) smooth. Since p was arbitrary, the result follows.

($\Leftarrow$): Straightforward.   □

We now turn our attention to maps. Let M and N be smooth manifolds, and let $F : M \longrightarrow N$ be a map. Since M and N are topological spaces, we understand what is meant by F being **continuous (on M)**. With Theorem 11.6.1 as motivation, we say that F is **smooth (on M)** if for every point p in M, there is a chart (U, φ) at p and a chart (V, ψ) at $F(p)$ such that $F(U) \subseteq V$ and the map $\psi \circ F \circ \varphi^{-1} : \varphi(U) \longrightarrow \mathbb{R}^n$ is (Euclidean) smooth. The condition $F(U) \subseteq V$ is included as part of the definition of smoothness to ensure that the next result holds.

Theorem 14.2.2. *Let M and N be smooth manifolds, and let $F : M \longrightarrow N$ be a map. If F is smooth, then it is continuous.*

Proof. Let p be a point in M. Since F is smooth, there are charts (U, φ) at p and (V, ψ) at $F(p)$ such that $F(U) \subseteq V$ and $\psi \circ F \circ \varphi^{-1} : \varphi(U) \longrightarrow \mathbb{R}^n$ is (Euclidean) smooth. By Theorem 10.1.7, $\psi \circ F \circ \varphi^{-1}$ is continuous, and according to property [C2] of Section 14.1, $\varphi : U \longrightarrow \varphi(U)$ and $\psi^{-1} : \psi(V) \longrightarrow V$ are homeomorphisms, hence continuous. It follows from Theorem 9.1.8 that

$$F = \psi^{-1} \circ (\psi \circ F \circ \varphi^{-1}) \circ \varphi : U \longrightarrow V$$

is continuous. Since p was arbitrary, the result follows.   □

Theorem 14.2.3 (Smoothness Criterion for Maps). *Let M and N be smooth manifolds, and let $F : M \longrightarrow N$ be a map. Then F is smooth if and only if (i) F is continuous, and (ii) for every chart (U, φ) on M and every chart (V, ψ) on N such that $F(U) \subseteq V$, the map $\psi \circ F \circ \varphi^{-1} : \varphi(U) \longrightarrow \mathbb{R}^n$ is (Euclidean) smooth.*

Proof. ($\Rightarrow$): By Theorem 14.2.2, F is continuous. Let (U, φ) and (V, ψ) be charts on M and N, respectively, such that $F(U) \subseteq V$, and let p be a point in U. By assumption, there are charts $(\widetilde{U}, \widetilde{\varphi})$ at p and $(\widetilde{V}, \widetilde{\psi})$ at $F(p)$ such that $F(\widetilde{U}) \subseteq \widetilde{V}$ and $\widetilde{\psi} \circ F \circ \widetilde{\varphi}^{-1} : \widetilde{\varphi}(\widetilde{U}) \longrightarrow \mathbb{R}^n$ is (Euclidean) smooth. Since U and $\widetilde{U}$ are open in M, so is $U \cap \widetilde{U}$, and by Theorem A.3(c), $(U \cap \widetilde{U}) \subseteq F^{-1}(V \cap \widetilde{V})$. Because $\widetilde{\varphi} \circ \varphi^{-1} : \varphi(U \cap \widetilde{U}) \longrightarrow \mathbb{R}^m$ and $\psi \circ \widetilde{\psi}^{-1} : \widetilde{\psi}(V \cap \widetilde{V}) \longrightarrow \mathbb{R}^n$ are transition maps, they are (Euclidean) smooth. It follows from Theorem 10.1.12 from

$$\psi \circ F \circ \varphi^{-1} = (\psi \circ \widetilde{\psi}^{-1}) \circ (\widetilde{\psi} \circ F \circ \widetilde{\varphi}^{-1}) \circ (\widetilde{\varphi} \circ \varphi^{-1}) : \varphi(U \cap \widetilde{U}) \longrightarrow \mathbb{R}^n$$

is (Euclidean) smooth. Since p was arbitrary, the result follows.

($\Leftarrow$): Let p be a point in M, and let (U, φ) and (V, ψ) be charts at p and $F(p)$, respectively. Since F is continuous, by Theorem 9.1.7, $F^{-1}(V)$ is open in M, and therefore, so is $U' = U \cap F^{-1}(V)$. Then $(\varphi|_{U'}, U')$ is a chart at p such that $F(U') \subseteq V$. The argument used in the proof of ($\Rightarrow$) shows that the map $\psi \circ F \circ \varphi^{-1} : \varphi(U') \longrightarrow \mathbb{R}^n$ is (Euclidean) smooth.   □

Theorem 14.2.4. *Let M, N, and P be smooth manifolds, and let $F : M \longrightarrow N$ and $G : N \longrightarrow P$ be maps. If F and G are smooth, then so is $G \circ F$.* □

Theorem 14.2.5. *Let V and W be vector spaces, let $A : V \longrightarrow W$ be a linear map, and suppose V and W have standard smooth structures. Then A is a smooth map.* □

We close this section with a brief look at two important methods of construction on smooth manifolds—bump functions and partitions of unity.

Theorem 14.2.6 (Bump Function). *Let M be a smooth manifold, let p be a point in M, and let U be a neighborhood of p in M. Then there is a function β in $C^\infty(M)$, called a **bump function at p**, such that:*
(a) $0 \leq \beta(p) \leq 1$ *for all p in M.*
(b) $\mathrm{supp}(\beta) \subset U$.
(c) $\beta = 1$ *on some neighborhood $\widetilde{U} \subseteq U$ of p in M.*
See Figure 14.2.1.

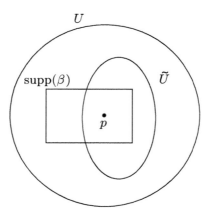

Figure 14.2.1. Bump function

Proof. We consider only the case $M = \mathbb{R}$ and $p = 0$. Let $[a, b]$ be a closed interval containing 0, and consider the smooth functions $f(t), g(t), \beta(t) : \mathbb{R} \longrightarrow \mathbb{R}$ given by

$$f(t) = \begin{cases} 0 & \text{if } t \leq 0 \\ e^{-1/t} & \text{if } t > 0, \end{cases}$$

$$g(t) = \frac{f(t)}{f(t) + f(1 - t)},$$

and

$$\beta(t) = 1 - g\left(\frac{x^2 - a^2}{b^2 - a^2}\right).$$

As depicted in Figure 14.2.2(a),

$$\begin{cases} g(t) = 0 & \text{if } t \leq 0 \\ 0 < g(t) < 1 & \text{if } t \in (0,1) \\ g(t) = 1 & \text{if } t \geq 1 \end{cases}$$

and

$$\begin{cases} \beta(t) = 0 & \text{if } t \leq -b \\ 0 < \beta(t) < 1 & \text{if } t \in (-b, -a) \\ \beta(t) = 1 & \text{if } t \in [-a, a] \\ 0 < \beta(t) < 1 & \text{if } t \in (a, b) \\ \beta(t) = 0 & \text{if } t \geq b, \end{cases}$$

so that $\text{supp}(\beta) = [-b, b]$. Taking $U = \mathbb{R}$ and $\widetilde{U} = (-a, a)$, for example, we find that β is a bump function at 0. □

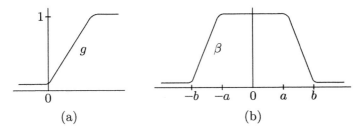

Figure 14.2.2. Diagram for Theorem 14.2.6

A glance at Figure 14.2.2(b) explains why a function such as β is called a bump function. Bump functions are often called upon to extend the domain of a smooth map, as in the proof of the next result.

Theorem 14.2.7 (Extension of Function). *Let M be a smooth manifold, let p be a point in M, let U be a neighborhood of p in M, and let f be a function in $C^\infty(U)$. Then there is a function $\widetilde{f}$ in $C^\infty(M)$ and a neighborhood $\widetilde{U} \subseteq U$ of p in M such that f and $\widetilde{f}$ agree on $\widetilde{U}$; that is, $f|_{\widetilde{U}} = \widetilde{f}|_{\widetilde{U}}$.*

Proof. Let β be a bump function in $C^\infty(M)$ that has support in U and constant value 1 on a neighborhood $\widetilde{U} \subseteq U$ of p in M. We define a function $\widetilde{f} : M \longrightarrow \mathbb{R}$ by

$$\widetilde{f}(q) = \begin{cases} \beta(q)f(q) & \text{if } q \in U \\ 0 & \text{if } q \in M \backslash U. \end{cases}$$

Because β and f are smooth on U, so is $\widetilde{f}$. Let q be a point in $M \backslash U$. Since β has support in U, q is in $M \backslash \text{supp}(\beta)$, and because $\text{supp}(\beta)$ is closed in M, $M \backslash \text{supp}(\beta)$ is open in M. Thus, $\widetilde{f} = 0$ on the neighborhood $M \backslash \text{supp}(\beta)$ of q in M. Since q was arbitrary, $\widetilde{f}$ is smooth on $M \backslash U$. This shows that $\widetilde{f}$ is smooth on M. Lastly, because $\beta = 1$ on $\widetilde{U}$, f and $\widetilde{f}$ agree on $\widetilde{U}$. □

Theorem 14.2.8 (Partition of Unity). *If M is a smooth manifold and $\mathcal{U} = \{U_\alpha : \alpha \in A\}$ is an open cover of M, then there is a family $\{\pi_\alpha \in C^\infty(M) : \alpha \in A\}$ of smooth functions on M, called a* **partition of unity on M subordinate to $\mathcal{U}$**, *such that:*

(a) *$0 \leq \pi_\alpha(p) \leq 1$ for all α in A and all p in M.*
(b) *$\operatorname{supp}(\pi_\alpha) \subset U_\alpha$ for all α in A.*
(c) *Every point p in M has a neighborhood U in M that intersects $\operatorname{supp}(\pi_\alpha)$ for only finitely many values of α.*
(d) *$\sum_{\alpha \in A} \pi_\alpha(p) = 1$ for all p in M.*   □

Although far from being intuitive, partitions of unity are indispensable for certain constructions in differential geometry, and we will see several such applications. The basic idea is to define the mathematical object of interest (for example, a function, vector field, or integral) on each set in the given open cover, and then form a weighted average using the π_α as weights to combine the individual contributions into a mathematical object defined on all of M. Because of part (c), there are no issues of convergence of infinite series.

14.3   Tangent Spaces

In the introduction to Part III, it was remarked that a crucial step in developing the theory of what we now call smooth manifolds is to devise a way of defining "tangent vector" when there is no ambient space. The definition created by differential geometers and provided here meets this challenge in an ingenious fashion. Framed in algebraic terms, it is both mathematically elegant and computationally convenient. However, it unfortunately lacks intuitive appeal compared to the methods adopted for surfaces, where tangent vectors were defined in terms of derivatives of smooth curves and could be thought of as "arrows". Later on we will see that the algebraic approach leads to a theory closely resembling that developed for surfaces, thereby lending the algebraic theory a certain geometric flavor.

Before proceeding, we need to establish some notation. In Chapter 10, coordinates on $\mathbb{R}^m$ were denoted by $(x^1, \ldots, x^m)$ or $(y^1, \ldots, y^m)$. In Chapter 11 and Chapter 12, it was necessary to clearly distinguish between coordinates on $\mathbb{R}^2$ and $\mathbb{R}^3$. For $\mathbb{R}^2$, the notation used was (r^1, r^2) or (r, s), and for $\mathbb{R}^3$, it was (x^1, x^2, x^3) or (x, y, z). In Chapter 13, coordinates on $\mathbb{R}^2$ and $\mathbb{R}^3$ were denoted by (x, y) and (x, y, z), respectively. The former choice was made because in the setting of graphs of functions, $\mathbb{R}^2$ was identified with the xy-plane in $\mathbb{R}^3$. That brings us to the present chapter, and beyond.

> **Henceforth, coordinate maps will be denoted by $(x^1, \ldots, x^m)$ or $(y^1, \ldots, y^m)$, except for standard coordinates on $\mathbb{R}^m$, which will be denoted by $(r^1, \ldots, r^m)$ or $(s^1, \ldots, s^m)$.**

Let M be a smooth manifold, and let p be a point in M. A **(tangent) vector at p** is defined to be a linear function

$$v : C^\infty(M) \longrightarrow \mathbb{R}$$

that satisfies the following product rule:

$$v(fg) = f(p)\,v(g) + g(p)\,v(f) \tag{14.3.1}$$

for all functions f, g in $C^\infty(M)$. The set of tangent vectors at p is denoted by $T_p(M)$ and called the **tangent space of M at p**. The **zero vector** in $T_p(M)$, denoted by 0, is the tangent vector that sends all functions in $C^\infty(M)$ to the real number 0. We make $T_p(M)$ into a vector space by defining operations as follows: for all vectors v, w in $T_p(M)$ and all real numbers c, let

$$(v + w)(f) = v(f) + w(f)$$

and

$$(cv)(f) = c\,v(f)$$

for all functions f in $C^\infty(M)$.

Let $(U, \varphi = (x^i))$ be a chart at p. The **partial derivative with respect to x^i at p** is the map

$$\left.\frac{\partial}{\partial x^i}\right|_p : C^\infty(M) \longrightarrow \mathbb{R}$$

defined by

$$\frac{\partial f}{\partial x^i}(p) = \frac{\partial(f \circ \varphi^{-1})}{\partial r^i}(\varphi(p)) \tag{14.3.2}$$

for all functions f in $C^\infty(M)$, where we denote

$$\left.\frac{\partial}{\partial x^i}\right|_p (f) \qquad \text{by} \qquad \frac{\partial f}{\partial x^i}(p). \tag{14.3.3}$$

The right-hand side of (14.3.2) is simply the ordinary (Euclidean) partial derivative of $f \circ \varphi^{-1}$ with respect to r^i at $\varphi(p)$. When $m = 1$, we denote

$$\frac{\partial f}{\partial x}(p) \qquad \text{by} \qquad \frac{df}{dx}(p).$$

Note that although the x^i have the domain U, the $(\partial/\partial x^i)|_p$ have the domain $C^\infty(M)$, as opposed to $C^\infty(U)$.

Theorem 14.3.1. *With the above setup:*
(a) *$(\partial/\partial x^i)|_p$ is a tangent vector at p, hence $(\partial/\partial x^i)|_p$ is in $T_p(M)$ for $i = 1, \dots, m$.*
(b)

$$\frac{\partial x^i}{\partial x^j}(p) = \delta^i_j$$

for $i, j = 1, \dots, m$, where δ^i_j is Kronecker's delta.

Proof. (a): This follows from the properties of Euclidean partial derivatives.
(b): Straightforward.   □

Theorem 14.3.2. *Let M be a smooth m-manifold, let p be a point in M, and let v be a vector in $T_p(M)$. Then:*
(a) *If f and g are functions in $C^\infty(M)$ that are equal on a neighborhood of p in M, then $v(f) = v(g)$.*
(b) *If h is a function in $C^\infty(M)$ that is constant on a neighborhood of p in M, then $v(h) = 0$.*

Proof. (a): Since v is linear, it suffices to show that if $f = 0$ on a neighborhood U of p in M, then $v(f) = 0$. A first observation is that

$$v(0_M) = v(0_M + 0_M) = v(0_M) + v(0_M),$$

hence $v(0_M) = 0$. Let β be a bump function at p that has support in U. Since $f = 0$ on U and $\text{supp}(\beta) \subset U$, we have $\beta f = 0_M$. It follows from $\beta(p) = 1$, $f(p) = 0$, and (14.3.1) that

$$0 = v(\beta f) = \beta(p)\, v(f) + f(p)\, v(\beta) = v(f).$$

(b): By part (a), it suffices to consider the case where h is constant on M. We have from (14.3.1) that

$$v(1_M) = v(1_M 1_M) = 1_M(p)\, v(1_M) + 1_M(p)\, v(1_M) = 2\, v(1_M),$$

hence $v(1_M) = 0$. Since $h = c 1_M$ for some real number c,

$$v(h) = v(c 1_M) = c\, v(1_M) = 0.$$ □

The next result is reminiscent of Theorem 11.4.1.

Theorem 14.3.3 (Open Set). *Let $(M, \mathfrak{A})$ be a smooth m-manifold, let U be an open set in M, and let*

$$\mathfrak{A}_U = \{(V \cap U, \psi|_{V \cap U}) : (V, \psi) \in \mathfrak{A}\}.$$

Then:
(a) *$\mathfrak{A}_U$ is a smooth atlas on U, hence $(U, \mathfrak{A}_U)$ is a smooth m-manifold, where U has the subspace topology induced by M.*
(b) *For all points p in U, $T_p(U)$ can be identified with $T_p(M)$, written*

$$T_p(U) = T_p(M).$$

Proof. (a): Straightforward.
(b): Using Theorem 14.3.4(c) and Theorem 14.7.3, an argument similar to that employed in the proof of Theorem 11.4.1(b) gives the result. Alternatively, consider the map $\iota : T_p(U) \longrightarrow T_p(M)$ defined by $\iota(v)(f) = v(f|_U)$ for all vectors v in $T_p(U)$ and all functions f in $C^\infty(M)$. It is easily shown that $\iota(v)$ is a vector in $T_p(M)$, so ι is well-defined. Furthermore, it can also be shown that ι is a linear isomorphism.   □

**Throughout, any open set in a smooth *m*-manifold
is viewed as a smooth *m*-manifold.**

Theorem 14.3.2(a) and Theorem 14.3.3(b) show that tangent vectors operate locally.

Let V be an m-dimensional vector space that we suppose has the standard smooth structure, so that V is a smooth m-manifold. For each vector v in V, the tangent space $T_v(V)$ is an m-dimensional vector space. In an obvious way, we identify $T_v(V)$ with V (viewed as a vector space), and write

$$T_v(V) = V. \tag{14.3.4}$$

Theorem 14.3.4 (Existence of Coordinate Basis). *Let M be a smooth m-manifold, let p be a point in M, and let $(U, \varphi = (x^i))$ be a chart at p. Then:*
(a)

$$\left(\left. \frac{\partial}{\partial x^1} \right|_p, \dots, \left. \frac{\partial}{\partial x^m} \right|_p \right)$$

*is a basis for $T_p(M)$, called the **coordinate basis at p** corresponding to (U, φ).*
(b) *For all vectors v in $T_p(M)$,*

$$v = \sum_i v(x^i) \left. \frac{\partial}{\partial x^i} \right|_p.$$

(c) *$T_p(M)$ has dimension m.*

Remark. Theorem 14.3.3(b) justifies having $T_p(M)$ appear in part (b) rather than $T_p(U)$, even though the x^i are functions in $C^\infty(U)$.

Proof. (b): We consider only the case $U = M$. Suppose without loss of generality that $\varphi(p) = (0, \dots, 0)$ and that $\varphi(M)$ is star-shaped about $(0, \dots, 0)$. Let $g : \varphi(M) \longrightarrow \mathbb{R}$ be a smooth function, and define a corresponding smooth function $g_i : \varphi(M) \longrightarrow \mathbb{R}$ by

$$g_i(r^1, \dots, r^m) = \int_0^1 \frac{\partial g}{\partial r^i}(tr^1, \dots, tr^m)\, dt$$

for all $(r^1, \dots, r^m)$ in $\varphi(M)$ for $i = 1, \dots, m$. For a given point $(r^1, \dots, r^m)$ in $\varphi(M)$, define a smooth function $h(t) : [0, 1] \longrightarrow \mathbb{R}$ by

$$h(t) = g(tr^1, \dots, tr^m).$$

This definition makes sense because $\varphi(M)$ is star-shaped about $(0, \dots, 0)$. It follows from the fundamental theorem of calculus that

$$\int_0^1 \frac{dh}{dt}(t)\, dt = h(1) - h(0) = g(r^1, \dots, r^m) - g(0, \dots, 0). \tag{14.3.5}$$

On the other hand, by Theorem 10.1.10,

$$\frac{dh}{dt}(t) = \sum_i r^i \frac{\partial g}{\partial r^i}(tr^1, \ldots, tr^m),$$

so

$$\int_0^1 \frac{dh}{dt}(t)\, dt = \sum_i \left(r^i \int_0^1 \frac{\partial g}{\partial r^i}(tr^1, \ldots, tr^m) \right) dt$$

$$= \sum_i r^i g_i(r^1, \ldots, r^m). \tag{14.3.6}$$

Combining (14.3.5) and (14.3.6) yields

$$g(r^1, \ldots, r^m) = g(0, \ldots, 0) + \sum_i r^i g_i(r^1, \ldots, r^m). \tag{14.3.7}$$

By assumption,

$$\left(x^1(p), \ldots, x^m(p) \right) = \varphi(p) = (0, \ldots, 0). \tag{14.3.8}$$

Let $q = \varphi^{-1}(r^1, \ldots, r^m)$, so that

$$\left(x^1(q), \ldots, x^m(q) \right) = \varphi(q) = (r^1, \ldots, r^m). \tag{14.3.9}$$

From (14.3.7)–(14.3.9), we obtain

$$g \circ \varphi(q) = g \circ \varphi(p) + \sum_i x^i(q)\,(g_i \circ \varphi)(q). \tag{14.3.10}$$

Let f be a function in $C^\infty(M)$. Setting $g = f \circ \varphi^{-1}$ and $f_i = g_i \circ \varphi$ in (14.3.10) gives

$$f(q) = f(p) + \sum_i x^i(q) f_i(q).$$

Since $(r^1, \ldots, r^m)$, hence q, was arbitrary,

$$f = f(p) + \sum_i x^i f_i.$$

For a vector v in $T_p(M)$, we have

$$v(f) = v\left(f(p) + \sum_i x^i f_i \right)$$

$$= \sum_i v(x^i f_i) \qquad\qquad\qquad \text{[Th 14.3.2(b)]}$$

$$= \sum_i x^i(p)\, v(f_i) + \sum_i f_i(p)\, v(x^i) \qquad \text{[(14.3.1)]}$$

$$= \sum_i f_i(p)\, v(x^i). \qquad\qquad\qquad \text{[(14.3.8)]}$$

$$\tag{14.3.11}$$

As a special case,

$$
\begin{aligned}
\left.\frac{\partial}{\partial x^i}\right|_p (f) &= \sum_j \left.\frac{\partial}{\partial x^i}\right|_p (x^j)\, f_j(p) & [\text{Th } 14.3.1(a)] \\
&= \sum_j \frac{\partial x^j}{\partial x^i}(p)\, f_j(p) & [(14.3.3)] \\
&= f_i(p) & [\text{Th } 14.3.1(b)]
\end{aligned}
$$

$$(14.3.12)$$

for $i = 1, \ldots, m$. It follows from (14.3.11) and (14.3.12) that

$$
v(f) = \left(\sum_i v(x^i) \left.\frac{\partial}{\partial x^i}\right|_p \right)(f).
$$

Since f was arbitrary,

$$
v = \sum_i v(x^i) \left.\frac{\partial}{\partial x^i}\right|_p .
$$

(a): We have from part (b) that the tangent vectors $(\partial/\partial x^i)|_p$ span $T_p(M)$. Suppose

$$
\sum_i a^i \left.\frac{\partial}{\partial x^i}\right|_p = 0
$$

for some real numbers a^i. Applying both sides of the preceding identity to x^j and using Theorem 14.3.1(b) gives

$$
0 = \sum_i a^i \frac{\partial x^j}{\partial x^i}(p) = a^j
$$

for $j = 1, \ldots, m$. Thus, the $(\partial/\partial x^i)|_p$ are linearly independent.

(c): This follows from part (a). $\square$

Theorem 14.3.5 (Change of Coordinate Basis). *Let M be a smooth m-manifold, let p be a point in M, let $(U, \varphi = (x^i))$ and $(\widetilde{U}, \widetilde{\varphi} = (\widetilde{x}^j))$ be charts at p, and let $\mathcal{X}_p = ((\partial/\partial x^1)|_p, \ldots, (\partial/\partial x^m)|_p)$ and $\widetilde{\mathcal{X}}_p = ((\partial/\partial \widetilde{x}^1)|_p, \ldots, (\partial/\partial \widetilde{x}^m)|_p)$ be the corresponding coordinate bases at p. Then:*
(a)

$$
\begin{aligned}
\left.\frac{\partial}{\partial x^j}\right|_p &= \sum_i \frac{\partial \widetilde{x}^i}{\partial x^j}(p) \left.\frac{\partial}{\partial \widetilde{x}^i}\right|_p \\
&= \sum_i \frac{\partial(\widetilde{x}^i \circ \varphi^{-1})}{\partial r^j}(\varphi(p)) \left.\frac{\partial}{\partial \widetilde{x}^i}\right|_p .
\end{aligned}
$$

(b)

$$\left[\mathrm{id}_{T_p(M)}\right]_{\mathcal{X}_p}^{\widetilde{\mathcal{X}}_p} = \begin{bmatrix} \dfrac{\partial \widetilde{x}^1}{\partial x^1}(p) & \cdots & \dfrac{\partial \widetilde{x}^1}{\partial x^m}(p) \\ \vdots & \ddots & \vdots \\ \dfrac{\partial \widetilde{x}^m}{\partial x^1}(p) & \cdots & \dfrac{\partial \widetilde{x}^m}{\partial x^m}(p) \end{bmatrix}$$

$$= \begin{bmatrix} \dfrac{\partial(\widetilde{x}^1 \circ \varphi^{-1})}{\partial r^1}(\varphi(p)) & \cdots & \dfrac{\partial(\widetilde{x}^1 \circ \varphi^{-1})}{\partial r^m}(\varphi(p)) \\ \vdots & \ddots & \vdots \\ \dfrac{\partial(\widetilde{x}^m \circ \varphi^{-1})}{\partial r^1}(\varphi(p)) & \cdots & \dfrac{\partial(\widetilde{x}^m \circ \varphi^{-1})}{\partial r^m}(\varphi(p)) \end{bmatrix}.$$

Remark. The second equalities in parts (a) and (b), both of which follow from (14.3.2), are included as a reminder that partial derivatives are ultimately computed using Euclidean methods.

Proof. (a): Let $\left[a_j^i\right] = \left[\mathrm{id}_{T_p(M)}\right]_{\mathcal{X}_p}^{\widetilde{\mathcal{X}}_p}$, so that from (2.2.6) and (2.2.7),

$$\left.\frac{\partial}{\partial x^j}\right|_p = \sum_k a_j^k \left.\frac{\partial}{\partial \widetilde{x}^k}\right|_p.$$

Applying both sides of the preceding identity to $\widetilde{x}^i$ and using Theorem 14.3.1(b) and (14.3.3) yields

$$\frac{\partial \widetilde{x}^i}{\partial x^j}(p) = \sum_k a_j^k \frac{\partial \widetilde{x}^i}{\partial \widetilde{x}^k}(p) = a_j^i.$$

(b): This follows from part (a). $\square$

Let M be a smooth m-manifold, let p be a point in M, and let $(U, \varphi = (x^i))$ be a chart at p. The **second order partial derivative with respect x^i and x^j at p** is the map

$$\left.\frac{\partial^2}{\partial x^i \partial x^j}\right|_p : C^\infty(M) \longrightarrow \mathbb{R}$$

defined by

$$\frac{\partial^2 f}{\partial x^i \partial x^j}(p) = \frac{\partial}{\partial x^i}\left(\frac{\partial f}{\partial x^j}\right)(p)$$

for all functions f in $C^\infty(M)$, where we denote

$$\left.\frac{\partial^2}{\partial x^i \partial x^j}\right|_p(f) \qquad \text{by} \qquad \frac{\partial^2 f}{\partial x^i \partial x^j}(p).$$

Theorem 14.3.6 (Equality of Mixed Partial Derivatives). *With the above setup,*

$$\frac{\partial^2 f}{\partial x^i \partial x^j}(p) = \frac{\partial^2 f}{\partial x^j \partial x^i}(p)$$

for $i, j = 1, \ldots, m$.

Proof. This follows from Theorem 10.1.6. $\square$

14.4   Differential of Maps

To define the differential map between two manifolds, we need a way to send vectors in one tangent space to vectors in another tangent space, and in a linear fashion. With the algebraic approach to vectors, this turns out to be surprisingly straightforward. Let M and N be smooth manifolds, let $F : M \longrightarrow N$ be a smooth map, and let p be a point in M. For each vector v in $T_p(M)$, define a map

$$d_p(F)(v) : C^\infty(N) \longrightarrow \mathbb{R}$$

by

$$d_p(F)(v)(g) = v(g \circ F) \tag{14.4.1}$$

for all functions g in $C^\infty(N)$. Since $g \circ F$ is in $C^\infty(M)$, the definition makes sense.

Theorem 14.4.1. *With the above setup, $d_p(F)(v)$ is a vector in $T_{F(p)}(N)$.*

Proof. Clearly, $d_p(F)(v)$ is linear. To show that $d_p(F)(v)$ satisfies (14.3.1), let g, h be functions in $C^\infty(N)$. Then

$$
\begin{aligned}
d_p(F)(v)(gh) &= v(gh \circ F) && [(14.4.1)] \\
&= v\big((g \circ F)(h \circ F)\big) \\
&= (g \circ F)(p) \cdot v(h \circ F) + (h \circ F)(p) \cdot v(g \circ F) && [(14.3.1)] \\
&= g\big(F(p)\big) \cdot d_p(F)(v)(h) + h\big(F(p)\big) \cdot d_p(F)(v)(g). && [(14.4.1)]
\end{aligned}
$$

Thus, $d_p(F)(v)$ is a tangent vector at $F(p)$. □

Continuing with above notation, the **differential of F at p** is the map

$$d_p(F) : T_p(M) \longrightarrow T_{F(p)}(N)$$

defined by the assignment

$$v \longmapsto d_p(F)(v)$$

for all vectors v in $T_p(M)$.

Theorem 14.4.2. *With the above setup, $d_p(F)$ is linear.*

Proof. Let v, w be vectors in $T_p(M)$, let g be a function in $C^\infty(N)$, and let c be a real number. Then

$$
\begin{aligned}
d_p(F)(cv + w)(g) &= (cv + w)(g \circ F) && [(14.4.1)] \\
&= c\,v(g \circ F) + w(g \circ F) \\
&= c\,d_p(F)(v)(g) + d_p(F)(w)(g) && [(14.4.1)] \\
&= \big(c\,d_p(F)(v) + d_p(F)(w)\big)(g).
\end{aligned}
$$

Since g was arbitrary, the result follows. □

The remaining results of this section give the basic properties of differential maps.

Theorem 14.4.3. *Let M be a smooth manifold, let $\mathrm{id}_M : M \longrightarrow M$ be the identity map, and let p be a point in M. Then*

$$d_p(\mathrm{id}_M) = \mathrm{id}_{T_p(M)}.$$

Proof. For a vector v in $T_p(M)$ and a function f in $C^\infty(M)$, we have from (14.4.1) that

$$d_p(\mathrm{id}_M)(v)(f) = v(f \circ \mathrm{id}_M) = v(f).$$

Since v and f were arbitrary, the result follows. $\square$

Theorem 14.4.4. *Let V and W be vector spaces, let $A : V \longrightarrow W$ be a linear map, and let v be a vector in V. Suppose V and W have standard smooth structures, and make the identifications $T_v(V) = V$ and $T_{A(v)}(W) = W$. Then*

$$d_v(A) = A.$$

Proof. Straightforward. $\square$

Theorem 14.4.5 (Chain Rule). *Let M, N, and P be smooth manifolds, let $F : M \longrightarrow N$ and $G : N \longrightarrow P$ be smooth maps, and let p be a point in M. Then*

$$d_p(G \circ F) = d_{F(p)}(G) \circ d_p(F).$$

Proof. For a vector v in $T_p(M)$ and a function f in $C^\infty(P)$, we have

$$
\begin{aligned}
d_p(G \circ F)(v)(f) &= v(f \circ G \circ F) & [(14.4.1)]\\
&= d_p(F)(v)(f \circ G) & [(14.4.1)]\\
&= d_{F(p)}(G)\big(d_p(F)(v)(f)\big) & [(14.4.1)]\\
&= \big(d_{F(p)}(G) \circ d_p(F)\big)(v)(f).
\end{aligned}
$$

Since v and f were arbitrary, the result follows. $\square$

The next result is a generalization of Theorem 14.3.5.

Theorem 14.4.6 (Matrix of Differential Map). *Let M be a smooth m-manifold, let N be a smooth n-manifold, and let $F : M \longrightarrow N$ be a smooth map. Let p be a point in M, let $(U, \varphi = (x^i))$ and $(V, (y^j))$ be charts at p and $F(p)$, respectively, and let $\mathcal{X}_p = \big((\partial/\partial x^1)|_p, \ldots, (\partial/\partial x^m)|_p\big)$ and $\mathcal{Y}_{F(p)} = \big((\partial/\partial y^1)|_{F(p)}, \ldots, (\partial/\partial y^n)|_{F(p)}\big)$ be the corresponding coordinate bases at p and $F(p)$, respectively. Then:*
(a)

$$
\begin{aligned}
d_p(F)\left(\left.\frac{\partial}{\partial x^j}\right|_p\right) &= \sum_i \frac{\partial(y^i \circ F)}{\partial x^j}(p) \left.\frac{\partial}{\partial y^i}\right|_{F(p)}\\
&= \sum_i \frac{\partial(y^i \circ F \circ \varphi^{-1})}{\partial r^j}(\varphi(p)) \left.\frac{\partial}{\partial y^i}\right|_{F(p)}.
\end{aligned}
$$

(b)

$$[d_p(F)]_{\mathcal{X}_p}^{\mathcal{Y}_{F(p)}} = \begin{bmatrix} \dfrac{\partial(y^1 \circ F)}{\partial x^1}(p) & \cdots & \dfrac{\partial(y^1 \circ F)}{\partial x^m}(p) \\ \vdots & \ddots & \vdots \\ \dfrac{\partial(y^n \circ F)}{\partial x^1}(p) & \cdots & \dfrac{\partial(y^n \circ F)}{\partial x^m}(p) \end{bmatrix}$$

$$= \begin{bmatrix} \dfrac{\partial(y^1 \circ F \circ \varphi^{-1})}{\partial r^1}(\varphi(p)) & \cdots & \dfrac{\partial(y^1 \circ F \circ \varphi^{-1})}{\partial r^m}(\varphi(p)) \\ \vdots & \ddots & \vdots \\ \dfrac{\partial(y^n \circ F \circ \varphi^{-1})}{\partial r^1}(\varphi(p)) & \cdots & \dfrac{\partial(y^n \circ F \circ \varphi^{-1})}{\partial r^m}(\varphi(p)) \end{bmatrix}.$$

Remark. The second equalities in parts (a) and (b), both of which follow from (14.3.2), are included as a reminder that partial derivatives are ultimately computed using Euclidean methods.

Proof. (a): Let $[a_j^k] = [d_p(F)]_{\mathcal{X}_p}^{\mathcal{Y}_{F(p)}}$, so that from (2.2.6) and (2.2.7),

$$d_p(F)\left(\left.\frac{\partial}{\partial x^j}\right|_p\right) = \sum_k a_j^k \left.\frac{\partial}{\partial y^k}\right|_{F(p)}.$$

Applying both sides of the preceding identity to y^i and using Theorem 14.3.1(b) and (14.3.3) yields

$$d_p(F)\left(\left.\frac{\partial}{\partial x^j}\right|_p\right)(y^i) = \sum_k a_j^k \frac{\partial y^i}{\partial y^k}(F(p)) = a_j^i.$$

On the other hand, we have from (14.3.3) and (14.4.1) that

$$d_p(F)\left(\left.\frac{\partial}{\partial x^j}\right|_p\right)(y^i) = \left(\left.\frac{\partial}{\partial x^j}\right|_p\right)(y^i \circ F) = \frac{\partial(y^i \circ F)}{\partial x^j}(p).$$

Combining the above identities gives the result.

(b): This follows from (2.2.6), (2.2.7), and part (a). □

14.5 Differential of Functions

Let M be a smooth manifold, let f be a function in $C^\infty(M)$, let p be a point in M, and let $(U, \varphi = (x^i))$ be a chart at p. Viewing $\mathbb{R}$ as a smooth 1-manifold, we have the differential map

$$d_p(f) : T_p(M) \longrightarrow T_{f(p)}(\mathbb{R}). \tag{14.5.1}$$

A covering chart for $\mathbb{R}$ is $(\mathbb{R}, \text{id}_{\mathbb{R}} = r)$, where r is the standard coordinate on $\mathbb{R}$. The corresponding coordinate basis at $f(p)$ is $((d/dr)|_{f(p)})$. Then Theorem

14.4.6(a) and $\mathrm{id}_{\mathbb{R}} = r$ give

$$d_p(f)\left(\left.\frac{\partial}{\partial x^i}\right|_p\right) = \frac{\partial(r \circ f)}{\partial x^i}(p) \left.\frac{d}{dr}\right|_{f(p)} = \frac{\partial f}{\partial x^i}(p) \left.\frac{d}{dr}\right|_{f(p)}. \tag{14.5.2}$$

Using (14.3.4), we identify $T_{f(p)}(\mathbb{R})$ with $\mathbb{R}$, and write $T_{f(p)}(\mathbb{R}) = \mathbb{R}$. To be consistent, we also identity the basis $((d/dr)|_{f(p)})$ for $T_{f(p)}(\mathbb{R})$ with the basis (1) for $\mathbb{R}$. Then (14.5.1) and (14.5.2) become

$$d_p(f) : T_p(M) \longrightarrow \mathbb{R} \tag{14.5.3}$$

and

$$d_p(f)\left(\left.\frac{\partial}{\partial x^i}\right|_p\right) = \frac{\partial f}{\partial x^i}(p) \tag{14.5.4}$$

for $i = 1, \ldots, m$. Let us denote the dual space

$$T_p(M)^* \qquad \text{by} \qquad T_p^*(M).$$

The usual identification of a vector space with its double dual gives

$$T_p^{**}(M) = T_p(M).$$

Theorem 14.5.1. *With the above setup, $d_p(f)$ is a covector in $T_p^*(M)$.*

Proof. This follows from Theorem 14.4.2 and (14.5.3). $\qquad\qquad\square$

Theorem 14.5.2 (Existence of Dual Coordinate Basis). *Let M be a smooth m-manifold, let p be a point in M, let $(U, (x^i))$ be a chart at p, and let $\mathcal{X}_p = ((\partial/\partial x^1)|_p, \ldots, (\partial/\partial x^m)|_p)$ be the corresponding coordinate basis at p. Then $(d_p(x^1), \ldots, d_p(x^m))$ is the dual basis of $\mathcal{X}_p$, called the **dual coordinate basis at p** corresponding to $(U, (x^i))$.*

Proof. Setting $f = x^j$ in (14.5.4) and using Theorem 14.3.1(b) gives the result. $\qquad\square$

With the next result, we recover an identity that is familiar from the differential calculus of several real variables, except that here "differentials" replace "infinitesimals".

Theorem 14.5.3. *Let M be a smooth m-manifold, and let f be a function in $C^\infty(M)$. Let p be a point in M, let $(U, (x^i))$ be a chart at p, and let $((\partial/\partial x^1)|_p, \ldots, (\partial/\partial x^m)|_p)$ and $(d_p(x^1), \ldots, d_p(x^m))$ be the corresponding coordinate and dual coordinate bases at p. Then, in local coordinates,*

$$d_p(f) = \sum_i \frac{\partial f}{\partial x^i}(p)\, d_p(x^i).$$

Proof. By Theorem 14.5.2, $d_p(f) = \sum_j a_j d_p(x^j)$ for some real numbers a_j. We have

$$\frac{\partial f}{\partial x^i}(p) = d_p(f)\left(\frac{\partial}{\partial x^i}\bigg|_p\right) \qquad\qquad [(14.5.4)]$$

$$= \left(\sum_j a_j d_p(x^j)\right)\left(\frac{\partial}{\partial x^i}\bigg|_p\right)$$

$$= \sum_j a_j\, d_p(x^j)\left(\frac{\partial}{\partial x^i}\bigg|_p\right)$$

$$= a_i \qquad\qquad [\text{Th } 14.5.2]$$

for $i = 1, \ldots, m$.   □

Theorem 14.5.4 (Change of Dual Coordinate Basis). *Let M be a smooth m-manifold, and let p be a point in M. Let $(U, (x^i))$ and $(\widetilde{U}, (\widetilde{x}^j))$ be charts at p, let $\mathcal{X}_p = ((\partial/\partial x^1)|_p, \ldots, (\partial/\partial x^m)|_p)$ and $\widetilde{\mathcal{X}}_p = ((\partial/\partial \widetilde{x}^1)|_p, \ldots, (\partial/\partial \widetilde{x}^m)|_p)$ be the corresponding coordinate bases at p, and let $\mathcal{X}_p^* = (d_p(x^1), \ldots, d_p(x^m))$ and $\widetilde{\mathcal{X}}_p^* = (d_p(\widetilde{x}^1), \ldots, d_p(\widetilde{x}^m))$ be the corresponding dual coordinate bases at p. Then:*
(a)

$$d_p(x^j) = \sum_i \frac{\partial x^j}{\partial \widetilde{x}^i}(p)\, d_p(\widetilde{x}^i)$$

for $j = 1, \ldots, m$.
(b)

$$\left[\mathrm{id}_{T_p^*(M)}\right]_{\mathcal{X}_p^*}^{\widetilde{\mathcal{X}}_p^*} = \begin{bmatrix} \dfrac{\partial x^1}{\partial \widetilde{x}^1}(p) & \cdots & \dfrac{\partial x^m}{\partial \widetilde{x}^1}(p) \\ \vdots & \ddots & \vdots \\ \dfrac{\partial x^1}{\partial \widetilde{x}^m}(p) & \cdots & \dfrac{\partial x^m}{\partial \widetilde{x}^m}(p) \end{bmatrix} = \left(\left[\mathrm{id}_{T_p(M)}\right]_{\widetilde{\mathcal{X}}_p}^{\mathcal{X}_p}\right)^{\mathrm{T}}.$$

Proof. (a): Setting $f = x^j$ and $x^i = \widetilde{x}^i$ in Theorem 14.5.3 gives the result.

(b): The first equality follows from (2.2.6), (2.2.7), and part (a). The second equality follows from Theorem 14.3.5(b).   □

Theorem 14.5.5. *Let M be a smooth manifold, let f be a function in $C^\infty(M)$, let p be a point in M, and let v be a vector in $T_p(M)$. Then*

$$d_p(f)(v) = v(f).$$

Proof. By Theorem 1.2.1(d) and Theorem 14.5.2,

$$v = \sum_i d_p(x^i)(v)\, \frac{\partial}{\partial x^i}\bigg|_p,$$

hence

$$v(f) = \left(\sum_i d_p(x^i)(v) \left. \frac{\partial}{\partial x^i} \right|_p \right)(f)$$

$$= \sum_i d_p(x^i)(v) \frac{\partial f}{\partial x^i}(p) \qquad [(14.3.3)]$$

$$= \left(\sum_i \frac{\partial f}{\partial x^i}(p) \, d_p(x^i) \right)(v)$$

$$= d_p(f)(v). \qquad [\text{Th } 14.5.3] \qquad \square$$

Theorem 14.5.6. *Let M be a smooth manifold, let p be a point in M, let f, g be functions in $C^\infty(M)$, and let c be a real number. Then:*
(a) $d_p(cf + g) = c\, d_p(f) + d_p(g)$.
(b) $d_p(fg) = f(p)\, d_p(g) + g(p)\, d_p(f)$.

Proof. For a vector v in $T_p(M)$, we have

$$d_p(cf + g)(v) = v(cf + g) \qquad [\text{Th } 14.5.5]$$

$$= c\, v(f) + v(g)$$

$$= c\, d_p(f)(v) + d_p(g)(v) \qquad [\text{Th } 14.5.5]$$

$$= \big(c\, d_p(f) + d_p(g) \big)(v)$$

and

$$d_p(fg)(v) = v(fg) \qquad [\text{Th } 14.5.5]$$

$$= f(p)\, v(g) + g(p)\, v(f) \qquad [(14.3.1)]$$

$$= f(p)\, d_p(g)(v) + g(p)\, d_p(f)(v) \qquad [\text{Th } 14.5.5]$$

$$= \big(f(p)\, d_p(g) + g(p)\, d_p(f) \big)(v).$$

Since v was arbitrary, the result follows. $\qquad \square$

Theorem 14.5.7. *Let $(e_1, \ldots, e_m)$ be the standard basis for $\mathbb{R}^m$, and let $(\xi^1, \ldots, \xi^m)$ be its dual basis. Let $(r^1, \ldots, r^m)$ be standard coordinates on $\mathbb{R}^m$, let p be a point in $\mathbb{R}^m$, let $((\partial/\partial r^1)|_p, \ldots, \partial/\partial r^m)|_p)$ be the corresponding coordinate basis at p, called the **standard coordinate basis at p**, and let $(d_p(r^1), \ldots, d_p(r^m))$ be the corresponding dual coordinate basis at p, called the **standard dual coordinate basis at p**. Then $((\partial/\partial r^1)|_p, \ldots, \partial/\partial r^m)|_p)$ can be identified with $(e_1, \ldots, e_m)$, and $(d_p(r^1), \ldots, d_p(r^m))$ can be identified with $(\xi^1, \ldots, \xi^m)$, written*

$$\left(\left. \frac{\partial}{\partial r^1} \right|_p, \ldots, \left. \frac{\partial}{\partial r^m} \right|_p \right) = (e_1, \ldots, e_m)$$

and

$$(d_p(r^1), \ldots, d_p(r^m)) = (\xi^1, \ldots, \xi^m).$$

Proof. It is sufficient to prove either of the above identities, the other following by taking the dual. In keeping with (14.3.4), we make the identification

$T_p(\mathbb{R}^m) = \mathbb{R}^m$. By definition, $(r^1, \ldots, r^m)$ is the coordinate map of the chart $(\mathbb{R}^m, \mathrm{id}_{\mathbb{R}^m} = (r^1, \ldots, r^m))$. Since r^i is linear, by Theorem 14.4.4, $d_p(r^i) = r^i$ for $i = 1, \ldots, m$. For a point q in $\mathbb{R}^m$,

$$\sum_i d_p(r^i)(q)\, e_i = \sum_i r^i(q) e_i = (r^1(q), \ldots, r^m(q)) = (r^1, \ldots, r^m)(q)$$

$$= \mathrm{id}_{\mathbb{R}^m}(q) = q = \sum_i \xi^i(q) e_i,$$

where the last equality follows from Theorem 1.2.1(d). Since q was arbitrary, $d_p(r^i) = \xi^i$ for $i = 1, \ldots, m$. $\qquad\square$

14.6 Immersions and Diffeomorphisms

In this brief section, we generalize the discussion of immersions and diffeomorphisms in Section 10.2 to the setting of smooth manifolds.

Let M and N be smooth manifolds, where $\dim(M) \le \dim(N)$, let $F : M \longrightarrow N$ be a smooth map, and let p be a point in M. We say that F is an **immersion at p** if the differential map $d_p(F) : T_p(M) \longrightarrow T_{F(p)}(N)$ is injective, and that F is an **immersion (on M)** if it is an immersion at every p in M.

Now suppose M and N have the same dimension, and let $G, H : M \longrightarrow N$ be smooth maps. We say that G is a **diffeomorphism**, and that M and N are **diffeomorphic**, if G is bijective and $G^{-1} : N \longrightarrow M$ is smooth. We say that H is a **local diffeomorphism at p** if there is a neighborhood U of p in M and a neighborhood V of $H(p)$ in N such that $H|_U : U \longrightarrow V$ is a diffeomorphism. Then H is said to be a **local diffeomorphism (on M)** if it is a local diffeomorphism at every p in M.

Theorem 14.6.1. *Let M and N be smooth manifolds with the same dimension, let $F : M \longrightarrow N$ be a diffeomorphism, and let p be a point in M. Then the differential map $d_p(F) : T_p(M) \longrightarrow T_{F(p)}(N)$ is a linear isomorphism, with inverse $d_p(F)^{-1} = d_{F(p)}(F^{-1})$.*

Proof. By Theorem 14.4.2, $d_p(F)$ is a linear map. We have from Theorem 14.4.3 and Theorem 14.4.5 that

$$\mathrm{id}_{T_p(M)} = d_p(\mathrm{id}_M) = d_p(F^{-1} \circ F) = d_{F(p)}(F^{-1}) \circ d_p(F).$$

Similarly, $\mathrm{id}_{T_{F(p)}(N)} = d_p(F) \circ d_{F(p)}(F^{-1})$. By Theorem A.4, $d_p(F)$ is bijective. $\qquad\square$

Theorem 14.6.2 (Inverse Map Theorem). *Let M and N be smooth manifolds with the same dimension, let $F : M \longrightarrow N$ be a smooth map, and let p be a point in M. Then F is a local diffeomorphism at p if and only if it is an immersion at p. Thus, F is a local diffeomorphism if and only if it is an immersion.* $\qquad\square$

14.7 Curves

The following definitions are borrowed more or less directly from Section 10.1 and Section 11.1.

A **(parametrized) curve** on a smooth manifold M is a map $\lambda : I \longrightarrow M$, where I is an interval in $\mathbb{R}$ that is either open, closed, half-open, or half-closed, and where the possibility that I is infinite is not excluded. Our focus will be on the case where I is a finite open interval, usually denoted by (a, b). Rather than provide a separate statement identifying the independent variable for the curve, most often denoted by t, and sometimes by u, it is helpful to incorporate this into the notation for λ, as in $\lambda(t) : I \longrightarrow M$. When I is a closed interval $[a, b]$ and λ is continuous, we say that λ **joins** $\lambda(a)$ to $\lambda(b)$. It is convenient to adopt the following convention.

**Henceforth, when required by the context,
the interval I is assumed to contain 0.**

Consider the curve $\lambda(t) : (a, b) \longrightarrow M$. Viewing (a, b) as a smooth 1-manifold, we say that λ is **smooth [on (a, b)]** if it is smooth as a map between smooth manifolds. Suppose λ is in fact smooth. For a given point t in (a, b), we have the differential map

$$d_t(\lambda) : T_t\big((a, b)\big) \longrightarrow T_{\lambda(t)}(M).$$

A covering chart for (a, b) is $\big((a, b), \mathrm{id}_{(a,b)} = r\big)$, where r is the standard coordinate on (a, b). The corresponding coordinate basis at t is $\big((d/dr)|_t\big)$. It follows that

$$d_t(\lambda)\left(\frac{d}{dr}\bigg|_t\right) \qquad \text{is a vector in} \qquad T_{\lambda(t)}(M) \tag{14.7.1}$$

for all t in (a, b). Thus, for all functions f in $C^\infty\big((a, b)\big)$,

$$\begin{aligned}
d_t(\lambda)\left(\frac{d}{dr}\bigg|_t\right)(f) &= \frac{d}{dr}\bigg|_t (f \circ \lambda) && [(14.4.1)] \\
&= \frac{d(f \circ \lambda)}{dr}(t). && [(14.3.3)]
\end{aligned} \tag{14.7.2}$$

In an effort to ensure that the notation adopted for smooth manifolds resembles as much as possible the notation from differential calculus, let us denote

$$d_t(\lambda)\left(\frac{d}{dr}\bigg|_t\right) \qquad \text{by} \qquad \frac{d\lambda}{dr}(t). \tag{14.7.3}$$

We continue to indulge in the usual (and sometimes confusing) practice of obscuring the difference between a variable and its value. With this understanding, (14.7.1) and (14.7.2) become:

$$\frac{d\lambda}{dt}(t) \qquad \text{is a vector in} \qquad T_{\lambda(t)}(M) \tag{14.7.4}$$

and

$$\frac{d\lambda}{dt}(t)(f) = \frac{d(f \circ \lambda)}{dt}(t). \tag{14.7.5}$$

We refer to $(d\lambda/dt)(t)$ as the **velocity of λ at t**.

Theorem 14.7.1. *Let M be a smooth m-manifold, let $\lambda(t) : (a,b) \longrightarrow M$ be a smooth curve, and let $(U, \varphi = (x^i))$ be a chart on M such that $\lambda((a,b)) \subset U$. Then:*
(a) *The smooth curve $\varphi \circ \lambda(t) : (a,b) \longrightarrow \mathbb{R}^m$ is given by*

$$\varphi \circ \lambda(t) = \left(x^1 \circ \lambda(t), \dots, x^m \circ \lambda(t)\right).$$

(b)

$$\frac{d\lambda}{dt}(t) = \sum_i \frac{d(x^i \circ \lambda)}{dt}(t) \left.\frac{\partial}{\partial x^i}\right|_{\lambda(t)}$$

for all t in (a,b). Thus, the components of $(d\lambda/dt)(t)$ with respect to the coordinate basis at $\lambda(t)$ are the components of $(d(\varphi \circ \lambda)/dt)(t)$ with respect to the standard basis for $\mathbb{R}^m$.

Proof. (a): We have

$$\varphi \circ \lambda(t) = (x^1, \dots, x^m) \circ \lambda(t) = \left(x^1 \circ \lambda(t), \dots, x^m \circ \lambda(t)\right).$$

(b): This follows from Theorem 14.4.6(a), (14.7.3), and part (a). $\qquad\square$

It was remarked in the introduction to Part III that the algebraic approach to defining tangent vectors gives rise to results that have something of the geometric flavor found in the theory of surfaces. We close this section with several such instances.

Theorem 14.7.2. *Let M be a smooth m-manifold, let p be a point in M, and let v be a vector in $T_p(M)$. Then there is a real number $\varepsilon > 0$ and a smooth curve $\lambda(t) : (-\varepsilon, \varepsilon) \longrightarrow M$ such that $\lambda(0) = p$ and $(d\lambda/dt)(0) = v$.*

Proof. Let $(U, \varphi = (x^i))$ be a chart at p such that $\varphi(p) = (0, \dots, 0)$, and, in local coordinates, let

$$v = \sum_i a^i \left.\frac{\partial}{\partial x^i}\right|_p.$$

Define a smooth curve $\lambda(t) : (-\varepsilon, \varepsilon) \longrightarrow M$ by $\lambda(t) = \varphi^{-1}(ta^1, \dots, ta^m)$, where ε is chosen small enough that $\lambda((-\varepsilon, \varepsilon)) \subset U$. Clearly, $\lambda(0) = p$. We have from Theorem 14.7.1(a) that $x^i \circ \lambda(t) = ta^i$, hence

$$a^i = \frac{d(x^i \circ \lambda)}{dt}(0)$$

for $i = 1, \dots, m$. It follows from the preceding identities and Theorem 14.7.1(b) that

$$\frac{d\lambda}{dt}(0) = \sum_i \frac{d(x^i \circ \lambda)}{dt}(0) \left.\frac{\partial}{\partial x^i}\right|_{\lambda(0)} = v. \qquad\square$$

Theorem 14.7.3. *If M is a smooth manifold and p is a point in M, then*

$$T_p(M) = \left\{ \frac{d\lambda}{dt}(t_0) : \lambda(t) : (a,b) \longrightarrow M \text{ is smooth, } \lambda(t_0) = p, \, t_0 \in (a,b) \right\}.$$

Proof. This follows from Theorem 14.7.2 and (14.7.4).   □

Let M be a smooth manifold, let p be a point in M, and let v be a vector in $T_p(M)$. We have from the preceding theorem that there is a smooth curve $\lambda(t) : (a,b) \longrightarrow M$ such that $\lambda(t_0) = p$ and $(d\lambda/dt)(t_0) = v$ for some t_0 in (a,b). Let X be a vector field in $\mathfrak{X}(M)$. Then X_p is a vector in $T_p(M)$, hence there corresponds such a smooth curve. We will make use of these observations frequently.

Theorem 14.7.4. *Let M and N be smooth manifolds, let $F : M \longrightarrow N$ be a smooth map, let p be a point in M, and let v be a vector in $T_p(M)$. Then:*
(a)

$$d_p(F)(v) = \frac{d(F \circ \lambda)}{dt}(t_0),$$

where $\lambda(t) : (a,b) \longrightarrow M$ is any smooth curve such that $\lambda(t_0) = p$ and $(d\lambda/dt)(t_0) = v$ for some t_0 in (a,b).
(b) *If $\psi(t) : (a,b) \longrightarrow M$ is a smooth curve, then*

$$d_{\psi(t)}(F)\left(\frac{d\psi}{dt}(t) \right) = \frac{d(F \circ \psi)}{dt}(t)$$

for all t in (a,b).

Remark. Since $F \circ \lambda, F \circ \psi : (a,b) \longrightarrow N$ are smooth curves, the assertions makes sense.

Proof. (a): For a function g in $C^\infty(N)$, we have

$$
\begin{aligned}
d_p(F)(v)(g) &= v(g \circ F) & [(14.4.1)] \\
&= \frac{d\lambda}{dt}(t_0)(g \circ F) & \\
&= \frac{d(g \circ F \circ \lambda)}{dt}(t_0) & [(14.7.5)] \\
&= \frac{d(F \circ \lambda)}{dt}(t_0)(g). & [(14.7.5)]
\end{aligned}
$$

Since g was arbitrary, the result follows.
   (b): This follows from part (a).   □

14.8   Submanifolds

Having defined smooth manifolds, it is natural to consider subsets with corresponding properties. Let $\overline{M}$ be a smooth $\overline{m}$-manifold, and let M be a subset of

$\overline{M}$ that is a smooth m-manifold in its own right. Without further assumptions, there is no reason to expect a connection between the topologies on $\overline{M}$ and M, and likewise for their smooth structures. We say that M is an **m-submanifold** of $\overline{M}$ if:

[**S1**] M has the subspace topology.

[**S2**] The inclusion map $\iota : M \longrightarrow \overline{M}$ is an immersion.

Suppose M is in fact an m-submanifold of $\overline{M}$. It follows from [S2] that $m \leq \overline{m}$. We say that M is a **hypersurface** of $\overline{M}$ if $\dim(M) = \dim(\overline{M}) - 1$. Since $\iota : M \longrightarrow \overline{M}$ is an immersion, for each point p in M, the differential map $d_p(\iota) : T_p(M) \longrightarrow T_p(\overline{M})$ is injective. Given a vector v in $T_p(M)$, the image vector $d_p(\iota)(v)$ behaves as follows: for all functions f in $C^\infty(M)$,

$$d_p(\iota)(v)(f) = v(f \circ \iota) = v(f|_M).$$

We adopt the established convention of identifying $T_p(M)$ with its image under $d_p(\iota)$. Thus, $T_p(M)$ is viewed as a vector subspace of $T_p(\overline{M})$, and we write

$$T_p(M) \subseteq T_p(\overline{M}). \tag{14.8.1}$$

Theorem 14.8.1. *A subset of a smooth manifold can be made into a submanifold in at most one way. Thus, a subset of a smooth manifold is either a submanifold or not.*   □

Theorem 14.8.2 (Regular Surface). *Any regular surface is a hypersurface of $\mathbb{R}^3$.*

Remark. According to Example 14.1.4 and Example 14.1.5, a regular surface is a smooth 2-manifold, and $\mathbb{R}^3$ is a smooth 3-manifold, so the assertion makes sense.

Proof. We need to verify that [S1] and [S2] are satisfied.

[S1]: By definition, M has the subspace topology.

[S2]: To prove that the inclusion map $\iota : M \longrightarrow \mathbb{R}^3$ is an immersion, we have to show that it is smooth and the differential map $d_p(\iota) : T_p(M) \longrightarrow T_p(\mathbb{R}^3)$ is injective for all p in M.

Smoothness. Let (U, φ) and (V, ψ) be charts on M and $\mathbb{R}^3$, respectively, such that $\iota(U) = U \subset V$. By Theorem 11.2.9, the map

$$\psi \circ \iota \circ \varphi^{-1} = \psi \circ \varphi^{-1} : \varphi(U) \longrightarrow \mathbb{R}^3$$

is smooth. Since (U, φ) and (V, ψ) were arbitrary, it follows from Theorem 14.2.3 that ι is smooth.

Injectivity. Let p be a point in M, and let $(U, \varphi = (\varphi^1, \varphi^2, \varphi^3))$ be a chart at p in the sense of Section 11.2. Then $(\varphi(U), \varphi^{-1})$ is a chart at p in the sense of Section 14.1. Let $\mathcal{X}_p$ be the corresponding coordinate basis at p, and let (r^1, r^2) and (s^1, s^2, s^3) be standard coordinates on $\mathbb{R}^2$ and $\mathbb{R}^3$, respectively.

Then $\left(\mathbb{R}^3, (s^1, s^2, s^3)\right)$ is a chart on $\mathbb{R}^3$. Let $\mathcal{Y}_p$ be the corresponding coordinate basis at $p = \iota(p)$. It follow from Theorem 14.4.6(b) that

$$[d_p(\iota)]_{\mathcal{X}_p}^{\mathcal{Y}_p} = \left[\frac{\partial(s^i \circ \iota \circ \varphi)}{\partial r^j}(\varphi^{-1}(p))\right] = \left[\frac{\partial \varphi^i}{\partial r^j}(\varphi^{-1}(p))\right],$$

and then from property [C1] of Section 11.2 and Theorem 11.2.2 that $d_p(\iota)$ is injective.   □

Reflecting on the above definition of a submanifold, [S1] seems like an obvious requirement, but the same cannot be said of [S2]. Theorem 14.8.2 gives some insight into [S2], but its rationale is far from transparent. Rather than search for a deeper understanding of [S2], we change course and provide an alternative perspective on submanifolds.

For a given integer $1 \le k \le m$, we define a type of projection map $\mathcal{P}_k : \mathbb{R}^m \longrightarrow \mathbb{R}^k \times \{0\}^{m-k}$ by

$$\mathcal{P}_k(r^1, \ldots, r^k, r^{k+1}, \ldots, r^m) = (r^1, \ldots, r^k, 0, \ldots, 0).$$

Let U be an open set in $\mathbb{R}^m$, and let S be a subset of U. We say that S is a **k-slice of U** if $S \subseteq \mathcal{P}_k(U)$.

Theorem 14.8.3 (Slice Criterion for Submanifolds). *Let $\overline{M}$ be a smooth $\overline{m}$-manifold, let M be a subset of $\overline{M}$ with the subspace topology, and let $1 \le m \le \overline{m}$. Then:*
(a) *M is an m-submanifold of $\overline{M}$ if and only if for every point p in M, there is a chart $\left(U, \varphi = (x^1, \ldots, x^{\overline{m}})\right)$ on $\overline{M}$ at p, called a **slice chart for M in $\overline{M}$**, such that $\varphi(M \cap U)$ is an m-slice of $\varphi(U)$.*
(b) *If M is an m-submanifold of $\overline{M}$, then corresponding to the slice chart $\left(U, \varphi = (x^1, \ldots, x^{\overline{m}})\right)$ on $\overline{M}$ at p, there is the chart $\left(M \cap U, (x^1|_{M \cap U}, \ldots, x^m|_{M \cap U})\right)$ on M at p such that*

$$\left(\frac{\partial}{\partial(x^1|_{M \cap U})}\bigg|_q, \ldots, \frac{\partial}{\partial(x^m|_{M \cap U})}\bigg|_q\right)$$

can be identified with

$$\left(\frac{\partial}{\partial x^1}\bigg|_q, \ldots, \frac{\partial}{\partial x^m}\bigg|_q\right)$$

for all q in $M \cap U$. Thus, $T_q(M)$ can be identified with the subspace of $T_q(\overline{M})$ spanned by $((\partial/\partial x^1)|_q, \ldots, (\partial/\partial x^m)|_q)$ for all q in $M \cap U$. □

The above notation is rather cumbersome. Henceforth, $x^i|_{M \cap U}$ will be abbreviated to x^i for $i = 1, \ldots, m$. In this revised notation, we denote

$$\left(M \cap U, (x^1|_{M \cap U}, \ldots, x^m|_{M \cap U})\right) \qquad \text{by} \qquad \left(M \cap U, (x^1, \ldots, x^m)\right),$$

and

$$\left(\frac{\partial}{\partial(x^1|_{M \cap U})}\bigg|_q, \ldots, \frac{\partial}{\partial(x^m|_{M \cap U})}\bigg|_q\right) \qquad \text{by} \qquad \left(\frac{\partial}{\partial x^1}\bigg|_q, \ldots, \frac{\partial}{\partial x^m}\bigg|_q\right).$$

We require two further results on submanifolds, both of which are straightforward consequences of Theorem 14.8.3.

We saw in Theorem 14.3.3(a) that an open set in a smooth manifold is itself a smooth manifold. The next result says that it is also a submanifold.

Theorem 14.8.4 (Open Submanifold). *If M is a smooth m-manifold and U is an open set in M, then U is an m-submanifold of M, called the **open submanifold** of M corresponding to U.* $\qquad\square$

> **Throughout, any open set in a smooth manifold is viewed as an open submanifold.**

Theorem 14.8.5 (Chart). *Let M be a smooth m-manifold, let p be a point in M, let (U, φ) be a chart at p, and let $\mathcal{X}_p = ((\partial/\partial x^1)|_p, \ldots, (\partial/\partial x^m)|_p)$ be the corresponding coordinate basis at p. Let $(r^1, \ldots, r^m)$ be standard coordinates on $\mathbb{R}^m$, and let $((\partial/\partial r^1)|_p, \ldots, (\partial/\partial r^m)|_p)$ be the standard coordinate basis at p. Viewing U and $\varphi(U)$ as open m-submanifolds of M and $\mathbb{R}^m$, respectively, make the identifications $T_p(U) = T_p(M)$ and $T_{\varphi(p)}(\varphi(U)) = \mathbb{R}^m$. Let $\mathcal{E} = (e_1, \ldots, e_m)$ be the standard basis for $\mathbb{R}^m$, and make the identification $((\partial/\partial r^1)|_p, \ldots, (\partial/\partial r^m)|_p) = \mathcal{E}$ (as given by Theorem 14.5.7). Then:*
(a) *$\varphi : U \longrightarrow \varphi(U)$ is a diffeomorphism.*
(b) *$\mathcal{X}_p$ is a basis for $T_p(U)$.*
(c)

$$\left[d_p(\varphi)\right]_{\mathcal{X}_p}^{\mathcal{E}} = I_m = \left[d_{\varphi(p)}(\varphi^{-1})\right]_{\mathcal{E}}^{\mathcal{X}_p}.$$

(d)

$$d_{\varphi(p)}(\varphi^{-1})(e_i) = \frac{\partial}{\partial x^i}\bigg|_p$$

for $i = 1, \ldots, m$.

Proof. (a): Straightforward.

(b): This follows from Theorem 14.3.3(b) and Theorem 14.3.4(a). Alternatively, setting $M = U$ and $\overline{M} = M$ in Theorem 14.8.3(b) gives the result.

(c): We have

$$\begin{aligned}
\frac{\partial(r^i \circ \varphi)}{\partial x^j}(p) &= \frac{\partial(r^i \circ \varphi \circ \varphi^{-1})}{\partial r^j}(\varphi(p)) && [(14.3.2)] \\
&= \frac{\partial r^i}{\partial r^j}(\varphi(p)) \\
&= \delta_j^i && [\text{Th } 14.3.1(b)]
\end{aligned}$$

for $i, j = 1, \ldots, m$. Then Theorem 14.4.6(b) yields $\left[d_p(\varphi)\right]_{\mathcal{X}_p}^{\mathcal{E}} = I_m$, which proves the first equality. We also have

$$\begin{aligned}
\left[d_{\varphi(p)}(\varphi^{-1})\right]_{\mathcal{E}}^{\mathcal{X}_p} &= \left[d_p(\varphi)^{-1}\right]_{\mathcal{E}}^{\mathcal{X}_p} && [\text{Th } 14.6.1, \text{ part (a)}] \\
&= \left(\left[d_p(\varphi)\right]_{\mathcal{X}_p}^{\mathcal{E}}\right)^{-1} && [\text{Th } 2.2.5(b)] \\
&= I_m, && [\text{first equality}]
\end{aligned}$$

which proves the second equality.

(d): This follows from (2.2.2), (2.2.3), and part (c).   □

14.9   Parametrized Surfaces

A **parametrized surface** on a smooth manifold M is a smooth map of the form

$$\sigma(r,s) : (a,b) \times (-\varepsilon, \varepsilon) \longrightarrow M,$$

where $\varepsilon > 0$ is a real number. For a given point r in (a,b), we define a smooth curve

$$\sigma_r(s) : (-\varepsilon, \varepsilon) \longrightarrow M$$

by

$$\sigma_r(s) = \sigma(r,s)$$

for all s in $(-\varepsilon, \varepsilon)$. Similarly, for a given point s in $(-\varepsilon, \varepsilon)$, we define a smooth curve

$$\sigma_s(r) : (a,b) \longrightarrow M$$

by

$$\sigma_s(r) = \sigma(r,s)$$

for all r in (a,b). In keeping with the terminology introduced in the context of surfaces of revolution, we refer to σ_r as the **latitude curve** (or **transverse curve**) corresponding to r, and to σ_s as the **longitude curve** corresponding to s. In most applications, we tend to think of σ as a family of longitude curves indexed by s.

Let us now consider the smooth curve σ_s from the perspective of Section 14.7. For a given point s in $(-\varepsilon, \varepsilon)$, and using (14.7.3), we denote

$$\frac{d\sigma_s}{dr}(r) \qquad \text{by} \qquad \frac{\partial \sigma}{\partial r}(r,s). \tag{14.9.1}$$

According to (14.7.4),

$$\frac{\partial \sigma}{\partial r}(r,s) \qquad \text{is a vector in} \qquad T_{\sigma(r,s)}(M).$$

We define $(\partial\sigma/\partial s)(r,s)$ similarly.

Example 14.9.1. By Theorem 11.4.3, a surface of revolution is a regular surface with a chart of the form $\big((a,b) \times (-\pi,\pi), \varphi = (t,\phi)\big)$, where

$$\varphi(t,\phi) = \big(\rho(t)\cos(\phi), \rho(t)\sin(\phi), h(t)\big)$$

for all (t,ϕ) in $(a,b) \times (-\pi,\pi)$. Thus, φ is a parametrized surface in the above sense, and φ_t and φ_ϕ are the latitude curve and longitude curve, respectively, as defined in Section 11.4.   ◇

Theorem 14.9.2. *Let M be a smooth m-manifold, let $\sigma(r,s) : (a,b) \times (-\varepsilon, \varepsilon) \longrightarrow M$ be a parametrized surface, and let $(U, (x^i))$ be a chart on M that intersects $\sigma((a,b) \times (-\varepsilon, \varepsilon))$. Then, in local coordinates,*

$$\frac{\partial \sigma}{\partial r}(r,s) = \sum_i \frac{\partial (x^i \circ \sigma)}{\partial r}(r,s) \left.\frac{\partial}{\partial x^i}\right|_{\sigma(r,s)}$$

and

$$\frac{\partial \sigma}{\partial s}(r,s) = \sum_i \frac{\partial (x^i \circ \sigma)}{\partial s}(r,s) \left.\frac{\partial}{\partial x^i}\right|_{\sigma(r,s)}$$

for all (r,s) in the intersection.

Remark. We observe that the $\partial(x^i \circ \sigma)/\partial r$ and $\partial(x^i \circ \sigma)/\partial s$ are usual (Euclidean) partial derivatives.

Proof. It follows from Theorem 14.7.1(b) and (14.9.1) that

$$\frac{\partial \sigma}{\partial r}(r,s) = \frac{d\sigma_s}{dr}(r) = \sum_i \frac{d(x^i \circ \sigma_s)}{dr}(r) \left.\frac{\partial}{\partial x^i}\right|_{\sigma_s(r)}$$

$$= \sum_i \frac{\partial (x^i \circ \sigma)}{\partial r}(r,s) \left.\frac{\partial}{\partial x^i}\right|_{\sigma(r,s)},$$

which gives the first identity. The second identity is demonstrated similarly. □

Chapter 15

Fields on Smooth Manifolds

In this chapter, we provide a generalization of vector fields to smooth manifolds and define a range of other types of "fields".

15.1 Vector Fields

Vector fields arise in a variety of contexts. In this section, we discuss vector fields on smooth manifolds, curves, parametrized surfaces, and submanifolds.

Smooth manifolds. Let M be a smooth manifold. A **vector field on M** is a map X that assigns to each point p in M a vector X_p in $T_p(M)$. As was the case for vector fields on regular surfaces, we sometimes use "$|_p$" notation as an alternative to "subscript p" notation, especially when other subscripts are involved. According to (14.3.1), X_p satisfies the product rule

$$X_p(fg) = f(p)\, X_p(g) + g(p)\, X_p(f) \tag{15.1.1}$$

for all functions f, g in $C^\infty(M)$.

Let f be a function in $C^\infty(M)$, and let

$$X(f) : M \longrightarrow \mathbb{R}$$

be the function defined by

$$X(f)(p) = X_p(f) \tag{15.1.2}$$

for all p in M. It follows from (15.1.1) that

$$X(fg) = fX(g) + gX(f). \tag{15.1.3}$$

We say that X is **smooth (on M)** if $X(f)$ is in $C^\infty(M)$ for all functions f in $C^\infty(M)$. The set of smooth vector fields on M is denoted by $\mathfrak{X}(M)$. We

Semi-Riemannian Geometry, First Edition. Stephen C. Newman.
© 2019 John Wiley & Sons, Inc. Published 2019 by John Wiley & Sons, Inc.

make $\mathfrak{X}(M)$ into both a vector space over $\mathbb{R}$ and a module over $C^\infty(M)$ by defining operations as follows: for all vector fields X, Y in $\mathfrak{X}(M)$, all functions f in $C^\infty(M)$, and all real numbers c, let

$$(X + Y)_p = X_p + Y_p,$$

$$(fX)_p = f(p)X_p,$$

and

$$(cX)_p = cX_p$$

for all p in M.

Looking back at the definition of a smooth vector field $X : U \longrightarrow \mathbb{R}^m$ between Euclidean spaces as presented in Section 10.3, we observe that for each point p in U, the vector X_p was taken to be in $\mathbb{R}^m$. With hindsight, it appears that we were implicitly identifying the tangent space $T_p(\mathbb{R}^m)$ with $\mathbb{R}^m$.

Theorem 15.1.1. *Let M be a smooth manifold, let X be a vector field in $\mathfrak{X}(M)$, and let p be a point in M. If $\lambda(t) : (a, b) \longrightarrow M$ is a smooth curve such that $\lambda(t_0) = p$ and $(d\lambda/dt)(t_0) = X_p$ for some t_0 in (a, b), then*

$$X(f)(p) = \frac{d(f \circ \lambda)}{dt}(t_0)$$

for all functions f in $C^\infty(M)$.

Proof. We have

$$X(f)(p) = X_p(f) \qquad\qquad [(15.1.2)]$$

$$= \frac{d\lambda}{dt}(t_0)(f)$$

$$= \frac{d(f \circ \lambda)}{dt}(t_0). \qquad [(14.7.5)] \qquad \square$$

The next result guarantees that for a given vector in a tangent space, there is always a smooth vector field with that vector as a value. Its proof (not given) relies on bump functions.

Theorem 15.1.2 (Smooth Extension of Vector). *Let M be a smooth manifold, let p be a point in M, and let v be a vector in $T_p(M)$. Then there is a vector field X in $\mathfrak{X}(M)$ such that $X_p = v$.* $\square$

Let M be a smooth m-manifold, and let U be an open set in M. Viewing U as an m-manifold, let $X_1, \ldots, X_m$ be vector fields in $\mathfrak{X}(U)$. The m-tuple $\mathcal{X} = (X_1, \ldots, X_m)$ is said to be a **frame on U** if $\mathcal{X}_p = (X_1|_p, \ldots, X_m|_p)$ is a basis for $T_p(U)$ for all p in U. We will see later in this section that for each point p in M, there is always a neighborhood of p on which there is a frame. However, there may not be a frame on all of M.

Curves. Let M be a smooth manifold, and let $\lambda : (a, b) \longrightarrow M$ be a smooth curve. A **vector field on λ** is a map J that assigns to each point t in (a, b)

a vector $J(t)$ in $T_{\lambda(t)}(M)$. We observe that there is no requirement that λ be injective, so the image of λ might self-intersect. As a consequence, there could be two (or more) distinct vectors assigned to a given point in $\lambda((a,b))$. This represents a distinct difference between a vector field on a curve and a vector field on a smooth manifold.

Let f be a function in $C^\infty(M)$, and consider the function

$$J(f) : (a,b) \longrightarrow \mathbb{R}$$

defined by

$$J(f)(t) = J(t)(f) \tag{15.1.4}$$

for all t in (a,b). We say that J is **smooth (on λ)** if $J(f)$ is in $C^\infty((a,b))$ for all functions f in $C^\infty(M)$. The set of smooth vector fields on λ is denoted by $\mathfrak{X}_M(\lambda)$. Recall from Section 14.7 that the velocity of λ at t is $(d\lambda/dt)(t)$. The **velocity of λ** is the vector field on λ defined by the assignment $t \longmapsto (d\lambda/dt)(t)$ for all t in (a,b). We say that λ is **regular** if its velocity is nowhere-vanishing; that is, $(d\lambda/dt)(t)$ is not the zero vector in $T_{\lambda(t)}(M)$ for any t in (a,b).

Theorem 15.1.3. *In the above notation, $d\lambda/dt$ is a vector field in $\mathfrak{X}_M(\lambda)$.*

Proof. Setting $J = d\lambda/dt$ in (15.1.4), we have from (14.7.5) that

$$\frac{d\lambda}{dt}(f)(t) = \frac{d\lambda}{dt}(t)(f) = \frac{d(f \circ \lambda)}{dt}(t)$$

for all t in (a,b). Since f and λ are smooth, by Theorem 14.2.4, so is $f \circ \lambda$. It follows that $(d\lambda/dt)(f)$ is in $C^\infty((a,b))$ for all functions f in $C^\infty(M)$. $\square$

Theorem 15.1.4. *Continuing with the above notation, if X is a vector field in $\mathfrak{X}(M)$, then $X \circ \lambda$ is a vector field in $\mathfrak{X}_M(\lambda)$.*

Remark. Clearly, $X \circ \lambda$ is a vector field on λ, so the assertion makes sense.

Proof. Setting $J = X \circ \lambda$ in (15.1.4), we have from (15.1.2) that

$$(X \circ \lambda)(f)(t) = (X \circ \lambda)(t)(f) = X_{\lambda(t)}(f) = X(f)(\lambda(t))$$
$$= (X(f) \circ \lambda)(t)$$

for all t in (a,b). Since X and f are smooth, by definition, so is $X(f)$, and because λ is smooth, by Theorem 14.2.4, so is $X(f) \circ \lambda$. Thus, $(X \circ \lambda)(f)$ is in $C^\infty((a,b))$ for all functions f in $C^\infty(M)$. $\square$

Depending on λ, not every vector field J in $\mathfrak{X}_M(\lambda)$ arises as the composition of λ with some vector field in $\mathfrak{X}(M)$. For example, suppose the image of λ self-intersects at the points t_1, t_2 in (a,b) and that $J(t_1) \neq J(t_2)$. Since every vector field in $\mathfrak{X}(M)$ assigns to each point p in M a distinct vector, there is no vector field X in $\mathfrak{X}(M)$ such that $J = X \circ \lambda$.

Let $J_1, \dots, J_m$ be vector fields in $\mathfrak{X}_M(\lambda)$. The m-tuple $\mathcal{J} = (J_1, \dots, J_m)$ is said to be a **frame on** $\boldsymbol{\lambda}$ if $\mathcal{J}(t) = \big(J_1(t), \dots, J_m(t)\big)$ is a basis for $T_{\lambda(t)}(M)$ for all t in (a, b).

Parametrized surfaces. Let M be a smooth manifold, and let $\sigma(r, s)$: $(a, b) \times (-\varepsilon, \varepsilon) \longrightarrow M$ be a parametrized surface. A **vector field on** $\boldsymbol{\sigma}$ is a map V that assigns to each point (r, s) in $(a, b) \times (-\varepsilon, \varepsilon)$ a vector $V(r, s)$ in $T_{\sigma(r,s)}(M)$. Once again, there is no requirement that σ be injective. Let f be a function in $C^\infty(M)$, and consider the function

$$V(f) : (a, b) \times (-\varepsilon, \varepsilon) \longrightarrow \mathbb{R}$$

defined by

$$V(f)(r, s) = V(r, s)(f) \tag{15.1.5}$$

for all (r, s) in $(a, b) \times (-\varepsilon, \varepsilon)$. We say that V is **smooth (on** $\boldsymbol{\sigma}$**)** if $V(f)$ is in $C^\infty\big((a, b) \times (-\varepsilon, \varepsilon)\big)$ for all functions f in $C^\infty(M)$. The set of smooth vector fields on σ is denoted by $\mathfrak{X}_M(\sigma)$.

Recalling the notation in (14.9.1), we define a vector field $\partial\sigma/\partial r$ on σ by the assignment $(r, s) \longmapsto (\partial\sigma/\partial r)(r, s)$, and likewise for $\partial\sigma/\partial s$.

Theorem 15.1.5. *With the above setup,* $\partial\sigma/\partial r$ *and* $\partial\sigma/\partial s$ *are vector fields in* $\mathfrak{X}_M(\sigma)$.

Proof. We have

$$\frac{\partial\sigma}{\partial r}(f)(r, s) = \frac{\partial\sigma}{\partial r}(r, s)(f) \qquad [(15.1.5)]$$

$$= \frac{d\sigma_s}{dr}(r)(f) \qquad [(14.9.1)]$$

$$= \frac{d(f \circ \sigma_s)}{dr}(r) \qquad [(14.7.5)]$$

$$= \frac{\partial(f \circ \sigma)}{\partial r}(r, s)$$

for all (r, s) in $(a, b) \times (-\varepsilon, \varepsilon)$. Since f and σ are smooth, by Theorem 14.2.4, so is $f \circ \sigma$, and therefore, so is $\partial(f \circ \sigma)/\partial r$. Thus, $(\partial\sigma/\partial r)(f)$ is in $C^\infty\big((a, b) \times (-\varepsilon, \varepsilon)\big)$ for all functions f in $C^\infty(M)$, and likewise for $(\partial\sigma/\partial s)(f)$. $\square$

Submanifolds. Let $\overline{M}$ be a smooth manifold, and let M be a submanifold. A **vector field** *along* $\boldsymbol{M}$ is a map V that assigns to each point p in M a vector V_p in $T_p(\overline{M})$. We note that V_p is required to be in $T_p(\overline{M})$ but not necessarily in $T_p(M)$. This explains the change in terminology to "along M" from "on M".

Let f be a function in $C^\infty(\overline{M})$, and consider the function

$$V(f) : M \longrightarrow \mathbb{R}$$

defined by

$$V(f)(p) = V_p(f)$$

for all p in M. We say that V is **smooth (along M)** if $V(f)$ is in $C^\infty(M)$ for all functions f in $C^\infty(\overline{M})$. The set of smooth vector fields along M is denoted by $\mathfrak{X}_{\overline{M}}(M)$. In particular, for each vector field X in $\mathfrak{X}(\overline{M})$, the restriction $X|_M$ is in $\mathfrak{X}_{\overline{M}}(M)$. With the usual definitions of addition and scalar multiplication, $\mathfrak{X}_{\overline{M}}(M)$ is a vector space over $\mathbb{R}$ and a module over $C^\infty(M)$. Furthermore, after making the appropriate identifications, $\mathfrak{X}(M)$ is a $C^\infty(M)$-submodule of $\mathfrak{X}_{\overline{M}}(M)$. According to (14.8.1), for each point p in M, $T_p(M)$ is a subspace of $T_p(\overline{M})$. We say that a vector field V in $\mathfrak{X}_{\overline{M}}(M)$ is **nowhere-tangent** to M if V_p is not in $T_p(M)$ for all p in M, or equivalently, if V_p is in $T_p(\overline{M}) \setminus T_p(M)$ for all p in M.

Let M be a smooth m-manifold, and let U be an open set in M. Recall from Theorem 14.8.4 that U is an open m-submanifold of M. Suppose U is the coordinate domain of a chart $(U,(x^i))$ on M. The **ith coordinate vector field** of $(U,(x^i))$ is the vector field

$$\frac{\partial}{\partial x^i} \qquad \text{in} \qquad \mathfrak{X}(U)$$

defined by the assignment

$$p \longmapsto \frac{\partial}{\partial x^i}\bigg|_p$$

for all p in U for $i = 1, \ldots, m$, where we denote

$$\frac{\partial}{\partial x^i}(p) \qquad \text{by} \qquad \frac{\partial}{\partial x^i}\bigg|_p.$$

Then $(\partial/\partial x^1, \ldots, \partial/\partial x^m)$ is a frame on U, called the **coordinate frame** corresponding to $(U,(x^i))$.

Let X be a (not necessarily smooth) vector field on M. Then $X|_U$ can be expressed as

$$X|_U = \sum_i \alpha^i \frac{\partial}{\partial x^i}, \qquad (15.1.6)$$

where the α^i are uniquely determined functions on U, called the **components of X** with respect to $(U,(x^i))$. For brevity, we denote

$$X|_U \qquad \text{by} \qquad X.$$

The right-hand side of (15.1.6) is said to express X in **local coordinates** with respect to $(U,(x^i))$. We often give the local coordinate expression of a vector field without mentioning the underlying chart. This should not introduce any confusion because the notation for the coordinate frame is imbedded in the notation used in (15.1.6), and the specifics of the coordinate domain are usually of no immediate interest.

Example 15.1.6 ($\mathbb{R}^m$). Let $(r^1, \ldots, r^m)$ be standard coordinates on $\mathbb{R}^m$, let $(\partial/\partial r^1, \ldots, \partial/\partial r^m)$ be the corresponding coordinate frame, called the **standard coordinate frame**, and let $(e_1, \ldots, e_m)$ be the standard basis for $\mathbb{R}^m$.

It follows from Theorem 14.5.7 that $(\partial/\partial r^1, \ldots, \partial/\partial r^m)$ can be identified with $(e_1, \ldots, e_m)$, written

$$\left(\frac{\partial}{\partial r^1}, \ldots, \frac{\partial}{\partial r^m}\right) = (e_1, \ldots, e_m). \qquad \Diamond$$

Theorem 15.1.7 (Smoothness Criterion for Vector Fields). *Let M be a smooth manifold, and let X be a (not necessarily smooth) vector field on M. Then X is in $\mathfrak{X}(M)$ if and only if for every chart $(U, (x^i))$ on M, the components of X are in $C^\infty(U)$.*

Proof. ($\Rightarrow$): Let $X = \sum_i \alpha^i(\partial/\partial x^i)$ be the local coordinate expression of X with respect to $(U, (x^i))$, and let p be a point in U. Since x^i is a function in $C^\infty(U)$, by Theorem 14.2.7, there is function $\tilde{x}^i$ in $C^\infty(M)$ and a neighborhood $\tilde{U} \subseteq U$ of p in M such that x^i and $\tilde{x}^i$ agree on $\tilde{U}$. By Theorem 14.3.1(b),

$$X(\tilde{x}^i) = \left(\sum_j \alpha^j \frac{\partial}{\partial x^j}\right)(\tilde{x}^i) = \sum_j \alpha^j \frac{\partial \tilde{x}^i}{\partial x^j} = \alpha^i$$

on $\tilde{U}$. Since X is in $\mathfrak{X}(M)$ and $\tilde{x}^i$ in $C^\infty(M)$, by definition, $X(\tilde{x}^i)$ in $C^\infty(M)$. It follows that α^i is smooth on $\tilde{U}$. Since p was arbitrary, α^i is smooth on U for $i = 1, \ldots, m$.

($\Leftarrow$): Let $X = \sum_i \alpha^i(\partial/\partial x^i)$ be the local coordinate expression of X with respect to $(U, (x^i))$. For a function f in $C^\infty(M)$, we have

$$X(f) = \sum_i \alpha^i \frac{\partial f}{\partial x^i}$$

on U. By assumption, the α^i are smooth on U, and therefore, so is $X(f)$. Since M is covered by the coordinate domains of its atlas, $X(f)$ is smooth on M. $\qquad \square$

Theorem 15.1.8 (Change of Coordinate Frame). *Let M be a smooth m-manifold, let $(U, (x^i))$ and $(\tilde{U}, (\tilde{x}^j))$ be overlapping charts on M, and let $(\partial/\partial x^1, \ldots, \partial/\partial x^m)$ and $(\partial/\partial \tilde{x}^1, \ldots, \partial/\partial \tilde{x}^m)$ be the corresponding coordinate frames. Then*

$$\frac{\partial}{\partial x^i} = \sum_j \frac{\partial \tilde{x}^j}{\partial x^i} \frac{\partial}{\partial \tilde{x}^j}$$

on $U \cap \tilde{U}$ for $i = 1, \ldots, m$.

Proof. This follows from Theorem 14.3.5(a). $\qquad \square$

15.2   Representation of Vector Fields

Let M be a smooth manifold. A linear map

$$\mathcal{D} : C^\infty(M) \longrightarrow C^\infty(M)$$

is said to be a **derivation [on $C^\infty(M)$]** if it satisfies the following product rule:

$$\mathcal{D}(fg) = f\mathcal{D}(g) + g\mathcal{D}(f) \tag{15.2.1}$$

for all functions f, g in $C^\infty(M)$; that is,

$$\mathcal{D}(fg)(p) = f(p)\,\mathcal{D}(g)(p) + g(p)\,\mathcal{D}(f)(p)$$

for all p in M. The set of derivations on $C^\infty(M)$ is denoted by $\mathrm{Der}(M)$. The **zero derivation** in $\mathrm{Der}(M)$, denoted by 0, is the derivation that sends all functions in $C^\infty(M)$ to the zero function in $C^\infty(M)$. We make $\mathrm{Der}(M)$ into both a vector space over $\mathbb{R}$ and a module over $C^\infty(M)$ by defining operations as follows: for all derivations $\mathcal{D}, \mathcal{E}$ in $\mathrm{Der}(M)$, all functions f, g in $C^\infty(M)$, and all real numbers c, let

$$(\mathcal{D} + \mathcal{E})(f)(p) = \mathcal{D}(f)(p) + \mathcal{E}(f)(p),$$

$$(f\mathcal{D})(g)(p) = f(p)\,\mathcal{D}(g)(p),$$

and

$$(c\mathcal{D})(f)(p) = c\,\mathcal{D}(f)(p)$$

for all p in M.

We see from (15.1.3) that a vector field on M can be thought of as derivation on $C^\infty(M)$. Pursuing this line of reasoning, let us consider the map

$$\mathfrak{F} : \mathfrak{X}(M) \longrightarrow \mathrm{Der}(M)$$

defined by

$$\mathfrak{F}(X)(f) = X(f)$$

for all vector fields X in $\mathfrak{X}(M)$ and all functions f in $C^\infty(M)$; that is,

$$\mathfrak{F}(X)(f)(p) = X(f)(p)$$

for all p in M, where the right-hand side of the above identity is given by (15.1.2).

Theorem 15.2.1 (Representation of Vector Fields). *If M is a smooth manifold, then $\mathfrak{F}$ is a $C^\infty(M)$-module isomorphism:*

$$\mathfrak{X}(M) \approx \mathrm{Der}(M).$$

Proof. It is easily shown that $\mathfrak{F}$ is a $C^\infty(M)$-module homomorphism. It remains to show that $\mathfrak{F}$ is bijective.

Injectivity. Suppose X is a vector field in $\mathfrak{X}(M)$ such that $\mathfrak{F}(X) = 0$; that is, $X(f) = 0$ for all functions f in $C^\infty(M)$. By definition, $X = 0$, so $\ker(\mathfrak{F}) = \{0\}$. It follows from Theorem B.5.3 that $\mathfrak{F}$ is injective.

Surjectivity. Let $\mathcal{D}$ be a derivation on $C^\infty(M)$. For each point p in M, define a function

$$X_\mathcal{D}|_p : C^\infty(M) \longrightarrow \mathbb{R}$$

by

$$X_{\mathcal{D}}|_p(f) = \mathcal{D}(f)(p) \tag{15.2.2}$$

for all functions f in $C^\infty(M)$. Since $\mathcal{D}$ is linear, so is $X_{\mathcal{D}}|_p$. For functions f, g in $C^\infty(M)$, we have from (15.2.1) that

$$\begin{aligned}
X_{\mathcal{D}}|_p(fg) &= \mathcal{D}(fg)(p) \\
&= f(p)\,\mathcal{D}(g)(p) + g(p)\,\mathcal{D}(f)(p) \\
&= f(p)\,X_{\mathcal{D}}|_p(g) + g(p)\,X_{\mathcal{D}}|_p(f),
\end{aligned}$$

hence $X_{\mathcal{D}}|_p$ satisfies (14.3.1). Thus, $X_{\mathcal{D}}|_p$ is a tangent vector at p. Let us define a vector field $X_{\mathcal{D}}$ on M by the assignment $p \longmapsto X_{\mathcal{D}}|_p$ for all p in M. Then

$$\begin{aligned}
X_{\mathcal{D}}(f)(p) &= X_{\mathcal{D}}|_p(f) && [(15.1.2)] \\
&= \mathcal{D}(f)(p) && [(15.2.2)]
\end{aligned}$$

for all p in M, so $X_{\mathcal{D}}(f) = \mathcal{D}(f)$. By definition, $\mathcal{D}(f)$ is in $C^\infty(M)$ for all functions f in $C^\infty(M)$, and therefore, so is $X_{\mathcal{D}}(f)$. Thus, by definition, $X_{\mathcal{D}}$ is a smooth vector field on M; that is, $X_{\mathcal{D}}$ is in $\mathfrak{X}(M)$. Since $\mathfrak{F}(X_{\mathcal{D}}) = \mathcal{D}$ and $\mathcal{D}$ was arbitrary, $\mathfrak{F}$ is surjective.

This shows that $\mathfrak{F}$ is a $C^\infty(M)$-module isomorphism. $\qquad\square$

From now on, we often (but not always) identify $\mathfrak{X}(M)$ with $\mathrm{Der}(M)$. However, we will continue to use the previous terminology and notation, and say, for example, that "X is a vector field in $\mathfrak{X}(M)$" rather than "X is a derivation in $\mathrm{Der}(M)$". It will usually be clear from the context whether the identification is being made, but sometimes, for emphasis, we make it explicit.

15.3   Lie Bracket

Let M be a smooth manifold, and let X and Y be vector fields in $\mathfrak{X}(M)$. The **Lie bracket of X and Y** is the map

$$[X, Y] : C^\infty(M) \longrightarrow C^\infty(M)$$

defined by

$$[X, Y](f) = X\big(Y(f)\big) - Y\big(X(f)\big) \tag{15.3.1}$$

for all functions f in $C^\infty(M)$. Observe that this definition employs the representation of vector fields given by Theorem 15.2.1. Reverting for the moment to the vector field formulation, we have from (15.1.2) that

$$X\big(Y(f)\big)(p) = X_p\big(Y(f)\big) \quad \text{and} \quad Y\big(X(f)\big)(p) = Y_p\big(X(f)\big),$$

so that

$$[X, Y](f)(p) = X_p\big(Y(f)\big) - Y_p\big(X(f)\big)$$

for all p in M.

Theorem 15.3.1. *With the above setup, $[X, Y]$ is a derivation on $C^\infty(M)$.*

Proof. Since Y is in $\mathfrak{X}(M)$ and f is in $C^\infty(M)$, by definition, $Y(f)$ is in $C^\infty(M)$. Furthermore, since X is in $\mathfrak{X}(M)$ and $Y(f)$ is in $C^\infty(M)$, by definition, $X\big(Y(f)\big)$ is in $C^\infty(M)$. Similarly, $Y\big(X(f)\big)$ is in $C^\infty(M)$. It follows that $[X, Y](f)$ is in $C^\infty(M)$. The remainder of the verification is straightforward.   □

Lie bracket is the map

$$[\cdot, \cdot] : \mathfrak{X}(M) \times \mathfrak{X}(M) \longrightarrow \mathfrak{X}(M)$$

defined by the assignment

$$(X, Y) \longmapsto [X, Y] = X \circ Y - Y \circ X$$

for all vector fields X, Y in $\mathfrak{X}(M)$.

Theorem 15.3.2. *Let M be a smooth manifold, let X, Y, Z be vector fields in $\mathfrak{X}(M)$, and let f, g be functions in $C^\infty(M)$. Then:*
(a) $[Y, X] = -[X, Y]$.
(b) $[X + Y, Z] = [X, Z] + [Y, Z]$.
(c) $[X, Y + Z] = [X, Y] + [X, Z]$.
(d) $[fX, gY] = fg[X, Y] + fX(g)Y - gY(f)X$.
(e) $\big[X, [Y, Z]\big] + \big[Y, [Z, X]\big] + \big[Z, [X, Y]\big] = 0$.   (Jacobi's identity)

Proof. (a)–(d): Straightforward.
(e): We have

$$\begin{aligned}
\big[X, [Y, Z]\big] &= X \circ [Y, Z] - [Y, Z] \circ X \\
&= X \circ (Y \circ Z - Z \circ Y) - (Y \circ Z - Z \circ Y) \circ X \\
&= X \circ Y \circ Z - X \circ Z \circ Y - Y \circ Z \circ X + Z \circ Y \circ X.
\end{aligned}$$

Likewise,

$$\begin{aligned}
\big[Y, [Z, X]\big] &= Y \circ Z \circ X - Y \circ X \circ Z - Z \circ X \circ Y + X \circ Z \circ Y \\
\big[Z, [X, Y]\big] &= Z \circ X \circ Y - Z \circ Y \circ X - X \circ Y \circ Z + Y \circ X \circ Z.
\end{aligned}$$

Summing the identities gives the result.   □

It was observed in Section 15.1 that $\mathfrak{X}(M)$ is a module over $C^\infty(M)$. It follows from Theorem 15.3.2 that $\mathfrak{X}(M)$ is also a Lie algebra over $\mathbb{R}$.

Theorem 15.3.3. *Let M be a smooth manifold, let X, Y be vector fields in $\mathfrak{X}(M)$, and, in local coordinates, let*

$$X = \sum_i \alpha^i \frac{\partial}{\partial x^i} \qquad and \qquad Y = \sum_j \beta^j \frac{\partial}{\partial x^j}.$$

Then:

(a)
$$[X,Y] = \sum_j \left\{ \sum_i \left(\alpha^i \frac{\partial \beta^j}{\partial x^i} - \beta^i \frac{\partial \alpha^j}{\partial x^i} \right) \right\} \frac{\partial}{\partial x^j}.$$

(b)
$$\left[\frac{\partial}{\partial x^i}, \frac{\partial}{\partial x^j} \right] = 0$$

for $i,j = 1,\ldots,m$.

Proof. (a): For a function f in $C^\infty(M)$, we have

$$(X \circ Y)(f) = X\left(Y(f)\right) = \left(\sum_i \alpha^i \frac{\partial}{\partial x^i} \right) \left(\sum_j \beta^j \frac{\partial f}{\partial x^j} \right) = \sum_{ij} \alpha^i \frac{\partial}{\partial x^i} \left(\beta^j \frac{\partial f}{\partial x^j} \right)$$

$$= \sum_{ij} \alpha^i \left(\frac{\partial \beta^j}{\partial x^i} \frac{\partial f}{\partial x^j} + \beta^j \frac{\partial^2 f}{\partial x^i \partial x^j} \right)$$

$$= \left(\sum_{ij} \alpha^i \frac{\partial \beta^j}{\partial x^i} \frac{\partial}{\partial x^j} + \sum_{ij} \alpha^i \beta^j \frac{\partial^2}{\partial x^i \partial x^j} \right)(f).$$

Since f was arbitrary,

$$X \circ Y = \sum_{ij} \alpha^i \frac{\partial \beta^j}{\partial x^i} \frac{\partial}{\partial x^j} + \sum_{ij} \alpha^i \beta^j \frac{\partial^2}{\partial x^i \partial x^j}.$$

Likewise,

$$Y \circ X = \sum_{ij} \beta^i \frac{\partial \alpha^j}{\partial x^i} \frac{\partial}{\partial x^j} + \sum_{ij} \beta^i \alpha^j \frac{\partial^2}{\partial x^i \partial x^j}.$$

The preceding two identities and Theorem 14.3.6 give the result.
    (b): This follows from part (a).   □

15.4   Covector Fields

Let M be smooth m-manifold. A **covector field on M** is a map ω that assigns to each point p in M a covector ω_p in $T_p^*(M)$. We say that ω **vanishes at p** if $\omega_p = 0$, is **nonvanishing at p** if $\omega_p \neq 0$, and is **nowhere-vanishing (on M)** if it is nonvanishing at every p in M.
    Let X be a vector field in $\mathfrak{X}(M)$, and let

$$\omega(X) : M \longrightarrow \mathbb{R}$$

be the function defined by

$$\omega(X)(p) = \omega_p(X_p) \tag{15.4.1}$$

for all p in M. We say that ω is **smooth (on M)** if the function $\omega(X)$ is in $C^\infty(M)$ for all vector fields X in $\mathfrak{X}(M)$. The set of smooth covector fields on

M is denoted by $\mathfrak{X}^*(M)$. We make $\mathfrak{X}^*(M)$ into both a vector space over $\mathbb{R}$ and a module over $C^\infty(M)$ by defining operations as follows: for all covector fields ω, ξ in $\mathfrak{X}^*(M)$, all functions f in $C^\infty(M)$, and all real numbers c, let

$$(\omega + \xi)_p = \omega_p + \xi_p,$$

$$(f\omega)_p = f(p)\omega_p,$$

and

$$(c\omega)_p = c\omega_p$$

for all p in M. With the identification $T_p^{**}(M) = T_p(M)$, we have from (5.1.3) that

$$\omega_p(X_p) = X_p(\omega_p)$$

for all p in M, which we express as

$$\omega(X) = X(\omega). \tag{15.4.2}$$

Let U be an open set in M. Viewing U as a smooth m-manifold, let $\omega^1, \ldots, \omega^m$ be covector fields in $\mathfrak{X}^*(U)$. The m-tuple $\Upsilon = (\omega^1, \ldots, \omega^m)$ is said to be a **dual frame on U** if $\Upsilon(p) = (\omega^1|_p, \ldots, \omega^m|_p)$ is a basis for $T_p^*(U)$ for all p in U. Given a frame $(X_1, \ldots, X_m)$ on U, there is a uniquely determined dual frame $(\omega^1, \ldots, \omega^m)$ on U defined as follows: $(\omega^1|_p, \ldots, \omega^m|_p)$ is the dual basis corresponding to $(X_1|_p, \ldots, X_m|_p)$ for all p in U. Conversely, given a dual frame on U, there is a uniquely determined frame on U defined in the obvious way.

Suppose U is the coordinate domain of a chart $(U, (x^i))$ on M. The **ith coordinate covector field** of $(U, (x^i))$ is the covector field

$$d(x^i) \qquad \text{in} \qquad \mathfrak{X}^*(U)$$

defined by the assignment

$$p \longmapsto d_p(x^i)$$

for all p in U for $i = 1, \ldots, m$. Then $(d(x^1), \ldots, d(x^m))$ is a dual frame on U, called the **dual coordinate frame** corresponding to $(U, (x^i))$.

Let ω be a (not necessarily smooth) covector field on M. Then $\omega|_U$ can be expressed as

$$\omega|_U = \sum_i \alpha^i d(x^i), \tag{15.4.3}$$

where the α^i are uniquely determined functions on U, called the **components of ω** with respect to $(U, (x^i))$. For brevity, we denote

$$\omega|_U \qquad \text{by} \qquad \omega.$$

The right-hand side of (15.4.3) is said to express ω in **local coordinates** with respect to $(U, (x^i))$.

Example 15.4.1 ($\mathbb{R}^m$). Let $(r^1, \ldots, r^m)$ be standard coordinates on $\mathbb{R}^m$, let $\big(d(r^1), \ldots, d(r^m)\big)$ be the corresponding dual coordinate frame, called the **standard dual coordinate frame**, and let $(\xi^1, \ldots, \xi^m)$ be the dual basis corresponding to the standard basis for $\mathbb{R}^m$. It follows from Theorem 14.5.7 that $\big(d(r^1), \ldots, d(r^m)\big)$ can be identified with $(\xi^1, \ldots, \xi^m)$, written

$$\big(d(r^1), \ldots, d(r^m)\big) = (\xi^1, \ldots, \xi^m). \qquad \Diamond$$

The next three results are the covector field counterparts to Theorem 15.1.2, Theorem 15.1.7, and Theorem 15.1.8.

Theorem 15.4.2 (Smoothness Criterion for Covector Fields). *Let M be a smooth manifold, and let ω be a (not necessarily smooth) covector field on M. Then ω is in $\mathfrak{X}^*(M)$ if and only if for every chart $(U, (x^i))$ on M, the components of ω are in $C^\infty(U)$.* $\qquad\square$

Theorem 15.4.3 (Change of Dual Coordinate Frame). *Let M be a smooth m-manifold, let $(U, (x^i))$ and $(\widetilde{U}, (\widetilde{x}^j))$ be overlapping charts on M, and let $\mathcal{X}^* = \big(d(x^1), \ldots, d(x)\big)$ and $\widetilde{\mathcal{X}}^* = \big(d(\widetilde{x}^1), \ldots, d(\widetilde{x}^m)\big)$ be the corresponding dual coordinate frames. Then*

$$d(x^j) = \sum_i \frac{\partial x^j}{\partial \widetilde{x}^i}\, d(\widetilde{x}^i)$$

on $U \cap \widetilde{U}$ for $j = 1, \ldots, m$.

Proof. This follows from Theorem 14.5.4(a). $\qquad\square$

Theorem 15.4.4 (Smooth Extension of Covector). *Let M be a smooth manifold, let p be a point in M, and let η be a covector in $T_p^*(M)$. Then there is a covector field ω in $\mathfrak{X}^*(M)$ such that $\omega_p = \eta$.* $\qquad\square$

Let M be a smooth manifold, and define a map

$$d : C^\infty(M) \longrightarrow \mathfrak{X}^*(M),$$

called the **exterior derivative**, by

$$d(f)_p(v) = d_p(f)(v) = v(f) \tag{15.4.4}$$

for all functions f in $C^\infty(M)$, all points p in M, and all vectors v in $T_p(M)$, where the second equality follows from Theorem 14.5.5. Part (a) of the next result shows that this definition makes sense.

Theorem 15.4.5. *Let M be a smooth manifold, let f be a function in $C^\infty(M)$, and let X be a vector field in $\mathfrak{X}(M)$. Then:*
(a) *$d(f)$ is a covector field in $\mathfrak{X}^*(M)$.*
(b) *$d(f)(X) = X(f)$.*

Proof. (b): By Theorem 14.5.1, $d(f)_p$ is a covector in $T_p^*(M)$, hence $d(f)$ is a covector field. For a point p in M, we have

$$
\begin{aligned}
d(f)(X)(p) &= d(f)_p(X_p) & &[(15.4.1)] \\
&= X_p(f) & &[(15.4.4)] \\
&= X(f)(p). & &[(15.1.2)]
\end{aligned}
$$

Since p was arbitrary, the result follows.

(a): Since X and f are smooth, by definition, so is $X(f)$. Because X was arbitrary, it follows from part (b) that, by definition, $d(f)$ is smooth.   □

Theorem 15.4.6. *Let M be a smooth manifold, let f, g be functions in $C^\infty(M)$, and let c be a real number. Then:*
(a) $d(cf + g) = c\,d(f) + d(g)$.
(b) $d(fg) = f d(g) + g d(f)$.

Proof. This follows from Theorem 14.5.6 and (15.4.4).   □

Theorem 15.4.7. *If M is a smooth manifold and f is a function in $C^\infty(M)$, then, in local coordinates,*

$$
d(f) = \sum_i \frac{\partial f}{\partial x^i}\, d(x^i).
$$

Proof. For a point p in M and a vector v in $T_p(M)$, we have

$$
\begin{aligned}
d(f)_p(v) &= d_p(f)(v) & &[(15.4.4)] \\
&= \sum_i \frac{\partial f}{\partial x^i}(p)\, d_p(x^i)(v) & &[\text{Th } 14.5.3] \\
&= \sum_i \frac{\partial f}{\partial x^i}(p)\, d(x^i)_p(v) & &[(15.4.4)] \\
&= \left(\sum_i \frac{\partial f}{\partial x^i}\, d(x^i) \right)\bigg|_p (v).
\end{aligned}
$$

Since p and v were arbitrary, the result follows.   □

15.5   Representation of Covector Fields

Theorem 15.5.1. *If M is a smooth manifold and ω is a covector field in $\mathfrak{X}^*(M)$, then ω is $C^\infty(M)$-linear. That is, for all vector fields X, Y in $\mathfrak{X}(M)$ and all functions f in $C^\infty(M)$:*
(a) $\omega(X + Y) = \omega(X) + \omega(Y)$
(b) $\omega(fX) = f\omega(X)$.

Proof. For a point p in M, we have

$$
\begin{aligned}
\omega(X+Y)(p) &= \omega_p\big((X+Y)_p\big) && [(15.4.1)] \\
&= \omega_p(X_p + Y_p) \\
&= \omega_p(X_p) + \omega_p(Y_p) \\
&= \omega(X)(p) + \omega(Y)(p) && [(15.4.1)] \\
&= \big(\omega(X) + \omega(Y)\big)(p)
\end{aligned}
$$

and

$$
\begin{aligned}
\omega(fX)(p) &= \omega_p\big((fX)_p\big) && [(15.4.1)] \\
&= \omega_p\big(f(p)X_p\big) \\
&= f(p)\,\omega_p(X_p) \\
&= (f\omega)_p(X_p) \\
&= \big(f\omega(X)\big)(p). && [(15.4.1)]
\end{aligned}
$$

Since p was arbitrary, the result follows. $\square$

Following Section B.5, we denote by

$$
\mathrm{Lin}_{C^\infty(M)}\big(\mathfrak{X}(M), C^\infty(M)\big)
$$

the $C^\infty(M)$-module of $C^\infty(M)$-linear maps from $\mathfrak{X}(M)$ to $C^\infty(M)$. Let us define a map

$$
\mathfrak{C} : \mathfrak{X}^*(M) \longrightarrow \mathrm{Lin}_{C^\infty(M)}\big(\mathfrak{X}(M), C^\infty(M)\big),
$$

called the **characterization map**, by

$$
\mathfrak{C}(\omega)(X) = \omega(X) \tag{15.5.1}
$$

for all covector fields ω in $\mathfrak{X}^*(M)$ and all vector fields X in $\mathfrak{X}(M)$, where the right-hand side of (15.5.1) is given by (15.4.1). It follows from Theorem 15.5.1 that $\mathfrak{C}(\omega)$ is a map in $\mathrm{Lin}_{C^\infty(M)}\big(\mathfrak{X}(M), C^\infty(M)\big)$, so the definition makes sense. At this point, $\mathfrak{C}(\omega)$ amounts to little more than notational shorthand for viewing the covector field ω in $\mathfrak{X}^*(M)$ from the perspective of (15.4.1): as a mechanism for turning vector fields into functions. The purpose of this formalism will become clear as we proceed.

We say that an $\mathbb{R}$-linear map $\mathcal{F} : \mathfrak{X}(M) \longrightarrow C^\infty(M)$ is **determined pointwise** if for all points p in M, we have $\mathcal{F}(X)(p) = \mathcal{F}(\widetilde{X})(p)$ whenever $X, \widetilde{X}$ are vector fields in $\mathfrak{X}(M)$ such that $X_p = \widetilde{X}_p$. Since $\mathcal{F}(X)(p) = \mathcal{F}(\widetilde{X})(p)$ is equivalent to $\mathcal{F}(X - \widetilde{X})(p) = 0$, and $X_p = \widetilde{X}_p$ is equivalent to $(X - \widetilde{X})_p = 0$, $\mathcal{F}$ is determined pointwise if and only if for every point p in M, $\mathcal{F}(Y)(p) = 0$ whenever Y is a vector field in $\mathfrak{X}(M)$ such that $Y_p = 0$.

Let ω be a covector field in $\mathfrak{X}^*(M)$, let p be a point in M, and let X be a vector field in $\mathfrak{X}(M)$ such that $X_p = 0$. It follows from (15.4.1) and (15.5.1) that $\mathfrak{C}(\omega)(X)(p) = 0$. Thus, $\mathfrak{C}(\omega)$ is a map in $\mathrm{Lin}_{C^\infty(M)}\big(\mathfrak{X}(M), C^\infty(M)\big)$ that is determined pointwise. Remarkably, as the next result shows, all maps in $\mathrm{Lin}_{C^\infty(M)}\big(\mathfrak{X}(M), C^\infty(M)\big)$ have this property.

Theorem 15.5.2. *If M is a smooth manifold and Υ is a map in* $\mathrm{Lin}_{C^\infty(M)}\big(\mathfrak{X}(M), C^\infty(M)\big)$, *then Υ is determined pointwise.*

Proof. Let p be a point in M, and let X be a vector field in $\mathfrak{X}(M)$ such that $X_p = 0$. We need to show that $\Upsilon(X)(p) = 0$. Only the case where M is covered by a single chart $(M, (x^i))$ is considered. In local coordinates, let $X = \sum_i \alpha^i(\partial/\partial x^i)$. This gives an expression for X on all of M, so we can apply Υ to obtain

$$\Upsilon(X) = \sum_i \alpha^i \Upsilon\left(\frac{\partial}{\partial x^i}\right),$$

hence

$$\Upsilon(X)(p) = \sum_i \alpha^i(p)\,\Upsilon\left(\frac{\partial}{\partial x^i}\right)(p).$$

Since $X_p = 0$, it follows that each $\alpha^i(p) = 0$, so $\Upsilon(X)(p) = 0$. $\qquad\square$

Theorem 15.5.3 (Covector Field Characterization Theorem). *If M is a smooth manifold, then $\mathfrak{C}$ is a $C^\infty(M)$-module isomorphism:*

$$\mathfrak{X}^*(M) \approx \mathrm{Lin}_{C^\infty(M)}\big(\mathfrak{X}(M), C^\infty(M)\big).$$

Proof. It is easily shown that $\mathfrak{C}$ is a $C^\infty(M)$-module homomorphism. We need to show that $\mathfrak{C}$ is bijective.

Injectivity. Suppose $\mathfrak{C}(\omega) = 0$ for some covector field ω in $\mathfrak{X}^*(M)$. It follows from (15.5.1) that $\omega(X) = 0$ for all vector fields X in $\mathfrak{X}(M)$. Let p be a point in M, and let v be a vector in $T_p(M)$. By Theorem 15.1.2, there is a vector field Y in $\mathfrak{X}(M)$ such that $Y_p = v$. Then $\omega(Y) = 0$, so (15.4.1) gives

$$\omega_p(v) = \omega_p(Y_p) = \omega(Y)(p) = 0.$$

Since p and v were arbitrary, $\omega = 0$. Thus, $\ker(\mathfrak{C}) = \{0\}$. By Theorem B.5.3, $\mathfrak{C}$ is injective.

Surjectivity. Let Υ be a map in $\mathrm{Lin}_{C^\infty(M)}\big(\mathfrak{X}(M), C^\infty(M)\big)$. We need to find a covector field $\widetilde{\Upsilon}$ in $\mathfrak{X}^*(M)$ such that $\mathfrak{C}(\widetilde{\Upsilon}) = \Upsilon$. Let p be a point in M, and define a map $\widetilde{\Upsilon}_p : T_p(M) \longrightarrow \mathbb{R}$ by

$$\widetilde{\Upsilon}_p(v) = \Upsilon(X)(p) \tag{15.5.2}$$

for all vectors v in $T_p(M)$, where X is any vector field in $\mathfrak{X}(M)$ such that $X_p = v$. By Theorem 15.1.2, such a vector field always exists. We have from Theorem 15.5.2 that $\widetilde{\Upsilon}_p$ is independent of the choice of X, so $\widetilde{\Upsilon}_p$ is well-defined. Let w be a vector in $T_p(M)$, let Y be a vector field in $\mathfrak{X}(M)$ such that $Y_p = w$, and let c be a real number. Then $cX + Y$ is a vector field in $\mathfrak{X}(M)$ and

$$(cX + Y)_p = cX_p + Y_p = cv + w,$$

hence

$$\widetilde{\Upsilon}_p(cv + w) = \Upsilon(cX + Y)(p) \qquad [(15.5.2)]$$
$$= c\Upsilon(X)(p) + \Upsilon(Y)(p)$$
$$= c\widetilde{\Upsilon}_p(v) + \widetilde{\Upsilon}_p(w), \qquad [(15.5.2)]$$

so $\widetilde{\Upsilon}_p$ is linear. Thus, $\widetilde{\Upsilon}_p$ is a covector in $T_p^*(M)$. Let us define a covector field $\widetilde{\Upsilon}$ on M by the assignment $p \longmapsto \widetilde{\Upsilon}_p$. We claim that $\widetilde{\Upsilon}$ is smooth; that is, the function $\widetilde{\Upsilon}(Z)$ is smooth for all vector fields Z in $\mathfrak{X}(M)$. From (15.4.1) and (15.5.2),

$$\widetilde{\Upsilon}(Z)(p) = \widetilde{\Upsilon}_p(Z_p) = \Upsilon(Z)(p)$$

for all p in M, hence

$$\widetilde{\Upsilon}(Z) = \Upsilon(Z). \qquad (15.5.3)$$

Since Υ is in $\mathrm{Lin}_{C^\infty(M)}\big(\mathfrak{X}(M), C^\infty(M)\big)$ and Z is in $\mathfrak{X}(M)$, by definition, $\Upsilon(Z)$ is a smooth function, and therefore, so is $\widetilde{\Upsilon}(Z)$. This proves the claim. It follows from (15.5.1) and (15.5.3) that $\mathfrak{C}(\widetilde{\Upsilon})(Z) = \Upsilon(Z)$. Since Z was arbitrary, $\mathfrak{C}(\widetilde{\Upsilon}) = \Upsilon$. Thus, $\mathfrak{C}$ is surjective. $\qquad\square$

It is useful to isolate an aspect of the proof of Theorem 15.5.3. Let Υ be a map in $\mathrm{Lin}_{C^\infty(M)}\big(\mathfrak{X}(M), C^\infty(M)\big)$. We showed that $\mathfrak{C}^{-1}(\Upsilon)$ is the covector field in $\mathfrak{X}^*(M)$ defined by

$$\mathfrak{C}^{-1}(\Upsilon)_p(v) = \Upsilon(X)(p) \qquad (15.5.4)$$

for all points p in M and all vectors v in $T_p(M)$, where X is any vector field in $\mathfrak{X}(M)$ such that $X_p = v$, the existence of which is guaranteed by Theorem 15.1.2.

Now that we have Theorem 15.5.3, we usually (but not always) view $\mathfrak{X}^*(M)$ as the vector space over $\mathbb{R}$ and module over $C^\infty(M)$ consisting of all $C^\infty(M)$-linear maps from $\mathfrak{X}(M)$ to $C^\infty(M)$. We will see a significant generalization of Theorem 15.5.3 in Section 15.7.

15.6   Tensor Fields

In this section, we generalize some of the material in Section 15.4.

Let M be a smooth m-manifold, and let $r, s \geq 0$ be integers. An (r, s)-**tensor field on M** is a map $\mathcal{A}$ that assigns to each point p in M an (r, s)-tensor $\mathcal{A}_p$ in $\mathcal{T}_s^r\big(T_p(M)\big)$. We also refer to $\mathcal{A}$ as an **r-contravariant-s-covariant tensor field** or simply a **tensor field**, and we define the **rank of $\mathcal{A}$** to be (r, s). When $s = 0$, $\mathcal{A}$ is said to be an **r-contravariant tensor field** or just a **contravariant tensor field**; and when $r = 0$, $\mathcal{A}$ is said to be an **s-covariant tensor field** or simply a **covariant tensor field**.

Let $\omega^1, \ldots, \omega^r$ be covector fields in $\mathfrak{X}^*(M)$, let $X_1, \ldots, X_s$ be vector fields in $\mathfrak{X}(M)$, and consider the function

$$\mathcal{A}(\omega^1, \ldots, \omega^r, X_1, \ldots, X_s) : M \longrightarrow \mathbb{R}$$

defined by

$$\mathcal{A}(\omega^1, \ldots, \omega^r, X_1, \ldots, X_s)(p) = \mathcal{A}_p(\omega^1|_p, \ldots, \omega^r|_p, X_1|_p, \ldots, X_s|_p) \quad (15.6.1)$$

for all p in M. We say that $\mathcal{A}$ is **smooth (on M)** if the function $\mathcal{A}(\omega^1, \ldots, \omega^r, X_1, \ldots, X_s)$ is in $C^\infty(M)$ for all covector fields $\omega^1, \ldots, \omega^r$ in $\mathfrak{X}^*(M)$ and all vector fields $X_1, \ldots, X_s$ in $\mathfrak{X}(M)$. The set of smooth (r, s)-tensor fields on M is denoted by $\mathcal{T}_s^r(M)$. In particular,

$$\mathcal{T}_1^0(M) = \mathfrak{X}^*(M) \qquad \text{and} \qquad \mathcal{T}_0^1(M) = \mathfrak{X}(M). \quad (15.6.2)$$

For completeness, we define

$$\mathcal{T}_0^0(M) = C^\infty(M). \quad (15.6.3)$$

It is instructive to compare identities (15.6.2) and (15.6.3) to identities (5.1.1) and (5.1.2). From now on, we avoid the following trivial case.

Throughout, unless stated otherwise, $(r, s) \neq (0, 0)$.

Defining operations on $\mathcal{T}_s^r(M)$ in a manner analogous to that described for vector fields and covector fields, we make $\mathcal{T}_s^r(M)$ into both a vector space over $\mathbb{R}$ and a module over $C^\infty(M)$.

Many of the definitions presented for smooth manifolds are expressed in a pointwise fashion (not to be confused with "determined pointwise") and ultimately rest on earlier definitions given in the context of vector spaces. For example, a tensor field on a smooth manifold is essentially a collection of tensors, one for each point in the smooth manifold. An important consequence of the pointwise approach is that earlier theorems presented for vectors spaces generalize immediately to smooth manifolds. We will say that the resulting smooth manifold theorem is the **manifold version** (abbreviated **mv**) of the earlier vector space theorem. Here is an example.

Theorem 15.6.1. *Let M be a smooth manifold, let $\mathcal{A}, \mathcal{A}_1, \mathcal{A}_2$ and $\mathcal{B}, \mathcal{B}_1, \mathcal{B}_2$ and $\mathcal{C}$ be tensor fields in $\mathcal{T}_s^r(M)$ and $\mathcal{T}_{s'}^{r'}(M)$ and $\mathcal{T}_{s''}^{r''}(M)$, respectively, and let f be a function in $C^\infty(M)$. Then:*
(a) *$(\mathcal{A}_1 + \mathcal{A}_2) \otimes \mathcal{B} = \mathcal{A}_1 \otimes \mathcal{B} + \mathcal{A}_2 \otimes \mathcal{B}$.*
(b) *$\mathcal{A} \otimes (\mathcal{B}_1 + \mathcal{B}_2) = \mathcal{A} \otimes \mathcal{B}_1 + \mathcal{A} \otimes \mathcal{B}_2$.*
(c) *$(f\mathcal{A}) \otimes \mathcal{B} = f(\mathcal{A} \otimes \mathcal{B}) = \mathcal{A} \otimes (f\mathcal{B})$.*
(d) *$(\mathcal{A} \otimes \mathcal{B}) \otimes \mathcal{C} = \mathcal{A} \otimes (\mathcal{B} \otimes \mathcal{C})$.*

Proof. This is the manifold version of Theorem 5.1.2. $\square$

Let M be a smooth m-manifold, and let $(U, (x^i))$ be a chart on M. Let $1 \leq i_1, \ldots, i_r \leq m$ and $1 \leq j_1, \ldots, j_s \leq m$ be integers, and consider the tensor field

$$\frac{\partial}{\partial x^{i_1}} \otimes \cdots \otimes \frac{\partial}{\partial x^{i_r}} \otimes d(x^{j_1}) \otimes \cdots \otimes d(x^{j_s}) \qquad \text{in} \qquad \mathcal{T}_s^r(U)$$

defined by the assignment

$$p \longmapsto \left.\frac{\partial}{\partial x^{i_1}}\right|_p \otimes \cdots \otimes \left.\frac{\partial}{\partial x^{i_r}}\right|_p \otimes d_p(x^{j_1}) \otimes \cdots \otimes d_p(x^{j_s})$$

for all p in U. Suppose $\mathcal{A}$ is a (not necessarily smooth) (r,s)-tensor field on M. Then $\mathcal{A}|_U$ can be expressed as

$$\mathcal{A}|_U = \sum_{\substack{1 \le i_1,\ldots,i_r \le m \\ 1 \le j_1,\ldots,j_s \le m}} \mathcal{A}^{i_1 \ldots i_r}_{j_1 \ldots j_s} \frac{\partial}{\partial x^{i_1}} \otimes \cdots \otimes \frac{\partial}{\partial x^{i_r}} \otimes d(x^{j_1}) \otimes \cdots \otimes d(x^{j_s}), \quad (15.6.4)$$

where the $\mathcal{A}^{i_1 \ldots i_r}_{j_1 \ldots j_s}$ are uniquely determined functions on U, called the **components of $\mathcal{A}$** with respect to $(U, (x^i))$. For brevity, we denote

$$\mathcal{A}|_U \qquad \text{by} \qquad \mathcal{A}.$$

The right-hand side of (15.6.4) is said to express $\mathcal{A}$ in **local coordinates** with respect to $(U, (x^i))$.

Theorem 15.6.2 (Smoothness Criterion for (r,s)-Tensor Fields). *With the above setup, $\mathcal{A}$ is in $T^r_s(M)$ if and only if for every chart $(U, (x^i))$ on M, the components of $\mathcal{A}$ are in $C^\infty(U)$.*   □

Theorem 15.6.3 (Change of Coordinate Frame). *Let M be a smooth m-manifold, let $\mathcal{A}$ be a tensor field in $T^r_s(M)$, let $(U, (x^i))$ and $(\widetilde{U}, (\widetilde{x}^i))$ be overlapping charts on M, and let $\mathcal{A}^{i_1 \ldots i_r}_{j_1 \ldots j_s}$ and $\widetilde{\mathcal{A}}^{i_1 \ldots i_r}_{j_1 \ldots j_s}$ be the components of $\mathcal{A}$ with respect to $(U, (x^i))$ and $(\widetilde{U}, (\widetilde{x}^i))$, respectively. Then*

$$\widetilde{\mathcal{A}}^{i_1 \ldots i_r}_{j_1 \ldots j_s} = \sum_{\substack{1 \le k_1,\ldots,k_r \le m \\ 1 \le l_1,\ldots,l_s \le m}} \frac{\partial \widetilde{x}^{i_1}}{\partial x^{k_1}} \cdots \frac{\partial \widetilde{x}^{i_r}}{\partial x^{k_r}} \frac{\partial x^{l_1}}{\partial \widetilde{x}^{j_1}} \cdots \frac{\partial x^{l_s}}{\partial \widetilde{x}^{j_s}} \mathcal{A}^{k_1 \ldots k_r}_{l_1 \ldots l_s}$$

on $U \cap \widetilde{U}$.

Proof. This is the manifold version of Theorem 5.1.4, but it is instructive to work through the details. We have from Theorem 15.1.8 and Theorem 15.4.3

that

$$\widetilde{\mathcal{A}}^{i_1\ldots i_r}_{j_1\ldots j_s} = \mathcal{A}\left(d(\widetilde{x}^{i_1}),\ldots,d(\widetilde{x}^{i_r}),\frac{\partial}{\partial\widetilde{x}^{j_1}},\ldots,\frac{\partial}{\partial\widetilde{x}^{j_s}}\right)$$

$$= \mathcal{A}\left(\sum_{k_1}\frac{\partial\widetilde{x}^{i_1}}{\partial x^{k_1}}d(x^{k_1}),\ldots,\sum_{k_r}\frac{\partial\widetilde{x}^{i_r}}{\partial x^{k_r}}d(x^{k_r}),\right.$$

$$\left.\sum_{l_1}\frac{\partial x^{l_1}}{\partial\widetilde{x}^{j_1}}\frac{\partial}{\partial x^{l_1}},\ldots,\sum_{l_s}\frac{\partial x^{l_s}}{\partial\widetilde{x}^{j_s}}\frac{\partial}{\partial x^{l_s}}\right)$$

$$= \sum_{\substack{1\leq k_1,\ldots,k_r\leq m\\1\leq l_1,\ldots,l_s\leq m}}\frac{\partial\widetilde{x}^{i_1}}{\partial x^{k_1}}\cdots\frac{\partial\widetilde{x}^{i_r}}{\partial x^{k_r}}\frac{\partial x^{l_1}}{\partial\widetilde{x}^{j_1}}\cdots\frac{\partial x^{l_s}}{\partial\widetilde{x}^{j_s}}$$

$$\cdot\mathcal{A}\left(d(x^{k_1}),\ldots,d(x^{k_r}),\frac{\partial}{\partial x^{l_1}},\ldots,\frac{\partial}{\partial x^{l_s}}\right)$$

$$= \sum_{\substack{1\leq k_1,\ldots,k_r\leq m\\1\leq l_1,\ldots,l_s\leq m}}\frac{\partial\widetilde{x}^{i_1}}{\partial x^{k_1}}\cdots\frac{\partial\widetilde{x}^{i_r}}{\partial x^{k_r}}\frac{\partial x^{l_1}}{\partial\widetilde{x}^{j_1}}\cdots\frac{\partial x^{l_s}}{\partial\widetilde{x}^{j_s}}\mathcal{A}^{k_1\ldots k_r}_{l_1\ldots l_s}. \qquad\square$$

15.7   Representation of Tensor Fields

In this section, we present generalizations of the definitions and results of Section 15.5.

Let M be a smooth manifold, and let $r, s \geq 0$ be integers. Following Section B.5, we denote by $\mathrm{Mult}_{C^\infty(M)}\big(\mathfrak{X}^*(M)^r \times \mathfrak{X}(M)^s, C^\infty(M)\big)$ the $C^\infty(M)$-module of $C^\infty(M)$-multilinear maps from $\mathfrak{X}^*(M)^r \times \mathfrak{X}(M)^s$ to $C^\infty(M)$. Let us define a map

$$\mathfrak{C}^r_s : \mathcal{T}^r_s(M) \longrightarrow \mathrm{Mult}_{C^\infty(M)}\big(\mathfrak{X}^*(M)^r \times \mathfrak{X}(M)^s, C^\infty(M)\big),$$

called the **characterization map**, by

$$\mathfrak{C}^r_s(\mathcal{A})(\omega^1,\ldots,\omega^r, X_1,\ldots,X_s) = \mathcal{A}(\omega^1,\ldots,\omega^r, X_1,\ldots,X_s) \qquad (15.7.1)$$

for all tensor fields $\mathcal{A}$ in $\mathcal{T}^r_s(M)$, all covector fields $\omega^1,\ldots,\omega^s$ in $\mathfrak{X}^*(M)$, and all vector fields $X_1,\ldots,X_s$ in $\mathfrak{X}(M)$, where the right-hand side of (15.7.1) is given by (15.6.1). It follows from a generalization of Theorem 15.5.1 that $\mathcal{A}$ is $C^\infty(M)$-multilinear. Thus, $\mathfrak{C}^r_s(\mathcal{A})$ is in $\mathrm{Mult}_{C^\infty(M)}\big(\mathfrak{X}^*(M)^r \times \mathfrak{X}(M)^s, C^\infty(M)\big)$, so the definition makes sense.

We say that an $\mathbb{R}$-linear map $\mathcal{F} : \mathfrak{X}^*(M)^r \times \mathfrak{X}(M)^s \longrightarrow C^\infty(M)$ is **determined pointwise** if for all points p in M,

$$\mathcal{F}(\omega^1,\ldots,\omega^s, X_1,\ldots,X_s)(p) = \mathcal{F}(\widetilde{\omega}^1,\ldots,\widetilde{\omega}^s, \widetilde{X}_1,\ldots,\widetilde{X}_s)(p)$$

whenever $\omega^i, \widetilde{\omega}^i$ are covector fields in $\mathfrak{X}^*(M)$ such that $\omega^i|_p = \widetilde{\omega}^i|_p$ for $i = 1,\ldots,r$, and $X_j, \widetilde{X}_j$ are vector fields in $\mathfrak{X}(M)$ such that $X_j|_p = \widetilde{X}_j|_p$ for $j = 1,\ldots,s$.

Theorem 15.7.1. *If M is a smooth manifold and Υ is a map in* $\mathrm{Mult}_{C^\infty(M)}\big(\mathfrak{X}^*(M)^r \times \mathfrak{X}(M)^s, C^\infty(M)\big)$, *then Υ is determined pointwise.* $\qquad\square$

Theorem 15.7.2 (Tensor Field Characterization Theorem). *If M is a smooth manifold, then $\mathfrak{C}_s^r$ is a $C^\infty(M)$-module isomorphism:*

$$\mathcal{T}_s^r(M) \approx \mathrm{Mult}_{C^\infty(M)}\big(\mathfrak{X}^*(M)^r \times \mathfrak{X}(M)^s, C^\infty(M)\big).$$ $\qquad\square$

Let Υ be a map in $\mathrm{Mult}_{C^\infty(M)}\big(\mathfrak{X}^*(M)^r \times \mathfrak{X}(M)^s, C^\infty(M)\big)$. Analogous to (15.5.4), $(\mathfrak{C}_s^r)^{-1}(\Upsilon)$ is the tensor field in $\mathcal{T}_s^r(M)$ defined by

$$(\mathfrak{C}_s^r)^{-1}(\Upsilon)_p(\eta^1, \ldots, \eta^r, v_1, \ldots, v_s) = \Upsilon(\omega^1, \ldots, \omega^r, X_1, \ldots, X_s)(p) \qquad (15.7.2)$$

for all points p in M, all covectors $\eta^1, \ldots, \eta^r$ in $T_p^*(M)$, and all vectors $v_1, \ldots, v_s$ in $T_p(M)$, where $\omega^1, \ldots, \omega^r$ are any covector fields in $\mathfrak{X}^*(M)$ such that $\omega^i|_p = \eta^i$ for $i = 1, \ldots, r$, and $X_1, \ldots, X_s$ are any vector fields in $\mathfrak{X}(M)$ such that $X_j|_p = v_j$ for $j = 1, \ldots, s$.

For the remainder of this section, we attempt to place the above technical material in a larger context.

Let $\mathcal{A}$ be a tensor field in $\mathcal{T}_s^r(M)$. For a given point p in M, $\mathcal{A}_p$ is a tensor in $\mathcal{T}_s^r(T_p(M))$, and for given covectors $\eta^1, \ldots, \eta^r$ in $T_p^*(M)$ and vectors $v_1, \ldots, v_s$ in $T_p(M)$, $\mathcal{A}_p(\eta^1, \ldots, \eta^r, v_1, \ldots, v_s)$ is its value in $\mathbb{R}$. Making the identification given by the isomorphism in Theorem 15.7.2, $\mathcal{A}$ can now viewed as a map in $\mathrm{Mult}_{C^\infty(M)}\big(\mathfrak{X}^*(M)^r \times \mathfrak{X}(M)^s, C^\infty(M)\big)$. For given covector fields $\omega^1, \ldots, \omega^r$ in $\mathfrak{X}^*(M)$ and vector fields $X_1, \ldots, X_s$ in $\mathfrak{X}(M)$, $\mathcal{A}(\omega^1, \ldots, \omega^r, X_1, \ldots, X_s)$ is a function in $C^\infty(M)$, and for a given point p in M, $\mathcal{A}(\omega^1, \ldots, \omega^r, X_1, \ldots, X_s)(p)$ is its value in $\mathbb{R}$. The innovation introduced by Theorem 15.7.2 is that we have gone from evaluating the tensor $\mathcal{A}_p$ at covectors and vectors to evaluating the function $\mathcal{A}$ at forms and vector fields.

Now that we have Theorem 15.7.2 at our disposal, we often (but not always) view $\mathcal{T}_s^r(M)$ as the vector space over $\mathbb{R}$ and module over $C^\infty(M)$ consisting of all $C^\infty(M)$-multilinear maps from $\mathfrak{X}^*(M)^r \times \mathfrak{X}(M)^s$ to $C^\infty(M)$. We will not be fastidious about whether "$\mathfrak{C}_s^r$" is included in the notation, allowing the context to make the situation clear and thereby providing a welcome simplification of notation.

An advantage of our new approach to tensor fields is the mechanism it provides for deciding whether a given map

$$\mathcal{F} : \mathfrak{X}^*(M)^r \times \mathfrak{X}(M)^s \longrightarrow C^\infty(M)$$

is (or at least can be identified with) a tensor field in $\mathcal{T}_s^r(M)$. According to Theorem 15.7.2, this identification can be made as long as $\mathcal{F}$ can be shown to be $C^\infty(M)$-multilinear. In practice, deciding if $\mathcal{F}$ is additive is usually straightforward. The challenge typically resides in determining whether functions in $C^\infty(M)$ can be "factored out" of $\mathcal{F}$. That is, if for all covector fields $\omega^1, \ldots, \omega^r$ in $\mathfrak{X}^*(M)$, all vector fields $X_1, \ldots, X_s$ in $\mathfrak{X}(M)$, and all functions f in $C^\infty(M)$, we have

$$\mathcal{F}(\omega^1, \ldots, f\omega^i, \ldots, \omega^r, X_1, \ldots, X_s) = f\mathcal{F}(\omega^1, \ldots, \omega^i, \ldots, \omega^r, X_1, \ldots, X_s)$$

for $i = 1, \ldots, r$, and

$$\mathcal{F}(\omega^1, \ldots, \omega^r, X_1, \ldots, fX_j, \ldots, X_s) = f\mathcal{F}(\omega^1, \ldots, \omega^r, X_1, \ldots, X_j, \ldots, X_s)$$

for $j = 1, \ldots, s$. We will encounter several instances of such computations in subsequent chapters.

Let us close this section with a few remarks on "representations". In Section 15.2, we showed that a vector field in $\mathfrak{X}(M)$ is equivalent to a type of map from $C^\infty(M)$ to $C^\infty(M)$. In Section 15.5, it was demonstrated that a covector field in $\mathfrak{X}^*(M)$ is equivalent to a type of map from $\mathfrak{X}(M)$ to $C^\infty(M)$. In this section, we showed (or at least asserted) that a tensor field in $\mathcal{T}_s^r(M)$ is equivalent to a type of map from $\mathfrak{X}^*(M)^r \times \mathfrak{X}(M)^s$ to $C^\infty(M)$. Loosely speaking, we have been involved in a campaign to represent "fields" as maps that produce "functions".

15.8   Differential Forms

Let M be a smooth m-manifold, and let $0 \le s \le m$ be an integer. A **differential s-form on M** is a map ω that assigns to each point p in M an s-covector ω_p in $\Lambda^s\big(T_p(M)\big)$. In the literature, a differential s-form is usually referred to as an **s-form** or simply a **form**. Observe that 1-forms and covector fields are the same thing. Let $X_1, \ldots, X_s$ be vector fields in $\mathfrak{X}(M)$, and define a function

$$\omega(X_1, \ldots, X_s) : M \longrightarrow \mathbb{R}$$

by

$$\omega(X_1, \ldots, X_s)(p) = \omega_p(X_1|_p, \ldots, X_s|_p)$$

for all p in M. We say that ω is **smooth (on M)** if the function $\omega(X_1, \ldots, X_s)$ is in $C^\infty(M)$ for all vector fields $X_1, \ldots, X_s$ in $\mathfrak{X}(M)$. The set of smooth s-forms on M is denoted by $\Lambda^s(M)$. Clearly, $\Lambda^s(M)$ is an $\mathbb{R}$-subspace and $C^\infty(M)$-submodule of $\mathcal{T}_s^0(M)$, and

$$\Lambda^1(M) = \mathfrak{X}^*(M) = \mathcal{T}_1^0(M). \tag{15.8.1}$$

For completeness, and to be consistent with (15.6.3), let us define

$$\Lambda^0(M) = C^\infty(M). \tag{15.8.2}$$

In view of Theorem 7.2.12(b), we set $\Lambda^s(M) = \{0\}$ for $s > m$.

Let ω and ξ be forms in $\Lambda^s(M)$ and $\Lambda^{s'}(M)$, respectively. We define a form $\omega \wedge \xi$ in $\Lambda^{s+s'}(M)$, called the **wedge product of ω and ξ**, by

$$(\omega \wedge \xi)_p = \omega_p \wedge \xi_p$$

for all p in M. In particular, for a function f in $C^\infty(M) = \Lambda^0(M)$, we have

$$f \wedge \omega = f\omega. \tag{15.8.3}$$

Let $\big(U, (x^i)\big)$ be a chart on M, let $1 \leq i_1 < \cdots < i_r \leq m$ be integers, and let

$$d(x^{i_1}) \wedge \cdots \wedge d(x^{i_s}) : U \longrightarrow \Lambda^s(U)$$

be the map defined by the assignment

$$p \longmapsto d_p(x^{i_1}) \wedge \cdots \wedge d_p(x^{i_s})$$

for all p in U. Suppose ω is a form in $\Lambda^s(M)$. Then $\omega|_U$ can be expressed as

$$\omega|_U = \sum_{1 \leq i_1 < \cdots < i_s \leq m} \alpha_{i_1,\dots,i_s} d(x^{i_1}) \wedge \cdots \wedge d(x^{i_s}), \qquad (15.8.4)$$

where the $\alpha_{i_1,\dots,i_s}$ are uniquely determined functions in $C^\infty(U)$, called the **components of ω** with respect to $\big(U, (x^i)\big)$. For brevity, we denote

$$\omega|_U \qquad \text{by} \qquad \omega.$$

The right-hand side of (15.8.4) is said to express ω in **local coordinates** with respect to $\big(U, (x^i)\big)$.

Theorem 15.8.1. *Let M be a smooth manifold, let $\omega, \omega^1, \omega^2$ and ξ, ξ^1, ξ^2 be forms in $\Lambda^s(M)$ and $\Lambda^{s'}(M)$, respectively, and let f be a function in $C^\infty(M)$. Then:*
(a) $(\omega^1 + \omega^2) \wedge \xi = \omega^1 \wedge \xi + \omega^2 \wedge \xi$.
(b) $\omega \wedge (\xi^1 + \xi^2) = \omega \wedge \xi^1 + \omega \wedge \xi^2$.
(c) $(f\omega) \wedge \xi = f(\omega \wedge \xi) = \omega \wedge (f\xi)$.

Proof. This is the manifold version of Theorem 7.2.2.   □

Theorem 15.8.2. *Let M be a smooth manifold, and let ω and ξ be forms in $\Lambda^s(M)$ and $\Lambda^{s'}(M)$, respectively. Then:*
(a) $\omega \wedge \xi = (-1)^{ss'} \xi \wedge \omega$.
(b) If $s = s' = 1$, then $\omega \wedge \xi = -\xi \wedge \omega$.
(c) If $s = 1$, then $\omega \wedge \omega = 0$.

Proof. This is the manifold version of Theorem 7.2.3.   □

Theorem 15.8.3. *Let M be a smooth manifold, let $\omega^1, \dots, \omega^s$ be covector fields in $\mathfrak{X}^*(M)$, and let $X_1, \dots, X_s$ be vector fields in $\mathfrak{X}(M)$. Then:*
(a)

$$\omega^1 \wedge \cdots \wedge \omega^s(X_1, \dots, X_s) = \det\left(\begin{bmatrix} \omega^1(X_1) & \cdots & \omega^1(X_s) \\ \vdots & \ddots & \vdots \\ \omega^s(X_1) & \cdots & \omega^s(X_s) \end{bmatrix} \right).$$

(b) If $\omega^i = \omega^j$ for some $1 \leq i < j \leq s$, then $\omega^1 \wedge \cdots \wedge \omega^s = 0$.

Proof. This is the manifold version of Theorem 7.2.6 combined with the manifold version of Theorem 7.2.8.   □

Theorem 15.8.4. *Let M be a smooth m-manifold, and let U be an open set in M. If $(X_1, \ldots, X_m)$ is a frame on U and $(\omega^1, \ldots, \omega^m)$ is its dual frame, then*

$$\omega^{j_1} \wedge \cdots \wedge \omega^{j_s}(X_{i_1}, \ldots, X_{i_s}) = \begin{cases} 1 & \text{if } (i_1, \ldots, i_s) = (j_1, \ldots, j_s) \\ 0 & \text{if } (i_1, \ldots, i_s) \neq (j_1, \ldots, j_s) \end{cases}$$

for all $1 \leq i_1 < \cdots < i_s \leq m$ and $1 \leq j_1 < \cdots < j_s \leq m$. In particular, if $(U, (x^i))$ is a chart on M, and $(\partial/\partial x^1, \ldots, \partial/\partial x^m)$ and $(d(x^1), \ldots, d(x^m))$ are the corresponding coordinate and dual coordinate frames, then

$$d(x^{j_1}) \wedge \cdots \wedge d(x^{j_s})\left(\frac{\partial}{\partial x^{i_1}}, \ldots, \frac{\partial}{\partial x^{i_s}}\right) = \begin{cases} 1 & \text{if } (i_1, \ldots, i_s) = (j_1, \ldots, j_s) \\ 0 & \text{if } (i_1, \ldots, i_s) \neq (j_1, \ldots, j_s) \end{cases}$$

for all $1 \leq i_1 < \cdots < i_s \leq m$ and $1 \leq j_1 < \cdots < j_s \leq m$.

Proof. This is the manifold version of Theorem 7.2.9. □

Theorem 15.8.5. *Let M be a smooth m-manifold, let $(U, (x^i))$ be a chart on M, and let $(d(x^1), \ldots, d(x^m))$ be the corresponding dual coordinate frame. If ω is a form in $\Lambda^m(U)$, then there is a uniquely determined function f in $C^\infty(U)$ such that $\omega = f\, d(x^1) \wedge \cdots \wedge d(x^m)$.*

Proof. Let $(\partial/\partial x^1, \ldots, \partial/\partial x^m)$ be the coordinate frame corresponding to $(U, (x^i))$, and, for brevity, let $\Omega = d(x^1) \wedge \cdots \wedge d(x^m)$. For each point p in U, by definition, ω_p is a covector in $T_p^*(U)$. It follows from Theorem 7.2.12(d) that $\omega_p = f(p)\,\Omega_p$ for some real number $f(p)$. The assignment $p \longmapsto f(p)$ defines a function $f : U \longrightarrow \mathbb{R}$ such that $\omega = f\Omega$. Since ω is in $\Lambda^m(U)$, by definition, $\omega(X_1, \ldots, X_m)$ is a function in $C^\infty(U)$ for all vector fields $X_1, \ldots, X_m$ in $\mathfrak{X}(U)$. In particular, this is so for $\partial/\partial x^1, \ldots, \partial/\partial x^m$. By Theorem 15.8.4,

$$f = f\Omega\left(\frac{\partial}{\partial x^1}, \ldots, \frac{\partial}{\partial x^m}\right) = \omega\left(\frac{\partial}{\partial x^1}, \ldots, \frac{\partial}{\partial x^m}\right),$$

hence f is in $C^\infty(U)$. The uniqueness of f follows from the preceding computations. □

15.9 Pushforward and Pullback of Functions

Let M be a smooth manifold, and let $F : M \longrightarrow N$ be a diffeomorphism. **Pushforward by F (for functions)** is the map

$$F_\bullet : C^\infty(M) \longrightarrow C^\infty(N)$$

defined by

$$F_\bullet(f) = f \circ F^{-1} \tag{15.9.1}$$

for all functions f in $C^\infty(M)$.

Theorem 15.9.1. *If M, N, and P are smooth manifolds, and $F : M \longrightarrow N$ and $G : N \longrightarrow P$ are diffeomorphisms, then*

$$(G \circ F)_\bullet = G_\bullet \circ F_\bullet.$$

Proof. For a function f in $C^\infty(M)$, we have

$$(G \circ F)_\bullet(f) = f \circ (G \circ F)^{-1} = (f \circ F^{-1}) \circ G^{-1} = F_\bullet(f) \circ G^{-1}$$
$$= G_\bullet\big(F_\bullet(f)\big) = (G_\bullet \circ F_\bullet)(f).$$

Since f was arbitrary, the result follows. $\square$

Let M and N be smooth manifolds, and let $F : M \longrightarrow N$ be a smooth map (but not necessarily a diffeomorphism). **Pullback by F (for functions)** is the map

$$F^\bullet : C^\infty(N) \longrightarrow C^\infty(M)$$

defined by

$$F^\bullet(g) = g \circ F \tag{15.9.2}$$

for all functions g in $C^\infty(N)$.

Theorem 15.9.2. *If M, N, and P are smooth manifolds, and $F : M \longrightarrow N$ and $G : N \longrightarrow P$ are smooth maps, then*

$$(G \circ F)^\bullet = F^\bullet \circ G^\bullet.$$

Proof. For a function g in $C^\infty(P)$, we have

$$(G \circ F)^\bullet(g) = (g \circ G) \circ F = G^\bullet(g) \circ F = F^\bullet\big(G^\bullet(g)\big) = (F^\bullet \circ G^\bullet)(g).$$

Since g was arbitrary, the result follows. $\square$

Theorem 15.9.3. *If M is a smooth manifold and $F : M \longrightarrow N$ is a diffeomorphism, then:*
(a) *$F^\bullet$ is bijective, with inverse $(F^\bullet)^{-1} = (F^{-1})^\bullet = F_\bullet$.*
(b) *$F_\bullet$ is bijective, with inverse $(F_\bullet)^{-1} = (F^{-1})_\bullet = F^\bullet$.*

Proof. (a): By Theorem 15.9.2,

$$\mathrm{id}_{C^\infty(M)} = (\mathrm{id}_M)^\bullet = (F^{-1} \circ F)^\bullet = F^\bullet \circ (F^{-1})^\bullet$$

and

$$\mathrm{id}_{C^\infty(N)} = (\mathrm{id}_N)^\bullet = (F \circ F^{-1})^\bullet = (F^{-1})^\bullet \circ F^\bullet.$$

It follows from Theorem A.4 that $F^\bullet$ is bijective. The preceding identities give the first equality. For a function f in $C^\infty(M)$, we have

$$(F^{-1})^\bullet(f) = f \circ F^{-1} = F_\bullet(f).$$

Since f was arbitrary, the second equality follows.
(b): This follows from part (a). $\square$

15.10   Pushforward and Pullback of Vector Fields

Let M and N be smooth manifolds, let $F : M \longrightarrow N$ be a smooth map, let X be a vector field in $\mathfrak{X}(M)$, and let p be a point in M. By definition, $d_p(F)(X_p)$ is a vector in $T_{F(p)}(N)$. Without further assumptions, the assignment $p \longmapsto d_p(F)(X_p)$ does not necessarily produce a vector field in $\mathfrak{X}(N)$. For example, if F is not surjective, there is no way to assign a vector to any point outside the image of F. Furthermore, if F is not injective, then there are distinct points p_1, p_2 in M such that $F(p_1) = F(p_2)$. When $d_{p_1}(F)(X_{p_1}) \neq d_{p_2}(F)(X_{p_2})$, there is no unambiguous way to assign a vector to $F(p_1)$. The way out of this dilemma is to assume, as we now do, that F is a diffeomorphism. In what follows, we use Theorem 15.2.1 to identify vector fields in $\mathfrak{X}(M)$ with derivations in $\mathrm{Der}(M)$.

The **pushforward of X by F** is the vector field $F_*(X)$ in $\mathfrak{X}(N)$ defined by

$$F_*(X) = F_\bullet \circ X \circ F^\bullet. \tag{15.10.1}$$

This definition makes sense because we have the maps $F^\bullet : C^\infty(N) \longrightarrow C^\infty(M)$, $X : C^\infty(M) \longrightarrow C^\infty(M)$, and $F_\bullet : C^\infty(M) \longrightarrow C^\infty(N)$, hence

$$F_*(X) : C^\infty(N) \longrightarrow C^\infty(N).$$

Pushforward by F (for vector fields) is the map

$$F_* : \mathfrak{X}(M) \longrightarrow \mathfrak{X}(N)$$

defined by the assignment

$$X \longmapsto F_*(X)$$

for all vector fields X in $\mathfrak{X}(M)$. **Pullback by F (for vector fields)** is the map

$$F^* : \mathfrak{X}(N) \longrightarrow \mathfrak{X}(M)$$

defined by

$$F^* = (F^{-1})_* \; ;$$

that is,

$$F^*(Y) = (F^{-1})_*(Y) \tag{15.10.2}$$

for all vector fields Y in $\mathfrak{X}(N)$. We call $F^*(Y)$ the **pullback of Y by F** and observe that it equals the pushforward of Y by F^{-1}.

**The notation F_* and F^* will be used only
when F is a diffeomorphism.**

The next result shows that the pushforward provides a response to the issue raised in the introduction.

Theorem 15.10.1. *Let M and N be smooth manifolds, let $F : M \longrightarrow N$ be a diffeomorphism, and let X be a vector field in $\mathfrak{X}(M)$. Then*

$$d_p(F)(X_p) = F_*(X)_{F(p)}$$

for all p in M.

Remark. Working through the definitions, we find that both $d_p(F)(X_p)$ and $F_*(X)_{F(p)}$ are vectors in $T_{F(p)}(N)$, so the assertion makes sense. In the literature, the notation F_{*p} or simply F_* is often used in place of $d_p(F)$. In fact, such notation is frequently adopted even when F is not a diffeomorphism.

Proof. For a function g in $C^\infty(N)$, we have

$$
\begin{aligned}
d_p(F)(X_p)(g) &= X_p(g \circ F) & [(14.4.1)] \\
&= X(g \circ F)(p) & [(15.1.2)] \\
&= X\big(F^\bullet(g)\big) \circ F^{-1}\big(F(p)\big) & [(15.9.2)] \\
&= F_\bullet\big(X(F^\bullet(g))\big)\big(F(p)\big) & [(15.9.1)] \\
&= (F_\bullet \circ X \circ F^\bullet)(g)\big(F(p)\big) & \\
&= F_*(X)(g)\big(F(p)\big) & [(15.10.1)] \\
&= F_*(X)_{F(p)}(g). & [(15.1.2)]
\end{aligned}
$$

Since g was arbitrary, the result follows.   □

Theorem 15.10.2. *Let M, N, and P be smooth manifolds, let $F : M \longrightarrow N$ and $G : N \longrightarrow P$ be diffeomorphisms, and let Y be a vector field in $\mathfrak{X}(N)$. Then:*
(a) $(G \circ F)_* = G_* \circ F_*$
(b) $(G \circ F)^* = F^* \circ G^*$
(c) $F^*(Y) = F^\bullet \circ Y \circ F_\bullet$.

Proof. (a): For a vector field X in $\mathfrak{X}(M)$, we have

$$
\begin{aligned}
(G \circ F)_*(X) &= (G \circ F)_\bullet \circ X \circ (G \circ F)^\bullet & [(15.10.1)] \\
&= G_\bullet \circ (F_\bullet \circ X \circ F^\bullet) \circ G^\bullet & [\text{Th } 15.9.1,\ \text{Th } 15.9.2] \\
&= G_\bullet \circ F_*(X) \circ G^\bullet & [(15.10.1)] \\
&= G_*\big(F_*(X)\big) & [(15.10.1)] \\
&= (G_* \circ F_*)(X). &
\end{aligned}
$$

Since X was arbitrary, the result follows.
   (b): For a vector field Z in $\mathfrak{X}(P)$, we have

$$
\begin{aligned}
(G \circ F)^*(Z) &= \big((G \circ F)^{-1}\big)_*(Z) & [(15.10.2)] \\
&= (F^{-1} \circ G^{-1})_*(Z) & \\
&= (F^{-1})_* \circ (G^{-1})_*(Z) & [\text{part (a)}] \\
&= (F^* \circ G^*)(Z). & [(15.10.2)]
\end{aligned}
$$

Since Z was arbitrary, the result follows.

   (c): We have

$$
\begin{aligned}
F^*(Y) &= (F^{-1})_*(Y) && [(15.10.2)] \\
&= (F^{-1})_\bullet \circ Y \circ (F^{-1})^\bullet && [(15.10.1)] \\
&= F^\bullet \circ Y \circ F_\bullet. && [\text{Th } 15.9.3] \qquad \square
\end{aligned}
$$

Theorem 15.10.3. *If M is a smooth manifold and $F : M \longrightarrow N$ is a diffeomorphism, then:*
(a) *F^* is bijective, with inverse $(F^*)^{-1} = (F^{-1})^* = F_*$.*
(b) *F_* is bijective, with inverse $(F_*)^{-1} = (F^{-1})_* = F^*$.*

Proof. The arguments are similar to those used in the proof of Theorem 15.9.3.
$$\square$$

Theorem 15.10.4. *Let M and N be smooth manifolds, let $F : M \longrightarrow N$ be a diffeomorphism, let Y be a vector field in $\mathfrak{X}(N)$, and let g be a function in $C^\infty(N)$. Then*

$$
F^\bullet(Y(g)) = F^*(Y)(F^\bullet(g)).
$$

Proof. We have

$$
\begin{aligned}
F^*(Y)(F^\bullet(g)) &= (F^\bullet \circ Y \circ F_\bullet)(F^\bullet(g)) && [\text{Th } 15.10.2(\text{c})] \\
&= F^\bullet(Y(g)). && [\text{Th } 15.9.3] \qquad \square
\end{aligned}
$$

15.11   Pullback of Covector Fields

Let M and N be smooth manifolds, let $F : M \longrightarrow N$ be a smooth map, and let p be a point in M. The corresponding differential map is $d_p(F) : T_p(M) \longrightarrow T_{F(p)}(N)$. According to (7.1.2), $\Lambda^1(T_p(M)) = T_p^*(M)$ and $\Lambda^1(T_{F(p)}(N)) = T_{F(p)}^*(N)$, so we have from (7.3.1) that the pullback by $d_p(F)$ for covectors is the map

$$
d_p(F)^* : T_{F(p)}^*(N) \longrightarrow T_p^*(M)
$$

defined by

$$
d_p(F)^*(\eta)(v) = \eta(d_p(F)(v)) \tag{15.11.1}
$$

for all covectors η in $T_{F(p)}^*(N)$ and all vectors v in $T_p(M)$. **Pullback by F (for covector fields)** is the linear map

$$
F^* : \mathfrak{X}^*(N) \longrightarrow \mathfrak{X}^*(M)
$$

defined by

$$
F^*(\omega)_p(v) = d_p(F)^*(\omega_{F(p)})(v) = \omega_{F(p)}(d_p(F)(v)) \tag{15.11.2}
$$

for all covector fields ω in $\mathfrak{X}(N)$, all points p in M, and all vectors v in $T_p(M)$, where the second equality follows from setting $\eta = \omega_{F(p)}$ in (15.11.1). We refer to $F^*(\omega)$ as the **pullback of ω by F**. An important observation is that unlike the situation with pullbacks of vector fields, pullbacks of covector fields do not require diffeomorphisms for their definition.

Theorem 15.11.1. *Let M, N, and P be smooth manifolds, let $F : M \longrightarrow N$ and $G : N \longrightarrow P$ be smooth maps, let ω, ξ be covector fields in $\mathfrak{X}^*(N)$, and let g be a function in $C^\infty(N)$. Then:*
(a) $F^*(\omega + \xi) = F^*(\omega) + F^*(\xi)$.
(b) $F^*(g\omega) = F^\bullet(g)F^*(\omega)$.
(c) $(G \circ F)^* = F^* \circ G^*$.

Proof. For a point p in M and a vector v in $T_p(M)$, we have

$$\big(F^*(\omega + \xi)\big)_p(v) = (\omega + \xi)_{F(p)}\big(d_p(F)(v)\big) \qquad\qquad [(15.11.2)]$$
$$= \omega_{F(p)}\big(d_p(F)(v)\big) + \xi_{F(p)}\big(d_p(F)(v)\big)$$
$$= F^*(\omega)_p(v) + F^*(\xi)_p(v) \qquad\qquad [(15.11.2)]$$
$$= \big(F^*(\omega) + F^*(\xi)\big)_p(v),$$

$$\big(F^*(g\omega)\big)_p(v) = (g\omega)_{F(p)}\big(d_p(F)(v)\big) \qquad\qquad [(15.11.2)]$$
$$= g\big(F(p)\big) \cdot \omega_{F(p)}\big(d_p(F)(v)\big)$$
$$= F^\bullet(g)(p) \cdot F^*(\omega)_p(v) \qquad\qquad [(15.9.2),\ (15.11.2)]$$
$$= \big(F^\bullet(g)F^*(\omega)\big)_p(v),$$

and

$$\big((G \circ F)^*(\xi)\big)_p(v) = \xi_{G\circ F(p)}\big(d_p(G \circ F)(v)\big) \qquad\qquad [(15.11.2)]$$
$$= \xi_{G(F(p))}\big(d_{F(p)}(G) \circ d_p(F)(v)\big) \qquad\qquad [\text{Th } 14.4.5]$$
$$= \xi_{G(F(p))}\big(d_{F(p)}(G)\big(d_p(F)(v)\big)\big)$$
$$= G^*(\xi)_{F(p)}\big(d_p(F)(v)\big) \qquad\qquad [(15.11.2)]$$
$$= \big(F^*(G^*(\xi))\big)_p(v) \qquad\qquad [(15.11.2)]$$
$$= \big(F^* \circ G^*(\xi)\big)_p(v).$$

Since p, v, and ξ were arbitrary, the result follows. $\square$

Theorem 15.11.2 (Commutativity of Pullback and Exterior Derivative of Functions). *If M and N are smooth manifolds and $F : M \longrightarrow N$ is a smooth map, then*

$$F^* \circ d = d \circ F^\bullet; \qquad\qquad (15.11.3)$$

that is,

$$F^*\big(d(g)\big) = d\big(F^\bullet(g)\big) \qquad\qquad (15.11.4)$$

for all functions g in $C^\infty(N)$.

Remark. For the left-hand side of (15.11.3), we have the maps $d : C^\infty(N) \longrightarrow$
$\mathfrak{X}^*(N)$ and $F^* : \mathfrak{X}^*(N) \longrightarrow \mathfrak{X}^*(M)$, and for the right-hand side, we have the
maps $F^\bullet : C^\infty(N) \longrightarrow C^\infty(M)$ and $d : C^\infty(M) \longrightarrow \mathfrak{X}^*(M)$, so the assertion
makes sense.

Proof. For a point p in M and a vector v in $T_p(M)$, we have

$$
\begin{aligned}
F^*\big(d(g)\big)_p(v) &= d(g)_{F(p)}\big(d_p(F)(v)\big) && [(15.11.2)] \\
&= d_{F(p)}(g)\big(d_p(F)(v)\big) && [(15.4.4)] \\
&= d_{F(p)}(g) \circ d_p(F)(v) \\
&= \big(d_p(g \circ F)\big)(v) && [\text{Th } 14.4.5] \\
&= d(g \circ F)_p(v) && [(15.4.4)] \\
&= d\big(F^\bullet(g)\big)_p(v). && [(15.9.2)]
\end{aligned}
$$

Since p, v, and g were arbitrary, the result follows.   □

Theorem 15.11.3 (Pullback of Covector Field). *Let M and N be smooth
manifolds, let $F : M \longrightarrow N$ be a smooth map, and let $(U, (x^i))$ and $(V, (y^j))$ be
charts on M and N, respectively, such that $U \cap F^{-1}(V)$ is nonempty. Let ω be
a covector field in $\mathfrak{X}^*(N)$, and, in local coordinates, let*

$$
\omega = \sum_j \alpha_j d(y^j).
$$

Then

$$
F^*(\omega) = \sum_i \left(\sum_j (\alpha_j \circ F) \frac{\partial(y^j \circ F)}{\partial x^i} \right) d(x^i).
$$

Proof. We have

$$
\begin{aligned}
F^*(\omega) &= F^* \left(\sum_j \alpha_j d(y^j) \right) \\
&= \sum_j F^\bullet(\alpha_j) \, F^*\big(d(y^j)\big) && [\text{Th } 15.11.1] \\
&= \sum_j F^\bullet(\alpha_j) \, d\big(F^\bullet(y^j)\big) && [\text{Th } 15.11.2] \\
&= \sum_j (\alpha_j \circ F) \, d(y^j \circ F), && [(15.9.2)]
\end{aligned}
$$

and from Theorem 15.4.7 that

$$
d(y^j \circ F) = \sum_i \frac{\partial(y^j \circ F)}{\partial x^i} d(x^i).
$$

Combining the above identities gives the result.   □

Example 15.11.4. Let $f(t) : \mathbb{R} \longrightarrow \mathbb{R}$ be the function given by $f(t) = t^2$, and let ω be the covector field in $\mathfrak{X}^*(\mathbb{R})$ defined by $\omega_r = \alpha(r)d(r)$, where α is a function in $C^\infty(\mathbb{R})$. By Theorem 15.11.3,

$$f^*(\omega)_t = \alpha\big(f(t)\big)\frac{df}{dt}(t)\,d(t) = 2t\,\alpha(t^2)\,d(t).$$

Alternatively, substituting $r = t^2$ and $d(r) = 2t\,d(t)$ in $\alpha(r)d(r)$ also gives $2t\,\alpha(t^2)\,d(t)$. This illustrates that the "pullback" operation is nothing other than the technique of "substitution" familiar from the differential calculus of one real variable. $\diamond$

Example 15.11.5. Let (r, s) and (x, y, z) be standard coordinates on $\mathbb{R}^2$ and $\mathbb{R}^3$, respectively, let $F : \mathbb{R}^3 \longrightarrow \mathbb{R}^2$ be the smooth map given by

$$F(x, y, z) = \big(x^2y, y\sin(z)\big),$$

and consider the covector field

$$\omega_{(r,s)} = s\,d(r) + r\,d(s)$$

in $\mathfrak{X}^*(\mathbb{R}^2)$. Setting $(U, (x^i)) = (\mathbb{R}^3, (x, y, z))$ and $(V, (y^j)) = (\mathbb{R}^2, (r, s))$ in Theorem 15.11.3 yields

$$
\begin{aligned}
F^*(\omega)_{(x,y,z)} &= \left((s \circ F)\frac{\partial(r \circ F)}{\partial x} + (r \circ F)\frac{\partial(s \circ F)}{\partial x}\right)d(x) \\
&+ \left((s \circ F)\frac{\partial(r \circ F)}{\partial y} + (r \circ F)\frac{\partial(s \circ F)}{\partial y}\right)d(y) \\
&+ \left((s \circ F)\frac{\partial(r \circ F)}{\partial z} + (r \circ F)\frac{\partial(s \circ F)}{\partial z}\right)d(z) \\
&= \left(y\sin(z)\frac{\partial(x^2y)}{\partial x} + x^2y\frac{\partial(y\sin(z))}{\partial x}\right)d(x) \\
&+ \left(y\sin(z)\frac{\partial(x^2y)}{\partial y} + x^2y\frac{\partial(y\sin(z))}{\partial y}\right)d(y) \\
&+ \left(y\sin(z)\frac{\partial(x^2y)}{\partial z} + x^2y\frac{\partial(y\sin(z))}{\partial z}\right)d(z) \\
&= 2xy^2\sin(z)\,d(x) + 2x^2y\sin(z)\,d(y) + x^2y^2\cos(z)\,d(z). \quad \diamond
\end{aligned}
$$

Example 15.11.6 (Polar Coordinates). Let (r, s) be standard coordinates on $\mathbb{R}^2$, and let (ρ, ϕ) be polar coordinates on

$$U = \{(\rho, \phi) \in \mathbb{R}^2 : \rho > 0, 0 \le \phi < 2\pi\}.$$

Let $F : U \longrightarrow \mathbb{R}^2$ be the smooth map given by

$$F(\rho, \phi) = \big(\rho\cos(\phi), \rho\sin(\phi)\big),$$

and consider the covector field

$$\omega_{(r,s)} = -s\,d(r) + r\,d(s)$$

in $\mathfrak{X}^*(\mathbb{R}^2)$. Setting $(U, (x^i)) = (U, (\rho, \phi))$ and $(V, (y^j)) = (\mathbb{R}^2, (r, s))$ in Theorem 15.11.3 yields

$$F^*(\omega)_{(\rho,\phi)} = \left((-s \circ F)\frac{\partial(r \circ F)}{\partial \rho} + (r \circ F)\frac{\partial(s \circ F)}{\partial \rho} \right) d(\rho)$$

$$+ \left((-s \circ F)\frac{\partial(r \circ F)}{\partial \phi} + (r \circ F)\frac{\partial(s \circ F)}{\partial \phi} \right) d(\phi)$$

$$= \left(-\rho\sin(\phi)\frac{\partial(\rho\cos(\phi))}{\partial \rho} + \rho\cos(\phi)\frac{\partial(\rho\sin(\phi))}{\partial \rho} \right) d(\rho)$$

$$+ \left(-\rho\sin(\phi)\frac{\partial(\rho\cos(\phi))}{\partial \phi} + \rho\cos(\phi)\frac{\partial(\rho\sin(\phi))}{\partial \phi} \right) d(\phi)$$

$$= \rho^2 d(\phi). \qquad \diamond$$

Example 15.11.7 (Polar Coordinates and Punctured Plane). Let (r, s) be standard coordinates on the set $\mathbb{R}^2\backslash\{(0,0)\}$, called the **punctured plane**, and let (ρ, ϕ) be polar coordinates on

$$U = \{(\rho, \phi) \in \mathbb{R}^2 : \rho > 0, 0 \leq \phi < 2\pi\}.$$

Let $F : U \longrightarrow \mathbb{R}^2$ be the smooth map given by

$$F(\rho, \phi) = (\rho\cos(\phi), \rho\sin(\phi)),$$

and consider the covector field

$$\omega_{(r,s)} = -\frac{s}{r^2 + s^2}\,d(r) + \frac{r}{r^2 + s^2}\,d(s)$$

in $\mathfrak{X}^*(\mathbb{R}^2\backslash\{(0,0)\})$.   Setting $(U, (x^i)) = (U, (\rho, \phi))$ and $(V, (y^j)) = (\mathbb{R}^2\backslash\{(0,0)\}, (r, s))$ in Theorem 15.11.3 yields

$$F^*(\omega)_{(\rho,\phi)} = \left[\left(-\frac{s}{r^2 + s^2} \circ F \right)\frac{\partial(r \circ F)}{\partial \rho} + \left(\frac{r}{r^2 + s^2} \circ F \right)\frac{\partial(s \circ F)}{\partial \rho} \right] d(\rho)$$

$$+ \left[\left(-\frac{s}{r^2 + s^2} \circ F \right)\frac{\partial(r \circ F)}{\partial \phi} + \left(\frac{r}{r^2 + s^2} \circ F \right)\frac{\partial(s \circ F)}{\partial \phi} \right] d(\phi)$$

$$= \left(-\frac{\rho\sin(\phi)}{[\rho\cos(\phi)]^2 + [\rho\sin(\phi)]^2}\frac{\partial(\rho\cos(\phi))}{\partial \rho} \right.$$

$$+ \left. \frac{\rho\cos(\phi)}{[\rho\cos(\phi)]^2 + [\rho\sin(\phi)]^2}\frac{\partial(\rho\sin(\phi))}{\partial \rho} \right) d(\rho)$$

$$+ \left(-\frac{\rho\sin(\phi)}{[\rho\cos(\phi)]^2 + [\rho\sin(\phi)]^2}\frac{\partial(\rho\cos(\phi))}{\partial \phi} \right.$$

$$+ \left. \frac{\rho\cos(\phi)}{[\rho\cos(\phi)]^2 + [\rho\sin(\phi)]^2}\frac{\partial(\rho\sin(\phi))}{\partial \phi} \right) d(\phi)$$

$$= d(\phi). \qquad \diamond$$

Theorem 15.11.8. *Let M be a smooth manifold, let $\lambda(t) : (a,b) \longrightarrow M$ be a smooth curve, let ω be a covector field in $\mathfrak{X}^*(M)$, and view (a,b) as a smooth 1-manifold. Then*

$$\lambda^*(\omega)_t = \omega_{\lambda(t)} \left(\frac{d\lambda}{dt}(t) \right) d(t)$$

for all t in (a,b).

Proof. Let p be a point in M, let $(U, (x^i))$ be a chart at p, and, in local coordinates, let

$$\omega = \sum_i \alpha_i d(x^i).$$

By Theorem 15.11.3,

$$\lambda^*(\omega)_t = \left(\sum_i \alpha_i(\lambda(t)) \frac{d(x^i \circ \lambda)}{dt}(t) \right) d(t). \qquad (15.11.5)$$

On the other hand, (15.4.4) gives

$$\omega_{\lambda(t)} = \sum_j \alpha_j(\lambda(t)) \, d(x^j)_{\lambda(t)} = \sum_j \alpha_j(\lambda(t)) \, d_{\lambda(t)}(x^j),$$

hence

$$\omega_{\lambda(t)} \left(\frac{\partial}{\partial x^i} \Big|_{\lambda(t)} \right) = \sum_j \left[\alpha_j(\lambda(t)) \, d_{\lambda(t)}(x^j) \left(\frac{\partial}{\partial x^i} \Big|_{\lambda(t)} \right) \right] = \alpha_i(\lambda(t))$$

for $i = 1, \ldots, m$. By Theorem 14.7.1(b),

$$\frac{d\lambda}{dt}(t) = \sum_i \frac{d(x^i \circ \lambda)}{dt}(t) \frac{\partial}{\partial x^i} \Big|_{\lambda(t)},$$

so

$$\omega_{\lambda(t)} \left(\frac{d\lambda}{dt}(t) \right) = \sum_{ij} \frac{d(x^i \circ \lambda)}{dt}(t) \, \omega_{\lambda(t)} \left(\frac{\partial}{\partial x^i} \Big|_{\lambda(t)} \right)$$

$$= \sum_{ij} \alpha_i(\lambda(t)) \frac{d(x^i \circ \lambda)}{dt}(t). \qquad (15.11.6)$$

Substituting (15.11.6) into (15.11.5) gives the result. $\square$

15.12   Pullback of Covariant Tensor Fields

Let M and N be smooth manifolds, let $F : M \longrightarrow N$ be a smooth map, and let p be a point in M. The corresponding differential map is $d_p(F) : T_p(M) \longrightarrow T_{F(p)}(N)$. According to (5.2.1), the pullback by $d_p(F)$ for covariant tensors is the family of linear maps

$$d_p(F)^* : \mathcal{T}_s^0(T_{F(p)}(N)) \longrightarrow \mathcal{T}_s^0(T_p(M))$$

defined for $s \geq 1$ by

$$d_p(F)^*(\mathcal{B})(v_1, \ldots, v_s) = \mathcal{B}\big(d_p(F)(v_1), \ldots, d_p(F)(v_s)\big) \qquad (15.12.1)$$

for all tensors $\mathcal{B}$ in $T_s^0(T_{F(p)}(N))$ and all vectors $v_1, \ldots, v_s$ in $T_p(M)$. **Pullback by F (for covariant tensor fields)** is the family of linear maps

$$F^* : T_s^0(N) \longrightarrow T_s^0(M)$$

defined for $s \geq 1$ by

$$\begin{aligned}
F^*(\mathcal{A})_p(v_1, \ldots, v_s) &= d_p(F)^*(\mathcal{A}_{F(p)})(v_1, \ldots, v_s) \\
&= \mathcal{A}_{F(p)}\big(d_p(F)(v_1), \ldots, d_p(F)(v_s)\big)
\end{aligned} \qquad (15.12.2)$$

for all tensor fields $\mathcal{A}$ in $T_s^0(N)$, all points p in M, and all vectors $v_1, \ldots, v_s$ in $T_p(M)$, where the second equality follows from setting $\mathcal{B} = \mathcal{A}_{F(p)}$ in (15.12.1). We refer to $F^*(\mathcal{A})$ as the **pullback of $\mathcal{A}$ by F**.

To give meaning to F^* when $s = 0$, recall from (15.6.3) that $T_0^0(N) = C^\infty(N)$. We therefore define

$$F^*(g)(p) = g\big(F(p)\big) = F^\bullet(g)(p)$$

for all functions g in $C^\infty(N)$ and all p in M; that is, we define

$$F^* = F^\bullet. \qquad (15.12.3)$$

Theorem 15.12.1. *Let M, N, and P be smooth manifolds, let $F : M \longrightarrow N$ and $G : N \longrightarrow P$ be smooth maps, let $\mathcal{A}, \mathcal{B}$ and $\mathcal{C}$ be tensor fields in $T_s^0(N)$ and $T_{s'}^0(N)$, respectively, and let g be a function in $C^\infty(N)$. Then:*
(a) $F^*(\mathcal{A} + \mathcal{B}) = F^*(\mathcal{A}) + F^*(\mathcal{B})$.
(b) $F^*(g\mathcal{A}) = F^*(g)F^*(\mathcal{A})$.
(c) $F^*(\mathcal{A} \otimes \mathcal{C}) = F^*(\mathcal{A}) \otimes F^*(\mathcal{C})$.
(d) $(G \circ F)^* = F^* \circ G^*$. $\qquad\qquad\qquad\qquad\qquad\qquad\qquad\qquad \square$

Since $g \otimes \mathcal{A} = g\mathcal{A}$, part (b) of Theorem 15.12.1 follows from part (c). Identity (15.12.3) has several implications: Theorem 15.9.2 follows from Theorem 15.12.1(d); the identity in Theorem 15.10.4 can be expressed as

$$F^*\big(Y(g)\big) = F^*(Y)\big(F^*(g)\big)$$

for all vector fields Y in $\mathfrak{X}(N)$ and all functions g in $C^\infty(N)$; Theorem 15.11.1(b) follows from Theorem 15.12.1(b); and (15.11.4) can be expressed as

$$F^*\big(d(g)\big) = d\big(F^*(g)\big)$$

for all functions g in $C^\infty(N)$.

Theorem 15.12.2 (Pullback of Covariant Tensor Field). *Let M be a smooth m-manifold, let N be a smooth n-manifold, let $F : M \longrightarrow N$ be a smooth map, and let $(U, (x^j))$ and $(V, (y^i))$ be charts on M and N, respectively, such $U \cap F^{-1}(V)$ is nonempty. Let $\mathcal{A}$ be a tensor field in $\mathcal{T}_s^0(N)$, and, in local coordinates, let*

$$\mathcal{A} = \sum_{1 \leq i_1, \ldots, i_s \leq n} \mathcal{A}_{i_1 \ldots i_s} d(y^{i_1}) \otimes \cdots \otimes d(y^{i_s}).$$

Then

$$F^*(\mathcal{A}) = \sum_{1 \leq j_1, \ldots, j_s \leq m} \left(\sum_{1 \leq i_1, \ldots, i_s \leq n} (\mathcal{A}_{i_1 \ldots i_s} \circ F) \frac{\partial(y^{i_1} \circ F)}{\partial x^{j_1}} \cdots \frac{\partial(y^{i_s} \circ F)}{\partial x^{j_s}} \right)$$
$$\cdot d(x^{j_1}) \otimes \cdots \otimes d(x^{j_s}).$$

Proof. We have from Theorem 15.12.1 that

$$F^*(\mathcal{A}) = F^* \left(\sum_{1 \leq i_1, \ldots, i_s \leq n} \mathcal{A}_{i_1 \ldots i_s} d(y^{i_1}) \otimes \cdots \otimes d(y^{i_s}) \right) \tag{15.12.4}$$
$$= \sum_{1 \leq i_1, \ldots, i_s \leq n} F^*(\mathcal{A}_{i_1 \ldots i_s}) \, F^*\big(d(y^{i_1}) \otimes \cdots \otimes d(y^{i_s})\big),$$

and from (15.9.2) and (15.12.3) that

$$F^*(\mathcal{A}_{i_1 \ldots i_s}) = \mathcal{A}_{i_1 \ldots i_s} \circ F. \tag{15.12.5}$$

We also have

$$F^* \big(d(y^{i_1}) \otimes \cdots \otimes d(y^{i_s})\big)$$
$$= F^* \big(d(y^{i_1})\big) \otimes \cdots \otimes F^* \big(d(y^{i_s})\big) \qquad [\text{Th } 15.12.1(\text{c})]$$
$$= d\big(F^*(y^{i_1})\big) \otimes \cdots \otimes d\big(F^*(y^{i_s})\big) \qquad [\text{Th } 15.11.2, \ (15.12.3)]$$
$$= d(y^{i_1} \circ F) \otimes \cdots \otimes d(y^{i_s} \circ F), \qquad [(15.9.2), \ (15.12.3)]$$

and from Theorem 15.4.7 that

$$d(y^{i_k} \circ F) = \sum_{j_k} \frac{\partial(y^{i_k} \circ F)}{\partial x^{j_k}} d(x^{j_k})$$

for $k = 1, \ldots, s$. Thus,

$$F^* \big(d(y^{i_1}) \otimes \cdots \otimes d(y^{i_s})\big)$$
$$= \left(\sum_{j_1} \frac{\partial(y^{i_1} \circ F)}{\partial x^{j_1}} d(x^{j_1}) \right) \otimes \cdots \otimes \left(\sum_{j_s} \frac{\partial(y^{i_s} \circ F)}{\partial x^{j_s}} d(x^{j_s}) \right) \tag{15.12.6}$$
$$= \sum_{1 \leq j_1, \ldots, j_s \leq m} \frac{\partial(y^{i_1} \circ F)}{\partial x^{j_1}} \cdots \frac{\partial(y^{i_s} \circ F)}{\partial x^{j_s}} d(x^{j_1}) \otimes \cdots \otimes d(x^{j_s}).$$

Substituting (15.12.5) and (15.12.6) into (15.12.4) and reversing summations gives the result. $\qquad \square$

Example 15.12.3 (Polar Coordinates). Let (r, s) be standard coordinates on $\mathbb{R}^2$, and let (ρ, ϕ) be polar coordinates on

$$U = \{(\rho, \phi) \in \mathbb{R}^2 : \rho > 0, 0 \le \phi < 2\pi\}.$$

Let $F : U \longrightarrow \mathbb{R}^2$ be the smooth map given by

$$F(\rho, \phi) = \big(\rho\cos(\phi), \rho\sin(\phi)\big),$$

and consider the tensor field

$$\mathfrak{e}_{(r,s)} = d(r) \otimes d(r) + d(s) \otimes d(s)$$

in $\mathcal{T}_2^0(\mathbb{R}^2)$. Setting $\big(U, (x^i)\big) = \big(U, (\rho, \phi)\big)$ and $\big(V, (y^j)\big) = \big(\mathbb{R}^2, (r, s)\big)$ in Theorem 15.12.2, and observing that $\mathfrak{e}_{11} \circ F = \mathfrak{e}_{22} \circ F = 1$ and $\mathfrak{e}_{12} \circ F = \mathfrak{e}_{21} \circ F = 0$, yields

$$F^*(\mathfrak{e})_{(\rho,\phi)}$$
$$= \left(\frac{\partial(r \circ F)}{\partial\rho} \frac{\partial(r \circ F)}{\partial\rho} + dfrac{\partial(s \circ F)}{\partial\rho}\partial\rho\frac{\partial(s \circ F)}{\partial\rho} \right) d(\rho) \otimes d(\rho)$$
$$+ \left(\frac{\partial(r \circ F)}{\partial\rho} \frac{\partial(r \circ F)}{\partial\phi} + \frac{\partial(s \circ F)}{\partial\rho} \frac{\partial(s \circ F)}{\partial\phi} \right) d(\rho) \otimes d(\phi)$$
$$+ \left(\frac{\partial(r \circ F)}{\partial\phi} \frac{\partial(r \circ F)}{\partial\rho} + \frac{\partial(s \circ F)}{\partial\phi} \frac{\partial(s \circ F)}{\partial\rho} \right) d(\phi) \otimes d(\rho)$$
$$+ \left(\frac{\partial(r \circ F)}{\partial\phi} \frac{\partial(r \circ F)}{\partial\phi} + \frac{\partial(s \circ F)}{\partial\phi} \frac{\partial(s \circ F)}{\partial\phi} \right) d(\phi) \otimes d(\phi)$$
$$= \left(\frac{\partial\big(\rho\cos(\phi)\big)}{\partial\rho} \frac{\partial\big(\rho\cos(\phi)\big)}{\partial\rho} + \frac{\partial\big(\rho\sin(\phi)\big)}{\partial\rho} \frac{\partial\big(\rho\sin(\phi)\big)}{\partial\rho} \right) d(\rho) \otimes d(\rho)$$
$$+ \left(\frac{\partial\big(\rho\cos(\phi)\big)}{\partial\rho} \frac{\partial\big(\rho\cos(\phi)\big)}{\partial\phi} + \frac{\partial\big(\rho\sin(\phi)\big)}{\partial\rho} \frac{\partial\big(\rho\sin(\phi)\big)}{\partial\phi} \right) d(\rho) \otimes d(\phi)$$
$$+ \left(\frac{\partial\big(\rho\cos(\phi)\big)}{\partial\phi} \frac{\partial\big(\rho\cos(\phi)\big)}{\partial\rho} + \frac{\partial\big(\rho\sin(\phi)\big)}{\partial\phi} \frac{\partial\big(\rho\sin(\phi)\big)}{\partial\rho} \right) d(\phi) \otimes d(\rho)$$
$$+ \left(\frac{\partial\big(\rho\cos(\phi)\big)}{\partial\phi} \frac{\partial\big(\rho\cos(\phi)\big)}{\partial\phi} + \frac{\partial\big(\rho\sin(\phi)\big)}{\partial\phi} \frac{\partial\big(\rho\sin(\phi)\big)}{\partial\phi} \right) d(\phi) \otimes d(\phi)$$
$$= [\cos^2(\phi) + \sin^2(\phi)]\, d(\rho) \otimes d(\rho)$$
$$+ [-\rho\cos(\phi)\sin(\phi) + \rho\sin(\phi)\cos(\phi)]\, d(\rho) \otimes d(\phi)$$
$$+ [-\rho\sin(\phi)\cos(\phi) + \rho\cos(\phi)\sin(\phi)]\, d(\phi) \otimes d(\rho)$$
$$+ [\rho^2\sin^2(\phi) + \rho^2\cos^2(\phi)]\, d(\phi) \otimes d(\phi)$$
$$= d(\rho) \otimes d(\rho) + \rho^2\, d(\phi) \otimes d(\phi). \qquad \diamondsuit$$

15.13   Pullback of Differential Forms

Let M and N be smooth manifolds, and let $F : M \longrightarrow N$ be a smooth map. In Section 15.12, we defined $F^* : \mathcal{T}_s^0(N) \longrightarrow \mathcal{T}_s^0(M)$, the pullback by F for covariant tensor fields. We seek a corresponding pullback for differential forms. An observation is that $\Lambda^s(M)$ is a subspace of $\mathcal{T}^0_s(M)$, $\Lambda^s(N)$ is a subspace of $\mathcal{T}^0_s(N)$, and $F^*\big(\Lambda^s(N)\big)$ is a subspace of $\Lambda^s(M)$, so we can proceed by restricting the maps defined in (15.12.1)–(15.12.3). **Pullback by F (for differential forms)** is the family of linear maps

$$F^* : \Lambda^s(N) \longrightarrow \Lambda^s(M)$$

defined for $s \geq 1$ by

$$F^*(\omega)_p(v_1, \ldots, v_s) = d_p(F)^*(\omega_{F(p)})(v_1, \ldots, v_s)$$
$$= \omega_{F(p)}\big(d_p(F)(v_1), \ldots, d_p(F)(v_s)\big)$$

for all differential forms ω in $\Lambda^s(N)$, all points p in M, and all vectors $v_1, \ldots, v_s$ in $T_p(M)$. We refer to $F^*(\omega)$ as the **pullback of ω by F**. As before, when $s = 0$, we define $F^* = F^\bullet$.

Theorem 15.13.1. *Let M, N, and P be smooth manifolds, let $F : M \longrightarrow N$ and $G : N \longrightarrow P$ be smooth maps, let ω, ξ, and ζ be forms in $\Lambda^s(N)$ and $\Lambda^{s'}(N)$, respectively, and let g be a function in $C^\infty(N)$. Then:*
(a) $F^*(\omega + \xi) = F^*(\omega) + F^*(\xi)$.
(b) $F^*(g\omega) = F^*(g)F^*(\omega)$.
(c) $F^*(\omega \wedge \zeta) = F^*(\omega) \wedge F^*(\zeta)$.
(d) $(G \circ F)^* = F^* \circ G^*$. $\square$

Theorem 15.13.2. *Let M be a smooth m-manifold, and let $f^1, \ldots, f^s$ be functions in $C^\infty(M)$, where $s \leq m$. Let $\big(U, (x^j)\big)$ be a chart on M, and let $(\partial/\partial x^1, \ldots, \partial/\partial x^m)$ and $\big(d(x^1), \ldots, d(x^m)\big)$ be the corresponding coordinate and dual coordinate frames. Then, in local coordinates,*

$$d(f^1) \wedge \cdots \wedge d(f^s)$$

$$= \sum_{1 \leq j_1 < \cdots < j_s \leq m} \det\left(\begin{bmatrix} \dfrac{\partial f^1}{\partial x^1} & \cdots & \dfrac{\partial f^1}{\partial x^m} \\ \vdots & \ddots & \vdots \\ \dfrac{\partial f^s}{\partial x^1} & \cdots & \dfrac{\partial f^s}{\partial x^m} \end{bmatrix}_{(j_1, \ldots, j_s)}\right) d(x^{j_1}) \wedge \cdots \wedge d(x^{j_s}).$$

Proof. In local coordinates, let

$$d(f^1) \wedge \cdots \wedge d(f^s) = \sum_{1 \leq j_1 < \cdots < j_s \leq m} \alpha_{j_1, \ldots, j_s} d(x^{j_1}) \wedge \cdots \wedge d(x^{j_s}). \qquad (15.13.1)$$

Evaluating both sides of the preceding identity at $(\partial/\partial x^{i_1}, \ldots, \partial/\partial x^{i_s})$ and using Theorem 15.8.4 yields

$$d(f^1) \wedge \cdots \wedge d(f^s)\left(\frac{\partial}{\partial x^{i_1}}, \ldots, \frac{\partial}{\partial x^{i_s}}\right)$$

$$= \sum_{1 \leq j_1 < \cdots < j_s \leq m} \alpha_{j_1, \ldots, j_s} d(x^{j_1}) \wedge \cdots \wedge d(x^{j_s})\left(\frac{\partial}{\partial x^{i_1}}, \ldots, \frac{\partial}{\partial x^{i_s}}\right) = \alpha_{i_1, \ldots, i_s}.$$

With the aid of (14.5.4) and Theorem 15.8.3(a), the left-hand side of the preceding identity can be expressed as

$$d(f^1) \wedge \cdots \wedge d(f^s)\left(\frac{\partial}{\partial x^{i_1}}, \ldots, \frac{\partial}{\partial x^{i_s}}\right)$$

$$= \det\left(\begin{bmatrix} d(f^1)\left(\dfrac{\partial}{\partial x^{i_1}}\right) & \cdots & d(f^1)\left(\dfrac{\partial}{\partial x^{i_s}}\right) \\ \vdots & \ddots & \vdots \\ d(f^s)\left(\dfrac{\partial}{\partial x^{i_1}}\right) & \cdots & d(f^s)\left(\dfrac{\partial}{\partial x^{i_s}}\right) \end{bmatrix}\right)$$

$$= \det\left(\begin{bmatrix} \dfrac{\partial f^1}{\partial x^{i_1}} & \cdots & \dfrac{\partial f^1}{\partial x^{i_s}} \\ \vdots & \ddots & \vdots \\ \dfrac{\partial f^s}{\partial x^{i_1}} & \cdots & \dfrac{\partial f^s}{\partial x^{i_s}} \end{bmatrix}\right) = \det\left(\begin{bmatrix} \dfrac{\partial f^1}{\partial x^1} & \cdots & \dfrac{\partial f^1}{\partial x^m} \\ \vdots & \ddots & \vdots \\ \dfrac{\partial f^s}{\partial x^1} & \cdots & \dfrac{\partial f^s}{\partial x^m} \end{bmatrix}_{(i_1,\ldots,i_s)}\right).$$

Thus,

$$\alpha_{j_1,\ldots,j_s} = \det\left(\begin{bmatrix} \dfrac{\partial f^1}{\partial x^1} & \cdots & \dfrac{\partial f^1}{\partial x^m} \\ \vdots & \ddots & \vdots \\ \dfrac{\partial f^s}{\partial x^1} & \cdots & \dfrac{\partial f^s}{\partial x^m} \end{bmatrix}_{(j_1,\ldots,j_s)}\right). \tag{15.13.2}$$

Substituting (15.13.2) into (15.13.1) gives the result.   □

Theorem 15.13.3 (Pullback of Differential Form). *Let M be a smooth m-manifold, let N be a smooth n-manifold, let $F : M \longrightarrow N$ be a smooth map, and let $(U,(x^i))$ and $(V,(y^j))$ be charts on M and N, respectively, such that $U \cap F^{-1}(V)$ is nonempty. Let ω be a form in $\Lambda^s(N)$, and, in local coordinates, let*

$$\omega = \sum_{1 \le j_1 < \cdots < j_s \le n} \alpha_{j_1 \ldots j_s} d(y^{j_1}) \wedge \cdots \wedge d(y^{j_s}),$$

where $s \le \min(m,n)$. Then:
(a)

$$F^*(\omega) = \sum_{1 \le i_1 < \cdots < i_s \le m}\left[\sum_{1 \le j_1 < \cdots < j_s \le n} (\alpha_{j_1 \ldots j_s} \circ F) \det\left((J_F)^{(j_1,\ldots,j_s)}_{(i_1,\ldots,i_s)}\right)\right]$$
$$\cdot d(x^{i_1}) \wedge \cdots \wedge d(x^{i_s}),$$

where

$$J_F = \begin{bmatrix} \dfrac{\partial(y^1 \circ F)}{\partial x^1} & \cdots & \dfrac{\partial(y^1 \circ F)}{\partial x^m} \\ \vdots & \ddots & \vdots \\ \dfrac{\partial(y^n \circ F)}{\partial x^1} & \cdots & \dfrac{\partial(y^n \circ F)}{\partial x^m} \end{bmatrix} \tag{15.13.3}$$

*is the corresponding **Jacobian matrix.***

(b) *If $s = m = n$ and $\omega = \alpha\, d(y^1) \wedge \cdots \wedge d(y^m)$, then*

$$F^*(\omega) = (\alpha \circ F) \det(J_F)\, d(x^1) \wedge \cdots \wedge d(x^m).$$

Proof. (a): We have from Theorem 15.13.1 that

$$
\begin{aligned}
F^*(\omega) &= F^*\left(\sum_{1 \le j_1 < \cdots < j_s \le n} \alpha_{j_1 \ldots j_s} d(y^{j_1}) \wedge \cdots \wedge d(y^{j_s}) \right) \\
&= \sum_{1 \le j_1 < \cdots < j_s \le n} F^*(\alpha_{j_1 \ldots j_s})\, F^*\big(d(y^{j_1}) \wedge \cdots \wedge d(y^{j_s})\big),
\end{aligned}
\tag{15.13.4}
$$

and from (15.9.2) and (15.12.3) that

$$F^*(\alpha_{j_1 \ldots j_s}) = \alpha_{j_1 \ldots j_s} \circ F^*.
\tag{15.13.5}$$

We also have

$$
\begin{aligned}
&F^*\big(d(y^{j_1}) \wedge \cdots \wedge d(y^{j_s})\big) \\
&\quad = F^*\big(d(y^{j_1})\big) \wedge \cdots \wedge F^*\big(d(y^{j_s})\big) && \text{[Th 15.13.1(c)]} \\
&\quad = d\big(F^*(y^{j_1})\big) \wedge \cdots \wedge d\big(F^*(y^{j_s})\big) && \text{[Th 15.11.2, (15.12.3)]} \\
&\quad = d(y^{j_1} \circ F) \wedge \cdots \wedge d(y^{j_s} \circ F), && [(15.9.2),\ (15.12.3)]
\end{aligned}
$$

and from Theorem 15.13.2 that

$$
\begin{aligned}
&d(y^{j_1} \circ F) \wedge \cdots \wedge d(y^{j_s} \circ F) \\
&= \sum_{1 \le i_1 < \cdots < i_s \le m} \det\left(\begin{bmatrix} \dfrac{\partial(y^{j_1} \circ F)}{\partial x^1} & \cdots & \dfrac{\partial(y^{j_1} \circ F)}{\partial x^m} \\ \vdots & \ddots & \vdots \\ \dfrac{\partial(y^{j_s} \circ F)}{\partial x^1} & \cdots & \dfrac{\partial(y^{j_s} \circ F)}{\partial x^m} \end{bmatrix}_{(i_1, \ldots, i_s)} \right) \\
&\qquad \cdot d(x^{i_1}) \wedge \cdots \wedge d(x^{i_s}) \\
&= \sum_{1 \le i_1 < \cdots < i_s \le m} \det\left(\begin{bmatrix} \dfrac{\partial(y^1 \circ F)}{\partial x^1} & \cdots & \dfrac{\partial(y^1 \circ F)}{\partial x^m} \\ \vdots & \ddots & \vdots \\ \dfrac{\partial(y^n \circ F)}{\partial x^1} & \cdots & \dfrac{\partial(y^n \circ F)}{\partial x^m} \end{bmatrix}^{(j_1, \ldots, j_s)}_{(i_1, \ldots, i_s)} \right) \\
&\qquad \cdot d(x^{i_1}) \wedge \cdots \wedge d(x^{i_s}) \\
&= \sum_{1 \le i_1 < \cdots < i_s \le m} \det\left((J_F)^{(j_1, \ldots, j_s)}_{(i_1, \ldots, i_s)} \right) d(x^{i_1}) \wedge \cdots \wedge d(x^{i_s}).
\end{aligned}
$$

Thus,

$$
\begin{aligned}
&F^*\big(d(y^{j_1}) \wedge \cdots \wedge d(y^{j_s})\big) \\
&\quad = \sum_{1 \le i_1 < \cdots < i_s \le m} \det\left((J_F)^{(j_1, \ldots, j_s)}_{(i_1, \ldots, i_s)} \right) d(x^{i_1}) \wedge \cdots \wedge d(x^{i_s}).
\end{aligned}
\tag{15.13.6}
$$

Substituting (15.13.5) and (15.13.6) into (15.13.4) and reversing summations gives the result.

(b): This follows from part (a). □

Example 15.13.4 (Polar Coordinates). Let (r, s) be standard coordinates on $\mathbb{R}^2$, and let (ρ, ϕ) be polar coordinates on

$$U = \{(\rho, \phi) \in \mathbb{R}^2 : \rho > 0, 0 \leq \phi < 2\pi\}.$$

Let $F : U \longrightarrow \mathbb{R}^2$ be the smooth map given by

$$F(\rho, \phi) = \big(\rho \cos(\phi), \rho \sin(\phi)\big),$$

and consider the form

$$\omega_{(r,s)} = d(r) \wedge d(s)$$

in $\Lambda^2(\mathbb{R}^2)$. Setting $(U, (x^i)) = (U, (\rho, \phi))$ and $(V, (y^j)) = (\mathbb{R}^2, (r, s))$ in Theorem 15.13.3(b) yields

$$F^*\big(d(r) \wedge d(s)\big) = \det\left(\begin{bmatrix} \dfrac{\partial(r \circ F)}{\partial \rho} & \dfrac{\partial(r \circ F)}{\partial \phi} \\[2mm] \dfrac{\partial(s \circ F)}{\partial \rho} & \dfrac{\partial(s \circ F)}{\partial \phi} \end{bmatrix}\right) d(\rho) \wedge d(\phi)$$

$$= \det\left(\begin{bmatrix} \dfrac{\partial\big(\rho \cos(\phi)\big)}{\partial \rho} & \dfrac{\partial\big(\rho \cos(\phi)\big)}{\partial \phi} \\[2mm] \dfrac{\partial\big(\rho \sin(\phi)\big)}{\partial \rho} & \dfrac{\partial\big(\rho \sin(\phi)\big)}{\partial \phi} \end{bmatrix}\right) d(\rho) \wedge d(\phi)$$

$$= \det\left(\begin{bmatrix} \cos(\phi) & -\rho \sin(\phi) \\ \sin(\phi) & \rho \cos(\phi) \end{bmatrix}\right) d(\rho) \wedge d(\phi)$$

$$= \rho \, d(\rho) \wedge d(\phi). \qquad \Diamond$$

15.14 Contraction of Tensor Fields

This brief section presents the manifold versions of several of the results in Section 5.4.

Theorem 15.14.1. *If M is a smooth manifold, then there is a unique $C^\infty(M)$-linear map*

$$C_1^1 : \mathcal{T}_1^1(M) \longrightarrow C^\infty(M),$$

called **(1, 1)-contraction,** *such that*

$$C_1^1(X \otimes \omega) = \omega(X)$$

for all vector fields X in $\mathfrak{X}(M)$ and all covector fields ω in $\mathfrak{X}^*(M)$. □

Theorem 15.14.2. *Let M be a smooth manifold, let $\mathcal{A}$ be a tensor field in $\mathcal{T}_1^1(M)$, and, in local coordinates, let*

$$\mathcal{A} = \sum_{ij} \mathcal{A}_j^i \frac{\partial}{\partial x^i} \otimes d(x^j).$$

Then

$$\mathsf{C}_1^1(\mathcal{A}) = \sum_i \mathcal{A}_i^i. \qquad \square$$

Theorem 15.14.3. *If M is a smooth manifold, and $1 \le k \le r$ and $1 \le l \le s$ are integers, then there is a unique $C^\infty(M)$-linear map*

$$\mathsf{C}_l^k : \mathcal{T}_s^r(M) \longrightarrow \mathcal{T}_{s-1}^{r-1}(M),$$

called (k,l)-contraction, such that

$$\mathsf{C}_l^k(X_1 \otimes \cdots \otimes X_r \otimes \omega^1 \otimes \cdots \otimes \omega^s)$$
$$= X_k(\omega^l)\, X_1 \otimes \cdots \otimes \widehat{X_k} \otimes \cdots \otimes X_r \otimes \omega^1 \otimes \cdots \otimes \widehat{\omega^l} \otimes \cdots \otimes \omega^s$$

for all vector fields $X_1, \ldots, X_r$ in $\mathfrak{X}(M)$ and all covector fields $\omega^1, \ldots, \omega^s$ in $\mathfrak{X}^*(M)$, where $\widehat{\ }$ indicates that an expression is omitted. $\square$

Theorem 15.14.4. *Let M be a smooth manifold, let $\mathcal{A}$ be a tensor field in $\mathcal{T}_s^r(M)$, let $\omega^1, \ldots, \omega^r$ be covector fields in $\mathfrak{X}^*(M)$, and let $X_1, \ldots, X_s$ be vector fields in $\mathfrak{X}(M)$. Then there is a contraction C on M such that*

$$\mathcal{A}(\omega^1, \ldots, \omega^r, X_1, \ldots, X_s) = \mathsf{C}(X_1 \otimes \cdots \otimes X_s \otimes \mathcal{A} \otimes \omega^1 \otimes \cdots \otimes \omega^r). \qquad \square$$

Chapter 16

Differentiation and Integration on Smooth Manifolds

16.1 Exterior Derivatives

Let M be a smooth m-manifold and recall that $C^\infty(M) = \Lambda^0(M)$ and $\mathfrak{X}^*(M) = \Lambda^1(M)$. The exterior derivative $d : C^\infty(M) \longrightarrow \mathfrak{X}^*(M)$ of Section 15.4 can therefore be expressed as

$$d : \Lambda^0(M) \longrightarrow \Lambda^1(M).$$

According to Theorem 15.4.6, d is linear and satisfies a type of product rule. In this section, we describe a generalization of d that has corresponding properties. Three equivalent approaches to this theory are available—axiomatic, local, and global.

Theorem 16.1.1 (Axioms for Exterior Derivative). *If M is a smooth m-manifold, then there is a unique family of linear maps*

$$d : \Lambda^s(M) \longrightarrow \Lambda^{s+1}(M)$$

*defined for $s \geq 0$, called the **exterior derivative**, such that:*
(a) *$d : \Lambda^0(M) \longrightarrow \Lambda^1(M)$ is the exterior derivative defined in Section 15.4.*
(b) *For all forms ω in $\Lambda^s(M)$ and ξ in $\Lambda^{s'}(M)$,*

$$d(\omega \wedge \xi) = d(\omega) \wedge \xi + (-1)^s \omega \wedge d(\xi).$$

(c) *$d^2 = 0$; that is, $d \circ d(\omega) = 0$ for all forms ω in $\Lambda^s(M)$.* $\square$

Semi-Riemannian Geometry, First Edition. Stephen C. Newman.
© 2019 John Wiley & Sons, Inc. Published 2019 by John Wiley & Sons, Inc.

We observe that when $s \geq m$, $\Lambda^m(M)$ is the zero vector space, in which case d is the zero map. For any function f in $C^\infty(M) = \Lambda^0(M)$ and any form ω in $\Lambda^s(M)$, we have from (15.8.3) that $f\omega = f \wedge \omega$, and then from Theorem 16.1.1(b) that

$$d(f\omega) = d(f \wedge \omega) = d(f) \wedge \omega + f \wedge d(\omega) = d(f) \wedge \omega + f d(\omega). \tag{16.1.1}$$

Theorem 16.1.2 (Local Coordinate Expression for Exterior Derivative). *Let M be a smooth m-manifold, let ω be a form in $\Lambda^s(M)$, and, in local coordinates, let*

$$\omega = \sum_{1 \leq i_1 < \cdots < i_s \leq m} \alpha_{i_1 \ldots i_s} d(x^{i_1}) \wedge \cdots \wedge d(x^{i_s}).$$

Then

$$d(\omega) = \sum_{1 \leq i_1 < \cdots < i_s \leq m} d(\alpha_{i_1 \ldots i_s}) \wedge d(x^{i_1}) \wedge \cdots \wedge d(x^{i_s}). \qquad \square$$

Theorem 16.1.2 will be used often, usually without attribution.

Theorem 16.1.3 (Global Expression for Exterior Derivative). *Let M be a smooth m-manifold, let ω be a form in $\Lambda^s(M)$, and let $X_1, \ldots, X_{s+1}$ be vector fields in $\mathfrak{X}(M)$. Then*

$$d(\omega)(X_1, \ldots, X_{s+1})$$

$$= \sum_{i=1}^{s+1} (-1)^{i-1} X_i \big(\omega(X_1, \ldots, \widehat{X_i}, \ldots, X_{s+1}) \big)$$

$$+ \sum_{1 \leq i < j \leq s+1} (-1)^{i+j} \omega([X_i, X_j], X_1, \ldots, \widehat{X_i}, \ldots, \widehat{X_j}, \ldots, X_{s+1}),$$

where $\widehat{}$ indicates that an expression is omitted. $\qquad \square$

Example 16.1.4. Setting $s = 1$ in Theorem 16.1.3 gives

$$d(\omega)(X_1, X_2) = X_1\big(\omega(X_2)\big) - X_2\big(\omega(X_1)\big) - \omega([X_1, X_2]). \qquad \diamondsuit$$

The next result is a generalization of Theorem 15.11.2.

Theorem 16.1.5 (Commutativity of Pullback and Exterior Derivative of Forms). *If M and N are smooth manifolds and $F : M \longrightarrow N$ is a smooth map, then*

$$F^* \circ d = d \circ F^*;$$

that is,

$$F^*\big(d(\omega)\big) = d\big(F^*(\omega)\big)$$

for all forms ω in $\Lambda^s(M)$. $\qquad \square$

The following three examples show that the exterior derivative is related to the classical curl, gradient, and divergence operators presented in Section 10.6.

Example 16.1.6 (Curl). Let $\alpha_1, \alpha_2, \alpha_3$ be functions in $C^\infty(\mathbb{R}^3)$, and define a covector field ω in $\mathfrak{X}^*(\mathbb{R}^3) = \Lambda^1(\mathbb{R}^3)$ by

$$\omega = \alpha_1 d(x) + \alpha_2 d(y) + \alpha_3 d(z).$$

Temporarily denoting (x, y, z) by (x^1, x^2, x^3), we have

$$d(\omega) = d\left(\sum_i \alpha_i d(x^i)\right) = \sum_i d(\alpha_i d(x^i)) = \sum_i [d(\alpha_i) \wedge d(x^i) + \alpha_i d^2(x^i)]$$

$$= \sum_i \left[\left(\sum_j \frac{\partial \alpha_i}{\partial x^j} d(x^j)\right) \wedge d(x^i)\right] = \sum_{ij} \frac{\partial \alpha_i}{\partial x^j} d(x^j) \wedge d(x^i)$$

$$= \sum_{1 \le i < j \le m} \frac{\partial \alpha_i}{\partial x^j} d(x^j) \wedge d(x^i) + \sum_{1 \le j < i \le m} \frac{\partial \alpha_i}{\partial x^j} d(x^j) \wedge d(x^i)$$

$$= -\sum_{1 \le i < j \le m} \frac{\partial \alpha_i}{\partial x^j} d(x^i) \wedge d(x^j) + \sum_{1 \le i < j \le m} \frac{\partial \alpha_j}{\partial x^i} d(x^i) \wedge d(x^j)$$

$$= \sum_{1 \le i < j \le m} \left(\frac{\partial \alpha_j}{\partial x^i} - \frac{\partial \alpha_i}{\partial x^j}\right) d(x^i) \wedge d(x^j),$$

where the third equality follows from (16.1.1), the fourth equality from Theorem 15.4.7 and Theorem 16.1.1(c), the sixth equality from Theorem 15.8.2(c), and the seventh equality from Theorem 15.8.2(b). Reverting to the earlier notation gives

$$d(\omega) = \left(\frac{\partial \alpha_2}{\partial x} - \frac{\partial \alpha_1}{\partial y}\right) d(x) \wedge d(y) + \left(\frac{\partial \alpha_3}{\partial x} - \frac{\partial \alpha_1}{\partial z}\right) d(x) \wedge d(z)$$

$$+ \left(\frac{\partial \alpha_3}{\partial y} - \frac{\partial \alpha_2}{\partial z}\right) d(y) \wedge d(z)$$

$$= \left(\frac{\partial \alpha_3}{\partial y} - \frac{\partial \alpha_2}{\partial z}\right) d(y) \wedge d(z) + \left(\frac{\partial \alpha_1}{\partial z} - \frac{\partial \alpha_3}{\partial x}\right) d(z) \wedge d(x) \qquad (16.1.2)$$

$$+ \left(\frac{\partial \alpha_2}{\partial x} - \frac{\partial \alpha_1}{\partial y}\right) d(x) \wedge d(y).$$

Corresponding to ω, let us define the smooth map $F = (\alpha_1, \alpha_2, \alpha_3) : \mathbb{R}^3 \longrightarrow \mathbb{R}^3$. Recall from Section 10.6 that

$$\text{curl}(F) = \left(\frac{\partial \alpha_3}{\partial y} - \frac{\partial \alpha_2}{\partial z}, \frac{\partial \alpha_1}{\partial z} - \frac{\partial \alpha_3}{\partial x}, \frac{\partial \alpha_2}{\partial x} - \frac{\partial \alpha_1}{\partial y}\right).$$

Thus, the components of $d(\omega)$ are the same as the components of $\text{curl}(F)$.   ◇

Example 16.1.7 (Gradient). Let f be a function in $C^\infty(M) = \Lambda^0(M)$. We have from Section 10.6 and (11.4.3) that

$$\text{grad}(f) = \left(\frac{\partial f}{\partial x}, \frac{\partial f}{\partial y}, \frac{\partial f}{\partial z}\right),$$

and from Theorem 15.4.7 that

$$d(f) = \frac{\partial f}{\partial x}d(x) + \frac{\partial f}{\partial y}d(y) + \frac{\partial f}{\partial z}d(z).$$

Setting $\omega = d(f)$ and $(\alpha_1, \alpha_2, \alpha_3) = \mathrm{grad}(f)$ in (16.1.2) gives

$$d^2(f) = \left(\frac{\partial^2 f}{\partial y \partial z} - \frac{\partial^2 f}{\partial z \partial y}\right)d(y) \wedge d(z) + \left(\frac{\partial^2 f}{\partial z \partial x} - \frac{\partial^2 f}{\partial x \partial z}\right)d(z) \wedge d(x)$$
$$+ \left(\frac{\partial^2 f}{\partial x \partial y} - \frac{\partial^2 f}{\partial y \partial x}\right)d(x) \wedge d(y).$$

It follows from the equality of mixed partial derivatives (Theorem 10.1.6) that $d^2(f) = 0$, as expected from Theorem 16.1.1(c). Looked at differently, $d^2 = 0$ is a way of expressing the equality of mixed partial derivatives in terms of the exterior derivative. In view of the connection between $d(\omega)$ and $\mathrm{curl}(F)$ noted in Example 16.1.6, we now see that $d^2(f) = 0$ is equivalent to $\mathrm{curl}(\mathrm{grad}(f)) = 0$, which is Theorem 10.6.1(c).   ◇

Example 16.1.8 (Divergence). Let $\alpha_1, \alpha_2, \alpha_3$ be functions in $C^\infty(\mathbb{R}^3)$, and define a form ξ in $\Lambda^2(\mathbb{R}^3)$ by

$$\xi = \alpha_1 d(y) \wedge d(z) + \alpha_2 d(z) \wedge d(x) + \alpha_3 d(x) \wedge d(y).$$

Computing as in Example 16.1.6 yields

$$d(\xi) = \left(\frac{\partial \alpha_1}{\partial x} + \frac{\partial \alpha_2}{\partial y} + \frac{\partial \alpha_3}{\partial z}\right)d(x) \wedge d(y) \wedge d(z). \tag{16.1.3}$$

Corresponding to ξ, let us define the smooth map $F = (\alpha_1, \alpha_2, \alpha_3) : \mathbb{R}^3 \longrightarrow \mathbb{R}^3$. Recall from Section 10.6 that

$$\mathrm{div}(F) = \frac{\partial \alpha_1}{\partial x} + \frac{\partial \alpha_2}{\partial y} + \frac{\partial \alpha_3}{\partial z}.$$

Thus, the single component of $d(\xi)$ equals $\mathrm{div}(F)$.

  With ω as in Example 16.1.6, we have from (16.1.2) that $d(\omega)$ is a form in $\Lambda^2(\mathbb{R}^3)$. Setting $\xi = d(\omega)$ in (16.1.3) yields

$$d^2(\omega) = \left[\frac{\partial}{\partial x}\left(\frac{\partial \alpha_3}{\partial y} - \frac{\partial \alpha_2}{\partial z}\right) + \frac{\partial}{\partial y}\left(\frac{\partial \alpha_1}{\partial z} - \frac{\partial \alpha_3}{\partial x}\right) + \frac{\partial}{\partial z}\left(\frac{\partial \alpha_2}{\partial x} - \frac{\partial \alpha_1}{\partial y}\right)\right]$$
$$\cdot d(x) \wedge d(y) \wedge d(z)$$
$$= \left[\left(\frac{\partial^2 \alpha_3}{\partial x \partial y} - \frac{\partial^2 \alpha_2}{\partial x \partial z}\right) + \left(\frac{\partial^2 \alpha_1}{\partial y \partial z} - \frac{\partial^2 \alpha_3}{\partial y \partial x}\right) + \left(\frac{\partial \alpha_2}{\partial z \partial x} - \frac{\partial^2 \alpha_1}{\partial z \partial y}\right)\right]$$
$$\cdot d(x) \wedge d(y) \wedge d(z).$$

It follows from the equality of mixed partial derivatives (Theorem 10.1.6) that $d^2(\omega) = 0$, as expected from Theorem 16.1.1(c). In view of the connection between $d(\omega)$ and $\mathrm{curl}(F)$ noted in Example 16.1.6, we now see that $d^2(\omega) = 0$ is equivalent to $\mathrm{div}(\mathrm{curl}(F)) = 0$, which is Theorem 10.6.1(d).   ◇

We close this section with two computations relevant to electrodynamics that will be needed in Section 22.2.

Example 16.1.9 (Electrodynamics). Let (x, y, z, t) be standard coordinates on $\mathbb{R}^4$, let $\alpha_1, \alpha_2, \alpha_3, \alpha_4$ be functions in $C^\infty(\mathbb{R}^4)$, and define a covector field ω in $\mathfrak{X}^*(\mathbb{R}^4)$ by

$$\omega = \alpha_1 d(x) + \alpha_2 d(y) + \alpha_3 d(z) + \alpha_4 d(t).$$

Then

$$d(\omega) = d(\alpha_1) \wedge d(x) + d(\alpha_2) \wedge d(y) + d(\alpha_3) \wedge d(z) + d(\alpha_4) \wedge d(t).$$

We have

$$\begin{aligned}
d(\alpha_1) \wedge d(x) &= \left(\frac{\partial \alpha_1}{\partial x} d(x) + \frac{\partial \alpha_1}{\partial y} d(y) + \frac{\partial \alpha_1}{\partial z} d(z) + \frac{\partial \alpha_1}{\partial t} d(t) \right) \wedge d(x) \\
&= \frac{\partial \alpha_1}{\partial y} d(y) \wedge d(x) + \frac{\partial \alpha_1}{\partial z} d(z) \wedge d(x) + \frac{\partial \alpha_1}{\partial t} d(t) \wedge d(x) \\
&= -\frac{\partial \alpha_1}{\partial y} d(x) \wedge d(y) - \frac{\partial \alpha_1}{\partial z} d(x) \wedge d(z) - \frac{\partial \alpha_1}{\partial t} d(x) \wedge d(t).
\end{aligned}$$

Likewise,

$$d(\alpha_2) \wedge d(y) = \frac{\partial \alpha_2}{\partial x} d(x) \wedge d(y) - \frac{\partial \alpha_2}{\partial z} d(y) \wedge d(z) - \frac{\partial \alpha_2}{\partial t} d(y) \wedge d(t)$$

$$d(\alpha_3) \wedge d(z) = \frac{\partial \alpha_3}{\partial x} d(x) \wedge d(z) + \frac{\partial \alpha_3}{\partial y} d(y) \wedge d(z) - \frac{\partial \alpha_3}{\partial t} d(z) \wedge d(t)$$

$$d(\alpha_4) \wedge d(t) = \frac{\partial \alpha_4}{\partial x} d(x) \wedge d(t) + \frac{\partial \alpha_4}{\partial y} d(y) \wedge d(t) + \frac{\partial \alpha_4}{\partial z} d(z) \wedge d(t).$$

Thus,

$$\begin{aligned}
d(\omega) =\ & \left(\frac{\partial \alpha_4}{\partial x} - \frac{\partial \alpha_1}{\partial t} \right) d(x) \wedge d(t) + \left(\frac{\partial \alpha_4}{\partial y} - \frac{\partial \alpha_2}{\partial t} \right) d(y) \wedge d(t) \\
&+ \left(\frac{\partial \alpha_4}{\partial z} - \frac{\partial \alpha_3}{\partial t} \right) d(z) \wedge d(t) \\
&+ \left(\frac{\partial \alpha_3}{\partial y} - \frac{\partial \alpha_2}{\partial z} \right) d(y) \wedge d(z) + \left(\frac{\partial \alpha_1}{\partial z} - \frac{\partial \alpha_3}{\partial x} \right) d(z) \wedge d(x) \\
&+ \left(\frac{\partial \alpha_2}{\partial x} - \frac{\partial \alpha_1}{\partial y} \right) d(x) \wedge d(y). \hspace{2cm} \Diamond
\end{aligned}$$

Example 16.1.10 (Electrodynamics). Let (x, y, z, t) be standard coordinates on $\mathbb{R}^4$, and let $\alpha_1, \alpha_2, \alpha_3, \beta_1, \beta_2, \beta_3$ be functions in $C^\infty(\mathbb{R}^4)$. Define a covector field ξ in $\mathfrak{X}^*(\mathbb{R}^4)$ and forms ζ, φ in $\Lambda^2(\mathbb{R}^4)$ by

$$\xi = \alpha_1 d(x) + \alpha_2 d(y) + \alpha_3 d(z),$$

$$\zeta = \beta_1 d(y) \wedge d(z) + \beta_2 d(z) \wedge d(x) + \beta_3 d(x) \wedge d(y),$$

and

$$\varphi = \xi \wedge d(t) + \zeta$$
$$= \alpha_1 d(x) \wedge d(t) + \alpha_2 d(y) \wedge d(t) + \alpha_3 d(z) \wedge d(t)$$
$$+ \beta_1 d(y) \wedge d(z) + \beta_2 d(z) \wedge d(x) + \beta_3 d(x) \wedge d(y).$$

We seek an alternative expression for

$$d(\varphi) = d(\alpha_1) \wedge d(x) \wedge d(t) + d(\alpha_2) \wedge d(y) \wedge d(t) + d(\alpha_3) \wedge d(z) \wedge d(t)$$
$$+ d(\beta_1) \wedge d(y) \wedge d(z) + d(\beta_2) \wedge d(z) \wedge d(x) + d(\beta_3) \wedge d(x) \wedge d(y).$$

We have

$$d(\alpha_1) \wedge d(x) \wedge d(t)$$
$$= \left(\frac{\partial \alpha_1}{\partial x} d(x) + \frac{\partial \alpha_1}{\partial y} d(y) + \frac{\partial \alpha_1}{\partial z} d(z) + \frac{\partial \alpha_1}{\partial t} d(t) \right) \wedge d(x) \wedge d(t)$$
$$= \frac{\partial \alpha_1}{\partial x} d(x) \wedge d(x) \wedge d(t) + \frac{\partial \alpha_1}{\partial y} d(y) \wedge d(x) \wedge d(t)$$
$$+ \frac{\partial \alpha_1}{\partial z} d(z) \wedge d(x) \wedge d(t) + \frac{\partial \alpha_1}{\partial t} d(t) \wedge d(x) \wedge d(t)$$
$$= -\frac{\partial \alpha_1}{\partial y} d(x) \wedge d(y) \wedge d(t) - \frac{\partial \alpha_1}{\partial z} d(x) \wedge d(z) \wedge d(t).$$

Likewise,

$$d(\alpha_2) \wedge d(t) \wedge d(y) = \frac{\partial \alpha_2}{\partial x} d(x) \wedge d(y) \wedge d(t) - \frac{\partial \alpha_2}{\partial z} d(y) \wedge d(z) \wedge d(t)$$

$$d(\alpha_3) \wedge d(t) \wedge d(z) = \frac{\partial \alpha_3}{\partial x} d(x) \wedge d(z) \wedge d(t) + \frac{\partial \alpha_3}{\partial y} d(y) \wedge d(z) \wedge d(t)$$

$$d(\beta_1) \wedge d(y) \wedge d(z) = \frac{\partial \beta_1}{\partial x} d(x) \wedge d(y) \wedge d(z) + \frac{\partial \beta_1}{\partial t} d(y) \wedge d(z) \wedge d(t)$$

$$d(\beta_2) \wedge d(z) \wedge d(x) = \frac{\partial \beta_2}{\partial y} d(x) \wedge d(y) \wedge d(z) - \frac{\partial \beta_2}{\partial t} d(x) \wedge d(z) \wedge d(t)$$

$$d(\beta_3) \wedge d(x) \wedge d(y) = \frac{\partial \beta_3}{\partial z} d(x) \wedge d(y) \wedge d(z) + \frac{\partial \beta_3}{\partial t} d(x) \wedge d(y) \wedge d(t).$$

Thus,

$$d(\varphi) = \left(\frac{\partial \beta_1}{\partial x} + \frac{\partial \beta_2}{\partial y} + \frac{\partial \beta_3}{\partial z} \right) d(x) \wedge d(y) \wedge d(z)$$
$$+ \left(\frac{\partial \alpha_2}{\partial x} - \frac{\partial \alpha_1}{\partial y} + \frac{\partial \beta_3}{\partial t} \right) d(x) \wedge d(y) \wedge d(t)$$
$$- \left(\frac{\partial \alpha_1}{\partial z} - \frac{\partial \alpha_3}{\partial x} + \frac{\partial \beta_2}{\partial t} \right) d(x) \wedge d(z) \wedge d(t)$$
$$+ \left(\frac{\partial \alpha_3}{\partial y} - \frac{\partial \alpha_2}{\partial z} + \frac{\partial \beta_1}{\partial t} \right) d(y) \wedge d(z) \wedge d(t). \qquad \diamond$$

16.2   Tensor Derivations

A **tensor derivation** on a smooth manifold M is a family D of linear maps

$$\mathsf{D}_s^r : \mathcal{T}_s^r(M) \longrightarrow \mathcal{T}_s^r(M)$$

defined for $r, s \geq 0$ such that for all tensor fields $\mathcal{A}$ in $\mathcal{T}_s^r(M)$ and $\mathcal{B}$ in $\mathcal{T}_{s'}^{r'}(M)$, and all (k, l)-contractions C_l^k:

[D1]

$$\mathsf{D}_{s+s'}^{r+r'}(\mathcal{A} \otimes \mathcal{B}) = \mathsf{D}_s^r(\mathcal{A}) \otimes \mathcal{B} + \mathcal{A} \otimes \mathsf{D}_{s'}^{r'}(\mathcal{B}).$$

[D2]

$$\mathsf{C}_l^k \circ \mathsf{D}_s^r(\mathcal{A}) = \mathsf{D}_{s-1}^{r-1} \circ \mathsf{C}_l^k(\mathcal{A}).$$

[D1] is a type of product rule for tensor fields, and [D2] says that "contraction commutes with tensor derivation".

Theorem 16.2.1. *Let M be a smooth manifold, let D be a tensor derivation on M, and let $\mathcal{A}_i$ be a tensor field in $\mathcal{T}_{s_i}^{r_i}(M)$ for $i = 1, \ldots, k$. Then*

$$\mathsf{D}_{s_1 + \cdots + s_k}^{r_1 + \cdots + r_k}(\mathcal{A}_1 \otimes \cdots \otimes \mathcal{A}_k) = \sum_{i=1}^k \mathcal{A}_1 \otimes \cdots \otimes \mathsf{D}_{s_i}^{r_i}(\mathcal{A}_i) \otimes \cdots \otimes \mathcal{A}_k.$$

Proof. This follows from [D1] using an inductive argument. $\qquad\square$

Theorem 16.2.2. *Let M be a smooth manifold, let D be a tensor derivation on M, let f be a function in $C^\infty(M)$, and let $\mathcal{A}$ be a tensor field in $\mathcal{T}_s^r(M)$. Then:*
(a) D_0^0 *is a vector field in $\mathfrak{X}(M)$.*
(b) $\mathsf{D}_s^r(f\mathcal{A}) = \mathsf{D}_0^0(f)\mathcal{A} + f\mathsf{D}_s^r(\mathcal{A}).$

Proof. (a): For functions f, g in $C^\infty(M) = \mathcal{T}_0^0(M)$, we have from [D1] that

$$\mathsf{D}_0^0(fg) = f\mathsf{D}_0^0(g) + g\mathsf{D}_0^0(f).$$

Thus, D_0^0 is a derivation on $C^\infty(M)$. The result now follows from the identification given by Theorem 15.2.1.

(b): Since f is a function in $C^\infty(M) = \mathcal{T}_0^0(M)$, so is $\mathsf{D}_0^0(f)$. By definition, $f \otimes \mathcal{A} = f\mathcal{A}$ and $\mathsf{D}_0^0(f) \otimes \mathcal{A} = \mathsf{D}_0^0(f)\mathcal{A}$. Then [D1] gives

$$\mathsf{D}_s^r(f\mathcal{A}) = \mathsf{D}_0^0(f) \otimes \mathcal{A} + f \otimes \mathsf{D}_s^r(\mathcal{A}) = \mathsf{D}_0^0(f)\mathcal{A} + f\mathsf{D}_s^r(\mathcal{A}). \qquad\square$$

Example 16.2.3. Let M be a smooth manifold, let D be a tensor derivation on M, and let $\mathcal{A}$ be a tensor field in $\mathcal{T}_1^1(M)$. We seek an explicit expression for $\mathsf{D}_1^1(\mathcal{A})$. Let $X = \mathsf{D}_0^0$ be the vector field in $\mathfrak{X}(M)$ given by Theorem 16.2.2(a), and let $\mathsf{C} = \mathsf{C}_1^1 \circ \mathsf{C}_2^2$. Using [D2] twice yields $\mathsf{D}_0^0 \circ \mathsf{C} = \mathsf{C} \circ \mathsf{D}_2^2$. For a covector field ω in $\mathfrak{X}^*(M)$ and a vector field Y in $\mathfrak{X}(M)$, we have

$$
\begin{aligned}
X\big(\mathcal{A}(\omega, Y)\big) &= \mathsf{D}_0^0\big(\mathcal{A}(\omega, Y)\big) \\
&= \mathsf{D}_0^0\big(\mathsf{C}(Y \otimes \mathcal{A} \otimes \omega)\big) \\
&= \mathsf{C}\big(\mathsf{D}_2^2(Y \otimes \mathcal{A} \otimes \omega)\big) \\
&= \mathsf{C}\big(\mathsf{D}_0^1(Y) \otimes \mathcal{A} \otimes \omega + Y \otimes \mathsf{D}_1^1(\mathcal{A}) \otimes \omega + Y \otimes \mathcal{A} \otimes \mathsf{D}_1^0(\omega)\big) \\
&= \mathsf{C}\big(\mathsf{D}_0^1(Y) \otimes \mathcal{A} \otimes \omega\big) + \mathsf{C}\big(Y \otimes \mathsf{D}_1^1(\mathcal{A}) \otimes \omega\big) + \mathsf{C}\big(Y \otimes \mathcal{A} \otimes \mathsf{D}_1^0(\omega)\big) \\
&= \mathcal{A}\big(\omega, \mathsf{D}_0^1(Y)\big) + \mathsf{D}_1^1(\mathcal{A})(\omega, Y) + \mathcal{A}\big(\mathsf{D}_1^0(\omega), Y\big),
\end{aligned}
$$

where the second and sixth equalities follow from the manifold version of Example 5.4.6, and the third equality from [D2]. Thus,

$$
\mathsf{D}_1^1(\mathcal{A})(\omega, Y) = X\big(\mathcal{A}(\omega, Y)\big) - \mathcal{A}\big(\mathsf{D}_1^0(\omega), Y\big) - \mathcal{A}\big(\omega, \mathsf{D}_0^1(Y)\big). \qquad \Diamond
$$

Observe that in the preceding computations, we implicitly used Theorem 15.7.2 to identify $\mathcal{A}$ with a map in $\mathrm{Mult}_{C^\infty(M)}\big(\mathfrak{X}^*(M) \times \mathfrak{X}(M), C^\infty(M)\big)$. The next result generalizes Example 16.2.3 using the same approach.

Theorem 16.2.4 (Global Expression for Tensor Derivation). *Let M be a smooth manifold, let D be a tensor derivation on M, and let $X = \mathsf{D}_0^0$ be the vector field in $\mathfrak{X}(M)$ given by Theorem 16.2.2(a). Let $\mathcal{A}$ be a tensor field in $\mathcal{T}_s^r(M)$, let $\omega^1, \ldots, \omega^r$ be covector fields in $\mathfrak{X}^*(M)$, and let $Y_1, \ldots, Y_s$ be vector fields in $\mathfrak{X}(M)$. Then*

$$
\begin{aligned}
\mathsf{D}_s^r&(\mathcal{A})(\omega^1, \ldots, \omega^r, Y_1, \ldots, Y_s) \\
&= X\big(\mathcal{A}(\omega^1, \ldots, \omega^r, Y_1, \ldots, Y_s)\big) \\
&\quad - \sum_{i=1}^r \mathcal{A}\big(\omega^1, \ldots, \omega^{i-1}, \mathsf{D}_1^0(\omega^i), \omega^{i+1}, \ldots, \omega^r, Y_1, \ldots, Y_s\big) \qquad (16.2.1) \\
&\quad - \sum_{j=1}^s \mathcal{A}\big(\omega^1, \ldots, \omega^r, Y_1, \ldots, Y_{j-1}, \mathsf{D}_0^1(Y_j), Y_{j+1}, \ldots, Y_s\big).
\end{aligned}
$$

Proof. The proof is an elaboration of Example 16.2.3. We have

$$
\begin{aligned}
X&\big(\mathcal{A}(\omega^1, \ldots, \omega^r, Y_1, \ldots, Y_s)\big) \\
&= \mathsf{D}_0^0\big(\mathcal{A}(\omega^1, \ldots, \omega^r, Y_1, \ldots, Y_s)\big) \\
&= \mathsf{D}_0^0\big(\mathsf{C}(Y_1 \otimes \cdots \otimes Y_s \otimes \mathcal{A} \otimes \omega^1 \otimes \cdots \otimes \omega^r)\big) \\
&= \mathsf{C}\big(\mathsf{D}_{2s}^{2r}(Y_1 \otimes \cdots \otimes Y_s \otimes \mathcal{A} \otimes \omega^1 \otimes \cdots \otimes \omega^r)\big)
\end{aligned}
$$

$$= \mathsf{C}\bigg(\sum_{j=1}^{s} Y_1 \otimes \cdots \otimes \mathsf{D}_0^1(Y_j) \otimes \cdots \otimes Y_s \otimes \mathcal{A} \otimes \omega^1 \otimes \cdots \otimes \omega^r$$

$$+ Y_1 \otimes \cdots \otimes Y_s \otimes \mathsf{D}_s^r(\mathcal{A}) \otimes \omega^1 \otimes \cdots \otimes \omega^r$$

$$+ \sum_{i=1}^{r} Y_1 \otimes \cdots \otimes Y_s \otimes \mathcal{A} \otimes \omega^1 \otimes \cdots \otimes \mathsf{D}_1^0(\omega^i) \otimes \cdots \cdots \otimes \omega^r \bigg)$$

$$= \sum_{j=1}^{s} \mathsf{C}\big(Y_1 \otimes \cdots \otimes \mathsf{D}_0^1(Y_j) \otimes \cdots \otimes Y_s \otimes \mathcal{A} \otimes \omega^1 \otimes \cdots \otimes \omega^r \big)$$

$$+ \mathsf{C}\big(Y_1 \otimes \cdots \otimes Y_s \otimes \mathsf{D}_s^r(\mathcal{A}) \otimes \omega^1 \otimes \cdots \otimes \omega^r \big)$$

$$+ \sum_{i=1}^{r} \mathsf{C}\big(Y_1 \otimes \cdots \otimes Y_s \otimes \mathcal{A} \otimes \omega^1 \otimes \cdots \otimes \mathsf{D}_1^0(\omega^i) \otimes \cdots \cdots \otimes \omega^r \big)$$

$$= \sum_{j=1}^{s} \mathcal{A}\big(\omega^1, \ldots, \omega^r, Y_1, \ldots, Y_{j-1}, \mathsf{D}_0^1(Y_j), Y_{j+1}, \ldots, Y_s \big)$$

$$+ \mathsf{D}_s^r(\mathcal{A})(\omega^1, \ldots, \omega^r, Y_1, \ldots, Y_s)$$

$$+ \sum_{i=1}^{r} \mathcal{A}\big(\omega^1, \ldots, \omega^{i-1}, \mathsf{D}_1^0(\omega^i), \omega^{i+1}, \ldots, \omega^r, Y_1, \ldots, Y_s \big),$$

where the second and sixth equalities follow from Theorem 15.14.4, and the third equality from [D2]. The result follows. $\qquad\square$

Example 16.2.5. For a covector field ω in $\mathfrak{X}^*(M) = \mathcal{T}_1^0(M)$ and a vector field Y in $\mathfrak{X}(M)$, setting $r = 0$ and $s = 1$ in Theorem 16.2.4 gives

$$\mathsf{D}_1^0(\omega)(Y) = X\big(\omega(Y) \big) - \omega\big(\mathsf{D}_0^1(Y) \big), \tag{16.2.2}$$

where $X = \mathsf{D}_0^0$. $\qquad\diamond$

Theorem 16.2.6. *A tensor derivation* D *on a smooth manifold* M *is uniquely determined by its values on* $C^\infty(M)$ *and* $\mathfrak{X}(M)$.

Proof. We see from Theorem 16.2.4 that D is uniquely determined by its values on $C^\infty(M)$, $\mathfrak{X}(M)$, and $\mathfrak{X}^*(M)$. However, (16.2.2) shows that the values of D on $\mathfrak{X}^*(M)$ are uniquely determined by its values on $\mathfrak{X}(M)$. $\qquad\square$

Theorem 16.2.7 (Existence of Tensor Derivation). *Let* M *be a smooth manifold, let* X *be a vector field in* $\mathfrak{X}(M)$, *and let*

$$\mathcal{D}_X : \mathfrak{X}(M) \longrightarrow \mathfrak{X}(M)$$

be a linear map such that

$$\mathcal{D}_X(fY) = X(f)Y + f\mathcal{D}_X(Y) \tag{16.2.3}$$

for all functions f *in* $C^\infty(M)$ *and all vector fields* Y *in* $\mathfrak{X}(M)$. *Then there is a unique tensor derivation* D *on* M *such that*

$$\mathsf{D}_0^0 = X \qquad and \qquad \mathsf{D}_0^1 = \mathcal{D}_X. \tag{16.2.4}$$

Proof. We need to construct a tensor derivation D on M starting with the building blocks $\mathsf{D}_0^0 = X$ and $\mathsf{D}_0^1 = \mathcal{D}_X$. Inevitably, (16.2.1) forces our hand as to the specifics of the construction. We illustrate the approach for the cases $(r, s) = (0, 0), (1, 0)$, and $(0, 1)$.

For $(r, s) = (0, 0)$, where $\mathcal{T}_0^0(M) = C^\infty(M)$, we have from Theorem 15.2.1 that D_0^0 satisfies [D1]. Since no contractions are possible, [D2] is trivially satisfied.

For $(r, s) = (1, 0)$, where $\mathcal{T}_0^1(M) = \mathfrak{X}(M)$, we have from (16.2.3) and (16.2.4) that

$$
\begin{aligned}
\mathsf{D}_{0+0}^{0+1}(f \otimes X) = \mathsf{D}_0^1(fX) &= \mathcal{D}_X(fX) = X(f)Y + f\mathcal{D}_X(Y) \\
&= \mathsf{D}_0^0(f)Y + f\mathsf{D}_0^1(Y) = \mathsf{D}_0^0(f) \otimes Y + f \otimes \mathsf{D}_0^1(Y),
\end{aligned}
$$

hence [D1] is satisfied. Again no contractions are possible, so [D2] is trivially satisfied.

For $(r, s) = (0, 1)$, where $\mathcal{T}_1^0(M) = \mathfrak{X}^*(M)$, we proceed as follows.

Existence. For a given covector field ω in $\mathfrak{X}^*(M)$, let us define a map

$$
\mathsf{D}_1^0(\omega) : \mathfrak{X}(M) \longrightarrow C^\infty(M)
$$

by

$$
\mathsf{D}_1^0(\omega)(Y) = X\big(\omega(Y)\big) - \omega\big(\mathcal{D}_X(Y)\big) \tag{16.2.5}
$$

for all vector fields Y in $\mathfrak{X}(M)$. This definition is based on (16.2.2), which follows from (16.2.1) and incorporates (16.2.4).

We claim that $\mathsf{D}_1^0(\omega)$ is in $\text{Lin}_{C^\infty(M)}\big(\mathfrak{X}(M), C^\infty(M)\big)$. Let Y, Z be vector fields in $\mathfrak{X}(M)$, and let f be a function in $C^\infty(M)$. We have from (16.2.5) that

$$
\begin{aligned}
\mathsf{D}_1^0(\omega)&(fY + Z) \\
&= X\big(\omega(fY + Z)\big) - \omega\big(\mathcal{D}_X(fY + Z)\big) \\
&= X\big(\omega(fY)\big) + X\big(\omega(Z)\big) - \omega\big(\mathcal{D}_X(fY)\big) - \omega\big(\mathcal{D}_X(Z)\big) \\
&= X\big(\omega(fY)\big) - \omega\big(\mathcal{D}_X(fY)\big) + \mathsf{D}_1^0(\omega)(Z).
\end{aligned} \tag{16.2.6}
$$

We also have

$$
\begin{aligned}
X\big(\omega(fY)\big) &= X\big(f\omega(Y)\big) && \text{[Th 15.5.1(b)]} \\
&= fX\big(\omega(Y)\big) + \omega(Y)X(f) && \text{[(15.1.3)]}
\end{aligned}
$$

and

$$
\begin{aligned}
\omega\big(\mathcal{D}_X(fY)\big) &= \omega\big(X(f)Y + f\mathcal{D}_X(Y)\big) && \text{[(16.2.3)]} \\
&= X(f)\omega(Y) + f\omega\big(\mathcal{D}_X(Y)\big), && \text{[Th 15.5.1]}
\end{aligned}
$$

hence

$$
\begin{aligned}
X\big(\omega(fY)\big) - \omega\big(\mathcal{D}_X(fY)\big) &= f\big[X\big(\omega(Y)\big) - \omega\big(\mathcal{D}_X(Y)\big)\big] \\
&= f\mathsf{D}_1^0(\omega)(Y).
\end{aligned} \tag{16.2.7}
$$

Combining (16.2.6) and (16.2.7) gives

$$\mathsf{D}_1^0(\omega)(fY + Z) = f\mathsf{D}_1^0(\omega)(Y) + \mathsf{D}_1^0(\omega)(Z),$$

which proves the claim.

The next step is to define a map

$$\mathsf{D}_1^0 : \mathfrak{X}^*(M) \longrightarrow \mathrm{Lin}_{C^\infty(M)}\big(\mathfrak{X}(M), C^\infty(M)\big) \qquad (16.2.8)$$

by the assignment

$$\omega \longmapsto \mathsf{D}_1^0(\omega)$$

for all covector fields ω in $\mathfrak{X}^*(M)$. We have $\mathfrak{X}^*(M) = \mathcal{T}_1^0(M)$, and by Theorem 15.5.3, $\mathrm{Lin}_{C^\infty(M)}\big(\mathfrak{X}(M), C^\infty(M)\big)$ can be identified with $\mathfrak{X}^*(M)$. Thus, (16.2.8) can be expressed as $\mathsf{D}_1^0 : \mathcal{T}_1^0(M) \longrightarrow \mathcal{T}_1^0(M)$, which is what we need for a tensor derivation corresponding to $(r, s) = (0, 1)$.

It remains to show that [D1] and [D2] are satisfied. Let f be a function in $C^\infty(M)$, let ω be a covector field in $\mathfrak{X}^*(M)$, and let Y be a vector field in $\mathfrak{X}(M)$. Then

$$
\begin{aligned}
\mathsf{D}_{0+1}^{0+0}&(f \otimes \omega)(Y) \\
&= \mathsf{D}_1^0(f\omega)(Y) \\
&= X\big(f\omega(Y)\big) - f\omega\big(\mathcal{D}_X(Y)\big) && [(16.2.5)] \\
&= X(f)\omega(Y) + fX\big(\omega(Y)\big) - f\omega\big(\mathcal{D}_X(Y)\big) && [(15.1.3)] \\
&= \mathsf{D}_0^0(f)\omega(Y) + f\mathsf{D}_1^0(\omega)(Y) && [(16.2.4),\ (16.2.5)] \\
&= \big(\mathsf{D}_0^0(f) \otimes \omega + f \otimes \mathsf{D}_1^0(\omega)\big)(Y).
\end{aligned}
$$

Since Y was arbitrary,

$$\mathsf{D}_{0+1}^{0+0}(f \otimes \omega) = \mathsf{D}_0^0(f) \otimes \omega + f \otimes \mathsf{D}_1^0(\omega),$$

hence [D1] is satisfied. For $(r, s) = (0, 1)$, no contractions are possible, so [D2] is trivially satisfied.

Uniqueness. It follows from (16.2.2) and (16.2.4) that (16.2.5) is the only option for defining D_1^0. $\qquad\qquad\square$

16.3   Form Derivations

A **form derivation** on a smooth manifold M is a family $\mathfrak{D}$ of linear maps

$$\mathfrak{D}_s : \Lambda^s(M) \longrightarrow \Lambda^s(M)$$

defined for $s \geq 0$ such that

$$\mathfrak{D}_{s+s'}(\omega \wedge \xi) = \mathfrak{D}_s(\omega) \wedge \xi + \omega \wedge \mathfrak{D}_{s'}(\xi) \qquad (16.3.1)$$

for all forms ω in $\Lambda^s(M)$ and ξ in $\Lambda^{s'}(M)$.

Theorem 16.3.1. *If M is a smooth manifold and $\mathfrak{D}$ is form derivation on M, then, in local coordinates,*

$$\mathfrak{D}_s\big(d(x^1) \wedge \cdots \wedge d(x^s)\big) = \sum_i d(x^1) \wedge \cdots \wedge \mathfrak{D}_1\big(d(x^i)\big) \wedge \cdots \wedge d(x^s).$$

Proof. The proof is by induction. For $s = 1$, the result is trivial. Let $s > 1$, and suppose the assertion is true for all indices $< s$. We have from (16.3.1) that

$$\mathfrak{D}_s\big(d(x^1) \wedge \cdots \wedge d(x^s)\big) = \mathfrak{D}_1\big(d(x^1)\big) \wedge d(x^2) \wedge \cdots \wedge d(x^s) \tag{16.3.2}$$
$$+ d(x^1) \wedge \mathfrak{D}_{s-1}\big(d(x^2) \wedge \cdots \wedge d(x^s)\big).$$

By the induction hypothesis,

$$\mathfrak{D}_{s-1}\big(d(x^2) \wedge \cdots \wedge d(x^s)\big) = \sum_{i=2}^{s} d(x^2) \wedge \cdots \wedge \mathfrak{D}_1\big(d(x^i)\big) \wedge \cdots \wedge d(x^s),$$

hence

$$d(x^1) \wedge \mathfrak{D}_{s-1}\big(d(x^2) \wedge \cdots \wedge d(x^s)\big)$$
$$= \sum_{i=2}^{s} d(x^1) \wedge d(x^2) \wedge \cdots \wedge \mathfrak{D}_1\big(d(x^i)\big) \wedge \cdots \wedge d(x^s). \tag{16.3.3}$$

Combining (16.3.2) and (16.3.3) gives the result for s. □

Theorem 16.3.2. *A form derivation $\mathfrak{D}$ on a smooth manifold M is uniquely determined by its values on $C^\infty(M)$ and $\{d(f) : f \in C^\infty(M)\}$.*

Proof. Let $\dim(M) = m$, let ω be a form in $\Lambda^s(M)$, and, in local coordinates, let

$$\omega = \sum_{1 \leq i_1 < \cdots < i_s \leq m} \alpha_{i_1 \ldots i_s} d(x^{i_1}) \wedge \cdots \wedge d(x^{i_s}).$$

Then

$$\mathfrak{D}_s(\omega) = \sum_{1 \leq i_1 < \cdots < i_s \leq m} \mathfrak{D}_{0+s}\big(\alpha_{i_1 \ldots i_s} d(x^{i_1}) \wedge \cdots \wedge d(x^{i_s})\big)$$

$$= \sum_{1 \leq i_1 < \cdots < i_s \leq m} \big[\mathfrak{D}_0(\alpha_{i_1 \ldots i_s}) \wedge d(x^{i_1}) \wedge \cdots \wedge d(x^{i_s})$$
$$+ \alpha_{i_1 \ldots i_s} \wedge \mathfrak{D}_s\big(d(x^{i_1}) \wedge \cdots \wedge d(x^{i_s})\big)\big]$$

$$= \sum_{1 \leq i_1 < \cdots < i_s \leq m} \Big[\mathfrak{D}_0(\alpha_{i_1 \ldots i_s}) d(x^{i_1}) \wedge \cdots \wedge d(x^{i_s})$$
$$+ \alpha_{i_1 \ldots i_s} \sum_{j=1}^{s} d(x^{i_1}) \wedge \cdots \wedge \mathfrak{D}_1\big(d(x^{i_j})\big) \wedge \cdots \wedge d(x^{i_s})\big)\Big],$$

where the second equality follows from (16.3.1), and the third equality from Theorem 16.3.1. The result follows. □

16.4   Lie Derivative

Let M be a smooth manifold, let X be a vector field in $\mathfrak{X}(M)$, and consider the map $\mathcal{L}_X : \mathfrak{X}(M) \longrightarrow \mathfrak{X}(M)$ defined by $\mathcal{L}_X(Y) = [X, Y]$ for all vector fields Y in $\mathfrak{X}(M)$. By parts (c) and (d) of Theorem 15.3.2, $\mathcal{L}_X$ is linear and satisfies $\mathcal{L}_X(fY) = X(f)Y + f\mathcal{L}_X(Y)$ for all functions f in $C^\infty(M)$ and all vector fields Y in $\mathfrak{X}(M)$. It follows from Theorem 16.2.7 that there is a unique tensor derivation

$$\mathcal{L}_X : T_s^r(M) \longrightarrow T_s^r(M)$$

on M, called the **Lie derivative with respect to** X, such that

$$\mathcal{L}_X(f) = X(f) \qquad \text{and} \qquad \mathcal{L}_X(Y) = [X, Y] \qquad (16.4.1)$$

for all functions f in $C^\infty(M)$ and all vector fields Y in $\mathfrak{X}(M)$, where $(\mathcal{L}_X)_s^r$ has been abbreviated to $\mathcal{L}_X$.

The Lie derivative enjoys a wealth of algebraic properties.

Theorem 16.4.1 (Vector Field Properties of Lie Derivative). *Let M be a smooth manifold, let X, Y, Z be vector fields in $\mathfrak{X}(M)$, and let f, g be functions in $C^\infty(M)$. Then:*
(a) $\mathcal{L}_X(Y) = -\mathcal{L}_Y(X)$.
(b) $\mathcal{L}_{X+Y}(Z) = \mathcal{L}_X(Z) + \mathcal{L}_Y(Z)$.
(c) $\mathcal{L}_X(Y + Z) = \mathcal{L}_X(Y) + \mathcal{L}_X(Z)$.
(d) $\mathcal{L}_{fX}(gY) = fg\mathcal{L}_X(Y) + fX(g)Y - gY(f)X$.
(e) $\mathcal{L}_X([Y, Z]) = [\mathcal{L}_X(Y), Z] + [Y, \mathcal{L}_X(Z)]$.

Proof. This follows from Theorem 15.3.2.   □

We note that Jacobi's identity in part (e) of Theorem 15.3.2 gives rise to a product rule for Lie brackets in part (e) of the above theorem. The next result contains a product rule for tensor products, and the one after that a product rule for wedge products.

Theorem 16.4.2 (Tensor Field Properties of Lie Derivative). *Let M be a smooth manifold, let X be a vector field in $\mathfrak{X}(M)$, let $\mathcal{A}, \mathcal{B}$ and $\mathcal{C}$ be tensor fields in $T_s^0(M)$ and $T_{s'}^0(M)$, respectively, and let f be a function in $C^\infty(M)$. Then:*
(a) $\mathcal{L}_X(\mathcal{A} + \mathcal{B}) = \mathcal{L}_X(\mathcal{A}) + \mathcal{L}_X(\mathcal{B})$.
(b) $\mathcal{L}_X(f\mathcal{A}) = X(f)\mathcal{A} + f\mathcal{L}_X(\mathcal{A})$.
(c) $\mathcal{L}_X(\mathcal{A} \otimes \mathcal{C}) = \mathcal{L}_X(\mathcal{A}) \otimes \mathcal{C} + \mathcal{A} \otimes \mathcal{L}_X(\mathcal{C})$.

Proof. This follows from the properties of $\mathcal{L}_X$ as a tensor derivation on M.   □

Theorem 16.4.3 (Form Derivation Property of Lie Derivative). *If M is a smooth manifold and X is a vector field in $\mathfrak{X}(M)$, then $\mathcal{L}_X$ is a form derivation; that is,*

$$\mathcal{L}_X(\omega \wedge \xi) = \mathcal{L}_X(\omega) \wedge \xi + \omega \wedge \mathcal{L}_X(\xi)$$

for all forms ω in $\Lambda^s(M)$ and ξ in $\Lambda^{s'}(M)$.   □

Theorem 16.4.4 (Global Expression for Lie Derivative). *Let M be a smooth manifold, and let X be a vector field in $\mathfrak{X}(M)$. Let $\mathcal{A}$ be a tensor field in $\mathcal{T}_s^r(M)$, let $\omega^1, \ldots, \omega^r$ be covector fields in $\mathfrak{X}^*(M)$, and let $Y_1, \ldots, Y_s$ be vector fields in $\mathfrak{X}(M)$. Then*

$$
\begin{aligned}
\mathcal{L}_X(\mathcal{A})&(\omega^1, \ldots, \omega^r, Y_1, \ldots, Y_s) \\
&= X\big(\mathcal{A}(\omega^1, \ldots, \omega^r, Y_1, \ldots, Y_s)\big) \\
&\quad - \sum_{i=1}^r \mathcal{A}\big(\omega^1, \ldots, \omega^{i-1}, \mathcal{L}_X(\omega^i), \omega^{i+1}, \ldots, \omega^r, Y_1, \ldots, Y_s\big) \\
&\quad - \sum_{j=1}^s \mathcal{A}\big(\omega^1, \ldots, \omega^r, Y_1, \ldots, Y_{j-1}, [X, Y_j], Y_{j+1}, \ldots, Y_s\big).
\end{aligned}
$$

Remark. We have from (16.4.1) that $X\big(\mathcal{A}(\omega^1, \ldots, \omega^r, Y_1, \ldots, Y_s)\big)$ in the above identity can be replaced with $\mathcal{L}_X\big(\mathcal{A}(\omega^1, \ldots, \omega^r, Y_1, \ldots, Y_s)\big)$.

Proof. Since $\mathcal{L}_X(Y_j) = [X, Y_j]$, the result follows from Theorem 16.2.4. □

Example 16.4.5. For a covector field ω in $\mathfrak{X}^*(M) = \mathcal{T}_1^0(M)$ and vector fields X, Y in $\mathfrak{X}(M)$, setting $r = 0$ and $s = 1$ in Theorem 16.4.4 gives

$$
\mathcal{L}_X(\omega)(Y) = X\big(\omega(Y)\big) - \omega([X, Y]). \qquad \diamond
$$

Theorem 16.4.6 (Commutativity of Lie Derivative and Exterior Derivative of Functions). *If M is a smooth manifold and X is a vector field in $\mathfrak{X}(M)$, then*

$$
\mathcal{L}_X \circ d = d \circ \mathcal{L}_X;
$$

that is,

$$
\mathcal{L}_X\big(d(f)\big) = d\big(\mathcal{L}_X(f)\big)
$$

for all functions f in $C^\infty(M)$.

Proof. For a vector field Y in $\mathfrak{X}(M)$, we have

$$
\begin{aligned}
\mathcal{L}_X\big(d(f)\big)(Y) &= X\big(d(f)(Y)\big) - d(f)([X, Y]) && \text{[Ex 16.4.5]} \\
&= X\big(Y(f)\big) - [X, Y](f) && \text{[Th 15.4.5(b)]} \\
&= Y\big(X(f)\big) && \text{[(15.3.1)]} \\
&= d\big(X(f)\big)(Y) && \text{[Th 15.4.5(b)]} \\
&= d\big(\mathcal{L}_X(f)\big)(Y). && \text{[(16.4.1)]}
\end{aligned}
$$

Since f and Y were arbitrary, the result follows. □

Theorem 16.4.7. *Let M be a smooth m-manifold, let X be a vector field in $\mathfrak{X}(M)$, and, in local coordinates, let*

$$
X = \sum_i \alpha^i \frac{\partial}{\partial x^i}.
$$

Then

$$\mathcal{L}_X\big(d(x^i)\big) = d(\alpha^i)$$

for $i = 1, \ldots, m$.

Proof. It follows from (16.4.1) that

$$\mathcal{L}_X(x^i) = X(x^i) = \left(\sum_j \alpha^j \frac{\partial}{\partial x^j}\right)(x^i) = \sum_j \alpha^j \frac{\partial x^i}{\partial x^j} = \alpha^i,$$

and then from Theorem 16.4.6 that

$$\mathcal{L}_X\big(d(x^i)\big) = d\big(\mathcal{L}_X(x^i)\big) = d(\alpha^i). \qquad \square$$

Theorem 16.4.8 (Local Coordinate Expression for Lie Derivative). *Let M be a smooth m-manifold, let X be a vector field in $\mathfrak{X}(M)$, and let $\mathcal{A}$ be a tensor field in $\mathcal{T}^r_s(M)$. In local coordinates, let*

$$X = \sum_{i=1}^m \alpha^i \frac{\partial}{\partial x^i},$$

and let $\mathcal{A}^{i_1 \ldots i_r}_{j_1 \ldots j_s}$ be the components of $\mathcal{A}$. Then the components of $\mathcal{L}_X(\mathcal{A})$ are

$$\mathcal{L}_X(\mathcal{A})^{i_1 \ldots i_r}_{j_1 \ldots j_s} = \sum_{k=1}^m \alpha^k \frac{\partial \mathcal{A}^{i_1 \ldots i_r}_{j_1 \ldots j_s}}{\partial x^k} - \sum_{l=1}^r \left(\sum_{k=1}^m \mathcal{A}^{i_1 \ldots i_{l-1} \, k \, i_{l+1} \ldots i_r}_{j_1 \ldots j_s} \frac{\partial \alpha^{i_l}}{\partial x^k}\right)$$

$$+ \sum_{l=1}^s \left(\sum_{k=1}^m \mathcal{A}^{i_1 \ldots i_r}_{j_1 \ldots j_{l-1} \, k \, j_{l+1} \ldots j_s} \frac{\partial \alpha^k}{\partial x^{j_l}}\right).$$

Proof. By Theorem 16.4.4,

$$\mathcal{L}_X(\mathcal{A})^{i_1 \ldots i_r}_{j_1 \ldots j_s}$$

$$= \mathcal{L}_X(\mathcal{A})\left(d(x^{i_1}), \ldots, d(x^{i_r}), \frac{\partial}{\partial x^{j_1}}, \ldots, \frac{\partial}{\partial x^{j_s}}\right)$$

$$= X\left(\mathcal{A}\left(d(x^{i_1}), \ldots, d(x^{i_r}), \frac{\partial}{\partial x^{j_1}}, \ldots, \frac{\partial}{\partial x^{j_s}}\right)\right) \qquad (16.4.2)$$

$$- \sum_{l=1}^r \mathcal{A}\left(d(x^{i_1}), \ldots, \mathcal{L}_X\big(d(x^{i_l})\big), \ldots, d(x^{i_r}), \frac{\partial}{\partial x^{j_1}}, \ldots, \frac{\partial}{\partial x^{j_s}}\right)$$

$$- \sum_{l=1}^s \mathcal{A}\left(d(x^{i_1}), \ldots, d(x^{i_r}), \frac{\partial}{\partial x^{i_1}}, \ldots, \left[X, \frac{\partial}{\partial x^{j_l}}\right], \ldots, \frac{\partial}{\partial x^{j_s}}\right).$$

We seek alternative expressions for the terms in the last three lines of (16.4.2). For the first term,

$$X\left(\mathcal{A}\left(d(x^{i_1}), \ldots, d(x^{i_r}), \frac{\partial}{\partial x^{j_1}}, \ldots, \frac{\partial}{\partial x^{j_s}}\right)\right) = X\big(\mathcal{A}^{i_1 \ldots i_r}_{j_1 \ldots j_s}\big)$$

$$= \left(\sum_{k=1}^m \alpha^k \frac{\partial}{\partial x^k}\right)\big(\mathcal{A}^{i_1 \ldots i_r}_{j_1 \ldots j_s}\big) = \sum_{k=1}^m \alpha^k \frac{\partial \mathcal{A}^{i_1 \ldots i_r}_{j_1 \ldots j_s}}{\partial x^k}. \qquad (16.4.3)$$

For the second term, Theorem 15.4.7 and Theorem 16.4.7 give

$$\mathcal{L}_X\big(d(x^{i_l})\big) = d(\alpha^{i_l}) = \sum_{k=1}^{m} \frac{\partial \alpha^{i_l}}{\partial x^k} d(x^k),$$

hence

$$\sum_{l=1}^{r} \mathcal{A}\bigg(d(x^{i_1}),\ldots,\mathcal{L}_X\big(d(x^{i_l})\big),\ldots,d(x^{i_r}),\frac{\partial}{\partial x^{j_1}},\ldots,\frac{\partial}{\partial x^{j_s}}\bigg)$$

$$= \sum_{l=1}^{r} \mathcal{A}\bigg(d(x^{i_1}),\ldots,\sum_{k=1}^{m} \frac{\partial \alpha^{i_l}}{\partial x^k} d(x^k),\ldots,d(x^{i_r}),\frac{\partial}{\partial x^{j_1}},\ldots,\frac{\partial}{\partial x^{j_s}}\bigg)$$

$$= \sum_{l=1}^{r} \bigg\{\sum_{k=1}^{m} \frac{\partial \alpha^{i_l}}{\partial x^k} \mathcal{A}\bigg(d(x^{i_1}),\ldots,d(x^k),\ldots,d(x^{i_r}),\frac{\partial}{\partial x^{j_1}},\ldots,\frac{\partial}{\partial x^{j_s}}\bigg)\bigg\}$$

$$= \sum_{l=1}^{r} \bigg(\sum_{k=1}^{m} \mathcal{A}^{i_1\ldots i_{l-1}\,k\,i_{l+1}\ldots i_r}_{j_1\ldots j_s} \frac{\partial \alpha^{i_l}}{\partial x^k}\bigg). \qquad (16.4.4)$$

For the third term,

$$\bigg[X,\frac{\partial}{\partial x^{j_l}}\bigg] = \sum_{k=1}^{m} \bigg[\alpha^k \frac{\partial}{\partial x^k},\frac{\partial}{\partial x^{j_l}}\bigg] \qquad \text{[Th 15.3.2(b)]}$$

$$= \sum_{k=1}^{m} \bigg(\alpha^k \bigg[\frac{\partial}{\partial x^k},\frac{\partial}{\partial x^{j_l}}\bigg] - \frac{\partial \alpha^k}{\partial x^{j_l}} \frac{\partial}{\partial x^k}\bigg) \qquad \text{[Th 15.3.2(d)]}$$

$$= -\sum_{k=1}^{m} \frac{\partial \alpha^k}{\partial x^{j_l}} \frac{\partial}{\partial x^k}, \qquad \text{[Th 15.3.3(b)]}$$

hence

$$\sum_{l=1}^{s} \mathcal{A}\bigg(d(x^{i_1}),\ldots,d(x^{i_r}),\frac{\partial}{\partial x^{j_1}},\ldots,\bigg[X,\frac{\partial}{\partial x^{j_l}}\bigg],\ldots,\frac{\partial}{\partial x^{j_s}}\bigg)$$

$$= \sum_{l=1}^{s} \mathcal{A}\bigg(d(x^{i_1}),\ldots,d(x^{i_r}),\frac{\partial}{\partial x^{j_1}},\ldots,-\sum_{k=1}^{m} \frac{\partial \alpha^k}{\partial x^{j_l}} \frac{\partial}{\partial x^k},\ldots,\frac{\partial}{\partial x^{j_s}}\bigg)$$

$$= -\sum_{l=1}^{s} \bigg\{\sum_{k=1}^{m} \frac{\partial \alpha^k}{\partial x^{j_l}} \mathcal{A}\bigg(d(x^{i_1}),\ldots,d(x^{i_r}),\frac{\partial}{\partial x^{j_1}},\ldots,\frac{\partial}{\partial x^k},\ldots,\frac{\partial}{\partial x^{j_s}}\bigg)\bigg\}$$

$$= -\sum_{l=1}^{s} \bigg(\sum_{k=1}^{m} \mathcal{A}^{i_1\ldots i_r}_{j_1\ldots j_{l-1}\,k\,j_{l+1}\ldots j_s} \frac{\partial \alpha^k}{\partial x^{j_l}}\bigg).$$

$$(16.4.5)$$

Substituting (16.4.3)–(16.4.5) into (16.4.2) gives the result.   □

Example 16.4.9. Let M be a smooth manifold, let X be a vector field in $\mathfrak{X}(M)$, let ω be a covector field in $\mathfrak{X}^*(M)$, and, in local coordinates, let

$$X = \sum_{i=1}^{m} \alpha^i \frac{\partial}{\partial x^i} \qquad \text{and} \qquad \omega = \sum_{j=1}^{m} \beta_j d(x^j).$$

By Theorem 16.4.8, the covector field $\mathcal{L}_X(\omega)$ in $\mathfrak{X}^*(M)$ has the components

$$\mathcal{L}_X(\omega)_j = \sum_{k=1}^{m} \alpha^k \frac{\partial \beta_j}{\partial x^k} + \sum_{k=1}^{m} \beta_k \frac{\partial \alpha^k}{\partial x^j}. \qquad \diamond$$

16.5   Interior Multiplication

The following definitions are the manifold versions of those appearing in Section 7.4. Let M be a smooth manifold, and let X be a vector field in $\mathfrak{X}(M)$. **Interior multiplication by X** is the family of linear maps

$$i_X : \Lambda^s(M) \longrightarrow \Lambda^{s-1}(M)$$

defined for $s \geq 2$ by

$$i_X(\omega)(Y_1, \ldots, Y_{s-1}) = \omega(X, Y_1, \ldots, Y_{s-1})$$

for all forms ω in $\Lambda^s(M)$ and all vector fields $Y_1, \ldots, Y_{s-1}$ in $\mathfrak{X}(M)$. Recalling from (15.8.1) and (15.8.2) that $\Lambda^1(M) = \mathfrak{X}^*(M)$ and $\Lambda^0(M) = C^\infty(M)$, we extend the preceding definition to $s = 1$ as follows:

$$i_X : \Lambda^1(M) \longrightarrow \Lambda^0(M)$$

is given by

$$i_X(\omega) = \omega(X) \qquad (16.5.1)$$

for all covector fields ω in $\mathfrak{X}^*(M)$. In particular, for any function f in $C^\infty(M)$, we have from Theorem 15.4.5(b) and (16.5.1) that

$$i_X\big(d(f)\big) = d(f)(X) = X(f). \qquad (16.5.2)$$

For $s = 0$, we trivially define $i_X = 0$; that is,

$$i_X(f) = 0 \qquad (16.5.3)$$

for all functions f in $C^\infty(M)$.

Theorem 16.5.1. *Let M be a smooth manifold, let X, Y be vector fields in $\mathfrak{X}(M)$, let $\omega, \xi,$ and ζ be forms in $\Lambda^s(M)$ and $\Lambda^{s'}(M)$, respectively, and let f be a function in $C^\infty(M)$. Then:*
(a) $i_{X+Y}(\omega) = i_X(\omega) + i_Y(\omega)$.
(b) $i_{fX}(\omega) = f\, i_X(\omega)$.
(c) $i_X(\omega + \xi) = i_X(\omega) + i_X(\xi)$.
(d) $i_X(f\omega) = f\, i_X(\omega)$.
(e) $i_X(\omega \wedge \zeta) = i_X(\omega) \wedge \zeta + (-1)^s \omega \wedge i_X(\zeta)$.

Proof. This is the manifold version of Theorem 7.4.1 combined with the manifold version of Theorem 7.4.3. $\square$

Example 16.5.2. For a vector field X in $\mathfrak{X}(M)$ and covector fields ω, ξ in $\Lambda^1(M) = \mathfrak{X}^*(M)$, we have from (15.8.3), (16.5.1), and Theorem 16.5.1(e) that

$$i_X(\omega \wedge \xi) = i_X(\omega) \wedge \xi - \omega \wedge i_X(\xi) = \omega(X) \wedge \xi - \omega \wedge \xi(X)$$
$$= \omega(X)\xi - \xi(X)\omega. \qquad \Diamond$$

We have defined two differential operators on smooth manifolds—the exterior derivative and the Lie derivative. Despite their evident differences in construction and properties, they are related by a remarkable identity that involves interior multiplication.

Theorem 16.5.3 (Cartan's Identity). *If M is a smooth manifold and X is a vector field in $\mathfrak{X}(M)$, then*

$$\mathcal{L}_X = d \circ i_X + i_X \circ d;$$

that is,

$$\mathcal{L}_X(\omega) = d(i_X(\omega)) + i_X(d(\omega))$$

for all forms ω in $\Lambda^s(M)$.

Proof. Let ω and ξ be forms in $\Lambda^s(M)$ and $\Lambda^{s'}(M)$, respectively. We have

$$(d \circ i_X + i_X \circ d)(\omega \wedge \xi) = d(i_X(\omega \wedge \xi)) + i_X(d(\omega \wedge \xi)),$$

$$
\begin{aligned}
d(i_X(\omega \wedge \xi)) &= d(i_X(\omega) \wedge \xi + (-1)^s \omega \wedge i_X(\xi)) && \text{[Th 16.5.1(e)]}\\
&= d(i_X(\omega) \wedge \xi) + (-1)^s d(\omega \wedge i_X(\xi))\\
&= d(i_X(\omega)) \wedge \xi + (-1)^{s-1} i_X(\omega) \wedge d(\xi)\\
&\quad + (-1)^s\big[d(\omega) \wedge i_X(\xi) + (-1)^s \omega \wedge d(i_X(\xi))\big] && \text{[Th 16.1.1(b)]}\\
&= (d \circ i_X)(\omega) \wedge \xi - (-1)^s i_X(\omega) \wedge d(\xi)\\
&\quad + (-1)^s d(\omega) \wedge i_X(\xi) + \omega \wedge (d \circ i_X)(\xi),
\end{aligned}
$$

and

$$
\begin{aligned}
i_X(d(\omega \wedge \xi)) &= i_X(d(\omega) \wedge \xi + (-1)^s \omega \wedge d(\xi)) && \text{[Th 16.1.1(b)]}\\
&= i_X(d(\omega) \wedge \xi) + (-1)^s i_X(\omega \wedge d(\xi))\\
&= i_X(d(\omega)) \wedge \xi + (-1)^{s+1} d(\omega) \wedge i_X(\xi)\\
&\quad + (-1)^s\big[i_X(\omega) \wedge d(\xi) + (-1)^s \omega \wedge i_X(d(\xi))\big] && \text{[Th 16.5.1(e)]}\\
&= (i_X \circ d)(\omega) \wedge \xi - (-1)^s d(\omega) \wedge i_X(\xi)\\
&\quad + (-1)^s i_X(\omega) \wedge d(\xi) + \omega \wedge (i_X \circ d)(\xi).
\end{aligned}
$$

Combining the above identities gives

$$
\begin{aligned}
(d \circ i_X + i_X \circ d)(\omega \wedge \xi) &= (d \circ i_X)(\omega) \wedge \xi + (i_X \circ d)(\omega) \wedge \xi\\
&\quad + \omega \wedge (d \circ i_X)(\xi) + \omega \wedge (i_X \circ d)(\xi)\\
&= (d \circ i_X + i_X \circ d)(\omega) \wedge \xi + \omega \wedge (d \circ i_X + i_X \circ d)(\xi).
\end{aligned}
$$

Thus, $d \circ i_X + i_X \circ d$ is a form derivation on M. By Theorem 16.4.3, the same is true of $\mathcal{L}_X$.

Let f be a function in $C^\infty(M)$. Then

$$
\begin{aligned}
(d \circ i_X + i_X \circ d)(f) &= d\big(i_X(f)\big) + i_X\big(d(f)\big) \\
&= X(f) && [(16.5.2),\ (16.5.3)] \\
&= \mathcal{L}_X(f) && [(16.4.1)]
\end{aligned}
$$

and

$$
\begin{aligned}
&(d \circ i_X + i_X \circ d)\big(d(f)\big) \\
&= d\big(i_X\big(d(f)\big)\big) + i_X\big(d^2(f)\big) \\
&= d\big(X(f)\big) && [\text{Th } 16.1.1(\text{c}),\ (16.5.2)] \\
&= d\big(\mathcal{L}_X(f)\big) && [(16.4.1)] \\
&= \mathcal{L}_X\big(d(f)\big). && [\text{Th } 16.4.6]
\end{aligned}
$$

Since f was arbitrary, $d \circ i_X + i_X \circ d$ and $\mathcal{L}_X$ agree on $C^\infty(M)$ and $\{d(f) : f \in C^\infty(M)\}$. The result now follows from Theorem 16.3.2.   □

The next result is a generalization of Theorem 16.4.6.

Theorem 16.5.4 (Commutativity of Lie Derivative and Exterior Derivative of Forms). *If M is a smooth manifold and X is a vector field in $\mathfrak{X}(M)$, then*

$$
\mathcal{L}_X \circ d = d \circ \mathcal{L}_X;
$$

that is,

$$
\mathcal{L}_X\big(d(\omega)\big) = d\big(\mathcal{L}_X(\omega)\big)
$$

for all forms ω in $\Lambda^s(M)$.

Proof. It follows from Theorem 16.1.1(c) and Theorem 16.5.3 that

$$
\begin{aligned}
\mathcal{L}_X \circ d &= (d \circ i_X + i_X \circ d) \circ d = d \circ i_X \circ d + i_X \circ d^2 \\
&= d \circ i_X \circ d
\end{aligned}
$$

and

$$
\begin{aligned}
d \circ \mathcal{L}_X &= d \circ (d \circ i_X + i_X \circ d) = d^2 \circ i_X + d \circ i_X \circ d \\
&= d \circ i_X \circ d.
\end{aligned}
$$

The result follows.   □

16.6   Orientation

The material in this section borrows heavily from Section 8.2 and Section 12.7.

Let M be a smooth manifold, let $(U, (x^i))$ and $(\widetilde{U}, (\widetilde{x}^i))$ be overlapping charts on M, and let $\mathcal{X}$ and $\widetilde{\mathcal{X}}$ be the corresponding coordinate frames. Recall from Section 8.2 that for a given point p in $U \cap \widetilde{U}$, the coordinate bases $\mathcal{X}_p$ and $\widetilde{\mathcal{X}}_p$ are said to be **consistent** if

$$\det\left(\left[\mathrm{id}_{T_p(M)} \right]_{\mathcal{X}_p}^{\widetilde{\mathcal{X}}_p} \right) > 0.$$

We say that $(U, (x^i))$ and $(\widetilde{U}, (\widetilde{x}^i))$ are **consistent** if $\mathcal{X}_p$ and $\widetilde{\mathcal{X}}_p$ are consistent for all p in $U \cap \widetilde{U}$. A smooth atlas $\mathfrak{A}$ for M is said to be **consistent** if every pair of overlapping charts in $\mathfrak{A}$ is consistent.

A **pointwise orientation of M** is a collection of orientations

$$\mathfrak{O} = \{\mathfrak{O}(p) : p \in M\}, \tag{16.6.1}$$

one for each tangent space of M. Without additional structure, the orientations might vary erratically from point to point. We say that $\mathfrak{O}$ is a **(smooth) orientation of M** if there is a consistent smooth atlas $\mathfrak{A}$ for M such that for each chart $(U, (x^i))$ in $\mathfrak{A}$ and corresponding coordinate frame $\mathcal{X}$,

$$\mathfrak{O}(p) = [\mathcal{X}_p] \tag{16.6.2}$$

all p in U, where we recall that $[\mathcal{X}_p]$ is the equivalence class of all bases for $T_p(M)$ (not just coordinate bases) that are consistent with $\mathcal{X}_p$. Let $(U, (x^i))$ and $(\widetilde{U}, (\widetilde{x}^i))$ be overlapping charts in $\mathfrak{A}$, let $\mathcal{X}$ and $\widetilde{\mathcal{X}}$ be the corresponding coordinate frames, and let p be a point in $U \cap \widetilde{U}$. Since the charts are consistent, $[\mathcal{X}_p] = [\widetilde{\mathcal{X}}_p]$, hence (16.6.2) is independent of the choice of chart in $\mathfrak{A}$ at p.

Suppose $\mathfrak{O}$ is in fact an orientation of M. The existence of a consistent smooth atlas $\mathfrak{A}$ for M ensures that the orientations in $\mathfrak{O}$ vary "smoothly" on M. We refer to the pair $(M, \mathfrak{O})$ as an **oriented smooth manifold**. The notation $(M, \mathfrak{A})$, and sometimes $(M, \mathfrak{A}, \mathfrak{O})$, is used as an alternative to $(M, \mathfrak{O})$. Each chart in $\mathfrak{A}$ is said to be **positively oriented (with respect to $\mathfrak{O}$)**. According to terminology introduced in Section 8.2, for each point p in M, we say that each basis in $\mathfrak{O}(p)$ is positively oriented [with respect to $\mathfrak{O}(p)$].

Let M be a smooth manifold, and let $\mathfrak{A}$ be a consistent smooth atlas for M. Then (16.6.1) and (16.6.2) can be used to define a corresponding orientation of M, called the **orientation induced by $\mathfrak{A}$**. The remarks above show that the orientation is well-defined. We say that a smooth manifold is **orientable** if it has a consistent smooth atlas. Evidently, the concept of an orientation of a smooth manifold and that of a consistent smooth atlas for a smooth manifold are closely related, almost to the extent of being indistinguishable.

Let $(M, \mathfrak{A}, \mathfrak{O})$ be an oriented smooth m-manifold, let $(U, (x^1, \ldots, x^m))$ be a chart in $\mathfrak{A}$, and let

$$\mathcal{X} = \left(\frac{\partial}{\partial x^1}, \frac{\partial}{\partial x^2}, \ldots, \frac{\partial}{\partial x^m} \right)$$

be the corresponding coordinate frame. Then

$$(U, (-x^1, \ldots, x^m))$$

is a chart on M, and the corresponding coordinate frame is

$$-\mathcal{X} = \left(-\frac{\partial}{\partial x^1}, \frac{\partial}{\partial x^2}, \ldots, \frac{\partial}{\partial x^m}\right).$$

It is easily shown using Theorem 14.3.5(b) that

$$-\mathfrak{A} = \{(U, (-x^1, \ldots, x^m)) : (U, (x^1, \ldots, x^m)) \in \mathfrak{A}\}$$

is a consistent smooth atlas for M. The orientation of M induced by $-\mathfrak{A}$ is

$$-\mathfrak{O} = \{-\mathfrak{O}(p) : p \in M\},$$

where

$$-\mathfrak{O}(p) = [-\mathcal{X}_p] = \left[\left(-\frac{\partial}{\partial x^1}\bigg|_p, \frac{\partial}{\partial x^2}\bigg|_p, \ldots, \frac{\partial}{\partial x^m}\bigg|_p\right)\right].$$

We say that the orientation $-\mathfrak{O}$ is the **opposite of** $\mathfrak{O}$.

A vector space always has precisely two orientations, but the situation is different for smooth manifolds. As we saw in Example 12.7.9, the Möbius band does not have a smooth unit normal vector field, and therefore, by Theorem 12.7.6, it fails to have an orientation. On the other hand, if a smooth manifold has an orientation, then it has at least two of them, namely, the given orientation and its opposite—but there could be more.

Defining the orientation of a smooth manifold in terms of an atlas ultimately rests on the way the bases for tangent spaces are oriented, which is geometrically appealing. However, this approach is not computationally convenient. We now present an algebraic alternative that eases this problem, but at the expense of greater abstraction.

Let M be a smooth m-manifold. We say that a form ϖ in $\Lambda^m(M)$ is an **orientation form (on M)** if it is nowhere-vanishing; or equivalently, if ϖ_p is a nonzero multicovector in $\Lambda^m(T_p(M))$ for all p in M; or equivalently, if ϖ_p is an orientation multicovector on $T_p(M)$ for all p in M. In the literature, such a form is often referred to as a **volume form**, but we reserve this terminology for a particular type of orientation form that makes it possible (in certain settings) to measure "volume". As we will soon see, there is no guarantee that a given smooth manifold has an orientation form.

Theorem 16.6.1. *Let M be a smooth manifold, and let ϖ, ϑ be orientation forms on M. Then:*
(a) *There is a uniquely determined nowhere-vanishing function f in $C^\infty(M)$ such that $\varpi = f\vartheta$.*
(b) *If M is connected as a topological space, then f is either strictly positive or strictly negative.*

Proof. (a): Let p be a point in M. By definition, ϖ_p and ϑ_p are covectors in $T_p^*(M)$, so we have from Theorem 7.2.12(d) that $\varpi_p = f(p)\vartheta_p$ for some real number $f(p)$. By definition, ϖ and ϑ are nowhere-vanishing, so $f(p) \neq 0$. The assignment $p \longmapsto f(p)$ defines a nowhere-vanishing function $f : M \longrightarrow \mathbb{R}$ such that $\varpi = f\vartheta$. Let $(U, (x^i))$ be a chart on M, and let $(d(x^1), \ldots, d(x^m))$ be the corresponding dual frame. It follows from Theorem 14.3.3(b) that $\varpi|_U$ and $\vartheta|_U$ are orientation forms on U, and then from Theorem 15.8.5 that $\varpi|_U = g\, d(x^1) \wedge \cdots \wedge d(x^m)$ and $\vartheta|_U = h\, d(x^1) \wedge \cdots \wedge d(x^m)$, where g and h are uniquely determined functions in $C^\infty(U)$. Since $\varpi|_U$ and $\vartheta|_U$ are nowhere-vanishing, so are g and h. It follows that $\varpi|_U = (g/h)\vartheta|_U$, hence $f|_U = g/h$ is a function in $C^\infty(U)$. Since p was arbitrary, f is in $C^\infty(M)$. The uniqueness of f follows from the preceding arguments.

(b): This follows from Theorem 9.1.20 and Theorem 10.1.1. $\square$

Let M be a smooth manifold, and let ϖ and ϑ be orientation forms on M. By Theorem 16.6.1(a), there is a uniquely determined nowhere-vanishing function f in $C^\infty(M)$ such that $\varpi = f\vartheta$. We say that ϖ and ϑ are **consistent (on M)**, and write $\varpi \sim \vartheta$, if f is strictly positive on M. It is easily shown that $\sim$ is an equivalence relation on the set of orientation forms on M. The equivalence class containing ϖ is denoted by $[\varpi]$. Since ϖ and $-\varpi$ are not consistent, $[\varpi]$ and $[-\varpi]$ are distinct equivalence classes. We emphasize again that there is no guarantee that a given smooth manifold has even one orientation form, so the preceding equivalence relation may be vacuous. On the other hand, if the equivalence relation does exist, there could be more than two equivalence classes.

Theorem 16.6.2. *Let M be a smooth m-manifold, let $(U, (x^i))$ and $(U, (\tilde{x}^i))$ be overlapping charts on M, let $\mathcal{X}$ and $\widetilde{\mathcal{X}}$ be the corresponding coordinate frames, and let $(d(x^1), \ldots, d(x^m))$ and $(d(\tilde{x}^1), \ldots, d(\tilde{x}^m))$ be the corresponding dual coordinate frames. Then $(U, (x^i))$ and $(U, (\tilde{x}^i))$ are consistent if and only if $d(x^1) \wedge \cdots \wedge d(x^m)$ and $d(\tilde{x}^1) \wedge \cdots \wedge d(\tilde{x}^m)$ are consistent on $U \cap \tilde{U}$.*

Proof. Let p be a point in $U \cap \tilde{U}$. Then $\mathcal{X}_p$ and $\widetilde{\mathcal{X}}_p$ are coordinate bases at p, and $(d_p(x^1), \ldots, d_p(x^m))$ and $(d_p(\tilde{x}^1), \ldots, d_p(\tilde{x}^m))$ are the corresponding dual coordinate bases. We have from Theorem 7.3.3(a) that

$$d_p(\tilde{x}^1) \wedge \cdots \wedge d_p(\tilde{x}^m) = f(p)\, d_p(x^1) \wedge \cdots \wedge d_p(x^m),$$

where

$$f(p) = \det\left(\left[\mathrm{id}_{T_p(M)}\right]_{\mathcal{X}_p}^{\widetilde{\mathcal{X}}_p}\right).$$

The assignment $p \longmapsto f(p)$ defines a function f in $C^\infty(U \cap \tilde{U})$ such that

$$d(\tilde{x}^1) \wedge \cdots \wedge d(\tilde{x}^m) = f\, d(x^1) \wedge \cdots \wedge d(x^m)$$

on $U \cap \tilde{U}$. By Theorem 2.5.3, f is nowhere-vanishing. The result follows. $\square$

Theorem 16.6.3. *Let M be a smooth manifold, and let ϖ be an orientation form on M. For each point p in M, let $\mathfrak{O}(p)$ be the orientation of $T_p(M)$ induced by the orientation multicovector ϖ_p (as given by Theorem 8.2.3). Then $\mathfrak{O} = \{\mathfrak{O}(p) : p \in M\}$ is an orientation of M, called the **orientation induced by ϖ**.*

Proof. We need to construct a consistent smooth atlas $\mathfrak{A}$ for M that has $\mathfrak{O}$ as its induced orientation. Let $\mathfrak{B}$ be a smooth atlas for M. Arguing as in the proof of Theorem 12.7.4, we assume without loss of generality that the coordinate domain of each chart in $\mathfrak{B}$ is a connected set in M. Let $\big(U, (x^1, x^2, \dots, x^m)\big)$ be a chart in $\mathfrak{B}$, and let $d(x^1) \wedge \cdots \wedge d(x^m)$ be the corresponding dual coordinate frame. Then $\varpi|_U$ and $d(x^1) \wedge \cdots \wedge d(x^m)$ are orientation forms on U. Viewing U as a smooth m-manifold, it follows from Theorem 16.6.1 that there is a uniquely determined function f in $C^\infty(U)$ such $\varpi|_U = f\, d(x^1) \wedge \cdots \wedge d(x^m)$, with f either strictly positive or strictly negative. In the former case, $\varpi|_U$ and $d(x^1) \wedge d(x^2) \wedge \cdots \wedge d(x^m)$ are consistent, and we include $\big(U, (x^1, x^2, \dots, x^m)\big)$ in $\mathfrak{A}$. In the latter case, $\varpi|_U$ and $d(-x^1) \wedge d(x^2) \wedge \cdots \wedge d(x^m)$ are consistent, and we include $\big(U, (-x^1, x^2, \dots, x^m)\big)$ in $\mathfrak{A}$. By construction, the dual coordinate frames corresponding to any pair of overlapping charts in $\mathfrak{A}$ are consistent with a restriction of ϖ, hence are consistent with each other. It follows from Theorem 16.6.2 that any pair of overlapping charts in $\mathfrak{A}$ is consistent. Thus, $\mathfrak{A}$ is a consistent smooth atlas for M.

Let p be a point in M, let $\big(U, (x^1, \dots, x^m)\big)$ be a chart in $\mathfrak{A}$ at p, and let $\mathcal{X} = (\partial/\partial x^1, \dots, \partial/\partial x^m)$ and $\big(d(x^1), \dots, d(x^m)\big)$ be the corresponding coordinate and dual coordinate frames. From the above construction, there is a uniquely determined strictly positive function f in $C^\infty(U)$ such that $\varpi|_U = f\, d(x^1) \wedge \cdots \wedge d(x^m)$. Then Theorem 7.2.9 gives

$$\varpi_p\left(\left.\frac{\partial}{\partial x^1}\right|_p, \dots, \left.\frac{\partial}{\partial x^m}\right|_p\right) = f(p)\, d_p(x^1) \wedge \cdots \wedge d_p(x^m)\left(\left.\frac{\partial}{\partial x^1}\right|_p, \dots, \left.\frac{\partial}{\partial x^m}\right|_p\right)$$
$$= f(p) > 0.$$

Thus, $\mathcal{X}_p$ is in $\mathfrak{O}(p)$, or equivalently, $\mathfrak{O}(p) = [\mathcal{X}_p]$. Since p and $\big(U, (x^1, \dots, x^m)\big)$ were arbitrary, $\mathfrak{O}$ is the orientation induced by $\mathfrak{A}$. $\qquad\square$

Theorem 16.6.4. *If $(M, \mathfrak{O})$ is an oriented smooth manifold, then there is an orientation form on M that induces $\mathfrak{O}$.*

Proof. Let $\mathfrak{A} = \big\{(U_\alpha, (x_\alpha^i)) : \alpha \in A\big\}$ be a consistent smooth atlas for M, and let $\mathcal{X}_\alpha = (\partial/\partial x_\alpha^1, \dots, \partial/\partial x_\alpha^m)$ and $\big(d(x_\alpha^1), \dots, d(x_\alpha^m)\big)$ be the coordinate and dual coordinate frames corresponding to $\big(U_\alpha, (x_\alpha^i)\big)$. Let $\varpi_\alpha = d(x_\alpha^1) \wedge \cdots \wedge d(x_\alpha^m)$, and let $\{\pi_\alpha : \alpha \in A\}$ be a partition of unity subordinate to the open cover $\{U_\alpha : \alpha \in A\}$ of M. It can be shown that there is a smooth extension of $\pi_\alpha \varpi_\alpha$ to all of M such that the extension, also denoted by $\pi_\alpha \varpi_\alpha$, has constant value 0 on $M \backslash \mathrm{supp}(\pi_\alpha)$. It can also be shown that $\Omega = \sum_\alpha \pi_\alpha \varpi_\alpha$ is a form in $\Lambda^m(M)$.

We claim that Ω is an orientation form that induces $\mathfrak{O}$. For a given point p in M, let $\big(U_\beta, (x_\beta^i)\big)$ be a chart in $\mathfrak{A}$ at p, and consider

$$\Omega_p\left(\left.\frac{\partial}{\partial x_\beta^1}\right|_p, \dots, \left.\frac{\partial}{\partial x_\beta^m}\right|_p\right) = \sum_\alpha \pi_\alpha(p)\, \varpi_\alpha|_p\left(\left.\frac{\partial}{\partial x_\beta^1}\right|_p, \dots, \left.\frac{\partial}{\partial x_\beta^m}\right|_p\right).$$

From the properties of partitions of unity, $\pi_\alpha(p) \geq 0$ for all α, and $\pi_\alpha(p) > 0$ for at least one α. Because $\mathcal{X}_\beta|_p$ is consistent with $\mathcal{X}_\alpha|_p$ for all α, by Theorem

8.2.2,

$$\varpi_\alpha|_p\left(\frac{\partial}{\partial x_\beta^1}\bigg|_p,\ldots,\frac{\partial}{\partial x_\beta^m}\bigg|_p\right) > 0$$

for all α, hence

$$\Omega_p\left(\frac{\partial}{\partial x_\beta^1}\bigg|_p,\ldots,\frac{\partial}{\partial x_\beta^m}\bigg|_p\right) > 0.$$

Since p was arbitrary, we have from the preceding inequality that Ω is nowhere-vanishing and is therefore an orientation form. Since $\left(U_\beta,(x_\beta^i)\right)$ was arbitrary, it also follows from the preceding inequality and Theorem 16.6.3 that Ω induces $\mathfrak{O}$. $\square$

Theorem 16.6.5 (Orientability Criterion for Smooth Manifolds). *A smooth manifold is orientable if and only if it has an orientation form.*

Proof. ($\Rightarrow$): This follows from Theorem 16.6.4.
        ($\Leftarrow$): This follows from Theorem 16.6.3. $\square$

Theorem 16.6.6. *Let* $(r^1,\ldots,r^m)$ *be standard coordinates on* $\mathbb{R}^m$, *and let* $\left(d(r^1),\ldots,d(r^m)\right)$ *be the corresponding standard dual coordinate frame. Then* $d(r^1)\wedge\cdots\wedge d(r^m)$ *induces the standard orientation of* $\mathbb{R}^m$.

Proof. According to Example 15.1.6, we have the identification $(\partial/\partial r^1,\ldots,\partial/\partial r^m) = (e_1,\ldots,e_m)$, where $(e_1,\ldots,e_m)$ is the standard basis for $\mathbb{R}^m$. By Theorem 15.8.4, $d(r^1)\wedge\cdots\wedge d(r^m)(e_1,\ldots,e_m) = 1$. Thus, $d(r^1)\wedge\cdots\wedge d(r^m)$ is an orientation form on $\mathbb{R}^m$. By Theorem 16.6.3, $[(e_1,\ldots,e_m)]$ is in the orientation of $\mathbb{R}^m$ induced by $d(r^1)\wedge\cdots\wedge d(r^m)$. $\square$

Theorem 16.6.7. *Let M be an orientable smooth manifold, let $\{[\varpi]\}$ be the set of equivalence classes of orientation forms on M, and let $\{\mathfrak{O}\}$ be the set of orientations of M. Then the map $\iota : \{[\varpi]\} \longrightarrow \{\mathfrak{O}\}$ defined by assigning $[\varpi]$ to the orientation induced by ϖ is bijective.*

Remark. It is easily shown that all orientation forms in the same equivalence class induce the same orientation, so ι is well-defined.

Proof. By Theorem 16.6.4, ι is surjective. Let ϖ and ϑ be orientation forms on M that induce the same orientation $\mathfrak{O}$. By Theorem 16.6.1(a), there is a uniquely determined nowhere-vanishing function f in $C^\infty(M)$ such that $\varpi = f\vartheta$. Let p be a point in M, and let $(v_1,\ldots,v_m)$ be a basis that is positively oriented with respect to $\mathfrak{O}(p)$. Then $\varpi_p(v_1,\ldots,v_m), \vartheta_p(v_1,\ldots,v_m) > 0$, hence

$$f(p) = \frac{\varpi_p(v_1,\ldots,v_m)}{\vartheta_p(v_1,\ldots,v_m)} > 0.$$

Since p was arbitrary, f is strictly positive, so ϖ and ϑ are consistent, or equivalently, $[\varpi] = [\vartheta]$. Thus, ι is injective. $\square$

It was remarked in the introduction to this section that if a smooth manifold has an orientation, then it has at least two of them. The next result gives a condition under which there are two orientations, and no more.

Theorem 16.6.8. *A connected orientable smooth manifold has precisely two orientations.*

Proof. Let M be a connected orientable smooth manifold. By Theorem 16.6.5, M has an orientation form ϖ. Each orientation form in $[\varpi]$ induces the same orientation, and each orientation form in $[-\varpi]$ induces its opposite. Let ϑ be another orientation form on M. We have from Theorem 16.6.1(a) that there is a uniquely determined nowhere-vanishing function f in $C^\infty(M)$ such that $\vartheta = f\varpi$, and then from Theorem 16.6.1(b) that ϑ is either in $[\varpi]$ or $[-\varpi]$. $\square$

It follows from Theorem 14.1.3 that a smooth manifold has countably many connected components, each of which is an open set in the smooth manifold. Let M be a smooth manifold that has a finite number of connected components $C_1, \ldots, C_k$. Suppose C_i, viewed as an open submanifold of M, is orientable for $i = 1, \ldots, k$. Then C_i can be given an orientation $\mathfrak{O}_i$ independently of the other connected components. Taken together, the $\mathfrak{O}_i$ give an orientation of M, which we denote by $(\mathfrak{O}_1, \ldots, \mathfrak{O}_k)$. It follows from Theorem 16.6.8 that by taking all combinations of signs in $(\pm\mathfrak{O}_1, \ldots, \pm\mathfrak{O}_k)$ we obtain the 2^k possible orientations of M.

Let $(\overline{M}, \overline{\mathfrak{D}})$ be an oriented smooth $\overline{m}$-manifold, where $\overline{m} \geq 2$, and let M be a submanifold of $\overline{M}$. In general, there is no mechanism for constructing an orientation of M from $\overline{\mathfrak{D}}$. However, when M is a hypersurface of $\overline{M}$, this may be possible, as the next result shows.

Theorem 16.6.9. *Let $(\overline{M}, \overline{\mathfrak{D}})$ be an oriented smooth $\overline{m}$-manifold, where $\overline{m} \geq 2$, let M be a hypersurface of $\overline{M}$, and suppose there is a nowhere-tangent vector field V in $\mathfrak{X}_{\overline{M}}(M)$. Then:*

(a) *There is a unique orientation $\mathfrak{O}_M$ of M, called the **orientation induced by** V, such that for all points p in M, $(v_1, \ldots, v_{\overline{m}-1})$ is a basis for $T_p(M)$ that is positively oriented with respect to $\mathfrak{O}_M(p)$ if and only if $(V_p, v_1, \ldots, v_{\overline{m}-1})$ is a basis for $T_p(\overline{M})$ that is positively oriented with respect to $\overline{\mathfrak{D}}(p)$.*

(b) *If ϖ is an orientation form on $\overline{M}$ that induces $\overline{\mathfrak{D}}$, then $i_V(\varpi)|_M$ is an orientation form on M that induces $\mathfrak{O}_M$.*

(c) *$\mathfrak{O}_M$ is independent of the choice of orientation form on $\overline{M}$ that induces $\overline{\mathfrak{D}}$.*

Proof. This is the manifold version of Theorem 8.2.5, but we provide a detailed proof.

(a), (b): Let ϖ be an orientation form on $\overline{M}$ that induces $\overline{\mathfrak{D}}$. According to Theorem 16.6.3, by definition, ϖ_q is an orientation multicovector on $T_q(\overline{M})$ that induces $\overline{\mathfrak{D}}(q)$ for all q in $\overline{M}$. It follows from Theorem 8.2.5(b) that $\left(i_V(\varpi)|_M\right)\big|_p = i_{V_p}(\varpi_p)|_{T_p(M)}$ is an orientation multicovector on $T_p(M)$ for all p in M. Thus, $i_V(\varpi)|_M$ is a nowhere-vanishing form on M. For vector fields $X_1, \ldots, X_{m-1}$ in $\mathfrak{X}(M)$, we have

$$i_V(\varpi)|_M(X_1, \ldots, X_{m-1}) = \varpi(V, X_1, \ldots, X_{m-1}).$$

Since ϖ is smooth and $X_1, \ldots, X_{m-1}$ were arbitrary, by definition, $i_V(\varpi)|_M$ is smooth. This shows that $i_V(\varpi)|_M$ is an orientation form on M. It follows from Theorem 8.2.5(a) and Theorem 16.6.3 that the orientation induced by $i_V(\varpi)|_M$ is the one described in the statement of part (a).

(c): This follows from part (a).   □

A couple of remarks on Theorem 16.6.9 are in order. First, although $\mathfrak{O}_M$ is independent of the choice of orientation form ϖ on $\overline{M}$ that induces $\overline{\mathfrak{O}}$, the same cannot be said for the choice of nowhere-tangent vector field V in $\mathfrak{X}_{\overline{M}}(M)$. In particular, the orientation induced by $-V$ is $-\mathfrak{O}_M$. Second, without further assumptions, there is no guarantee that a nowhere-tangent vector field in $\mathfrak{X}_{\overline{M}}(M)$ exists in the first place, an issue we return to in Theorem 19.11.1.

Let $(M, \mathfrak{O})$ and $(\widetilde{M}, \widetilde{\mathfrak{O}})$ be oriented smooth m-manifolds, and let $F : M \longrightarrow \widetilde{M}$ be a diffeomorphism. We say that F is **orientation-preserving** if $d_p(F) : T_p(M) \longrightarrow T_{F(p)}(\widetilde{M})$ is orientation-preserving as a linear map for all p in M.

Theorem 16.6.10. *Let U and V be open sets in $\mathbb{R}^m$, and let $F : U \longrightarrow V$ be a diffeomorphism. Then F is orientation-preserving if and only if $\det(J_F(p)) > 0$ for all p in U, where $J_F(p)$ is the Jacobian matrix of F at p.*

Proof. The $\mathcal{E}$ be the standard basis for $\mathbb{R}^m$. From (10.1.3), we have $[d_p(F)]_{\mathcal{E}}^{\mathcal{E}} = J_F(p)$, hence $\det(d_p(F)) = \det(J_F(p))$. The result now follows from Theorem 8.2.8.   □

16.7   Integration of Differential Forms

In this section, we introduce the theory of integration on smooth manifolds. It may come as a surprise to learn that in this setting the objects to be integrated are not functions but rather differential forms. In order for integration of functions to be possible, we need a type of smooth manifold that has structure beyond what has been considered up to now. This topic is covered in Section 19.10.

Let M be a smooth m-manifold, and let ω be a form in $\Lambda^m(M)$. The **support of ω** is denoted by $\mathrm{supp}(\omega)$ and defined to be the closure in M of the set of points at which ω is nonvanishing:

$$\mathrm{supp}(\omega) = \mathrm{cl}_M(\{p \in M : \omega_p \neq 0\}).$$

We say that ω has **compact support** if $\mathrm{supp}(\omega)$ is compact in M. Let U be an open set in M. It is said that ω has **support in U** if $\mathrm{supp}(\omega) \subseteq U$, and that ω has **compact support in U** if $\mathrm{supp}(\omega) \subseteq U$ and $\mathrm{supp}(\omega)$ is compact in M.

We begin by defining the integral of differential forms in the setting of Euclidean spaces and then generalize to smooth manifolds.

Let V be an open set in $\mathbb{R}^m$, and let ω be a form in $\Lambda^m(V)$ that has compact support. By Theorem 15.8.5, ω can be expressed as

$$\omega = f\, d(r^1) \wedge \cdots \wedge d(r^m), \tag{16.7.1}$$

where f is a uniquely determined function in $C^\infty(V)$. Since f is smooth, by Theorem 10.1.1, it is continuous, and from

$$\mathrm{supp}(\omega) = \mathrm{cl}_M\left(\{p \in M : \omega_p \neq 0\}\right) = \mathrm{cl}_M\left(\{p \in M : f(p) \neq 0\}\right)$$
$$= \mathrm{supp}(f),$$

it follows that f has compact support. We have from (10.5.3) and the associated discussion that the integral of f over V exists and is denoted by $\int_V f \, dr^1 \cdots dr^m$. The **integral of ω over V** is denoted by $\int_V \omega$ and defined by

$$\int_V \omega = \int_V f \, dr^1 \cdots dr^m. \tag{16.7.2}$$

We can express the preceding identity more suggestively as

$$\int_V f \, d(r^1) \wedge \cdots \wedge d(r^m) = \int_V f \, dr^1 \cdots dr^m.$$

The next result shows that the integral is independent of the way the form is parametrized.

Theorem 16.7.1 (Diffeomorphism Invariance of Integral). *Let U and V be open sets in $\mathbb{R}^m$, let $F : U \longrightarrow V$ be an orientation-preserving or orientation-reversing diffeomorphism, and let ω be a form in $\Lambda^m(V)$ that has compact support. Then*

$$\int_V \omega = \pm \int_U F^*(\omega),$$

where the positive (negative) sign is chosen if F is orientation-preserving (orientation-reversing).

Proof. Let $(r^1, \ldots, r^m)$ and $(s^1, \ldots, s^m)$ be standard coordinates on U and V, respectively, and let $\omega = f \, d(s^1) \wedge \cdots \wedge d(s^m)$, where f is a uniquely determined function in $C^\infty(V)$. If F is orientation-preserving, then

$$\int_V \omega = \int_V f \, ds^1 \cdots ds^m \qquad [(16.7.2)]$$

$$= \int_U (f \circ F) \, |\det(J_F)| \, dr^1 \cdots dr^m \qquad [\text{Th 10.5.3}]$$

$$= \int_U (f \circ F) \, \det(J_F) \, dr^1 \cdots dr^m \qquad [\text{Th 16.6.10}]$$

$$= \int_U (f \circ F) \, \det(J_F) \, d(r^1) \wedge \cdots \wedge d(r^m) \qquad [(16.7.2)]$$

$$= \int_U F^*(\omega). \qquad [\text{Th 15.13.3(b)}]$$

The proof when F is orientation-reversing is the same, except that a negative sign is introduced when the absolute value bars are removed. $\qquad\square$

We now generalize to smooth manifolds. Let $(M, \mathfrak{A}, \mathfrak{O})$ be an oriented smooth m-manifold, and let ω be a form in $\Lambda^m(M)$ that has compact support. We first consider the case where ω has compact support in the coordinate domain of a single chart $(U, \varphi = (x^i))$ in $\mathfrak{A}$. Viewing $\varphi(U)$ as a smooth manifold, it can be shown that the form $(\varphi^{-1})^*(\omega)$ in $\Lambda^m(\varphi(U))$ has compact support in the open set $\varphi(U)$ in $\mathbb{R}^m$. In keeping with (16.7.2), the **integral of ω over U** is denoted by $\int_U \omega$ and defined by

$$\int_U \omega = \int_{\varphi(U)} (\varphi^{-1})^*(\omega). \qquad (16.7.3)$$

It can be shown that $\int_U \omega$ does not depend on the choice of chart with coordinate domain containing $\mathrm{supp}(\omega)$.

Let us look more closely at the right-hand side of (16.7.3). By Theorem 15.8.5, ω can be expressed in local coordinates as

$$\omega = f \, d(x^1) \wedge \cdots \wedge d(x^m),$$

where f is a uniquely determined function in $C^\infty(U)$. It follows from Theorem 14.8.5(a) that $\varphi^{-1} : \varphi(U) \longrightarrow U$ is smooth. Since

$$r^i \circ \varphi = r^i \circ (x^1, \ldots, x^m) = x^i,$$

hence $r^i = x^i \circ \varphi^{-1}$, we have from (15.13.3) that the corresponding Jacobian matrix is

$$J_{\varphi^{-1}} = \left[\frac{\partial(x^i \circ \varphi^{-1})}{\partial r^j} \right] = I_m.$$

Then Theorem 15.13.3(b) gives

$$(\varphi^{-1})^*(\omega) = (f \circ \varphi^{-1}) \, d(r^1) \wedge \cdots \wedge d(r^m).$$

Using (10.5.3), (16.7.1), and (16.7.2), the right-hand side of (16.7.3) can now be expressed as

$$\int_{\varphi(U)} (\varphi^{-1})^*(\omega) = \int_D f \circ \varphi^{-1} \, dr^1 \cdots dr^m,$$

where D is any compact domain of integration in $\mathbb{R}^m$ such that $\mathrm{supp}((\varphi^{-1})^*(\omega)) \subseteq D \subset \varphi(U)$.

Using a partition of unity argument, we now address the case where ω does not have compact support in the coordinate domain of a single chart in $\mathfrak{A}$. It is clear that the union of the coordinate domains of all charts in $\mathfrak{A}$ comprises an open cover of $\mathrm{supp}(\omega)$. Since ω has compact support, $\mathrm{supp}(\omega)$ has a finite open cover $\mathcal{U} = \{U_i : i = 1, \ldots, k\}$ consisting of the coordinate domains of a finite number of charts in $\mathfrak{A}$. Let $\{\pi_i : i = 1, \ldots, k\}$ be a partition of unity subordinate to $\mathcal{U}$. It can be shown that the form $\pi_i \omega$ in $\Lambda^m(M)$ has compact support in U_i. Thus, definition (16.7.3) applies and each integral $\int_{U_i} \pi_i \omega$ exists. The **integral of ω over M** is denoted by $\int_M \omega$ and defined by

$$\int_M \omega = \sum_{i=1}^{k} \int_{U_i} \pi_i \omega.$$

It can be shown that $\int_M \omega$ does not depend on the choice of open cover or the choice of partition of unity.

Integration of differential forms has the properties desired of an integral.

Theorem 16.7.2. *Let $(M, \mathfrak{O})$ be an oriented smooth m-manifold, let ω, ξ be forms in $\Lambda^m(M)$ that have compact support, and let c be a real number. Then:*
(a)

$$\int_M (c\omega + \xi) = c \int_M \omega + \int_M \xi.$$

(b)

$$\int_{-M} \omega = -\int_M \omega,$$

where $-M$ denotes $(M, -\mathfrak{O})$.   □

16.8   Line Integrals

Due to its endpoints, a closed interval $[a, b]$ in $\mathbb{R}$ is not a smooth 1-manifold. However, as discussed in Section 17.1, $[a, b]$ is what we will later refer to as a smooth 1-manifold with boundary. For present purposes, it is convenient to ignore the distinction and treat $[a, b]$ as a smooth 1-manifold.

Let M be a smooth manifold, and let $\lambda(t) : [a, b] \longrightarrow M$ be a (not necessarily smooth) curve. We say that λ is **piecewise smooth** if there is a finite series of real numbers $a = a_1 < a_2 < \cdots < a_{k+1} = b$ such that the restriction of λ to $[a_i, a_{i+1}]$ is a smooth curve for $i = 1, \ldots, k$. Suppose λ is in fact piecewise smooth, and let ω be a covector field in $\mathfrak{X}^*(M)$. Motivated by (16.7.3), the **line integral of ω over λ** is denoted by $\int_\lambda \omega$ and defined by

$$\int_\lambda \omega = \sum_{i=1}^k \int_{[a_i, a_{i+1}]} \lambda^*(\omega), \tag{16.8.1}$$

where the ith integrand on the right-hand side is interpreted as the restriction of $\lambda^*(\omega)$ to $[a_i, a_{i+1}]$. When λ is smooth, (16.8.1) simplifies to

$$\int_\lambda \omega = \int_{[a,b]} \lambda^*(\omega). \tag{16.8.2}$$

Clearly, $[a_i, a_{i+1}]$ is a domain of integration in $\mathbb{R}$, and by Theorem 9.4.2, it is compact for $i = 1, \ldots, k$. It follows from Theorem 15.11.8, (16.7.2), and (16.8.1) that

$$\int_\lambda \omega = \sum_{i=1}^k \int_{a_i}^{a_{i+1}} \omega_{\lambda(t)} \left(\frac{d\lambda}{dt}(t) \right) dt. \tag{16.8.3}$$

When λ is smooth, (16.8.3) simplifies to

$$\int_\lambda \omega = \int_a^b \omega_{\lambda(t)} \left(\frac{d\lambda}{dt}(t) \right) dt. \tag{16.8.4}$$

Let $g(u) : [c, d] \longrightarrow [a, b]$ be a diffeomorphism, and let $c_i = g^{-1}(a_i)$ for $i = 1, \ldots, k + 1$. According to Theorem 10.2.4(c), g is either strictly increasing or strictly decreasing, so we have either $c = c_1 < c_2 < \cdots < c_{k+1} = d$ or $c = c_{k+1} < c_k < \cdots < c_1 = d$. From (16.8.1), the line integral of ω over $\lambda \circ g$ is

$$\int_{\lambda \circ g} \omega = \sum_{i=1}^{k} \int_{g^{-1}([a_i, a_{i+1}])} (\lambda \circ g)^*(\omega), \qquad (16.8.5)$$

where

$$g^{-1}([a_i, a_{i+1}]) = \begin{cases} [c_i, c_{i+1}] & \text{if } g \text{ is strictly increasing} \\ [c_{k+2-i}, c_{k+1-i}] & \text{if } g \text{ is strictly decreasing} \end{cases}$$

for $i = 1, \ldots, k$. When λ is smooth, (16.8.5) simplifies to

$$\int_{\lambda \circ g} \omega = \int_{[c,d]} (\lambda \circ g)^*(\omega). \qquad (16.8.6)$$

The next result is the line integral counterpart of Theorem 16.7.1.

Theorem 16.8.1 (Diffeomorphism Invariance of Line Integral). *Let M be a smooth manifold, let ω be a covector field in $\mathfrak{X}^*(M)$, let $\lambda(t) : [a, b] \longrightarrow M$ be a piecewise smooth curve, and let $g(u) : [c, d] \longrightarrow [a, b]$ be a diffeomorphism. Then*

$$\int_{\lambda \circ g} \omega = \pm \int_{\lambda} \omega,$$

where the positive (negative) sign is chosen if g is strictly increasing (strictly decreasing).

Proof. We consider only the case where λ is smooth. Then

$$\begin{aligned} \int_{\lambda} \omega &= \int_{[a,b]} \lambda^*(\omega) && [(16.8.2)] \\ &= \pm \int_{[c,d]} g^*(\lambda^*(\omega)) && [\text{Th } 16.7.1] \\ &= \pm \int_{[c,d]} (\lambda \circ g)^*(\omega) && [\text{Th } 15.9.2, (15.12.3)] \\ &= \pm \int_{\lambda \circ g} \omega. && [(16.8.6)] \qquad \square \end{aligned}$$

Theorem 16.8.1 says that, up to a sign determined by the "direction" of the reparametrizing diffeomorphism, the line integral of ω is independent of the way λ is parametrized.

The next result can be viewed as a generalization of the fundamental theorem of calculus.

Theorem 16.8.2 (Fundamental Theorem of Line Integrals). *Let M be a smooth manifold, let f be a function in $C^\infty(M)$, and let $\lambda(t) : [a, b] \longrightarrow M$ be a piecewise smooth curve. Then*

$$\int_\lambda d(f) = f\big(\lambda(b)\big) - f\big(\lambda(a)\big).$$

Proof. We consider only the case where λ is smooth. Then

$$\int_\lambda d(f) = \int_a^b d(f)_{\lambda(t)}\left(\frac{d\lambda}{dt}(t)\right) dt \qquad [(16.8.4)]$$

$$= \int_a^b d_{\lambda(t)}(f)\left(\frac{d\lambda}{dt}(t)\right) dt \qquad [(15.4.4)]$$

$$= \int_a^b \frac{d(f \circ \lambda)}{dt}(t)\, dt \qquad [\text{Th } 14.7.4(\text{b})]$$

$$= f\big(\lambda(b)\big) - f\big(\lambda(a)\big),$$

where the last equality follows from the fundamental theorem of calculus. $\qquad \square$

16.9   Closed and Exact Covector Fields

Let M be a smooth manifold, and let ω be a covector field in $\mathfrak{X}^*(M)$. We say that ω is **closed** if $d(\omega) = 0$, **exact** if there is a function f in $C^\infty(M)$ such that $\omega = d(f)$, **conservative** if $\int_\lambda \omega = 0$ for all piecewise smooth curves $\lambda(t) : [a, b] \longrightarrow M$ such that $\lambda(a) = \lambda(b)$, and **path-independent** if $\int_\mu \omega = \int_\psi \omega$ for all piecewise smooth curves $\mu(t) : [a, b] \longrightarrow M$ and $\psi(u) : [c, d] \longrightarrow M$ such that $\mu(a) = \psi(c)$ and $\mu(b) = \psi(d)$. In this section, we explore the various ways in which these properties are related.

Theorem 16.9.1. *Let M be a smooth m-manifold, let ω be a covector field in $\mathfrak{X}^*(M)$, and, in local coordinates, let*

$$\omega = \sum_i \alpha^i d(x^i).$$

Then ω is closed if and only if for every such local coordinate expression,

$$\frac{\partial \alpha^i}{\partial x^j} = \frac{\partial \alpha^j}{\partial x^i}$$

for $i, j = 1, \ldots, m$.

Proof. Computing as in Example 16.1.6 gives

$$d(\omega) = \sum_{1 \le i < j \le m} \left(\frac{\partial \alpha^j}{\partial x^i} - \frac{\partial \alpha^i}{\partial x^j}\right) d(x^i) \wedge d(x^j),$$

from which the result follows. $\qquad \square$

Theorem 16.9.2. *If M is a smooth manifold and ω is a covector field in $\mathfrak{X}^*(M)$, then the following are equivalent:*
(a) *ω is exact.*
(b) *ω is conservative.*
(c) *ω is path-independent.*

Proof. (a) $\Rightarrow$ (b): Let f be a function in $C^\infty(M)$ such that $\omega = d(f)$, and let $\lambda(t) : [a, b] \longrightarrow M$ be a piecewise smooth curve such that $\lambda(a) = \lambda(b)$. By Theorem 16.8.2,

$$\int_\lambda \omega = \int_\lambda d(f) = f\big(\lambda(b)\big) - f\big(\lambda(a)\big) = 0.$$

(b) $\Rightarrow$ (c): Let $\mu(t) : [a, b] \longrightarrow M$ and $\psi(u) : [c, d] \longrightarrow M$ be smooth curves such that $\mu(a) = \psi(c)$ and $\mu(b) = \psi(d)$. Without loss of generality, we assume that $[a, b] = [-1, 0]$ and $[c, d] = [0, 1]$, so that $\mu(-1) = \psi(0)$ and $\mu(0) = \psi(1)$. Define the smooth curve $g(s) : [0, 1] \longrightarrow [0, 1]$ by $g(s) = 1 - s$, and the piecewise smooth curve $\lambda(s) : [-1, 1] \longrightarrow M$ by

$$\lambda(s) = \begin{cases} \mu(s) & \text{if } s \in [-1, 0] \\ \psi \circ g(s) & \text{if } s \in [0, 1]. \end{cases}$$

The image of λ is a path starting and ending at $\mu(-1) = \psi(0)$, and passing through $\mu(0) = \psi(1)$. We have

$$0 = \int_\lambda \omega = \int_\mu \omega + \int_{\psi \circ g} \omega = \int_\mu \omega - \int_\psi \omega,$$

where the last equality comes from Theorem 16.8.1. The result follows.

(c) $\Rightarrow$ (a): Only a sketch of the proof is provided. The construction to follow can be carried out on each connected component of M, so we assume without loss of generality that M is connected. Let p_0 be a given point in M. Since M is connected, it can be shown that for an arbitrary point p in M, there is a piecewise smooth curve $\mu(t) : [a, b] \longrightarrow M$ such that $\mu(a) = p_0$ and $\mu(b) = p$. Let

$$f(p) = \int_\mu \omega.$$

By assumption, line integrals of ω are path-independent, so the real number $f(p)$ does not depend on the choice of curve joining p_0 and p. Thus, the assignment $p \longmapsto f(p)$ defines a function $f : M \longrightarrow \mathbb{R}$. It can be shown that f is smooth.

We claim that $d(f) = \omega$. Let $\lambda_1(t) : [a, b] \longrightarrow M$ and $\lambda_2(t) : [b, c] \longrightarrow M$ be piecewise smooth curves such that $\lambda_1(a) = p_0$ and $\lambda_1(b) = \lambda_2(b)$. Define a piecewise smooth curve $\lambda(t) : [a, c] \longrightarrow M$ by

$$\lambda(t) = \begin{cases} \lambda_1(t) & \text{if } t \in [a, b] \\ \lambda_2(t) & \text{if } t \in [b, c], \end{cases}$$

so that $\lambda(a) = p_0$ and $\lambda(b) = \lambda_2(b)$. Then

$$f\big(\lambda_2(c)\big) = \int_\lambda \omega = \int_{\lambda_1} \omega + \int_{\lambda_2} \omega = f\big(\lambda_1(b)\big) + \int_{\lambda_2} \omega$$
$$= f\big(\lambda_2(b)\big) + \int_{\lambda_2} \omega,$$

hence

$$\int_{\lambda_2} \omega = f\big(\lambda_2(c)\big) - f\big(\lambda_2(b)\big) = \int_{\lambda_2} d(f),$$

where the last equality follows from Theorem 16.8.2. Thus,

$$\int_{\lambda_2} [\omega - d(f)] = 0.$$

It can be shown that since λ_2 was arbitrary, $\omega - d(f)$ must be the zero form; that is, $\omega = d(f)$. $\qquad\square$

Theorem 16.9.3. *Let M be a smooth manifold, and let ω be a covector field in $\mathfrak{X}^*(M)$. If ω is exact, then it is closed.*

Proof. This follows from Theorem 16.1.1(c). $\qquad\square$

In light of Theorem 16.9.3, there are three possibilities for a covector field with respect to being exact or not, and closed or not: (i) exact (hence closed), (ii) closed and not exact, and (iii) not closed (hence not exact). The following examples illustrate each of these cases.

Example 16.9.4. Let ω be the covector field in $\mathfrak{X}^*(\mathbb{R}^2)$ given by

$$\omega_{(r,s)} = s\,d(r) + r\,d(s).$$

Since $\omega = d(rs)$, it is exact. $\qquad\diamond$

Example 16.9.5 (Punctured Plane). Let ϕ be the standard coordinate on $[0, 2\pi]$, and let (r, s) be standard coordinates on $\mathbb{R}^2\backslash\{(0,0)\}$, the punctured plane. Let $\lambda(\phi) : [0, 2\pi] \longrightarrow \mathbb{R}^2\backslash\{(0,0)\}$ be the smooth curve given by $\lambda(\phi) = \big(\cos(\phi), \sin(\phi)\big)$, and consider the form

$$\omega_{(r,s)} = -\frac{s}{r^2 + s^2}\,d(r) + \frac{r}{r^2 + s^2}\,d(s)$$

in $\Lambda^2(\mathbb{R}^2\backslash\{(0,0)\})$. Using Theorem 16.9.1, it is easily shown that ω is closed. Setting $(U, (x^i)) = ([0, 2\pi], (\phi))$ and $(V, (y^j)) = (\mathbb{R}^2, (r, s))$ in a generalization of Theorem 15.11.3 yields

$$\lambda^*(\omega)(\phi)$$
$$= \left[\left(-\frac{s}{r^2 + s^2} \circ \lambda\right)\frac{\partial(r \circ \lambda)}{\partial \phi} + \left(\frac{r}{r^2 + s^2} \circ \lambda\right)\frac{\partial(s \circ \lambda)}{\partial \phi}\right]d(\phi)$$
$$= \left(-\frac{\sin(\phi)}{\cos^2(\phi) + \sin^2(\phi)}\frac{\partial(\cos(\phi))}{\partial \phi} + \frac{\cos(\phi)}{\cos^2(\phi) + \sin^2(\phi)}\frac{\partial(\sin(\phi))}{\partial \phi}\right)d(\phi)$$
$$= d(\phi).$$

Then (16.8.2) gives

$$\int_\lambda \omega = \int_{[0,2\pi]} \lambda^*(\omega) = \int_0^{2\pi} d\phi = 2\pi.$$

Thus, ω is not conservative. By Theorem 16.9.2, it is not exact. ◇

Example 16.9.6. Let ω be the covector field in $\mathfrak{X}^*(\mathbb{R}^2)$ given by

$$\omega_{(r,s)} = rs\,d(r) + rs\,d(s).$$

It follows from Theorem 16.9.1 that ω is not closed. ◇

Let M be a smooth manifold, and let ω be a form in $\Lambda^s(M)$. Generalizing earlier definitions, we say that ω is **closed** if $d(\omega) = 0$, and **exact** if there is a form ξ in $\Lambda^{s-1}(M)$ such that $\omega = d(\xi)$.

Theorem 16.9.3 has an immediate generalization.

Theorem 16.9.7. *Let M be a smooth manifold, and let ω be a form in $\Lambda^s(U)$. If ω is exact, then it is closed.*

Proof. This follows from Theorem 16.1.1(c). ☐

Recall that a subset S of $\mathbb{R}^m$ is said to be star-shaped about a point p_0 in S if $tp + (1-t)p_0$ is in S for all p in S and all t in $[0,1]$; that is, for all p in S, the line segment joining p_0 to p is contained in S.

Theorem 16.9.8 (Poincaré's Lemma). *Let U be a star-shaped open set in $\mathbb{R}^m$, and let ω be a form in $\Lambda^s(U)$. If ω is closed, then it is exact.*

Proof. We seek a form ξ in $\Lambda^{s-1}(U)$ such that $d(\xi) = \omega$. Assuming without loss of generality that U is star-shaped about $(0,\ldots,0)$ in $\mathbb{R}^m$, let us view U as a smooth m-manifold with the covering chart $\big(U, \mathrm{id}_U = (r^i)\big)$ and corresponding coordinate and dual coordinate frames $(\partial/\partial r^1, \ldots, \partial/\partial r^m)$ and $\big(d(r^1), \ldots, d(r^m)\big)$. Any form in $\Lambda^s(U)$ can be expressed uniquely as the sum of forms of the type $f\,d(r^{i_1}) \wedge \cdots \wedge d(r^{i_s})$, where f is a function in $C^\infty(U)$ and $1 \le i_1 < \cdots < i_s \le m$. We define a family of linear maps

$$h : \Lambda^s(U) \longrightarrow \Lambda^s(U)$$

for $s \ge 1$ by specifying h on such forms and then extending to all of $\Lambda^s(U)$ by linearity:

$$h\big(f(r^1,\ldots,r^m)\,d(r^{i_1}) \wedge \cdots \wedge d(r^{i_s})\big)$$
$$= \left(\int_0^1 t^{s-1} f(tr^1,\ldots,tr^m)\,dt\right) d(r^{i_1}) \wedge \cdots \wedge d(r^{i_s})$$

for all $(r^1,\ldots,r^m)$ in U. Since U is star-shaped about $(0,\ldots,0)$, $f(tr^1,\ldots,tr^m)$ is defined for all t in $[0,1]$. The vector field R in $\mathfrak{X}(U)$ defined by

$$R = \sum_i r^i \frac{\partial}{\partial r^i}$$

is called the **radial vector field**.

 Claim 1. $h \circ \mathcal{L}_R = \mathrm{id}_{\Lambda^s(U)}$.

We have

$$\mathcal{L}_R\big(f\,d(r^{i_1}) \wedge \cdots \wedge d(r^{i_s})\big) = \mathcal{L}_R(f)\,d(r^{i_1}) \wedge \cdots \wedge d(r^{i_s})$$
$$+ f\,\mathcal{L}_R\big(d(r^{i_1}) \wedge \cdots \wedge d(r^{i_s})\big),$$

and from (16.4.1),

$$\mathcal{L}_R(f) = R(f) = \sum_i r^i \frac{\partial f}{\partial r^i}.$$

According to Theorem 16.4.3, $\mathcal{L}_R$ is a form derivation. It follows from Theorem 16.4.7 that $\mathcal{L}_R\big(d(r^i)\big) = d(r^i)$, and then from Theorem 16.3.1 that

$$\mathcal{L}_R\big(d(r^{i_1}) \wedge \cdots \wedge d(r^{i_s})\big) = s\,d(r^{i_1}) \wedge \cdots \wedge d(r^{i_s}).$$

Combining the above identities yields

$$\mathcal{L}_R\big(f\,d(r^{i_1}) \wedge \cdots \wedge d(r^{i_s})\big) = \bigg(\sum_i r^i \frac{\partial f}{\partial r^i} + sf\bigg) d(r^{i_1}) \wedge \cdots \wedge d(r^{i_s}).$$

Then

$$h\Big(\mathcal{L}_R\big(f(r^1,\ldots,r^m)\,d(r^{i_1}) \wedge \cdots \wedge d(r^{i_s})\big)\Big)$$

$$= \int_0^1 t^{s-1}\bigg(\sum_i tr^i \frac{\partial f}{\partial r^i}(tr^1,\ldots,tr^m) + sf(tr^1,\ldots,tr^m)\bigg) dt$$
$$\cdot d(r^{i_1}) \wedge \cdots \wedge d(r^{i_s})$$

$$= \int_0^1 \bigg(\sum_i t^s r^i \frac{\partial f}{\partial r^i}(tr^1,\ldots,tr^m) + st^{s-1} f(tr^1,\ldots,tr^m)\bigg) dt$$
$$\cdot d(r^{i_1}) \wedge \cdots \wedge d(r^{i_s})$$

$$= \bigg(\int_0^1 \frac{d}{dt}\big(t^s f(tr^1,\ldots,tr^m)\big)\,dt\bigg) d(r^{i_1}) \wedge \cdots \wedge d(r^{i_s})$$

$$= \big(t^s f(tr^1,\ldots,tr^m)\big\vert_{t=0}^{t=1}\big) d(r^{i_1}) \wedge \cdots \wedge d(r^{i_s})$$

$$= f(r^1,\ldots,r^m)\,d(r^{i_1}) \wedge \cdots \wedge d(r^{i_s}),$$

where the fourth equality follows from the fundamental theorem of calculus. This proves the claim.

Claim 2. $d \circ h = h \circ d$.
We have

$$d\left(\int_0^1 t^{s-1} f(tr^1, \dots, tr^m)\, dt\right)$$

$$= \sum_j \frac{\partial}{\partial r^j}\left(\int_0^1 t^{s-1} f(tr^1, \dots, tr^m)\, dt\right) d(r^j) \qquad \text{[Th 15.4.7]}$$

$$= \sum_j \left(\int_0^1 \frac{\partial}{\partial r^j}\left(t^{s-1} f(tr^1, \dots, tr^m)\right) dt\right) d(r^j) \qquad \text{[Th 10.5.5]}$$

$$= \sum_j \left(\int_0^1 t^s \frac{\partial f}{\partial r^j}(tr^1, \dots, tr^m)\, dt\right) d(r^j),$$

hence

$$h\left(d\big(f(r^1, \dots, r^m)\, d(r^{i_1}) \wedge \cdots \wedge d(r^{i_s})\big)\right)$$

$$= h\left(d(f)_{(r^1, \dots, r^m)} \wedge d(r^{i_1}) \wedge \cdots \wedge d(r^{i_s})\right)$$

$$= h\left(\left[\sum_j \frac{\partial f}{\partial r^j}(r^1, \dots, r^m)\, d(r^j)\right] \wedge d(r^{i_1}) \wedge \cdots \wedge d(r^{i_s})\right)$$

$$= \sum_j h\left(\frac{\partial f}{\partial r^j}(r^1, \dots, r^m)\, d(r^j) \wedge d(r^{i_1}) \wedge \cdots \wedge d(r^{i_s})\right)$$

$$= \sum_j \left[\left(\int_0^1 t^s \frac{\partial f}{\partial r^j}(tr^1, \dots, tr^m)\, dt\right) d(r^j) \wedge d(r^{i_1}) \wedge \cdots \wedge d(r^{i_s})\right]$$

$$= \left[\sum_j \frac{\partial}{\partial r^j}\left(\int_0^1 t^{s-1} f(tr^1, \dots, tr^m)\, dt\right) d(r^j)\right] \wedge d(r^{i_1}) \wedge \cdots \wedge d(r^{i_s})$$

$$= d\left(\int_0^1 t^{s-1} f(tr^1, \dots, tr^m)\, dt\right) \wedge d(r^{i_1}) \wedge \cdots \wedge d(r^{i_s})$$

$$= d\left(\left[\int_0^1 t^{s-1} f(tr^1, \dots, tr^m)\, dt\right] d(r^{i_1}) \wedge \cdots \wedge d(r^{i_s})\right)$$

$$= d\left(h\big(f(r^1, \dots, r^m)\, d(r^{i_1}) \wedge \cdots \wedge d(r^{i_s})\big)\right),$$

where the first and seventh equalities follow from Theorem 16.1.2, the second and sixth equalities from Theorem 15.4.7, and the fifth equality from Theorem 10.5.5. This proves the claim.

Then

$$\omega = h \circ \mathcal{L}_R(\omega) \qquad\qquad \text{[Claim 1]}$$

$$= (h \circ d \circ i_R + h \circ i_R \circ d)(\omega) \qquad \text{[Th 16.5.3]}$$

$$= h \circ d \circ i_R(\omega) \qquad\qquad [\omega \text{ is closed}]$$

$$= d \circ h \circ i_R(\omega) \qquad\qquad \text{[Claim 2]}$$

$$= d\big(h \circ i_R(\omega)\big).$$

We take $\xi = h\big(i_R(\omega)\big)$.   □

Theorem 16.9.8 explains why the form in Example 16.9.6 is not exact: the punctured plane is not star-shaped about the origin.

16.10   Flows

We saw in Section 15.1 that the velocity of a smooth curve on a smooth manifold is a vector field on the curve. In this section, we turn this observation around and ask if a given smooth vector field on a smooth manifold can be realized as a family of velocities of smooth curves.

Let M be a smooth manifold, let X be a vector field in $\mathfrak{X}(M)$, and let p be a point in M. An **integral curve of X with starting point p** is a smooth curve $\lambda(t) : (a, b) \longrightarrow M$ such that

$$\lambda(0) = p \qquad \text{and} \qquad \frac{d\lambda}{dt}(t) = X_{\lambda(t)}$$

for all t in (a, b). We say that λ is **maximal** if it cannot be extended to a smooth curve satisfying these properties on a larger open interval.

Theorem 16.10.1 (Existence of Maximal Integral Curve). *Let M be a smooth manifold, let X be a vector field in $\mathfrak{X}(M)$, and let p be a point in M. Then there is a unique maximal integral curve*

$$\Phi_p(t) : (a_p, b_p) \longrightarrow M$$

of X with starting point p.

Remark. By definition,

$$\Phi_p(0) = p \qquad \text{and} \qquad \frac{d\Phi_p}{dt}(t) = X_{\Phi_p(t)} \qquad\qquad (16.10.1)$$

for all t in (a_p, b_p).

Proof. We consider only the case where $\Phi_p\big((a_p, b_p)\big) \subset U$ for some chart $(U, (x^i))$ on M. It follows from Theorem 14.7.1(b) that, in local coordinates, Φ_p satisfies

$$\frac{d\Phi_p}{dt}(t) = \sum_i \frac{d(x^i \circ \Phi_p)}{dt}(t) \frac{\partial}{\partial x^i}\bigg|_{\Phi_p(t)}$$

for all t in (a_p, b_p). In local coordinates, let $X = \sum_i \alpha^i (\partial/\partial x^i)$, so that

$$X_{\Phi_p(t)} = \sum_i \alpha^i \big(\Phi_p(t)\big) \frac{\partial}{\partial x^i}\bigg|_{\Phi_p(t)}.$$

The second identity in (16.10.1) can now be expressed as

$$\sum_i \frac{d(x^i \circ \Phi_p)}{dt}(t) \frac{\partial}{\partial x^i}\bigg|_{\Phi_p(t)} = \sum_i \alpha^i \big(\Phi_p(t)\big) \frac{\partial}{\partial x^i}\bigg|_{\Phi_p(t)},$$

which is equivalent to the system of ordinary differential equations

$$\frac{d(x^i \circ \Phi_p)}{dt}(t) = \alpha^i\big(\Phi_p(t)\big) \tag{16.10.2}$$

for $i = 1, \ldots, m$. The result now follows from the theory of differential equations. $\square$

Example 16.10.2. Let (r, s) be standard coordinates on $\mathbb{R}^2$, and consider the vector field X in $\mathfrak{X}(\mathbb{R}^2)$ given by $X = (-s, r)$. Let $\Phi_{(1,0)}(t) : \mathbb{R} \longrightarrow \mathbb{R}^2$ be an integral curve of X with starting point $(1, 0)$. The system of differential equations corresponding to (16.10.2) is

$$\frac{d(r \circ \Phi_{(1,0)})}{dt}(t) = -s \circ \Phi_{(1,0)}(t) \quad \text{and} \quad \frac{d(s \circ \Phi_{(1,0)})}{dt}(t) = r \circ \Phi_{(1,0)}(t),$$

with initial condition $\Phi_{(1,0)}(0) = (1, 0)$. Writing $\Phi_{(1,0)} = \big(\Phi^1_{(1,0)}, \Phi^2_{(1,0)}\big)$, the system becomes

$$\frac{d(\Phi^1_{(1,0)})}{dt}(t) = -\Phi^2_{(1,0)}(t) \quad \text{and} \quad \frac{d(\Phi^2_{(1,0)})}{dt}(t) = \Phi^1_{(1,0)}(t),$$

with initial conditions $\Phi^1_{(1,0)}(0) = 1$ and $\Phi^2_{(1,0)}(0) = 0$. The obvious solution is $\Phi_{(1,0)}(t) = \big(\cos(t), \sin(t)\big)$, which is clearly maximal. Figure 16.10.1 depicts the vector field as arrows and the integral curves as broken circles. $\Diamond$

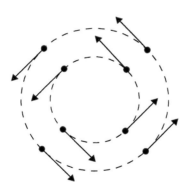

Figure 16.10.1. Diagram for Example 16.10.2

Continuing with the above setup, let $\Phi_p(t) : (a_p, b_p) \longrightarrow M$ be the maximal integral curve of X with starting point p. The **maximal flow domain** is defined by

$$D = \{(t, p) : t \in (a_p, b_p), p \in M\},$$

and the **maximal flow of X** is the map $\Phi : D \longrightarrow M$ defined by

$$\Phi(t, p) = \Phi_p(t)$$

for all (t, p) in D. For each t in $\mathbb{R}$, let

$$M_t = \{p \in M : t \in (a_p, b_p)\},$$

and consider the map

$$\Phi_t : M_t \longrightarrow M$$

defined by

$$\Phi_t(p) = \Phi(t, p)$$

for all p in M_t. We note that for sufficiently large t, M_t might be empty.

In physical terms, M_t is the set of points p in M for which a "particle" starting at p is able to "flow" for t units under the "action" of Φ. It is helpful to visualize Φ_t as a map that causes a portion of the vector field to "flow" bodily across the underlying smooth manifold for a "distance" of t units. We observe that with the above definitions,

$$\Phi_p(t) = \Phi(t, p) = \Phi_t(p). \qquad (16.10.3)$$

Theorem 16.10.3 (Fundamental Theorem on Flows). *Let M be a smooth manifold, let X be a vector field in $\mathfrak{X}(M)$, and let $\Phi(t, p) : D \longrightarrow M$ be the maximal flow of X. Then:*
(a) *D is an open set in $\mathbb{R} \times M$.*
(b) *Φ is smooth.*
(c) *Φ satisfies the **group laws**: for all p in M,*

$$\Phi_p(0) = p,$$

and for all s in (a_p, b_p) and t in $(a_{\Phi_p(s)}, b_{\Phi_p(s)})$ such that $s + t$ is in (a_p, b_p),

$$\Phi_{\Phi_p(s)}(t) = \Phi_p(s + t).$$

(d) *For each t in $\mathbb{R}$ such that M_t is nonempty, M_t is an open set in M and $\Phi_t : M_t \longrightarrow M_{-t}$ is a diffeomorphism, with inverse $(\Phi_t)^{-1} = \Phi_{-t}$.* ◻

Theorem 16.10.3 says that the maximal integral curves corresponding to a vector field combine to form a smooth map defined on an open set, and that the "flow" of particles referred to above takes place in a "diffeomorphic" fashion.

We close this section with a pair of results that demonstrate the surprisingly close connection between maximal flows and Lie derivatives.

Theorem 16.10.4 (Lie Derivative of Function). *Let M be a smooth manifold, let X be a vector field in $\mathfrak{X}(M)$, let $\Phi : D \longrightarrow M$ be the maximal flow of X, and let f be a function in $C^\infty(M)$. Then*

$$\mathcal{L}_X(f)(p) = \lim_{t \to 0} \frac{(\Phi_t)^\bullet(f)(p) - f(p)}{t}$$

for all p in M.

Proof. We have

$$
\begin{aligned}
\mathcal{L}_X(f)(p) &= X(f)(p) && [(16.4.1)] \\
&= X_p(f) && [(15.1.2)] \\
&= d_p(f)(X_p) && [\text{Th } 14.5.5] \\
&= d_{\Phi_p(0)}(f)\left(\frac{d\Phi_p}{dt}(0)\right) && [(16.10.1)] \\
&= \frac{d(f \circ \Phi_p)}{dt}(0) && [\text{Th } 14.7.4(b)] \\
&= \lim_{t \to 0} \frac{(f \circ \Phi_p)(t) - (f \circ \Phi_p)(0)}{t} \\
&= \lim_{t \to 0} \frac{(f \circ \Phi_t)(p) - f(p)}{t} && [(16.10.3)] \\
&= \lim_{t \to 0} \frac{(\Phi_t)^\bullet(f)(p) - f(p)}{t}. && [(15.9.2)] \qquad \square
\end{aligned}
$$

Theorem 16.10.5 (Lie Derivative of Vector Field). *Let M be a smooth manifold, let X, Y be vector fields in $\mathfrak{X}(M)$, and let $\Phi : D \longrightarrow M$ be the maximal flow of X. Then*

$$
\begin{aligned}
\mathcal{L}_X(Y)_p &= \lim_{t \to 0} \frac{d_{\Phi_t(p)}(\Phi_{-t})(Y_{\Phi_t(p)}) - Y_p}{t} \\
&= \lim_{t \to 0} \frac{(\Phi_t)^*(Y)_p - Y_p}{t} \qquad\qquad (16.10.4) \\
&= \lim_{t \to 0} \frac{Y_p - (\Phi_t)_*(Y)_p}{t}
\end{aligned}
$$

for all p in M.

Proof. We begin by showing that the three right-hand sides of (16.10.4) are equal. It follows from

$$
\begin{aligned}
d_{\Phi_t(p)}(\Phi_{-t})(Y_{\Phi_t(p)}) &= (\Phi_{-t})_*(Y)_p && [\text{Th } 15.10.1, \text{ Th } 16.10.3(d)] \\
&= (\Phi_t)^*(Y)_p && [(15.10.2), \text{ Th } 16.10.3(d)]
\end{aligned}
$$

that the first and second right-hand sides are equal, and also that

$$
\begin{aligned}
\lim_{t \to 0} \frac{Y_p - (\Phi_t)_*(Y)_p}{t} &= \lim_{t \to 0} \frac{Y_p - (\Phi_{-t})_*(Y)_p}{t} \\
&= \lim_{t \to 0} \frac{(\Phi_{-t})_*(Y)_p - Y_p}{-t} \\
&= \lim_{t \to 0} \frac{(\Phi_t)^*(Y)_p - Y_p}{t},
\end{aligned}
$$

hence the second and third right-hand sides are equal.

For a sufficiently small real number $\varepsilon > 0$, define a smooth function $g_t(p)$: $(-\varepsilon, \varepsilon) \times M \longrightarrow \mathbb{R}$ by

$$g_t(p) = (\Phi_t)^{\bullet}(f)(p) - f(p) = f \circ \Phi_t - f, \tag{16.10.5}$$

and observe that $g_0(p) = 0$. Consider the smooth function $h_t(p)$: $(-\varepsilon, \varepsilon) \times M \longrightarrow \mathbb{R}$ defined by

$$h_t(p) = \int_0^1 \frac{\partial g_{st}}{\partial s}(p) \, ds.$$

For $t \neq 0$, make a change of variable, and use the fundamental theorem of calculus to obtain

$$h_t(p) = \frac{1}{t} \int_0^t \frac{dg_r}{dr}(p) \, dr = \frac{1}{t} \left(g_r(p) \Big|_{r=0}^{r=t} \right) = \frac{g_t(p)}{t}. \tag{16.10.6}$$

Then

$$\begin{aligned}
h_0(p) &= \lim_{t \to 0} h_t(p) \\
&= \mathcal{L}_X(f)(p) \qquad \text{[Th 16.10.4]} \\
&= X(f)(p) \qquad \text{[(16.4.1)]}
\end{aligned}$$

for all p in M, hence

$$h_0 = X(f). \tag{16.10.7}$$

We have

$$\begin{aligned}
(\Phi_t)_*(Y)_p(f) &= (\Phi_t)_*(Y)_{\Phi_t((\Phi_{-t})(p))}(f) & \text{[Th 16.10.3(d)]} \\
&= d_{\Phi_{-t}(p)}(\Phi_t)(Y_{\Phi_{-t}(p)})(f) & \text{[Th 15.10.1]} \\
&= Y_{\Phi_{-t}(p)}(f \circ \Phi_t) & \text{[(14.4.1)]} \\
&= Y_{\Phi_{-t}(p)}(f + th_t) & \text{[(16.10.5), (16.10.6)]} \\
&= Y_{\Phi_{-t}(p)}(f) + t\, Y_{\Phi_{-t}(p)}(h_t) \\
&= Y(f)(\Phi_{-t}(p)) + t\, Y(h_t)(\Phi_{-t}(p)), & \text{[(15.1.2)]}
\end{aligned}$$

so

$$\begin{aligned}
&\left(Y_p - (\Phi_t)_*(Y)_p\right)(f) \\
&= Y_p(f) - (\Phi_t)_*(Y)_p(f) \\
&= Y(f)(p) - Y(f)(\Phi_{-t}(p)) - t\, Y(h_t)(\Phi_{-t}(p)). \qquad \text{[(15.1.2)]}
\end{aligned}$$

Thus,

$$\begin{aligned}
&\lim_{t \to 0} \frac{\left(Y_p - (\Phi_t)_*(Y)_p\right)(f)}{t} \\
&= \lim_{t \to 0} \frac{Y(f)(p) - Y(f)(\Phi_{-t}(p))}{t} - \lim_{t \to 0} Y(h_t)(\Phi_{-t}(p)).
\end{aligned} \tag{16.10.8}$$

The two terms on the right-hand side of (16.10.8) are

$$\lim_{t \to 0} \frac{Y(f)(p) - Y(f)\big(\Phi_{-t}(p)\big)}{t}$$
$$= \lim_{t \to 0} \frac{Y(f)\big(\Phi_{-t}(p)\big) - Y(f)(p)}{-t}$$
$$= \lim_{t \to 0} \frac{Y(f)\big(\Phi_{t}(p)\big) - Y(f)(p)}{t} \qquad\qquad\qquad (16.10.9)$$
$$= \lim_{t \to 0} \frac{(\Phi_{t})^{\bullet}\big(Y(f)\big)(p) - Y(f)(p)}{t} \qquad [(15.9.2)]$$
$$= \mathcal{L}_X\big(Y(f)\big)(p) \qquad\qquad\qquad [\text{Th } 16.10.4]$$
$$= X\big(Y(f)\big)(p) \qquad\qquad\qquad [(16.4.1)]$$

and

$$\lim_{t \to 0} Y(h_t)\big(\Phi_{-t}(p)\big) = Y(h_0)\big(\Phi_0(p)\big)$$
$$= Y\big(X(f)\big)(p). \qquad [(16.10.7)] \qquad (16.10.10)$$

Substituting (16.10.9) and (16.10.10) into (16.10.8) gives

$$\lim_{t \to 0} \frac{\big(Y_p - (\Phi_t)_*(Y)_p\big)(f)}{t} = X\big(Y(f)\big)(p) - Y\big(X(f)\big)(p)$$
$$= [X,Y](f)(p)$$
$$= [X,Y]_p(f) \qquad\qquad [(15.1.2)]$$
$$= \mathcal{L}_X(Y)_p(f). \qquad\qquad [(16.4.1)]$$

Since f was arbitrary,

$$\lim_{t \to 0} \frac{(\Phi_t)^*(Y)_p - Y_p}{t} = \mathcal{L}_X(Y)_p. \qquad\qquad \square$$

The second right-hand side of (16.10.4) allows us to think of $\mathcal{L}_X(Y)_p$ as a type of "directional derivative" of Y in the "direction" X_p, in the following sense. Ordinarily there is no connection between vectors in $T_p(M)$ and those in other tangent spaces of M. The maximal flow of X establishes a link between $T_p(M)$ and those tangent spaces of M that correspond to points that lie on the maximal integral curve with starting point p. To find $\mathcal{L}_X(Y)_p$, the diffeomorphism Φ_t is used to compute the pullback of Y at p, yielding a vector $(\Phi_t)^*(Y)_p$ in $T_p(M)$, which is then compared with Y_p by taking a limit. This makes sense because all computations are performed in $T_p(M)$.

In Section 16.4, the Lie derivative was formulated as a tensor derivation. In particular, the Lie derivative of a vector field was defined in terms of the Lie bracket. It is more usual in the literature for the Lie derivative of a vector field to be defined using the first equality in (16.10.4). Theorem 16.10.5 shows that the two approaches are equivalent.

Chapter 17

Smooth Manifolds with Boundary

It is not unusual in practice to encounter what would otherwise be a smooth manifold except for the presence of some type of "boundary". In this chapter, we introduce smooth manifolds with boundary and prove one of most important results in differential geometry—Stokes's theorem.

17.1 Smooth Manifolds with Boundary

The **closed upper half-space** of $\mathbb{R}^m$, defined by

$$\mathbb{H}^m = \{(x^1, \ldots, x^m) \in \mathbb{R}^m : x^m \geq 0\},$$

is the model for what we later call a smooth m-manifold with boundary. It is easily shown that

$$\mathrm{int}_{\mathbb{R}^m}(\mathbb{H}^m) = \{(x^1, \ldots, x^m) \in \mathbb{R}^m : x^m > 0\}$$

$$\mathrm{ext}_{\mathbb{R}^m}(\mathbb{H}^m) = \{(x^1, \ldots, x^m) \in \mathbb{R}^m : x^m < 0\}$$

$$\mathrm{bd}_{\mathbb{R}^m}(\mathbb{H}^m) = \{(x^1, \ldots, x^m) \in \mathbb{R}^m : x^m = 0\}.$$

For example, $\mathbb{H}^3 = \{(x, y, z) \in \mathbb{R}^3 : z \geq 0\}$ is the upper half of $\mathbb{R}^3$ including the xy-plane, $\mathrm{int}_{\mathbb{R}^3}(\mathbb{H}^3)$ is the upper half of $\mathbb{R}^3$ excluding the xy-plane, $\mathrm{ext}_{\mathbb{R}^3}(\mathbb{H}^3)$ is the lower half of $\mathbb{R}^3$ excluding the xy-plane, and $\mathrm{bd}_{\mathbb{R}^3}(\mathbb{H}^3)$ is the xy-plane.

> **Throughout, $\mathbb{H}^m$ is assumed to have the subspace topology induced by $\mathbb{R}^m$.**

Semi-Riemannian Geometry, First Edition. Stephen C. Newman.
© 2019 John Wiley & Sons, Inc. Published 2019 by John Wiley & Sons, Inc.

Our first goal is to use $\mathbb{H}^m$ to broaden our earlier notion of "chart". Let M be a topological space, and consider the following modification of the definition of chart given in Section 14.1. A **chart on M** is a pair (U, φ), where U is an open set in M and $\varphi : U \longrightarrow \mathbb{H}^m$ is a map such that:

[**C1**] $\varphi(U)$ is an open set in $\mathbb{H}^m$.

[**C2**] $\varphi : U \longrightarrow \varphi(U)$ is a homeomorphism.

The difference from the earlier formulation is that $\mathbb{H}^m$ appears in [C1] instead of $\mathbb{R}^m$. We refer to m as the **dimension** of the chart. For each point p in $\varphi(U)$, (U, φ) is said to be a **chart at p**. As before, an **atlas** for M is a collection $\mathfrak{A} = \{(U_\alpha, \varphi_\alpha) : \alpha \in A\}$ of charts on M such that the U_α form an open cover of M.

A **topological m-manifold with boundary** is a pair $(M, \mathfrak{A})$, where M is a topological space and $\mathfrak{A}$ is an atlas for M such that:

[**T1**] Each chart in $\mathfrak{A}$ has dimension m.

[**T2**] The topology of M has a countable basis.

[**T3**] For every pair of distinct points p_1, p_2 in M, there are disjoint open sets U_1 and U_2 in M such that p_1 is in U_1 and p_2 is in U_2.

Let M be a topological m-manifold with boundary, and let (U, φ) be a chart on M. We say that (U, φ) is an **interior chart** if $\varphi(U)$ is contained in $\mathrm{int}_{\mathbb{R}^m}(\mathbb{H}^m)$, and a **boundary chart** if $\varphi(U)$ intersects $\mathrm{bd}_{\mathbb{R}^m}(\mathbb{H}^m)$. Evidently, any chart on M is either an interior chart or a boundary chart. Since $\mathrm{int}_{\mathbb{R}^m}(\mathbb{H}^m)$ is an open set in $\mathbb{R}^m$, it follows from Theorem 9.1.5 that if (U, φ) is an interior chart, then it is a chart in the sense of Section 14.1. On the other hand, if $\varphi(U)$ intersects $\mathrm{bd}_{\mathbb{R}^m}(\mathbb{H}^m)$, we have something entirely new.

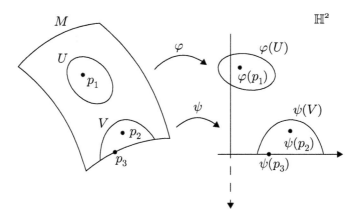

Figure 17.1.1. Interior chart and boundary chart

Let p be a point in M. We say that p is an **interior point** of M if there is a chart (U, φ) at p such that $\varphi(p)$ is in $\mathrm{int}_{\mathbb{R}^m}(\mathbb{H}^m)$, and that p is a **boundary point** of M if there is a chart (V, ψ) at p such that $\psi(p)$ is in $\mathrm{bd}_{\mathbb{R}^m}(\mathbb{H}^m)$. See Figure 17.1.1, where p_1 and p_2 are interior points, and p_3 is a boundary point. We observe that a chart at an interior point is either an interior chart or a boundary chart, but a chart at a boundary point is necessarily a boundary chart. Since any point p in M is contained in at least one chart, it is either an interior point (with respect to that chart) or a boundary point (with respect to that chart). This raises the question of whether p can be an interior point with respect to one chart, and at the same time a boundary point with respect to another chart. It is a crucial fact that this cannot happen.

Theorem 17.1.1 (Topological Invariance of Boundary). *Each point in a topological manifold with boundary is either an interior point or a boundary point, but not both.*   □

Let M be a topological manifold with boundary, and let p be a point in M. A consequence of Theorem 17.1.1 is that determining whether p is an interior point or a boundary point is independent of the choice of chart at p. The set of interior points of M is denoted by M° and called the **(manifold) interior of M**, and the set of boundary points of M is denoted by ∂M and called the **(manifold) boundary of M**. Thus, M is the disjoint union of M° and ∂M.

With the notion of chart in hand, the other core definitions presented in Section 14.1—transition map, tangent space, smooth atlas, and so on—transfer to the present setting virtually word for word. However, the boundary of $\mathbb{H}^m$ must be taken into account. To that end, when smoothness is being considered, we adopt the "extended" definition of smoothness introduced in Section 10.1.

A topological manifold with boundary $(M, \mathfrak{A})$ is said to be a **smooth m-manifold with boundary** if $\mathfrak{A}$ is a smooth atlas for M.

Example 17.1.2 ($\mathbb{H}^m$). The obvious example of a smooth m-manifold with boundary is the topological space $\mathbb{H}^m$ paired with the smooth atlas consisting of the single chart $\left(\mathbb{H}^m, \mathrm{id}_{\mathbb{H}^m} = (r^1, \ldots, r^m)\right)$. The interior and boundary of $\mathbb{H}^m$ are

$$(\mathbb{H}^m)^\circ = \mathrm{int}_{\mathbb{R}^m}(\mathbb{H}^m) \qquad \text{and} \qquad \partial \mathbb{H}^m = \mathrm{bd}_{\mathbb{R}^m}(\mathbb{H}^m),$$

respectively. Mimicking the situation with $\mathbb{R}^m$, we refer to $(r^1, \ldots, r^m)$ as **standard coordinates** on $\mathbb{H}^m$, to $(\partial/\partial r^1, \ldots, \partial/\partial r^m)$ and $\left(d(r^1), \ldots, d(r^m)\right)$ as the corresponding **standard coordinate frame** and **standard dual coordinate frame** of $\mathbb{H}^m$, respectively, and to the orientation of $\mathbb{H}^m$ induced by $d(r^1) \wedge \cdots \wedge d(r^m)$ as the **standard orientation**.   ◇

Example 17.1.3 (Closed Interval). A closed interval $[a, b]$ in $\mathbb{R}$ is a smooth 1-manifold with boundary, where $[a, b]^\circ = (a, b)$ and $\partial[a, b] = \{a, b\}$. This example was anticipated in Section 16.8.   ◇

Let M be a smooth manifold with boundary, let (U, φ) be a boundary chart on M, and let p be a point in $\partial M \cap U$. By definition, p is a boundary point, so we have from Theorem 17.1.1 that $\varphi(p)$ is in $\partial \mathbb{H}^m$. Thus, a coordinate map sends boundary points to boundary points. More generally, we have the following result.

Theorem 17.1.4 (Diffeomorphism Invariance of Boundary). *If M and N are smooth manifolds with boundary and $F : M \longrightarrow N$ is a diffeomorphism, then $F(\partial M) = \partial N$, hence $F(M^\circ) = F(N^\circ)$.* $\square$

There is nothing in the definition of a smooth manifold with boundary that requires the boundary to be nonempty. If the boundary is in fact empty, then the smooth atlas is comprised entirely of interior charts. In that case, the resulting smooth structure is that of a smooth manifold as defined in Section 14.1. For this reason, a smooth manifold with boundary where the boundary is empty will be referred to simply as a **smooth manifold**. On the other hand, we establish the following convention.

Henceforth, a smooth manifold with boundary is assumed to have a nonempty boundary.

The definition of submanifold given in Section 14.8 generalizes immediately to the present setting. A submanifold of either a smooth manifold or a smooth manifold with boundary might or might not have a boundary. We use the terms **submanifold** and **submanifold with boundary** as necessary to clarify the situation. For example, $\mathbb{H}^m$ is an m-submanifold with boundary of the smooth m-manifold $\mathbb{R}^m$. When dealing with either a smooth manifold or a smooth manifold with boundary, we employ the term **chart** in a generic sense to refer to either an interior chart or a boundary chart, reserving the latter terminology for situations where the difference needs to be made explicit.

The distinction between a manifold boundary and a topological boundary deserves emphasis. Consider, for example, the unit open disk D in $\mathbb{R}^2$. Viewed as an open 2-submanifold of the smooth 2-manifold $\mathbb{R}^2$, D is a smooth 2-manifold in its own right, and as such has no boundary. On the other hand, viewed as a subset of the topological space $\mathbb{R}^2$, D has the topological boundary consisting of the unit circle.

According to the next result, the boundary of a smooth manifold with boundary automatically has a smooth structure.

Theorem 17.1.5. *If M is a smooth m-manifold with boundary, where $m \geq 2$, then ∂M is an $(m-1)$-submanifold of M, hence a hypersurface of M.* $\square$

As an example, $\partial \mathbb{H}^m$ is a hypersurface of $\mathbb{H}^m$. It is sometimes convenient to identify the smooth $(m-1)$-manifolds $\partial \mathbb{H}^m$ and $\mathbb{R}^{m-1}$.

17.2   Inward-Pointing and Outward-Pointing Vectors

For a boundary point of a smooth manifold with boundary, there are two associated tangent spaces, and they need to be distinguished. Let M be a smooth m-manifold with boundary, and let p be a point in M. It follows from the definitions given above that $T_p(M)$ is an m-dimensional vector space, and this is so whether p is an interior point or a boundary point. However, when p is a boundary point, we have from Theorem 17.1.5 that ∂M is a hypersurface of M, hence $T_p(\partial M)$ is an $(m-1)$-dimensional vector space. It follows from (14.8.1) and Theorem 17.1.5 that $T_p(\partial M)$ is (or at least can be identified with) a subspace of $T_p(M)$.

Theorem 17.2.1. *Let M be a smooth m-manifold with boundary, let p be a boundary point, and let v be a vector in $T_p(M)$. Let $(U, \varphi = (x^i))$ and $(\tilde{U}, \tilde{\varphi} = (\tilde{x}^i))$ be (boundary) charts at p, and, in local coordinates, let*

$$v = \sum_{i=1}^{m} a^i \frac{\partial}{\partial x^i}\bigg|_p \qquad and \qquad v = \sum_{i=1}^{m} \tilde{a}^i \frac{\partial}{\partial \tilde{x}^i}\bigg|_p.$$

Then either a^m and $\tilde{a}^m$ are both 0, or they are both nonzero with the same sign.

Proof. We assume without loss of generality that $\varphi(p) = (0, \ldots, 0)$. Consider the transition map $\tilde{\varphi} \circ \varphi^{-1} : \varphi(W) \longrightarrow \tilde{\varphi}(W)$, where $W = U \cap \tilde{U}$ and

$$\tilde{\varphi} \circ \varphi^{-1} = (\tilde{x}^1 \circ \varphi^{-1}, \ldots, \tilde{x}^m \circ \varphi^{-1}).$$

By Theorem 17.1.4, $\tilde{\varphi} \circ \varphi^{-1}$ sends boundary (interior) points of $\mathbb{H}^m$ to boundary (interior) points of $\mathbb{H}^m$. Thus,

$$\tilde{x}^m \circ \varphi^{-1}(r^1, \ldots, r^{m-1}, 0) = 0 \qquad (17.2.1)$$

for all $(r^1, \ldots, r^{m-1}, 0)$ in $\varphi(W) \cap \partial \mathbb{H}^m$, and

$$\tilde{x}^m \circ \varphi^{-1}(r^1, \ldots, r^{m-1}, r^m) > 0 \qquad (17.2.2)$$

for all $(r^1, \ldots, r^{m-1}, r^m)$ in $\varphi(W) \cap (\mathbb{H}^m)^\circ$, where we note that $r^m > 0$. It follows from (14.3.2) and (17.2.1) that

$$\frac{\partial \tilde{x}^m}{\partial x^i}(p) = \frac{\partial(\tilde{x}^m \circ \varphi^{-1})}{\partial r^i}(0, \ldots, 0, 0) = 0 \qquad (17.2.3)$$

for $i = 1, \ldots, m-1$, and from (14.3.2), (17.2.1), and (17.2.2) that

$$\begin{aligned}
\frac{\partial \tilde{x}^m}{\partial x^m}(p) &= \frac{\partial(\tilde{x}^m \circ \varphi^{-1})}{\partial r^m}(0, \ldots, 0, 0) \\
&= \lim_{r \to 0^+} \frac{\tilde{x}^m \circ \varphi^{-1}(0, \ldots, 0, r) - \tilde{x}^m \circ \varphi^{-1}(0, \ldots, 0, 0)}{r} \qquad (17.2.4) \\
&= \lim_{r \to 0^+} \frac{\tilde{x}^m \circ \varphi^{-1}(0, \ldots, 0, r)}{r} \geq 0.
\end{aligned}$$

Let $\mathcal{X}$ and $\widetilde{\mathcal{X}}$ be the coordinate frames corresponding to $(U, (x^i))$ and $(\widetilde{U}, (\widetilde{x}^i))$, respectively. Then Theorem 14.3.5(b) and (17.2.3) give

$$\left[\mathrm{id}_{T_p(M)}\right]_{\mathcal{X}_p}^{\widetilde{\mathcal{X}}_p} = \begin{bmatrix} \dfrac{\partial \widetilde{x}^1}{\partial x^1}(p) & \cdots & \dfrac{\partial \widetilde{x}^1}{\partial x^{m-1}}(p) & \dfrac{\partial \widetilde{x}^1}{\partial x^m}(p) \\ \vdots & \ddots & \vdots & \vdots \\ \dfrac{\partial \widetilde{x}^{m-1}}{\partial x^1}(p) & \cdots & \dfrac{\partial \widetilde{x}^{m-1}}{\partial x^{m-1}}(p) & \dfrac{\partial \widetilde{x}^{m-1}}{\partial x^m}(p) \\ 0 & \cdots & 0 & \dfrac{\partial \widetilde{x}^m}{\partial x^m}(p) \end{bmatrix}, \tag{17.2.5}$$

hence

$$\det\left(\left[\mathrm{id}_{T_p(M)}\right]_{\mathcal{X}_p}^{\widetilde{\mathcal{X}}_p}\right) = \dfrac{\partial \widetilde{x}^m}{\partial x^m}(p)\, \det\left(\begin{bmatrix} \dfrac{\partial \widetilde{x}^1}{\partial x^1}(p) & \cdots & \dfrac{\partial \widetilde{x}^1}{\partial x^{m-1}}(p) \\ \vdots & \ddots & \vdots \\ \dfrac{\partial \widetilde{x}^{m-1}}{\partial x^1}(p) & \cdots & \dfrac{\partial \widetilde{x}^{m-1}}{\partial x^{m-1}}(p) \end{bmatrix}\right).$$

By Theorem 2.5.3, the left-hand side, hence the right-hand side, of the preceding identity is nonzero. Then (17.2.4) yields

$$\dfrac{\partial \widetilde{x}^m}{\partial x^m}(p) > 0. \tag{17.2.6}$$

We have from (2.2.5) that

$$\begin{bmatrix} \widetilde{a}^1 \\ \vdots \\ \widetilde{a}^m \end{bmatrix} = \left[\mathrm{id}_{T_p(M)}\right]_{\mathcal{X}(p)}^{\widetilde{\mathcal{X}}(p)} \begin{bmatrix} a^1 \\ \vdots \\ a^m \end{bmatrix},$$

and then from (17.2.5) that

$$\widetilde{a}^m = a^m \dfrac{\partial \widetilde{x}^m}{\partial x^m}(p). \tag{17.2.7}$$

The result now follows from (17.2.6) and (17.2.7).   □

Continuing with the above notation, we say that v is **inward-pointing** if $a^m > 0$, and **outward-pointing** if $a^m < 0$. According to Theorem 17.2.1, this determination is independent of the choice of chart used to express v in local coordinates. We need to consider the case $a^m = 0$. It follows from Theorem 14.8.3(b) and Theorem 17.1.5 that there is a (boundary) chart on M at p that is a slice chart for ∂M in M. If $(U, \varphi = (x^1, \ldots, x^m))$ is such a chart, then $((\partial/\partial x^1)|_p, \ldots, (\partial/\partial x^{m-1})|_p)$ is a basis for $T_p(\partial M)$. Thus, $a^m = 0$ if and only if v is in $T_p(\partial M)$. This observation along with Theorem 17.2.1 proves the following result.

Theorem 17.2.2. *Let M be a smooth manifold with boundary, and let p be a boundary point in M. Then $T_p(M)$ is the disjoint union of the set of inward-pointing tangent vectors at p, the set of outward-pointing tangent vectors at p, and $T_p(\partial M)$.* $\qquad\square$

See Figure 17.2.1, where v_1 is inward-pointing, v_2 is outward pointing, and v_3 is in $T_p(\partial M)$.

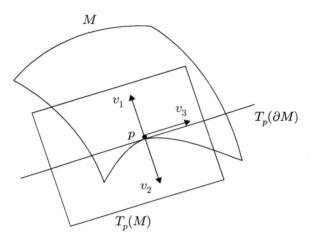

Figure 17.2.1. Inward-pointing and outward-pointing tangent vectors

Recall from Section 15.1 that $\mathfrak{X}_M(\partial M)$ is the set of smooth vector fields along ∂M. Let V be a vector field in $\mathfrak{X}_M(\partial M)$. By definition, V_p is in $T_p(M)$ for all p in ∂M. We say that V is **inward-pointing (outward-pointing)** if V_p is inward-pointing (outward-pointing) for all p in ∂M. Suppose V is in fact either inward-pointing or outward-pointing. According to the definition given in Section 15.1, V is nowhere-tangent to ∂M; or equivalently, V_p is in $T_p(M)\backslash T_p(\partial M)$ for all p in ∂M.

Example 17.2.3 ($\mathbb{H}^2$). Let (r, s) be standard coordinates on $\mathbb{H}^2$, and let $(\partial/\partial r, \partial/\partial s)$ be the corresponding standard coordinate frame. From Example 15.1.6, we have the identification $(\partial/\partial r, \partial/\partial s) = (e_1, e_2)$. It follows that $\partial/\partial s$ and $-\partial/\partial s$ are inward-pointing and outward-pointing vector fields in $\mathfrak{X}_{\mathbb{H}^2}(\partial \mathbb{H}^2)$, respectively. $\qquad\Diamond$

Example 17.2.4 (Hemisphere). Let (x, y, z) be standard coordinates on $\mathbb{R}^3$, and let $(\partial/\partial x, \partial/\partial y, \partial/\partial z)$ be the corresponding standard coordinate frame. According to Example 15.1.6, we have the identification $(\partial/\partial x, \partial/\partial y, \partial/\partial z) = (e_1, e_2, e_3)$. Consider the hemisphere-shaped smooth 2-manifold with boundary

$$\mathcal{S}^2 \cap \mathbb{H}^3 = \{(x, y, z) \in \mathbb{R}^3 : x^2 + y^2 + z^2 = 1, z \geq 0\},$$

where the boundary

$$\partial(\mathcal{S}^2 \cap \mathbb{H}^3) = \{(x, y, z) \in \mathbb{R}^3 : x^2 + y^2 = 1, z = 0\}$$

is the unit circle in the xy-plane of $\mathbb{R}^3$. Then $\partial/\partial z$ and $-\partial/\partial z$ are inward-pointing and outward-pointing vector fields in $\mathfrak{X}_{\mathcal{S}^2 \cap \mathbb{H}^3}\big(\partial(\mathcal{S}^2 \cap \mathbb{H}^3)\big)$, respectively.

$\Diamond$

The "construction" in the preceding example generalizes to an arbitrary smooth manifold with boundary, as we now show.

Theorem 17.2.5 (Existence of Outward-Pointing Vector Field). *If M is a smooth manifold with boundary, then there is an outward-pointing vector field in $\mathfrak{X}_M(\partial M)$.*

Proof. Only a sketch of the proof is provided. Let $\mathfrak{A} = \big\{ (U_\alpha, (x_\alpha^1, \ldots, x_\alpha^m)) : \alpha \in A \big\}$ be a smooth atlas for M, and let $(\partial/\partial x_\alpha^1, \ldots, \partial/\partial x_\alpha^m)$ be the coordinate frame corresponding to $(U_\alpha, (x_\alpha^1, \ldots, x_\alpha^m))$. For brevity, let $V_\alpha = -(\partial/\partial x_\alpha^m)|_{\partial M \cap U_\alpha}$. Then V_α is an outward-pointing vector field in $\mathfrak{X}_M(\partial M \cap U_\alpha)$. Let $\{\pi_\alpha : \alpha \in A\}$ be a partition of unity subordinate to the open cover $\{\partial M \cap U_\alpha : \alpha \in A\}$ of ∂M. It can be shown that there is a smooth extension of $\pi_\alpha V_\alpha$ to all of ∂M such that the extension, also denoted by $\pi_\alpha V_\alpha$, has constant value 0 on $\partial M \backslash \mathrm{supp}(\pi_\alpha)$. It can also be shown that $V = \sum_\alpha \pi_\alpha V_\alpha$ is a vector field in $\mathfrak{X}_M(\partial M)$.

We claim that V is outward-pointing. Let p be a point in ∂M, let $(U, (x^1, \ldots, x^m))$ be a (boundary) chart at p, and let $((\partial/\partial x^1)|_p, \ldots, ((\partial/\partial x^m)|_p)$ be the corresponding coordinate basis at p. For each α in A,

$$V_\alpha|_p = \sum_i^m \beta_\alpha^i(p) \frac{\partial}{\partial x^i}\bigg|_p$$

for some real numbers $\beta_\alpha^1(p), \ldots, \beta_\alpha^m(p)$, hence

$$V_p = \sum_\alpha \pi_\alpha(p) V_\alpha|_p = \sum_i^m \bigg(\sum_\alpha \pi_\alpha(p) \beta_\alpha^i(p) \bigg) \frac{\partial}{\partial x^i}\bigg|_p.$$

From the properties of partitions of unity, $\pi_\alpha(p) \geq 0$ for all α, and $\pi_\alpha(p) > 0$ for at least one α. Since $V_\alpha|_p$ is outward-pointing, we have $\beta_\alpha^m(p) < 0$ for all α. It follows that $\sum_\alpha \pi_\alpha(p)\beta_\alpha^m(p) < 0$, hence V_p is outward-pointing. Since p was arbitrary, V is outward-pointing. $\qquad\square$

17.3 Orientation of Boundaries

Let $(M, \mathfrak{O})$ be an oriented smooth m-manifold with boundary, where $m \geq 2$. By Theorem 17.2.5, there is an outward-pointing vector field V in $\mathfrak{X}_M(\partial M)$. Let $\mathfrak{O}_{\partial M}$ be the orientation of ∂M induced by V, as given by Theorem 16.6.9. We refer to $\mathfrak{O}_{\partial M}$ as the **Stokes orientation** of ∂M. It was remarked in connection with Theorem 16.6.9 that the orientation of a hypersurface induced by a nowhere-tangent vector field is not independent of the choice of nowhere-tangent vector field. Here the hypersurface is a boundary and the nowhere-tangent vector field is outward-pointing. This added specificity has the following implication.

Theorem 17.3.1. *If* $(M, \mathfrak{O})$ *is an oriented smooth m-manifold with boundary, where* $m \geq 2$, *then the Stokes orientation of* ∂M *is independent of the choice of outward-pointing vector field in* $\mathfrak{X}_M(\partial M)$.

Proof. Let V and W be outward-pointing vector fields in $\mathfrak{X}_M(\partial M)$. By Theorem 16.6.9(b), $i_V(\varpi)|_{\partial M}$ and $i_W(\varpi)|_{\partial M}$ are orientation forms on ∂M. We have from Theorem 16.6.1(a) that there is a uniquely determined nowhere-vanishing function f in $C^\infty(\partial M)$ such that $i_W(\varpi)|_{\partial M} = f\, i_V(\varpi)|_{\partial M}$. It suffices to show that $i_V(\varpi)|_{\partial M}$ and $i_W(\varpi)|_{\partial M}$ are consistent, meaning that f is strictly positive on ∂M. Let p be a point in ∂M, let $(U, (x^1, \ldots, x^m))$ be a (boundary) chart at p, let $((\partial/\partial x^1)|_p, \ldots, (\partial/\partial x^m)|_p)$ be the corresponding coordinate basis at p, and, in local coordinates, let

$$V_p = \sum_{i=1}^m a^i \frac{\partial}{\partial x^i}\Big|_p \qquad \text{and} \qquad W_p = \sum_{i=1}^m b^i \frac{\partial}{\partial x^i}\Big|_p. \tag{17.3.1}$$

Arguments similar to those used to prove Theorem 1.1.2 and Theorem 1.1.4 show that

$$\mathcal{V}_p = \left(V_p, \frac{\partial}{\partial x^1}\Big|_p, \ldots, \frac{\partial}{\partial x^{m-1}}\Big|_p \right)$$

and

$$\mathcal{W}_p = \left(W_p, \frac{\partial}{\partial x^1}\Big|_p, \ldots, \frac{\partial}{\partial x^{m-1}}\Big|_p \right)$$

are bases for $T_p(M)$. Then

$$\left(i_V(\varpi)|_{\partial M} \right)\Big|_p \left(\frac{\partial}{\partial x^1}\Big|_p, \ldots, \frac{\partial}{\partial x^{m-1}}\Big|_p \right)$$
$$= \varpi_p \left(V_p, \frac{\partial}{\partial x^1}\Big|_p, \ldots, \frac{\partial}{\partial x^{m-1}}\Big|_p \right)$$
$$= \det\left(\left[\mathrm{id}_{T_p(M)} \right]_{\mathcal{V}_p}^{\mathcal{W}_p} \right) \varpi_p \left(W_p, \frac{\partial}{\partial x^1}\Big|_p, \ldots, \frac{\partial}{\partial x^{m-1}}\Big|_p \right)$$
$$= \det\left(\left[\mathrm{id}_{T_p(M)} \right]_{\mathcal{V}_p}^{\mathcal{W}_p} \right) \left(i_W(\varpi)|_{\partial M} \right)\Big|_p \left(\frac{\partial}{\partial x^1}\Big|_p, \ldots, \frac{\partial}{\partial x^{m-1}}\Big|_p \right),$$

where the second equality follows from Theorem 7.3.3(b). Thus,

$$f(p) = \det\left(\left[\mathrm{id}_{T_p(M)} \right]_{\mathcal{V}_p}^{\mathcal{W}_p} \right).$$

We have from (17.3.1) that

$$V_p = \frac{b^m}{a^m} W_p + \sum_{i=1}^{m-1} \frac{a^m b^i - b^m a^i}{a^m} \frac{\partial}{\partial x^i}\Big|_p,$$

and then from (2.2.6) and (2.2.7) that

$$
\left[\mathrm{id}_{T_p(M)}\right]_{\mathcal{V}_p}^{\mathcal{W}_p} = \left[\begin{array}{cc} \left[\begin{array}{c} \dfrac{b^m}{a^m} \\[2mm] \dfrac{a^m b^2 - b^m a^2}{a^m} \\ \vdots \\ \dfrac{a^m b^{m-1} - b^m a^{m-1}}{a^m} \end{array}\right] & \begin{array}{c} O_{1\times(m-1)} \\[6mm] I_{m-1} \end{array} \end{array}\right],
$$

hence $f(p) = b^m/a^m$. Because V_p and W_p are outward-pointing, by definition, $a^m, b^m < 0$, so $f(p) > 0$. Since p was arbitrary, the result follows. $\qquad\square$

Theorem 17.3.2 (Stokes Orientation of $\partial\mathbb{H}^m$). *Let $(r^1, \ldots, r^m)$ be standard coordinates on $\mathbb{H}^m$, let $(\partial/\partial r^1, \ldots, \partial/\partial r^m)$ and $(d(r^1), \ldots, d(r^m))$ be the corresponding standard coordinate and dual coordinate frames, and suppose $\mathbb{H}^m$ has the standard orientation. Then the frame $((-1)^m \partial/\partial r^1, \ldots, \partial/\partial r^{m-1})$ on $\partial\mathbb{H}^m$ has the Stokes orientation.*

Proof. Recall from Example 17.1.2 that, by definition, the standard orientation of $\mathbb{H}^m$ is induced by $\Omega_{\mathbb{H}^m} = d(r^1) \wedge \cdots \wedge d(r^m)$. Since $-\partial/\partial r^m$ is an outward-pointing vector field in $\mathfrak{X}_{\mathbb{H}^m}(\partial\mathbb{H}^m)$, it follows from Theorem 16.6.9(b) and Theorem 17.3.1 that $i_{-\partial/\partial r^m}(\Omega_{\mathbb{H}^m})|_{\partial\mathbb{H}^m}$ is an orientation form on $\partial\mathbb{H}^m$ that induces the Stokes orientation. Let $X_1, \ldots, X_{m-1}$ be vector fields in $\mathfrak{X}(\partial\mathbb{H}^m)$. By Theorem 15.8.3(a),

$$
i_{-\partial/\partial r^m}(\Omega_{\mathbb{H}^m})|_{\partial\mathbb{H}^m}(X_1, \ldots, X_{m-1})
$$

$$
= \Omega_{\mathbb{H}^m}\left(-\frac{\partial}{\partial r^m}, X_1, \ldots, X_{m-1}\right)
$$

$$
= \det\left(\begin{bmatrix} 0 & d(r^1)(X_1) & \cdots & d(r^1)(X_{m-1}) \\ \vdots & \vdots & \ddots & \vdots \\ 0 & d(r^{m-1})(X_1) & \cdots & d(r^{m-1})(X_{m-1}) \\ -1 & d(r^m)(X_1) & \cdots & d(r^m)(X_{m-1}) \end{bmatrix}\right)
$$

$$
= (-1)^m \det\left(\begin{bmatrix} d(r^1)(X_1) & \cdots & d(r^1)(X_{m-1}) \\ \vdots & \ddots & \vdots \\ d(r^{m-1})(X_1) & \cdots & d(r^{m-1})(X_{m-1}) \end{bmatrix}\right)
$$

$$
= (-1)^m d(r^1) \wedge \cdots \wedge d(r^{m-1})(X_1, \ldots, X_{m-1}).
$$

Since $X_1, \ldots, X_{m-1}$ were arbitrary,

$$
i_{-\partial/\partial r^m}(\Omega_{\mathbb{H}^m})|_{\partial\mathbb{H}^m} = (-1)^m d(r^1) \wedge \cdots \wedge d(r^{m-1}).
$$

In keeping with a remark at the end of Section 17.1, we identify $\partial\mathbb{H}^m$ with $\mathbb{R}^{m-1}$. Then $(r^1, \ldots, r^{m-1})$ are standard coordinates on $\partial\mathbb{H}^m$, and $(\partial/\partial r^1, \ldots,$

$\partial/\partial r^{m-1})$ is the corresponding standard coordinate frame. By Theorem 15.8.4,

$$i_{-\partial/\partial r^m}(\Omega_{\mathbb{H}^m})|_{\partial\mathbb{H}^m}\left(\frac{\partial}{\partial r^1},\ldots,\frac{\partial}{\partial r^{m-1}}\right)=(-1)^m,$$

hence

$$i_{-\partial/\partial r^m}(\Omega_{\mathbb{H}^m})|_{\partial\mathbb{H}^m}\left((-1)^m\frac{\partial}{\partial r^1},\ldots,\frac{\partial}{\partial r^{m-1}}\right)=1.$$

Since $i_{-\partial/\partial r^m}(\Omega_{\mathbb{H}^m})|_{\partial\mathbb{H}^m}$ induces the Stokes orientation, it follows from Theorem 16.6.3 that $((-1)^m\partial/\partial r^1,\ldots,\partial/\partial r^{m-1})$ has the Stokes orientation.   □

17.4   Stokes's Theorem

In this section, we prove one of the central results of differential geometry—Stokes's theorem. In one sweeping statement it unifies numerous core theorems of vector calculus, as illustrated in Section 21.2.

Although details will not be presented here, the theory of integration of differential forms on smooth manifolds given in Section 16.7 can be extended to include the integration of differential forms on smooth manifolds with boundary.

Theorem 17.4.1 (Stokes's Theorem). *Let $(M,\mathfrak{O})$ be an oriented smooth m-manifold with boundary, and suppose ∂M has the Stokes orientation. If ω is a form in $\Lambda^{m-1}(M)$ that has compact support, then*

$$\int_M d(\omega)=\int_{\partial M}\omega\,,\qquad\qquad(17.4.1)$$

where the integrand on the right-hand side is interpreted as $\omega|_{\partial M}$.

Proof. We consider only the case $M=\mathbb{H}^m$. Let $(r^1,\ldots,r^m)$ be standard coordinates on $\mathbb{H}^m$, and let $(d(r^1),\ldots,d(r^m))$ be the corresponding standard dual coordinate frame. Suppose $\mathbb{H}^m$ has the standard orientation and that $\partial\mathbb{H}^m$ has the corresponding Stokes orientation.

Step 1. Compute $\int_{\mathbb{H}^m}d(\omega)$, the left-hand side of (17.4.1).

Since ω has compact support, it follows from Theorem 9.4.2 that there are real numbers $c_1,\ldots,c_m>0$ such that $\text{supp}(\omega)$ is properly contained in the closed cell

$$C=[-c_1,c_1]\times\cdots\times[-c_{m-1},c_{m-1}]\times[0,c_m].$$

From the manifold version of Theorem 7.2.12(e), we have in local coordinates that

$$\omega=\sum_{i=1}^m\alpha_i d(r^1)\wedge\cdots\wedge\widehat{d(r^i)}\wedge\cdots\wedge d(r^m),\qquad\qquad(17.4.2)$$

hence

$$d(\omega) = \sum_{i=1}^{m} d(\alpha_i) \wedge d(r^1) \wedge \cdots \wedge \widehat{d(r^i)} \wedge \cdots \wedge d(r^m)$$

$$= \sum_{i=1}^{m} \left[\left(\sum_{j} \frac{\partial \alpha_i}{\partial r^j} d(r^j) \right) \wedge d(r^1) \wedge \cdots \wedge \widehat{d(r^i)} \wedge \cdots \wedge d(r^m) \right]$$

$$= \sum_{i=1}^{m} \left(\sum_{j} \frac{\partial \alpha_i}{\partial r^j} d(r^j) \wedge d(r^1) \wedge \cdots \wedge \widehat{d(r^i)} \wedge \cdots \wedge d(r^m) \right)$$

$$= \sum_{i=1}^{m} \left(\frac{\partial \alpha_i}{\partial r^i} d(r^i) \wedge d(r^1) \wedge \cdots \wedge \widehat{d(r^i)} \wedge \cdots \wedge d(r^m) \right)$$

$$= \left(\sum_{i=1}^{m} (-1)^{i-1} \frac{\partial \alpha_i}{\partial r^i} \right) d(r^1) \wedge \cdots \wedge d(r^m),$$

where the first equality follows from Theorem 16.1.2, the second equality from Theorem 15.4.7, the fourth equality from Theorem 15.8.3(b), and the last equality from Theorem 15.8.2(b), where $\widehat{}$ indicates that an expression is omitted. Then

$$\int_{\mathbb{H}^m} d(\omega)$$

$$= \int_{C} \left(\sum_{i=1}^{m} (-1)^{i-1} \frac{\partial \alpha_i}{\partial r^i} \right) d(r^1) \wedge \cdots \wedge d(r^m)$$

$$= \sum_{i=1}^{m} (-1)^{i-1} \int_{C} \frac{\partial \alpha_i}{\partial r^i} d(r^1) \wedge \cdots \wedge d(r^m)$$

$$= \sum_{i=1}^{m} (-1)^{i-1} \int_{0}^{c_m} \int_{-c_{m-1}}^{c_{m-1}} \cdots \int_{-c_1}^{c_1} \frac{\partial \alpha_i}{\partial r^i} (r^1, \ldots, r^m) \, dr^1 \cdots dr^m$$

$$= \sum_{i=1}^{m-1} (-1)^{i-1} \int_{0}^{c_m} \int_{-c_{m-1}}^{c_{m-1}} \cdots \widehat{\int_{-c_i}^{c_i}} \cdots \int_{-c_1}^{c_1} \left(\int_{-c_i}^{c_i} \frac{\partial \alpha_i}{\partial r^i} (r^1, \ldots, r^m) \, dr^i \right)$$

$$\quad dr^1 \cdots \widehat{dr^i} \cdots dr^m$$

$$+ (-1)^{m-1} \int_{-c_{m-1}}^{c_{m-1}} \cdots \int_{-c_1}^{c_1} \left(\int_{0}^{c_m} \frac{\partial \alpha_m}{\partial r^m} (r^1, \ldots, r^m) \, dr^m \right) dr^1 \cdots dr^{m-1}$$

$$= \sum_{i=1}^{m-1} (-1)^{i-1} \int_{0}^{c_m} \int_{-c_{m-1}}^{c_{m-1}} \cdots \widehat{\int_{-c_i}^{c_i}} \cdots \int_{-c_1}^{c_1} \alpha_i(r^1, \ldots, r^m) \Big|_{r^i = -c_i}^{r^i = c_i}$$

$$\quad dr^1 \cdots \widehat{dr^i} \cdots dr^m$$

$$+ (-1)^{m-1} \int_{-c_{m-1}}^{c_{m-1}} \cdots \int_{-c_1}^{c_1} \alpha_m(r^1, \ldots, r^m) \Big|_{r^m = 0}^{r^m = c_m} dr^1 \cdots dr^{m-1},$$

where the third equality follows from (16.7.2), the fourth equality from Theorem 10.5.4, and last equality from the fundamental theorem of calculus. The choice of c_i ensures that $\alpha_i(r^1, \ldots, r^m) = 0$ when $r^i = \pm c_i$ for $i = 1, \ldots, m-1$, and that $\alpha_m(r^1, \ldots, r^{m-1}, c_m) = 0$. Thus,

$$\int_{\mathbb{H}^m} d(\omega) = (-1)^m \int_{-c_{m-1}}^{c_{m-1}} \cdots \int_{-c_1}^{c_1} \alpha_m(r^1, \ldots, r^{m-1}, 0)\, dr^1 \cdots dr^{m-1}. \quad (17.4.3)$$

Step 2. Compute $\int_{\partial\mathbb{H}^m} \omega$, the right-hand side of (17.4.1).
We have from (17.4.2) that

$$\int_{\partial\mathbb{H}^m} \omega = \int_{C\cap\partial\mathbb{H}^m} \left(\sum_{i=1}^{m} \alpha_i\, d(r^1) \wedge \cdots \wedge \widehat{d(r^i)} \wedge \cdots \wedge d(r^m) \right)$$

$$= \sum_{i=1}^{m} \int_{C\cap\partial\mathbb{H}^m} \alpha_i\, d(r^1) \wedge \cdots \wedge \widehat{d(r^i)} \wedge \cdots \wedge d(r^m).$$

Since r^m vanishes on $\partial\mathbb{H}^m$, so does $d(r^m)$, hence

$$\int_{\partial\mathbb{H}^m} \omega = \int_{C\cap\partial\mathbb{H}^m} \alpha_m\, d(r^1) \wedge \cdots \wedge d(r^{m-1}).$$

By Theorem 17.3.2, $\big((-1)^m \partial/\partial r^1, \ldots, \partial/\partial r^{m-1}\big)$ has the Stokes orientation, so (16.7.2) gives

$$\int_{\partial\mathbb{H}^m} \omega = (-1)^m \int_{C\cap\partial\mathbb{H}^m} \alpha_m(r^1, \ldots, r^{m-1}, 0)\, dr^1 \cdots dr^{m-1}$$

$$= (-1)^m \int_{-c_{m-1}}^{c_{m-1}} \cdots \int_{-c_1}^{c_1} \alpha_m(r^1, \ldots, r^{m-1}, 0)\, dr^1 \cdots dr^{m-1}. \quad (17.4.4)$$

The result now follows from (17.4.3) and (17.4.4).   □

We will see several applications of Stokes's theorem in Section 21.2.

Chapter 18

Smooth Manifolds with a Connection

In the introduction to Part III, we set out the task of developing a theory of differential geometry built upon our earlier study of curves and surfaces, but without having to involve an ambient space. Implicit in this undertaking was the aim of recovering, to the extent possible, the results presented for curves and surfaces. Chapters 14–17 have met significant parts of this agenda. Noticeably absent, however, is a discussion of "covariant derivative" and "metric". We remedy the first of these deficits in this chapter by adding "connection" to our discussion of smooth manifolds.

18.1   Covariant Derivatives

Let M be a smooth manifold. A **connection** on M is a map

$$\nabla : \mathfrak{X}(M) \times \mathfrak{X}(M) \longrightarrow \mathfrak{X}(M)$$

such that for all vector fields X, Y, Z in $\mathfrak{X}(M)$ and all functions f in $C^\infty(M)$:

[$\nabla 1$] $\nabla(X + Y, Z) = \nabla(X, Z) + \nabla(Y, Z)$.
[$\nabla 2$] $\nabla(fX, Y) = f\nabla(X, Y)$.
[$\nabla 3$] $\nabla(X, Y + Z) = \nabla(X, Y) + \nabla(X, Z)$.
[$\nabla 4$] $\nabla(X, fY) = X(f)Y + f\nabla(X, Y)$.

We refer to the pair (M, ∇) as a **smooth manifold with a connection**. It is possible for a given smooth manifold to have more than one connection. Thus, a connection is not a fundamental constituent of the manifold in the same way that, for example, the boundary forms part of a smooth manifold with boundary.

Semi-Riemannian Geometry, First Edition. Stephen C. Newman.

Example 18.1.1. There are two obvious examples of connections. Let U be an open set in $\mathbb{R}^m$. Viewing U as a smooth m-manifold, let

$$D : \mathfrak{X}(U) \times \mathfrak{X}(U) \longrightarrow \mathfrak{X}(U)$$

be the map defined by

$$D(X,Y) = D_X(Y)$$

for all vector fields X,Y in $\mathfrak{X}(U)$, where D_X is the Euclidean derivative with respect to X. It follows from Theorem 10.3.1 that D is a connection on U.

Similarly, let $(M,\mathfrak{g})$ be a regular surface in $\mathbb{R}^3_\nu$. According to Example 14.1.5, M is a smooth 2-manifold. Let

$$\nabla : \mathfrak{X}(M) \times \mathfrak{X}(M) \longrightarrow \mathfrak{X}(M)$$

be the map defined by

$$\nabla(X,Y) = \nabla_X(Y)$$

for all vector fields X,Y in $\mathfrak{X}(M)$, where ∇_X is the covariant derivative with respect to X. We have from Theorem 12.4.1 that ∇ is a connection on M. $\Diamond$

In each of the preceding examples, we started with a "derivative" and used it to define a connection. Given a smooth manifold with a connection, we can reverse that process.

Let (M,∇) be a smooth manifold with a connection, and let X be a vector field in $\mathfrak{X}(M)$. We define a map

$$\nabla_X : \mathfrak{X}(M) \longrightarrow \mathfrak{X}(M)$$

by

$$\nabla_X(Y) = \nabla(X,Y)$$

for all vector fields Y in $\mathfrak{X}(M)$. It follows from [$\nabla 3$] and [$\nabla 4$] that ∇_X is $\mathbb{R}^m$-linear, and from [$\nabla 4$] that

$$\nabla_X(fY) = X(f)Y + f\nabla_X(Y)$$

for all functions f in $C^\infty(M)$ and all vector fields Y in $\mathfrak{X}(M)$. By Theorem 16.2.7, there corresponds a unique tensor derivation

$$\nabla_X : \mathcal{T}^r_s(M) \longrightarrow \mathcal{T}^r_s(M)$$

on M, called the **covariant derivative with respect to X**, where $(\nabla_X)^r_s$ has been abbreviated to ∇_X. In particular, we have

$$\nabla_X(f) = X(f) \tag{18.1.1}$$

for all functions f in $C^\infty(M)$.

Theorem 18.1.2 (Vector Field Properties of Covariant Derivative). *Let (M,∇) be a smooth manifold with a connection, let X,Y,Z be vector fields in $\mathfrak{X}(M)$, and let f be a function in $C^\infty(M)$. Then:*

(a) $\nabla_{X+Y}(Z) = \nabla_X(Z) + \nabla_Y(Z)$.
(b) $\nabla_{fX}(Y) = f\nabla_X(Y)$.
(c) $\nabla_X(Y+Z) = \nabla_X(Y) + \nabla_X(Z)$.
(d) $\nabla_X(fY) = X(f)Y + f\nabla_X(Y)$.

Proof. This follows immediately from properties [∇1]–[∇4]. □

Theorem 18.1.3 (Tensor Field Properties of Covariant Derivative).
Let (M, ∇) be a smooth manifold with a connection, let X be a vector field in $\mathfrak{X}(M)$, and let $\mathcal{A}, \mathcal{B}$ and $\mathcal{C}$ be tensor fields in $T^r_s(M)$ and $T^{r'}_{s'}(M)$, respectively. Then:
(a) $\nabla_X(\mathcal{A} + \mathcal{B}) = \nabla_X(\mathcal{A}) + \nabla_X(\mathcal{B})$.
(b) $\nabla_X(\mathcal{A} \otimes \mathcal{C}) = \nabla_X(\mathcal{A}) \otimes \mathcal{C} + \mathcal{A} \otimes \nabla_X(\mathcal{C})$.

Proof. This follows from the defining properties of tensor derivations given in Section 16.2. □

Theorem 18.1.4 (Global Expression for Covariant Derivative). *Let (M, ∇) be a smooth manifold with a connection, let $\mathcal{A}$ be a tensor field in $T^r_s(M)$, let $X, Y_1, \ldots, Y_s$ be vector fields in $\mathfrak{X}(M)$, and let $\omega^1, \ldots, \omega^r$ be covector fields in $\mathfrak{X}^*(M)$. Then*

$$
\begin{aligned}
&\nabla_X(\mathcal{A})(\omega^1, \ldots, \omega^r, Y_1, \ldots, Y_s) \\
&= X\big(\mathcal{A}(\omega^1, \ldots, \omega^r, Y_1, \ldots, Y_s)\big) \\
&\quad - \sum_{i=1}^{r} \mathcal{A}(\omega^1, \ldots, \omega^{i-1}, \nabla_X(\omega^i), \omega^{i+1}, \ldots, \omega^r, Y_1, \ldots, Y_s) \\
&\quad - \sum_{j=1}^{s} \mathcal{A}(\omega^1, \ldots, \omega^r, Y_1, \ldots, Y_{j-1}, \nabla_X(Y_j), Y_{j+1}, \ldots, Y_s).
\end{aligned}
$$

Proof. This follows from Theorem 16.2.4. □

Example 18.1.5. Let (M, ∇) be a smooth manifold with a connection, let $\mathcal{A}$ be a tensor field in $T^0_2(M)$, let X, Y, Z be vector fields in $\mathfrak{X}(M)$, and let ω be a covector field in $\mathfrak{X}^*(M) = T^0_1(M)$. We have from (18.1.1) and Theorem 18.1.4 that

$$\nabla_X(\omega)(Y) = X\big(\omega(Y)\big) - \omega\big(\nabla_X(Y)\big) = \nabla_X\big(\omega(Y)\big) - \omega\big(\nabla_X(Y)\big) \tag{18.1.2}$$

and

$$
\begin{aligned}
\nabla_X(\mathcal{A})(Y, Z) &= X\big(\mathcal{A}(Y, Z)\big) - \mathcal{A}\big(\nabla_X(Y), Z\big) - \mathcal{A}\big(Y, \nabla_X(Z)\big) \\
&= \nabla_X\big(\mathcal{A}(Y, Z)\big) - \mathcal{A}\big(\nabla_X(Y), Z\big) - \mathcal{A}\big(Y, \nabla_X(Z)\big).
\end{aligned} \tag{18.1.3}
$$

It is easily shown that if $\mathcal{A}$ is symmetric in its two arguments, then so is $\nabla_X(\mathcal{A})$.

Let (M, ∇) be a smooth manifold with a connection, and let X be a vector field in $\mathfrak{X}(M)$. It can be shown that ∇_X is determined pointwise in the following sense: for all points p in M and all vector fields Y in $\mathfrak{X}(M)$,

$$\nabla_X(Y)_p = \nabla_{\widetilde{X}}(Y)_p$$

for all vector fields $X, \widetilde{X}$ in $\mathfrak{X}(M)$ such that $X_p = \widetilde{X}_p$. For each point p in M and vector v in $T_p(M)$, we define a vector $\nabla_v(Y)$ in $T_p(M)$ by

$$\nabla_v(Y) = \nabla_X(Y)_p, \tag{18.1.4}$$

where X is any vector field in $\mathfrak{X}(M)$ such that $X_p = v$. By Theorem 15.1.2, such an X always exists. Since ∇_X is determined pointwise, $\nabla_v(Y)$ is independent of the choice of vector field and is therefore well-defined. It follows from (18.1.4) that for all vector fields X, Y in $\mathfrak{X}(M)$,

$$\nabla_{X_p}(Y) = \nabla_X(Y)_p \tag{18.1.5}$$

for all p in M.

By definition, $\mathrm{Lin}\big(\mathfrak{X}(M), \mathfrak{X}(M)\big)$ is the vector space of $\mathbb{R}$-linear maps from the vector space $\mathfrak{X}(M)$ to itself. We make $\mathrm{Lin}\big(\mathfrak{X}(M), \mathfrak{X}(M)\big)$ into a ring over $\mathbb{R}$ by defining multiplication to be composition of maps. Let Θ, Υ, Ψ be three such maps, and define the bracket of Θ and Υ by

$$[\Theta, \Upsilon] = \Theta \circ \Upsilon - \Upsilon \circ \Theta.$$

It is easily shown that with this definition, $\mathrm{Lin}\big(\mathfrak{X}(M), \mathfrak{X}(M)\big)$ is a Lie algebra over $\mathbb{R}$. In particular, Jacobi's identity is satisfied:

$$[[\Theta, \Upsilon], \Psi] + [[\Upsilon, \Psi], \Theta] + [[\Psi, \Theta], \Upsilon] = 0.$$

For vector fields X, Y, Z in $\mathfrak{X}(M)$, we have from parts (c) and (d) of Theorem 18.1.2 that $\nabla_X, \nabla_Y, \nabla_Z$ are maps in $\mathrm{Lin}\big(\mathfrak{X}(M), \mathfrak{X}(M)\big)$, hence

$$\big[[\nabla_X, \nabla_Y], \nabla_Z\big] + \big[[\nabla_Y, \nabla_Z], \nabla_X\big] + \big[[\nabla_Z, \nabla_X], \nabla_Y\big] = 0. \tag{18.1.6}$$

18.2   Christoffel Symbols

In our study of regular surfaces in $\mathbb{R}^3_\nu$, we defined Christoffel symbols and used them to construct the covariant derivative, which we now think of as equivalent to a connection (see Example 18.1.1). Our approach to smooth manifolds with a connection proceeds in reverse: we start with a connection and then use it to define Christoffel symbols.

Let (M, ∇) be a smooth m-manifold with a connection, let X be a vector field in $\mathfrak{X}(M)$, let $\big(U, (x^i)\big)$ be a chart on M, and let $\mathcal{X} = (\partial/\partial x^1, \ldots, \partial/\partial x^m)$ be the corresponding coordinate frame. It can be shown that ∇ restricts unambiguously to the smooth m-manifold U. For simplicity of notation, let us denote the restriction

$$\nabla|_U \qquad \text{by} \qquad \nabla.$$

Then (U, ∇) is a smooth m-manifold with a connection, $(U, (x^i))$ is a covering chart on U, and $\mathcal{X}$ is the corresponding coordinate frame. We can therefore express $\nabla_{\partial/\partial x^i}(\partial/\partial x^j)$ as

$$\nabla_{\partial/\partial x^i}\left(\frac{\partial}{\partial x^j}\right) = \sum_{k=1}^{m} \Gamma_{ij}^{k} \frac{\partial}{\partial x^k} \tag{18.2.1}$$

for $i, j = 1, \ldots, m$, where the Γ_{ij}^{k} are uniquely determined functions in $C^\infty(U)$, called the **Christoffel symbols**.

Theorem 18.2.1 (Local Coordinate Expression for Covariant Derivative). *Let (M, ∇) be a smooth m-manifold with a connection, let X, Y be vector fields in $\mathfrak{X}(M)$, let ω be a covector field in $\mathfrak{X}^*(M)$, and, in local coordinates, let*

$$X = \sum_i \alpha^i \frac{\partial}{\partial x^i}, \qquad Y = \sum_j \beta^j \frac{\partial}{\partial x^j}, \qquad and \qquad \omega = \sum_j \gamma_j d(x^j).$$

Let

$$\alpha_{;j}^{i} = \frac{\partial \alpha^i}{\partial x^j} + \sum_k \Gamma_{jk}^{i} \alpha^k \tag{18.2.2}$$

and

$$\gamma_{i;j} = \frac{\partial \gamma_i}{\partial x^j} - \sum_k \Gamma_{ji}^{k} \gamma_k \tag{18.2.3}$$

for $i, j = 1, \ldots, m$. Then:
(a)

$$\nabla_X(Y) = \sum_j \left(\sum_i \alpha^i \beta_{;i}^{j} \right) \frac{\partial}{\partial x^j}.$$

(b)

$$\nabla_X(\omega) = \sum_j \left(\sum_i \alpha^i \gamma_{j;i} \right) d(x^j).$$

Proof. (a): Theorem 18.1.2 is used repeatedly in what follows. We have

$$\nabla_{\partial/\partial x^i}(Y) = \sum_j \nabla_{\partial/\partial x^i}\left(\beta^j \frac{\partial}{\partial x^j}\right)$$

$$= \sum_j \left[\nabla_{\partial/\partial x^i}(\beta^j)\frac{\partial}{\partial x^j} + \beta^j \nabla_{\partial/\partial x^i}\left(\frac{\partial}{\partial x^j}\right) \right]$$

$$= \sum_j \left(\frac{\partial \beta^j}{\partial x^i}\frac{\partial}{\partial x^j} + \beta^j \sum_k \Gamma_{ij}^{k}\frac{\partial}{\partial x^k} \right)$$

$$= \sum_j \frac{\partial \beta^j}{\partial x^i}\frac{\partial}{\partial x^j} + \sum_{jk} \beta^j \Gamma_{ij}^{k}\frac{\partial}{\partial x^k} = \sum_j \frac{\partial \beta^j}{\partial x^i}\frac{\partial}{\partial x^j} + \sum_{jk} \beta^k \Gamma_{ik}^{j}\frac{\partial}{\partial x^j}$$

$$= \sum_j \left(\frac{\partial \beta^j}{\partial x^i} + \sum_k \beta^k \Gamma_{ik}^{j} \right)\frac{\partial}{\partial x^j} = \sum_j \beta_{;i}^{j}\frac{\partial}{\partial x^j},$$

hence

$$\nabla_X(Y) = \sum_i \alpha^i \nabla_{\partial/\partial x^i}(Y) = \sum_j \left(\sum_i \alpha^i \beta^j_{;i}\right)\frac{\partial}{\partial x^j}.$$

(b): The components of $\nabla_{\partial/\partial x^i}\left(d(x^j)\right)$ are

$$\nabla_{\partial/\partial x^i}\left(d(x^j)\right)\left(\frac{\partial}{\partial x^k}\right)$$

$$= \frac{\partial}{\partial x^i}\left(d(x^j)\left(\frac{\partial}{\partial x^k}\right)\right) - d(x^j)\left(\nabla_{\partial/\partial x^i}\left(\frac{\partial}{\partial x^k}\right)\right) \qquad [(18.1.2)]$$

$$= -d(x^j)\left(\sum_l \Gamma^l_{ik}\frac{\partial}{\partial x^l}\right) \qquad\qquad\qquad [(18.2.1)]$$

$$= -\sum_l \Gamma^l_{ik} d(x^j)\left(\frac{\partial}{\partial x^l}\right) = -\Gamma^j_{ik},$$

hence

$$\nabla_{\partial/\partial x^i}\left(d(x^j)\right) = -\sum_k \Gamma^j_{ik} d(x^k). \qquad (18.2.4)$$

Then

$$\nabla_{\partial/\partial x^i}(\omega) = \nabla_{\partial/\partial x^i}\left(\sum_j \gamma_j d(x^j)\right) = \sum_j \nabla_{\partial/\partial x^i}(\gamma_j d(x^j))$$

$$= \sum_j \left(\nabla_{\partial/\partial x^i}(\gamma_j)d(x^j) + \gamma_j \nabla_{\partial/\partial x^i}\left(d(x^j)\right)\right)$$

$$= \sum_j \left(\frac{\partial \gamma_j}{\partial x^i}d(x^j) - \gamma_j \sum_k \Gamma^j_{ik} d(x^k)\right) \qquad [(18.2.4)]$$

$$= \sum_j \frac{\partial \gamma_j}{\partial x^i}d(x^j) - \sum_{jk} \gamma_j \Gamma^j_{ik} d(x^k)$$

$$= \sum_j \frac{\partial \gamma_j}{\partial x^i}d(x^j) - \sum_{jk} \gamma_k \Gamma^k_{ij} d(x^j)$$

$$= \sum_j \left(\frac{\partial \gamma_j}{\partial x^i} - \sum_k \gamma_k \Gamma^k_{ij}\right)d(x^j) = \sum_j \gamma_{j;i} d(x^j),$$

so

$$\nabla_X(\omega) = \sum_i \alpha^i \nabla_{\partial/\partial x^i}(\omega) = \sum_i \alpha^i \left(\sum_j \gamma_{j;i} d(x^j)\right)$$

$$= \sum_j \left(\sum_i \alpha^i \gamma_{j;i}\right)d(x^j). \qquad \square$$

We observe that the expressions for $\nabla_X(Y)$ and $\nabla_X(\omega)$ in parts (a) and (b) of Theorem 18.2.1 involve the α^i but not their partial derivatives. In order to compute $\nabla_X(Y)$ and $\nabla_X(\omega)$ at a given point, all we need to know about X is its value at that point. This is consistent with ∇_X being determined pointwise, as was remarked in conjunction with (18.1.4).

In the notation of Theorem 18.2.1, we have as special cases

$$\nabla_{\partial/\partial x^i}(Y) = \sum_j \beta^j_{;i} \frac{\partial}{\partial x^j} \qquad \text{and} \qquad \nabla_{\partial/\partial x^i}(\omega) = \sum_j \gamma_{j;i}\, d(x^j).$$

These formulas tell us that to take the covariant derivatives of Y and ω with respect to $\partial/\partial x^i$, we simply add a subscript "$;i$" to each of the components in the local coordinate expressions.

Theorem 18.2.2 (Local Coordinate Expression for Covariant Derivative). *Let (M, ∇) be a smooth m-manifold with a connection, let $\mathcal{A}$ be a tensor field in $\mathcal{T}^r_s(M)$, and, in local coordinates, let $\mathcal{A}^{i_1 \ldots i_r}_{j_1 \ldots j_s}$ be the components of $\mathcal{A}$. Then the components of $\nabla_{\partial/\partial x^k}(\mathcal{A})$ are*

$$\nabla_{\partial/\partial x^k}(\mathcal{A})^{i_1 \ldots i_r}_{j_1 \ldots j_s} = \frac{\partial \mathcal{A}^{i_1 \ldots i_r}_{j_1 \ldots j_s}}{\partial x^k} + \sum_{l=1}^{r}\left(\sum_{p=1}^{m} \Gamma^{i_l}_{kp} \mathcal{A}^{i_1 \ldots i_{l-1}\, p\, i_{l+1} \ldots i_r}_{j_1 \ldots j_s} \right)$$

$$- \sum_{l=1}^{s}\left(\sum_{p=1}^{m} \Gamma^{p}_{kj_l} \mathcal{A}^{i_1 \ldots i_r}_{j_1 \ldots j_{l-1}\, p\, j_{l+1} \ldots j_s} \right).$$

Proof. By Theorem 18.1.4,

$$\nabla_{\partial/\partial x^k}(\mathcal{A})^{i_1 \ldots i_r}_{j_1 \ldots j_s}$$

$$= \nabla_{\partial/\partial x^k}(\mathcal{A})\left(d(x^{i_1}), \ldots, d(x^{i_r}), \frac{\partial}{\partial x^{j_1}}, \ldots, \frac{\partial}{\partial x^{j_s}} \right)$$

$$= \frac{\partial}{\partial x^k}\left(\mathcal{A}\left(d(x^{i_1}), \ldots, d(x^{i_r}), \frac{\partial}{\partial x^{j_1}}, \ldots, \frac{\partial}{\partial x^{j_s}} \right) \right)$$

$$- \sum_{l=1}^{r} \mathcal{A}\left(d(x^{i_1}), \ldots, \nabla_{\partial/\partial x^k}\big(d(x^{i_l})\big), \ldots, d(x^{i_r}), \frac{\partial}{\partial x^{j_1}}, \ldots, \frac{\partial}{\partial x^{j_s}} \right)$$

$$- \sum_{l=1}^{s} \mathcal{A}\left(d(x^{i_1}), \ldots, d(x^{i_r}), \frac{\partial}{\partial x^{i_1}}, \ldots, \nabla_{\partial/\partial x^k}\left(\frac{\partial}{\partial x^{j_l}} \right), \ldots, \frac{\partial}{\partial x^{j_s}} \right).$$

We have

$$\frac{\partial}{\partial x^k}\left(\mathcal{A}\left(d(x^{i_1}), \ldots, d(x^{i_r}), \frac{\partial}{\partial x^{j_1}}, \ldots, \frac{\partial}{\partial x^{j_s}} \right) \right) = \frac{\partial \mathcal{A}^{i_1 \ldots i_r}_{j_1 \ldots j_s}}{\partial x^k}$$

and

$$\sum_{l=1}^{r} \mathcal{A}\left(d(x^{i_1}), \ldots, \nabla_{\partial/\partial x^k}\left(d(x^{i_l})\right), \ldots, d(x^{i_r}), \frac{\partial}{\partial x^{j_1}}, \ldots, \frac{\partial}{\partial x^{j_s}}\right)$$

$$= \sum_{l=1}^{r} \mathcal{A}\left(d(x^{i_1}), \ldots, -\sum_{p=1}^{m} \Gamma_{kp}^{i_l} d(x^p), \ldots, d(x^{i_r}), \frac{\partial}{\partial x^{j_1}}, \ldots, \frac{\partial}{\partial x^{j_s}}\right)$$

$$= -\sum_{l=1}^{r}\sum_{p=1}^{m} \Gamma_{kp}^{i_l} \mathcal{A}\left(d(x^{i_1}), \ldots, d(x^p), \ldots, d(x^{i_r}), \frac{\partial}{\partial x^{j_1}}, \ldots, \frac{\partial}{\partial x^{j_s}}\right)$$

$$= -\sum_{l=1}^{r}\left(\sum_{p=1}^{m} \Gamma_{kp}^{i_l} \mathcal{A}_{j_1 \ldots j_s}^{i_1 \ldots i_{l-1} \, p \, i_{l+1} \ldots i_r}\right),$$

where the first equality follows from (18.2.4). We also have

$$\sum_{l=1}^{s} \mathcal{A}\left(d(x^{i_1}), \ldots, d(x^{i_r}), \frac{\partial}{\partial x^{j_1}}, \ldots, \nabla_{\partial/\partial x^k}\left(\frac{\partial}{\partial x^{j_l}}\right), \ldots, \frac{\partial}{\partial x^{j_s}}\right)$$

$$= \sum_{l=1}^{s} \mathcal{A}\left(d(x^{i_1}), \ldots, d(x^{i_r}), \frac{\partial}{\partial x^{j_1}}, \ldots, \sum_{p=1}^{m} \Gamma_{kj_l}^{p} \frac{\partial}{\partial x^p}, \ldots, \frac{\partial}{\partial x^{j_s}}\right)$$

$$= \sum_{l=1}^{s}\sum_{p=1}^{m} \Gamma_{kj_l}^{p} \mathcal{A}\left(d(x^{i_1}), \ldots, d(x^{i_r}), \frac{\partial}{\partial x^{j_1}}, \ldots, \frac{\partial}{\partial x^p}, \ldots, \frac{\partial}{\partial x^{j_s}}\right)$$

$$\sum_{l=1}^{s}\left(\sum_{p=1}^{m} \Gamma_{kj_l}^{p} \mathcal{A}_{j_1 \ldots j_{l-1} \, p \, j_{l+1} \ldots j_s}^{i_1 \ldots i_r}\right),$$

where the first equality follows from (18.2.1). Combining the above identities gives the result. □

Theorem 18.2.3. *Let (M, ∇) be a smooth m-manifold with a connection, let $(U, (x^i))$ and $(\widetilde{U}, (\widetilde{x}^j))$ be overlapping charts on M, let $(\partial/\partial x^1, \ldots, \partial/\partial x^m)$ and $(\partial/\partial \widetilde{x}^1, \ldots, \partial/\partial \widetilde{x}^m)$ be the corresponding coordinate frames, and let Γ_{ij}^k and $\widetilde{\Gamma}_{pq}^r$ be the corresponding Christoffel symbols. Then*

$$\widetilde{\Gamma}_{pq}^r = \sum_{ijk} \frac{\partial x^i}{\partial \widetilde{x}^p} \frac{\partial x^j}{\partial \widetilde{x}^q} \frac{\partial \widetilde{x}^r}{\partial x^k} \Gamma_{ij}^k + \sum_{l} \frac{\partial^2 x^l}{\partial \widetilde{x}^p \partial \widetilde{x}^q} \frac{\partial \widetilde{x}^r}{\partial x^l}$$

on $U \cap \widetilde{U}$ for $p, q, r = 1, \ldots, m$.

Proof. We have

$$\nabla_{\partial/\partial\widetilde{x}^p}\left(\frac{\partial}{\partial\widetilde{x}^q}\right) = \sum_l \widetilde{\Gamma}^l_{pq}\frac{\partial}{\partial\widetilde{x}^l} \qquad [(18.2.1)]$$

$$= \sum_l \widetilde{\Gamma}^l_{pq}\left(\sum_k \frac{\partial x^k}{\partial\widetilde{x}^l}\frac{\partial}{\partial x^k}\right) \qquad [\text{Th } 15.1.8] \qquad (18.2.5)$$

$$= \sum_k \left(\sum_l \widetilde{\Gamma}^l_{pq}\frac{\partial x^k}{\partial\widetilde{x}^l}\right)\frac{\partial}{\partial x^k}.$$

On the other hand,

$$\nabla_{\partial/\partial\widetilde{x}^p}\left(\frac{\partial}{\partial\widetilde{x}^q}\right) = \nabla_{\sum_i(\partial x^i/\partial\widetilde{x}^p)(\partial/\partial x^i)}\left(\sum_j \frac{\partial x^j}{\partial\widetilde{x}^q}\frac{\partial}{\partial x^j}\right)$$

$$= \sum_{ij}\frac{\partial x^i}{\partial\widetilde{x}^p}\nabla_{\partial/\partial x^i}\left(\frac{\partial x^j}{\partial\widetilde{x}^q}\frac{\partial}{\partial x^j}\right)$$

$$= \sum_{ij}\frac{\partial x^i}{\partial\widetilde{x}^p}\left[\nabla_{\partial/\partial x^i}\left(\frac{\partial x^j}{\partial\widetilde{x}^q}\right)\frac{\partial}{\partial x^j} + \frac{\partial x^j}{\partial\widetilde{x}^q}\nabla_{\partial/\partial x^i}\left(\frac{\partial}{\partial x^j}\right)\right] \qquad (18.2.6)$$

$$= \sum_{ij}\frac{\partial x^i}{\partial\widetilde{x}^p}\left[\nabla_{\partial/\partial x^i}\left(\frac{\partial x^j}{\partial\widetilde{x}^q}\right)\frac{\partial}{\partial x^j} + \frac{\partial x^j}{\partial\widetilde{x}^q}\sum_k \Gamma^k_{ij}\frac{\partial}{\partial x^k}\right],$$

where the first equality follows from Theorem 15.1.8, and the last equality from (18.2.1). We also have

$$\nabla_{\partial/\partial x^i}\left(\frac{\partial x^j}{\partial\widetilde{x}^q}\right) = \nabla_{\sum_k(\partial\widetilde{x}^k/\partial x^i)(\partial/\partial\widetilde{x}^k)}\left(\frac{\partial x^j}{\partial\widetilde{x}^q}\right)$$

$$= \sum_k \frac{\partial\widetilde{x}^k}{\partial x^i}\nabla_{\partial/\partial\widetilde{x}^k}\left(\frac{\partial x^j}{\partial\widetilde{x}^q}\right) \qquad (18.2.7)$$

$$= \sum_k \frac{\partial\widetilde{x}^k}{\partial x^i}\frac{\partial^2 x^j}{\partial\widetilde{x}^k\partial\widetilde{x}^q},$$

where the first equality follows from Theorem 15.1.8. Combining (18.2.6) and (18.2.7) yields

$$\nabla_{\partial/\partial\widetilde{x}^p}\left(\frac{\partial}{\partial\widetilde{x}^q}\right)$$

$$= \sum_{ij}\frac{\partial x^i}{\partial\widetilde{x}^p}\left[\left(\sum_k \frac{\partial\widetilde{x}^k}{\partial x^i}\frac{\partial^2 x^j}{\partial\widetilde{x}^k\partial\widetilde{x}^q}\right)\frac{\partial}{\partial x^j} + \frac{\partial x^j}{\partial\widetilde{x}^q}\sum_k \Gamma^k_{ij}\frac{\partial}{\partial x^k}\right] \qquad (18.2.8)$$

$$= \sum_{ijk}\frac{\partial x^i}{\partial\widetilde{x}^p}\frac{\partial x^j}{\partial\widetilde{x}^q}\Gamma^k_{ij}\frac{\partial}{\partial x^k} + \sum_{ijk}\frac{\partial x^i}{\partial\widetilde{x}^p}\frac{\partial\widetilde{x}^k}{\partial x^i}\frac{\partial^2 x^j}{\partial\widetilde{x}^k\partial\widetilde{x}^q}\frac{\partial}{\partial x^j}.$$

The second term in the last line of (18.2.8) can be expressed as

$$\sum_{ijk} \frac{\partial x^i}{\partial \widetilde{x}^p} \frac{\partial \widetilde{x}^k}{\partial x^i} \frac{\partial^2 x^j}{\partial \widetilde{x}^k \partial \widetilde{x}^q} \frac{\partial}{\partial x^j} = \sum_{jk} \left[\left(\sum_i \frac{\partial \widetilde{x}^k}{\partial x^i} \frac{\partial x^i}{\partial \widetilde{x}^p} \right) \frac{\partial^2 x^j}{\partial \widetilde{x}^k \partial \widetilde{x}^q} \frac{\partial}{\partial x^j} \right]$$

$$= \sum_{jk} \frac{\partial \widetilde{x}^k}{\partial \widetilde{x}^p} \frac{\partial^2 x^j}{\partial \widetilde{x}^k \partial \widetilde{x}^q} \frac{\partial}{\partial x^j} = \sum_j \frac{\partial^2 x^j}{\partial \widetilde{x}^p \partial \widetilde{x}^q} \frac{\partial}{\partial x^j}$$

$$= \sum_k \frac{\partial^2 x^k}{\partial \widetilde{x}^p \partial \widetilde{x}^q} \frac{\partial}{\partial x^k},$$

so (18.2.8) becomes

$$\nabla_{\partial/\partial \widetilde{x}^p} \left(\frac{\partial}{\partial \widetilde{x}^q} \right) = \sum_k \left(\sum_{ij} \frac{\partial x^i}{\partial \widetilde{x}^p} \frac{\partial x^j}{\partial \widetilde{x}^q} \Gamma^k_{ij} + \frac{\partial^2 x^k}{\partial \widetilde{x}^p \partial \widetilde{x}^q} \right) \frac{\partial}{\partial x^k}. \tag{18.2.9}$$

Comparing (18.2.5) and (18.2.9) shows that

$$\sum_{ij} \frac{\partial x^i}{\partial \widetilde{x}^p} \frac{\partial x^j}{\partial \widetilde{x}^q} \Gamma^k_{ij} + \frac{\partial^2 x^k}{\partial \widetilde{x}^p \partial \widetilde{x}^q} = \sum_l \widetilde{\Gamma}^l_{pq} \frac{\partial x^k}{\partial \widetilde{x}^l}.$$

Multiplying both sides of the preceding identity by $\partial \widetilde{x}^r / \partial x^k$ and summing over k yields

$$\sum_{ijk} \frac{\partial x^i}{\partial \widetilde{x}^p} \frac{\partial x^j}{\partial \widetilde{x}^q} \frac{\partial \widetilde{x}^r}{\partial x^k} \Gamma^k_{ij} + \sum_k \frac{\partial^2 x^k}{\partial \widetilde{x}^p \partial \widetilde{x}^q} \frac{\partial \widetilde{x}^r}{\partial x^k} = \sum_{kl} \widetilde{\Gamma}^l_{pq} \frac{\partial \widetilde{x}^r}{\partial x^k} \frac{\partial x^k}{\partial \widetilde{x}^l}$$

$$= \sum_l \widetilde{\Gamma}^l_{pq} \left(\sum_k \frac{\partial \widetilde{x}^r}{\partial x^k} \frac{\partial x^k}{\partial \widetilde{x}^l} \right) = \sum_l \widetilde{\Gamma}^l_{pq} \frac{\partial \widetilde{x}^r}{\partial \widetilde{x}^l} = \widetilde{\Gamma}^r_{pq}. \qquad \square$$

We see from Theorem 15.6.3 and Theorem 18.2.3 that, despite appearances, the Christoffel symbols are not the components of a tensor. In classical terminology it is said that the Christoffel symbols "do not transform" like the components of a tensor.

18.3   Covariant Derivative on Curves

The definition of covariant derivative given in Section 18.1 refers to an entire smooth manifold. We now discuss a version of covariant derivative that applies only to smooth curves on a smooth manifold.

Theorem 18.3.1 (Existence of Covariant Derivative on a Curve). *If* (M, ∇) *is a smooth manifold with a connection and* $\lambda(t) : (a, b) \longrightarrow M$ *is a smooth curve, then there is a unique linear map*

$$\frac{\nabla}{dt} : \mathfrak{X}_M(\lambda) \longrightarrow \mathfrak{X}_M(\lambda),$$

*called the **covariant derivative on** λ, such that for all vector fields J, K in $\mathfrak{X}_M(\lambda)$, all functions f in $C^\infty(M)$, all vector fields X in $\mathfrak{X}(M)$, and all points t in (a, b):*

(a)
$$\frac{\nabla(J + K)}{dt}(t) = \frac{\nabla J}{dt}(t) + \frac{\nabla K}{dt}(t).$$

(b)
$$\frac{\nabla(fJ)}{dt}(t) = \frac{df}{dt}(t)\, J(t) + f(t)\, \frac{\nabla J}{dt}(t).$$

(c)
$$\frac{\nabla(X \circ \lambda)}{dt}(t) = \nabla_{(d\lambda/dt)(t)}(X).$$

Remark. The notation $\nabla_{(d\lambda/dt)(t)}(X)$ in part (c) is justified by the discussion surrounding (18.1.4). $\qquad\square$

Theorem 18.3.2 (Local Coordinate Expression for Covariant Derivative on a Curve). *Let (M, ∇) be a smooth manifold with a connection, let $\lambda(t) : (a, b) \longrightarrow M$ be a smooth curve, and let J be a vector field in $\mathfrak{X}_M(\lambda)$. Let $(U, (x^i))$ be a chart on M such that $\lambda((a, b)) \subset U$, and, in local coordinates, let*

$$J(t) = \sum_i \alpha^i(t)\, \frac{\partial}{\partial x^i}\bigg|_{\lambda(t)}.$$

Then

$$\frac{\nabla J}{dt}(t) = \sum_k \left(\frac{d\alpha^k}{dt}(t) + \sum_{ij} \frac{d(x^i \circ \lambda)}{dt}(t)\, \alpha^j(t)\, \Gamma^k_{ij}\big(\lambda(t)\big) \right) \frac{\partial}{\partial x^k}\bigg|_{\lambda(t)}$$

for all t in (a, b).

Proof. Theorem 18.3.1 is used repeatedly in what follows. We have

$$\frac{\nabla}{dt}\left(\frac{\partial}{\partial x^j}\bigg|_{\lambda(t)} \right) = \nabla_{(d\lambda/dt)(t)}\left(\frac{\partial}{\partial x^j} \right),$$

so

$$\begin{aligned}
\frac{\nabla J}{dt}(t) &= \sum_j \frac{\nabla}{dt}\left(\alpha^j(t)\, \frac{\partial}{\partial x^j}\bigg|_{\lambda(t)} \right) \\
&= \sum_j \frac{d\alpha^j}{dt}(t)\, \frac{\partial}{\partial x^j}\bigg|_{\lambda(t)} + \sum_j \alpha^j(t)\, \nabla_{(d\lambda/dt)(t)}\left(\frac{\partial}{\partial x^j} \right).
\end{aligned} \tag{18.3.1}$$

By Theorem 14.7.1(b),

$$\frac{d\lambda}{dt}(t) = \sum_i \frac{d(x^i \circ \lambda)}{dt}(t)\, \frac{\partial}{\partial x^i}\bigg|_{\lambda(t)},$$

hence

$$\nabla_{(d\lambda/dt)(t)}\left(\frac{\partial}{\partial x^j}\right)$$

$$= \sum_i \frac{d(x^i \circ \lambda)}{dt}(t)\, \nabla_{(\partial/\partial x^i)(\lambda(t))}\left(\frac{\partial}{\partial x^j}\right)$$

$$= \sum_i \frac{d(x^i \circ \lambda)}{dt}(t)\left(\sum_k \Gamma_{ij}^k\big(\lambda(t)\big)\left.\frac{\partial}{\partial x^k}\right|_{\lambda(t)}\right) \qquad [(18.1.5),\ (18.2.1)]$$

$$= \sum_{ik} \frac{d(x^i \circ \lambda)}{dt}(t)\, \Gamma_{ij}^k\big(\lambda(t)\big)\left.\frac{\partial}{\partial x^k}\right|_{\lambda(t)}.$$

The second term in the last line of (18.3.1) can therefore be expressed as

$$\sum_j \alpha^j(t)\, \nabla_{(d\lambda/dt)(t)}\left(\frac{\partial}{\partial x^j}\right) = \sum_j \alpha^j(t)\left(\sum_{ik} \frac{d(x^i \circ \lambda)}{dt}(t)\, \Gamma_{ij}^k\big(\lambda(t)\big)\left.\frac{\partial}{\partial x^k}\right|_{\lambda(t)}\right)$$

$$= \sum_k\left(\sum_{ij} \frac{d(x^i \circ \lambda)}{dt}(t)\, \alpha^j(t)\, \Gamma_{ij}^k\big(\lambda(t)\big)\right)\left.\frac{\partial}{\partial x^k}\right|_{\lambda(t)}.$$

Then (18.3.1) becomes

$$\frac{\nabla J}{dt}(t)$$

$$= \sum_k \frac{d\alpha^k}{dt}(t)\left.\frac{\partial}{\partial x^k}\right|_{\lambda(t)} + \sum_k\left(\sum_{ij} \frac{d(x^i \circ \lambda)}{dt}(t)\, \alpha^j(t)\, \Gamma_{ij}^k\big(\lambda(t)\big)\right)\left.\frac{\partial}{\partial x^k}\right|_{\lambda(t)}$$

$$= \sum_k\left(\frac{d\alpha^k}{dt}(t) + \sum_{ij} \frac{d(x^i \circ \lambda)}{dt}(t)\, \alpha^j(t)\, \Gamma_{ij}^k\big(\lambda(t)\big)\right)\left.\frac{\partial}{\partial x^k}\right|_{\lambda(t)}. \qquad \square$$

Theorem 18.3.3. *Let (M, ∇) be a smooth manifold with a connection, let $\lambda(t) : (a, b) \longrightarrow M$ be a smooth curve, let J be a vector field in $\mathfrak{X}_M(\lambda)$, and let $g(u) : (c, d) \longrightarrow (a, b)$ be a diffeomorphism. Then, for all u in (c, d):*
(a)

$$\frac{\nabla(J \circ g)}{du}(u) = \frac{dg}{du}(u)\, \frac{\nabla J}{dt}\big(g(u)\big).$$

(b)

$$\frac{\nabla}{du}\left(\frac{d(\lambda \circ g)}{du}\right)(u) = \frac{d^2 g}{du^2}(u)\, \frac{d\lambda}{dt}\big(g(u)\big) + \left(\frac{dg}{du}(u)\right)^2 \frac{\nabla}{dt}\left(\frac{d\lambda}{dt}\right)\big(g(u)\big).$$

Remark. ∇/dt and ∇/du denote covariant differentiation on λ and $\lambda \circ g$, respectively.

Proof. We consider only the case where $(U, (x^i))$ is a chart on M such that $\lambda(a, b) \subset U$.

(a): In local coordinates, let

$$J(t) = \sum_k \alpha^k(t) \left. \frac{\partial}{\partial x^k} \right|_{\lambda(t)},$$

so that

$$J \circ g(u) = \sum_k (\alpha^k \circ g)(u) \left. \frac{\partial}{\partial x^k} \right|_{\lambda \circ g(u)}.$$

Then

$$
\begin{aligned}
\frac{\nabla(J \circ g)}{du}(u) &= \sum_k \left(\frac{d(\alpha^k \circ g)}{du}(u) \right. \\
&\quad \left. + \sum_{ij} \frac{d(x^i \circ \lambda \circ g)}{du}(u) \, (\alpha^j \circ g)(u) \, \Gamma_{ij}^k \big(\lambda \circ g(u)\big) \right) \left. \frac{\partial}{\partial x^k} \right|_{\lambda \circ g(u)} \\
&= \sum_k \left(\frac{d\alpha^k}{dt}(g(u)) \, \frac{dg}{du}(u) \right. \\
&\quad \left. + \sum_{ij} \frac{d(x^i \circ \lambda)}{dt}(g(u)) \, \frac{dg}{du}(u) \, \alpha^j(g(u)) \, \Gamma_{ij}^k \big(\lambda(g(u))\big) \right) \left. \frac{\partial}{\partial x^k} \right|_{\lambda(g(u))} \\
&= \frac{dg}{du}(u) \sum_k \left(\frac{d\alpha^k}{dt}(g(u)) \right. \\
&\quad \left. + \sum_{ij} \frac{d(x^i \circ \lambda)}{dt}(g(u)) \, \alpha^j(g(u)) \, \Gamma_{ij}^k \big(\lambda(g(u))\big) \right) \left. \frac{\partial}{\partial x^k} \right|_{\lambda(g(u))} \\
&= \frac{dg}{du}(u) \, \frac{\nabla J}{dt}(g(u)),
\end{aligned}
$$

where the first and last equalities follow from Theorem 18.3.2, and the second equality from Theorem 14.4.5.

(b): We have

$$
\begin{aligned}
\frac{\nabla}{du} \left(\frac{d(\lambda \circ g)}{du} \right)(u) &= \frac{\nabla}{du} \left(\left(\frac{dg}{du} \right) \left(\frac{d\lambda}{dt} \circ g \right) \right)(u) \\
&= \frac{d^2 g}{du^2}(u) \, \frac{d\lambda}{dt}(g(u)) + \frac{dg}{du}(u) \, \frac{\nabla}{du} \left(\frac{d\lambda}{dt} \circ g \right)(u) \\
&= \frac{d^2 g}{du^2}(u) \, \frac{d\lambda}{dt}(g(u)) + \left(\frac{dg}{du}(u) \right)^2 \frac{\nabla}{dt} \left(\frac{d\lambda}{dt} \right)(g(u)),
\end{aligned}
$$

where the first equality follows from Theorem 14.4.5, the second equality from Theorem 18.3.1(b), and the last equality from part (a). $\qquad \square$

The above discussion is devoted to covariant derivatives on smooth curves. We close this section with an application to parametrized surfaces.

Let M be a smooth manifold, let $\sigma(r,s) : (a,b) \times (-\varepsilon, \varepsilon) \longrightarrow M$ be a parametrized surface, and let V be a vector field in $\mathfrak{X}_M(\sigma)$. For a given point s in $(-\varepsilon, \varepsilon)$, we define a vector field V_s in $\mathfrak{X}_M(\sigma_s)$ by $V_s(r) = V(r,s)$, where we recall from Section 14.9 that the longitude curve $\sigma_s(r) : (a,b) \longrightarrow M$ is defined by $\sigma_s(r) = \sigma(r,s)$. The **partial covariant derivative on σ with respect to r** is the map

$$\frac{\nabla}{\partial r} : \mathfrak{X}_M(\sigma) \longrightarrow \mathfrak{X}_M(\sigma)$$

defined by

$$\frac{\nabla V}{\partial r}(r,s) = \frac{\nabla V_s}{dr}(r) \tag{18.3.2}$$

for all vector fields V in $\mathfrak{X}_M(\sigma)$. The partial covariant derivative on σ with respect to s is defined similarly. As an example, we have from Theorem 15.1.5 that the vectors fields $\partial\sigma/\partial r$ and $\partial\sigma/\partial s$ are in $\mathfrak{X}_M(\sigma)$. The partial covariant derivatives on σ with respect to r are

$$\frac{\nabla}{\partial r}\left(\frac{\partial\sigma}{\partial r}\right) \quad \text{and} \quad \frac{\nabla}{\partial r}\left(\frac{\partial\sigma}{\partial s}\right), \tag{18.3.3}$$

and the partial covariant derivatives on σ with respect to s are defined in a corresponding manner.

18.4   Total Covariant Derivatives

Let (M, ∇) be a smooth manifold with a connection. **Total covariant derivative** (or **total covariant differential**) is the family of linear maps

$$\nabla : \mathcal{T}_s^r(M) \longrightarrow \mathcal{T}_{s+1}^r(M)$$

defined for $r, s \geq 0$ by

$$\nabla(\mathcal{A})(\omega^1, \ldots, \omega^r, X, Y_1, \ldots, Y_s) = \nabla_X(\mathcal{A})(\omega^1, \ldots, \omega^r, Y_1, \ldots, Y_s) \tag{18.4.1}$$

for all tensor fields $\mathcal{A}$ in $\mathcal{T}_s^r(M)$, all covector fields $\omega^1, \ldots, \omega^r$ in $\mathfrak{X}^*(M)$, and all vector fields $X, Y_1, \ldots, Y_s$ in $\mathfrak{X}(M)$. An alternative definition in the literature (not adopted here) is to place X last in the sequence of vector fields rather than first:

$$\nabla(\mathcal{A})(\omega^1, \ldots, \omega^r, Y_1, \ldots, Y_s, X) = \nabla_X(\mathcal{A})(\omega^1, \ldots, \omega^r, Y_1, \ldots, Y_s).$$

We say that $\mathcal{A}$ is **parallel** (with respect to ∇) if $\nabla(\mathcal{A}) = 0$. Let us denote

$$\nabla \circ \nabla \quad \text{by} \quad \nabla^2.$$

Theorem 18.4.1. *With the above setup, $\mathcal{A}$ is parallel with respect to ∇ if and only if $\nabla_X(\mathcal{A}) = 0$ for all vector fields X in $\mathfrak{X}(M)$.*

Proof. This follows from (18.4.1). $\qquad\qquad\qquad\qquad\qquad\qquad\qquad\qquad\square$

Let $(U, (x^i))$ be a chart on M, and, in local coordinates, let $\mathcal{A}$ have the components $\mathcal{A}^{i_1 \ldots i_r}_{j_1 \ldots j_s}$. The components of $\nabla(\mathcal{A})$ are denoted by

$$\mathcal{A}^{i_1 \ldots i_r}_{j_1 \ldots j_s; k}$$

(not by $\nabla(\mathcal{A})^{i_1 \ldots i_r}_{k j_1 \ldots j_s}$, as might be expected from (18.4.1) and previous notation conventions).

Theorem 18.4.2. *With the above setup,*

$$\mathcal{A}^{i_1 \ldots i_r}_{j_1 \ldots j_s; k} = \frac{\partial \mathcal{A}^{i_1 \ldots i_r}_{j_1 \ldots j_s}}{\partial x^k} + \sum_{l=1}^{r} \left(\sum_{p=1}^{m} \Gamma^{i_l}_{kp} \mathcal{A}^{i_1 \ldots i_{l-1} \, p \, i_{l+1} \ldots i_r}_{j_1 \ldots j_s} \right)$$

$$- \sum_{l=1}^{s} \left(\sum_{p=1}^{m} \Gamma^{p}_{k j_l} \mathcal{A}^{i_1 \ldots i_r}_{j_1 \ldots j_{l-1} \, p \, j_{l+1} \ldots j_s} \right). \tag{18.4.2}$$

Proof. By definition, $\mathcal{A}^{i_1 \ldots i_r}_{j_1 \ldots j_s; k} = \nabla_{\partial / \partial x^k} (\mathcal{A})^{i_1 \ldots i_r}_{j_1 \ldots j_s}$, so the result follows from Theorem 18.2.2. □

We observe that the notation introduced in (18.2.2) and (18.2.3) is consistent with (18.4.2).

Theorem 18.4.3. *If (M, ∇) is a smooth manifold with a connection and f is a function in $C^\infty(M)$, then*

$$\nabla(f) = d(f).$$

Proof. For a vector field X in $\mathfrak{X}(M)$, we have

$$\begin{aligned}
\nabla(f)(X) &= \nabla_X(f) && [(18.4.1)] \\
&= X(f) && [(18.1.1)] \\
&= d(f)(X). && [\text{Th } 15.4.5(b)]
\end{aligned}$$

Since X was arbitrary, the result follows. □

Theorem 18.4.4. *Let (M, ∇) be a smooth manifold with a connection, let X, Y be vector fields in $\mathfrak{X}(M)$, and let ω be a covector field in $\mathfrak{X}^*(M)$. Then:*
(a)
$$\nabla(X)(\omega, Y) = \omega(\nabla_Y(X)).$$

(b) *If $X = \sum_i \alpha^i (\partial / \partial x^i)$ in local coordinates, then*

$$\nabla(X) = \sum_{ij} \alpha^i_{;j} \frac{\partial}{\partial x^i} \otimes d(x^j).$$

Proof. (a): We have

$$\begin{aligned}
\nabla(X)(\omega, Y) &= \nabla_Y(X)(\omega) && [(18.4.1)] \\
&= \omega(\nabla_Y(X)). && [(15.4.2)]
\end{aligned}$$

(b): The components of $\nabla(X)$ are

$$\nabla(X)\left(d(x^i), \frac{\partial}{\partial x^j}\right) = d(x^i)\left(\nabla_{\partial/\partial x^j}(X)\right) \qquad \text{[part (a)]}$$

$$= d(x^i)\left(\sum_k \alpha^k_{;j} \frac{\partial}{\partial x^k}\right) \qquad \text{[Th 18.2.1(a)]}$$

$$= \sum_k \alpha^k_{;j}\, d(x^i)\left(\frac{\partial}{\partial x^k}\right) = \alpha^i_{;j},$$

from which the result follows.   □

Theorem 18.4.5. *Let (M, ∇) be a smooth manifold with a connection, let X, Y be vector fields in $\mathfrak{X}(M)$, and let ω be a covector field in $\mathfrak{X}^*(M)$. Then:*
(a)
$$\nabla(\omega)(X, Y) = X\big(\omega(Y)\big) - \omega\big(\nabla_X(Y)\big).$$
(b) *If $\omega = \sum_k \gamma_k d(x^k)$ in local coordinates, then*
$$\nabla(\omega) = \sum_{ij} \gamma_{j;i}\, d(x^i) \otimes d(x^j).$$

Proof. (a): We have

$$\nabla(\omega)(X, Y) = \nabla_X(\omega)(Y) \qquad \text{[(18.4.1)]}$$
$$= X\big(\omega(Y)\big) - \omega\big(\nabla_X(Y)\big). \qquad \text{[(18.1.2)]}$$

(b): The components of $\nabla(\omega)$ are

$$\nabla(\omega)\left(\frac{\partial}{\partial x^i}, \frac{\partial}{\partial x^j}\right) = \nabla_{\partial/\partial x^i}(\omega)\left(\frac{\partial}{\partial x^j}\right) \qquad \text{[(18.4.1)]}$$

$$= \left(\sum_k \gamma_{k;i} d(x^k)\right)\left(\frac{\partial}{\partial x^j}\right) \qquad \text{[Th 18.2.1(b)]}$$

$$= \sum_k \gamma_{k;i}\, d(x^k)\left(\frac{\partial}{\partial x^j}\right) = \gamma_{j;i},$$

from which the result follows.   □

Theorem 18.4.6. *Let (M, ∇) be a smooth manifold with a connection, let $\mathcal{A}$ be a tensor field in $T^0_2(M)$, and let X, Y, Z be vector fields in $\mathfrak{X}(M)$. Then*

$$\nabla(\mathcal{A})(X, Y, Z) = X\big(\mathcal{A}(Y, Z)\big) - \mathcal{A}\big(\nabla_X(Y), Z\big) - \mathcal{A}\big(Y, \nabla_X(Z)\big).$$

Proof. We have

$$\nabla(\mathcal{A})(X, Y, Z)$$
$$= \nabla_X(\mathcal{A})(Y, Z) \qquad \text{[(18.4.1)]}$$
$$= X\big(\mathcal{A}(Y, Z)\big) - \mathcal{A}\big(\nabla_X(Y), Z\big) - \mathcal{A}\big(Y, \nabla_X(Z)\big). \qquad \text{[(18.1.3)]} \qquad □$$

Theorem 18.4.7. *Let* (M, ∇) *be a smooth manifold with a connection, let f be a function in $C^\infty(M)$, and let X, Y be vector fields in $\mathfrak{X}(M)$. Then*

$$\nabla^2(f)(X, Y) = X(Y(f)) - \nabla_X(Y)(f).$$

Proof. We have

$$
\begin{aligned}
\nabla^2(f)(X, Y) &= \nabla(\nabla(f))(X, Y) \\
&= \nabla(d(f))(X, Y) && \text{[Th 18.4.3]} \\
&= X(d(f)(Y)) - d(f)(\nabla_X(Y)) && \text{[Th 18.4.5(a)]} \\
&= X(Y(f)) - \nabla_X(Y)(f). && \text{[Th 15.4.5(b)]} \qquad \square
\end{aligned}
$$

Adopting the notation of Theorem 18.4.7 and using (18.1.1) three times, we obtain

$$
\begin{aligned}
\nabla^2(f)(X, Y) &= X(Y(f)) - \nabla_X(Y)(f) \\
&= \nabla_X(\nabla_Y(f)) - \nabla_{\nabla_X(Y)}(f).
\end{aligned}
\tag{18.4.3}
$$

This shows that in the context of regular surfaces in $\mathbb{R}^3_\nu$, $\nabla^2(f)(X, Y)$ is precisely $\nabla^2_{X,Y}(f)$, the second order covariant derivative of f with respect to X and Y, as given by (12.4.3).

18.5 Parallel Translation

Let (M, ∇) be a smooth manifold with a connection, and let $\lambda : (a, b) \longrightarrow M$ be a smooth curve. We say that a vector field J in $\mathfrak{X}_M(\lambda)$ is **parallel** (with respect to ∇) if

$$\frac{\nabla J}{dt}(t) = 0$$

for all t in (a, b).

Theorem 18.5.1 (Existence of Parallel Vector Field on a Curve). *Let (M, ∇) be a smooth m-manifold with a connection, let $\lambda(t) : (a, b) \longrightarrow M$ be a smooth curve, let t_0 be a point in (a, b), and let v be a vector in $T_{\lambda(t_0)}(M)$. Then:*
(a) *There is a unique parallel vector field J in $\mathfrak{X}_M(\lambda)$ such that $J(t_0) = v$.*
(b) *Let $(U, (x^i))$ be a chart on M such that $\lambda((a, b)) \subset U$, and, in local coordinates, let*

$$J(t) = \sum_i \alpha^i(t) \left.\frac{\partial}{\partial x^i}\right|_{\lambda(t)}.\tag{18.5.1}$$

Then the α^i satisfy the following system of differential equations, with initial condition $J(t_0) = v$:

$$\frac{d\alpha^k}{dt}(t) + \sum_{ij} \frac{d(x^i \circ \lambda)}{dt}(t)\, \alpha^j(t)\, \Gamma^k_{ij}(\lambda(t)) = 0$$

for all t in (a, b) for $k = 1, \ldots, m$.

Proof. By Theorem 18.3.2, the identity $(\nabla J/dt)(t) = 0$ for all t in (a, b) is equivalent to the above system of ordinary differential equations. The result now follows from the theory of differential equations. $\square$

Continuing with the notation of Theorem 18.5.1, let us denote

$$J(t) \qquad \text{by} \qquad \Pi_{\lambda(t_0)}^{\lambda(t)}(v).$$

Then (18.5.1) becomes

$$\Pi_{\lambda(t_0)}^{\lambda(t)}(v) = \sum_i \alpha^i(t) \frac{\partial}{\partial x^i}\bigg|_{\lambda(t)}.$$

Fixing t_0 and t in (a, b) and allowing v to vary, the assignment

$$v \longmapsto \Pi_{\lambda(t_0)}^{\lambda(t)}(v)$$

for all vectors v in $T_{\lambda(t_0)}(M)$ defines a map

$$\Pi_{\lambda(t_0)}^{\lambda(t)} : T_{\lambda(t_0)}(M) \longrightarrow T_{\lambda(t)}(M)$$

called **parallel translation on λ** from $\lambda(t_0)$ to $\lambda(t)$. We say that $\Pi_{\lambda(t_0)}^{\lambda(t)}(v)$ is the **parallel translate of v**. Let us observe that the initial condition of Theorem 18.5.1 can now be expressed as $\Pi_{\lambda(t_0)}^{\lambda(t_0)}(v) = v$.

Example 18.5.2 (Smooth 2-Manifolds). For smooth 2-manifolds with a connection, the differential equations in Theorem 18.5.1 are simply

$$\frac{d\alpha^1}{dt}(t) + \frac{d(x^1 \circ \lambda)}{dt}(t)\big[\alpha^1(t)\,\Gamma_{11}^1(\lambda(t)) + \alpha^2(t)\,\Gamma_{12}^1(\lambda(t))\big]$$
$$+ \frac{d(x^2 \circ \lambda)}{dt}(t)\big[\alpha^1(t)\,\Gamma_{21}^1(\lambda(t)) + \alpha^2(t)\,\Gamma_{22}^1(\lambda(t))\big] = 0$$

and

$$\frac{d\alpha^2}{dt}(t) + \frac{d(x^1 \circ \lambda)}{dt}(t)\big[\alpha^1(t)\,\Gamma_{11}^2(\lambda(t)) + \alpha^2(t)\,\Gamma_{12}^2(\lambda(t))\big]$$
$$+ \frac{d(x^2 \circ \lambda)}{dt}(t)\big[\alpha^1(t)\,\Gamma_{21}^2(\lambda(t)) + \alpha^2(t)\,\Gamma_{22}^2(\lambda(t))\big] = 0.$$

In particular, these equations apply to regular surfaces with the connection defined in Example 18.1.1. $\Diamond$

Example 18.5.3 (Pln). Let us view the regular surface Pln, discussed in Section 13.1, as a smooth 2-manifold with the connection defined in Example 18.1.1. Let $\lambda(t) : (a, b) \longrightarrow$ Pln be a smooth curve, let t_0 be a point in (a, b), and let v be a vector in Pln. The Christoffel symbols have constant value 0, so $\Pi_{\lambda(t_0)}^{\lambda(t)}(v) = (\alpha^1(t), \alpha^2(t))$. From Example 18.5.2, the differential equations are $(d\alpha^k/dt)(t) = 0$ for $k = 1, 2$, with the initial condition $\Pi_{\lambda(t_0)}^{\lambda(t_0)}(v) = v$. It follows that the α^k are independent of t, so $\Pi_{\lambda(t_0)}^{\lambda(t)}(v) = v$ for all t in (a, b). This shows that parallel translation of a vector from one point to another in Pln is independent of the choice of smooth curve joining the two points. In particular, parallel translation of a vector along the entire trajectory of a closed smooth curve in Pln, such as a circle, takes the vector to itself.   ◇

Example 18.5.4 ($\mathcal{S}^2$). Let us view the regular surface $\mathcal{S}^2$, discussed in Section 13.4, as a smooth 2-manifold with the connection defined in Example 18.1.1. Let

$$\theta(t) : (a, b) \longrightarrow (0, \pi) \qquad \text{and} \qquad \phi(t) : (a, b) \longrightarrow (-\pi, \pi)$$

be diffeomorphisms, and consider the smooth curve $\lambda(t) : (a, b) \longrightarrow \mathcal{S}^2$ given by

$$\lambda(t) = \varphi\big(\theta(t), \phi(t)\big) = \Big(\sin\big(\theta(t)\big)\cos\big(\phi(t)\big), \sin\big(\theta(t)\big)\sin\big(\phi(t)\big), \cos\big(\theta(t)\big)\Big).$$

The coordinate basis at $\lambda(t)$ is $(H_1|_{(\theta(t),\phi(t))}, H_2|_{(\theta(t),\phi(t))})$, where

$$H_1|_{(\theta(t),\phi(t))} = \Big(\cos\big(\theta(t)\big)\cos\big(\phi(t)\big), \cos\big(\theta(t)\big)\sin\big(\phi(t)\big), -\sin\big(\theta(t)\big)\Big)$$

$$H_2|_{(\theta(t),\phi(t))} = \Big(-\sin\big(\theta(t)\big)\sin\big(\phi(t)\big), \sin\big(\theta(t)\big)\cos\big(\phi(t)\big), 0\Big).$$

Setting

$$v = \frac{d\lambda}{dt}(t_0) \tag{18.5.2}$$

in Theorem 18.5.1 and recalling the results on Christoffel symbols in Section 13.4, the differential equations in Example 18.5.2 become

$$\frac{d\alpha^1}{dt}(t) - \cos\big(\theta(t)\big)\sin\big(\theta(t)\big)\frac{d\phi}{dt}(t)\,\alpha^2(t) = 0$$

$$\frac{d\alpha^2}{dt}(t) + \cot\big(\theta(t)\big)\left(\frac{d\phi}{dt}(t)\,\alpha^1(t) + \frac{d\theta}{dt}(t)\,\alpha^2(t)\right) = 0, \tag{18.5.3}$$

where we have used

$$\big(x^1 \circ \lambda(t), x^2 \circ \lambda(t)\big) = (x^1, x^2) \circ \lambda(t) = \varphi^{-1} \circ \lambda(t) = \big(\theta(t), \phi(t)\big).$$

For a given point θ_0 in $(0, \pi)$, let θ be the function that has constant value θ_0, and let ϕ be the identity function. As well, let us denote t by ϕ, so that $\phi_0 = t_0$ and ϕ is in $(-\pi, \pi)$. Setting $c_0 = \cos(\theta_0)$ and $s_0 = \sin(\theta_0)$, we have

$$\lambda(\phi) = \big(s_0 \cos(\phi), s_0 \sin(\phi), c_0\big) \tag{18.5.4}$$

$$\frac{d\lambda}{d\phi}(\phi) = \big(-s_0 \sin(\phi), s_0 \cos(\phi), 0\big) \tag{18.5.5}$$

$$H_1|_{(\theta_0,\phi)} = \big(c_0 \cos(\phi), c_0 \sin(\phi), -s_0\big)$$

$$H_2|_{(\theta_0,\phi)} = \big(-s_0 \sin(\phi), s_0 \cos(\phi), 0\big) \tag{18.5.6}$$

$$\frac{d\alpha^1}{d\phi}(\phi) - c_0 s_0 \alpha^2(\phi) = 0$$

$$\frac{d\alpha^2}{d\phi}(\phi) + \frac{c_0}{s_0} \alpha^1(\phi) = 0, \tag{18.5.7}$$

where we assume $s_0 \neq 0$, or equivalently, $\theta_0 \neq \pi/2$. Observe that the image of λ falls on the latitude curve of $\mathcal{S}^2$ corresponding to θ_0. See Figure 13.4.1.

Now set $\phi_0 = 0$. Then (18.5.2) and (18.5.5) give $v = (0, s_0, 0)$. We seek an explicit expression for

$$\Pi_{\lambda(0)}^{\lambda(\phi)}\big((0, s_0, 0)\big) = \alpha^1(\phi) H_1|_{(\theta_0,\phi)} + \alpha^2(\phi) H_2|_{(\theta_0,\phi)}, \tag{18.5.8}$$

subject to the initial condition

$$(0, s_0, 0) = \Pi_{\lambda(0)}^{\lambda(0)}\big((0, s_0, 0)\big) = \alpha^1(0) H_1|_{(\theta_0,0)} + \alpha^2(0) H_2|_{(\theta_0,0)}. \tag{18.5.9}$$

The solution of (18.5.7) is

$$\alpha^1(\phi) = A\cos(c_0\phi) - B\sin(c_0\phi)$$

$$\alpha^2(\phi) = -\frac{1}{s_0}[A\sin(c_0\phi) + B\cos(c_0\phi)] \tag{18.5.10}$$

for some real numbers A, B. Setting $\phi = 0$ in (18.5.6) and (18.5.10) gives

$$H_1|_{(\theta_0,0)} = (c_0, 0, -s_0) \qquad H_2|_{(\theta_0,0)} = (0, s_0, 0) \tag{18.5.11}$$

and

$$\alpha^1(0) = A \qquad \alpha^2(0) = -\frac{B}{s_0}. \tag{18.5.12}$$

It follows from (18.5.9), (18.5.11), and (18.5.12) that

$$(0, s_0, 0) = A(c_0, 0, -s_0) - \frac{B}{s_0}(0, s_0, 0) = (Ac_0, -B, -As_0),$$

hence $A = 0$ and $B = -s_0$. We then have from (18.5.10) that

$$\alpha^1(\phi) = s_0 \sin(c_0\phi) \qquad \alpha^2(\phi) = \cos(c_0\phi). \tag{18.5.13}$$

Finally, substituting (18.5.6) and (18.5.13) into (18.5.8) yields

$$\Pi_{\lambda(0)}^{\lambda(\phi)}\big((0, s_0, 0)\big) = s_0 \sin(c_0\phi)\big(c_0\cos(\phi), c_0\sin(\phi), -s_0\big)$$
$$+ \cos(c_0\phi)\big(-s_0\sin(\phi), s_0\cos(\phi), 0\big).$$

For purposes of investigating the properties of parallel translation of $v = (0, s_0, 0)$, there is no harm in extending the domain of ϕ to all of $\mathbb{R}$. In particular, we have for $\phi = 2\pi$ that

$$\Pi_{\lambda(0)}^{\lambda(2\pi)}\big((0, s_0, 0)\big) = s_0 \sin(2\pi c_0)(c_0, 0, -s_0) + \cos(2\pi c_0)(0, s_0, 0).$$

It is easily shown that $\Pi_{\lambda(0)}^{\lambda(2\pi)}((0, s_0, 0)) = (0, s_0, 0)$ if and only if $\theta_0 = \pi/2$; that is, if and only if the image of λ lies on the "equator". We therefore have the surprising finding that parallel translation of v along the entire trajectory of a latitude curve other than the equator does not send v to itself. This is a manifestation of the "curvature" of the sphere.

The intersection of S^2 and a plane through the origin is called a **great circle** of S^2, the equator being the obvious example. The parametrization we selected at the outset features the equator, but this was a matter of convenience. Due to the symmetry of S^2, any great circle would do.   ◇

Theorem 18.5.5 (Diffeomorphism Invariance of Parallelism of a Curve). *Let (M, ∇) be a smooth manifold with a connection, let $\lambda(t) : (a, b) \longrightarrow M$ be a smooth curve, and let $g(u) : (c, d) \longrightarrow (a, b)$ be a diffeomorphism. Then the vector field J in $\mathfrak{X}_M(\lambda)$ is parallel if and only if the vector field $J \circ g$ in $\mathfrak{X}_M(\lambda \circ g)$ is parallel.*

Proof. This follows from Theorem 10.2.4(a) and Theorem 18.3.3(a).   □

Theorem 18.5.6. *Let (M, ∇) be a smooth manifold with a connection, let $\lambda : (a, b) \longrightarrow M$ be a smooth curve, and let t_0 and t be points in (a, b). Then $\Pi_{\lambda(t_0)}^{\lambda(t)} : T_{\lambda(t_0)}(M) \longrightarrow T_{\lambda(t)}(M)$ is a linear isomorphism, and its inverse*

$$\left(\Pi_{\lambda(t_0)}^{\lambda(t)}\right)^{-1} : T_{\lambda(t)}(M) \longrightarrow T_{\lambda(t_0)}(M)$$

is given by

$$\left(\Pi_{\lambda(t_0)}^{\lambda(t)}\right)^{-1}(v) = \Pi_{\lambda(t)}^{\lambda(t_0)}(v)$$

for all vectors v in $T_{\lambda(t)}(M)$.

Proof. Let v, w be vectors in $T_{\lambda(t_0)}(M)$. By Theorem 18.5.1, there are parallel vector fields J, K in $\mathfrak{X}_M(\lambda)$ such that $J(t_0) = v$ and $K(t_0) = w$. For a real number c, we have

$$\frac{\nabla(cJ + K)}{dt}(t) = c\frac{\nabla J}{dt}(t) + \frac{\nabla K}{dt}(t) = 0$$

for all t in (a, b). Thus, $cJ + K$ is a parallel vector field in $\mathfrak{X}_M(\lambda)$ such that

$$(cJ + K)(t_0) = cJ(t_0) + K(t_0) = cv + w.$$

It follows from the uniqueness property in Theorem 18.5.1 that

$$\Pi_{\lambda(t_0)}^{\lambda(t)}(cv + w) = (cJ + K)(t) = cJ(t) + K(t)$$
$$= c\Pi_{\lambda(t_0)}^{\lambda(t)}(v) + \Pi_{\lambda(t_0)}^{\lambda(t)}(w).$$

Thus, $\Pi_{\lambda(t_0)}^{\lambda(t)}$ is linear.

Let v and w be vectors in $T_{\lambda(t_0)}(M)$ and $T_{\lambda(t)}(M)$, respectively. It follows from the properties of ordinary differential equations that w is the parallel translate of v if and only if v is the parallel translate of w. Thus, $\Pi_{\lambda(t_0)}^{\lambda(t)}$ is a bijective map, and therefore a linear isomorphism, with inverse $\left(\Pi_{\lambda(t_0)}^{\lambda(t)}\right)^{-1} = \Pi_{\lambda(t)}^{\lambda(t_0)}$.   □

Theorem 18.5.7 (Existence of Parallel Frame on a Curve). *Let* (M, ∇) *be a smooth m-manifold with a connection, let* $\lambda(t) : (a, b) \longrightarrow M$ *be a smooth curve, let* t_0 *be a point in* (a, b), *and let* $(v_1, \ldots, v_m)$ *be a basis for* $T_{\lambda(t_0)}(M)$. *Then there is a unique frame* $\mathcal{J}$ *on* λ *consisting of parallel vector fields in* $\mathfrak{X}_M(\lambda)$ *such that* $\mathcal{J}(t_0) = (v_1, \ldots, v_m)$ *and* $\mathcal{J}(t)$ *is a basis for* $T_{\lambda(t)}(M)$ *for all* t *in* (a, b).

Proof. According to Theorem 18.5.1, there is a unique parallel vector field J_i in $\mathfrak{X}_M(\lambda)$ such that $J_i(t_0) = v_i$ for $i = 1, \ldots, m$. By definition, $J_i(t) = \Pi^{\lambda(t)}_{\lambda(t_0)}(v_i)$ for all t in (a, b). Suppose there is a point t in (a, b) such that $\sum_i a^i J_i(t) = 0$ for some real numbers $a^1, \ldots, a^m$. Using Theorem 18.5.6 to apply $\left(\Pi^{\lambda(t)}_{\lambda(t_0)}\right)^{-1}$ to both sides of the preceding identity yields $\sum_i a^i v_i = 0$, hence $a^i = 0$ for $i = 1, \ldots, m$. Thus, $J_1(t), \ldots, J_m(t)$ are linearly independent vectors in $T_{\lambda(t)}(M)$. By Theorem 14.3.4(c), $T_{\lambda(t)}(M)$ has dimension m, so $\mathcal{J}(t)$ is a basis for $T_{\lambda(t)}(M)$ for all t in (a, b). $\qquad \square$

At the beginning of this section, covariant differentiation along a curve was used to define parallel translation. The next result shows that parallel translation can be used to define covariant differentiation along a curve. In this sense, covariant differentiation and parallel translation are equivalent concepts.

Theorem 18.5.8. *Let* (M, ∇) *be a smooth manifold with a connection, let* $\lambda(t) : (a, b) \longrightarrow M$ *be a smooth curve, let* t_0 *be a point in* (a, b), *and let* K *be a vector field in* $\mathfrak{X}_M(\lambda)$. *Then*

$$\frac{\nabla K}{dt}(t_0) = \lim_{t \to t_0} \frac{\Pi^{\lambda(t_0)}_{\lambda(t)}\big(K(t)\big) - K(t_0)}{t}.$$

Proof. Let $\mathcal{J} = (J_1, \ldots, J_m)$ be the frame given by Theorem 18.5.7. Then $K(t)$ can be expressed as

$$K(t) = \sum_i \alpha^i(t) J_i(t), \tag{18.5.14}$$

where the α^i are uniquely determined functions in $C^\infty\big((a, b)\big)$. In particular

$$K(t_0) = \sum_i \alpha^i(t_0) v_i.$$

Since each J_i is a parallel vector field in $\mathfrak{X}_M(\lambda)$, we have $(\nabla J_i / dt)(t) = 0$, hence

$$\frac{\nabla K}{dt}(t) = \sum_i \left(\frac{d\alpha^i}{dt}(t) J_i(t) + \alpha^i(t) \frac{\nabla J_i}{dt}(t) \right) = \sum_i \frac{d\alpha^i}{dt}(t) J_i(t),$$

so

$$\frac{\nabla K}{dt}(t_0) = \sum_i \frac{d\alpha^i}{dt}(t_0) J_i(t_0) = \sum_i \frac{d\alpha^i}{dt}(t_0) v_i. \tag{18.5.15}$$

By definition, $J_i(t) = \Pi^{\lambda(t)}_{\lambda(t_0)}(v_i)$ for all t in (a, b). Using Theorem 18.5.6 to apply $\left(\Pi^{\lambda(t)}_{\lambda(t_0)}\right)^{-1}$ to both sides of (18.5.14) yields

$$\Pi^{\lambda(t_0)}_{\lambda(t)}\left(K(t)\right) = \sum_i \alpha^i(t)v_i.$$

Thus,

$$\lim_{t \to t_0} \frac{\Pi^{\lambda(t_0)}_{\lambda(t)}\left(K(t)\right) - K(t_0)}{t} = \sum_i \left(\lim_{t \to t_0} \frac{\alpha^i(t) - \alpha^i(t_0)}{t}\right)v_i$$

$$= \sum_i \frac{d\alpha^i}{dt}(t_0)v_i. \qquad (18.5.16)$$

The result now follows from (18.5.15) and (18.5.16).   □

18.6   Torsion Tensors

Let (M, ∇) be a smooth manifold with a connection. The **torsion tensor (field)** corresponding to ∇ is the map

$$T : \mathfrak{X}(M) \times \mathfrak{X}(M) \longrightarrow \mathfrak{X}(M)$$

defined by

$$T(X, Y) = \nabla_X(Y) - \nabla_Y(X) - [X, Y] \qquad (18.6.1)$$

for all vector fields X, Y in $\mathfrak{X}(M)$. We say that ∇ is **symmetric** (or **torsion-free**) if $T = 0$; that is,

$$\nabla_X(Y) - \nabla_Y(X) = [X, Y]$$

for all vector fields X, Y in $\mathfrak{X}(M)$.

Since T takes values in $\mathfrak{X}(M)$, referring to T as a "tensor field", although usual in the literature, is not consistent with Theorem 15.7.2. We take up this matter in Section 19.5.

Example 18.6.1 (Regular Surface in $\mathbb{R}^3_\nu$). Viewing a regular surface as a smooth 2-manifold with the connection defined in Example 18.1.1, it follows from (12.6.1) that the connection is symmetric.   ◇

Theorem 18.6.2. *If (M, ∇) is a smooth manifold with a connection and X, Y are vector fields in $\mathfrak{X}(M)$, then:*
(a) $T(Y, X) = -T(X, Y)$
(b) T is $C^\infty(M)$-bilinear.

Proof. (a): This follows from Theorem 15.3.2(a).

(b): Using Theorem 15.3.2 and Theorem 18.1.2, it is easily shown that T is additive in both arguments. For functions f, g in $C^\infty(M)$, we have from the same two theorems that

$$
\begin{aligned}
T(fX, gY) &= \nabla_{fX}(gY) - \nabla_{gY}(fX) - [fX, gY] \\
&= \big(fX(g)Y + fg\nabla_X(Y)\big) - \big(gY(f)X + fg\nabla_Y(X)\big) \\
&\quad - \big(fg[X, Y] + fX(g)Y - gY(f)X\big) \\
&= fg\big(\nabla_X(Y) - \nabla_Y(X) - [X, Y]\big) \\
&= fg\,T(X, Y),
\end{aligned}
$$

so T is $C^\infty(M)$-bilinear. $\square$

Let (M, ∇) be a smooth m-manifold with a connection, and let $(U, (x^i))$ be a chart on M. In local coordinates, $T(\partial/\partial x^i, \partial/\partial x^j)$ can be expressed as

$$
T\left(\frac{\partial}{\partial x^i}, \frac{\partial}{\partial x^j}\right) = \sum_k T_{ij}^k \frac{\partial}{\partial x^k},
$$

where the T_{ij}^k are uniquely determined functions in $C^\infty(U)$.

Theorem 18.6.3. *With the above setup:*
(a) $T_{ij}^k = \Gamma_{ij}^k - \Gamma_{ji}^k$ *for* $i, j, k = 1, \ldots, m$.
(b) ∇ *is symmetric if and only if for all such local coordinate expressions,* $\Gamma_{ij}^k = \Gamma_{ji}^k$ *for* $i, j, k = 1, \ldots, m$.

Proof. (a): We have from Theorem 15.3.3(b) and (18.2.1) that

$$
\begin{aligned}
T\left(\frac{\partial}{\partial x^i}, \frac{\partial}{\partial x^j}\right) &= \nabla_{\partial/\partial x_i}\left(\frac{\partial}{\partial x^j}\right) - \nabla_{\partial/\partial x_j}\left(\frac{\partial}{\partial x^i}\right) - \left[\frac{\partial}{\partial x^i}, \frac{\partial}{\partial x^j}\right] \\
&= \sum_k (\Gamma_{ij}^k - \Gamma_{ji}^k)\frac{\partial}{\partial x^k}.
\end{aligned}
$$

(b): From Theorem 18.6.2(b), T is $\mathbb{R}$-bilinear, so symmetry is determined by the behavior of T on coordinates frames. By part (a), $T_{ij}^k = 0$ if and only if $\Gamma_{ij}^k = \Gamma_{ji}^k$. The result follows. $\square$

Theorem 18.6.4. *If (M, ∇) is a smooth manifold with a symmetric connection and $\sigma(r, s) : (a, b) \times (-\varepsilon, \varepsilon) \longrightarrow M$ is a parametrized surface, then*

$$
\frac{\nabla}{\partial r}\left(\frac{\partial \sigma}{\partial s}\right) = \frac{\nabla}{\partial s}\left(\frac{\partial \sigma}{\partial r}\right),
$$

where $\partial\sigma/\partial r$ and $\partial\sigma/\partial s$ are given by (14.9.1) and Theorem 15.1.5, and $\nabla/\partial r$ and $\nabla/\partial s$ are given by (18.3.3).

Proof. Let $(U, (x^i))$ be a chart on M that intersects $\sigma\big((a, b) \times (-\varepsilon, \varepsilon)\big)$. By Theorem 14.9.2,

$$\frac{\partial \sigma}{\partial r} = \sum_i \frac{\partial (x^i \circ \sigma)}{\partial r} \left(\frac{\partial}{\partial x^i} \circ \sigma \right) \tag{18.6.2}$$

and

$$\frac{\partial \sigma}{\partial s} = \sum_i \frac{\partial (x^i \circ \sigma)}{\partial s} \left(\frac{\partial}{\partial x^i} \circ \sigma \right). \tag{18.6.3}$$

We have from Theorem 18.3.1(c) and (18.6.3) that

$$\frac{\nabla}{\partial r} \left(\frac{\partial \sigma}{\partial s} \right) = \sum_i \frac{\nabla}{\partial r} \left(\frac{\partial (x^i \circ \sigma)}{\partial s} \left(\frac{\partial}{\partial x^i} \circ \sigma \right) \right)$$

$$= \sum_i \frac{\nabla}{\partial r} \left(\frac{\partial (x^i \circ \sigma)}{\partial s} \right) \left(\frac{\partial}{\partial x^i} \circ \sigma \right) + \sum_i \frac{\partial (x^i \circ \sigma)}{\partial s} \frac{\nabla}{\partial r} \left(\frac{\partial}{\partial x^i} \circ \sigma \right)$$

$$= \sum_k \frac{\partial^2 (x^k \circ \sigma)}{\partial r \partial s} \left(\frac{\partial}{\partial x^k} \circ \sigma \right) + \sum_i \frac{\partial (x^i \circ \sigma)}{\partial s} \nabla_{\partial \sigma / \partial r} \left(\frac{\partial}{\partial x^i} \right),$$

and from (18.2.1) and (18.6.2) that

$$\sum_i \frac{\partial (x^i \circ \sigma)}{\partial s} \nabla_{\partial \sigma / \partial r} \left(\frac{\partial}{\partial x^i} \right)$$

$$= \sum_i \frac{\partial (x^i \circ \sigma)}{\partial s} \left[\sum_j \frac{\partial (x^j \circ \sigma)}{\partial r} \nabla_{(\partial / \partial x^j) \circ \sigma} \left(\frac{\partial}{\partial x^i} \right) \right]$$

$$= \sum_{ij} \left[\frac{\partial (x^j \circ \sigma)}{\partial r} \frac{\partial (x^i \circ \sigma)}{\partial s} \nabla_{(\partial / \partial x^j) \circ \sigma} \left(\frac{\partial}{\partial x^i} \right) \right]$$

$$= \sum_{ij} \left\{ \frac{\partial (x^j \circ \sigma)}{\partial r} \frac{\partial (x^i \circ \sigma)}{\partial s} \left[\sum_k (\Gamma^k_{ji} \circ \sigma) \left(\frac{\partial}{\partial x^k} \circ \sigma \right) \right] \right\}$$

$$= \sum_k \left[\sum_{ij} \frac{\partial (x^j \circ \sigma)}{\partial r} \frac{\partial (x^i \circ \sigma)}{\partial s} (\Gamma^k_{ji} \circ \sigma) \right] \left(\frac{\partial}{\partial x^k} \circ \sigma \right),$$

hence

$$\frac{\nabla}{\partial r} \left(\frac{\partial \sigma}{\partial s} \right) = \sum_k \left[\frac{\partial^2 (x^k \circ \sigma)}{\partial r \partial s} \right.$$

$$+ \sum_{ij} \left(\frac{\partial (x^j \circ \sigma)}{\partial r} \frac{\partial (x^i \circ \sigma)}{\partial s} (\Gamma^k_{ji} \circ \sigma) \right) \right] \left(\frac{\partial}{\partial x^k} \circ \sigma \right).$$

Reversing the roles of r and s in the preceding identity and using Theorem 10.1.6 yields

$$
\frac{\nabla}{\partial s}\left(\frac{\partial \sigma}{\partial r}\right) = \sum_k \left[\frac{\partial^2 (x^k \circ \sigma)}{\partial s \partial r}\right.
$$
$$
\left. + \sum_{ij}\left(\frac{\partial (x^j \circ \sigma)}{\partial s}\frac{\partial (x^i \circ \sigma)}{\partial r}(\Gamma^k_{ji} \circ \sigma)\right)\right]\left(\frac{\partial}{\partial x^k}\circ \sigma\right)
$$
$$
= \sum_k \left[\frac{\partial^2 (x^k \circ \sigma)}{\partial r \partial s}\right.
$$
$$
\left. + \sum_{ij}\left(\frac{\partial (x^j \circ \sigma)}{\partial r}\frac{\partial (x^i \circ \sigma)}{\partial s}(\Gamma^k_{ij} \circ \sigma)\right)\right]\left(\frac{\partial}{\partial x^k}\circ \sigma\right).
$$

Comparing the last two identities and invoking Theorem 18.6.3(b) gives the result.   □

18.7   Curvature Tensors

In this section, we generalize to smooth manifolds with a connection the Riemann curvature tensor on regular surfaces in $\mathbb{R}^3_\nu$.

Let (M, ∇) be a smooth manifold with a connection. The **curvature tensor (field)** is the map

$$
R : \mathfrak{X}(M)^3 \longrightarrow \mathfrak{X}(M)
$$

defined by

$$
R(X,Y)Z = \nabla_X\left(\nabla_Y(Z)\right) - \nabla_Y\left(\nabla_X(Z)\right) - \nabla_{[X,Y]}(Z) \tag{18.7.1}
$$

for all vector fields X, Y, Z in $\mathfrak{X}(M)$. Since R takes values in $\mathfrak{X}(M)$, referring to R as a "tensor field", although ubiquitous in the literature, presents the same issues of terminology encountered with the torsion tensor in Section 18.6. This concern will be addressed in Section 19.5.

Theorem 18.7.1. *If (M, ∇) is a smooth manifold with a connection and X, Y, Z are vector fields in $\mathfrak{X}(M)$, then:*
(a) $R(Y, X)Z = -R(X, Y)Z.$
(b) R *is $C^\infty(M)$-multilinear.*

Proof. (a): This follows from Theorem 15.3.2(a).

(b): Using Theorem 15.3.2 and Theorem 18.1.2, it is easily shown that R is additive in each argument. For a function f in $C^\infty(M)$, we have from Theorem

15.3.2(d) and Theorem 18.1.2 that

$$
\begin{aligned}
R(fX,Y)Z &= \nabla_{fX}\big(\nabla_Y(Z)\big) - \nabla_Y\big(\nabla_{fX}(Z)\big) - \nabla_{[fX,Y]}(Z) \\
&= f\nabla_X\big(\nabla_Y(Z)\big) - \nabla_Y\big(f\nabla_X(Z)\big) - \nabla_{f[X,Y]-Y(f)X}(Z) \\
&= f\nabla_X\big(\nabla_Y(Z)\big) - \big\{Y(f)\nabla_X(Z) + f\nabla_Y\big(\nabla_X(Z)\big)\big\} \\
&\quad - \big(f\nabla_{[X,Y]}(Z) - Y(f)\nabla_X(Z)\big) \\
&= f\big(\nabla_X\big(\nabla_Y(Z)\big) - \nabla_Y\big(\nabla_X(Z)\big) - \nabla_{[X,Y]}(Z)\big) \\
&= fR(X,Y)Z.
\end{aligned}
$$

Likewise,
$$
R(X,fY)Z = fR(X,Y)Z.
$$

We also have

$$
R(X,Y)(fZ) = \nabla_X\big(\nabla_Y(fZ)\big) - \nabla_Y\big(\nabla_X(fZ)\big) - \nabla_{[X,Y]}(fZ),
$$

$$
\begin{aligned}
\nabla_X\big(\nabla_Y(fZ)\big) &= \nabla_X\big(Y(f)Z + f\nabla_Y(Z)\big) = \nabla_X\big(Y(f)Z\big) + \nabla_X\big(f\nabla_Y(Z)\big) \\
&= X\big(Y(f)\big)Z + Y(f)\nabla_X(Z) + X(f)\nabla_Y(Z) + f\nabla_X\big(\nabla_Y(Z)\big),
\end{aligned}
$$

$$
\nabla_Y\big(\nabla_X(fZ)\big) = Y\big(X(f)\big)Z + X(f)\nabla_Y(Z) + Y(f)\nabla_X(Z) + f\nabla_Y\big(\nabla_X(Z)\big),
$$

and
$$
\begin{aligned}
\nabla_{[X,Y]}(fZ) &= [X,Y](f)Z + f\nabla_{[X,Y]}(Z) \\
&= X\big(Y(f)\big)Z - Y\big(X(f)\big)Z + f\nabla_{[X,Y]}(Z),
\end{aligned}
$$

hence
$$
\begin{aligned}
R(X,Y)(fZ) &= f\big\{\nabla_X\big(\nabla_Y(Z)\big) - \nabla_Y\big(\nabla_X(Z)\big) - \nabla_{[X,Y]}(Z)\big\} \\
&= fR(X,Y)(Z).
\end{aligned}
$$

Thus, R is $C^\infty(M)$-multilinear. $\qquad\square$

Let (M,∇) be a smooth m-manifold with a connection, and let $(U,(x^i))$ be a chart on M. In local coordinates, $R(\partial/\partial x^i,\partial/\partial x^j)\partial/\partial x^k$ can be expressed as

$$
R\left(\frac{\partial}{\partial x^i},\frac{\partial}{\partial x^j}\right)\frac{\partial}{\partial x^k} = \sum_l R^l_{ijk}\frac{\partial}{\partial x^l}, \tag{18.7.2}
$$

where the R^l_{ijk} are uniquely determined functions in $C^\infty(U)$.

Theorem 18.7.2 (Local Coordinate Expression for Curvature Tensor).
With the above setup,

$$
R^l_{ijk} = \frac{\partial \Gamma^l_{jk}}{\partial x^i} - \frac{\partial \Gamma^l_{ik}}{\partial x^j} + \sum_n (\Gamma^l_{in}\Gamma^n_{jk} - \Gamma^l_{jn}\Gamma^n_{ik})
$$

for $i,j,k,l = 1,\ldots,m$.

Proof. We have from (18.7.1) and (18.7.2) that

$$\sum_l R^l_{ijk} \frac{\partial}{\partial x^l} = R\left(\frac{\partial}{\partial x^i}, \frac{\partial}{\partial x^j}\right) \frac{\partial}{\partial x^k}$$

$$= \nabla_{\partial/\partial x^i}\left(\nabla_{\partial/\partial x^j}\left(\frac{\partial}{\partial x^k}\right)\right) - \nabla_{\partial/\partial x^j}\left(\nabla_{\partial/\partial x^i}\left(\frac{\partial}{\partial x^k}\right)\right)$$

$$- \nabla_{[\partial/\partial x^i, \partial/\partial x^j]}\left(\frac{\partial}{\partial x^k}\right).$$

Then (18.2.1) gives

$$\nabla_{\partial/\partial x^i}\left(\nabla_{\partial/\partial x^j}\left(\frac{\partial}{\partial x^k}\right)\right) = \nabla_{\partial/\partial x^i}\left(\sum_l \Gamma^l_{jk} \frac{\partial}{\partial x^l}\right) = \sum_l \nabla_{\partial/\partial x^i}\left(\Gamma^l_{jk} \frac{\partial}{\partial x^l}\right)$$

$$= \sum_l \left[\frac{\partial \Gamma^l_{jk}}{\partial x^i} \frac{\partial}{\partial x^l} + \Gamma^l_{jk} \nabla_{\partial/\partial x^i}\left(\frac{\partial}{\partial x^l}\right)\right]$$

$$= \sum_l \left(\frac{\partial \Gamma^l_{jk}}{\partial x^i} \frac{\partial}{\partial x^l} + \Gamma^l_{jk} \sum_n \Gamma^n_{il} \frac{\partial}{\partial x^n}\right)$$

$$= \sum_l \frac{\partial \Gamma^l_{jk}}{\partial x^i} \frac{\partial}{\partial x^l} + \sum_{ln} \Gamma^l_{jk} \Gamma^n_{il} \frac{\partial}{\partial x^n}$$

$$= \sum_l \frac{\partial \Gamma^l_{jk}}{\partial x^i} \frac{\partial}{\partial x^l} + \sum_{ln} \Gamma^n_{jk} \Gamma^l_{in} \frac{\partial}{\partial x^l}$$

$$= \left(\sum_l \frac{\partial \Gamma^l_{jk}}{\partial x^i} + \sum_{ln} \Gamma^n_{jk} \Gamma^l_{in}\right) \frac{\partial}{\partial x^l}$$

and

$$\nabla_{\partial/\partial x^j}\left(\nabla_{\partial/\partial x^i}\left(\frac{\partial}{\partial x^k}\right)\right) = \left(\sum_l \frac{\partial \Gamma^l_{ik}}{\partial x^j} + \sum_{ln} \Gamma^n_{ik} \Gamma^l_{jn}\right) \frac{\partial}{\partial x^l},$$

and Theorem 15.3.3(b) yields

$$\nabla_{[\partial/\partial x^i, \partial/\partial x^j]}\left(\frac{\partial}{\partial x^k}\right) = 0.$$

Combining the above identities gives the result.   □

In the literature, the indexing of R^l_{ijk} is sometimes defined by

$$R\left(\frac{\partial}{\partial x^j}, \frac{\partial}{\partial x^k}\right) \frac{\partial}{\partial x^i} = \sum_l R^l_{ijk} \frac{\partial}{\partial x^l}.$$

This changes the indexing of certain identities, but has no substantive implications.

Let M be a smooth manifold (with or without a connection), let $\mathcal{F} : \mathfrak{X}(M)^3 \longrightarrow \mathfrak{X}(M)$ be a map, and define

$$\sum_{(X,Y,Z)} \mathcal{F}(X,Y,Z) = \mathcal{F}(X,Y,Z) + \mathcal{F}(Y,Z,X) + \mathcal{F}(Z,X,Y)$$

for all vector fields X, Y, Z in $\mathfrak{X}(M)$. Thus, $\sum_{(X,Y,Z)} \mathcal{F}(X,Y,Z)$ is a sum of terms obtained by permuting X, Y, Z cyclically. Clearly,

$$\sum_{(X,Y,Z)} \mathcal{F}(X,Y,Z) = \sum_{(X,Y,Z)} \mathcal{F}(Y,Z,X) = \sum_{(X,Y,Z)} \mathcal{F}(Z,X,Y). \qquad (18.7.3)$$

For example, Jacobi's identity for vector fields in $\mathfrak{X}(M)$, given by Theorem 15.3.2(e), can be expressed as

$$\sum_{(X,Y,Z)} \big[X,[Y,Z]\big] = 0, \qquad (18.7.4)$$

and Jacobi's identity for covariant derivatives in $\mathrm{Lin}\big(\mathfrak{X}(M),\mathfrak{X}(M)\big)$, defined by (18.1.6), becomes

$$\sum_{(X,Y,Z)} \big[\nabla_X,[\nabla_Y,\nabla_Z]\big] = 0. \qquad (18.7.5)$$

For a map $\mathcal{G} : \mathfrak{X}(M)^k \longrightarrow \mathfrak{X}(M)$, where $k > 3$, we use the above notation, but apply the cyclic permutation only to the three arguments specified by the 3-tuple under the summation sign, leaving the other arguments in place.

Theorem 18.7.3 (First Bianchi Identity). *Let (M, ∇) be a smooth manifold with a connection, and let X, Y, Z be vector fields in $\mathfrak{X}(M)$. Then:*
(a)

$$\sum_{(X,Y,Z)} R(X,Y)Z = \sum_{(X,Y,Z)} \nabla_X\big(T(Y,Z)\big) + \sum_{(X,Y,Z)} T(X,[Y,Z]),$$

where T is the torsion tensor.
(b) *If ∇ is symmetric, then*

$$\sum_{(X,Y,Z)} R(X,Y)Z = 0.$$

Proof. (a): We have from (18.6.1) that

$$T(X,[Y,Z]) = \nabla_X([Y,Z]) - \nabla_{[Y,Z]}(X) - \big[X,[Y,Z]\big],$$

hence

$$\sum_{(X,Y,Z)} T(X,[Y,Z])$$

$$= \sum_{(X,Y,Z)} \nabla_X([Y,Z]) - \sum_{(X,Y,Z)} \nabla_{[Y,Z]}(X) - \sum_{(X,Y,Z)} \big[X,[Y,Z]\big] \qquad (18.7.6)$$

$$= \sum_{(X,Y,Z)} \nabla_X([Y,Z]) - \sum_{(X,Y,Z)} \nabla_{[X,Y]}(Z),$$

where the last equality follows from (18.7.3) and (18.7.4). We also have from (18.6.1) and (18.7.1) that

$$
\begin{aligned}
R(X,Y)Z \\
&= \nabla_X\big(\nabla_Y(Z)\big) - \nabla_Y\big(\nabla_X(Z)\big) - \nabla_{[X,Y]}(Z) \\
&= \nabla_X\big(T(Y,Z) + \nabla_Z(Y) + [Y,Z]\big) - \nabla_Y\big(\nabla_X(Z)\big) - \nabla_{[X,Y]}(Z) \\
&= \nabla_X\big(T(Y,Z)\big) + \nabla_X\big(\nabla_Z(Y)\big) + \nabla_X([Y,Z]) - \nabla_Y\big(\nabla_X(Z)\big) - \nabla_{[X,Y]}(Z).
\end{aligned}
$$

Then (18.7.3) gives

$$
\begin{aligned}
\sum_{(X,Y,Z)} R(X,Y)Z \\
&= \sum_{(X,Y,Z)} \nabla_X\big(T(Y,Z)\big) + \sum_{(X,Y,Z)} \nabla_X\big(\nabla_Z(Y)\big) + \sum_{(X,Y,Z)} \nabla_X([Y,Z]) \\
&\quad - \sum_{(X,Y,Z)} \nabla_Y\big(\nabla_X(Z)\big) - \sum_{(X,Y,Z)} \nabla_{[X,Y]}(Z) \\
&= \sum_{(X,Y,Z)} \nabla_X\big(T(Y,Z)\big) + \sum_{(X,Y,Z)} \nabla_X([Y,Z]) - \sum_{(X,Y,Z)} \nabla_{[X,Y]}(Z).
\end{aligned}
\tag{18.7.7}
$$

The result now follows from (18.7.6) and (18.7.7).

(b): This follows from part (a). $\square$

Theorem 18.7.4 (Second Bianchi Identity). *If (M,∇) is a smooth manifold with a connection and X,Y,Z,W are vector fields in $\mathfrak{X}(M)$, then*

$$
\sum_{(X,Y,Z)} \nabla_X\big(R(Y,Z)W\big) = \sum_{(X,Y,Z)} R(X,Y)\nabla_Z(W) + \sum_{(X,Y,Z)} R([X,Y],Z)W.
$$

Proof. Recalling the discussion on the Lie algebra $\mathrm{Lin}\big(\mathfrak{X}(M),\mathfrak{X}(M)\big)$ at the end of Section 18.1, $R(X,Y)Z$ can be expressed as

$$
\begin{aligned}
R(X,Y)Z &= \nabla_X\big(\nabla_Y(Z)\big) - \nabla_Y\big(\nabla_X(Z)\big) - \nabla_{[X,Y]}(Z) \\
&= \big(\nabla_X \circ \nabla_Y - \nabla_Y \circ \nabla_X - \nabla_{[X,Y]}\big)(Z) \\
&= \big([\nabla_X, \nabla_Y] - \nabla_{[X,Y]}\big)(Z),
\end{aligned}
$$

hence

$$
\begin{aligned}
\nabla_Z\big(R(X,Y)W\big) &= \nabla_Z\big(([\nabla_X,\nabla_Y] - \nabla_{[X,Y]})(W)\big) \\
&= \big(\nabla_Z \circ [\nabla_X,\nabla_Y] - \nabla_Z \circ \nabla_{[X,Y]}\big)(W),
\end{aligned}
$$

$$
\begin{aligned}
R(X,Y)\nabla_Z(W) &= \big([\nabla_X,\nabla_Y] - \nabla_{[X,Y]}\big)\big(\nabla_Z(W)\big) \\
&= \big([\nabla_X,\nabla_Y] \circ \nabla_Z - \nabla_{[X,Y]} \circ \nabla_Z\big)(W),
\end{aligned}
$$

and

$$
R([X,Y],Z)W = \big([\nabla_{[X,Y]}, \nabla_Z] - \nabla_{[[X,Y],Z]}\big)(W).
$$

Combining the preceding three identities yields

$$
\begin{aligned}
&\nabla_Z\big(R(X,Y)W\big) - R(X,Y)\nabla_Z(W) - R([X,Y],Z)W \\
&= \big((\nabla_Z \circ [\nabla_X,\nabla_Y] - \nabla_Z \circ \nabla_{[X,Y]}) - ([\nabla_X,\nabla_Y] \circ \nabla_Z - \nabla_{[X,Y]} \circ \nabla_Z) \\
&\quad - ([\nabla_{[X,Y]},\nabla_Z] - \nabla_{[[X,Y],Z]})\big)(W) \\
&= \big((\nabla_Z \circ [\nabla_X,\nabla_Y] - [\nabla_X,\nabla_Y] \circ \nabla_Z) - (\nabla_Z \circ \nabla_{[X,Y]} - \nabla_{[X,Y]} \circ \nabla_Z) \\
&\quad + ([\nabla_Z,\nabla_{[X,Y]}] - \nabla_{[Z,[X,Y]]})\big)(W) \\
&= \big([\nabla_Z,[\nabla_X,\nabla_Y]] - [\nabla_Z,\nabla_{[X,Y]}] + [\nabla_Z,\nabla_{[X,Y]}] - \nabla_{[Z,[X,Y]]}\big)(W) \\
&= [\nabla_Z,[\nabla_X,\nabla_Y]](W) - \nabla_{[Z,[X,Y]]}(W).
\end{aligned}
$$

Then (18.7.3)–(18.7.5) give

$$
\begin{aligned}
&\sum_{(X,Y,Z)} \nabla_X\big(R(Y,Z)W\big) - \sum_{(X,Y,Z)} R(X,Y)\nabla_Z(W) - \sum_{(X,Y,Z)} R([X,Y],Z)W \\
&= \Big(\sum_{(X,Y,Z)} [\nabla_X,[\nabla_Y,\nabla_Z]]\Big)(W) - \big(\nabla_{\sum_{(X,Y,Z)}[X,[Y,Z]]}\big)(W) = 0. \qquad \square
\end{aligned}
$$

Theorem 18.7.5. *Let (M,∇) be a smooth manifold with a connection, let $\sigma(r,s): (a,b) \times (-\varepsilon,\varepsilon) \longrightarrow M$ be a parametrized surface, and let V be a vector field in $\mathfrak{X}_M(\sigma)$. Then*

$$
R\left(\frac{\partial\sigma}{\partial r},\frac{\partial\sigma}{\partial s}\right)V = \frac{\nabla}{\partial r}\left(\frac{\nabla V}{\partial s}\right) - \frac{\nabla}{\partial s}\left(\frac{\nabla V}{\partial r}\right). \tag{18.7.8}
$$

Proof. The proof is a lengthy computation. We seek alternative expressions for both sides of (18.7.8). Let $(U,(x^i))$ be a chart on M that intersects $\sigma\big((a,b) \times (-\varepsilon,\varepsilon)\big)$. By Theorem 14.9.2,

$$
\frac{\partial\sigma}{\partial r} = \sum_i \frac{\partial(x^i \circ \sigma)}{\partial r}\left(\frac{\partial}{\partial x^i} \circ \sigma\right) \tag{18.7.9}
$$

and

$$
\frac{\partial\sigma}{\partial s} = \sum_j \frac{\partial(x^j \circ \sigma)}{\partial s}\left(\frac{\partial}{\partial x^j} \circ \sigma\right). \tag{18.7.10}
$$

We first consider the case $V = (\partial/\partial x^k) \circ \sigma$. For the left-hand side of (18.7.8), we have

$$
R\left(\frac{\partial\sigma}{\partial r}, \frac{\partial\sigma}{\partial s}\right)\left(\frac{\partial}{\partial x^k} \circ \sigma\right)
$$

$$
= R\left(\sum_i \frac{\partial(x^i \circ \sigma)}{\partial r}\left(\frac{\partial}{\partial x^i} \circ \sigma\right), \sum_j \frac{\partial(x^j \circ \sigma)}{\partial s}\left(\frac{\partial}{\partial x^j} \circ \sigma\right)\right)\left(\frac{\partial}{\partial x^k} \circ \sigma\right)
$$

$$
= \sum_{ij} \frac{\partial(x^i \circ \sigma)}{\partial r}\frac{\partial(x^j \circ \sigma)}{\partial s} R\left(\frac{\partial}{\partial x^i} \circ \sigma, \frac{\partial}{\partial x^j} \circ \sigma\right)\left(\frac{\partial}{\partial x^k} \circ \sigma\right)
$$

$$
= \sum_{ij} \frac{\partial(x^i \circ \sigma)}{\partial r}\frac{\partial(x^j \circ \sigma)}{\partial s}\left[\nabla_{(\partial/\partial x^i)\circ\sigma}\left(\nabla_{\partial/\partial x^j}\left(\frac{\partial}{\partial x^k}\right)\right)\right.
$$

$$
\left. - \nabla_{(\partial/\partial x^j)\circ\sigma}\left(\nabla_{\partial/\partial x^i}\left(\frac{\partial}{\partial x^k}\right)\right)\right],
$$

$$(18.7.11)$$

where the last equality follows from Theorem 15.3.3(b) and (18.7.1).

For the right-hand side of (18.7.8), Theorem 18.3.1(c) and (18.7.10) give

$$
\frac{\nabla}{\partial s}\left(\frac{\partial}{\partial x^k} \circ \sigma\right) = \nabla_{\partial\sigma/\partial s}\left(\frac{\partial}{\partial x^k}\right) = \sum_j \frac{\partial(x^j \circ \sigma)}{\partial s}\nabla_{(\partial/\partial x^j)\circ\sigma}\left(\frac{\partial}{\partial x^k}\right),
$$

hence

$$
\frac{\nabla}{\partial r}\left(\frac{\nabla}{\partial s}\left(\frac{\partial}{\partial x^k} \circ \sigma\right)\right) = \sum_j \frac{\nabla}{\partial r}\left(\frac{\partial(x^j \circ \sigma)}{\partial s}\nabla_{(\partial/\partial x^j)\circ\sigma}\left(\frac{\partial}{\partial x^k}\right)\right)
$$

$$
= \sum_j \frac{\nabla}{\partial r}\left(\frac{\partial(x^j \circ \sigma)}{\partial s}\right)\nabla_{(\partial/\partial x^j)\circ\sigma}\left(\frac{\partial}{\partial x^k}\right)
$$

$$
+ \sum_j \frac{\partial(x^j \circ \sigma)}{\partial s}\nabla_{\partial\sigma/\partial r}\left(\nabla_{\partial/\partial x^j}\left(\frac{\partial}{\partial x^k}\right)\right) \qquad (18.7.12)
$$

$$
= \sum_j \frac{\partial^2 x^j}{\partial r\partial s}\nabla_{(\partial/\partial x^j)\circ\sigma}\left(\frac{\partial}{\partial x^k}\right)
$$

$$
+ \sum_j \frac{\partial(x^j \circ \sigma)}{\partial s}\nabla_{\partial\sigma/\partial r}\left(\nabla_{\partial/\partial x^j}\left(\frac{\partial}{\partial x^k}\right)\right).
$$

Using (18.7.9), the term in the last row of (18.7.12) can be expressed as

$$\sum_j \frac{\partial(x^j \circ \sigma)}{\partial s} \nabla_{\partial \sigma/\partial r} \left(\nabla_{\partial/\partial x^j} \left(\frac{\partial}{\partial x^k} \right) \right)$$

$$= \sum_j \frac{\partial(x^j \circ \sigma)}{\partial s} \left[\sum_i \frac{\partial(x^i \circ \sigma)}{\partial r} \nabla_{(\partial/\partial x^i)\circ\sigma} \left(\nabla_{\partial/\partial x^j} \left(\frac{\partial}{\partial x^k} \right) \right) \right]$$

$$= \sum_{ij} \frac{\partial(x^i \circ \sigma)}{\partial r} \frac{\partial(x^j \circ \sigma)}{\partial s} \nabla_{(\partial/\partial x^i)\circ\sigma} \left(\nabla_{(\partial/\partial x^j)\circ\sigma} \left(\frac{\partial}{\partial x^k} \right) \right).$$

Then (18.7.12) becomes

$$\frac{\nabla}{\partial r} \left(\frac{\nabla}{\partial s} \left(\frac{\partial}{\partial x^k} \circ \sigma \right) \right) = \sum_j \frac{\partial^2 x^j}{\partial r \partial s} \nabla_{(\partial/\partial x^j)\circ\sigma} \left(\frac{\partial}{\partial x^k} \right)$$

$$+ \sum_{ij} \frac{\partial(x^i \circ \sigma)}{\partial r} \frac{\partial(x^j \circ \sigma)}{\partial s} \nabla_{(\partial/\partial x^i)\circ\sigma} \left(\nabla_{\partial/\partial x^j} \left(\frac{\partial}{\partial x^k} \right) \right).$$

Interchanging r and s in the preceding identity and using Theorem 10.1.6 yields

$$\frac{\nabla}{\partial s} \left(\frac{\nabla}{\partial r} \left(\frac{\partial}{\partial x^k} \circ \sigma \right) \right) = \sum_j \frac{\partial^2 x^j}{\partial s \partial r} \nabla_{(\partial/\partial x^j)\circ\sigma} \left(\frac{\partial}{\partial x^k} \right)$$

$$+ \sum_{ij} \frac{\partial(x^i \circ \sigma)}{\partial s} \frac{\partial(x^j \circ \sigma)}{\partial r} \nabla_{(\partial/\partial x^i)\circ\sigma} \left(\nabla_{\partial/\partial x^j} \left(\frac{\partial}{\partial x^k} \right) \right)$$

$$= \sum_j \frac{\partial^2 x^j}{\partial r \partial s} \nabla_{(\partial/\partial x^j)\circ\sigma} \left(\frac{\partial}{\partial x^k} \right)$$

$$+ \sum_{ij} \frac{\partial(x^i \circ \sigma)}{\partial r} \frac{\partial(x^j \circ \sigma)}{\partial s} \nabla_{(\partial/\partial x^j)\circ\sigma} \left(\nabla_{\partial/\partial x^i} \left(\frac{\partial}{\partial x^k} \right) \right).$$

Subtracting the preceding two identities gives

$$\frac{\nabla}{\partial r} \left(\frac{\nabla}{\partial s} \left(\frac{\partial}{\partial x^k} \circ \sigma \right) \right) - \frac{\nabla}{\partial s} \left(\frac{\nabla}{\partial r} \left(\frac{\partial}{\partial x^k} \circ \sigma \right) \right)$$

$$= \sum_{ij} \frac{\partial(x^i \circ \sigma)}{\partial r} \frac{\partial(x^j \circ \sigma)}{\partial s} \left[\nabla_{(\partial/\partial x^i)\circ\sigma} \left(\nabla_{\partial/\partial x^j} \left(\frac{\partial}{\partial x^k} \right) \right) \right. \qquad (18.7.13)$$

$$\left. - \nabla_{(\partial/\partial x^j)\circ\sigma} \left(\nabla_{\partial/\partial x^i} \left(\frac{\partial}{\partial x^k} \right) \right) \right],$$

which is the desired expression for the right-hand side of (18.7.8).

From (18.7.11) and (18.7.13), we obtain

$$R\left(\frac{\partial\sigma}{\partial r},\frac{\partial\sigma}{\partial s}\right)\left(\frac{\partial}{\partial x^k}\circ\sigma\right)$$

$$=\frac{\nabla}{\partial r}\left(\frac{\nabla}{\partial s}\left(\frac{\partial}{\partial x^k}\circ\sigma\right)\right)-\frac{\nabla}{\partial s}\left(\frac{\nabla}{\partial r}\left(\frac{\partial}{\partial x^k}\circ\sigma\right)\right). \tag{18.7.14}$$

This proves the assertion for $V=(\partial/\partial x^k)\circ\sigma$.

More generally, in local coordinates, consider

$$V=\sum_k\alpha^k\left(\frac{\partial}{\partial x^k}\circ\sigma\right).$$

Then

$$R\left(\frac{\partial\sigma}{\partial r},\frac{\partial\sigma}{\partial s}\right)V$$

$$=R\left(\frac{\partial\sigma}{\partial r},\frac{\partial\sigma}{\partial s}\right)\left(\sum_k\alpha^k\left(\frac{\partial}{\partial x^k}\circ\sigma\right)\right)$$

$$=\sum_k\alpha^k R\left(\frac{\partial\sigma}{\partial r},\frac{\partial\sigma}{\partial s}\right)\left(\frac{\partial}{\partial x^k}\circ\sigma\right) \tag{18.7.15}$$

$$=\sum_k\alpha^k\left[\frac{\nabla}{\partial r}\left(\frac{\nabla}{\partial s}\left(\frac{\partial}{\partial x^k}\circ\sigma\right)\right)-\frac{\nabla}{\partial s}\left(\frac{\nabla}{\partial r}\left(\frac{\partial}{\partial x^k}\circ\sigma\right)\right)\right],$$

where the last equality follows from (18.7.14). We also have

$$\frac{\nabla V}{\partial s}=\sum_k\frac{\nabla}{\partial s}\left(\alpha^k\left(\frac{\partial}{\partial x^k}\circ\sigma\right)\right)$$

$$=\sum_k\frac{\partial\alpha^k}{\partial s}\left(\frac{\partial}{\partial x^k}\circ\sigma\right)+\sum_k\alpha^k\frac{\nabla}{\partial s}\left(\frac{\partial}{\partial x^k}\circ\sigma\right),$$

hence

$$\frac{\nabla}{\partial r}\left(\frac{\nabla V}{\partial s}\right)=\sum_k\frac{\nabla}{\partial r}\left(\frac{\partial\alpha^k}{\partial s}\left(\frac{\partial}{\partial x^k}\circ\sigma\right)\right)+\sum_k\frac{\nabla}{\partial r}\left(\alpha^k\frac{\nabla}{\partial s}\left(\frac{\partial}{\partial x^k}\circ\sigma\right)\right)$$

$$=\sum_k\frac{\partial^2\alpha^k}{\partial r\partial s}\left(\frac{\partial}{\partial x^k}\circ\sigma\right)+\sum_k\frac{\partial\alpha^k}{\partial s}\frac{\nabla}{\partial r}\left(\frac{\partial}{\partial x^k}\circ\sigma\right)$$

$$+\sum_k\frac{\partial\alpha^k}{\partial r}\frac{\nabla}{\partial s}\left(\frac{\partial}{\partial x^k}\circ\sigma\right)+\sum_k\alpha^k\frac{\nabla}{\partial r}\left(\frac{\nabla}{\partial s}\left(\frac{\partial}{\partial x^k}\circ\sigma\right)\right).$$

Interchanging r and s in the preceding identity and using Theorem 10.1.6 yields

$$\frac{\nabla}{\partial s}\left(\frac{\nabla V}{\partial r}\right) = \sum_k \frac{\partial^2 \alpha^k}{\partial r \partial s}\left(\frac{\partial}{\partial x^k} \circ \sigma\right) + \sum_k \frac{\partial \alpha^k}{\partial r}\frac{\nabla}{\partial s}\left(\frac{\partial}{\partial x^k} \circ \sigma\right)$$

$$+ \sum_k \frac{\partial \alpha^k}{\partial s}\frac{\nabla}{\partial r}\left(\frac{\partial}{\partial x^k} \circ \sigma\right) + \sum_k \alpha^k \frac{\nabla}{\partial s}\left(\frac{\nabla}{\partial r}\left(\frac{\partial}{\partial x^k} \circ \sigma\right)\right).$$

Subtracting the preceding two identities gives

$$\frac{\nabla}{\partial r}\left(\frac{\nabla V}{\partial s}\right) - \frac{\nabla}{\partial s}\left(\frac{\nabla V}{\partial r}\right)$$

$$= \sum_k \alpha^k \left[\frac{\nabla}{\partial r}\left(\frac{\nabla}{\partial s}\left(\frac{\partial}{\partial x^k} \circ \sigma\right)\right) - \frac{\nabla}{\partial s}\left(\frac{\nabla}{\partial r}\left(\frac{\partial}{\partial x^k} \circ \sigma\right)\right)\right]. \tag{18.7.16}$$

The result now follows from (18.7.15) and (18.7.16). $\square$

18.8   Geodesics

In this section, we introduce a type of "straight line" for smooth manifolds with a connection. A cardinal feature of straight lines in Euclidean space is that they give the shortest distance between distinct points. This is meaningless in the present setting because there is no notion of distance on a smooth manifold with a connection. However, in Newtonian physics, a particle with zero acceleration traces a straight line in Euclidean space. We make this observation our point of departure.

Let (M, ∇) be a smooth manifold with a connection, and let $\lambda(t) : (a, b) \longrightarrow M$ be a smooth curve. Recall from Theorem 15.1.3 that the **velocity of λ** is the vector field $d\lambda/dt$ in $\mathfrak{X}_M(\lambda)$, and that λ is said to be **regular** if its velocity is nowhere-vanishing. The **acceleration of λ** is defined to be the covariant derivative of the velocity, that is, the vector field $(\nabla/dt)(d\lambda/dt)$ in $\mathfrak{X}_M(\lambda)$. We say that λ is a **geodesic** if it has zero acceleration; that is,

$$\frac{\nabla}{dt}\left(\frac{d\lambda}{dt}\right)(t) = 0$$

for all t in (a, b); or equivalently, if $d\lambda/dt$ is a parallel vector field in $\mathfrak{X}_M(\lambda)$.

As an example, consider the straight line in $\mathbb{R}^3$ parametrized by $\lambda(t) = (t, t, t) : \mathbb{R} \longrightarrow \mathbb{R}^3$. The velocity and accelerations are $(d\lambda/dt)(t) = (1, 1, 1)$ and $(\nabla/dt)(d\lambda/dt)(t) = (0, 0, 0)$ for all t in $\mathbb{R}^3$, hence λ is a geodesic. Now consider the alternative parametrization $\widetilde{\lambda}(t) = (t^3, t^3, t^3) : \mathbb{R} \longrightarrow \mathbb{R}^3$. Then $(d\widetilde{\lambda}/dt)(t) = (3t^2, 3t^2, 3t^2)$ and $(\nabla/dt)(d\widetilde{\lambda}/dt)(t) = (6t, 6t, 6t)$, so $\widetilde{\lambda}$ is not a geodesic. This illustrates that deciding whether a smooth curve is a geodesic depends not only on its "geometry" but also on its parametrization.

Theorem 18.8.1 (Geodesic Equations). *Let* (M, ∇) *be a smooth m-manifold with a connection, and let* $\gamma(t) : (a, b) \longrightarrow M$ *be a smooth curve. Then* γ *is a geodesic if and only if for every chart* $(U, (x^i))$ *on* M *such that* $(a, b) \cap \gamma^{-1}(U)$ *is nonempty,* γ *satisfies the following system of differential equations, called the* **geodesic equations:**

$$\frac{d^2(x^k \circ \gamma)}{dt^2}(t) + \sum_{ij} \frac{d(x^i \circ \gamma)}{dt}(t) \frac{d(x^j \circ \gamma)}{dt}(t) \Gamma_{ij}^k(\gamma(t)) = 0$$

for all t *in* $(a, b) \cap \gamma^{-1}(U)$ *for* $k = 1, \ldots, m$.

Proof. By Theorem 14.7.1(b),

$$\frac{d\gamma}{dt}(t) = \sum_i \frac{d(x^i \circ \gamma)}{dt}(t) \frac{\partial}{\partial x^i}\bigg|_{\gamma(t)}$$

for all t in (a, b). Setting $J = d\gamma/dt$ and $\alpha^i = d(x^i \circ \lambda)/dt$ in Theorem 18.5.1 gives the above system of ordinary differential equations. The result now follows from the theory of differential equations. $\qquad\qquad\square$

Theorem 18.8.2 (Existence of Maximal Geodesic). *Let* (M, ∇) *be a smooth manifold with a connection, let* p *be a point in* M, *and let* v *be a vector in* $T_p(M)$. *Then there is a geodesic*

$$\gamma_{p,v}(t) : (a, b) \longrightarrow M,$$

called the **maximal geodesic with starting point** p **and initial velocity** v, *such that:*
(a)
$$\gamma_{p,v}(0) = p \qquad and \qquad \frac{d\gamma_{p,v}}{dt}(0) = v.$$

(b) *If* $\gamma(t) : (c, d) \longrightarrow M$ *is another geodesic such that* $\gamma(0) = p$ *and* $(d\gamma/dt)(0) = v$, *then* $(c, d) \subseteq (a, b)$ *and* γ *is the restriction of* $\gamma_{p,v}$ *to* (c, d). $\qquad\square$

Theorem 18.8.3. *Let* (M, ∇) *be a smooth manifold with a connection, let* p *be a point in* M, *and let* $\gamma_{p,0}(t) : (a, b) \longrightarrow M$ *be the maximal geodesic with starting point* p *and initial velocity* 0, *where* 0 *is the zero vector in* $T_p(M)$. *Then* $\gamma_{p,0}$ *has constant value* p.

Proof. Let $\gamma(t) : (a, b) \longrightarrow M$ be the smooth curve with constant value p on (a, b). Clearly, $(\nabla/dt)(d\gamma/dt)(t) = 0$ for all t in (a, b), so γ is a geodesic. Since $\gamma(0) = p$ and $(d\gamma/dt)(t) = 0$ for all t in (a, b), it follows from Theorem 18.8.2(b) that γ is the restriction of $\gamma_{p,0}$ to (a, b). $\qquad\square$

Theorem 18.8.4. *Let* (M, ∇) *be a smooth manifold with a connection, and let* $\gamma(t) : (a, b) \longrightarrow M$ *be a geodesic. If* γ *is nonconstant, then it is regular.*

Proof. We prove the logically equivalent assertion: if γ is not regular, then it is constant. Suppose $(d\gamma/dt)(t_0) = 0$ for some point t_0 in (a, b). Let $g(u)$: $(a - t_0, b - t_0) \longrightarrow (a, b)$ be the diffeomorphism given by $g(u) = u + t_0$. Since g is bijective, as u varies over (c, d), $g(u)$ varies over (a, b). Because γ is a geodesic, it follows from Theorem 18.3.3(b) that

$$\frac{\nabla}{du}\left(\frac{d(\gamma \circ g)}{du}\right)(u) = \frac{d^2 g}{du^2}(u)\frac{d\gamma}{dt}(g(u)) = 0$$

for all u in (c, d). Thus, $\gamma \circ g$ is a geodesic. Let $p = \gamma(t_0)$, so that $\gamma \circ g(0) = p$. By Theorem 14.4.5,

$$\frac{d(\gamma \circ g)}{du}(0) = \frac{dg}{du}(0)\frac{d\gamma}{dt}(g(0)) = \frac{d\gamma}{dt}(t_0) = 0.$$

We have from Theorem 18.8.2(b) that $\gamma \circ g$ is the restriction of $\gamma_{p,0}$ to $(a - t_0, b - t_0)$. By Theorem 18.8.3, $\gamma_{p,0}$ has constant value p, and therefore, so does γ. $\square$

Theorem 18.8.5. *Let* (M, ∇) *be a smooth manifold with a connection, let* $\gamma(t) : (a, b) \longrightarrow M$ *be a nonconstant geodesic, and let* $g(u) : (c, d) \longrightarrow (a, b)$ *be a diffeomorphism. Then* $\gamma \circ g(u) : (c, d) \longrightarrow (a, b)$ *is a geodesic if and only if* g *is linear.*

Proof. Since γ is a geodesic, by Theorem 18.3.3(b),

$$\frac{\nabla}{du}\left(\frac{d(\gamma \circ g)}{du}\right)(u) = \frac{d^2 g}{du^2}(u)\frac{d\gamma}{dt}(g(u)),$$

and because γ is nonconstant, by Theorem 18.8.4, $(d\gamma/dt)(g(u)) \neq 0$ for all u in (c, d). Since g is bijective, as u varies over (c, d), $g(u)$ varies over (a, b). Thus,

$$\gamma \circ g \text{ is a geodesic}$$
$$\Leftrightarrow \quad \frac{\nabla}{du}\left(\frac{d(\gamma \circ g)}{du}\right) = 0$$
$$\Leftrightarrow \quad \frac{d^2 g}{du^2} = 0$$
$$\Leftrightarrow \quad g \text{ is linear.} \qquad \square$$

Theorem 18.8.6. *Let* (M, ∇) *be a smooth manifold with a connection, let* p *be a point in* M, *and let* v *be a vector in* $T_p(M)$. *Let* $\gamma_{p,v}(t) : (a, b) \longrightarrow M$ *be the maximal geodesic with starting point* p *and initial velocity* v, *and let* $\varepsilon > 0$ *be a real number. Then the maximal geodesic with starting point* p *and initial velocity* εv *is* $\gamma_{p,\varepsilon v}(t) : (a/\varepsilon, b/\varepsilon) \longrightarrow M$, *where*

$$\gamma_{p,\varepsilon v}(t) = \gamma_{p,v}(\varepsilon t)$$

for all t *in* $(a/\varepsilon, b/\varepsilon)$.

Proof. Consider the diffeomorphism $g(u) : (a/\varepsilon, b/\varepsilon) \longrightarrow (a, b)$ given by $g(u) = \varepsilon u$. By Theorem 18.8.5, $\gamma_{p,v} \circ g(u) : (a/\varepsilon, b/\varepsilon) \longrightarrow M$ is a geodesic. Since $\gamma_{p,v}$ is maximal, so is $\gamma_{p,v} \circ g$, otherwise $\gamma_{p,v}$ could be extended. We have from Theorem 18.8.2(a) that $\gamma_{p,v} \circ g(0) = \gamma_{p,v}(0) = p$, and from Theorem 14.4.5 and Theorem 18.8.2(a) that

$$\frac{d(\gamma_{p,v} \circ g)}{du}(0) = \frac{dg}{du}(0)\frac{d\gamma_{p,v}}{dt}(g(0)) = \varepsilon\frac{d\gamma_{p,v}}{dt}(0) = \varepsilon v.$$

It follows from Theorem 18.8.2 that $\gamma_{p,\varepsilon v} = \gamma_{p,v} \circ g$. Thus, $\gamma_{p,\varepsilon v}(u) = \gamma_{p,v}(\varepsilon u)$ for all u in $(a/\varepsilon, b/\varepsilon)$. $\qquad\square$

Example 18.8.7 (Smooth 2-Manifold). For smooth 2-manifolds with a symmetric connection, the geodesic equations in Theorem 18.8.1 simplify to

$$\frac{d^2(x^1 \circ \lambda)}{dt^2}(t) + \left(\frac{d(x^1 \circ \lambda)}{dt}(t)\right)^2 \Gamma^1_{11}(\lambda(t))$$

$$+ 2\frac{d(x^1 \circ \lambda)}{dt}(t)\frac{d(x^2 \circ \lambda)}{dt}(t)\,\Gamma^1_{12}(\lambda(t))$$

$$+ \left(\frac{d(x^2 \circ \lambda)}{dt}(t)\right)^2 \Gamma^1_{22}(\lambda(t)) = 0$$

and

$$\frac{d^2(x^2 \circ \lambda)}{dt^2}(t) + \left(\frac{d(x^1 \circ \lambda)}{dt}(t)\right)^2 \Gamma^2_{11}(\lambda(t))$$

$$+ 2\frac{d(x^1 \circ \lambda)}{dt}(t)\frac{d(x^2 \circ \lambda)}{dt}(t)\,\Gamma^2_{12}(\lambda(t))$$

$$+ \left(\frac{d(x^2 \circ \lambda)}{dt}(t)\right)^2 \Gamma^2_{22}(\lambda(t)) = 0.$$

Viewing a regular surface as a smooth 2-manifold with the connection defined in Example 18.1.1, we have from Example 18.6.1 that the connection is symmetric. Thus, the above geodesic equations apply. $\qquad\diamond$

Example 18.8.8 (Pln). Let us view the regular surface Pln, discussed in Section 13.1, as a smooth 2-manifold with the connection defined in Example 18.1.1. Suppose $\gamma(t) = (\gamma^1, \gamma^2) : \mathbb{R} \longrightarrow \mathbb{R}^2$ is a geodesic. Since the Christoffel symbols have constant value 0, by Theorem 18.8.1, the geodesic equations are $(d^2\gamma^k/dt^2)(t) = 0$ for $k = 1, 2$. These have the solutions $\gamma^k(t) = a^k + b^k t$ for any choice of real numbers a^1, b^1, a^2, b^2. Thus,

$$\gamma(t) = (a^1, a^2) + t(b^1, b^2).$$

Not surprisingly, the image of γ is a straight line. This demonstrates that the maximal geodesics in Pln with a given starting point are the straight lines through that point. $\qquad\diamond$

Example 18.8.9 (Cyl). Let us view the regular surface Cyl, discussed in Section 13.2, as a smooth 2-manifold with the connection defined in Example 18.1.1. Suppose $\gamma(u) : \mathbb{R} \longrightarrow$ Cyl is a geodesic parametrized by

$$\gamma(u) = \big(\cos(\phi(u)), \sin(\phi(u)), t(u)\big).$$

We have from Example 18.8.7 that the geodesic equations are

$$\frac{d^2 t}{du^2}(u) = 0 \quad \text{and} \quad \frac{d^2 \phi}{du^2}(u) = 0,$$

These have the solutions $t(u) = a^1 + b^1 u$ and $\phi(u) = a^2 + b^2 u$ for any choice of real numbers a^1, b^1, a^2, b^2. Thus,

$$\gamma(u) = \big(\cos(a^1 + b^1 u), \sin(a^1 + b^1 u), a^2 + b^2 u\big).$$

If $b^1 = 0$, then the image of γ is a straight line parallel to the z-axis; and if $b^2 = 0$, then the image of γ is a circle lying in a plane parallel to the xy-plane. Otherwise, the image of γ is a helix. This shows that the maximal geodesics in Cyl with a given starting point are the straight lines, circles, and helices through that point. Whether it is one or another is determined by the initial velocity of the maximal geodesic.   ◇

Example 18.8.10 ($\mathcal{S}^2$). Let us view the regular surface $\mathcal{S}^2$, discussed in Section 13.4, as a smooth 2-manifold with the connection defined in Example 18.1.1. Continuing with Example 18.5.4, suppose $\gamma(t) : (a, b) \longrightarrow \mathcal{S}^2$ is a geodesic. Setting $\alpha^1 = d\theta/dt$ and $\alpha^2 = d\phi/dt$ in (18.5.3) gives the geodesic equations

$$\frac{d^2\theta}{dt^2}(t) - \cos(\theta(t)) \sin(\theta(t)) \left(\frac{d\phi}{dt}(t)\right)^2 = 0$$

$$\frac{d\phi^2}{dt^2}(t) + 2\cot(\theta(t)) \frac{d\theta}{dt}(t) \frac{d\phi}{dt}(t) = 0. \tag{18.8.1}$$

For ease of computation, we view $\mathcal{S}^2$ as what we will later refer to as a Riemannian 2-manifold. In the present context, this means treating $\mathcal{S}^2$ as a regular surface in $\mathbb{R}^3_0$ and permitting computations that involve the Euclidean inner product.

Setting

$$\mu = (\mu^1, \mu^2) = (\theta, \phi) \tag{18.8.2}$$

in (12.5.2), we have from Theorem 12.5.2(d) that

$$\left\langle \frac{d\gamma}{dt}(t), \frac{d\gamma}{dt}(t) \right\rangle = \left(\frac{d\theta}{dt}(t)\right)^2 + \sin^2(\theta(t)) \left(\frac{d\phi}{dt}(t)\right)^2.$$

By Theorem 19.9.1(a), γ has constant speed, say, c. The preceding identity then becomes

$$\left(\frac{d\theta}{dt}(t)\right)^2 + \sin^2(\theta(t)) \left(\frac{d\phi}{dt}(t)\right)^2 = c^2. \tag{18.8.3}$$

Using (18.8.1) and (18.8.3), a lengthy but straightforward computation yields

$$\frac{d^2\gamma}{dt^2}(t) + c^2\gamma(t) = 0. \tag{18.8.4}$$

Alternatively, we have from Theorem 12.5.2(c), the Christoffel symbol results in Section 13.4, and (18.8.2) that

$$\begin{aligned}
\frac{d^2\gamma}{dt^2}(t) &= \left[\frac{d^2\theta}{dt}(t) - \cos(\theta(t))\sin(\theta(t))\left(\frac{d\phi}{dt}(t)\right)^2\right]H_1|_{(\theta(t),\phi(t))} \\
&+ \left[\frac{d^2\phi}{dt}(t) + 2\cot(\theta(t))\frac{d\theta}{dt}(t)\frac{d\phi}{dt}(t)\right]H_2|_{(\theta(t),\phi(t))} \\
&- \left[\left(\frac{d\theta}{dt}(t)\right)^2 + \sin^2(\theta(t))\left(\frac{d\phi}{dt}(t)\right)^2\right]G_{(\theta(t),\phi(t))}.
\end{aligned}$$

Substituting from (18.8.1)–(18.8.3) gives $(d^2\gamma/dt^2)(t) + c^2 G_{(\theta(t),\phi(t))} = 0$. It follows from Theorem 12.7.3(a) and a remark at the end of Section 13.4 that

$$G_{(\theta(t),\phi(t))} = \mathcal{N}_{\varphi(\theta(t),\phi(t))} = \varphi(\theta(t),\phi(t)) = \gamma(t),$$

so we again arrive at (18.8.4).

The solution of (18.8.4) is

$$\gamma(t) = \cos(ct)v + \sin(ct)w, \tag{18.8.5}$$

for some vectors v, w in $\mathbb{R}^3$ as yet to be determined. If $c = 0$, then γ has constant value $\gamma(0) = v$. We exclude this case and suppose $\pi/(2c)$ is in (a, b). Then

$$\gamma(0) = v, \qquad \gamma\left(\frac{\pi}{2c}\right) = w, \qquad \text{and} \qquad \frac{d\gamma}{dt}(0) = cw.$$

It follows from Theorem 18.8.2 that γ is the restriction of $\gamma_{v,cw}$ to (a, b), and from (18.8.5) that the image of γ lies in the intersection of $\mathcal{S}^2$ and the subspace of $\mathbb{R}^3_0$ spanned by v and w. Thus, the image of γ is all or a portion of a great circle. More generally, the maximal geodesics in $\mathcal{S}^2$ with a given starting point are the great circles through that point. $\Diamond$

18.9 Radial Geodesics and Exponential Maps

Let (M, ∇) be a smooth manifold with a connection, let p be a point in M, and suppose $T_p(M)$ has the standard smooth structure. Let $\mathcal{E}_p$ be the set of vectors v in $T_p(M)$ such that $\gamma_{p,v}(1)$ is defined; that is, 1 is in the domain of $\gamma_{p,v}$, where $\gamma_{p,v}$ is the maximal geodesic with starting point p and initial velocity v.

Theorem 18.9.1. *With the above setup:*
(a) $\mathcal{E}_p$ *is an open set in* $T_p(M)$.
(b) $\mathcal{E}_p$ *is star-shaped about the zero vector in* $T_p(M)$.

(c) *If v is a vector in $\mathcal{E}_p$, then $\gamma_{p,v}$ is defined on $[0,1]$.*

Proof. We consider only parts (b) and (c).

(b): Let v be a vector in $T_p(M)$, and let $0 < \varepsilon \leq 1$. If $\gamma_{p,v}$ has domain (a, b), then, by Theorem 18.8.6, $\gamma_{p,\varepsilon v}$ has domain $(a/\varepsilon, b/\varepsilon)$, which contains (a, b). It follows that if $\gamma_{p,v}(1)$ is defined, then so is $\gamma_{p,\varepsilon v}(1)$; that is, if v is in $\mathcal{E}_p$, then so is εv.

(c): We have from part (b) that $\gamma_{p,tv}(1)$ is defined for all t in $[0,1]$, and from Theorem 18.8.6 that $\gamma_{p,v}(t) = \gamma_{p,tv}(1)$. The result follows. $\qquad\square$

Continuing with the above setup, let v be a vector in $\mathcal{E}_p$. Using Theorem 18.9.1(c), we restrict $\gamma_{p,v}$ to $[0,1]$ and obtain the geodesic

$$\rho_{p,v}(t) = \gamma_{p,v}|_{[0,1]} : [0,1] \longrightarrow M, \tag{18.9.1}$$

called the **radial geodesic with starting point p and initial velocity v**. It follows from Theorem 18.8.2(a) that $\rho_{p,v}(0) = p$. By varying v, we obtain a family of radial geodesics with starting point p.

Example 18.9.2 (Pln and $\mathcal{S}^2$). Let p be a point in Pln. We have from Example 18.8.8 that the radial geodesics with starting point p are line segments in Pln radiating from p. It follows that $\mathcal{E}_p = T_p(\text{Pln})$. Let q be another point in Pln. As is clear from the underlying Euclidean geometry, there is a unique radial geodesic starting at p and ending at q.

Now let p be a point in $\mathcal{S}^2$. We have from Example 18.8.10 that the radial geodesics with starting point p are portions (or all) of great circles on $\mathcal{S}^2$ emanating from p. It follows that $\mathcal{E}_p = T_p(\mathcal{S}^2)$. Let q be another point in $\mathcal{S}^2$. There is a unique great circle passing through p and q, obtained by intersecting $\mathcal{S}^2$ with the plane passing through p, q, and $(0,0,0)$. This gives two radial geodesics that start at p and end at q. $\qquad\diamond$

For the remainder of this section, we work toward a condition that ensures a uniqueness condition for radial geodesics.

Let (M, ∇) be a smooth m-manifold with a connection, and let p be a point in M. The **exponential map at p** is the smooth map

$$\exp_p : \mathcal{E}_p \longrightarrow M$$

defined by

$$\exp_p(v) = \gamma_{p,v}(1) \tag{18.9.2}$$

for all vectors v in $\mathcal{E}_p$.

Theorem 18.9.3. *Let (M, ∇) be a smooth manifold with a connection, let p be a point in M, and let v be a vector in $\mathcal{E}_p$. Let $\gamma_{p,v}(t) : (a, b) \longrightarrow M$ be the maximal geodesic with starting point p and initial velocity v, and let $\rho_{p,v}(t) : [0,1] \longrightarrow M$ be the radial geodesic with starting point p and initial velocity v. Then*

$$\gamma_{p,v}(t) = \exp_p(tv) \qquad \text{for all } t \text{ in } (a, b)$$

and

$$\rho_{p,v}(t) = \exp_p(tv) \qquad \text{for all } t \text{ in } [0,1]. \tag{18.9.3}$$

Proof. We have

$$t \text{ is in } (a, b)$$
$$\Leftrightarrow \quad \gamma_{p,v}(t) \text{ is defined}$$
$$\Leftrightarrow \quad \gamma_{p,tv}(1) \text{ is defined} \qquad \text{[Th 18.8.6]}$$
$$\Leftrightarrow \quad \exp_p(tv) \text{ is defined.} \qquad \text{[(18.9.2)]}$$

Thus, when any one of $\gamma_{p,v}(t)$, $\gamma_{p,tv}(1)$, or $\exp_p(tv)$ is defined, they all are, in which case equality holds:

$$\gamma_{p,v}(t) = \gamma_{p,tv}(1) = \exp_p(tv).$$

This proves the first equality. The second equality follows from (18.9.1). $\square$

Roughly speaking, Theorem 18.9.3 says that the exponential function sends lines (or portions of lines) passing through the zero vector of the tangent space at a point to geodesics in the smooth manifold that pass through the point. See Figure 18.9.1.

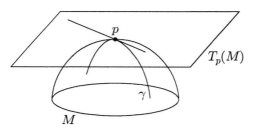

Figure 18.9.1. Exponential map

By Theorem 14.8.4 and Theorem 18.9.1(a), $\mathcal{E}_p$ is an open m-submanifold of $T_p(M)$, and from Theorem 14.3.3(b), we have the identification $T_v(\mathcal{E}_p) = T_v(T_p(M))$ for each vector v in $\mathcal{E}_p$. The differential map of $\exp_p : \mathcal{E}_p \longrightarrow M$ at v can therefore be expressed as

$$d_v(\exp_p) : T_v(T_p(M)) \longrightarrow T_{\exp_p(v)}(M). \tag{18.9.4}$$

See Figure 18.9.2, and, for context, also see Figure 11.6.1. Let w be a vector in $T_p(M)$, and consider the smooth curve $\mu_v(t) : (-\varepsilon, \varepsilon) \longrightarrow T_p(M)$ defined by $\mu_v(t) = v + tw$. According to (14.7.4), $(d\mu_v/dt)(t)$ is a vector in $T_{\mu_v(t)}(T_p(M))$. In the literature, $(d\mu_v/dt)(0)$ is sometimes denoted by w_v. Thus, w_v is a vector in $T_v(T_p(M))$, and it follows from (18.9.4) that $d_v(\exp_p)(w_v)$ is a vector in $T_{\exp_p(v)}(M)$. In geometric terms, we can think of $T_v(T_p(M))$ as the translation of $T_p(M)$ to v, with w_v the corresponding translation of w. It is sometimes convenient to denote w_v simply by w and allow the underlying translation to be understood. As a further simplification, when $v = 0$, we use (14.3.4) to identify $T_0(T_p(M))$ with $T_p(M)$, and write

$$T_0(T_p(M)) = T_p(M).$$

Since $\exp_p(0) = p$, we have from (18.9.4) that for $v = 0$ the differential map can be expressed as

$$d_0(\exp_p) : T_p(M) \longrightarrow T_p(M).$$

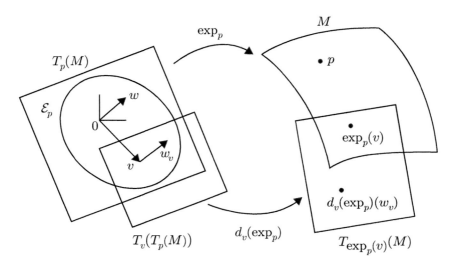

Figure 18.9.2. Differential map of exponential map

Theorem 18.9.4. *If (M, ∇) is a smooth manifold with a connection and p is a point in M, then $d_0(\exp_p)$ is the identity map on $T_p(M)$.*

Proof. Let v be a vector in $T_p(M)$, and let $\gamma_{p,v}(t) : (a, b) \longrightarrow M$ be the maximal geodesic with starting point p and initial velocity v. Consider the smooth curve $\lambda(t) : (a, b) \longrightarrow \mathcal{E}_p$ defined by $\lambda(t) = tv$. By Theorem 18.9.3, $\gamma_{p,v} = \exp_p \circ \lambda$. Since $\lambda(0) = 0$ and $(d\lambda/dt)(0) = v$, we have from Theorem 14.7.4(a) and Theorem 18.8.2(b) that

$$d_0(\exp_p)(v) = \frac{d(\exp_p \circ \lambda)}{dt}(0) = \frac{d\gamma_{p,v}}{dt}(0) = v. \qquad \square$$

Let (M, ∇) be a smooth manifold with a connection, and let p be a point in M. We say that a neighborhood V of p in M is **normal** if there is a star-shaped neighborhood $\widetilde{V} \subseteq \mathcal{E}_p$ of the zero vector in $T_p(M)$ such that $\exp_p|_{\widetilde{V}} : \widetilde{V} \longrightarrow V$ is a diffeomorphism. For brevity, we often denote

$$\exp_p|_{\widetilde{V}} : \widetilde{V} \longrightarrow V \qquad \text{by} \qquad \exp_p : \widetilde{V} \longrightarrow V.$$

Theorem 18.9.5 (Existence of Normal Neighborhood). *Every point in a smooth manifold with a connection has a normal neighborhood.*

Proof. Let (M, ∇) be a smooth manifold with a connection, and let p be a point in M. By Theorem 14.6.2 and Theorem 18.9.4, there is a neighborhood $\widetilde{W} \subseteq \mathcal{E}_p$ of the zero vector in $T_p(M)$ and a neighborhood W of p in M such that $\exp_p|_{\widetilde{W}} : \widetilde{W} \longrightarrow W$ is a diffeomorphism. Suppose $T_p(M)$ has the standard smooth structure, and let $(T_p(M), A)$ be a (covering) chart on $T_p(M)$, where $A : T_p(M) \longrightarrow \mathbb{R}^m$ is any linear isomorphism. By Theorem 9.1.14, $A(\widetilde{W})$ is an open set in $\mathbb{R}^m$. Let $B \subseteq A(\widetilde{W})$ be an open ball in $\mathbb{R}^m$ centered at $(0, \ldots, 0)$, and observe that B is star-shaped about $(0, \ldots, 0)$. Let v be a vector in $A^{-1}(B)$. Then $A(tv) = tA(v)$ is in B for all t in $[0, 1]$, hence tv is in $A^{-1}(B)$ for all t in $[0, 1]$. Thus, $A^{-1}(B) \subseteq \mathcal{E}_p$ is star-shaped about the zero vector in $T_p(M)$. Setting $\widetilde{V} = A^{-1}(B)$, we find that $V = \exp_p(A^{-1}(B))$ is a normal neighborhood of p in M. $\square$

As we now show, radial geodesics on a normal neighborhood have a uniqueness property.

Theorem 18.9.6 (Uniqueness of Radial Geodesic). *Let (M, ∇) be a smooth manifold with a connection, let p be a point in M, let V be a normal neighborhood of p in M, and let $\exp_p : \widetilde{V} \longrightarrow V$ be the corresponding diffeomorphism. Let q be a point in V, let $v = \exp_p^{-1}(q)$, and let $\rho_{p,v} : [0, 1] \longrightarrow M$ be the radial geodesic with starting point p and initial velocity v. Then $\rho_{p,v}$ is the unique radial geodesic such that $\rho_{p,v}(0) = p$, $\rho_{p,v}(1) = q$, and $\rho_{p,v}([0, 1]) \subset V$.*

Remark. Since q is in V and $\exp_p^{-1}(V) = \widetilde{V} \subseteq \mathcal{E}_p$, we see that v is in $\mathcal{E}_p$, hence $\rho_{p,v}$ is defined, so assertion makes sense.

Proof. As was observed in connection with (18.9.1), $\rho_{p,v}(0) = p$, and we have from (18.9.3) that $\rho_{p,v}(1) = \exp_p(v) = q$. By assumption, $\widetilde{V}$ is star-shaped about the zero vector in $T_p(M)$ and $\exp_p(\widetilde{V}) = V$, so it also follows from (18.9.3) that $\rho_{p,v}(t)$ is in V for all t in $[0, 1]$. Thus, $\rho_{p,v}$ satisfies the three specified properties. To show uniqueness, suppose $\rho_{r,w}$ is another radial geodesic satisfying these properties; that is, (i) $\rho_{r,w}(0) = p$, (ii) $\rho_{r,w}(1) = q$, and (iii) $\rho_{r,w}([0, 1]) \subset V$. We need to show that $r = p$ and $v = w$. It follows from Theorem 18.8.2(a), (18.9.1), and (i) that $r = \rho_{r,w}(0) = p$. We have from $r = p$ and (iii) that $\rho_{p,w}([0, 1]) \subset V$, hence $\rho_{p,w}(1)$ is defined, and then from (18.9.3) that $\rho_{p,w}(1) = \exp_p(w)$. Then (18.9.3), $r = p$ and (ii) give

$$\exp_p(v) = \rho_{p,v}(1) = q = \rho_{r,w}(1) = \rho_{p,w}(1) = \exp_p(w).$$

Since $\exp_p|_{\widetilde{V}}$ is a diffeomorphism, $v = w$. Thus, $\rho_{r,w} = \rho_{p,v}$. $\square$

Example 18.9.7 (Pln and $\mathcal{S}^2$). We continue with Example 18.9.2. It is clear that Pln is a normal neighborhood of any of its points. However, the same is not true of $\mathcal{S}^2$, because $\mathcal{S}^2$ is not homeomorphic (let alone diffeomorphic) to an open set in the Euclidean plane. However, an appropriate subset of $\mathcal{S}^2$, such as a hemisphere, is a normal neighborhood of any of its points. $\Diamond$

18.10   Normal Coordinates

Let (M, ∇) be a smooth m-manifold with a connection, let p be a point in M, and let $(\widetilde{e}_1, \ldots, \widetilde{e}_m)$ be a basis for $T_p(M)$. The map

$$\ell_p : \mathbb{R}^m \longrightarrow T_p(M)$$

defined by

$$\ell_p(r^1, \ldots, r^m) = \sum_i r^i \widetilde{e}_i \qquad (18.10.1)$$

is a linear isomorphism. Let V be a normal neighborhood of p in M, let $\exp_p : \widetilde{V} \longrightarrow V$ be the corresponding diffeomorphism, and consider the smooth map

$$\ell_p^{-1} \circ \exp_p^{-1} : V \longrightarrow \ell_p^{-1}(\widetilde{V}).$$

Then

$$\left(V, \ell_p^{-1} \circ \exp_p^{-1} = (x^1, \ldots, x^m)\right) \qquad (18.10.2)$$

is a chart at p, called **normal coordinates at p**. Let $\mathcal{X}_p = \left((\partial/\partial x^1)|_p, \ldots, (\partial/\partial x^m)|_p\right)$ and $\left(d_p(x^1), \ldots, d_p(x^m)\right)$ be the corresponding coordinate and dual coordinate bases at p.

Theorem 18.10.1. *With the above setup:*
(a)

$$\widetilde{e}_i = \left.\frac{\partial}{\partial x^i}\right|_p$$

for $i = 1, \ldots, m$.
(b)

$$x^i \circ \exp_p = d_p(x^i)$$

for $i = 1, \ldots, m$.
(c)

$$\exp_p^{-1}(q) = \sum_i x^i(q) \left.\frac{\partial}{\partial x^i}\right|_p$$

for all q in V.

Proof. (a): Let $\mathcal{E} = (e_1, \ldots, e_m)$ be the standard basis for $\mathbb{R}^m$, and let $\varphi = \ell_p^{-1} \circ \exp_p^{-1}$. We have

$$
\begin{aligned}
d_{\varphi(p)}(\varphi^{-1}) &= d_{\varphi(p)}(\exp_p \circ \ell_p) \\
&= d_{\ell_p(\varphi(p))}(\exp_p) \circ d_{\varphi(p)}(\ell_p) \qquad \text{[Th 14.4.5]} \\
&= d_0(\exp_p) \circ \ell_p \qquad \text{[Th 14.4.4]} \\
&= \ell_p. \qquad \text{[Th 18.9.4]}
\end{aligned}
$$

Then Theorem 14.8.5(d) and (18.10.1) give

$$\left.\frac{\partial}{\partial x^i}\right|_p = \ell_p(e_i) = \widetilde{e}_i.$$

(c): We have from (18.10.2) that $\exp_p^{-1} = \ell_p \circ (x^1, \ldots, x^m)$, so

$$\exp_p^{-1}(q) = \ell_p(x^1(q), \ldots, x^m(q))$$

$$= \sum_i x^i(q)\widetilde{e}^i \qquad\qquad [(18.10.1)]$$

$$= \sum_i x^i(q)\frac{\partial}{\partial x^i}\Big|_p. \qquad\qquad [\text{part (a)}]$$

(b): Applying $d_p(x^i)$ to both sides of the identity in part (c) yields

$$d_p(x^i) \circ \exp_p^{-1}(q) = \sum_j x^j(q)\, d_p(x^i)\left(\frac{\partial}{\partial x^j}\Big|_p\right) = x^i(q).$$

Since q was arbitrary, $d_p(x^i) \circ \exp_p^{-1} = x^i$, from which the result follows. $\square$

In light of Theorem 18.10.1(a), we can replace (18.10.1) with

$$\ell_p(r^1, \ldots, r^m) = \sum_i r^i \frac{\partial}{\partial x^i}\Big|_p. \qquad\qquad (18.10.3)$$

Theorem 18.10.2. *If (M, ∇) is a smooth m-manifold with a symmetric connection and p is a point in M, then, in normal coordinates at p,*

$$\Gamma_{ij}^k(p) = 0$$

for $i, j, k = 1, \ldots, m$.

Proof. Let $(r^1, \ldots, r^m)$ be a vector in $\ell_p^{-1}(\widetilde{V})$, let $\widetilde{v} = \ell_p(r^1, \ldots, r^m)$, and consider the maximal geodesic $\gamma_{p,\widetilde{v}}$ with starting point p and initial velocity $\widetilde{v}$. We have

$$x^i \circ \gamma_{p,\widetilde{v}}(t) = x^i \circ \exp_p(t\widetilde{v}) \qquad\qquad [\text{Th 18.9.3}]$$

$$= d_p(x^i)\left(t\sum_j r^j \frac{\partial}{\partial x^i}\Big|_p\right) \qquad [\text{Th 18.10.1(b), (18.10.3)}]$$

$$= t\sum_j r^j\, d_p(x^i)\left(\frac{\partial}{\partial x^j}\Big|_p\right) \qquad [\text{Th 14.4.2}]$$

$$= tr^i$$

for $i = 1, \ldots, m$. Since $\gamma_{p,\widetilde{v}}(t)$ satisfies the geodesic equations in Theorem 18.8.1,

$$\sum_{ij} r^i r^j \Gamma_{ij}^k(\gamma_{p,\widetilde{v}}(t)) = 0$$

for $i, j = 1, \ldots, m$. Because $\gamma_{p,\widetilde{v}}(0) = p$, we have

$$\sum_{ij} r^i r^j \Gamma_{ij}^k(p) = 0$$

for $i, j = 1, \ldots, m$. Observe that the $\Gamma_{ij}^k(p)$ are independent of the choice of $(r^1, \ldots, r^m)$. Setting

$$(r^1, \ldots, r^i, \ldots, r^m) = (0, \ldots, 1, \ldots, 0)$$

yields

$$\Gamma_{ii}^k(p) = 0 \qquad (18.10.4)$$

for $i = 1, \ldots, m$; and setting

$$(r^1, \ldots, r^i, \ldots, r^j, \ldots, r^m) = (0, \ldots, 1, \ldots, 1, \ldots, 0)$$

gives

$$0 = \Gamma_{ii}^k(p) + \Gamma_{ij}^k(p) + \Gamma_{ji}^k(p) + \Gamma_{jj}^k(p) = 2\,\Gamma_{ij}^k(p)$$

for $i \neq j = 1, \ldots, m$, where the last equality follows from Theorem 18.6.3(b) and (18.10.4). $\square$

18.11   Jacobi Fields

On intuitive grounds, it seems that the way geodesics radiate from a given point on a smooth manifold, and the rapidity with which they spread out, should be related to the "curvature" of the smooth manifold at that point. This line of reasoning leads to a consideration of Jacobi fields.

Let (M, ∇) be a smooth manifold with a connection, and let $\gamma(r) : (a, b) \longrightarrow M$ be a geodesic. We say that the vector field J in $\mathfrak{X}_M(\gamma)$ is a **Jacobi field** if it satisfies **Jacobi's equation**:

$$R\left(J, \frac{d\gamma}{dr}\right)\frac{d\gamma}{dr} + \frac{\nabla^2 J}{dr^2} = 0, \qquad (18.11.1)$$

where we denote

$$\frac{\nabla}{dr} \circ \frac{\nabla}{dr} \quad \text{by} \quad \frac{\nabla^2}{dr^2}.$$

By Theorem 18.7.1(a), Jacobi's equation is equivalent to

$$R\left(\frac{d\gamma}{dr}, J\right)\frac{d\gamma}{dr} = \frac{\nabla^2 J}{dr^2}.$$

Theorem 18.11.1 (Existence of Jacobi Field). *Let (M, ∇) be a smooth manifold with a connection, and let p be a point in M. Let $\gamma(r) : (a, b) \longrightarrow M$ be a geodesic, with $\gamma(0) = p$, and let v, w be vectors in $T_p(M)$. Then there is a unique Jacobi field J in $\mathfrak{X}_M(\gamma)$ such that $J(0) = v$ and $(\nabla J/\partial r)(0) = w$.*

Proof. It can be shown that Jacobi's equation is equivalent to a system of ordinary differential equations. The result now follows from the theory of differential equations. $\square$

Let (M, ∇) be a smooth manifold with a connection, and let

$$\sigma(r, s) : (a, b) \times (-\varepsilon, \varepsilon) \longrightarrow M$$

be a parametrized surface. Recall from Section 14.9 that for a given point s in $(-\varepsilon, \varepsilon)$, the corresponding longitude curve $\sigma_s(r) : (a, b) \longrightarrow M$ is defined by $\sigma_s(r) = \sigma(r, s)$. Let $\gamma(r) : (a, b) \longrightarrow M$ be a geodesic. We say that σ is a **geodesic variation** of γ if $\sigma_0 = \gamma$ and each σ_s is a geodesic.

Suppose σ is in fact a geodesic variation of γ. Then

$$\frac{d\gamma}{dr}(r) = \frac{\partial\sigma}{\partial r}(r, 0) \tag{18.11.2}$$

for all r in (a, b). The **variation field** of σ is the vector field J in $\mathfrak{X}_M(\gamma)$ defined by

$$J(r) = \frac{\partial\sigma}{\partial s}(r, 0) \tag{18.11.3}$$

for all r in (a, b). Since σ_s is a geodesic and $\partial\sigma/\partial r$ is its velocity, we have

$$\frac{\nabla}{\partial r}\left(\frac{\partial\sigma}{\partial r}\right)(r, s) = 0 \tag{18.11.4}$$

for all (r, s) in $(a, b) \times (-\varepsilon, \varepsilon)$. See Figure 18.11.1.

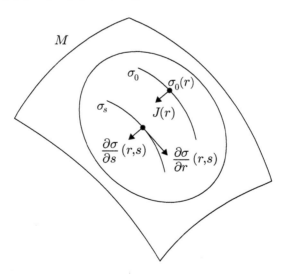

Figure 18.11.1. Geodesic variation

Example 18.11.2 ($\mathcal{S}^2$). In the notation of Section 13.4, consider the parametrized surface

$$\varphi(\theta, \phi) : (0, \pi) \times (-\pi, \pi) \longrightarrow \mathcal{S}^2$$

given by
$$\varphi(\theta, \phi) = \big(\sin(\theta)\cos(\phi), \sin(\theta)\sin(\phi), \cos(\theta)\big).$$

The image of the smooth curve

$$\gamma(\theta) = \varphi(\theta, 0) = \big(\sin(\theta), 0, \cos(\theta)\big) : (0, \pi) \longrightarrow \mathcal{S}^2$$

is a portion of the unit circle in the xz-plane centered at the origin, which is a great circle of $\mathcal{S}^2$. It follows from Example 18.8.10 that γ is a geodesic. Thus, φ is a geodesic variation of γ. ◊

Theorem 18.11.3. *If (M, ∇) is a smooth manifold with a symmetric connection and γ is a geodesic, then the variation field of any geodesic variation of γ is a Jacobi field in $\mathfrak{X}_M(\gamma)$.*

Proof. Let $\sigma(r, s) : (a, b) \times (-\varepsilon, \varepsilon) \longrightarrow M$ be a geodesic variation of γ, and let J be its variation field. Then

$$0 = \frac{\nabla}{\partial s}\left(\frac{\nabla}{\partial r}\left(\frac{\partial \sigma}{\partial r}\right)\right) \qquad [(18.11.4)]$$

$$= \frac{\nabla}{\partial r}\left(\frac{\nabla}{\partial s}\left(\frac{\partial \sigma}{\partial r}\right)\right) + R\left(\frac{\partial \sigma}{\partial s}, \frac{\partial \sigma}{\partial r}\right)\frac{\partial \sigma}{\partial r} \qquad \text{Th 18.7.5]}$$

$$= \frac{\nabla}{\partial r}\left(\frac{\nabla}{\partial r}\left(\frac{\partial \sigma}{\partial s}\right)\right) + R\left(\frac{\partial \sigma}{\partial s}, \frac{\partial \sigma}{\partial r}\right)\frac{\partial \sigma}{\partial r} \qquad [\text{Th 18.6.4}]$$

$$= \frac{\nabla^2}{\partial r^2}\left(\frac{\partial \sigma}{\partial s}\right) + R\left(\frac{\partial \sigma}{\partial s}, \frac{\partial \sigma}{\partial r}\right)\frac{\partial \sigma}{\partial r}.$$

Setting $s = 0$ and using (18.11.2) and (18.11.3) gives (18.11.1). □

Theorem 18.11.4. *Let (M, ∇) be a smooth manifold with a connection, let p be a point in M, and let $\psi(s) : (-\varepsilon, \varepsilon) \longrightarrow T_p(M)$ be a smooth curve. Consider the parametrized surface $\sigma(r, s) : [0, 1] \times (-\varepsilon, \varepsilon) \longrightarrow M$ given by*

$$\sigma(r, s) = \exp_p\big(r\psi(s)\big),$$

where it is assumed that $r\psi(s)$ is in $\mathcal{E}_p$ for all (r, s) in $[0, 1] \times (-\varepsilon, \varepsilon)$. Then:
(a) *For each point s in $(-\varepsilon, \varepsilon)$, σ_s is the radial geodesic $\rho_{p,\psi(s)}$ with starting point p and initial velocity $\psi(s)$.*
(b) *σ is a geodesic variation of σ_0.*
(c)

$$\frac{\partial \sigma}{\partial r}(r, s) = d_{r\psi(s)}(\exp_p)\big(\psi(s)\big).$$

(d)

$$\frac{\partial \sigma}{\partial s}(r, s) = r\, d_{r\psi(s)}(\exp_p)\left(\frac{d\psi}{ds}(s)\right).$$

Proof. (a): By Theorem 18.9.3,

$$\sigma_s(r) = \sigma(r, s) = \exp_p\big(r\psi(s)\big) = \rho_{p,\psi(s)}(r)$$

for all r in $[0, 1]$.

(b): This follows from part (a).

(c), (d): We define a smooth map

$$\widetilde{\sigma}(r, s) : [0, 1] \times (-\varepsilon, \varepsilon) \longrightarrow T_p(M)$$

by

$$\widetilde{\sigma}(r, s) = r\psi(s),$$

so that $\sigma = \exp_p \circ \widetilde{\sigma}$. Then

$$\frac{\partial \widetilde{\sigma}}{\partial r}(r, s) = \psi(s) \qquad \text{and} \qquad \frac{\partial \widetilde{\sigma}}{\partial s}(r, s) = r\frac{d\psi}{ds}(s),$$

hence

$$
\begin{aligned}
\frac{\partial \sigma}{\partial r}(r, s) &= \frac{\partial (\exp_p \circ \widetilde{\sigma})}{\partial r}(r, s) \\
&= d_{\widetilde{\sigma}(r,s)}(\exp_p)\left(\frac{\partial \widetilde{\sigma}}{\partial r}(r, s)\right) \qquad \text{[Th 14.7.4(b)]} \\
&= d_{r\psi(s)}(\exp_p)\big(\psi(s)\big).
\end{aligned}
$$

Similarly,

$$
\begin{aligned}
\frac{\partial \sigma}{\partial s}(r, s) &= \frac{\partial (\exp_p \circ \widetilde{\sigma})}{\partial s}(r, s) \\
&= d_{\widetilde{\sigma}(r,s)}(\exp_p)\left(\frac{\partial \widetilde{\sigma}}{\partial s}(r, s)\right) \qquad \text{[Th 14.7.4(b)]} \\
&= d_{r\psi(s)}(\exp_p)\left(r\frac{d\psi}{ds}(s)\right) \\
&= r\, d_{r\psi(s)}(\exp_p)\left(\frac{d\psi}{ds}(s)\right). \qquad \text{[Th 14.4.2]} \qquad \square
\end{aligned}
$$

Theorem 18.11.5. *Let (M, ∇) be a smooth manifold with a connection, let p be a point in M, and let v, w be vectors in $T_p(M)$. Let $\rho_{p,v}(r) = \exp_p(rv) :$ $[0, 1] \longrightarrow M$ be the radial geodesic with starting point p and initial velocity v. Let $\sigma(r, s) : [0, 1] \times (-\varepsilon, \varepsilon) \longrightarrow M$ be the geodesic variation of $\rho_{p,v}(r)$ defined by*

$$\sigma(r, s) = \exp_p\big(r(v + sw)\big),$$

where it is assumed that $r(v + sw)$ is in $\mathcal{E}_p$ for all (r, s) in $[0, 1] \times (-\varepsilon, \varepsilon)$. Then:

(a) *The variation field J of σ is given by*

$$J(r) = r\, d_{rv}(\exp_p)(w)$$

for all r in $[0, 1]$.

(b) *If ∇ is symmetric, then J is the unique Jacobi field in $\mathfrak{X}_M(\rho_{p,v})$ such that $J(0) = 0$ and $(\nabla J/dr)(0) = w$.*

Remark. That $\rho_{p,v}(r) = \exp_p(rv)$ follows from Theorem 18.9.3. By Theorem 18.11.4(b), σ is a geodesic variation of $\rho_{p,v}$. We have from (18.9.4) that the differential map of $\exp_p : \mathcal{E}_p \longrightarrow M$ at rv is

$$d_{rv}(\exp_p) : T_{rv}(T_p(M)) \longrightarrow T_{\exp_p(rv)}(M).$$

It would be more precise in part (a) to write $J(r) = r\, d_{rv}(\exp_p)(w_{rv})$. Stating part (a) as above amounts to making the identification $T_{rv}(T_p(M)) = T_p(M)$.

Proof. (a): Setting $\psi(s) = v + sw$ in Theorem 18.11.4(d) yields

$$\frac{\partial \sigma}{\partial s}(r, s) = r\, d_{r(v+sw)}(\exp_p)(w).$$

Setting $s = 0$ in the preceding identity and using (18.11.3) gives the result.

(b): Since ∇ is symmetric, it follows from Theorem 18.11.3 and part (a) that J is a Jacobi field in $\mathfrak{X}_M(\rho_{p,v})$. From the identity in part (a), we obtain $J(0) = 0$ and

$$\frac{\nabla J}{dr}(r) = d_{rv}(\exp_p)(w) + r\, \frac{\nabla}{dr}\big(d_{rv}(\exp_p)(w)\big).$$

By Theorem 18.9.4, $d_0(\exp_p)$ is the identity diffeomorphism. Setting $r = 0$ in the preceding identity yields $(\nabla J/dr)(0) = w$. The uniqueness property of J now follows from Theorem 18.11.1. $\qquad\square$

Example 18.11.6. Let (M, ∇) be a smooth manifold with a connection, and let $\gamma(r) : [0, 1] \longrightarrow M$ be a (radial) geodesic. Consider the vector fields J and K in $\mathfrak{X}_M(\gamma)$ defined by

$$J(r) = \frac{d\gamma}{dr}(r) \qquad \text{and} \qquad K(r) = r\frac{d\gamma}{dr}(r).$$

We have

$$\frac{\nabla}{dr}\left(\frac{d\gamma}{dr}\right) = 0 \qquad \text{and} \qquad R \circ \gamma\left(\frac{d\gamma}{dr}, \frac{d\gamma}{dr}\right)\frac{d\gamma}{dr} = 0,$$

where the second identity follows from Theorem 18.7.1(a). It is now easily shown that J and K are Jacobi fields. $\qquad\Diamond$

Chapter 19

Semi-Riemannian Manifolds

Looking back at the material presented on smooth manifolds and smooth manifolds with a connection, we see that many of the concepts and results motivated by our study of regular surfaces and regular surfaces in $\mathbb{R}^3_\nu$ have been recovered—but not all. Missing from our recent efforts has been any discussion of "length" and "area". To address this shortfall, it is necessary to endow smooth manifolds with additional structure, which entails a consideration of semi-Riemannian manifolds.

19.1 Semi-Riemannian Manifolds

The following material builds on portions of the discussion in Section 12.2. Let M be a smooth m-manifold, and let $\mathfrak{g}$ be a symmetric tensor field in $\mathcal{T}^0_2(M)$, so that $\mathfrak{g}_p$ is a symmetric tensor in $\mathcal{T}^0_2\big(T_p(M)\big)$ for all p in M. We say that $\mathfrak{g}$ is a **metric** on M if:

[G1] $\mathfrak{g}_p$ is nondegenerate on $T_p(M)$ for all p in M.

[G2] $\mathrm{ind}(\mathfrak{g}_p)$ is independent of p in M.

When [G1] is satisfied, $\mathfrak{g}_p$ is a scalar product on $T_p(M)$ for all p in M; or equivalently, $\big(T_p(M), \mathfrak{g}_p\big)$ is a scalar product space for all p in M.

Suppose $\mathfrak{g}$ is in fact a metric. In that case, the pair $(M, \mathfrak{g})$ is said to be a **semi-Riemannian m-manifold**. We ascribe to $\mathfrak{g}$ those properties of $\mathfrak{g}_p$ that are independent of p. Accordingly, $\mathfrak{g}$ is said to be bilinear, symmetric, nondegenerate, and so on. The common value of the $\mathrm{ind}(\mathfrak{g}_p)$ is denoted by ν and called the **index of $\mathfrak{g}$** or the **index of M**. The common signature of the $\mathfrak{g}_p$ is denoted by $(\varepsilon_1, \ldots, \varepsilon_m)$ and called the **signature of $\mathfrak{g}$** or the **signature of M**. For brevity, we usually denote

$$\mathfrak{g}_p(\cdot, \cdot) \qquad \text{by} \qquad \langle \cdot, \cdot \rangle.$$

Semi-Riemannian Geometry, First Edition. Stephen C. Newman.
© 2019 John Wiley & Sons, Inc. Published 2019 by John Wiley & Sons, Inc.

When $\nu = 0$, we say that M is a **Riemannian m-manifold**, and when $m \geq 2$ and $\nu = 1$, M is said to be a **Lorentz m-manifold**. A **semi-Riemannian m-manifold with a connection** is a triple $(M, \mathfrak{g}, \nabla)$, where $(M, \mathfrak{g})$ is a semi-Riemannian m-manifold and ∇ is a connection on M. For the time being, we do not assume that $\mathfrak{g}$ and ∇ are related.

We have already encountered several semi-Riemannian manifolds, although we were not in a position to use such terminology at the time. For example, semi-Euclidean (m, ν)-space $\mathbb{R}_\nu^m = (\mathbb{R}^m, \mathfrak{s})$ is a semi-Riemannian m-manifold of index ν. In particular, Euclidean m-space $\mathbb{R}_0^m = (\mathbb{R}^m, \mathfrak{e})$ is a Riemannian m-manifold, and Minkowski m-space $\mathbb{R}_1^m = (\mathbb{R}^m, \mathfrak{m})$ is a Lorentz m-manifold. With tensor notation now at our disposal, we can express $\mathfrak{s}$ in local coordinates as

$$
\mathfrak{s} = \begin{cases} \sum_{i=1}^m d(x^i) \otimes d(x^i) & \text{if } \nu = 0 \\ \sum_{i=1}^{m-\nu} d(x^i) \otimes d(x^i) - \sum_{i=m-\nu+1}^m d(x^i) \otimes d(x^i) & \text{if } 1 \leq \nu \leq m-1 \\ -\sum_{i=1}^m d(x^i) \otimes d(x^i) & \text{if } \nu = m. \end{cases}
$$

To our list of semi-Riemannian manifolds, let us add the unit sphere $\mathcal{S}^2$ and hyperbolic space $\mathcal{H}^2$, which are Riemannian 2-manifolds, and the pseudosphere $\mathcal{P}^2$, which is a Lorentz 2-manifold. In fact, each of the regular surfaces in $\mathbb{R}_0^3$ described in Sections 13.1–13.8 is a Riemannian 2-manifold. We now provide a less obvious example of a Riemannian 2-manifold.

Example 19.1.1. Let $\mathbb{P}^2 = \{(r, s) \in \mathbb{R}^2 : s > 0\}$, which is an open set in the smooth 2-manifold $\mathbb{R}^2$. It follows from Theorem 14.3.3(b), (14.3.4), and Theorem 14.8.4 that $\mathbb{P}^2$ is a 2-submanifold of $\mathbb{R}^2$, and that $T_{(r,s)}(\mathbb{P}^2)$ can be identified with $\mathbb{R}^2$. We define a tensor field $\mathfrak{p}$ in $\mathcal{T}_2^0(\mathbb{P}^2)$ by

$$
\mathfrak{p}_{(r,s)}(v, w) = \frac{1}{s^2} \langle v, w \rangle
$$

for all points (r, s) in $\mathbb{P}^2$ and all vectors v, w in $T_{(r,s)}(\mathbb{P}^2)$, where $\langle \cdot, \cdot \rangle$ is the Euclidean inner product. It is easily shown that $\mathfrak{p}_{(r,s)}$ is an inner product for all (r, s) in $\mathbb{P}^2$. Thus, $(\mathbb{P}^2, \mathfrak{p})$ is a Riemannian 2-manifold, called the **Poincaré half-plane**. $\diamond$

Theorem 19.1.2 (Open Set). *If $(M, \mathfrak{g})$ is a semi-Riemannian m-manifold of index ν and U is an open set in M, then $(U, \mathfrak{g}|_U)$ is a semi-Riemannian m-submanifold of $(M, \mathfrak{g})$ of index ν.*

Proof. By Theorem 14.8.4, U is an m-submanifold of M. From Theorem 14.3.3(b), we have the identification $T_p(U) = T_p(M)$ for all p in U. It follows that $(\mathfrak{g}|_U)_p$ is nondegenerate on $T_p(U)$ for all p in U, and $\text{ind}((\mathfrak{g}|_U)_p) = \text{ind}(\mathfrak{g}_p) = \nu$ for all p in U. Thus, $\mathfrak{g}|_U$ satisfies [G1] and [G2]. $\square$

The remainder of this section presents definitions that will play a role later on.

Let $(M, \mathfrak{g})$ be a semi-Riemannian m-manifold, and let X, Y be vector fields in $\mathfrak{X}(M)$. We say that X is a **unit vector field** if X_p is a unit vector for all p in M, and that X and Y are **orthogonal vector fields** if X_p and Y_p are orthogonal for all p in M.

Let us define a function

$$\langle X, Y \rangle : M \longrightarrow \mathbb{R}$$

in $C^\infty(M)$ by the assignment

$$p \longmapsto \langle X_p, Y_p \rangle$$

for all p in M. Similarly, for a smooth curve $\lambda : (a, b) \longrightarrow M$ and vector fields J, K in $\mathfrak{X}_M(\lambda)$, we define a function

$$\langle J, K \rangle : (a, b) \longrightarrow \mathbb{R}$$

in $C^\infty((a, b))$ by the assignment

$$t \longmapsto \langle J(t), K(t) \rangle$$

for all t in (a, b).

Let $(U, (x^i))$ be a chart on M, and let $\mathcal{X} = (\partial/\partial x^1, \ldots, \partial/\partial x^m)$ be the corresponding coordinate frame. We define functions $\mathfrak{g}_{ij}$ in $C^\infty(U)$ by

$$\mathfrak{g}_{ij}(p) = \left\langle \left. \frac{\partial}{\partial x^i} \right|_p, \left. \frac{\partial}{\partial x^j} \right|_p \right\rangle \tag{19.1.1}$$

for all p in U for $i, j = 1, \ldots, m$. The **matrix of $\mathfrak{g}$** with respect to $\mathcal{X}$ is denoted by $\mathfrak{g}_{\mathcal{X}}$ and defined by

$$\mathfrak{g}_{\mathcal{X}}(p) = \begin{bmatrix} \mathfrak{g}_{11}(p) & \cdots & \mathfrak{g}_{1m}(p) \\ \vdots & \ddots & \vdots \\ \mathfrak{g}_{m1}(p) & \cdots & \mathfrak{g}_{mm}(p) \end{bmatrix}$$

for all p in U. Recall from Section 3.1 that the matrix of $\mathfrak{g}_p$ with respect to $\mathcal{X}_p$ is $(\mathfrak{g}_p)_{\mathcal{X}_p} = [\mathfrak{g}_{ij}(p)]$. Thus, as a matter of notation,

$$\mathfrak{g}_{\mathcal{X}}(p) = (\mathfrak{g}_p)_{\mathcal{X}_p}.$$

The **inverse matrix of $\mathfrak{g}$** with respect to $\mathcal{X}$ is denoted by $\mathfrak{g}_{\mathcal{X}}^{-1}$ and defined by

$$\mathfrak{g}_{\mathcal{X}}^{-1}(p) = \mathfrak{g}_{\mathcal{X}}(p)^{-1}$$

for all p in U. It is usual to express the entries of $\mathfrak{g}_{\mathcal{X}}^{-1}(p)$ with superscripts:

$$\mathfrak{g}_{\mathcal{X}}^{-1}(p) = \begin{bmatrix} \mathfrak{g}_{11}(p) & \cdots & \mathfrak{g}_{1m}(p) \\ \vdots & \ddots & \vdots \\ \mathfrak{g}_{m1}(p) & \cdots & \mathfrak{g}_{mm}(p) \end{bmatrix}^{-1} = \begin{bmatrix} \mathfrak{g}^{11}(p) & \cdots & \mathfrak{g}^{1m}(p) \\ \vdots & \ddots & \vdots \\ \mathfrak{g}^{m1}(p) & \cdots & \mathfrak{g}^{mm}(p) \end{bmatrix}.$$

The assignment $p \longmapsto \mathfrak{g}^{ij}(p)$ defines functions $\mathfrak{g}^{ij}$ in $C^\infty(U)$ for $i, j = 1, \ldots, m$. Since $[\mathfrak{g}_{ij}]$ and $[\mathfrak{g}^{ij}]$ are symmetric matrices, the functions $\mathfrak{g}_{ij}$ and $\mathfrak{g}^{ij}$ are symmetric in i, j.

Theorem 19.1.3. *Let $(M, \mathfrak{g})$ be a semi-Riemannian m-manifold, let p be a point in M, let $(U, (x^i))$ and $(\widetilde{U}, (\widetilde{x}^i))$ be charts at p, and let $\mathcal{X}_p$ and $\widetilde{\mathcal{X}}_p$ be the corresponding coordinate bases at p. Then*

$$\mathfrak{g}_{\widetilde{\mathcal{X}}}(p) = \left(\left[\mathrm{id}_{T_p(M)} \right]_{\widetilde{\mathcal{X}}_p}^{\mathcal{X}_p} \right)^{\mathrm{T}} \mathfrak{g}_{\mathcal{X}}(p) \left[\mathrm{id}_{T_p(M)} \right]_{\widetilde{\mathcal{X}}_p}^{\mathcal{X}_p},$$

where

$$\left[\mathrm{id}_{T_p(M)} \right]_{\widetilde{\mathcal{X}}_p}^{\mathcal{X}_p} = \begin{bmatrix} \dfrac{\partial x^1}{\partial \widetilde{x}^1}(p) & \cdots & \dfrac{\partial x^1}{\partial \widetilde{x}^m}(p) \\ \vdots & \ddots & \vdots \\ \dfrac{\partial x^m}{\partial \widetilde{x}^1}(p) & \cdots & \dfrac{\partial x^m}{\partial \widetilde{x}^m}(p) \end{bmatrix}.$$

Proof. This follows from Theorem 3.1.3 and Theorem 14.3.5(b). $\qquad\qquad\square$

Let $(\overline{M}, \overline{\mathfrak{g}})$ be a semi-Riemannian manifold, and let M be a submanifold of $\overline{M}$. We say that $(M, \overline{\mathfrak{g}}|_M)$ is a **semi-Riemannian submanifold** of $(\overline{M}, \overline{\mathfrak{g}})$ if $\overline{\mathfrak{g}}|_M$ is a metric on M. In that case, by definition: (i) $(\overline{\mathfrak{g}}|_M)|_p$ is nondegenerate on $T_p(M)$ for all p in M, and (ii) $\mathrm{ind}((\overline{\mathfrak{g}}|_M)|_p)$ is independent of p in M. The common value of the $\mathrm{ind}((\overline{\mathfrak{g}}|_M)|_p)$ is denoted by $\mathrm{ind}(\overline{\mathfrak{g}}|_M)$ and called the **index of** $\overline{\mathfrak{g}}|_M$ or the **index of** M. The common signature of the $(\overline{\mathfrak{g}}|_M)|_p$ is called the **signature of** $\overline{\mathfrak{g}}|_M$ or the **signature of** M. We say that $(M, \overline{\mathfrak{g}}|_M)$ is a **Riemannian (Lorentz) submanifold** if $(M, \overline{\mathfrak{g}}|_M)$, viewed as a semi-Riemannian manifold in its own right, is a Riemannian (Lorentz) manifold. When $(M, \overline{\mathfrak{g}}|_M)$ is a Riemannian (Lorentz) submanifold of $\overline{M}$ and also a hypersurface, we refer to $(M, \overline{\mathfrak{g}}|_M)$ as a **Riemannian (Lorentz) hypersurface** of $\overline{M}$. Observe that if $(\overline{M}, \overline{\mathfrak{g}})$ is a Riemannian manifold, then $(M, \overline{\mathfrak{g}}|_M)$ is automatically a Riemannian submanifold. On the other hand, if $(\overline{M}, \overline{\mathfrak{g}})$ is a Lorentz manifold and $(M, \overline{\mathfrak{g}}|_M)$ is a semi-Riemannian submanifold, then the latter could be either a Riemannian or a Lorentz submanifold. To illustrate, we have from Theorem 12.2.12(d) and Theorem 14.8.2 that $\mathcal{S}^2$ is a Riemannian hypersurface of the Riemannian 3-manifold $\mathbb{R}_0^3$, while $\mathcal{H}^2$ is a Riemannian hypersurface and $\mathcal{P}^2$ is a Lorentz hypersurface of the Lorentz 3-manifold $\mathbb{R}_1^3$.

Suppose $(M, \overline{\mathfrak{g}}|_M)$ is a semi-Riemannian hypersurface of $(\overline{M}, \overline{\mathfrak{g}})$, and let p be a point in M. According to (14.8.1), $T_p(M)$ can be viewed as a vector subspace of $T_p(\overline{M})$. It follows from Theorem 4.1.3 that $T_p(\overline{M}) = T_p(M) \oplus T_p(M)^\perp$, and then from Theorem 1.1.18 that $T_p(M)^\perp$ is 1-dimensional. We say that a vector v in $T_p(\overline{M})$ is **normal at** p if v is in $T_p(M)^\perp$. If v is also a unit vector, it is said to be **unit normal at** p. In that case, since v is nonzero, we have $T_p(M)^\perp = \mathbb{R}v$. Let V be a (not necessarily smooth) vector field along M. We say that V is a **normal vector field** if V_p is normal at p for all p in M. When V is both a unit vector field and a normal vector field, it is said to be a **unit normal vector field**.

Theorems 12.2.2–12.2.5 generalize in a straightforward fashion to the setting of a semi-Riemannian hypersurface of a semi-Riemannian manifold. It follows from the generalization of Theorem 12.2.5 that there is a (not necessarily smooth) unit normal vector field V along M such that $\langle V_p, V_p \rangle$ is independent of p in M. The common value of the $\langle V_p, V_p \rangle$ is denoted by ϵ_M and called the **sign of** M. Thus,

$$\epsilon_M = \langle V_p, V_p \rangle$$

for all p in M. We have from the generalization of Theorem 12.2.2(b) that

$$\epsilon_M = (-1)^{\nu - \mathrm{ind}(\overline{\mathfrak{g}}|_M)}.$$

19.2   Curves

Let $(M, \mathfrak{g})$ be a semi-Riemannian manifold, let $\lambda(t) : (a, b) \longrightarrow M$ be a smooth curve, and let $g(u) : (c, d) \longrightarrow (a, b)$ be a diffeomorphism. The smooth curve $\lambda \circ g(u) : (c, d) \longrightarrow \mathbb{R}^3$ is said to be a **smooth reparametrization of λ**. Recall from Section 15.1 that the **velocity of λ** is the smooth curve $(d\lambda/dt)(t) : (a, b) \longrightarrow M$, and that λ is said to be **regular** if its velocity is nowhere-vanishing. The **speed of λ** is the (not necessarily smooth) curve $\|(d\lambda/dt)(t)\| : (a, b) \longrightarrow M$. When $\nu = 0$, λ is regular if and only if its speed is nowhere-vanishing. We say that λ has **constant speed** if there is a real number c such that $\|(d\lambda/dt)(t)\| = c$ for all t in (a, b). It is said that λ is **spacelike** (resp., **timelike, lightlike**) if $(d\lambda/dt)(t)$ is spacelike (resp., timelike, lightlike) for all t in (a, b). Also, λ is said to be **future-directed (past-directed)** if $(d\lambda/dt)(t)$ is future-directed (past-directed) for all t in (a, b).

Let $\lambda(t) : [a, b] \longrightarrow M$ be a smooth curve on a closed interval. The **length of λ** is defined by

$$L(\lambda) = \int_a^b \left\| \frac{d\lambda}{dt}(t) \right\| dt. \tag{19.2.1}$$

Theorem 19.2.1 (Diffeomorphism Invariance of Length). *Let $(M, \mathfrak{g})$ be a semi-Riemannian manifold, let $\lambda(t) : [a, b] \longrightarrow M$ be a smooth curve, and let $g(u) : [c, d] \longrightarrow [a, b]$ be a diffeomorphism. Then $L(\lambda) = L(\lambda \circ g)$.*

Proof. The proof is the same as that given for Theorem 12.1.1.   □

19.3   Fundamental Theorem of Semi-Riemannian Manifolds

Let $(M, \mathfrak{g}, \nabla)$ be a semi-Riemannian manifold with a connection. Without further assumptions, there is no reason to expect $\mathfrak{g}$ and ∇ to be related. We say that ∇ is **compatible with $\mathfrak{g}$** if

$$\begin{aligned} X(\langle Y, Z \rangle) &= \langle \nabla_X(Y), Z \rangle + \langle Y, \nabla_X(Z) \rangle \\ &= \nabla_X(\langle Y, Z \rangle) \end{aligned} \tag{19.3.1}$$

for all vector fields X, Y, Z in $\mathfrak{X}(M)$, where the second equality follows from (18.1.1). According to definitions given in Section 18.4, $\nabla(\mathfrak{g})$ is a tensor field in $\mathcal{T}_3^0(M)$, and $\mathfrak{g}$ is **parallel (with respect to ∇)** provided $\nabla(\mathfrak{g}) = 0$.

Theorem 19.3.1. *Let* $(M, \mathfrak{g}, \nabla)$ *be a semi-Riemannian manifold with a connection. Then the following are equivalent:*

(a) ∇ *is compatible with* $\mathfrak{g}$.
(b) $\mathfrak{g}$ *is parallel with respect to* ∇.
(c) *If* $\lambda(t) : (a, b) \longrightarrow M$ *is a regular smooth curve and* J, K *are vector fields in* $\mathfrak{X}_M(\lambda)$, *then*

$$\frac{d(\langle J, K \rangle)}{dt}(t) = \left\langle \frac{\nabla J}{dt}(t), K(t) \right\rangle + \left\langle J(t), \frac{\nabla K}{dt}(t) \right\rangle$$

for all t *in* (a, b).
(d) *If* $\lambda(t) : (a, b) \longrightarrow M$ *is a smooth curve and* J, K *are parallel vector fields in* $\mathfrak{X}_M(\lambda)$, *then the function* $\langle J, K \rangle(t) : (a, b) \longrightarrow \mathbb{R}$ *is constant.*

Proof. (a) $\Leftrightarrow$ (b): By Theorem 18.4.6,

$$\nabla(\mathfrak{g})(X, Y, Z) = X(\langle Y, Z \rangle) - \langle \nabla_X(Y), Z \rangle - \langle Y, \nabla_X(Z) \rangle$$

for all vector fields X, Y, Z in $\mathfrak{X}(M)$, from which the result follows.

(a) $\Rightarrow$ (c): Let $\lambda(t) : (a, b) \longrightarrow M$ be a regular smooth curve, let J, K be vector fields in $\mathfrak{X}_M(\lambda)$, let t_0 be a point in (a, b), and let $p = \lambda(t_0)$. By Theorem 15.1.2, there is a vector field X in $\mathfrak{X}(M)$ such that $X_p = (d\lambda/dt)(t_0)$. Since λ is regular, it can be shown that there is an interval $(c, d) \subseteq (a, b)$ containing t_0 and vector fields Y, Z in $\mathfrak{X}(M)$ such that $J|_{(c,d)} = Y \circ \lambda|_{(c,d)}$ and $K|_{(c,d)} = Z \circ \lambda|_{(c,d)}$. Then

$$\begin{aligned}
\frac{d(\langle J, K \rangle)}{dt}(t_0) &= \frac{d(\langle Y \circ \lambda, Z \circ \lambda \rangle)}{dt}(t_0) = \frac{d(\langle Y, Z \rangle \circ \lambda)}{dt}(t_0) \\
&= \frac{d\lambda}{dt}(t_0)(\langle Y, Z \rangle) = X_p(\langle Y, Z \rangle) = X(\langle Y, Z \rangle)(p) \\
&= \langle \nabla_X(Y)_p, Z_p \rangle + \langle Y_p, \nabla_X(Z)_p \rangle \\
&= \langle \nabla_{X_p}(Y), Z_p \rangle + \langle Y_p, \nabla_{X_p}(Z) \rangle \\
&= \langle \nabla_{(d\lambda/dt)(t_0)}(Y), Z_p \rangle + \langle Y_p, \nabla_{(d\lambda/dt)(t_0)}(Z) \rangle \\
&= \left\langle \frac{\nabla(Y \circ \lambda)}{dt}(t_0), Z_p \right\rangle + \left\langle Y_p, \frac{\nabla(Z \circ \lambda)}{dt}(t_0) \right\rangle \\
&= \left\langle \frac{\nabla J}{dt}(t_0), K(t_0) \right\rangle + \left\langle J(t_0), \frac{\nabla K}{dt}(t_0) \right\rangle,
\end{aligned}$$

where the third equality follows from (14.7.5), the fifth equality from (15.1.2), the sixth equality is by assumption, the seventh equality follows from (18.1.5), and the ninth equality from Theorem 18.3.1(c). Since t_0 was arbitrary, the result follows.

(c) $\Rightarrow$ (a): Let X, Y, Z be vector fields in $\mathfrak{X}(M)$, let p be a point in M, and let $\lambda(t) : (a, b) \longrightarrow M$ be a smooth curve such that $\lambda(t_0) = p$ and $(d\lambda/dt)(t_0) = X_p$

for some t_0 in (a, b). Then

$$
\begin{aligned}
X(\langle Y, Z \rangle)(p) = X_p(\langle Y, Z \rangle) &= \frac{d\lambda}{dt}(t_0)(\langle Y, Z \rangle) \\
&= \frac{d(\langle Y, Z \rangle \circ \lambda)}{dt}(t_0) = \frac{d(\langle Y \circ \lambda, Z \circ \lambda \rangle)}{dt}(t_0) \\
&= \left\langle \frac{\nabla(Y \circ \lambda)}{dt}(t_0), Z|_{\lambda(t_0)} \right\rangle + \left\langle Y|_{\lambda(t_0)}, \frac{\nabla(Z \circ \lambda)}{dt}(t_0) \right\rangle \\
&= \langle \nabla_{(d\lambda/dt)(t_0)}(Y), Z_p \rangle + \langle Y_p, \nabla_{(d\lambda/dt)(t_0)}(Z) \rangle \\
&= \langle \nabla_{X_p}(Y), Z_p \rangle + \langle Y_p, \nabla_{X_p}(Z) \rangle \\
&= \langle \nabla_X(Y)_p, Z_p \rangle + \langle Y_p, \nabla_X(Z)_p \rangle \\
&= \langle \nabla_X(Y), Z \rangle(p) + \langle Y, \nabla_X(Z) \rangle(p),
\end{aligned}
$$

where the first equality follows from (15.1.2), the third equality from (14.7.5), the fifth equality is by assumption, the sixth equality follows from Theorem 18.3.1(c), and the eighth equality from (18.1.5). Since p was arbitrary, the result follows.

(c) $\Rightarrow$ (d): By assumption, $\nabla J/dt = 0$ and $\nabla K/dt = 0$, so the identity in part (c) becomes $(d(\langle J, K \rangle)/dt)(t) = 0$ for all t in (a, b), hence the function $\langle J, K \rangle$ is constant.

(d) $\Rightarrow$ (c): Let $(e_1, \ldots, e_m)$ be an orthonormal basis for $T_{\lambda(0)}(M)$. Theorem 18.5.1 gives the parallel translate $E_i(t) = \Pi_{\lambda(0)}^{\lambda(t)}(e_i)$ of e_i for all t in (a, b), where $E_i(0) = e_i$ for $i = 1, \ldots, m$. By assumption, the function $\langle E_i, E_j \rangle$ is constant, so $\langle E_i(t), E_j(t) \rangle = \langle e_i, e_j \rangle$. Thus, $(E_1(t), \ldots, E_m(t))$ is an orthonormal basis for $T_{\lambda(t)}(M)$ for all t in (a, b). Expressing J and K as

$$
J = \sum_i \alpha^i E_i \qquad \text{and} \qquad K = \sum_j \beta^j E_j,
$$

where the α^i, β^j are uniquely determined functions in $C^\infty((a, b))$, we find that

$$
\langle J, K \rangle = \sum_{ij} \langle E_i, E_j \rangle \alpha_i \beta_j = \sum_i \langle e_i, e_i \rangle \alpha_i \beta_i,
$$

hence

$$
\frac{d(\langle J, K \rangle)}{dt} = \sum_i \langle e_i, e_i \rangle \left(\frac{d\alpha^i}{dt} \beta^i + \alpha^i \frac{d\beta^i}{dt} \right). \tag{19.3.2}
$$

Since the E_i are parallel,

$$
\frac{\nabla J}{dt} = \sum_i \frac{\nabla}{dt}(\alpha^i E_i) = \sum_i \frac{d\alpha^i}{dt} E_i + \sum_i \alpha^i \frac{\nabla E_i}{dt} = \sum_i \frac{d\alpha^i}{dt} E_i.
$$

Likewise,

$$
\frac{\nabla K}{dt} = \sum_j \frac{d\beta^j}{dt} E_j.
$$

Thus,

$$\left\langle \frac{\nabla J}{dt}, K \right\rangle = \left\langle \sum_i \frac{d\alpha^i}{dt} E_i, \sum_j \beta^j E_j \right\rangle = \sum_{ij} \langle E_i, E_j \rangle \frac{d\alpha^i}{dt} \beta^j$$

$$= \sum_i \langle e_i, e_i \rangle \frac{d\alpha^i}{dt} \beta^i. \tag{19.3.3}$$

Similarly,

$$\left\langle \frac{\nabla K}{dt}, J \right\rangle = \sum_i \langle e_i, e_i \rangle \frac{d\beta^i}{dt} \alpha_i. \tag{19.3.4}$$

Combining (19.3.2)–(19.3.4) gives the result.   □

The next result is sometimes called "the miracle of semi-Riemannian geometry".

Theorem 19.3.2 (Fundamental Theorem of Semi-Riemannian Manifolds). *If $(M, \mathfrak{g})$ is a semi-Riemannian manifold, then there is a unique connection on M, called the **Levi-Civita connection**, that is symmetric and compatible with $\mathfrak{g}$.*

Proof. Uniqueness. Suppose such a connection ∇ exists. Since it is symmetric,

$$\nabla_X(Z) = \nabla_Z(X) + [X, Z]$$
$$\nabla_Y(X) = \nabla_X(Y) + [Y, X]$$
$$\nabla_Z(Y) = \nabla_Y(Z) + [Z, Y],$$

and because it is compatible with $\mathfrak{g}$,

$$X(\langle Y, Z \rangle) = \langle \nabla_X(Y), Z \rangle + \langle Y, \nabla_X(Z) \rangle$$
$$Y(\langle Z, X \rangle) = \langle \nabla_Y(Z), X \rangle + \langle Z, \nabla_Y(X) \rangle$$
$$Z(\langle X, Y \rangle) = \langle \nabla_Z(X), Y \rangle + \langle X, \nabla_Z(Y) \rangle.$$

Combining the above identities gives

$$X(\langle Y, Z \rangle) = \langle \nabla_X(Y), Z \rangle + \langle Y, \nabla_Z(X) \rangle + \langle Y, [X, Z] \rangle$$
$$Y(\langle Z, X \rangle) = \langle \nabla_Y(Z), X \rangle + \langle Z, \nabla_X(Y) \rangle + \langle Z, [Y, X] \rangle$$
$$Z(\langle X, Y \rangle) = \langle \nabla_Z(X), Y \rangle + \langle X, \nabla_Y(Z) \rangle + \langle X, [Z, Y] \rangle.$$

Adding the first two identities in the above display and subtracting the third yields

$$X(\langle Y, Z \rangle) + Y(\langle Z, X \rangle) - Z(\langle X, Y \rangle)$$
$$= 2\langle \nabla_X(Y), Z \rangle - \langle X, [Z, Y] \rangle + \langle Y, [X, Z] \rangle + \langle Z, [Y, X] \rangle$$
$$= 2\langle \nabla_X(Y), Z \rangle + \langle X, [Y, Z] \rangle - \langle Y, [Z, X] \rangle - \langle Z, [X, Y] \rangle,$$

hence

$$2\langle \nabla_X(Y), Z \rangle = X(\langle Y, Z \rangle) + Y(\langle Z, X \rangle) - Z(\langle X, Y \rangle)$$
$$- \langle X, [Y, Z] \rangle + \langle Y, [Z, X] \rangle + \langle Z, [X, Y] \rangle,$$

which is known as **Koszul's formula.** Suppose $\widetilde{\nabla}$ is another connection satisfying Koszul's formula. The right-hand side of Koszul's formula does not depend on the choice of connection, so $\langle \nabla_X(Y) - \widetilde{\nabla}_X(Y), Z \rangle = 0$ for all vector fields X, Y, Z in $\mathfrak{X}(M)$. Since $\mathfrak{g}$ is nondegenerate on M, $\nabla_X(Y) = \widetilde{\nabla}_X(Y)$ for all vector fields X, Y in $\mathfrak{X}(M)$. Thus, $\nabla = \widetilde{\nabla}$.

Existence. Let $\mathfrak{K}(X, Y, Z)$ be the right-hand side of Koszul's formula. Temporarily treating X and Y as fixed, we define a map $\overline{\mathfrak{K}} : \mathfrak{X}(M) \longrightarrow C^\infty(M)$ by the assignment $Z \longmapsto \mathfrak{K}(X, Y, Z)$ for all Z in $\mathfrak{X}(M)$. It is easily shown that $\overline{\mathfrak{K}}$ is $C^\infty(M)$-linear, so $\overline{\mathfrak{K}}$ is in $\text{Lin}_{C^\infty(M)}\big(\mathfrak{X}(M), C^\infty(M)\big)$. It follows from Theorem 15.5.3 that $\overline{\mathfrak{K}}$ can be identified with a covector field in $\mathfrak{X}^*(M)$, and then from Theorem 19.4.2 that there is a unique vector field $\nabla(X, Y)$ in $\mathfrak{X}(M)$ such that $2\langle \nabla(X, Y), Z \rangle = \mathfrak{K}(X, Y, Z)$ for all vector fields Z in $\mathfrak{X}(M)$. This defines a map $\nabla : \mathfrak{X}(M) \times \mathfrak{X}(M) \longrightarrow \mathfrak{X}(M)$. We need to check that the criteria [$\nabla 1$]–[$\nabla 4$] for a connection are met. Clearly, [$\nabla 1$] and [$\nabla 3$] are satisfied, and computations similar to those showing that $\overline{\mathfrak{K}}$ is $C^\infty(M)$-linear demonstrate that [$\nabla 2$] and [$\nabla 4$] are satisfied. $\qquad \square$

Example 19.3.3 (Regular Surface in $\mathbb{R}^3_\nu$). Viewing a regular surface $(M, \mathfrak{g})$ in $\mathbb{R}^3_\nu$ as a smooth 2-manifold with the connection ∇ defined in Example 18.1.1, we have from (12.6.1) that ∇ is symmetric, and from Theorem 12.4.1(e) that it is compatible with $\mathfrak{g}$. By Theorem 19.3.2, ∇ is the Levi-Civita connection on M corresponding to $\mathfrak{g}$. $\qquad \diamond$

> **Throughout, unless stated otherwise, the connection on a semi-Riemannian manifold is the Levi-Civita connection.**

With this convention, when we say that $(M, \mathfrak{g})$ is a semi-Riemannian manifold, it is implicit that there is a connection on M, and it is the Levi-Civita connection. On the other hand, the notation $(M, \mathfrak{g}, \nabla)$ indicates that $(M, \mathfrak{g})$ is a semi-Riemannian manifold with a connection, but not necessarily the Levi-Civita connection.

Christoffel symbols for a regular surface in $\mathbb{R}^3_\nu$ were introduced in Section 12.3, and for a smooth manifold with a connection in Section 18.2. As the next result shows, with a metric at our disposal, we can recover the identity for Christoffel symbols given in Theorem 12.3.3(b).

Theorem 19.3.4 (Local Coordinate Expression for Christoffel Symbols). *If $(M, \mathfrak{g})$ is a semi-Riemannian manifold, then, in local coordinates,*

$$\Gamma^k_{ij} = \frac{1}{2} \sum_l \mathfrak{g}^{kl} \left(\frac{\partial \mathfrak{g}_{jl}}{\partial x^i} + \frac{\partial \mathfrak{g}_{il}}{\partial x^j} - \frac{\partial \mathfrak{g}_{ij}}{\partial x^l} \right)$$

for $i, j, k = 1, \ldots, m$.

Proof. We have

$$2 \sum_n \Gamma_{ij}^n \mathfrak{g}_{ln}$$

$$= 2 \sum_n \Gamma_{ij}^n \left\langle \frac{\partial}{\partial x^n}, \frac{\partial}{\partial x^l} \right\rangle = 2 \left\langle \sum_n \Gamma_{ij}^n \frac{\partial}{\partial x^n}, \frac{\partial}{\partial x^l} \right\rangle = 2 \left\langle \nabla_{\partial/\partial x^i} \left(\frac{\partial}{\partial x^j} \right), \frac{\partial}{\partial x^l} \right\rangle$$

$$= \frac{\partial}{\partial x^i} \left(\left\langle \frac{\partial}{\partial x^j}, \frac{\partial}{\partial x^l} \right\rangle \right) + \frac{\partial}{\partial x^j} \left(\left\langle \frac{\partial}{\partial x^l}, \frac{\partial}{\partial x^i} \right\rangle \right) - \frac{\partial}{\partial x^l} \left(\left\langle \frac{\partial}{\partial x^i}, \frac{\partial}{\partial x^j} \right\rangle \right)$$

$$- \left\langle \frac{\partial}{\partial x^i}, \left[\frac{\partial}{\partial x^j}, \frac{\partial}{\partial x^l} \right] \right\rangle + \left\langle \frac{\partial}{\partial x^j}, \left[\frac{\partial}{\partial x^l}, \frac{\partial}{\partial x^i} \right] \right\rangle + \left\langle \frac{\partial}{\partial x^l}, \left[\frac{\partial}{\partial x^i}, \frac{\partial}{\partial x^j} \right] \right\rangle$$

$$= \frac{\partial \mathfrak{g}_{jl}}{\partial x^i} + \frac{\partial \mathfrak{g}_{il}}{\partial x^j} - \frac{\partial \mathfrak{g}_{ij}}{\partial x^l},$$

where the third equality follows from (18.2.1), the fourth equality from Koszul's formula, and the last equality from Theorem 15.3.3(b). Thus,

$$\frac{1}{2} \left(\frac{\partial \mathfrak{g}_{jl}}{\partial x^i} + \frac{\partial \mathfrak{g}_{il}}{\partial x^j} - \frac{\partial \mathfrak{g}_{ij}}{\partial x^l} \right) = \sum_n \mathfrak{g}_{ln} \Gamma_{ij}^n.$$

Multiplying both sides of the preceding identity by $\mathfrak{g}^{kl}$ and summing over l gives

$$\frac{1}{2} \sum_l \mathfrak{g}^{kl} \left(\frac{\partial \mathfrak{g}_{jl}}{\partial r^i} + \frac{\partial \mathfrak{g}_{il}}{\partial r^j} - \frac{\partial \mathfrak{g}_{ij}}{\partial r^l} \right) = \sum_l \mathfrak{g}^{kl} \left(\sum_n \mathfrak{g}_{ln} \Gamma_{ij}^n \right)$$

$$= \sum_n \left(\sum_l \mathfrak{g}^{kl} \mathfrak{g}_{ln} \right) \Gamma_{ij}^n = \sum_n \delta_n^k \Gamma_{ij}^n = \Gamma_{ij}^k. \qquad \square$$

Theorem 19.3.5 (Standard Coordinates). *If $(\varepsilon_1, \ldots, \varepsilon_m)$ is the signature of $\mathbb{R}_\nu^m = (\mathbb{R}^m, \mathfrak{s})$, then, in standard coordinates,*

$$\mathfrak{s}_{ij}(p) = \varepsilon_i \delta_{ij} \qquad and \qquad \Gamma_{ij}^k(p) = 0$$

for all p in $\mathbb{R}_\nu^m$ for $i, j, k = 1, \ldots, m$.

Proof. Let $(e_1, \ldots, e_m)$ be the standard basis for $\mathbb{R}_\nu^m$. We have from (4.2.1), Theorem 14.5.7, and (19.1.1) that

$$\mathfrak{s}_{ij}(p) = \left\langle \left. \frac{\partial}{\partial r^i} \right|_p, \left. \frac{\partial}{\partial r^j} \right|_p \right\rangle = \langle e_i, e_j \rangle = \varepsilon_i \delta_{ij},$$

which proves the first identity. The second identity follows from the first identity and Theorem 19.3.4. $\qquad \square$

The next result shows that in normal coordinates at a given point, a semi-Riemannian m-manifold of index ν behaves like $\mathbb{R}_\nu^m$ with respect to the metric and the Christoffel symbols *at that point.*

Theorem 19.3.6 (Normal Coordinates). *If $(M, \mathfrak{g})$ is a semi-Riemannian m-manifold with signature $(\varepsilon_1, \ldots, \varepsilon_m)$ and p is a point in M, then, in normal coordinates at p,*

$$\mathfrak{g}_{ij}(p) = \varepsilon_i \delta_{ij} \qquad and \qquad \Gamma^k_{ij}(p) = 0$$

for $i, j, k = 1, \ldots, m$.

Proof. Continuing with the setup at the beginning of Section 18.10, suppose $(\tilde{e}_1, \ldots, \tilde{e}_m)$ is an orthonormal basis for $T_p(M)$. From Theorem 18.10.1(a) and (19.1.1),

$$\mathfrak{g}_{ij}(p) = \left\langle \frac{\partial}{\partial x^i}\Big|_p, \frac{\partial}{\partial x^j}\Big|_p \right\rangle = \langle \tilde{e}_i, \tilde{e}_j \rangle.$$

By definition, $T_p(M)$ and M have the same signature, hence $\langle \tilde{e}_i, \tilde{e}_j \rangle = \varepsilon_i \delta_{ij}$. This proves the first identity. The second identity is a repeat of Theorem 18.10.2. $\qquad\square$

Let $(M, \mathfrak{g})$ be a semi-Riemannian manifold, let U be an open set in M, and let $\mathcal{X} = (X_1, \ldots, X_m)$ be a frame on U. We say that $\mathcal{X}$ is **orthonormal** if $\mathcal{X}_p = (X_1|_p, \ldots, X_m|_p)$ is an orthonormal basis for $T_p(M)$ for all p in U. When M is oriented, $\mathcal{X}$ is said to be **positively oriented** if $\mathcal{X}_p$ is positively oriented for all p in U.

Theorem 19.3.7 (Existence of Orthonormal Frame). *Every point in a semi-Riemannian manifold has a neighborhood on which there is an orthonormal frame. If the semi-Riemannian manifold is oriented, then the orthonormal frame can be chosen to be positively oriented.*

Proof. Let $(M, \mathfrak{g})$ be a semi-Riemannian manifold, let p be a point in M, and let $(e_1, \ldots, e_m)$ be an orthonormal basis for $T_p(M)$. Let V be a normal neighborhood of p in M, let $\exp_p : \tilde{V} \longrightarrow V$ be the corresponding diffeomorphism, and let $\gamma(t) : [0, 1] \longrightarrow M$ be a radial geodesic with starting point p. By Theorem 18.5.7, there is a unique frame $(E_{1\gamma}, \ldots, E_{m\gamma})$ on γ consisting of parallel vector fields in $\mathfrak{X}_M(\gamma)$ such that $(E_{1\gamma}(0), \ldots, E_{m\gamma}(0)) = (e_1, \ldots, e_m)$ and $(E_{1\gamma}(t), \ldots, E_{m\gamma}(t))$ is a basis for $T_{\gamma(t)}(M)$ for all t in $[0, 1]$. It follows from Theorem 19.3.1 that $\langle E_{i\gamma}(t), E_{j\gamma}(t) \rangle = \langle e_i, e_j \rangle$ for all t in $[0, 1]$ for $i, j = 1, \ldots, m$. Thus, each $(E_{1\gamma}(t), \ldots, E_{m\gamma}(t))$ is an orthonormal basis. It can be shown that combining these vector fields across all radial geodesics with starting point p yields an orthonormal frame $(E_1, \ldots, E_m)$ on V.

Now suppose M is oriented. Replacing e_1 with $-e_1$ if necessary, we assume without loss of generality that $(E_1|_p, \ldots, E_m|_p) = (e_1, \ldots, e_m)$ is a positively oriented orthonormal basis for $T_p(M)$ and that the above construction has been revised accordingly. By Theorem 16.6.5, there is an orientation form ϖ that induces the orientation of M. Then $\varpi_p(E_1|_p, \ldots, E_m|_p) > 0$, hence $f = \varpi|_V(E_1, \ldots, E_m)$ is a function in $C^\infty(V)$ such that $f(p) > 0$. Since f is smooth, by Theorem 14.2.2, it is continuous. Let $\varepsilon > 0$ be a real number such that the open interval $I = \left(f(p) - \varepsilon, f(p) + \varepsilon \right)$ consists of positive real numbers. By Theorem 9.1.5 and Theorem 9.1.7, $W = f^{-1}(I)$ is a neighborhood of p in M on which f is positive-valued. It follows that $(E_1, \ldots, E_m)$ is positively oriented on W.   □

We call upon Theorem 19.3.7 often, but usually without attribution.

Theorem 19.3.8. *Let $(M, \mathfrak{g})$ be a semi-Riemannian manifold, let X be a vector field in $\mathfrak{X}(M)$, and let $\mathcal{A}$ be a tensor field in $T^r_s(M)$. Then*

$$\nabla_X(\mathfrak{g} \otimes \mathcal{A}) = \mathfrak{g} \otimes \nabla_X(\mathcal{A}).$$

Proof. We have

$$\begin{aligned}
\nabla_X(\mathfrak{g} \otimes \mathcal{A}) &= \nabla_X(\mathfrak{g}) \otimes \mathcal{A} + \mathfrak{g} \otimes \nabla_X(\mathcal{A}) && \text{[Th 18.1.3(b)]}\\
&= \mathfrak{g} \otimes \nabla_X(\mathcal{A}). && \text{[Th 18.4.1, Th 19.3.1]}
\end{aligned}$$
                                                                                 □

19.4   Flat Maps and Sharp Maps

In this section, we present manifold versions of the flat maps and sharp maps defined in Section 3.3, Section 6.2, and Section 6.3.

Let $(M, \mathfrak{g})$ be a semi-Riemannian manifold. For a given vector field X in $\mathfrak{X}(M)$, we define a map

$$X^{\mathsf{F}} : \mathfrak{X}(M) \longrightarrow C^\infty(M)$$

by

$$X^{\mathsf{F}}(Y) = \langle X, Y \rangle \tag{19.4.1}$$

for all vector fields Y in $\mathfrak{X}(M)$.

Theorem 19.4.1. *With the above setup, X^{F} is a covector field in $\mathfrak{X}^*(M)$.*

Proof. Let Y, Z be vector fields in $\mathfrak{X}(M)$, and let f be a function in $C^\infty(M)$. Then

$$X^{\mathsf{F}}(Y + Z) = \langle X, Y + Z \rangle = \langle X, Y \rangle + \langle X, Z \rangle = X^{\mathsf{F}}(Y) + X^{\mathsf{F}}(Z)$$

and

$$(fX)^{\mathsf{F}}(Y) = \langle fX, Y \rangle = f \langle X, Y \rangle = f\, X^{\mathsf{F}}(Y),$$

so X^{F} is a map in $\mathrm{Lin}_{C^\infty(M)}\big(\mathfrak{X}(M), C^\infty(M)\big)$. The result now follows from Theorem 15.5.3.   □

The **flat map** on M is the map

$$\mathsf{F} : \mathfrak{X}(M) \longrightarrow \mathfrak{X}^*(M)$$

defined by the assignment

$$X \longmapsto X^{\mathsf{F}}$$

for all vector fields X in $\mathfrak{X}(M)$. Thus, as a matter of notation,

$$\mathsf{F}(X) = X^{\mathsf{F}}.$$

Theorem 19.4.2. *With the above setup,* F *is a* $C^\infty(M)$*-module isomorphism:*

$$\mathfrak{X}(M) \approx \mathfrak{X}^*(M).$$

Proof. It is easily shown that F is a $C^\infty(M)$-module homomorphism. We need to show that F is bijective.

Injectivity. Since $\mathfrak{g}$ is nondegenerate, $\ker(\mathsf{F}) = 0$. It follows from Theorem B.5.3 that F is injective.

Surjectivity. Let ω be a covector field in $\mathfrak{X}^*(M)$, let $(U, (x^i))$ be a chart on M, and, in local coordinates, let $\omega|_U = \sum_i \alpha_i d(x^i)$. Define a vector field $X|_U$ in $\mathfrak{X}(U)$ by

$$X|_U = \sum_{ij} \mathfrak{g}^{ij} \alpha_i \frac{\partial}{\partial x^j}.$$

Then

$$(X|_U)^{\mathsf{F}}\left(\frac{\partial}{\partial x^i}\right) = \left\langle X|_U, \frac{\partial}{\partial x^i}\right\rangle = \left\langle \sum_{jk} \mathfrak{g}^{jk}\alpha_j \frac{\partial}{\partial x^k}, \frac{\partial}{\partial x^i}\right\rangle = \sum_{jk} \alpha_j \mathfrak{g}^{jk} \mathfrak{g}_{ki}$$

$$= \sum_j \alpha_j \left(\sum_k \mathfrak{g}^{jk}\mathfrak{g}_{ki}\right) = \sum_j \delta_i^j \alpha_j = \alpha_i$$

for $i = 1, \ldots, m$, hence

$$\omega|_U = \sum_i (X|_U)^{\mathsf{F}}\left(\frac{\partial}{\partial x^i}\right) d(x^i)$$

It follows from the manifold version of Theorem 1.2.1(e) that $(X|_U)^{\mathsf{F}} = \omega|_U$. Let $(\widetilde{U}, \widetilde{\varphi})$ be a chart on M that overlaps with (U, φ), and, in local coordinates, let $\omega|_{\widetilde{U}} = \sum_i \widetilde{\alpha}_i d(\widetilde{x}^i)$. We have

$$\frac{\partial}{\partial \widetilde{x}^i} = \sum_q \frac{\partial x^q}{\partial \widetilde{x}^i}\frac{\partial}{\partial x^q}, \qquad \widetilde{\alpha}_j = \sum_p \frac{\partial x^p}{\partial \widetilde{x}^j}\alpha_p, \qquad \text{and} \qquad \widetilde{\mathfrak{g}}^{ij} = \sum_{kl} \frac{\partial \widetilde{x}^i}{\partial x^k}\mathfrak{g}^{kl}\frac{\partial \widetilde{x}^j}{\partial x^l},$$

where the first identity follows from Theorem 15.1.8, the second identity from Theorem 15.6.3, and the third identity from the manifold version of Theorem 2.2.5(b), Theorem 14.3.5(b), and Theorem 19.1.3. Then

$$X|_{\widetilde{U}} = \sum_{ij} \widetilde{\mathfrak{g}}^{ij}\widetilde{\alpha}_j \frac{\partial}{\partial \widetilde{x}^i} = \sum_{ijklpq}\left(\frac{\partial \widetilde{x}^i}{\partial x^k}\mathfrak{g}^{kl}\frac{\partial \widetilde{x}^j}{\partial x^l}\right)\left(\frac{\partial x^p}{\partial \widetilde{x}^j}\alpha_p\right)\left(\frac{\partial x^q}{\partial \widetilde{x}^i}\frac{\partial}{\partial x^q}\right)$$

$$= \sum_{klpq}\left[\mathfrak{g}^{kl}\alpha_p\left(\sum_i \frac{\partial \widetilde{x}^i}{\partial x^k}\frac{\partial x^q}{\partial \widetilde{x}^i}\right)\left(\sum_j \frac{\partial \widetilde{x}^j}{\partial x^l}\frac{\partial x^p}{\partial \widetilde{x}^j}\right)\frac{\partial}{\partial x^q}\right]$$

$$= \sum_{klpq} \delta_k^q \delta_l^p \mathfrak{g}^{kl}\alpha_p \frac{\partial}{\partial x^q}$$

$$= \sum_{kl} \mathfrak{g}^{kl}\alpha_k \frac{\partial}{\partial x^l} = X|_U,$$

so there is agreement on overlaps. Taken together, the $X|_U$, one for each point in M, define a vector field X in $\mathfrak{X}(M)$ such that $X^{\mathsf{F}} = \omega$. Thus, F is surjective. $\qquad\square$

The **sharp map**

$$\mathsf{S} : \mathfrak{X}^*(M) \longrightarrow \mathfrak{X}(M)$$

on M is defined to be the inverse of F; that is,

$$\mathsf{S} = \mathsf{F}^{-1}.$$

Let $1 \leq k \leq r$ and $1 \leq l \leq s+1$ be integers (so that $r \geq 1$ and $s \geq 0$). The (k, l)**-flat map** is denoted by

$$\mathsf{F}_l^k : \mathcal{T}_s^r(M) \longrightarrow \mathcal{T}_{s+1}^{r-1}(M)$$

and defined (using Theorem 15.7.2) for all tensor fields $\mathcal{A}$ in $\mathcal{T}_s^r(M)$, all covector fields $\omega^1, \ldots, \omega^{r-1}$ in $\mathfrak{X}^*(M)$, and all vector fields $X_1, \ldots, X_{s+1}$ in $\mathfrak{X}(M)$ as follows, where $\frown$ indicates that an expression is omitted:

[F1] For $r = 1$ (so that $k = 1$):

$$\mathsf{F}_l^1(\mathcal{A})(X_1, \ldots, X_{s+1}) = \mathcal{A}(X_l^{\mathsf{F}}, X_1, \ldots, \widehat{X_l}, \ldots, X_{s+1}).$$

[F2] For $r \geq 2$ and $1 \leq k \leq r-1$:

$$\mathsf{F}_l^k(\mathcal{A})(\omega^1, \ldots, \omega^{r-1}, X_1, \ldots, X_{s+1})$$
$$= \mathcal{A}(\omega^1, \ldots, \omega^{k-1}, X_l^{\mathsf{F}}, \omega^k, \ldots, \omega^{r-1}, X_1, \ldots, \widehat{X_l}, \ldots, X_{s+1}).$$

[F3] For $r \geq 2$ and $k = r$:

$$\mathsf{F}_l^r(\mathcal{A})(\omega^1, \ldots, \omega^{r-1}, X_1, \ldots, X_{s+1})$$
$$= \mathcal{A}(\omega^1, \ldots, \omega^{r-1}, X_l^{\mathsf{F}}, X_1, \ldots, \widehat{X_l}, \ldots, X_{s+1}).$$

Similarly, let $1 \leq k \leq r+1$ and $1 \leq l \leq s$ be integers (so that $r \geq 0$ and $s \geq 1$). The (k, l)**-sharp map** is denoted by

$$\mathsf{S}_l^k : \mathcal{T}_s^r(M) \longrightarrow \mathcal{T}_{s-1}^{r+1}(M)$$

and defined (using Theorem 15.7.2) for all tensor fields $\mathcal{A}$ in $\mathcal{T}_s^r(M)$, all covector fields $\omega^1, \ldots, \omega^{r+1}$ in $\mathfrak{X}^*(M)$, and all vector fields $X_1, \ldots, X_{s-1}$ in $\mathfrak{X}(M)$ as follows:

[S1] For $s = 1$ (so that $l = 1$):

$$\mathsf{S}_1^k(\mathcal{A})(\omega^1, \ldots, \omega^{r+1}) = \mathcal{A}(\omega^1, \ldots, \widehat{\omega^k}, \ldots, \omega^{r+1}, \omega^{k\mathsf{S}}).$$

[S2] For $s \geq 2$ and $1 \leq l \leq s-1$:

$$\mathsf{S}_l^k(\mathcal{A})(\omega^1, \ldots, \omega^{r+1}, X_1, \ldots, X_{s-1})$$
$$= \mathcal{A}(\omega^1, \ldots, \widehat{\omega^k}, \ldots, \omega^{r+1}, X_1, \ldots, X_{l-1}, \omega^{k\mathsf{S}}, X_l, \ldots, X_{s-1}).$$

[S3] For $s \geq 2$ and $l = s$:

$$\mathsf{S}_s^k(\mathcal{A})(\omega^1, \ldots, \omega^{r+1}, X_1, \ldots, X_{s-1})$$
$$= \mathcal{A}(\omega^1, \ldots, \widehat{\omega^k}, \ldots, \omega^{r+1}, X_1, \ldots, X_{s-1}, \omega^k \mathsf{S}).$$

Theorem 19.4.3. *With the above setup, F_l^k and S_l^k are $C^\infty(M)$-module isomorphisms that are inverses.*

Proof. It is easily shown that F_l^k and S_l^k are $C^\infty(M)$-module homomorphisms. That they are $C^\infty(M)$-module isomorphisms and inverses follows from the manifold version of Theorem 6.3.7 and Theorem 19.4.2, respectively. $\qquad\square$

19.5 Representation of Tensor Fields

The representation of tensor fields on smooth manifolds was considered in Section 15.7. Let us now return to this topic with the additional structure afforded by semi-Riemannian manifolds. We introduce smooth manifold counterparts of the representation map and scalar product map presented in Section 5.3 and Section 6.4, respectively, and link them using the characterization map defined in Section 15.5 and Section 15.7. The notation is admittedly horrendous, but the ideas are essentially those introduced in the much less complicated setting of scalar product spaces.

Let $(M, \mathfrak{g})$ be a semi-Riemannian manifold. Following Section B.5, we denote by

$$\mathrm{Mult}_{C^\infty(M)}\big(\mathfrak{X}(M)^s, \mathfrak{X}(M)\big)$$

the $C^\infty(M)$-module of $C^\infty(M)$-multilinear maps from $\mathfrak{X}(M)^s$ to $\mathfrak{X}(M)$, and by

$$\mathrm{Mult}_{C^\infty(M)}\big(\mathfrak{X}^*(M)^r \times \mathfrak{X}(M)^s, C^\infty(M)\big)$$

the $C^\infty(M)$-module of $C^\infty(M)$-multilinear maps from $\mathfrak{X}^*(M)^r \times \mathfrak{X}(M)^s$ to $C^\infty(M)$.

Let us define a map

$$\mathfrak{S}_s : \mathrm{Mult}_{C^\infty(M)}\big(\mathfrak{X}(M)^s, \mathfrak{X}(M)\big) \longrightarrow \mathrm{Mult}_{C^\infty(M)}\big(\mathfrak{X}(M)^{s+1}, C^\infty(M)\big),$$

called the **scalar product map**, by

$$\mathfrak{S}_s(\Psi)(X_1, \ldots, X_{s+1}) = \langle X_1, \Psi(X_2, \ldots, X_{s+1}) \rangle$$

for all maps Ψ in $\mathrm{Mult}_{C^\infty(M)}\big(\mathfrak{X}(M)^s, \mathfrak{X}(M)\big)$ and all vector fields $X_1, \ldots, X_{s+1}$ in $\mathfrak{X}(M)$. We observe that $\mathfrak{S}_s$ is the smooth manifold counterpart of the map defined by (6.4.1).

Theorem 19.5.1. *If $(M, \mathfrak{g})$ is a semi-Riemannian manifold, then $\mathfrak{S}_s$ is a $C^\infty(M)$-module isomorphism:*

$$\mathrm{Mult}_{C^\infty(M)}\big(\mathfrak{X}(M)^s, \mathfrak{X}(M)\big) \approx \mathrm{Mult}_{C^\infty(M)}\big(\mathfrak{X}(M)^{s+1}, C^\infty(M)\big).$$

Proof. It is easily shown that $\mathfrak{S}_s$ is a $C^\infty(M)$-module homomorphism. We need to show that $\mathfrak{S}_s$ is bijective.

Injectivity. Suppose Ψ is a map in $\mathrm{Mult}_{C^\infty(M)}\big(\mathfrak{X}(M)^s, \mathfrak{X}(M)\big)$ such that $\mathfrak{S}_s(\Psi) = 0$. Then $\langle X_1|_p, \Psi(X_2, \ldots, X_{s+1})_p \rangle = 0$ for all vector fields $X_1, \ldots,$ X_{s+1} in $\mathfrak{X}(M)$ and all p in M. Since $\mathfrak{g}$ is nondegenerate, we have $\Psi(X_2, \ldots, X_{s+1})_p = 0$ for all vector fields $X_2, \ldots, X_{s+1}$ in $\mathfrak{X}(M)$ and all p in M, hence $\Psi(X_2, \ldots, X_{s+1}) = 0$ for all vector fields $X_2, \ldots, X_{s+1}$ in $\mathfrak{X}(M)$. Thus, $\Psi = 0$, so $\ker(\mathfrak{S}_s) = \{0\}$. By Theorem B.5.3, $\mathfrak{S}_s$ is injective.

Surjectivity. Let Υ be a map in $\mathrm{Mult}_{C^\infty(M)}\big(\mathfrak{X}(M)^{s+1}, C^\infty(M)\big)$. We need to find a map $\widetilde{\Upsilon}$ in $\mathrm{Mult}_{C^\infty(M)}\big(\mathfrak{X}(M)^s, \mathfrak{X}(M)\big)$ such that $\mathfrak{S}_s(\widetilde{\Upsilon}) = \Upsilon$. According to Theorem 19.4.2, defining a covector field in $\mathfrak{X}^*(M)$ is equivalent to defining a vector field in $\mathfrak{X}(M)$. For given vector fields $X_2, \ldots, X_{s+1}$ in $\mathfrak{X}(M)$, define a covector field $\widetilde{\Upsilon}(X_2, \ldots, X_{s+1})^\mathsf{F}$ in $\mathfrak{X}^*(M)$ by

$$\widetilde{\Upsilon}(X_2, \ldots, X_{s+1})^\mathsf{F}(X_1) = \Upsilon(X_1, \ldots, X_{s+1})$$

for all vector fields X_1 in $\mathfrak{X}(M)$. Then

$$
\begin{aligned}
\mathfrak{S}_s(\widetilde{\Upsilon})(X_1, \ldots, X_{s+1}) &= \langle X_1, \widetilde{\Upsilon}(X_2, \ldots, X_{s+1}) \rangle = \langle \widetilde{\Upsilon}(X_2, \ldots, X_{s+1}), X_1 \rangle \\
&= \widetilde{\Upsilon}(X_2, \ldots, X_{s+1})^\mathsf{F}(X_1) = \Upsilon(X_1, \ldots, X_{s+1}).
\end{aligned}
$$

Since $X_1, \ldots, X_{s+1}$ were arbitrary, $\mathfrak{S}_s(\widetilde{\Upsilon}) = \Upsilon$. Thus, $\mathfrak{S}_s$ is surjective. $\qquad\square$

Let us now define a map

$$\mathfrak{R}_s : \mathrm{Mult}_{C^\infty(M)}\big(\mathfrak{X}(M)^s, \mathfrak{X}(M)\big) \longrightarrow \mathrm{Mult}_{C^\infty(M)}\big(\mathfrak{X}^*(M) \times \mathfrak{X}(M)^s, C^\infty(M)\big),$$

called the **representation map,** by

$$\mathfrak{R}_s(\Psi)(\omega, X_1, \ldots, X_s) = \omega\big(\Psi(X_1, \ldots, X_s)\big) \qquad (19.5.1)$$

for all maps Ψ in $\mathrm{Mult}_{C^\infty(M)}\big(\mathfrak{X}(M)^s, \mathfrak{X}(M)\big)$, all covector fields ω in $\mathfrak{X}^*(M)$, and all vector fields $X_1, \ldots, X_s$ in $\mathfrak{X}(M)$. We observe that $\mathfrak{R}_s$ is the smooth manifold counterpart of the map defined by (5.3.1). When $s = 1$, we denote

$$\mathfrak{R}_1 \qquad \text{by} \qquad \mathfrak{R}$$

and obtain

$$\mathfrak{R} : \mathrm{Lin}_{C^\infty(M)}\big(\mathfrak{X}(M), \mathfrak{X}(M)\big) \longrightarrow \mathrm{Mult}_{C^\infty(M)}\big(\mathfrak{X}^*(M) \times \mathfrak{X}(M), C^\infty(M)\big).$$

Then

$$\mathfrak{R}(\Psi)(\omega, X) = \omega\big(\Psi(X)\big) \qquad (19.5.2)$$

for all maps Ψ in $\mathrm{Lin}_{C^\infty(M)}\big(\mathfrak{X}(M), \mathfrak{X}(M)\big)$, all covector fields ω in $\mathfrak{X}^*(M)$, and all vector fields X in $\mathfrak{X}(M)$.

Theorem 19.5.2. *If $(M, \mathfrak{g})$ is a semi-Riemannian manifold, then $\mathfrak{R}_s$ is a $C^\infty(M)$-module isomorphism:*

$$\mathrm{Mult}_{C^\infty(M)}\big(\mathfrak{X}(M)^s, \mathfrak{X}(M)\big) \approx \mathrm{Mult}_{C^\infty(M)}\big(\mathfrak{X}^*(M) \times \mathfrak{X}(M)^s, C^\infty(M)\big).$$

Proof. Recall the characterization map $\mathfrak{C}_s^r$ defined in Section 15.7. Let us define the "sharp" map

$$\mathfrak{P}_1^1 : \mathrm{Mult}_{C^\infty(M)}\big(\mathfrak{X}(M)^{s+1}, C^\infty(M)\big) \longrightarrow \mathrm{Mult}_{C^\infty(M)}\big(\mathfrak{X}^*(M) \times \mathfrak{X}(M)^s, C^\infty(M)\big)$$

by

$$\mathfrak{P}_1^1 = \mathfrak{C}_s^1 \circ \mathsf{S}_1^1 \circ (\mathfrak{C}_{s+1}^0)^{-1}. \tag{19.5.3}$$

By Theorem 15.7.2 and Theorem 19.4.3, we have that $\mathfrak{C}_s^1$, $\mathfrak{C}_{s+1}^0$, and S_1^1 are $C^\infty(M)$-module isomorphisms, and therefore, so is $\mathfrak{P}_1^1$.

We claim that

$$\mathfrak{R}_s = \mathfrak{P}_1^1 \circ \mathfrak{S}_s, \tag{19.5.4}$$

which is the manifold version of Theorem 6.4.1(d). Once this established, the result follows from Theorem 19.5.1. Let Υ be a map in $\mathrm{Mult}_{C^\infty(M)}\big(\mathfrak{X}(M)^{s+1}, C^\infty(M)\big)$. Making the identifications inherent in Theorem 15.7.2, it can be shown that

$$\mathfrak{P}_1^1(\Upsilon)(\omega, X_1, \ldots, X_s) = \Upsilon(\omega^\mathsf{S}, X_1, \ldots, X_s)$$

for all covector fields ω in $\mathfrak{X}^*(M)$ and all vector fields $X_1, \ldots, X_s$ in $\mathfrak{X}(M)$. For a map Ψ in $\mathrm{Mult}_{C^\infty(M)}\big(\mathfrak{X}(M)^s, \mathfrak{X}(M)\big)$, we have

$$
\begin{aligned}
\mathfrak{R}_s(\Psi)(\omega, X_1, \ldots, X_s) &= \omega\big(\Psi(X_1, \ldots, X_s)\big) = (\omega^\mathsf{S})^\mathsf{F}\big(\Psi(X_1, \ldots, X_s)\big) \\
&= \langle \omega^\mathsf{S}, \Psi(X_1, \ldots, X_s) \rangle = \mathfrak{S}_s(\Psi)(\omega^\mathsf{S}, X_1, \ldots, X_s) \\
&= \mathfrak{P}_1^1 \circ \mathfrak{S}_s(\Psi)(\omega, X_1, \ldots, X_s).
\end{aligned}
$$

Since Ψ, ω, and $X_1, \ldots, X_s$ were arbitrary, this proves the claim. $\square$

Theorem 19.5.3. *With the above setup, Figure 19.5.1 is a commutative diagram of $C^\infty(M)$-module isomorphisms. That is, any two $C^\infty(M)$-modules are connected by the $C^\infty(M)$-module isomorphism formed by taking the composite of intervening $C^\infty(M)$-module isomorphisms.*

Proof. It follows from (19.5.3) and (19.5.4) that $\mathfrak{R}_s = \mathfrak{C}_s^1 \circ \mathsf{S}_1^1 \circ (\mathfrak{C}_{s+1}^0)^{-1} \circ \mathfrak{S}_s$. $\square$

Theorem 19.5.4 (Representation of Covariant Tensor Fields). *If $(M, \mathfrak{g})$ is a semi-Riemannian manifold, then*

$$\mathcal{T}_{s+1}^0(M) \approx \mathrm{Mult}_{C^\infty(M)}\big(\mathfrak{X}(M)^s, \mathfrak{X}(M)\big) \approx \mathcal{T}_s^1(M).$$

Proof. This is the manifold version of Theorem 6.4.2 and follows from Theorem 19.5.3. $\square$

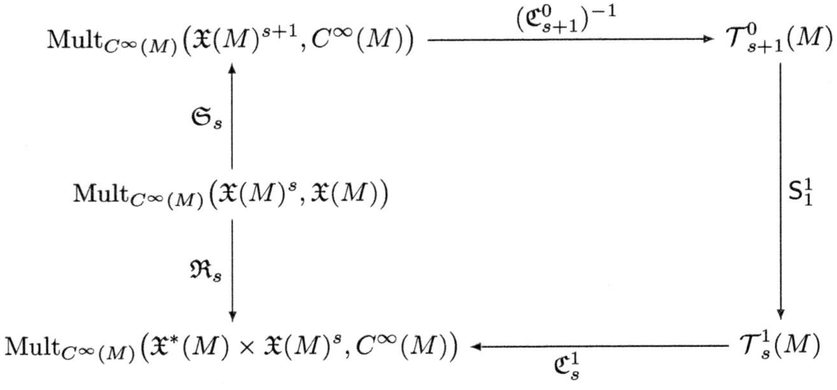

Figure 19.5.1. Commutative diagram for Theorem 19.5.3

Our efforts in this section, and much of what we previously developed in the area of "representations", have ultimately been directed at obtaining the following result. It provides the long-awaited justification for using the term "tensor" when referring to the torsion tensor and curvature tensor.

Theorem 19.5.5 (Representation of T and R). *Let $(M, \mathfrak{g})$ be a semi-Riemannian manifold, and let T and R be the torsion tensor and curvature tensor, respectively. Then:*

(a) *T is a map in $\mathrm{Mult}_{C^\infty(M)}\big(\mathfrak{X}(M)^2, \mathfrak{X}(M)\big)$ and can be identified with a tensor field in $\mathcal{T}_2^1(M)$.*

(b) *R is a map in $\mathrm{Mult}_{C^\infty(M)}\big(\mathfrak{X}(M)^3, \mathfrak{X}(M)\big)$ and can be identified with a tensor field in $\mathcal{T}_3^1(M)$.*

Proof. (a): This follows from Theorem 18.6.2(b) and Theorem 19.5.4.

(b): This follows from Theorem 18.7.1(b) and Theorem 19.5.4.   □

19.6   Contraction of Tensor Fields

The contraction of tensor fields on smooth manifolds was considered in Section 15.14. We now return to this topic armed with the properties of semi-Riemannian manifolds.

Theorem 19.6.1 (Orthonormal Frame Expression for Ordinary Contraction). *Let $(M, \mathfrak{g})$ be a semi-Riemannian m-manifold with signature $(\varepsilon_1, \ldots, \varepsilon_m)$, let $(E_1, \ldots, E_m)$ be an orthonormal frame on an open set U in M, and let $\mathcal{A}$ be a tensor field in $\mathcal{T}_s^1(M)$. Then:*

(a) *For integers $s \geq 2$ and $1 \leq l \leq s - 1$, the tensor field $\mathsf{C}_l^1(\mathcal{A})$ in $\mathcal{T}_{s-1}^0(M)$ is given on U by*

$$\mathsf{C}_l^1(\mathcal{A})(X_1, \ldots, X_{s-1})$$
$$= \sum_i \varepsilon_i \langle \mathfrak{R}_s^{-1} \circ \mathfrak{C}_s^1(\mathcal{A})(X_1, \ldots, X_{l-1}, E_i, X_l, \ldots, X_{s-1}), E_i \rangle$$

for all vector fields $X_1, \ldots, X_{s-1}$ in $\mathfrak{X}(M)$, where $\mathfrak{R}_s$ and $\mathfrak{C}_s^1$ are given by (19.5.1) and (15.7.1), respectively.

(b) *For $s = 1$, the function $\mathsf{C}_1^1(\mathcal{A})$ in $C^\infty(M)$ is given by*

$$\mathsf{C}_1^1(\mathcal{A}) = \sum_i \varepsilon_i \langle \mathfrak{R}^{-1} \circ \mathfrak{C}_1^1(\mathcal{A})(E_i), E_i \rangle,$$

where $\mathfrak{R}$ is given by (19.5.2).

Remark. According to Figure 19.5.1, $\mathfrak{R}_s^{-1} \circ \mathfrak{C}_s^1(\mathcal{A})$ is in $\mathrm{Mult}_{C^\infty(M)}(\mathfrak{X}(M)^s, \mathfrak{X}(M))$, so the scalar product in part (a) makes sense.

Proof. This is the manifold version of Theorem 6.1.1.   □

Let $(M, \mathfrak{g})$ be a semi-Riemannian manifold. Corresponding to the (k, l)-flat map F_l^k and the (k, l)-sharp map S_l^k introduced in Section 19.4, we now define manifold versions of the metric contraction maps described in Section 6.5.

For integers $r \geq 2$, $s \geq 0$, and $1 \leq k < l \leq r$, and motivated by (6.5.1), the (k, l)-**contravariant metric contraction** is the map

$$\mathsf{C}^{kl} : \mathcal{T}_s^r(M) \longrightarrow \mathcal{T}_s^{r-2}(M)$$

defined by

$$\mathsf{C}^{kl} = \mathsf{C}_1^k \circ \mathsf{F}_1^l. \tag{19.6.1}$$

Corresponding to (6.5.2), we have

$$\mathsf{C}^{k,k+1} = \mathsf{C}_1^k \circ \mathsf{F}_1^k.$$

For integers $r \geq 0$, $s \geq 2$, and $1 \leq k < l \leq s$, and motivated by (6.5.3), the (k, l)-**covariant metric contraction** is the map

$$\mathsf{C}_{kl} : \mathcal{T}_s^r(M) \longrightarrow \mathcal{T}_{s-2}^r(M)$$

defined by

$$\mathsf{C}_{kl} = \mathsf{C}_k^1 \circ \mathsf{S}_l^1. \tag{19.6.2}$$

Corresponding to (6.5.4), we have

$$\mathsf{C}_{k,k+1} = \mathsf{C}_k^1 \circ \mathsf{S}_k^1. \tag{19.6.3}$$

In order to avoid confusion between metric contractions and the contractions defined in Section 15.14, we sometimes refer to the latter as **ordinary contractions**.

Theorem 19.6.2 (Orthonormal Frame Expression for Metric Contraction). *Let $(M, \mathfrak{g})$ be a semi-Riemannian manifold with signature $(\varepsilon_1, \ldots, \varepsilon_m)$, let $(E_1, \ldots, E_m)$ be an orthonormal frame on an open set U in M, and let $\mathcal{A}$ be a tensor field in $\mathcal{T}_s^0(M)$. Then:*
(a) *For integers $s \geq 3$ and $1 \leq k < l \leq s$, the tensor field $\mathsf{C}_{kl}(\mathcal{A})$ in $\mathcal{T}_{s-2}^0(M)$ is given on U by*

$$\mathsf{C}_{kl}(\mathcal{A})(X_1, \ldots, X_{s-2})$$
$$= \sum_i \varepsilon_i \mathcal{A}(X_1, \ldots, X_{k-1}, E_i, X_k, \ldots, X_{l-2}, E_i, X_{l-1}, \ldots, X_{s-2})$$

for all vector fields $X_1, \ldots, X_{s-2}$ in $\mathfrak{X}(M)$.
(b) *For $s = 2$, the function $\mathsf{C}_{12}(\mathcal{A})$ in $C^\infty(M)$ is given by*

$$\mathsf{C}_{12}(\mathcal{A}) = \sum_i \varepsilon_i \mathcal{A}(E_i, E_i).$$

Remark. Observe that in part (a), the E_i appear in the kth and lth positions.

Proof. This is the manifold version of Theorem 6.5.1. $\qquad\qquad\square$

Theorem 19.6.3. *If $(M, \mathfrak{g})$ is a semi-Riemannian manifold, then covariant derivatives and total covariant derivatives commute with flat maps, sharp maps, ordinary contractions, and metric contractions.*

Proof. We illustrate the proof with a few representative cases. Let X be a vector field in $\mathfrak{X}(M)$, and let $\mathcal{A}$ be a tensor field in $\mathcal{T}_s^r(M)$. Since ∇_X is a tensor derivation, property [D2] in Section 16.2 gives

$$\nabla_X \circ \mathsf{C}_l^k = \mathsf{C}_l^k \circ \nabla_X. \qquad\qquad (19.6.4)$$

We have

$$
\begin{aligned}
\nabla_X \circ \mathsf{F}_1^k(\mathcal{A}) &= \nabla_X \big(\mathsf{C}_1^k (\mathfrak{g} \otimes \mathcal{A}) \big) && \text{[mv Th 6.2.6]} \\
&= \mathsf{C}_1^k \big(\nabla_X (\mathfrak{g} \otimes \mathcal{A}) \big) && \text{[(19.6.4)]} \\
&= \mathsf{C}_1^k \big(\mathfrak{g} \otimes \nabla_X (\mathcal{A}) \big) && \text{[Th 19.3.8]} \\
&= \mathsf{F}_1^k \big(\nabla_X (\mathcal{A}) \big) && \text{[mv Th 6.2.6]} \\
&= \mathsf{F}_1^k \circ \nabla_X (\mathcal{A}),
\end{aligned}
$$

and since $\mathcal{A}$ was arbitrary,

$$\nabla_X \circ \mathsf{F}_1^k = \mathsf{F}_1^k \circ \nabla_X. \qquad\qquad (19.6.5)$$

We also have

$$
\begin{aligned}
\nabla_X \circ \mathsf{C}^{kl} &= \nabla_X \circ \mathsf{C}_1^k \circ \mathsf{F}_1^l && \text{[(19.6.1)]} \\
&= \mathsf{C}_1^k \circ \nabla_X \circ \mathsf{F}_1^l && \text{[(19.6.4)]} \\
&= \mathsf{C}_1^k \circ \mathsf{F}_1^l \circ \nabla_X && \text{[(19.6.5)]} \\
&= \mathsf{C}^{kl} \circ \nabla_X. && \text{[(19.6.1)]} \qquad\square
\end{aligned}
$$

19.7   Isometries

Linear isometries on scalar product spaces were introduced in Section 4.4. We now describe their counterpart for semi-Riemannian manifolds.

Let $(M, \mathfrak{g})$ and $(\widetilde{M}, \widetilde{\mathfrak{g}})$ be semi-Riemannian manifolds with the same dimension, and let $F, G : M \longrightarrow \widetilde{M}$ be smooth maps. We say that F is an **isometry**, and that M and $\widetilde{M}$ are **isometric**, if F is a diffeomorphism and $F^*(\widetilde{\mathfrak{g}}) = \mathfrak{g}$, where, by definition, the latter condition is equivalent to $F^*(\widetilde{\mathfrak{g}})_p = \mathfrak{g}_p$ for all p in M. We say that G is a **local isometry**, and that M is **locally isometric** to $\widetilde{M}$, if for every point p in M, there is a neighborhood U of p in M and a neighborhood $\widetilde{U}$ of $G(p)$ in $\widetilde{M}$ such that $G|_U : U \longrightarrow \widetilde{U}$ is an isometry. We have from Theorem 19.1.2 that $(U, \mathfrak{g}|_U)$ and $(\widetilde{U}, \widetilde{\mathfrak{g}}|_{\widetilde{U}})$ are semi-Riemannian manifolds with the same dimension and index, so the preceding definition makes sense. Since every diffeomorphism is a local diffeomorphism, it follows that every isometry is a local isometry.

Theorem 19.7.1. *Let $(M, \mathfrak{g})$ and $(\widetilde{M}, \widetilde{\mathfrak{g}})$ be semi-Riemannian manifolds with the same dimension, and let $F : M \longrightarrow \widetilde{M}$ be a smooth map. Then the following are equivalent:*
(a) *F is a local isometry.*
(b) *$F^*(\widetilde{\mathfrak{g}}) = \mathfrak{g}$.*
(c) *The differential map $d_p(F) : T_p(M) \longrightarrow T_{F(p)}(\widetilde{M})$ is a linear isometry for all p in M.*

Proof. (b) ⇔ (c): For a point p in M,

$$F^*(\widetilde{\mathfrak{g}})_p = \mathfrak{g}_p$$
$$\Leftrightarrow \quad F^*(\widetilde{\mathfrak{g}})_p(v, w) = \mathfrak{g}_p(v, w) \text{ for all } v, w \text{ in } T_p(M)$$
$$\Leftrightarrow \quad \widetilde{\mathfrak{g}}_{F(p)}\big(d_p(F)(v), d_p(F)(w)\big) = \mathfrak{g}_p(v, w) \text{ for all } v, w \text{ in } T_p(M)$$
$$\Leftrightarrow \quad d_p(F) \text{ is a linear isometry,}$$

where the second equivalence follows from (15.12.2). Since p was arbitrary, the result follows.

(a) ⇒ (b): Straightforward.
(b) ⇒ (a): We have

$$F^*(\widetilde{\mathfrak{g}}) = \mathfrak{g}$$

⇒ $d_p(F)$ is a linear isometry for all p in M	[(b) ⇒ (c)]
⇒ $d_p(F)$ is a linear isomorphism for all p in M	[Th 4.4.1(a)]
⇒ $d_p(F)$ is injective for all p in M	[Th 1.1.14]
⇒ F is an immersion	[by definition]
⇒ F is a local diffeomorphism,	[Th 14.6.2]

(19.7.1)

from which the result follows.   □

Theorem 19.7.2. *Let $(M, \mathfrak{g})$ and $(\widetilde{M}, \widetilde{\mathfrak{g}})$ be semi-Riemannian manifolds with the same dimension, and let $F : M \longrightarrow \widetilde{M}$ be a smooth map. Then F is an isometry if and only if it is a bijective local isometry.*

Proof. ($\Rightarrow$): By definition, a diffeomorphism (hence an isometry) is bijective. As remarked above, an isometry is a local isometry.

($\Leftarrow$): Since F is a local isometry, by Theorem 19.7.1, $F^*(\widetilde{\mathfrak{g}}) = \mathfrak{g}$. It follows from (19.7.1) that F is a local diffeomorphism. Since a bijective local diffeomorphism is a diffeomorphism, F is a diffeomorphism. $\qquad\square$

Let $(M, \mathfrak{g})$ be a semi-Riemannian m-manifold of index ν, let p be a point in M, and let $(\widetilde{e}_1, \ldots, \widetilde{e}_m)$ be an orthonormal basis for $T_p(M)$. Replacing $\mathbb{R}^m$ with $\mathbb{R}^m_\nu$ in (18.10.1), we obtain the linear isomorphism

$$\ell_p : \mathbb{R}^m_\nu \longrightarrow T_p(M)$$

defined by

$$\ell_p(r^1, \ldots, r^m) = \sum_i r^i \widetilde{e}_i. \tag{19.7.2}$$

Using Theorem 14.1.6 and Theorem 14.2.5, let us now view $\mathbb{R}^m_\nu$ and $\big(T_p(M), \mathfrak{g}_p\big)$ as semi-Riemannian m-manifolds of index ν, and ℓ_p as a diffeomorphism.

Theorem 19.7.3. *With the above setup, ℓ_p is an isometry (between semi-Riemannian m-manifolds of index ν).*

Proof. Let q be a point in $\mathbb{R}^m_\nu$, and make the identifications $T_q(\mathbb{R}^m_\nu) = \mathbb{R}^m_\nu$ and $T_q\big(T_p(M)\big) = T_p(M)$. Since ℓ_p is linear, we have from Theorem 14.4.4 that $d_q(\ell_p) = \ell_p$. Let $(e_1, \ldots, e_m)$ be the standard basis for $\mathbb{R}^m_\nu$. Because $\mathbb{R}^m_\nu$ and $T_p(M)$ have the same index, hence the same signature,

$$\big\|d_q(\ell_p)(r^1, \ldots, r^m)\big\| = \big\|\ell_p(r^1, \ldots, r^m)\big\| = \sqrt{\Big|\Big\langle \sum_i r^i \widetilde{e}_i, \sum_j r^j \widetilde{e}_j \Big\rangle\Big|}$$

$$= \sqrt{\Big|\sum_i \langle \widetilde{e}_i, \widetilde{e}_i \rangle (r^i)^2 \Big|} = \sqrt{\Big|\sum_i \langle e_i, e_i \rangle (r^i)^2 \Big|}$$

$$= \sqrt{\Big|\Big\langle \sum_i r^i e_i, \sum_j r^j e_j \Big\rangle\Big|} = \big\|(r^1, \ldots, r^m)\big\|,$$

so $d_q(\ell_p)$ is a linear isometry for all q in $\mathbb{R}^m_\nu$. By Theorem 19.7.1, ℓ_p is a local isometry. Since ℓ_p is a diffeomorphism, and therefore bijective, it follows from Theorem 19.7.2 that ℓ_p is an isometry. $\qquad\square$

Theorem 19.7.4. *Let $(M, \mathfrak{g})$ be a semi-Riemannian manifold, let p be a point in M, and let $\psi(s) : (-\varepsilon, \varepsilon) \longrightarrow T_p(M)$ be a smooth curve. Consider the parametrized surface $\sigma(r, s) : [0, 1] \times (-\varepsilon, \varepsilon) \longrightarrow M$ given by*

$$\sigma(r, s) = \exp_p\big(r\psi(s)\big),$$

where it is assumed that $r\psi(s)$ is in $\mathcal{E}_p$ for all (r,s) in $[0,1] \times (-\varepsilon, \varepsilon)$. Then, for all (r,s) in $[0,1] \times (-\varepsilon, \varepsilon)$:

(a)

$$\left\langle \frac{\partial \sigma}{\partial r}(r,s), \frac{\partial \sigma}{\partial r}(r,s) \right\rangle = \langle \psi(s), \psi(s) \rangle.$$

(b)

$$\left\langle \frac{\partial \sigma}{\partial r}(r,s), \frac{\partial \sigma}{\partial s}(r,s) \right\rangle = r \left\langle \frac{d\psi}{ds}(s), \psi(s) \right\rangle.$$

Proof. (a): Recall from Section 14.9 that for a given point s in $(-\varepsilon, \varepsilon)$, the longitude curve $\sigma_s(r) : [0,1] \longrightarrow M$ is defined by $\sigma_s(r) = \sigma(r,s)$. By Theorem 18.11.4(a), σ_s is a geodesic with initial velocity $\psi(s)$; that is, $(d\sigma_s/dr)(0) = \psi(s)$. It follows from Theorem 19.9.1(a) that the function $\langle d\sigma_s/dr, d\sigma_s/dr \rangle : [0,1] \longrightarrow M$ is constant, hence

$$\left\langle \frac{\partial \sigma}{\partial r}(r,s), \frac{\partial \sigma}{\partial r}(r,s) \right\rangle = \left\langle \frac{d\sigma_s}{dr}(r), \frac{d\sigma_s}{dr}(r) \right\rangle = \left\langle \frac{d\sigma_s}{dr}(0), \frac{d\sigma_s}{dr}(0) \right\rangle$$
$$= \langle \psi(s), \psi(s) \rangle.$$

(b): By Theorem 19.3.1,

$$\frac{\partial}{\partial r} \left(\left\langle \frac{\partial \sigma}{\partial r}, \frac{\partial \sigma}{\partial s} \right\rangle \right) = \left\langle \frac{\nabla}{\partial r} \left(\frac{\partial \sigma}{\partial r} \right), \frac{\partial \sigma}{\partial s} \right\rangle + \left\langle \frac{\partial \sigma}{\partial r}, \frac{\nabla}{\partial r} \left(\frac{\partial \sigma}{\partial s} \right) \right\rangle.$$

Since σ_s is a geodesic,

$$\frac{\nabla}{\partial r} \left(\frac{\partial \sigma}{\partial r} \right) = 0,$$

and according to Theorem 18.6.4,

$$\left\langle \frac{\partial \sigma}{\partial r}, \frac{\nabla}{\partial r} \left(\frac{\partial \sigma}{\partial s} \right) \right\rangle = \left\langle \frac{\partial \sigma}{\partial r}, \frac{\nabla}{\partial s} \left(\frac{\partial \sigma}{\partial r} \right) \right\rangle.$$

Combining the preceding three identities gives

$$\frac{\partial}{\partial r} \left(\left\langle \frac{\partial \sigma}{\partial r}, \frac{\partial \sigma}{\partial s} \right\rangle \right) = \left\langle \frac{\partial \sigma}{\partial r}, \frac{\nabla}{\partial s} \left(\frac{\partial \sigma}{\partial r} \right) \right\rangle.$$

We have

$$2 \left\langle \frac{\partial \sigma}{\partial r}, \frac{\nabla}{\partial s} \left(\frac{\partial \sigma}{\partial r} \right) \right\rangle = \frac{\partial}{\partial s} \left(\left\langle \frac{\partial \sigma}{\partial r}, \frac{\partial \sigma}{\partial r} \right\rangle \right) \qquad \text{[Th 19.3.1]}$$
$$= \frac{d}{ds} (\langle \psi, \psi \rangle) \qquad \text{[part (a)]}$$
$$= 2 \left\langle \frac{d\psi}{ds}, \psi \right\rangle. \qquad \text{[Th 19.3.1]}$$

Then the preceding two identities give

$$\frac{\partial}{\partial r}\left(\left\langle \frac{\partial \sigma}{\partial r}(r,s), \frac{\partial \sigma}{\partial s}(r,s)\right\rangle\right) = \left\langle \frac{d\psi}{ds}(s), \psi(s)\right\rangle,$$

where we observe that the right-hand side is independent of r. From the differential calculus of one real variable,

$$\left\langle \frac{\partial \sigma}{\partial r}(r,s), \frac{\partial \sigma}{\partial s}(r,s)\right\rangle = c + r\left\langle \frac{d\psi}{ds}(s), \psi(s)\right\rangle \tag{19.7.3}$$

for some real number c. We have from Theorem 18.9.4 and Theorem 18.11.4(c) that

$$\frac{\partial \sigma}{\partial r}(0,s) = d_0(\exp_p)(\psi(s)) = \psi(s),$$

and from Theorem 18.11.4(d) that

$$\frac{\partial \sigma}{\partial s}(0,s) = 0.$$

Setting $r = 0$ in (19.7.3) and using the preceding two identities yields $c = 0$. $\square$

Let $(M, \mathfrak{g})$ be a semi-Riemannian manifold, and let p be a point in M. Let v and w be vectors in $\mathcal{E}_p$, and recall from (18.9.4) the differential map

$$d_v(\exp_p) : T_v\big(T_p(M)\big) \longrightarrow T_{\exp_p(v)}(M).$$

The next result shows that $d_v(\exp_p)$ can be viewed as a "linear isometry along radial geodesics".

Theorem 19.7.5 (Gauss's Lemma). *With the above setup,*

$$\langle v, w\rangle = \langle d_v(\exp_p)(v), d_v(\exp_p)(w)\rangle.$$

Proof. Let $\psi(s) : (-\varepsilon, \varepsilon) \longrightarrow T_p(M)$ be a smooth curve such that $\psi(0) = v$ and $(d\psi/ds)(0) = w$, for example, $\psi(s) = v + sw$. Consider the parametrized surface $\sigma(r,s) : [0,1] \times (-\varepsilon, \varepsilon) \longrightarrow M$ defined by $\sigma(r,s) = \exp_p(r\psi(s))$, where we assume that $r\psi(s)$ is in $\mathcal{E}_p$ for all (r,s) in $[0,1] \times (-\varepsilon, \varepsilon)$. It follows from parts (c) and (d) of Theorem 18.11.4 that

$$\frac{\partial \sigma}{\partial r}(1,0) = d_v(\exp_p)(v) \qquad \text{and} \qquad \frac{\partial \sigma}{\partial s}(1,0) = d_v(\exp_p)(w),$$

and then from Theorem 19.7.4(b) that

$$\langle d_v(\exp_p)(v), d_v(\exp_p)(w)\rangle = \left\langle \frac{d\psi}{ds}(0), \psi(0)\right\rangle = \langle w, v\rangle. \qquad \square$$

19.8 Riemann Curvature Tensor

Let $(M, \mathfrak{g})$ be a semi-Riemannian manifold. In this setting, the curvature tensor (field) R, originally introduced in the context of smooth manifolds with a connection, is called the **Riemann curvature tensor (field)**. We define a map

$$\mathcal{R} : \mathfrak{X}(M)^4 \longrightarrow C^\infty(M),$$

also called the Riemann curvature tensor (field) by

$$\mathcal{R}(X, Y, Z, W) = \langle R(X, Y)Z, W \rangle$$

for all vector fields X, Y, Z, W in $\mathfrak{X}(M)$.

Theorem 19.8.1 (Representation of $\mathcal{R}$). *If $(M, \mathfrak{g})$ is a semi-Riemannian manifold, then $\mathcal{R}$ is a map in* $\text{Mult}_{C^\infty(M)}(\mathfrak{X}(M)^4, C^\infty(M))$ *and can be identified with a tensor field in* $\mathcal{T}_4^0(M)$.

Proof. It follows from Theorem 18.7.1(b) that $\mathcal{R}$ is $C^\infty(M)$-linear in its first three arguments, and clearly it is $C^\infty(M)$-linear in its fourth. This proves the first assertion. The second assertion now follows from Theorem 15.7.2. $\square$

Working through the details of Figure 19.5.1, Theorem 19.5.5(b), and Theorem 19.8.1, we find that $\mathcal{R}$ can be identified with R. It is usual to think of $\mathcal{R}$ as being obtained from R by "lowering an index". This explains the practice of referring to both $\mathcal{R}$ and R as the Riemann curvature tensor.

Theorem 19.8.2. *If $(M, \mathfrak{g})$ is a semi-Riemannian m-manifold with a connection, then, in local coordinates, for $i, j, k, l = 1, \ldots, m$:*
(a) *The components of $\mathcal{R}$ with respect to the coordinate frame are*

$$\mathcal{R}_{ijkl} = \sum_n \mathfrak{g}_{ln} R_{ijk}^n.$$

(b)

$$R_{ijk}^l = \sum_n \mathfrak{g}^{ln} \mathcal{R}_{ijkn}.$$

Proof. (a): It follows from (18.7.2) that

$$\mathcal{R}_{ijkl} = \mathcal{R}\left(\frac{\partial}{\partial x^i}, \frac{\partial}{\partial x^j}, \frac{\partial}{\partial x^k}, \frac{\partial}{\partial x^l}\right) = \left\langle R\left(\frac{\partial}{\partial x^i}, \frac{\partial}{\partial x^j}\right)\frac{\partial}{\partial x^k}, \frac{\partial}{\partial x^l}\right\rangle$$

$$= \left\langle \sum_n R_{ijk}^n \frac{\partial}{\partial x^n}, \frac{\partial}{\partial x^l}\right\rangle = \sum_n \mathfrak{g}_{ln} R_{ijk}^n.$$

(b): By part (a),

$$\sum_n \mathfrak{g}^{ln} \mathcal{R}_{ijkn} = \sum_n \mathfrak{g}^{ln}\left(\sum_p \mathfrak{g}_{np} R_{ijk}^p\right) = \sum_p \left(\sum_n \mathfrak{g}^{ln} \mathfrak{g}_{np}\right) R_{ijk}^p$$

$$= \sum_p \delta_p^l R_{ijk}^p = R_{ijk}^l. \qquad \square$$

Theorem 19.8.3 (First Bianchi Identity). *Let* $(M, \mathfrak{g}, \nabla)$ *be a semi-Riemannian manifold with a connection, and let* X, Y, Z, W *be vector fields in* $\mathfrak{X}(M)$. *Then:*

(a)

$$
\sum_{(X,Y,Z)} \mathcal{R}(X, Y, Z, W) = \sum_{(X,Y,Z)} \langle \nabla_X(T(Y, Z)), W \rangle
$$

$$
+ \sum_{(X,Y,Z)} \langle T(X, [Y, Z]), W \rangle,
$$

where T is the torsion tensor.

(b) *If ∇ is symmetric (in particular, if ∇ is the Levi-Civita connection), then*

$$
\sum_{(X,Y,Z)} \mathcal{R}(X, Y, Z, W) = 0.
$$

Proof. This follows from Theorem 18.7.3. $\qquad\square$

Theorem 19.8.4 (Second Bianchi Identity). *Let* $(M, \mathfrak{g}, \nabla)$ *be a semi-Riemannian manifold with a connection, and let* X, Y, Z, V, W *be vector fields in* $\mathfrak{X}(M)$. *Then:*

(a)

$$
\sum_{(X,Y,Z)} \nabla_X(\mathcal{R})(Y, Z, V, W) = - \sum_{(X,Y,Z)} \mathcal{R}(T(X, Y), Z, V, W), \quad (19.8.1)
$$

where T is the torsion tensor.

(b) *If ∇ is symmetric (in particular, if ∇ is the Levi-Civita connection), then*

$$
\sum_{(X,Y,Z)} \nabla_X(\mathcal{R})(Y, Z, V, W) = 0.
$$

Proof. (a): We make repeated use of (18.7.3). By Theorem 18.1.4,

$$
\nabla_X(\mathcal{R})(Y, Z, V, W) = X(\mathcal{R}(Y, Z, V, W)) - \mathcal{R}(\nabla_X(Y), Z, V, W)
$$
$$
- \mathcal{R}(Y, \nabla_X(Z), V, W) - \mathcal{R}(Y, Z, \nabla_X(V), W)
$$
$$
- \mathcal{R}(Y, Z, V, \nabla_X(W)),
$$

so the left-hand side of (19.8.1) can be expressed as

$$
\sum_{(X,Y,Z)} \nabla_X(\mathcal{R})(Y,Z,V,W)
$$

$$
= \sum_{(X,Y,Z)} X\big(\mathcal{R}(Y,Z,V,W)\big) - \sum_{(X,Y,Z)} \mathcal{R}(\nabla_X(Y),Z,V,W)
$$

$$
- \sum_{(X,Y,Z)} \mathcal{R}(Y,\nabla_X(Z),V,W) - \sum_{(X,Y,Z)} \mathcal{R}(Y,Z,\nabla_X(V),W)
$$

$$
- \sum_{(X,Y,Z)} \mathcal{R}(Y,Z,V,\nabla_X(W)) \tag{19.8.2}
$$

$$
= \sum_{(X,Y,Z)} X\big(\mathcal{R}(Y,Z,V,W)\big) - \sum_{(X,Y,Z)} \mathcal{R}(\nabla_X(Y),Z,V,W)
$$

$$
- \sum_{(X,Y,Z)} \mathcal{R}(X,\nabla_Z(Y),V,W) - \sum_{(X,Y,Z)} \mathcal{R}(X,Y,\nabla_Z(V),W)
$$

$$
- \sum_{(X,Y,Z)} \mathcal{R}(X,Y,V,\nabla_Z(W)).
$$

We have from (18.6.1) and Theorem 18.7.1(a) that

$$
R\big(T(X,Y),Z\big)V = R\big(\nabla_X(Y) - \nabla_Y(X) - [X,Y],Z\big)V
$$
$$
= R\big(\nabla_X(Y),Z\big)V - R\big(\nabla_Y(X),Z\big)V - R\big([X,Y],Z\big)V
$$
$$
= R\big(\nabla_X(Y),Z\big)V + R\big(Z,\nabla_Y(X)\big)V - R\big([X,Y],Z\big)V,
$$

hence the sum in the right-hand side of (19.8.1) can be expressed as

$$
\sum_{(X,Y,Z)} \mathcal{R}\big(T(X,Y),Z,V,W\big)
$$

$$
= \sum_{(X,Y,Z)} \mathcal{R}\big(\nabla_X(Y),Z,V,W\big) + \sum_{(X,Y,Z)} \mathcal{R}\big(Z,\nabla_Y(X),V,W\big)
$$

$$
- \sum_{(X,Y,Z)} \mathcal{R}\big([X,Y],Z,V,W\big) \tag{19.8.3}
$$

$$
= \sum_{(X,Y,Z)} \mathcal{R}\big(\nabla_X(Y),Z,V,W\big) + \sum_{(X,Y,Z)} \mathcal{R}\big(X,\nabla_Z(Y),V,W\big)
$$

$$
- \sum_{(X,Y,Z)} \mathcal{R}\big([X,Y],Z,V,W\big).
$$

We need to show that the sum of terms in the last three lines of (19.8.2) and the sum of terms in the last two lines of (19.8.3) differ by a sign. The sums have two terms in common that differ by a sign, so we need to demonstrate that

$$
\begin{aligned}
\sum_{(X,Y,Z)} & \mathcal{R}([X,Y],Z,V,W) \\
= & \sum_{(X,Y,Z)} X\big(\mathcal{R}(Y,Z,V,W)\big) - \sum_{(X,Y,Z)} \mathcal{R}(X,Y,\nabla_Z(V),W) \\
& - \sum_{(X,Y,Z)} \mathcal{R}(X,Y,V,\nabla_Z(W)).
\end{aligned} \tag{19.8.4}
$$

Using (19.3.1) gives

$$
\begin{aligned}
X\big(\mathcal{R}(Y,Z,V,W)\big) &= X\big(\langle R(Y,Z)V,W\rangle\big) \\
&= \langle \nabla_X\big(R(Y,Z)V\big),W\rangle + \langle R(Y,Z)V,\nabla_X(W)\rangle \\
&= \langle \nabla_X\big(R(Y,Z)V\big),W\rangle + \mathcal{R}(Y,Z,V,\nabla_X(W)),
\end{aligned}
$$

hence

$$
\begin{aligned}
\sum_{(X,Y,Z)} & X\big(\mathcal{R}(Y,Z,V,W)\big) \\
= & \sum_{(X,Y,Z)} \langle \nabla_X\big(R(Y,Z)V\big),W\rangle + \sum_{(X,Y,Z)} \mathcal{R}(Y,Z,V,\nabla_X(W)) \\
= & \sum_{(X,Y,Z)} \langle \nabla_X\big(R(Y,Z)V\big),W\rangle + \sum_{(X,Y,Z)} \mathcal{R}(X,Y,V,\nabla_Z(W)).
\end{aligned} \tag{19.8.5}
$$

By Theorem 18.7.4,

$$
\begin{aligned}
\sum_{(X,Y,Z)} & \langle \nabla_X\big(R(Y,Z)V\big),W\rangle \\
= & \sum_{(X,Y,Z)} \mathcal{R}(X,Y,\nabla_Z(V),W) + \sum_{(X,Y,Z)} \mathcal{R}([X,Y],Z,V,W).
\end{aligned} \tag{19.8.6}
$$

Substituting (19.8.6) into (19.8.5) yields

$$
\begin{aligned}
\sum_{(X,Y,Z)} & X\big(\mathcal{R}(Y,Z,V,W)\big) \\
= & \sum_{(X,Y,Z)} \mathcal{R}(X,Y,\nabla_Z(V),W) + \sum_{(X,Y,Z)} \mathcal{R}([X,Y],Z,V,W) \\
& + \sum_{(X,Y,Z)} \mathcal{R}(X,Y,V,\nabla_Z(W)),
\end{aligned}
$$

which is equivalent to (19.8.4).

(b): This follows from part (a). $\square$

Theorem 19.8.5 (Symmetries of Riemann Curvature Tensor). *If* $(M, \mathfrak{g})$
is a semi-Riemannian manifold and X, Y, Z, W *are vector fields in* $\mathfrak{X}(M)$, *then*
$\mathcal{R}$ *satisfies the following symmetries:*
[S1] $\mathcal{R}(X, Y, Z, W) = -\mathcal{R}(Y, X, Z, W)$.
[S2] $\mathcal{R}(X, Y, Z, W) = -\mathcal{R}(X, Y, W, Z)$.
[S3] $\mathcal{R}(X, Y, Z, W) = \mathcal{R}(Z, W, X, Y)$.
[S4] $\mathcal{R}(X, Y, Z, W) + \mathcal{R}(Y, Z, X, W) + \mathcal{R}(Z, X, Y, W) = 0$.

Proof. [S1]: This follows from Theorem 18.7.1(a).
    [S2]: Since

$$\mathcal{R}(X, Y, Z + W, Z + W)$$
$$= \mathcal{R}(X, Y, Z, Z) + \mathcal{R}(X, Y, Z, W) + \mathcal{R}(X, Y, W, Z) + \mathcal{R}(X, Y, W, W),$$

it suffices to show that $\mathcal{R}(X, Y, Z, Z) = 0$ for all vector fields X, Y, Z in $\mathfrak{X}(M)$.
Using (19.3.1) gives

$$\langle \nabla_X (\nabla_Y(Z)), Z \rangle = X(\langle \nabla_Y(Z), Z \rangle) - \langle \nabla_Y(Z), \nabla_X(Z) \rangle$$

$$\langle \nabla_Y (\nabla_X(Z)), Z \rangle = Y(\langle \nabla_X(Z), Z \rangle) - \langle \nabla_X(Z), \nabla_Y(Z) \rangle$$

$$\langle \nabla_Y(Z), Z \rangle = \frac{1}{2} Y(\langle Z, Z \rangle)$$

$$\langle \nabla_X(Z), Z \rangle = \frac{1}{2} X(\langle Z, Z \rangle)$$

$$\langle \nabla_{[X,Y]}(Z), Z \rangle = \frac{1}{2} [X, Y](\langle Z, Z \rangle),$$

hence

$$\mathcal{R}(X, Y, Z, Z) = \langle R(X, Y)Z, Z \rangle$$
$$= \langle \nabla_X (\nabla_Y(Z)) - \nabla_Y (\nabla_X(Z)) - \nabla_{[X,Y]}(Z), Z \rangle$$
$$= \langle \nabla_X (\nabla_Y(Z)), Z \rangle - \langle \nabla_Y (\nabla_X(Z)), Z \rangle - \langle \nabla_{[X,Y]}(Z), Z \rangle$$
$$= X(\langle \nabla_Y(Z), Z \rangle) - Y(\langle \nabla_X(Z), Z \rangle) - \langle \nabla_{[X,Y]}(Z), Z \rangle$$
$$= \frac{1}{2} \{ X(Y(\langle Z, Z \rangle)) - Y(X(\langle Z, Z \rangle)) - [X, Y](\langle Z, Z \rangle) \} = 0.$$

    [S4]: This follows from Theorem 19.8.3(b).
    [S3]: This follows from [S1], [S2], [S4], and Theorem 6.6.1.   □

Example 19.8.6. The symmetries in Theorem 19.8.5 greatly simplify computations involving $\mathcal{R}$, especially when some of the vector fields are equal. For example, setting $X = Y$ in [S1] gives $\mathcal{R}(X, X, Z, W) = 0$. For a semi-Riemannian
2-manifold, we have the following tables:

(i,j,k,l)	$\mathcal{R}_{ijkl}$	(i,j,k,l)	$\mathcal{R}_{ijkl}$
$(1,1,1,1)$	0	$(2,1,1,1)$	0
$(1,1,1,2)$	0	$(2,1,1,2)$	$\mathcal{R}_{1221}$
$(1,1,2,1)$	0	$(2,1,2,1)$	$-\mathcal{R}_{1221}$
$(1,1,2,2)$	0	$(2,1,2,2)$	0
$(1,2,1,1)$	0	$(2,2,1,1)$	0
$(1,2,1,2)$	$-\mathcal{R}_{1221}$	$(2,2,1,2)$	0
$(1,2,2,1)$	$\mathcal{R}_{1221}$	$(2,2,2,1)$	0
$(1,2,2,2)$	0	$(2,2,2,2)$	0

Thus, fully 12 of the 16 values of $\mathcal{R}_{ijkl}$ have constant value 0, and the four remaining values equal $\pm\mathcal{R}_{1221}$.   ◇

Theorem 19.8.7. *If R and $\mathcal{R}$ are the Riemann curvature tensors on $\mathbb{R}^m_\nu$, then R has constant (vector) value 0, and $\mathcal{R}$ has constant (real number) value 0.*

Proof. We have from Theorem 19.3.5 that in standard coordinates the Christoffel symbols in $\mathbb{R}^m_\nu$ have constant value 0, and then from Theorem 18.7.2 that R has constant value 0. The corresponding result for $\mathcal{R}$ follows from Theorem 19.8.2(a).   □

Theorem 19.8.8. *Let $(M,\mathfrak{g})$ and $(\widetilde{M},\widetilde{\mathfrak{g}})$ be semi-Riemannian manifolds with the same dimension, and let $\mathcal{R}$ and $\widetilde{\mathcal{R}}$ be the corresponding Riemann curvature tensors. If $F : M \longrightarrow \widetilde{M}$ is an isometry, then $F^*(\widetilde{\mathcal{R}}) = \mathcal{R}$.*

Proof. Let p be a point in M, and let $\big(U,(x^i)\big)$ be a chart at p. Since F is a diffeomorphism, $\big(\widetilde{U},(\widetilde{x}^i)\big) = \big(F(U),(x^i \circ F^{-1})\big)$ is a chart at $F(p)$. By Theorem 14.4.6(a),

$$d_p(F)\left(\frac{\partial}{\partial x^i}\bigg|_p\right) = \frac{\partial}{\partial \widetilde{x}^i}\bigg|_{F(p)},$$

hence

$$\widetilde{\mathfrak{g}}_{ij}\big(F(p)\big) = \widetilde{\mathfrak{g}}_{F(p)}\left(\frac{\partial}{\partial \widetilde{x}^i}\bigg|_{F(p)}, \frac{\partial}{\partial \widetilde{x}^j}\bigg|_{F(p)}\right) \qquad\qquad [(19.1.1)]$$

$$= \widetilde{\mathfrak{g}}_{F(p)}\left(d_p(F)\left(\frac{\partial}{\partial x^i}\bigg|_p\right), d_p(F)\left(\frac{\partial}{\partial x^j}\bigg|_p\right)\right)$$

$$= F^*(\widetilde{\mathfrak{g}})_p\left(\frac{\partial}{\partial x^i}\bigg|_p, \frac{\partial}{\partial x^j}\bigg|_p\right) \qquad\qquad [(15.12.2)]$$

$$= \mathfrak{g}_p\left(\frac{\partial}{\partial x^i}\bigg|_p, \frac{\partial}{\partial x^j}\bigg|_p\right) \qquad\qquad [F^*(\widetilde{\mathfrak{g}}) = \mathfrak{g}]$$

$$= \mathfrak{g}_{ij}(p) \qquad\qquad [(19.1.1)]$$

for $i, j = 1, \ldots, m$. We then have from Theorem 18.7.2, Theorem 19.3.4, and Theorem 19.8.2(a) that $\widetilde{\mathcal{R}}_{ijkl}\big(F(p)\big) = \mathcal{R}_{ijkl}(p)$ for $i, j, k, l = 1, \ldots, m$. Since p was arbitrary, the result follows.   □

Let $(M, \mathfrak{g})$ be a semi-Riemannian m-manifold of index ν. We say that M is **flat** if for every point p in M, there is an isometry $F_p : (V, \mathfrak{g}|_V) \longrightarrow (U, \mathfrak{s}|_U)$, where V is a neighborhood of p in M, and U is an open set in $\mathbb{R}^m_\nu = (\mathbb{R}^m, \mathfrak{s})$. According to Theorem 19.1.2, open sets in semi-Riemannian manifolds are semi-Riemannian manifolds in their own right, so this definition makes sense. The preceding conditions are not the same as requiring M to be locally isometric to $\mathbb{R}^m_\nu$, because there is no guarantee that the F_p combine to give a smooth map from M to $\mathbb{R}^m_\nu$.

The Riemann curvature tensor (in either of its manifestations) is a complicated mathematical object. The next result shows that in at least one instance, it has something explicitly geometric to say about "curvature".

Theorem 19.8.9 (Riemann's Theorem). *Let $(M, \mathfrak{g})$ be a semi-Riemannian manifold, and let $\mathcal{R}$ be the Riemann curvature tensor. Then $\mathcal{R}$ has constant value 0 if and only if M is flat.*

Proof. Suppose M has dimension m and index ν.

($\Rightarrow$): Let p be a point in M, let V be a normal neighborhood of p in M, and let $\exp_p|_{\widetilde{V}} : \widetilde{V} \longrightarrow V$ be the corresponding diffeomorphism. Since $\big(T_p(M), \mathfrak{g}_p\big)$ and $(M, \mathfrak{g})$ are semi-Riemannian m-manifolds of index ν, by Theorem 19.1.2, so are $\widetilde{V}$ and V. We claim that $\exp_p|_{\widetilde{V}}$ is an isometry.

Let v, w_1, w_2 be vectors in $\widetilde{V} \subseteq \mathcal{E}_p$, and let $\rho_{p,v}(r) = \exp_p|_{\widetilde{V}}(rv) : [0, 1] \longrightarrow M$ be the radial geodesic with starting point p and initial velocity v. By Theorem 18.5.1, there are parallel vector fields J_i in $\mathfrak{X}_M(\rho_{p,v})$ such that $J_i(0) = w_i$ for $i = 1, 2$. Let us define vector fields $\widetilde{J}_i$ in $\mathfrak{X}_M(\rho_{p,v})$ by $\widetilde{J}_i(r) = r J_i(r)$. Since $\nabla J_i/dr$ has constant value 0, we have

$$\frac{\nabla \widetilde{J}_i}{dr}(r) = J_i(r) + r\frac{\nabla J_i}{dr}(r) = J_i(r),$$

hence

$$\frac{\nabla^2 \widetilde{J}_i}{dr^2}(r) = \frac{\nabla J_i}{dr}(r) = 0$$

for all r in $[0, 1]$. By Theorem 19.8.2, the Riemann curvature tensor $\mathcal{R}$ has constant (real number) value 0 if and only if the Riemann curvature tensor R has constant (vector) value 0. It follows that $\widetilde{J}_i$ satisfies Jacobi's equation (18.11.1) and is therefore a Jacobi field. Let us observe that $\widetilde{J}_i(0) = 0$ and $(\nabla \widetilde{J}_i/dr)(0) = w_i$. By Theorem 18.11.1, $\widetilde{J}_i$ is the unique Jacobi field in $\mathfrak{X}_M(\rho_{p,v})$ satisfying the preceding two identities.

Recall Theorem 18.11.4(b), and let $\sigma_i(r, s) : [0, 1] \times (-\varepsilon, \varepsilon) \longrightarrow M$ be the geodesic variation of $\rho_{p,v}(r)$ defined by

$$\sigma_i(r, s) = \exp_p|_{\widetilde{V}}\big(r(v + sw_i)\big),$$

where it is assumed that $r(v + sw_i)$ is in $\mathcal{E}_p$ for all (r,s) in $[0,1] \times (-\varepsilon, \varepsilon)$. It follows from Theorem 18.11.5 that $\tilde{J}_i$ is the variation field of σ_i, and $\tilde{J}_i(r) = r\, d_{rv}(\exp_p|_{\tilde{V}})(w_i)$ for all r in $[0,1]$. Thus,

$$J_i(1) = \tilde{J}_i(1) = d_v(\exp_p|_{\tilde{V}})(w_i)$$

for $i = 1,2$. By Theorem 19.3.1, the function $\langle J_1, J_2 \rangle$ is independent of r, so

$$\langle w_1, w_2 \rangle = \langle J_1(0), J_2(0) \rangle = \langle J_1(1), J_2(1) \rangle$$
$$= \langle d_v(\exp_p|_{\tilde{V}})(w_1), d_v(\exp_p|_{\tilde{V}})(w_2) \rangle.$$

Since v, w_1, w_2 were arbitrary, $d_v(\exp_p|_{\tilde{V}}) : T_v(\tilde{V}) \longrightarrow T_{\exp_p|_{\tilde{V}}(v)}(V)$ is a linear isometry for all v in $\tilde{V}$. By Theorem 19.7.1, $\exp_p|_{\tilde{V}}$ is a local isometry. Because $\exp_p|_{\tilde{V}}$ is a diffeomorphism, and therefore bijective, it follows from Theorem 19.7.2 that $\exp_p|_{\tilde{V}}$ is an isometry. This proves the claim.

With the map ℓ_p defined as in (19.7.2), let $U = \ell_p^{-1}(\tilde{V})$. We have from Theorem 9.1.7 and Theorem 14.2.2 that U is open in $\mathbb{R}_\nu^m$. It follows from Theorem 19.7.3 that $\ell_p|_U : U \longrightarrow \tilde{V}$ is an isometry. The composite of isometries is an isometry, as is the inverse of an isometry, hence $(\exp_p|_{\tilde{V}} \circ \ell_p|_U)^{-1} : V \longrightarrow U$ is an isometry. Since p was arbitrary, the result follows.

($\Leftarrow$): This follows from Theorem 19.8.7 and Theorem 19.8.8. $\square$

19.9   Geodesics

In Section 18.8, a geodesic was defined to be a curve with zero acceleration. We now add to our knowledge of geodesics using the metric properties of semi-Riemannian manifolds. In particular, we find a relationship between geodesics and length.

Theorem 19.9.1. *If $(M, \mathfrak{g})$ is a semi-Riemannian manifold and $\gamma(t) : (a,b) \longrightarrow M$ is a geodesic, then:*
(a) *$\langle d\gamma/dt, d\gamma/dt \rangle : (a,b) \longrightarrow M$ is a constant function, hence γ has constant speed.*
(b) *γ is either spacelike, timelike, or lightlike.*
(c) *γ has constant speed 0 if and only if it is lightlike.*

Proof. (a): Since γ is a geodesic, by definition, $d\gamma/dt$ is a parallel vector field in $\mathfrak{X}_M(\gamma)$. The result now follows from Theorem 19.3.1.
   (b): This follows from part (a).
   (c): Straightforward. $\square$

Theorem 19.9.2. *If p and q are distinct points in $\mathbb{R}_\nu^m$ and $a < b$ are real numbers, then:*
(a) *The smooth curve $\gamma(t) : [a,b] \longrightarrow \mathbb{R}_\nu^m$ defined by*

$$\gamma(t) = p + \frac{t-a}{b-a}(q-p)$$

is the unique geodesic such that $\gamma(a) = p$ and $\gamma(b) = q$.

(b) $L(\gamma) = \|q - p\|$.

Proof. (a): Let $\gamma = (\gamma^1, \ldots, \gamma^m)$. By Theorem 18.8.1 and Theorem 19.3.5, the geodesic equations in standard coordinates are $(d^2\gamma^i/dt)(t) = 0$ for $i = 1, \ldots, m$. Thus, any geodesic in $\mathbb{R}^m_\nu$ is linear. The result now follows from the conditions $\gamma(a) = p$ and $\gamma(b) = q$.

(b): We have from (19.2.1) and part (a) that

$$L(\gamma) = \int_a^b \left\| \frac{d\gamma}{dt}(t) \right\| dt = \frac{1}{b-a} \int_a^b \|q - p\| \, dt = \|q - p\| . \qquad \square$$

The next result demonstrates in a rigorous fashion the well-known property of Euclidean space that the straight line segment joining any pair of distinct points is the shortest path between them.

Theorem 19.9.3 (Length Minimizing Property of Geodesics in $\mathbb{R}^m_0$). *Let p and q be distinct points in $\mathbb{R}^m_0$, let $a < b$ be real numbers, and let $\gamma(t)$: $[a, b] \longrightarrow \mathbb{R}^m_0$ be the unique geodesic such that $\gamma(a) = p$ and $\gamma(b) = q$ (as given by Theorem 19.9.2). If $\lambda(t) : [a, b] \longrightarrow \mathbb{R}^m_0$ is a smooth curve such that $\lambda(a) = p$ and $\gamma(b) = q$, then $L(\gamma) \leq L(\lambda)$.*

Proof. As in the proof of Theorem 4.2.4, we extend the unit vector $e_m = (q - p)/\|q - p\|$ to an orthonormal basis $\mathcal{E} = (e_1, \ldots, e_m)$ for $\mathbb{R}^m_0$. Then λ can be expressed as

$$\lambda(t) = \sum_i \alpha^i(t) \, e_i,$$

where the α^i are uniquely determined functions in $C^\infty([a, b])$. We have

$$\frac{d\lambda}{dt}(t) = \sum_i \frac{d\alpha^i}{dt}(t) \, e_i,$$

hence

$$\left\| \frac{d\lambda}{dt}(t) \right\| = \sqrt{\left(\frac{d\alpha^1}{dt}(t) \right)^2 + \cdots + \left(\frac{d\alpha^m}{dt}(t) \right)^2} \geq \left| \frac{d\alpha^m}{dt}(t) \right|. \qquad (19.9.1)$$

Since $\mathcal{E}$ is a basis and

$$\|q - p\| \, e_m = q - p = \lambda(b) - \lambda(a) = \sum_i [\alpha^i(b) - \alpha^i(a)] e_i,$$

we also have

$$\alpha^m(b) - \alpha^m(a) = \|q - p\|. \qquad (19.9.2)$$

Then

$$L(\lambda) = \int_a^b \left\| \frac{d\lambda}{dt}(t) \right\| dt \qquad [(19.2.1)]$$

$$\geq \int_a^b \left| \frac{d\alpha^m}{dt}(t) \right| dt \qquad [(19.9.1)]$$

$$\geq \left| \int_a^b \frac{d\alpha^m}{dt}(t)\, dt \right|$$

$$= |\alpha^m(b) - \alpha^m(a)| \qquad [\text{FTC}]$$

$$= \|q - p\| \qquad [(19.9.2)]$$

$$= L(\gamma), \qquad [\text{Th } 19.9.2(b)]$$

where FTC stands for the fundamental theorem of calculus. $\qquad\square$

Theorem 19.9.4. *Let* $\lambda(t) : [a, b] \longrightarrow \mathbb{R}_1^m$ *be a future-directed timelike smooth curve. Then:*
(a) $\lambda(b) - \lambda(a)$ *is in* $\mathcal{T}_m^+$.
(b) *If* $\lambda(a)$ *is in* $\mathcal{T}_m^+$, *then* $\lambda(t)$ *is in* $\mathcal{T}_m^+$ *for all* t *in* $[a, b]$.

Proof. (b): Define a smooth function $f(t) : [a, b] \longrightarrow \mathbb{R}$ by $f(t) = \langle \lambda(t), \lambda(t) \rangle$, so that

$$\frac{df}{dt}(t) = 2\left\langle \frac{d\lambda}{dt}(t), \lambda(t) \right\rangle.$$

Since $\lambda(a)$ and $(d\lambda/dt)(t)$ are in $\mathcal{T}_m^+$ for all t in $[a, b]$, we have $f(a) < 0$ and $(df/dt)(a) < 0$. Thus, f starts off negative and is strictly decreasing, at least initially. In fact, f continues to decrease with increasing t as long as $\lambda(t)$ remains in $\mathcal{T}_m^+$. It is clear from the geometry of time cones in $\mathbb{R}_1^m$ that $\lambda(t)$ fails to be in $\mathcal{T}_m^+$ for all t in $[a, b]$ if and only if there is a point t_0 in $[a, b]$ such that $\lambda(t_0)$ is in Λ_m^+ or $\lambda(t_0) = (0, \ldots, 0)$; in either case, $f(t_0) = 0$. Suppose such a t_0 exists. Since f is smooth, by Theorem 10.1.1, it is continuous. It follows from basic properties of the real numbers and the continuity of f that there is a smallest such t_0. Then f is negative-valued and strictly decreasing on $[a, t_0)$, with $f(t_0) = 0$, which means that f is discontinuous at t_0. This contradiction shows that no such t_0 exists.

(a): Consider the smooth curve $\lambda'(t) : [a, b] \longrightarrow \mathbb{R}_1^m$ given by $\lambda'(t) = \lambda(t) - \lambda(a)$. Since $(d\lambda'/dt)(t) = (d\lambda/dt)(t)$ for all t in $[a, b]$, λ' is future-directed and timelike, so

$$\frac{d\lambda'}{dt}(a) = \lim_{\varepsilon \to 0^+} \frac{\lambda'(a+\varepsilon) - \lambda'(a)}{\varepsilon} = \lim_{\varepsilon \to 0^+} \frac{\lambda'(a+\varepsilon)}{\varepsilon}$$

is in $\mathcal{T}_m^+$. For a sufficiently small real number $\varepsilon_0 > 0$, $\lambda'(a+\varepsilon_0)/\varepsilon_0$ is $\mathcal{T}_m^+$, and then, by Theorem 4.9.3(a), so is $\lambda'(a+\varepsilon_0)$. Thus, the smooth curve $\lambda''(t) = \lambda'|_{[a+\varepsilon_0, b]}(t) : [a+\varepsilon_0, b] \longrightarrow \mathbb{R}_1^m$ is future-directed and timelike, and $\lambda''(a+\varepsilon_0)$ is in $\mathcal{T}_m^+$. By part (b), $\lambda(b) - \lambda(a) = \lambda''(b)$ is in $\mathcal{T}_m^+$. $\qquad\square$

Theorem 19.9.3 and the next result provide a dramatic illustration of how different the geometries of $\mathbb{R}_0^m$ and $\mathbb{R}_1^m$ are.

Theorem 19.9.5 (Length Maximizing Property of Geodesics in $\mathbb{R}_1^m$).
Let p and q be distinct points in $\mathbb{R}_1^m$, let $a < b$ be real numbers, and let $\gamma(t)$:
$[a, b] \longrightarrow \mathbb{R}_0^m$ be the unique geodesic such that $\gamma(a) = p$ and $\gamma(b) = q$ (as
given by Theorem 19.9.2). If there is a future-directed timelike smooth curve
$\lambda(t) : [a, b] \longrightarrow \mathbb{R}_1^m$ such that $\lambda(a) = p$ and $\lambda(b) = q$, then:
(a) γ is future-directed and timelike.
(b) $L(\gamma) \geq L(\lambda)$.

Proof. It follows from Theorem 19.9.4(a) that $q - p$ is in $\mathcal{T}_m^+$, and then from
Theorem 4.9.3(a) that so are $(q - p)/(b - a)$ and $e_m = (q - p)/\|q - p\|$.
    (a): By Theorem 19.9.2(a), $(d\gamma/dt)(t) = (q-p)/(b-a)$ for all t in $[a, b]$, and
as we just observed, $(q - p)/(b - a)$ is in $\mathcal{T}_m^+$.
    (b): The proof is a modification of that given for Theorem 19.9.3. As in the
proof of Theorem 4.2.4, we extend the unit vector e_m to an orthonormal basis
$\mathcal{E} = (e_1, \ldots, e_m)$ for $\mathbb{R}_1^m$. Then λ can be expressed as

$$\lambda(t) = \sum_i \alpha^i(t)\, e_i,$$

where the α^i are uniquely determined functions in $C^\infty([a, b])$, so

$$\frac{d\lambda}{dt}(t) = \sum_i \frac{d\alpha^i}{dt}(t)\, e_i. \tag{19.9.3}$$

Since $(d\lambda/dt)(t)$ is timelike,

$$0 < -\left\langle \frac{d\lambda}{dt}(t), \frac{d\lambda}{dt}(t) \right\rangle = \left(\frac{d\alpha^m}{dt}(t)\right)^2 - \left[\left(\frac{d\alpha^1}{dt}(t)\right)^2 + \cdots + \left(\frac{d\alpha^{m-1}}{dt}(t)\right)^2\right]$$

$$\leq \left(\frac{d\alpha^m}{dt}(t)\right)^2,$$

hence

$$\left\|\frac{d\lambda}{dt}(t)\right\| = \sqrt{\left|\left\langle \frac{d\lambda}{dt}(t), \frac{d\lambda}{dt}(t) \right\rangle\right|} \leq \left|\frac{d\alpha^m}{dt}(t)\right|.$$

Furthermore, because $(d\lambda/dt)(t)$ is future-directed, it follows from Theorem
4.9.7 and (19.9.3) that $(d\alpha^m/dt)(t) > 0$, so

$$\left\|\frac{d\lambda}{dt}(t)\right\| \leq \frac{d\alpha^m}{dt}(t). \tag{19.9.4}$$

Since $\mathcal{E}$ is a basis and

$$\|q - p\|\, e_m = q - p = \lambda(b) - \lambda(a) = \sum_i [\alpha^i(b) - \alpha^i(a)]e_i,$$

we have

$$\alpha^m(b) - \alpha^m(a) = \|q - p\|. \tag{19.9.5}$$

Then

$$L(\lambda) = \int_a^b \left\| \frac{d\lambda}{dt}(t) \right\| dt \qquad [(19.2.1)]$$

$$\leq \int_a^b \frac{d\alpha^m}{dt}(t)\, dt \qquad [(19.9.4)]$$

$$= \alpha^m(b) - \alpha^m(a) \qquad [\text{FTC}]$$

$$= \|q - p\| \qquad [(19.9.5)]$$

$$= L(\gamma), \qquad [\text{Th } 19.9.2(b)]$$

where FTC stands for the fundamental theorem of calculus.   □

A generalization of Theorem 19.9.3 to arbitrary Riemannian manifolds is presented in Theorem 21.1.1. There is also a corresponding generalization of Theorem 19.9.5 to arbitrary Lorentz manifolds, but it is not included here.

19.10   Volume Forms

In Section 16.7, we discussed the integration of differential forms on smooth manifolds. The structure of semi-Riemannian manifolds makes it possible to extend that theory to the integration of smooth functions.

Theorem 19.10.1 (Existence of Volume Form). *If* $(M, \mathfrak{g}, \mathfrak{O})$ *is an oriented semi-Riemannian m-manifold or an oriented semi-Riemannian m-manifold with boundary, then there is a unique orientation form* Ω_M *on* M, *called the* **volume form***, such that: (i)* Ω_M *induces* $\mathfrak{O}$, *and (ii) if* U *is an open set in* M *and* $(E_1, \ldots, E_m)$ *is a positively oriented orthonormal frame on* U, *then the function* $\Omega_M|_U(E_1, \ldots, E_m)$ *has constant value 1 on* U.

Proof. This is the manifold version of Theorem 8.3.4.   □

Theorem 19.10.2 (Local Coordinate Expression for Volume Form). *Let* $(M, \mathfrak{g}, \mathfrak{O})$ *be an oriented semi-Riemannian m-manifold or an oriented semi-Riemannian m-manifold with boundary, let* $(U, (x^i))$ *be a positively oriented chart on* M, *and let* $\mathcal{X}$ *and* $(d(x^1), \ldots, d(x^m))$ *be the corresponding coordinate and dual coordinate frames, respectively. Then*

$$\Omega_M|_U = \sqrt{|\det(\mathfrak{g}_{\mathcal{X}})|}\, d(x^1) \wedge \cdots \wedge d(x^m).$$

Proof. This is the manifold version of Theorem 8.3.5.   □

Let $(M, \mathfrak{g}, \mathfrak{O})$ be an oriented semi-Riemannian m-manifold, let Ω_M be its volume form, and let f be a function in $C^\infty(M)$ that has compact support. Then $f\,\Omega_M$ is a form in $\Lambda^m(M)$ that has compact support. The **integral of** f **over** M is defined by

$$\int_M f\,\Omega_M.$$

The **volume of M** is obtained by taking f to be the function in $C^\infty(M)$ with constant value 1:

$$\mathrm{vol}(M) = \int_M \Omega_M.$$

Example 19.10.3 (Area of Regular Surface in $\mathbb{R}_0^3$). Let $(M, \mathfrak{g}, \mathfrak{O})$ be a compact oriented regular surface in $\mathbb{R}_0^3$, let (U, φ) be a positively oriented chart on M, and let $\mathcal{H} = (H_1, H_2)$ be the corresponding coordinate frame. The term "chart" is used here in the sense of Section 11.2: U is an open set in $\mathbb{R}^2$ and $\varphi : U \longrightarrow \mathbb{R}_0^3$ is a smooth map. In the present setting, we refer to "volume" as "area". Working through the definitions in Section 16.7 and using Theorem 19.10.2, we find that

$$\mathrm{area}\big(\varphi(U)\big) = \iint_U \sqrt{\det(\mathfrak{g}_\mathcal{H})}\, dr^1 dr^2.$$

We have from Theorem 4.6.3 that $\det(\mathfrak{g}_\mathcal{H}) > 0$ on U, so the above integral is well-defined. By Theorem 8.4.6(b), $\|H_1 \times H_2\| = \sqrt{\det(\mathfrak{g}_\mathcal{H})}$, hence

$$\mathrm{area}\big(\varphi(U)\big) = \iint_U \|H_1 \times H_2\|\, dr^1 dr^2.$$

This is the definition of area familiar from the integral calculus of two real variables.   ◇

Example 19.10.4 ($\mathcal{S}^2$). Let us view $(\mathcal{S}^2, \mathfrak{e}|_{\mathcal{S}^2})$ as a Riemannian 2-submanifold of the Riemannian 3-manifold $\mathbb{R}_0^3 = (\mathbb{R}^3, \mathfrak{e})$. Recall the results of Section 13.4, and consider the chart $(U, \varphi = (\theta, \phi))$ on $\mathcal{S}^2$, where $U = (0, \pi) \times (-\pi, \pi)$ and $\varphi : U \longrightarrow \mathcal{S}^2$ is given by

$$\varphi(\theta, \phi) = \big(\sin(\theta)\cos(\phi), \sin(\theta)\sin(\phi), \cos(\theta)\big).$$

From

$$\mathfrak{g}_\mathcal{H}(\theta, \phi) = \begin{bmatrix} 1 & 0 \\ 0 & \sin^2(\theta) \end{bmatrix}$$

and Theorem 19.10.2, the volume form is given by

$$\Omega_{\mathcal{S}^2}|_{(\theta,\phi)} = \sqrt{\det\big(\mathfrak{g}_\mathcal{H}(\theta, \phi)\big)}\, d(\theta) \wedge d(\phi) = \sin(\theta)\, d(\theta) \wedge d(\phi).$$

Since $\varphi(U)$ equals $\mathcal{S}^2$ except for a portion of a great circle (which has an area of 0), the area of $\mathcal{S}^2$ is

$$\mathrm{area}(\mathcal{S}^2) = \iint_U \Omega_{\mathcal{S}^2}|_{(\theta,\phi)}\, d\theta d\phi = \int_{-\pi}^{\pi}\int_0^{\pi} \sin(\theta)\, d\theta d\phi = 4\pi.   ◇$$

19.11   Orientation of Hypersurfaces

In this section, we summarize and extend some of the earlier results on the orientation of hypersurfaces.

Let $(\overline{M}, \overline{\mathfrak{g}}, \overline{\mathfrak{O}})$ be an oriented semi-Riemannian $\overline{m}$-manifold, where $\overline{m} \geq 2$, and let $\Omega_{\overline{M}}$ be its volume form. Let $(M, \overline{\mathfrak{g}}|_M)$ be a semi-Riemannian hypersurface with boundary of $(\overline{M}, \overline{\mathfrak{g}}, \overline{\mathfrak{O}})$. We have from Theorem 17.1.5 that ∂M is a hypersurface of M. Suppose $(\partial M, \overline{\mathfrak{g}}|_{\partial M})$ is a semi-Riemannian hypersurface of $(M, \overline{\mathfrak{g}}|_M)$. (We note that if $(\overline{M}, \overline{\mathfrak{g}}, \overline{\mathfrak{O}})$ is Riemannian, then $(M, \overline{\mathfrak{g}}|_M)$ and $(\partial M, \overline{\mathfrak{g}}|_{\partial M})$ are automatically Riemannian.) Our goal is to use the orientation of $\overline{M}$ to obtain an orientation of M, and in turn to use the orientation of M to find an orientation of ∂M.

We already have experience with this type of undertaking. To construct an orientation of M, suppose there is a nowhere-tangent vector field V in $\mathfrak{X}_{\overline{M}}(M)$, and let $\mathfrak{O}_M$ be the orientation of M induced by V. Then $(M, \overline{\mathfrak{g}}|_M, \mathfrak{O}_M)$ is an oriented semi-Riemannian hypersurface with boundary of $(\overline{M}, \overline{\mathfrak{g}}, \overline{\mathfrak{O}})$. Let Ω_M be its volume form. It follows from Theorem 16.6.9 that: (i) for all points p in M, $(v_1, \ldots, v_{\overline{m}-1})$ is a basis for $T_p(M)$ that is positively oriented with respect to $\mathfrak{O}_M(p)$ if and only if $(V_p, v_1, \ldots, v_{\overline{m}-1})$ is a basis for $T_p(\overline{M})$ that is positively oriented with respect to $\overline{\mathfrak{O}}(p)$, and (ii) $i_V(\Omega_{\overline{M}})|_M$ is an orientation form on M that induces $\mathfrak{O}_M$. According to Theorem 19.11.2(a), $i_V(\Omega_{\overline{M}})|_M = \Omega_M$.

By Theorem 17.2.5, there is an outward-pointing vector field W in $\mathfrak{X}_M(\partial M)$. Let $\mathfrak{O}_{\partial M}$ be the (Stokes) orientation of ∂M induced by W. Then $(\partial M, \overline{\mathfrak{g}}|_{\partial M}, \mathfrak{O}_{\partial M})$ is an oriented semi-Riemannian hypersurface of $(M, \overline{\mathfrak{g}}|_M, \mathfrak{O}_M)$. Let $\Omega_{\partial M}$ be its volume form. It follows from Theorem 16.6.9 that: (iii) for all points q in ∂M, $(w_1, \ldots, w_{\overline{m}-2})$ is a basis for $T_q(\partial M)$ that is positively oriented with respect to $\mathfrak{O}_{\partial M}(q)$ if and only if $(W_q, w_1, \ldots, w_{\overline{m}-2})$ is a basis for $T_q(M)$ that is positively oriented with respect to $\mathfrak{O}_M(q)$, and (iv) $i_W(\Omega_M)|_{\partial M}$ is an orientation form on ∂M that induces $\mathfrak{O}_{\partial M}$. Once again, according to Theorem 19.11.2(a), $i_W(\Omega_M)|_{\partial M} = \Omega_{\partial M}$. Combining (i) and (iii) gives (v): for all points q in ∂M, $(w_1, \ldots, w_{\overline{m}-2})$ is a basis for $T_q(\partial M)$ that is positively oriented with respect to $\mathfrak{O}_{\partial M}(q)$ if and only if $(V_q, W_q, w_1, \ldots, w_{\overline{m}-2})$ is a basis for $T_q(\overline{M})$ that is positively oriented with respect to $\overline{\mathfrak{O}}(q)$.

The key to the above construction is the existence of a nowhere-tangent vector field in $\mathfrak{X}_{\overline{M}}(M)$. Unfortunately, there is no guarantee that such a vector field exists. This is evident from Example 12.7.9 when we view the Möbius band as a hypersurface of the Riemannian 3-manifold $\mathbb{R}_0^3$. The Möbius band has no shortage of nowhere-tangent vector fields, but none of them is smooth.

Let us note that in the above discussion, we did not take advantage of the metric properties of $\overline{M}$, M, or ∂M. This has special relevance to the existence of nowhere-tangent vector fields: in the context of semi-Riemannian manifolds, the premier nowhere-tangent vector field is a unit normal vector field.

The next result is a generalization of Theorem 12.7.6.

Theorem 19.11.1 (Orientability Criterion for Hypersurfaces). *Let $(\overline{M}, \overline{\mathfrak{g}}, \overline{\mathfrak{O}})$ be an oriented semi-Riemannian $\overline{m}$-manifold or an oriented semi-Riemannian $\overline{m}$-manifold with boundary, where $\overline{m} \geq 2$, and let $(M, \overline{\mathfrak{g}}|_M)$ be a semi-Riemannian hypersurface of $\overline{M}$ or a semi-Riemannian hypersurface with boundary of $\overline{M}$. Then M is orientable if and only if there is a unit normal vector field in $\mathfrak{X}_{\overline{M}}(M)$.*

Proof. ($\Rightarrow$): By Theorem 14.8.3, M has a smooth atlas $\mathfrak{B}$ consisting of charts of the form $\left(M \cap U, (x^1, \ldots, x^{\overline{m}-1})\right)$, where $\left(U, (x^1, \ldots, x^{\overline{m}})\right)$ is a slice chart for M in $\overline{M}$. See Figure 19.11.1(a), where $\overline{M}$ and M are represented by the "surface" and its "boundary", respectively; and corresponding to a given point p in the "boundary", $\partial/\partial x^1$ represents a vector in $T_p(M)$.

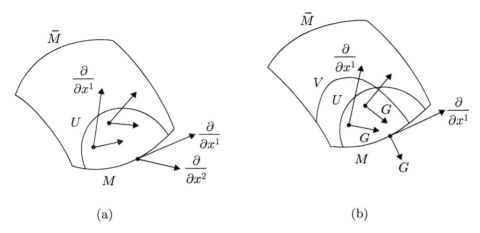

Figure 19.11.1. Diagram for Theorem 19.11.1

Since M is orientable, it follows from Theorem 16.6.5 that M has an orientation form. Arguing as in the proof of Theorem 16.6.3, we obtain from $\mathfrak{B}$ a consistent smooth atlas $\mathfrak{A}$ for M. Let $\left(M \cap U, (x^1, \ldots, x^{\overline{m}-1})\right)$ be a (positively oriented) chart in $\mathfrak{A}$, and let $\mathcal{X} = (\partial/\partial x^1, \ldots, \partial/\partial x^{\overline{m}-1})$ be the corresponding (positively oriented) coordinate frame on $M \cap U$. By Theorem 14.8.3(b), $((\partial/\partial x^1)|_p, \ldots, (\partial/\partial x^{\overline{m}-1})|_p)$ is a (positively oriented) basis for $T_p(M)$ for all p in $M \cap U$. It follows from Theorem 19.3.7 that there is a positively oriented orthonormal frame Ξ on an open set V in $\overline{M}$ that intersects $M \cap U$. By definition, Ξ_p is a positively oriented orthonormal basis for $T_p(\overline{M})$ for all p in V.

For each point p in $U \cap V$, consider the vector

$$G_p = \epsilon_M \frac{\left.\dfrac{\partial}{\partial x^1}\right|_p \times \cdots \times \left.\dfrac{\partial}{\partial x^{\overline{m}-1}}\right|_p}{\left\| \left.\dfrac{\partial}{\partial x^1}\right|_p \times \cdots \times \left.\dfrac{\partial}{\partial x^{\overline{m}-1}}\right|_p \right\|},$$

where the vector product is computed with respect to Ξ_p. By Theorem 19.1.2, $(U \cap V, \mathfrak{g}|_{U \cap V})$ is a semi-Riemannian submanifold of $(M, \mathfrak{g})$, hence $\mathfrak{g}|_{U \cap V}$ satisfies properties [G1] and [G2] of Section 19.1. Arguments similar to those used in the proof of parts (a) and (c) of Theorem 12.2.11 show that the assignment $p \longmapsto G_p$ defines a unit vector field G in $\mathfrak{X}(U \cap V)$. Since the construction of G is based on slice charts for M in $\overline{M}$, it follows from Theorem 14.8.3(b) that G_p is a vector in $T_p(M)^\perp$ for all p in $M \cap U \cap V$. Thus, $G|_{M \cap U \cap V}$ is a unit normal vector field in $\mathfrak{X}_{\overline{M}}(M \cap U \cap V)$. See Figure 19.11.1(b), where G is the "vector product of $\partial/\partial x^1$", and therefore orthogonal to $\partial/\partial x^1$.

Let $\left(M \cap \widetilde{U}, (\widetilde{x}^1, \ldots, \widetilde{x}^{\overline{m}-1})\right)$ be another (positively oriented) chart in $\mathfrak{A}$, let $\widetilde{\mathcal{X}} = (\partial/\partial \widetilde{x}^1, \ldots, \partial/\partial \widetilde{x}^{\overline{m}-1})$ be the corresponding (positively oriented) coordinate frame on $M \cap \widetilde{U}$, let $\widetilde{\Xi}$ be a positively oriented orthonormal frame on a neighborhood $\widetilde{V}$ in $\overline{M}$ that intersects $M \cap \widetilde{U}$, and let $\widetilde{G}|_{M \cap \widetilde{U} \cap \widetilde{V}}$ be the corresponding unit normal vector field in $\mathfrak{X}_{\overline{M}}(M \cap \widetilde{U} \cap \widetilde{V})$. Suppose $M \cap U \cap V$ intersects $M \cap \widetilde{U} \cap \widetilde{V}$. Since $\mathcal{X}$ and $\widetilde{\mathcal{X}}$ are positively oriented, they are consistent, and for the same reason, so are Ξ and $\widetilde{\Xi}$. It follows from Theorem 8.4.11(b) and Theorem 8.4.13(b) that $G_p = \widetilde{G}_p$ for all p in $M \cap (U \cap V) \cap (\widetilde{U} \cap \widetilde{V})$. We can therefore assign to each point p in M a vector $\mathcal{N}_p$ that is unit normal at p, and do so in such a way that there is agreement whenever coordinate domains and the domains of orthonormal frames overlap with M. This defines a unit normal vector field $\mathcal{N}$ along M. By construction, $\mathcal{N}$ is smooth on a neighborhood of every point in M, so it is smooth on M. Thus, $\mathcal{N}$ is a unit normal vector field in $\mathfrak{X}_{\overline{M}}(M)$.

($\Leftarrow$): Let V be a unit normal vector field in $\mathfrak{X}_{\overline{M}}(M)$. Since V is a nowhere-tangent vector field in $\mathfrak{X}_{\overline{M}}(M)$, it follows from Theorem 16.6.9 that V induces an orientation of M. By definition, there is a consistent smooth atlas for M that induces the orientation, and therefore, by definition, M is orientable. $\square$

Theorem 19.11.2. *Let $(\overline{M}, \overline{\mathfrak{g}}, \overline{\mathfrak{D}})$ be an oriented semi-Riemannian $\overline{m}$-manifold or an oriented semi-Riemannian $\overline{m}$-manifold with boundary, where $\overline{m} \geq 2$, and let $\Omega_{\overline{M}}$ be its volume form. Let $(M, \overline{\mathfrak{g}}|_M)$ be an orientable semi-Riemannian hypersurface of $\overline{M}$ or an orientable semi-Riemannian hypersurface with boundary of $\overline{M}$. Let V be a unit normal vector field in $\mathfrak{X}_{\overline{M}}(M)$ (as given by Theorem 19.11.1), and let $\mathfrak{D}_M$ be the orientation of M induced by $i_V(\Omega_{\overline{M}})|_M$ (as given by Theorem 16.6.9), so that $(M, \overline{\mathfrak{g}}|_M, \mathfrak{D}_M)$ is an oriented semi-Riemannian hypersurface of $\overline{M}$ or an oriented semi-Riemannian hypersurface with boundary of $\overline{M}$. Let Ω_M be its volume form. Then:*
(a)
$$i_V(\Omega_{\overline{M}})|_M = \Omega_M.$$
(b) *More generally, if X is a vector field in $\mathfrak{X}_{\overline{M}}(M)$, then*
$$i_X(\Omega_{\overline{M}})|_M = \epsilon_M \langle X, V \rangle \, \Omega_M.$$

Proof. This is the manifold version of Theorem 8.3.6. $\square$

We noted above that for smooth manifolds with boundary, Theorem 17.2.5 ensures the existence of an outward-pointing smooth vector field along the boundary. In the more specialized setting of semi-Riemannian manifolds, the outward-pointing smooth vector field can be upgraded to a unit normal smooth vector field, as we now show.

Theorem 19.11.3 (Existence of Outward-Pointing Vector Field). *Let $(M, \mathfrak{g})$ be a semi-Riemannian m-manifold with boundary, where $m \geq 2$, and* *suppose* $(\partial M, \mathfrak{g}|_{\partial M})$ *is a semi-Riemannian hypersurface of* $(M, \mathfrak{g})$. *Then there is a unique outward-pointing unit normal vector field in* $\mathfrak{X}_M(\partial M)$.

Remark. By Theorem 17.1.5, ∂M is a hypersurface of M, so the additional assumption here is that $\mathfrak{g}|_{\partial M}$ is a metric.

Proof. Uniqueness. Let p be a point in ∂M. According to (14.8.1), $T_p(\partial M)$ is a subspace of $T_p(M)$. We have from Theorem 4.1.3 that $T_p(M) = T_p(\partial M) \oplus T_p(\partial M)^\perp$, and then from Theorem 1.1.18 that $T_p(\partial M)^\perp$ is 1-dimensional. It follows that ∂M has precisely two unit normal vectors at p: one outward-pointing, which we denote by $\mathcal{N}_p$, and the other inward-pointing. The assignment $p \longmapsto \mathcal{N}_p$ defines the unique outward-pointing unit normal vector field $\mathcal{N}$ along ∂M. The question is whether $\mathcal{N}$ is smooth.

Existence. Emulating the proof of Theorem 19.11.1, let $\big(\partial M \cap U, (x^1, \ldots, x^{m-1})\big)$ be a chart on ∂M, where $\big(U, (x^1, \ldots, x^m)\big)$ is a slice chart for ∂M in M, and let $\mathcal{X} = \big(\partial/\partial x^1, \ldots, \partial/\partial x^{m-1}\big)$ be the corresponding coordinate frame. Let Ξ be an orthonormal frame on an open set V in M that intersects $\partial M \cap U$, and let $W = U \cap V$. Without loss of generality, we replace U and V with W in the preceding definitions, but otherwise retain the previous notation. Also without loss of generality, we assume W is a connected set in M. For each point p in W, consider the vector

$$G_p = \epsilon_{\partial M} \frac{\dfrac{\partial}{\partial x^1}\bigg|_p \times \cdots \times \dfrac{\partial}{\partial x^{m-1}}\bigg|_p}{\left\| \dfrac{\partial}{\partial x^1}\bigg|_p \times \cdots \times \dfrac{\partial}{\partial x^{m-1}}\bigg|_p \right\|},$$

where the vector product is computed with respect to Ξ_p. Arguments similar to those used in the proof of parts (a) and (c) of Theorem 12.2.11 show that the assignment $p \longmapsto G_p$ defines a unit vector field G in $\mathfrak{X}(W)$. Since the construction of G is based on slice charts for ∂M in M, it follows from Theorem 14.8.3(b) that G_p is a (unit) vector in $T_p(\partial M)^\perp$ for all p in $\partial M \cap W$. Thus, $G|_{\partial M \cap W}$ is a unit normal vector field in $\mathfrak{X}_M(\partial M \cap W)$. (Notice that unlike the proof of Theorem 19.11.1, we do not make any assumptions about the orientability of M or ∂M.)

Let

$$G = \sum_{i=1}^m \alpha^i \frac{\partial}{\partial x^i}$$

be the local coordinate expression of G with respect to $\big(W, (x^1, \ldots, x^m)\big)$, and let p be a point in W. By Theorem 8.4.10(b), G_p is not in the span of $\mathcal{X}_p$, so $\alpha^m(p) \neq 0$ for all p in W; that is, α is nowhere-vanishing on W. It follows from Theorem 9.1.20 that α^m is either strictly positive on W or strictly negative on W. Thus, $G|_{\partial M \cap W}$ is either inward-pointing on $\partial M \cap W$ or outward-pointing on $\partial M \cap W$; or equivalently, either $G|_{\partial M \cap W}$ or $-G|_{\partial M \cap W}$ is outward-pointing on $\partial M \cap W$. It follows from the uniqueness of $\mathcal{N}$ that either $\mathcal{N}|_{\partial M \cap W} = G|_{\partial M \cap W}$

or $\mathcal{N}|_{\partial M \cap W} = -G|_{\partial M \cap W}$. This shows that $\mathcal{N}$ is smooth on a neighborhood in ∂M of every point in ∂M. Thus, $\mathcal{N}$ is smooth on ∂M.   □

Theorem 19.11.4. *Let $(M, \mathfrak{e}|_M)$ be an orientable Riemannian hypersurface with boundary of $\mathbb{R}_0^3 = (\mathbb{R}^3, \mathfrak{e})$, let $V = (V^1, V^2, V^3)$ be a unit normal vector field in $\mathfrak{X}_{\mathbb{R}^3}(M)$ (as given by Theorem 19.11.1), and let $\mathfrak{O}_M$ be the orientation of M induced by V, so that $(M, \mathfrak{e}|_M, \mathfrak{O}_M)$ is an oriented Riemannian hypersurface with boundary of $\mathbb{R}_0^3$. Let Ω_M be its volume form. Let W be the unique outward-pointing unit normal vector field in $\mathfrak{X}_M(\partial M)$ (as given by Theorem 19.11.3), and let $\mathfrak{O}_{\partial M}$ be the (Stokes) orientation induced by W, so that $(\partial M, \mathfrak{e}|_{\partial M}, \mathfrak{O}_{\partial M})$ is an oriented Riemannian hypersurface of $(M, \mathfrak{e}|_M, \mathfrak{O}_M)$. Let $\Omega_{\partial M}$ be its volume form, and let $U = V \times W$. Then:*
(a) *For all points p in M and all vectors v, w in $T_p(M)$,*

$$\Omega_M|_p(v, w) = \langle V_p, v \times w \rangle$$
$$= \big(V^1(p)\, d(y) \wedge d(z) + V^2(p)\, d(z) \wedge d(x)$$
$$+ V^3(p)\, d(x) \wedge d(y)\big)(v, w).$$

(b) *For all points q in ∂M and all vectors u in $T_q(\partial M)$,*

$$\Omega_{\partial M}|_q(u) = \langle U_q, u \rangle.$$

(c) *For all points q in ∂M, (U_q, V_q, W_q) and (V_q, W_q, U_q) are orthonormal bases for $\mathbb{R}_0^3$ that have the standard orientation.*
(d) *U is the unique unit vector field in $\mathfrak{X}(\partial M)$ that is positively oriented with respect to $\mathfrak{O}_{\partial M}$.*
(e) *The function $\Omega_{\partial M}(U)$ in $C^\infty(\partial M)$ has constant value 1.*

Proof. (a): Let $v = (a^1, a^2, a^3)$ and $w = (b^1, b^2, b^3)$, and let $\Omega_{\mathbb{R}_0^3} = d(x) \wedge d(y) \wedge d(z)$ be the volume form on $\mathbb{R}_0^3$, where, as usual, $\mathbb{R}_0^3$ is assumed to have the standard orientation. Then

$$\Omega_M|_p(v, w) = \big(i_V(\Omega_{\mathbb{R}_0^3})|_M\big)\big|_p(v, w) = \Omega_{\mathbb{R}_0^3}|_p(V_p, v, w)$$
$$= d(x) \wedge d(y) \wedge d(z)(V_p, v, w)$$
$$= \langle V_p, v \times w \rangle = \det\left(\begin{bmatrix} V^1(p) & a^1 & b^1 \\ V^2(p) & a^2 & b^2 \\ V^3(p) & a^3 & b^3 \end{bmatrix}\right)$$
$$= V^1(p) \det\left(\begin{bmatrix} a^2 & b^2 \\ a^3 & b^3 \end{bmatrix}\right) - V^2(p) \det\left(\begin{bmatrix} a^1 & b^1 \\ a^3 & b^3 \end{bmatrix}\right)$$
$$+ V^3(p) \det\left(\begin{bmatrix} a^1 & b^1 \\ a^2 & b^2 \end{bmatrix}\right)$$
$$= \big(V^1(p)\, d(y) \wedge d(z) + V^2(p)\, d(z) \wedge d(x)$$
$$+ V^3(p)\, d(x) \wedge d(y)\big)(v, w),$$

where the first equality follows from Theorem 19.11.2(a), the fourth and fifth equalities from Theorem 8.4.5(a), and the last equality from Theorem 7.2.8.

(b): We have

$$\Omega_{\partial M}|_q(u) = \big(i_W(\Omega_M)|_{\partial M}\big)\big|_q(u) = \Omega_M|_q(W_q, u)$$
$$= \langle V_q, W_q \times u \rangle = \langle V_q \times W_q, u \rangle = \langle U_q, u \rangle,$$

where the first equality follows from Theorem 19.11.2(a), the third equality from part (a), and the fourth equality from Theorem 8.4.5(a).

(c): By definition, V_q is a unit vector in $T_q(M)^\perp$, W_q is a unit vector in $T_q(M)$, and $U_q = V_q \times W_q$. It follows from Theorem 8.4.6(b) that U_q is a unit vector. Thus, (U_q, V_q, W_q), hence (V_q, W_q, U_q), is an orthonormal basis for $\mathbb{R}_0^3$. By Theorem 8.4.10(b), (U_q, V_q, W_q) has the standard orientation, so (V_q, U_q, W_q) has the opposite orientation, hence (V_q, W_q, U_q) has the standard orientation. See Figure 19.11.2.

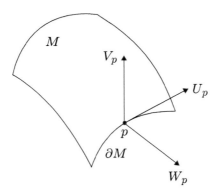

Figure 19.11.2. Diagram for Theorem 19.11.4

(d): Let u be a nonzero vector in $T_q(\partial M)$. It follows from item (v) in the introduction to this section that (u) is a basis for $T_q(\partial M)$ that is positively oriented with respect to $\mathfrak{O}_{\partial M}(q)$ if and only if (V_q, W_q, u) is a basis for $\mathbb{R}_0^3$ that has the standard orientation. We therefore have from part (c) that (U_q) is a basis for $T_q(\partial M)$ that is positively oriented with respect to $\mathfrak{O}_{\partial M}(q)$. Since q was arbitrary, the result follows.

(e): By part (b),

$$\Omega_{\partial M}|_q(U_q) = \langle U_q, U_q \rangle = 1$$

for all q in ∂M. The result follows.   □

We note that, taken together, parts (d) and (e) of the preceding theorem are consistent with Theorem 19.10.1. As an example of part (a), the outward-pointing unit normal vector field for the unit sphere S^2 is given by $V_{(x,y,z)} = (x, y, z)$, so

$$\Omega_{S^2}|_{(x,y,z)} = x\,d(y) \wedge d(z) + y\,d(z) \wedge d(x) + z\,d(x) \wedge d(y).$$

19.12   Induced Connections

In this section, we place the Euclidean derivative with respect to a vector field (Section 10.3) and the induced Euclidean derivative with respect to a vector field (Section 12.3) in the larger context of semi-Riemannian manifolds.

Let $(\overline{M}, \overline{\mathfrak{g}}, \overline{\nabla})$ be a semi-Riemannian $\overline{m}$-manifold, and let $(M, \mathfrak{g}, \nabla)$ be a semi-Riemannian m-submanifold, where $\overline{\nabla}$ and ∇ are the Levi-Civita connections. By definition, $\overline{\nabla}$ and ∇ are maps

$$\overline{\nabla} : \mathfrak{X}(\overline{M}) \times \mathfrak{X}(\overline{M}) \longrightarrow \mathfrak{X}(\overline{M}) \tag{19.12.1}$$

and

$$\nabla : \mathfrak{X}(M) \times \mathfrak{X}(M) \longrightarrow \mathfrak{X}(M),$$

where $\mathfrak{X}(\overline{M})$ is the $C^\infty(\overline{M})$-module of smooth vector fields on $\overline{M}$, and $\mathfrak{X}(M)$ is the $C^\infty(M)$-module of smooth vector fields on M. Recall from Section 15.1 that $\mathfrak{X}_{\overline{M}}(M)$ is the $C^\infty(M)$-module of smooth vector fields along M, and that $\mathfrak{X}(M)$ is a $C^\infty(M)$-submodule of $\mathfrak{X}_{\overline{M}}(M)$.

For each point p in M, we have by definition and (14.8.1) that $\overline{\mathfrak{g}}_p$ is nondegenerate on the subspace $T_p(M)$ of $T_p(\overline{M})$. By Theorem 4.1.3,

$$T_p(\overline{M}) = T_p(M) \oplus T_p(M)^\perp.$$

For brevity, let us denote the projection maps $\mathcal{P}_{T_p(M)}$ and $\mathcal{P}_{T_p(M)^\perp}$ by $\tan_p$ and nor_p, respectively, so that

$$\tan_p : T_p(\overline{M}) \longrightarrow T_p(M) \qquad \text{and} \qquad \mathrm{nor}_p : T_p(\overline{M}) \longrightarrow T_p(M)^\perp.$$

For each vector v in $T_p(\overline{M})$, $\tan_p(v)$ is a tangent vector at p, and $\mathrm{nor}_p(v)$ is a normal vector at p. We define the corresponding maps

$$\tan : \mathfrak{X}_{\overline{M}}(M) \longrightarrow \mathfrak{X}(M) \qquad \text{and} \qquad \mathrm{nor} : \mathfrak{X}_{\overline{M}}(M) \longrightarrow X(M)^\perp$$

by

$$\tan(V)_p = \tan_p(V_p) \qquad \text{and} \qquad \mathrm{nor}(V)_p = \mathrm{nor}_p(V_p)$$

for all points p in M and all vector fields V in $\mathfrak{X}_{\overline{M}}(M)$. Thus,

$$V = \tan(V) + \mathrm{nor}(V). \tag{19.12.2}$$

Let X and V be vector fields in $\mathfrak{X}(M)$ and $\mathfrak{X}_{\overline{M}}(M)$, respectively. As it stands, $\overline{\nabla}(X, V)$ is not meaningful because X and V are not in $\mathfrak{X}(\overline{M})$. It can be shown that for each p in M, there is a chart $(\overline{U}, \overline{\varphi})$ on $\overline{M}$ at p and vector fields $\overline{X}$ and $\overline{V}$ in $\mathfrak{X}(\overline{U})$ such that X and $\overline{X}$ agree on $\overline{U} \cap M$, and likewise for V and $\overline{V}$. Let us define $\overline{\nabla}(X, V)$ to be the restriction of $\overline{\nabla}(\overline{X}, \overline{V})$ to $\overline{U} \cap M$. In this way, we obtain a map

$$\overline{\nabla} : \mathfrak{X}(M) \times \mathfrak{X}_{\overline{M}}(M) \longrightarrow \mathfrak{X}_{\overline{M}}(M), \tag{19.12.3}$$

called the **induced connection on M**. Using the notation $\overline{\nabla}$ in both (19.12.1) and (19.12.3) is intentional and meant to emphasize the relatedness of the two maps. For a given vector field X in $\mathfrak{X}(M)$, the **induced covariant derivative with respect to X** is the map

$$\overline{\nabla}_X : \mathfrak{X}_{\overline{M}}(M) \longrightarrow \mathfrak{X}_{\overline{M}}(M)$$

defined by

$$\overline{\nabla}_X(V) = \overline{\nabla}(X, V)$$

for all V in $\mathfrak{X}_{\overline{M}}(M)$. Corresponding to Theorem 12.3.1, we have the following result.

Theorem 19.12.1. *With the above setup, let X, Y and V, W be vector fields in $\mathfrak{X}(M)$ and $\mathfrak{X}_{\overline{M}}(M)$, respectively, and let f be a function in $C^\infty(M)$. Then:*
(a) $\overline{\nabla}_{X+Y}(V) = \overline{\nabla}_X(V) + \overline{\nabla}_Y(V)$.
(b) $\overline{\nabla}_{fX}(V) = f\,\overline{\nabla}_X(V)$.
(c) $\overline{\nabla}_X(V + W) = \overline{\nabla}_X(V) + \overline{\nabla}_X(W)$.
(d) $\overline{\nabla}_X(fV) = X(f)V + f\,\overline{\nabla}_X(V)$.
(e) $\overline{\nabla}_X(\langle V, W\rangle) = \langle\overline{\nabla}_X(V), W\rangle + \langle V, \overline{\nabla}_X(W)\rangle$. $\qquad\square$

Theorem 19.12.2. *With the above setup, if X, Y are vector fields in $\mathfrak{X}(M)$, then*

$$\nabla_X(Y) = \tan(\overline{\nabla}_X(Y)). \qquad\square$$

Theorem 19.12.3. *With the above setup, define a function*

$$\mathfrak{h} : \mathfrak{X}(M) \times \mathfrak{X}(M) \longrightarrow \mathfrak{X}(M)^\perp,$$

*called the **shape tensor**, by*

$$\mathfrak{h}(X, Y) = \text{nor}(\overline{\nabla}_X(Y))$$

for all vector fields X, Y in $\mathfrak{X}(M)$. Then $\mathfrak{h}$ is $C^\infty(M)$-bilinear and symmetric. $\qquad\square$

We have from (19.12.2) and the preceding two theorems that

$$\overline{\nabla}_X(Y) = \nabla_X(Y) + \mathfrak{h}(X, Y),$$

which is the counterpart of (12.9.4).

For clarity in what follows, let us temporarily denote the above semi-Riemannian connection $\overline{\nabla}$ and induced semi-Riemannian connection $\overline{\nabla}$ by $\overline{\nabla}_1$ and $\overline{\nabla}_2$, respectively. Let us similarly denote the Euclidean connection D and the induced Euclidean connection D by D_1 and D_2, respectively. The following table summarizes notation. In the above discussion of semi-Riemannian connections, we began with $\overline{\nabla}_1$ and ∇, and then constructed $\overline{\nabla}_2$. By contrast, in the case of the Euclidean connection, we began with D_1 and D_2, and then used Theorem 19.12.2 as a definition to construct ∇.

Manifold	$(\overline{M}, \overline{\mathfrak{g}}, \overline{\nabla}_1)$	$\mathbb{R}^3_0 = (\mathbb{R}^3, \mathfrak{e}, D_1)$
Submanifold	$(M, \mathfrak{g}, \nabla)$	$(M, \mathfrak{g}, \nabla)$
Induced connection	$\overline{\nabla}_2$	D_2

Chapter 20

Differential Operators on Semi-Riemannian Manifolds

20.1 Hodge Star

Let $(M, \mathfrak{g}, \mathfrak{O})$ be an oriented semi-Riemannian m-manifold of index ν, and let Ω_M be its volume form. **Hodge star** is the family of linear maps

$$\star : \Lambda^s(M) \longrightarrow \Lambda^{m-s}(M)$$

defined for $0 \leq s \leq m$ that is the manifold version of Hodge star for scalar product spaces, as discussed in Section 8.5. Thus, for each form ω in $\Lambda^s(M)$, we have a unique form $\star(\omega)$ in $\Lambda^{m-s}(M)$ such that

$$\omega \wedge \xi = \langle \star(\omega), \xi \rangle^\Lambda \, \Omega_M$$

for all forms ξ in $\Lambda^{m-s}(M)$.

Theorem 20.1.1. *Let $(M, \mathfrak{g}, \mathfrak{O})$ be an oriented semi-Riemannian m-manifold of index ν, let Ω_M be its volume form, and let ω, ξ be forms in $\Lambda^s(M)$. Then:*
(a) $\star(1_M) = (-1)^\nu \Omega_M$.
(b) $\star(\Omega_M) = 1_M$.
(c) $\star^2(\omega) = (-1)^{s(m-s)+\nu} \omega$.
(d) $\omega \wedge \star(\xi) = (-1)^\nu \langle \omega, \xi \rangle^\Lambda \, \Omega_V$.

Proof. This is the manifold version of Theorem 8.5.3. $\qquad\square$

Theorem 20.1.2. *Let $(M, \mathfrak{g}, \mathfrak{O})$ be an oriented semi-Riemannian m-manifold of index ν, let Ω_M be its volume form, and let X be a vector field in $\mathfrak{X}(M)$. Then*

$$i_X(\Omega_M) = (-1)^\nu \star(X^\mathsf{F}).$$

Semi-Riemannian Geometry, First Edition. Stephen C. Newman.
© 2019 John Wiley & Sons, Inc. Published 2019 by John Wiley & Sons, Inc.

Proof. This is the manifold version of Theorem 8.5.7. □

Example 20.1.3 (Electrodynamics). Let (x, y, z, t) be standard coordinates on $\mathbb{R}_1^4$, let $\alpha_1, \alpha_2, \alpha_3, \alpha_4$ be functions in $C^\infty(\mathbb{R}_1^4)$, and define a covector field ω in $\mathfrak{X}^*(\mathbb{R}_1^4)$ by

$$\omega = \alpha_1 d(x) + \alpha_2 d(y) + \alpha_3 d(z) + \alpha_4 d(t).$$

Using Example 8.5.6, we have

$$\begin{aligned}
\star(\omega) &= \alpha_1 \star d(x) + \alpha_2 \star d(y) + \alpha_3 \star d(z) + \alpha_4 \star d(t) \\
&= -\alpha_1 d(y) \wedge d(z) \wedge d(t) + \alpha_2 d(x) \wedge d(z) \wedge d(t) \\
&\quad - \alpha_3 d(x) \wedge d(y) \wedge d(t) - \alpha_4 d(x) \wedge d(y) \wedge d(z),
\end{aligned}$$

hence

$$\begin{aligned}
d\star(\omega) &= -d(\alpha_1) \wedge d(y) \wedge d(z) \wedge d(t) + d(\alpha_2) \wedge d(x) \wedge d(z) \wedge d(t) \\
&\quad - d(\alpha_3) \wedge d(x) \wedge d(y) \wedge d(t) - d(\alpha_4) \wedge d(x) \wedge d(y) \wedge d(z).
\end{aligned}$$

We also have

$$\begin{aligned}
&d(\alpha_1) \wedge d(y) \wedge d(z) \wedge d(t) \\
&= \left(\frac{\partial \alpha_1}{\partial x} d(x) + \frac{\partial \alpha_1}{\partial y} d(y) + \frac{\partial \alpha_1}{\partial z} d(z) + \frac{\partial \alpha_1}{\partial t} d(t) \right) \wedge d(y) \wedge d(z) \wedge d(t) \\
&= \frac{\partial \alpha_1}{\partial x} d(x) \wedge d(y) \wedge d(z) \wedge d(t) + \frac{\partial \alpha_1}{\partial y} d(y) \wedge d(y) \wedge d(z) \wedge d(t) \\
&\quad + \frac{\partial \alpha_1}{\partial z} d(z) \wedge d(y) \wedge d(z) \wedge d(t) + \frac{\partial \alpha_1}{\partial t} d(t) \wedge d(y) \wedge d(z) \wedge d(t) \\
&= \frac{\partial \alpha_1}{\partial x} d(x) \wedge d(y) \wedge d(z) \wedge d(t);
\end{aligned}$$

likewise,

$$d(\alpha_2) \wedge d(x) \wedge d(z) \wedge d(t) = -\frac{\partial \alpha_2}{\partial y} d(x) \wedge d(y) \wedge d(z) \wedge d(t)$$

$$d(\alpha_3) \wedge d(x) \wedge d(y) \wedge d(t) = \frac{\partial \alpha_3}{\partial z} d(x) \wedge d(y) \wedge d(z) \wedge d(t)$$

$$d(\alpha_4) \wedge d(x) \wedge d(y) \wedge d(z) = -\frac{\partial \alpha_4}{\partial t} d(x) \wedge d(y) \wedge d(z) \wedge d(t).$$

Thus,

$$d\star(\omega) = \left(-\frac{\partial \alpha_1}{\partial x} - \frac{\partial \alpha_2}{\partial y} - \frac{\partial \alpha_3}{\partial z} + \frac{\partial \alpha_4}{\partial t} \right) d(x) \wedge d(y) \wedge d(z) \wedge d(t). \quad \diamond$$

20.2   Codifferential

Let $(M, \mathfrak{g}, \mathfrak{O})$ be an oriented semi-Riemannian m-manifold of index ν. **Codifferential** is the family of linear maps

$$\delta : \Lambda^s(M) \longrightarrow \Lambda^{s-1}(M)$$

defined for $1 \leq s \leq m$ by

$$\delta = (-1)^{m(s+1)+1+\nu} \star d \star ;$$

that is,

$$\delta(\omega) = (-1)^{m(s+1)+1+\nu} \star d \star (\omega) \qquad (20.2.1)$$

for all forms ω in $\Lambda^s(M)$. Let us denote

$$\delta \circ \delta \qquad \text{by} \qquad \delta^2.$$

Recall from (15.8.1) and (15.8.2) that $\Lambda^1(M) = \mathfrak{X}^*(M)$ and $\Lambda^0(M) = C^\infty(M)$. Thus, for $s = 1$, we have $\delta : \mathfrak{X}^*(M) \longrightarrow C^\infty(M)$. For $s = 0$, let us define $\delta = 0$; that is,

$$\delta(f) = 0 \qquad (20.2.2)$$

for all functions f in $C^\infty(M)$.

Theorem 20.2.1. *If $(M, \mathfrak{g}, \mathfrak{O})$ is an oriented semi-Riemannian m-manifold and ω is a form in $\Lambda^s(M)$, then:*
(a) $\delta^2 = 0$.
(b) $\star\delta(\omega) = (-1)^s d\star(\omega)$.
(c) $\delta\star(\omega) = (-1)^{s+1} \star d(\omega)$.

Proof. Suppose M has index ν. Since $d\star(\omega)$ is a form in $\Lambda^{m-s+1}(M)$, by Theorem 20.1.1(c),

$$\star^2\big(d\star(\omega)\big) = (-1)^{(m-s+1)(s-1)+\nu} d\star(\omega). \qquad (20.2.3)$$

(a): We have

$$
\begin{aligned}
\delta^2(\omega) &= (\star d\star)(\star d\star)(\omega) = \star d\big(\star^2\big(d\star(\omega)\big)\big) \\
&= (-1)^{(m-s+1)(s-1)+\nu} \star d^2 \star(\omega) && [(20.2.3)] \\
&= 0. && [\text{Th } 16.1.1(\text{c})]
\end{aligned}
$$

(b): We have

$$
\begin{aligned}
\star\delta(\omega) &= \star\big(\delta(\omega)\big) \\
&= (-1)^{m(s+1)+1+\nu} \star^2\big(d\star(\omega)\big) && [(20.2.1)] \\
&= (-1)^{[m(s+1)+1+\nu]+[(m-s+1)(s-1)+\nu]} d\star(\omega). && [(20.2.3)]
\end{aligned}
$$

Since

$$[m(s+1)+1+\nu]+[(m-s+1)(s-1)+\nu] = 2(ms+s+\nu)-s^2,$$

and because s^2 is even (odd) precisely when s is even (odd),

$$(-1)^{[m(s+1)+1+\nu]+[(m-s+1)(s-1)+\nu]} = (-1)^s.$$

The result follows.

(c): Since $\star(\omega)$ is a form in $\Lambda^{m-s}(M)$,

$$
\begin{aligned}
\delta\big(\star(\omega)\big) &= (-1)^{m(m-s+1)+1+\nu}\star d\star\big(\star(\omega)\big) && [(20.2.1)]\\
&= (-1)^{m(m-s+1)+1+\nu}\star d\big(\star^2(\omega)\big)\\
&= (-1)^{[m(m-s+1)+1+\nu]+[s(m-s)+\nu]}\star d(\omega). && [\text{Th }20.1.1(\text{c})]
\end{aligned}
$$

Since

$$
[m(m-s+1)+1+\nu]+[s(m-s)+\nu] = m(m+1)+2\nu+1-s^2,
$$

because $m(m+1)$ and 2ν are even, and since s^2 is even (odd) precisely when s is even (odd),

$$
(-1)^{[m(m-s+1)+1+\nu]+[s(m-s)+\nu]} = (-1)^{s+1}.
$$

The result follows.   □

Example 20.2.2 (Electrodynamics). We continue with Example 16.1.10, except that $\mathbb{R}^4$ is replaced with $\mathbb{R}^4_1$. The covector field ξ and forms ζ and φ, now viewed as elements of $\mathfrak{X}^*(\mathbb{R}^4_1)$ and $\Lambda^2(\mathbb{R}^4_1)$, respectively, are given by

$$
\xi = \alpha_1 d(x) + \alpha_2 d(y) + \alpha_3 d(z),
$$

$$
\zeta = \beta_1 d(y)\wedge d(z) + \beta_2 d(z)\wedge d(x) + \beta_3 d(x)\wedge d(y),
$$

and

$$
\begin{aligned}
\varphi &= \xi\wedge d(t) + \zeta\\
&= \alpha_1 d(x)\wedge d(t) + \alpha_2 d(y)\wedge d(t) + \alpha_3 d(z)\wedge d(t)\\
&\quad + \beta_1 d(y)\wedge d(z) + \beta_2 d(z)\wedge d(x) + \beta_3 d(x)\wedge d(y).
\end{aligned}
$$

We seek an expression for $\delta(\varphi)$. Setting $m=4$, $\nu=1$, and $s=2$ in (20.2.1) gives $\delta = \star d\star$. Using Example 8.5.6 yields

$$
\begin{aligned}
\star(\varphi) &= \alpha_1\star\big(d(x)\wedge d(t)\big) + \alpha_2\star\big(d(y)\wedge d(t)\big) + \alpha_3\star\big(d(z)\wedge d(t)\big)\\
&\quad + \beta_1\star\big(d(y)\wedge d(z)\big) + \beta_2\star\big(d(z)\wedge d(x)\big) + \beta_3\star\big(d(x)\wedge d(y)\big)\\
&= \alpha_1 d(y)\wedge d(z) + \alpha_2 d(z)\wedge d(x) + \alpha_3 d(x)\wedge d(y)\\
&\quad - [\beta_1 d(x)\wedge d(t) + \beta_2 d(y)\wedge d(t) + \beta_3 d(z)\wedge d(t)],
\end{aligned}
$$

hence

$$
\begin{aligned}
d\star(\varphi) &= d(\alpha_1)\wedge d(y)\wedge d(z) + d(\alpha_2)\wedge d(z)\wedge d(x) + d(\alpha_3)\wedge d(x)\wedge d(y)\\
&\quad - [d(\beta_1)\wedge d(x)\wedge d(t) + d(\beta_2)\wedge d(y)\wedge d(t) + d(\beta_3)\wedge d(z)\wedge d(t)].
\end{aligned}
$$

We also have

$$d(\alpha_1) \wedge d(y) \wedge d(z)$$
$$= \left(\frac{\partial \alpha_1}{\partial x} d(x) + \frac{\partial \alpha_1}{\partial y} d(y) + \frac{\partial \alpha_1}{\partial z} d(z) + \frac{\partial \alpha_1}{\partial t} d(t) \right) \wedge d(y) \wedge d(z)$$
$$= \frac{\partial \alpha_1}{\partial x} d(x) \wedge d(y) \wedge d(z) + \frac{\partial \alpha_1}{\partial y} d(y) \wedge d(y) \wedge d(z)$$
$$+ \frac{\partial \alpha_1}{\partial z} d(z) \wedge d(y) \wedge d(z) + \frac{\partial \alpha_1}{\partial t} d(t) \wedge d(y) \wedge d(z)$$
$$= \frac{\partial \alpha_1}{\partial x} d(x) \wedge d(y) \wedge d(z) + \frac{\partial \alpha_1}{\partial t} d(y) \wedge d(z) \wedge d(t);$$

likewise,

$$d(\alpha_2) \wedge d(z) \wedge d(x) = \frac{\partial \alpha_2}{\partial y} d(x) \wedge d(y) \wedge d(z) - \frac{\partial \alpha_2}{\partial t} d(x) \wedge d(z) \wedge d(t)$$

$$d(\alpha_3) \wedge d(x) \wedge d(y) = \frac{\partial \alpha_3}{\partial z} d(x) \wedge d(y) \wedge d(z) + \frac{\partial \alpha_3}{\partial t} d(x) \wedge d(y) \wedge d(t)$$

$$d(\beta_1) \wedge d(x) \wedge d(t) = -\frac{\partial \beta_1}{\partial y} d(x) \wedge d(y) \wedge d(t) - \frac{\partial \beta_1}{\partial z} d(x) \wedge d(z) \wedge d(t)$$

$$d(\beta_2) \wedge d(y) \wedge d(t) = \frac{\partial \beta_2}{\partial x} d(x) \wedge d(y) \wedge d(t) - \frac{\partial \beta_2}{\partial z} d(y) \wedge d(z) \wedge d(t)$$

$$d(\beta_3) \wedge d(z) \wedge d(t) = \frac{\partial \beta_3}{\partial x} d(x) \wedge d(z) \wedge d(t) + \frac{\partial \beta_3}{\partial y} d(y) \wedge d(z) \wedge d(t).$$

Thus,

$$d\star(\varphi) = \left(\frac{\partial \alpha_1}{\partial x} + \frac{\partial \alpha_2}{\partial y} + \frac{\partial \alpha_3}{\partial z} \right) d(x) \wedge d(y) \wedge d(z)$$
$$- \left(\frac{\partial \beta_2}{\partial x} - \frac{\partial \beta_1}{\partial y} - \frac{\partial \alpha_3}{\partial t} \right) d(x) \wedge d(y) \wedge d(t)$$
$$+ \left(\frac{\partial \beta_1}{\partial z} - \frac{\partial \beta_3}{\partial x} - \frac{\partial \alpha_2}{\partial t} \right) d(x) \wedge d(z) \wedge d(t)$$
$$- \left(\frac{\partial \beta_3}{\partial y} - \frac{\partial \beta_2}{\partial z} - \frac{\partial \alpha_1}{\partial t} \right) d(y) \wedge d(z) \wedge d(t).$$

Using Example 8.5.6 again, we obtain

$$\delta(\varphi) = \star d \star (\varphi)$$

$$= \left(\frac{\partial \alpha_1}{\partial x} + \frac{\partial \alpha_2}{\partial y} + \frac{\partial \alpha_3}{\partial z} \right) \star \big(d(x) \wedge d(y) \wedge d(z) \big)$$

$$- \left(\frac{\partial \beta_2}{\partial x} - \frac{\partial \beta_1}{\partial y} - \frac{\partial \alpha_3}{\partial t} \right) \star \big(d(x) \wedge d(y) \wedge d(t) \big)$$

$$+ \left(\frac{\partial \beta_1}{\partial z} - \frac{\partial \beta_3}{\partial x} - \frac{\partial \alpha_2}{\partial t} \right) \star \big(d(x) \wedge d(z) \wedge d(t) \big)$$

$$- \left(\frac{\partial \beta_3}{\partial y} - \frac{\partial \beta_2}{\partial z} - \frac{\partial \alpha_1}{\partial t} \right) \star \big(d(y) \wedge d(z) \wedge d(t) \big)$$

$$= \left(\frac{\partial \beta_3}{\partial y} - \frac{\partial \beta_2}{\partial z} - \frac{\partial \alpha_1}{\partial t} \right) d(x) + \left(\frac{\partial \beta_1}{\partial z} - \frac{\partial \beta_3}{\partial x} - \frac{\partial \alpha_2}{\partial t} \right) d(y)$$

$$+ \left(\frac{\partial \beta_2}{\partial x} - \frac{\partial \beta_1}{\partial y} - \frac{\partial \alpha_3}{\partial t} \right) d(z) - \left(\frac{\partial \alpha_1}{\partial x} + \frac{\partial \alpha_2}{\partial y} + \frac{\partial \alpha_3}{\partial z} \right) d(t). \qquad \Diamond$$

20.3   Gradient

Let $(M, \mathfrak{g})$ be a semi-Riemannian manifold. **Gradient** is the map

$$\mathrm{Grad} : C^\infty(M) \longrightarrow \mathfrak{X}(M)$$

defined by

$$\mathrm{Grad}(f) = d(f)^{\mathsf{S}}, \tag{20.3.1}$$

or equivalently, by

$$\mathrm{Grad}(f)^{\mathsf{F}} = d(f) \tag{20.3.2}$$

for all functions f in $C^\infty(M)$. We refer to $\mathrm{Grad}(f)$ as the **gradient of f**.

Theorem 20.3.1. *Let $(M, \mathfrak{g})$ be a semi-Riemannian manifold, let X be a vector field in $\mathfrak{X}(M)$, and let f be a function in $C^\infty(M)$. Then:*

$$\langle \mathrm{Grad}(f), X \rangle = d(f)(X) = X(f).$$

Proof. We have

$$\begin{aligned}
\langle \mathrm{Grad}(f), X \rangle &= \mathrm{Grad}(f)^{\mathsf{F}}(X) &&[(19.4.1)] \\
&= d(f)(X) &&[(20.3.2)] \\
&= X(f). &&[\text{Th } 15.4.5(\mathrm{b})] \qquad \square
\end{aligned}$$

Theorem 20.3.2 (Local Coordinate Expression for Gradient). *If $(M, \mathfrak{g})$ is a semi-Riemannian m-manifold and f is a function in $C^\infty(M)$, then, in local coordinates, the components of $\mathrm{Grad}(f)$ are*

$$\mathrm{Grad}(f)^i = \sum_j \mathfrak{g}^{ij} \frac{\partial f}{\partial x^j}$$

for $i = 1, \ldots, m$.

Proof. By definition,

$$\text{Grad}(f) = \sum_i \text{Grad}(f)^i \frac{\partial}{\partial x^i}.$$

It follows from Theorem 20.3.1 that

$$\frac{\partial f}{\partial x^j} = \left\langle \frac{\partial}{\partial x^j}, \text{Grad}(f) \right\rangle = \left\langle \frac{\partial}{\partial x^j}, \sum_k \text{Grad}(f)^k \frac{\partial}{\partial x^k} \right\rangle = \sum_k \mathfrak{g}_{jk} \text{Grad}(f)^k,$$

hence

$$\sum_j \mathfrak{g}^{ij} \frac{\partial f}{\partial x^j} = \sum_j \mathfrak{g}^{ij} \left(\sum_k \mathfrak{g}_{jk} \text{Grad}(f)^k \right) = \sum_k \left(\sum_j \mathfrak{g}^{ij} \mathfrak{g}_{jk} \right) \text{Grad}(f)^k$$

$$= \sum_k \delta_k^i \text{Grad}(f)^k = \text{Grad}(f)^i. \qquad \square$$

Theorem 20.3.3. *If $(M, \mathfrak{g})$ is a semi-Riemannian m-manifold and f, g are functions in $C^\infty(M)$, then*

$$\text{Grad}(fg) = f\,\text{Grad}(g) + g\,\text{Grad}(f).$$

Proof. By Theorem 19.4.3, S is $C^*(M)$-linear. The result now follows from Theorem 15.4.6(b) and (20.3.1). $\qquad \square$

The preceding result has an obvious resemblance to Theorem 10.6.1(a).

Example 20.3.4 ($\mathbb{R}_\nu^m$). Let $(r^1, \ldots, r^m)$, $(\partial/\partial r^1, \ldots, \partial/\partial r^m)$, and $(\varepsilon_1, \ldots, \varepsilon_m)$ be the standard coordinates, corresponding standard coordinate frame, and signature of $\mathbb{R}_\nu^m = (\mathbb{R}^m, \mathfrak{s})$, respectively. Since $\mathfrak{s}^{ij} = \mathfrak{s}_{ij} = \varepsilon_i \delta_{ij}$, we have from Theorem 20.3.2 that $\text{Grad}(f)^i = \varepsilon_i (\partial f / \partial r^i)$, hence

$$\text{Grad}(f) = \left(\varepsilon_1 \frac{\partial f}{\partial r^1}, \ldots, \varepsilon_m \frac{\partial f}{\partial r^m} \right).$$

In $\mathbb{R}_0^3$, with standard coordinates (x, y, z), this becomes

$$\text{Grad}(f) = \left(\frac{\partial f}{\partial x}, \frac{\partial f}{\partial y}, \frac{\partial f}{\partial z} \right) = \text{grad}(f),$$

and in $\mathbb{R}_1^4$, with standard coordinates (x, y, z, t), it becomes

$$\text{Grad}(f) = \left(\frac{\partial f}{\partial x}, \frac{\partial f}{\partial y}, \frac{\partial f}{\partial z}, -\frac{\partial f}{\partial t} \right). \qquad \Diamond$$

20.4   Divergence of Vector Fields

Let $(M, \mathfrak{g})$ be a semi-Riemannian manifold. **Divergence of vector fields is the map**

$$\mathrm{Div} : \mathfrak{X}(M) \longrightarrow C^\infty(M)$$

defined by

$$\mathrm{Div}(X) = \mathsf{C}_1^1 \circ \nabla(X) \tag{20.4.1}$$

for all vector fields X in $\mathfrak{X}(M)$. Since $\nabla(X)$ is a tensor field in $T_1^1(M)$, $\mathsf{C}_1^1(\nabla(X))$ is a function in $T_0^0(M) = C^\infty(M)$, so the definition makes sense. We refer to $\mathrm{Div}(X)$ as the **divergence of X**.

Theorem 20.4.1 (Local Coordinate Expression for Divergence). *Let $(M, \mathfrak{g})$ be a semi-Riemannian manifold with signature $(\varepsilon_1, \ldots, \varepsilon_m)$, let X be a vector field in $\mathfrak{X}(M)$, and, in local coordinates, let*

$$X = \sum_i \alpha^i \frac{\partial}{\partial x^i}.$$

Then

$$\mathrm{Div}(X) = \sum_i \alpha^i_{;i}.$$

Proof. We have from Theorem 15.14.1, Theorem 18.4.4(b), and (20.4.1) that

$$\mathrm{Div}(X) = \mathsf{C}_1^1 \left(\sum_{ij} \alpha^i_{;j} \frac{\partial}{\partial x^i} \otimes d(x^j) \right) = \sum_{ij} \alpha^i_{;j} \mathsf{C}_1^1 \left(\frac{\partial}{\partial x^i} \otimes d(x^j) \right)$$

$$= \sum_{ij} \frac{\partial}{\partial x^i} \big(d(x^j) \big) \alpha^i_{;j} = \sum_i \alpha^i_{;i}. \qquad \square$$

Theorem 20.4.2 (Orthonormal Frame Expression for Divergence). *Let $(M, \mathfrak{g})$ be a semi-Riemannian manifold with signature $(\varepsilon_1, \ldots, \varepsilon_m)$, let X be a vector field in $\mathfrak{X}(M)$, and let $(E_1, \ldots, E_m)$ be an orthonormal frame on an open set U in M. Then*

$$\mathrm{Div}(X) = \sum_i \varepsilon_i \langle \nabla_{E_i}(X), E_i \rangle = - \sum_i \varepsilon_i \langle [X, E_i], E_i \rangle$$

on U.

Proof. Consider the map $\nabla_{\bullet}(X)$ in $\mathrm{Lin}_{C^\infty(M)}\big(\mathfrak{X}(M), \mathfrak{X}(M)\big)$ defined by

$$\nabla_{\bullet}(X)(Y) = \nabla_Y(X) \tag{20.4.2}$$

for all vector fields Y in $\mathfrak{X}(M)$. Recall the maps $\mathfrak{C}_1^1$ and $\mathfrak{R}$ given by (15.7.1) and (19.5.2), respectively. We claim that

$$\mathfrak{R}^{-1} \circ \mathfrak{C}_1^1 \big(\nabla(X) \big) = \nabla_{\bullet}(X).$$

Since $\nabla(X)$ is in $\mathcal{T}_1^1(M)$, it follows from Theorem 15.7.2 that $\mathfrak{R}^{-1} \circ \mathsf{C}_1^1(\nabla(X))$ is in $\mathrm{Lin}_{C^\infty(M)}(\mathfrak{X}(M), \mathfrak{X}(M))$, so the assertion makes sense. Let ω be a covector field in $\mathfrak{X}^*(M)$. Then

$$
\begin{aligned}
\mathsf{C}_1^1(\nabla(X))(\omega, Y) &= \nabla(X)(\omega, Y) && [(15.7.1)] \\
&= \nabla_Y(X)(\omega) && [(18.4.1)] \\
&= \omega(\nabla_Y(X)) && [(15.4.2)] \\
&= \omega(\nabla_\bullet(X)(Y)) && [(20.4.2)] \\
&= \mathfrak{R}(\nabla_\bullet(X))(\omega, Y). && [(19.5.2)]
\end{aligned}
$$

Since ω and Y were arbitrary, $\mathsf{C}_1^1(\nabla(X)) = \mathfrak{R}(\nabla_\bullet(X))$, which proves the claim.
It follows that

$$
\mathfrak{R}^{-1} \circ \mathsf{C}_1^1(\nabla(X))(E_i) = \nabla_{E_i}(X),
$$

and then from Theorem 19.6.1(b) and (20.4.1) that

$$
\mathrm{Div}(X) = \mathsf{C}_1^1(\nabla(X)) = \sum_i \varepsilon_i \langle \nabla_{E_i}(X), E_i \rangle.
$$

This proves the first equality. Since the Levi-Civita connection ∇ is symmetric,

$$
[X, E_i] = \nabla_X(E_i) - \nabla_{E_i}(X),
$$

and because $\langle E_i, E_i \rangle = \pm 1$ and ∇ is compatible with $\mathfrak{g}$,

$$
0 = X(\langle E_i, E_i \rangle) = 2\langle \nabla_X(E_i), E_i \rangle.
$$

Thus,

$$
\langle [X, E_i], E_i \rangle = \langle \nabla_X(E_i), E_i \rangle - \langle \nabla_{E_i}(X), E_i \rangle = -\langle \nabla_{E_i}(X), E_i \rangle.
$$

The second equality now follows from the first. $\square$

Theorem 20.4.3. *Let $(M, \mathfrak{g}, \mathfrak{O})$ be an oriented semi-Riemannian manifold, let Ω_M be its volume form, and let X be a vector field in $\mathfrak{X}(M)$. Then:*
(a)
$$
\mathrm{Div}(X)\,\Omega_M = \mathcal{L}_X(\Omega_M) = d(i_X(\Omega_M)).
$$
(b)
$$
\mathrm{Div}(X) = -\delta(X^{\mathsf{F}}).
$$

Proof. (a): Let p be a point in M. By Theorem 19.3.7, there is a positively oriented orthonormal frame $(E_1, \ldots, E_m)$ on a neighborhood U of p. Since $\Omega_M|_U$ is a form in $\Lambda^m(U)$, so is $\mathcal{L}_X(\Omega_M|_U)$. It follows from Theorem 3.3.3, Theorem 15.8.5, and Theorem 19.10.2 that there is a uniquely determined function f in $C^\infty(U)$ such that $\mathcal{L}_X(\Omega_M|_U) = f\,\Omega_M|_U$. We have

$$
[X, E_i] = \sum_j \alpha^{ij} E_j \tag{20.4.3}
$$

for $i = 1, \ldots, m$, where the α^{ij} are uniquely determined functions in $C^\infty(U)$. Then

$$\langle [X, E_i], E_i \rangle = \sum_j \langle E_i, E_j \rangle \alpha^{ij} = \varepsilon_i \alpha^{ii},$$

where $(\varepsilon_1, \ldots, \varepsilon_m)$ is the signature of M. It follows from Theorem 20.4.2 that

$$\text{Div}(X)|_U = - \sum_i \alpha^{ii}.$$

Then

$$
\begin{aligned}
f &= f\, \Omega_M|_U(E_1, \ldots, E_m) && \text{[Th 19.10.1]} \\
&= \mathcal{L}_X(\Omega_M|_U)(E_1, \ldots, E_m) \\
&= X\big(\Omega_M|_U(E_1, \ldots, E_m)\big) \\
&\quad - \sum_i \Omega_M|_U(E_1, \ldots, E_{i-1}, [X, E_i], E_{i+1}, \ldots, E_m) && \text{[Th 16.4.4]} \\
&= - \sum_i \Omega_M|_U(E_1, \ldots, E_{i-1}, [X, E_i], E_{i+1}, \ldots, E_m) && \text{[Th 19.10.1]} \\
&= - \sum_{ij} \alpha^{ij} \Omega_M|_U(E_1, \ldots, E_{i-1}, E_j, E_{i+1}, \ldots, E_m) && \text{[(20.4.3)]} \\
&= - \sum_i \alpha^{ii} \Omega_M|_U(E_1, \ldots, E_{i-1}, E_i, E_{i+1}, \ldots, E_m) && \text{[mv Th 7.1.2]} \\
&= - \sum_i \alpha^{ii}. && \text{[Th 19.10.1]}
\end{aligned}
$$

Thus, $\text{Div}(X)|_U = f$. Since p was arbitrary, the first equality follows. For the second equality, we have

$$
\begin{aligned}
\text{Div}(X)\, \Omega_M &= \mathcal{L}_X(\Omega_M) && \text{[first equality]} \\
&= d \circ i_X(\Omega_M) + i_X \circ d(\Omega_M) && \text{[Th 16.5.3]} \\
&= d \circ i_X(\Omega_M). && \text{[mv Th 7.2.12(b)]}
\end{aligned}
$$

(b): Since X^{F} is a covector field in $\mathfrak{X}^*(M) = \Lambda^1(M)$, we have from (20.2.1) that $\delta = (-1)^{\nu+1} \star d\, \star$, where ν is the index of M. Then

$$
\begin{aligned}
\text{Div}(X) &= \text{Div}(X) \star(\Omega_M) && \text{[Th 20.1.1(b)]} \\
&= \star\big(\text{Div}(X)\, \Omega_M\big) \\
&= \star d\big(i_X(\Omega_M)\big) && \text{[part (a)]} \\
&= (-1)^\nu \star d \star(X^{\mathsf{F}}) && \text{[Th 20.1.2]} \\
&= -\delta(X^{\mathsf{F}}). && \square
\end{aligned}
$$

Theorem 20.4.4. *Let $(M, \mathfrak{g}, \mathfrak{O})$ be an oriented semi-Riemannian manifold, let f be a function in $C^\infty(M)$, and let X be a vector field in $\mathfrak{X}(M)$. Then*

$$\text{Div}(fX) = f\, \text{Div}(X) + \langle \text{Grad}(f), X \rangle.$$

Proof. Setting $\omega = d(f)$ and $\xi = \Omega_M$ in Theorem 16.5.1(e) gives

$$d(f) \wedge i_X(\Omega_M) = i_X\big(d(f)\big) \wedge \Omega_M - i_X\big(d(f) \wedge \Omega_M\big)$$
$$= X(f)\,\Omega_M \qquad\qquad\qquad [(16.5.2),\ \text{Th } 7.2.12(b)]$$
$$= \langle \text{Grad}(f), X \rangle\,\Omega_M, \qquad\qquad [\text{Th } 20.3.1]$$

hence

$$\text{Div}(fX)\,\Omega_M = d\big(i_{fX}(\Omega_M)\big) \qquad\qquad\quad [\text{Th } 20.4.3(a)]$$
$$= d\big(f\,i_X(\Omega_M)\big) \qquad\qquad\qquad\;\; [\text{Th } 16.5.1(b)]$$
$$= d(f) \wedge i_X(\Omega_M) + f\,d\big(i_X(\Omega_M)\big) \qquad [(16.1.1)]$$
$$= [\langle \text{Grad}(f), X \rangle + f\,\text{Div}(X)]\Omega_M. \qquad [\text{Th } 20.4.3(a)]$$

Since (Ω_M) is a basis for $\Lambda^m(M)$, the result follows. $\square$

The preceding result has a clear resemblance to Theorem 10.6.1(b).

Theorem 20.4.5 (Divergence Theorem for Semi-Riemannian Manifolds). *Let $(M, \mathfrak{g}, \mathfrak{D})$ be a compact oriented semi-Riemannian manifold with boundary, let Ω_M be its volume form, and let X be a vector field in $\mathfrak{X}(M)$. Then*

$$\int_M \text{Div}(X)\,\Omega_M = \int_{\partial M} i_X(\Omega_M).$$

Proof. We have

$$\int_M \text{Div}(X)\,\Omega_M = \int_M d\big(i_X(\Omega_M)\big) \qquad [\text{Th } 20.4.3(a)]$$
$$= \int_{\partial M} i_X(\Omega_M). \qquad\qquad [\text{Th } 17.4.1] \qquad \square$$

Example 20.4.6 ($\mathbb{R}^m_\nu$). Let $(r^1, \ldots, r^m)$, $(\partial/\partial r^1, \ldots, \partial/\partial r^m)$, and $(\varepsilon_1, \ldots, \varepsilon_m)$ be the standard coordinates, corresponding standard coordinate frame, and signature of $\mathbb{R}^m_\nu$, respectively. From Example 15.1.6, $(\partial/\partial r^1, \ldots, \partial/\partial r^m) = (e_1, \ldots, e_m)$, where $(e_1, \ldots, e_m)$ is the standard basis for $\mathbb{R}^m_\nu$. Let

$$X = (\alpha^1, \ldots, \alpha^m) = \sum_i \alpha^i e_i = \sum_i \alpha^i \frac{\partial}{\partial r^i}$$

be a vector field in $\mathfrak{X}(\mathbb{R}^m_\nu)$. It follows from (18.2.2), Theorem 19.3.5, and Theorem 20.4.1 that

$$\text{Div}(X) = \sum_i \frac{\partial \alpha^i}{\partial r^i}.$$

In $\mathbb{R}^3_0$, with standard coordinates (x, y, z), this becomes

$$\text{Div}(X) = \frac{\partial \alpha^1}{\partial x} + \frac{\partial \alpha^2}{\partial y} + \frac{\partial \alpha^3}{\partial z} = \text{div}(X),$$

and in $\mathbb{R}^4_1$, with standard coordinates (x, y, z, t), it becomes

$$\text{Div}(X) = \frac{\partial \alpha^1}{\partial x} + \frac{\partial \alpha^2}{\partial y} + \frac{\partial \alpha^3}{\partial z} + \frac{\partial \alpha^4}{\partial t}. \qquad \diamond$$

20.5   Curl

Let $(M, \mathfrak{g}, \mathfrak{O})$ be an oriented semi-Riemannian 3-manifold of index ν, and let Ω_M be its volume form. **Curl** is the map

$$\mathrm{Curl} : \mathfrak{X}(M) \longrightarrow \mathfrak{X}(M)$$

defined by

$$\mathrm{Curl}(X)^{\mathsf{F}} = \star d(X^{\mathsf{F}}) \qquad (20.5.1)$$

for all vector fields X in $\mathfrak{X}(M)$. Since X^{F} is a covector field in $\mathfrak{X}^*(M) = \Lambda^1(M)$, it follows that $d(X^{\mathsf{F}})$ is a form in $\Lambda^2(M)$, hence $\star d(X^{\mathsf{F}})$ is a covector field in $\Lambda^1(M)$, so the definition makes sense. We refer to $\mathrm{Curl}(X)$ as the **curl of** X.

Theorem 20.5.1. *With the above setup,*

$$i_{\mathrm{Curl}(X)}(\Omega_M) = d(X^{\mathsf{F}}).$$

Proof. We have

$$
\begin{aligned}
i_{\mathrm{Curl}(X)}(\Omega_M) &= (-1)^{\nu} \star \big(\mathrm{Curl}(X)^{\mathsf{F}} \big) && \text{[Th 20.1.2]} \\
&= (-1)^{\nu} \star^2 d(X^{\mathsf{F}}). && \text{[(20.5.1)]}
\end{aligned}
$$

Since $d(X^{\mathsf{F}})$ is a form in $\Lambda^2(M)$, setting $m = 3$ and $s = 2$ in Theorem 20.1.1(c) yields

$$\star^2 \big(d(X^{\mathsf{F}}) \big) = (-1)^{\nu} d(X^{\mathsf{F}}).$$

The result follows.   □

Despite the unfamiliar appearance of definition (20.5.1), the following example shows that the Curl and classical curl operators are identical in $\mathbb{R}_0^3$.

Example 20.5.2. Let (x, y, z) and $\big(d(x), d(y), d(z) \big)$ be the standard coordinates and corresponding standard dual coordinate frame on $\mathbb{R}_0^3$, respectively, and let $X = (\alpha^1, \alpha^2, \alpha^3)$ be a vector field in $\mathfrak{X}(\mathbb{R}_0^3)$. From the manifold version of Theorem 3.3.2,

$$X^{\mathsf{F}} = \alpha^1 d(x) + \alpha^2 d(y) + \alpha^3 d(z).$$

Then (16.1.2) gives

$$
\begin{aligned}
d(X^{\mathsf{F}}) = {} & \left(\frac{\partial \alpha^3}{\partial y} - \frac{\partial \alpha^2}{\partial z} \right) d(y) \wedge d(z) + \left(\frac{\partial \alpha^1}{\partial z} - \frac{\partial \alpha^3}{\partial x} \right) d(z) \wedge d(x) \\
& + \left(\frac{\partial \alpha^2}{\partial x} - \frac{\partial \alpha^1}{\partial y} \right) d(x) \wedge d(y),
\end{aligned}
$$

hence

$$\mathrm{Curl}(X)^{\mathsf{F}} = \star d(X^{\mathsf{F}})$$

$$= \left(\frac{\partial \alpha^3}{\partial y} - \frac{\partial \alpha^2}{\partial z}\right) \star (d(y) \wedge d(z)) + \left(\frac{\partial \alpha^1}{\partial z} - \frac{\partial \alpha^3}{\partial x}\right) \star (d(z) \wedge d(x))$$

$$+ \left(\frac{\partial \alpha^2}{\partial x} - \frac{\partial \alpha^1}{\partial y}\right) \star (d(x) \wedge d(y))$$

$$= \left(\frac{\partial \alpha^3}{\partial y} - \frac{\partial \alpha^2}{\partial z}\right) d(x) + \left(\frac{\partial \alpha^1}{\partial z} - \frac{\partial \alpha^3}{\partial x}\right) d(y)$$

$$+ \left(\frac{\partial \alpha^2}{\partial x} - \frac{\partial \alpha^1}{\partial y}\right) d(z)$$

$$= \left(\frac{\partial \alpha^3}{\partial y} - \frac{\partial \alpha^2}{\partial z}, \frac{\partial \alpha^1}{\partial z} - \frac{\partial \alpha^3}{\partial x}, \frac{\partial \alpha^2}{\partial x} - \frac{\partial \alpha^1}{\partial y}\right)^{\mathsf{F}}$$

$$= \mathrm{curl}(X)^{\mathsf{F}},$$

where the third equality follows from Example 8.5.4, the fourth equality from the manifold version of Theorem 3.3.2, and the last equality from Table 10.6.1. Thus, $\mathrm{Curl}(X) = \mathrm{curl}(X)$.   ◇

20.6   Hesse Operator

Let $(M, \mathfrak{g})$ be a semi-Riemannian manifold. The **Hesse operator** is the map

$$\mathrm{Hess} : C^\infty(M) \longrightarrow \mathcal{T}_2^0(M)$$

defined by

$$\mathrm{Hess}(f) = \nabla^2(f) \tag{20.6.1}$$

for all functions f in $C^\infty(M)$, where we denote

$$\nabla \circ \nabla \quad \text{by} \quad \nabla^2.$$

We refer to $\mathrm{Hess}(f)$ as the **Hessian of f**.

Theorem 20.6.1. *Let $(M, \mathfrak{g})$ be a semi-Riemannian m-manifold, let f be a function in $C^\infty(M)$, and let X, Y be vector fields in $\mathfrak{X}(M)$. Then:*
(a)

$$\mathrm{Hess}(f)(X, Y) = X\left(Y(f)\right) - \nabla_X(Y)(f)$$
$$= \nabla_X\left(\nabla_Y(f)\right) - \nabla_{\nabla_X(Y)}(f)$$
$$= \left\langle \nabla_X\left(\mathrm{Grad}(f)\right), Y \right\rangle.$$

(b) *$\mathrm{Hess}(f)$ is symmetric.*

Proof. (a): The first two equalities follow from (18.4.3). We have

$$
\begin{aligned}
X\big(Y(f)\big) &= X\big(\langle \mathrm{Grad}(f), Y \rangle\big) & \text{[Th 20.3.1]} \\
&= \big\langle \nabla_X\big(\mathrm{Grad}(f)\big), Y \big\rangle + \big\langle \mathrm{Grad}(f), \nabla_X(Y) \big\rangle & \text{[(19.3.1)]} \\
&= \big\langle \nabla_X\big(\mathrm{Grad}(f)\big), Y \big\rangle + \nabla_X(Y)(f). & \text{[Th 20.3.1]}
\end{aligned}
$$

The third equality now follows from the first.

(b): Since the Levi-Civita connection ∇ is symmetric,

$$
\nabla_X(Y)(f) - \nabla_Y(X)(f) = [X,Y](f) = X\big(Y(f)\big) - Y\big(X(f)\big),
$$

so

$$
\begin{aligned}
\mathrm{Hess}(f)(X,Y) &= X\big(Y(f)\big) - \nabla_X(Y)(f) & \text{[part(a)]} \\
&= Y\big(X(f)\big) - \nabla_Y(X)(f) \\
&= \mathrm{Hess}(f)(Y,X). & \text{[part(a)]} \qquad \square
\end{aligned}
$$

Theorem 20.6.2 (Local Coordinate Expression for Hesse Operator). *If $(M, \mathfrak{g})$ is a semi-Riemannian m-manifold and f is a function in $C^\infty(M)$, then, in local coordinates, the components of $\mathrm{Hess}(f)$ are*

$$
\mathrm{Hess}(f)_{ij} = \frac{\partial^2 f}{\partial x^i \partial x^j} - \sum_k \Gamma_{ij}^k \frac{\partial f}{\partial x^k}
$$

for $i, j = 1, \ldots, m$.

Proof. We have

$$
\begin{aligned}
\mathrm{Hess}(f)_{ij} &= \mathrm{Hess}(f)\left(\frac{\partial}{\partial x^i}, \frac{\partial}{\partial x^j} \right) \\
&= \frac{\partial}{\partial x^i}\left(\frac{\partial f}{\partial x^j} \right) - \nabla_{\partial/\partial x^i}\left(\frac{\partial}{\partial x^j} \right)(f) & \text{[Th 20.6.1(a)]} \\
&= \frac{\partial^2 f}{\partial x^i \partial x^j} - \sum_k \Gamma_{ij}^k \frac{\partial f}{\partial x^k}. & \text{[(18.2.1)]} \qquad \square
\end{aligned}
$$

Example 20.6.3 ($\mathbb{R}_\nu^m$). Let $(r^1, \ldots, r^m)$ and $(\partial/\partial r^1, \ldots, \partial/\partial r^m)$ be the standard coordinates and corresponding standard coordinate frame on $\mathbb{R}_\nu^m$, respectively. It follows from Theorem 19.3.5 and Theorem 20.6.2 that

$$
\mathrm{Hess}(f)_{ij} = \frac{\partial^2 f}{\partial r^i \partial r^j}
$$

for $i, j = 1, \ldots, m$. $\diamond$

20.7   Laplace Operator

Let $(M, \mathfrak{g})$ be a semi-Riemannian manifold. The **Laplace operator** is the map

$$\text{Lap} : C^\infty(M) \longrightarrow C^\infty(M)$$

defined by

$$\text{Lap}(f) = \text{Div} \circ \text{Grad}(f) \qquad (20.7.1)$$

for all functions f in $C^\infty(M)$. We refer to $\text{Lap}(f)$ as the **Laplacian of f**.

Theorem 20.7.1. *If $(M, \mathfrak{g})$ is a semi-Riemannian manifold and f is a function in $C^\infty(M)$, then*

$$\text{Lap}(f) = \mathsf{C}_{12} \circ \text{Hess}(f).$$

Proof. We have

$$
\begin{aligned}
\text{Lap}(f) &= \text{Div} \circ \text{Grad}(f) \\
&= \mathsf{C}_1^1 \circ \nabla \circ \mathsf{S}_1^1(d(f)) && [\text{mv Ex 6.3.1, (20.3.1), (20.4.1)}] \\
&= \mathsf{C}_1^1 \circ \mathsf{S}_1^1 \circ \nabla(\nabla(f)) && [\text{Th 18.4.3, Th 19.6.3}] \\
&= \mathsf{C}_1^1 \circ \mathsf{S}_1^1 \circ \nabla^2(f) \\
&= \mathsf{C}_{12} \circ \text{Hess}(f). && [(19.6.3), (20.6.1)] \qquad \square
\end{aligned}
$$

Theorem 20.7.2 (Orthonormal Frame Expression for Laplace Operator). *Let $(M, \mathfrak{g})$ be a semi-Riemannian manifold with signature $(\varepsilon_1, \ldots, \varepsilon_m)$, let f be a function in $C^\infty(M)$, and let $(E_1, \ldots, E_m)$ be an orthonormal frame on an open set U in M. Then*

$$\text{Lap}(f) = \sum_i \varepsilon_i \, \text{Hess}(f)(E_i, E_i)$$

on U.

Proof. This follows from Theorem 19.6.2(b) and Theorem 20.7.1. $\qquad \square$

Example 20.7.3 ($\mathbb{R}_\nu^m$). Let $(r^1, \ldots, r^m)$, $(\partial/\partial r^1, \ldots, \partial/\partial r^m)$, and $(\varepsilon_1, \ldots, \varepsilon_m)$ be the standard coordinates, corresponding standard coordinate frame, and signature of $\mathbb{R}_\nu^m$, respectively. We have from Example 20.6.3 that

$$\text{Hess}(f)\left(\frac{\partial}{\partial r^i}, \frac{\partial}{\partial r^j}\right) = \text{Hess}(f)_{ij} = \frac{\partial^2 f}{\partial r^i \partial r^j},$$

and from Example 15.1.6 that $(\partial/\partial r^1, \ldots, \partial/\partial r^m) = (e_1, \ldots, e_m)$, where $(e_1, \ldots, e_m)$ is the standard basis for $\mathbb{R}_\nu^m$. Thus, $(\partial/\partial r^1, \ldots, \partial/\partial r^m)$ is an orthonormal frame on $\mathbb{R}_\nu^m$. Then Theorem 20.7.2 gives

$$\text{Lap}(f) = \sum_i \varepsilon_i \frac{\partial^2 f}{\partial (r^i)^2}.$$

In $\mathbb{R}_0^3$, with standard coordinates (x, y, z), this becomes

$$\text{Lap}(f) = \frac{\partial^2 f}{\partial x^2} + \frac{\partial^2 f}{\partial y^2} + \frac{\partial^2 f}{\partial z^2} = \text{lap}(f),$$

and in $\mathbb{R}_1^4$, with standard coordinates (x, y, z, t), it is

$$\text{Lap}(f) = \frac{\partial^2 f}{\partial x^2} + \frac{\partial^2 f}{\partial y^2} + \frac{\partial^2 f}{\partial z^2} - \frac{\partial^2 f}{\partial t^2}.$$

In the physics literature, Lap in the context of $\mathbb{R}_1^4$ is usually denoted by $\square$ (sometimes by $\square^2$) and called the **d'Alembert operator**. $\diamond$

20.8   Laplace–de Rham Operator

Let $(M, \mathfrak{g}, \mathfrak{O})$ be an oriented semi-Riemannian manifold. The **Laplace–de Rham operator** is the family of linear maps

$$\Delta : \Lambda^s(M) \longrightarrow \Lambda^s(M)$$

defined for $1 \leq s \leq m$ by

$$\Delta = d\delta + \delta d; \qquad (20.8.1)$$

that is,

$$\Delta(\omega) = d\delta(\omega) + \delta d(\omega)$$

for all forms ω in $\Lambda^s(M)$. Recall from Theorem 16.1.1(c) that $d^2 = 0$, and from Theorem 20.2.1(a) that $\delta^2 = 0$. Thus,

$$(d + \delta) \circ (d + \delta) = d^2 + d\delta + \delta d + \delta^2 = \Delta,$$

which can be expressed as

$$\Delta = (d + \delta)^2.$$

The next result shows that, except for a sign, the Laplace–de Rham operator is a generalization of the Laplace operator.

Theorem 20.8.1. *If $(M, \mathfrak{g}, \mathfrak{O})$ is an oriented semi-Riemannian manifold and f is a function in $C^\infty(M)$, then*

$$\text{Lap}(f) = -\Delta(f).$$

Proof. We have

$$
\begin{aligned}
\text{Lap}(f) &= \text{Div}\big(\text{Grad}(f)\big) &&[(20.7.1)]\\
&= -\delta\big(\text{Grad}(f)^\mathsf{F}\big) &&[\text{Th } 20.4.3(\text{b})]\\
&= -\delta d(f) &&[(20.3.2)]\\
&= -(d\delta + \delta d)(f) &&[(20.2.2)]\\
&= -\Delta(f). &&[(20.8.1)]
\end{aligned}
$$

$\square$

20.9   Divergence of Symmetric 2-Covariant Tensor Fields

Let $(M, \mathfrak{g})$ be a semi-Riemannian manifold, and let $\Sigma(M)$ be the subspace of $T_2^0(M)$ consisting of symmetric (2-covariant) tensor fields. For example, $\mathfrak{g}$ is in $\Sigma(M)$. **Divergence of symmetric 2-covariant tensor fields** is the map

$$\mathrm{Div} : \Sigma(M) \longrightarrow \mathfrak{X}^*(M)$$

defined by

$$\mathrm{Div}(\mathcal{A}) = \mathsf{C}_{13} \circ \nabla(\mathcal{A})$$

for all tensor fields $\mathcal{A}$ in $\Sigma(M)$. Since $\nabla(\mathcal{A})$ is in $T_3^0(M)$, $\mathsf{C}_{13}(\nabla(\mathcal{A}))$ is in $T_1^0(M) = \mathfrak{X}^*(M)$, so the definition makes sense. We note in passing that by Theorem 19.3.1, $\nabla(\mathfrak{g}) = 0$, hence $\mathrm{Div}(\mathfrak{g}) = 0$.

Theorem 20.9.1 (Orthonormal Frame Expression for Divergence). *Let* $(M, \mathfrak{g})$ *be a semi-Riemannian manifold with signature* $(\varepsilon_1, \ldots, \varepsilon_m)$, *let* $(E_1, \ldots, E_m)$ *be an orthonormal frame on an open set* U *in* M, *let* $\mathcal{A}$ *be a symmetric tensor field in* $T_2^0(M)$, *and let* X *be a vector field in* $\mathfrak{X}(M)$. *Then*

$$\mathrm{Div}(\mathcal{A})(X) = \sum_i \varepsilon_i \nabla_{E_i}(\mathcal{A})(X, E_i) = \sum_i \varepsilon_i \nabla_{E_i}(\mathcal{A})(E_i, X)$$

on U.

Proof. We have

$$\begin{aligned}
\mathrm{Div}(\mathcal{A})(X) &= \mathsf{C}_{13}\big(\nabla(\mathcal{A})\big)(X) \\
&= \sum_i \varepsilon_i \nabla(\mathcal{A})(E_i, X, E_i) \qquad \text{[Th 19.6.2(a)]} \\
&= \sum_i \varepsilon_i \nabla_{E_i}(\mathcal{A})(X, E_i), \qquad \text{[(18.4.1)]}
\end{aligned}$$

which proves the first equality. It is easily shown using (18.1.3) that since $\mathcal{A}$ is symmetric, so is $\nabla_{E_i}(\mathcal{A})$. The second equality now follows from the first. $\qquad \square$

Theorem 20.9.2. *If* $(M, \mathfrak{g})$ *is a semi-Riemannian manifold and* f *is a function in* $C^\infty(M)$, *then*

$$\mathrm{Div}(f\mathfrak{g}) = d(f).$$

Proof. Let $(\varepsilon_1, \ldots, \varepsilon_m)$ be the signature of M, and let p be a point in M. By Theorem 19.3.7, there is an orthonormal frame $(E_1, \ldots, E_m)$ on a neighborhood U of p. Let X be a vector field in $\mathfrak{X}(M)$, and let $X = \sum_i \alpha^i E_i$ on U, where the α^i are uniquely determined functions in $C^\infty(U)$. From the manifold version of Theorem 4.2.7,

$$X = \sum_i \varepsilon_i \langle X, E_i \rangle E_i. \tag{20.9.1}$$

By Theorem 19.3.8,

$$\nabla_{E_i}(f\mathfrak{g}) = \nabla_{E_i}(\mathfrak{g} \otimes f) = \mathfrak{g} \otimes \nabla_{E_i}(f) = \nabla_{E_i}(f)\,\mathfrak{g},$$

hence

$$\nabla_{E_i}(f\mathfrak{g})(X, E_i) = \nabla_{E_i}(f)\,\mathfrak{g}(X, E_i) = \langle X, E_i \rangle\,\nabla_{E_i}(f). \qquad (20.9.2)$$

Then

$$
\begin{aligned}
d(f)(X) &= X(f) && \text{[Th 15.4.5(b)]}\\
&= \sum_i \varepsilon_i \langle X, E_i \rangle\,\nabla_{E_i}(f) && \text{[(18.1.1), (20.9.1)]}\\
&= \sum_i \varepsilon_i \nabla_{E_i}(f\mathfrak{g})(X, E_i) && \text{[(20.9.2)]}\\
&= \mathrm{Div}(f\mathfrak{g})(X). && \text{[Th 20.9.1]}
\end{aligned}
$$

Since X and p were arbitrary, the result follows.   □

Chapter 21

Riemannian Manifolds

In this chapter, we narrow the scope from semi-Riemannian manifolds to Riemannian manifolds and examine selected earlier topics in that more specialized setting.

21.1   Geodesics and Curvature on Riemannian Manifolds

Here is the promised generalization of Theorem 19.9.3 to Riemannian manifolds.

Theorem 21.1.1 (Length Minimizing Property of Geodesics on Riemannian Manifolds). *Let $(M, \mathfrak{g})$ be a Riemannian manifold, let p be a point in M, and let V be a normal neighborhood of p in M. Let q be a point in V, let $v = \exp_p^{-1}(q)$, and let $\rho_{p,v} : [0,1] \longrightarrow M$ be the radial geodesic with starting point p and initial velocity v. If $\lambda(s) : [0,1] \longrightarrow V$ is a smooth curve such that $\lambda(0) = p$, $\lambda(1) = q$, and $\lambda([0,1]) \subset V$, then $L(\rho_{p,v}) \leq L(\lambda)$.*

Proof. We have from Theorem 18.9.6 that $\rho_{p,v}$ is the unique radial geodesic such that $\rho_{p,v}(0) = p$, $\rho_{p,v}(1) = q$, and $\rho_{p,v}([0,1]) \subset V$. By Theorem 19.9.1(a), $\rho_{p,v}$ has constant speed

$$\left\| \frac{d\rho_{p,v}}{dr}(r) \right\| = \left\| \frac{d\rho_{p,v}}{dr}(0) \right\| = \|v\|,$$

so (19.2.1) gives

$$L(\rho_{p,v}) = \int_0^1 \left\| \frac{d\rho_{p,v}}{dr}(r) \right\| dr = \|v\|. \tag{21.1.1}$$

Semi-Riemannian Geometry, First Edition. Stephen C. Newman.
© 2019 John Wiley & Sons, Inc. Published 2019 by John Wiley & Sons, Inc.

Let $\exp_p : \tilde{V} \longrightarrow V$ be the diffeomorphism corresponding to V, and consider the smooth curve $\tilde{\lambda}(s) : (0,1] \longrightarrow \tilde{V}$ defined by

$$\tilde{\lambda}(s) = \exp_p^{-1} \circ \lambda(s).$$

Since $\tilde{\lambda}(s) = 0$ if and only if $\lambda(s) = p$, in which case λ has a loop joining p to itself, we assume without loss of generality that $\tilde{\lambda}(s) \neq 0$ for all s in $(0,1]$. Then $\tilde{\lambda}$ can be parametrized by

$$\tilde{\lambda}(s) = r(s)\psi(s),$$

where the smooth function $r(s) : [0,1] \longrightarrow \mathbb{R}$ and smooth map $\psi(s) : (0,1] \longrightarrow T_p(M)$ are given by

$$r(s) = \left\| \tilde{\lambda}(s) \right\| \qquad \text{and} \qquad \psi(s) = \frac{\tilde{\lambda}(s)}{\left\| \tilde{\lambda}(s) \right\|}.$$

We have from (21.1.1) that

$$r(1) = \left\| \exp_p^{-1}(\lambda(1)) \right\| = \left\| \exp_p^{-1}(q) \right\| = \|v\| = L(\rho_{p,v}). \tag{21.1.2}$$

Since M is Riemannian (as opposed to semi-Riemannian), we also have

$$\langle \psi(s), \psi(s) \rangle = 1. \tag{21.1.3}$$

Consider the parametrized surface $\sigma(r,s) : [0,1] \times (0,1] \longrightarrow M$ defined by

$$\sigma(r,s) = \exp_p(r\psi(s)),$$

where we assume that $r\psi(s)$ is in $\mathcal{E}_p$ for all (r,s) in $[0,1] \times (0,1]$. Then

$$\lambda(s) = \exp_p \circ \tilde{\lambda}(s) = \exp_p(r(s)\psi(s)) = \sigma(r(s),s) = \sigma \circ \mu(s),$$

where $\mu(s) = (r(s), s)$. Let $(U, (x^i))$ be a chart on M that intersects $\sigma([0,1] \times (0,1])$. Then

$$\frac{d\lambda}{ds}(s) = \frac{d(\sigma \circ \mu)}{ds}(s) = \sum_i \frac{d(x^i \circ \sigma \circ \mu)}{ds}(s) \frac{\partial}{\partial x^i}\bigg|_{\sigma \circ \mu(s)}$$

$$= \sum_i \left[\frac{dr}{ds}(s) \frac{\partial(x^i \circ \sigma)}{\partial r}(\mu(s)) + \frac{\partial(x^i \circ \sigma)}{\partial s}(\mu(s)) \right] \frac{\partial}{\partial x^i}\bigg|_{\sigma \circ \mu(s)}$$

$$= \frac{dr}{ds}(s) \sum_i \frac{\partial(x^i \circ \sigma)}{\partial r}(r(s), s) \frac{\partial}{\partial x^i}\bigg|_{\sigma(r(s),s)}$$

$$+ \sum_i \frac{\partial(x^i \circ \sigma)}{\partial s}(r(s), s) \frac{\partial}{\partial x^i}\bigg|_{\sigma(r(s),s)}$$

$$= \frac{dr}{ds}(s) \frac{\partial\sigma}{\partial r}(r(s), s) + \frac{\partial\sigma}{\partial s}(r(s), s),$$

where the second equality follows from Theorem 14.7.1(b), the third equality from Theorem 10.1.10, and the last equality from Theorem 14.9.2. Then

$$
\begin{aligned}
\left\langle \frac{d\lambda}{ds}(s), \frac{d\lambda}{ds}(s) \right\rangle = {}& \left(\frac{dr}{ds}(s) \right)^2 \left\langle \frac{\partial \sigma}{\partial r}(r(s), s), \frac{\partial \sigma}{\partial r}(r(s), s) \right\rangle \\
& + 2\frac{dr}{ds}(s) \left\langle \frac{\partial \sigma}{\partial r}(r(s), s), \frac{\partial \sigma}{\partial s}(r(s), s) \right\rangle \\
& + \left\langle \frac{\partial \sigma}{\partial s}(r(s), s), \frac{\partial \sigma}{\partial s}(r(s), s) \right\rangle.
\end{aligned}
\tag{21.1.4}
$$

We seek alternative expressions for the inner products in the first and second lines of (21.1.4). For the first line, part (a) of Theorem 19.7.4 and (21.1.3) give

$$
\left\langle \frac{\partial \sigma}{\partial r}(r(s), s), \frac{\partial \sigma}{\partial r}(r(s), s) \right\rangle = 1.
\tag{21.1.5}
$$

For the second line, we have from parts (c) and (d) of Theorem 18.11.4 that

$$
\frac{\partial \sigma}{\partial r}(r(s), s) = d_{r(s)\psi(s)}(\exp_p)(\psi(s))
$$

and

$$
\frac{\partial \sigma}{\partial s}(r(s), s) = r(s)\, d_{r(s)\psi(s)}(\exp_p)\left(\frac{d\psi}{ds}(s) \right),
$$

hence

$$
\begin{aligned}
& \left\langle \frac{\partial \sigma}{\partial r}(r(s), s), \frac{\partial \sigma}{\partial s}(r(s), s) \right\rangle \\
& = \left\langle d_{r(s)\psi(s)}(\exp_p)(r(s)\psi(s)), d_{r(s)\psi(s)}(\exp_p)\left(\frac{d\psi}{ds}(s) \right) \right\rangle \\
& = r(s)\left\langle \psi(s), \frac{d\psi}{ds}(s) \right\rangle,
\end{aligned}
\tag{21.1.6}
$$

where the first equality follows from $d_{r(s)\psi(s)}(\exp_p)$ being linear, and the second equality from Theorem 19.7.5. We have from (21.1.3) that

$$
2\left\langle \frac{d\psi}{ds}(s), \psi(s) \right\rangle = \frac{d}{dt}(\langle \psi(s), \psi(s) \rangle) = 0.
\tag{21.1.7}
$$

Then (21.1.6) and (21.1.7) give

$$
\left\langle \frac{\partial \sigma}{\partial r}(r(s), s), \frac{\partial \sigma}{\partial s}(r(s), s) \right\rangle = 0.
\tag{21.1.8}
$$

Substituting (21.1.5) and (21.1.8) into (21.1.4) yields

$$
\left\langle \frac{d\lambda}{ds}(s), \frac{d\lambda}{ds}(s) \right\rangle = \left(\frac{dr}{ds}(s) \right)^2 + \left\langle \frac{\partial \sigma}{\partial s}(r(s), s), \frac{\partial \sigma}{\partial s}(r(s), s) \right\rangle,
$$

which, since M is Riemannian, is equivalent to

$$\left\|\frac{d\lambda}{ds}(s)\right\|^2 = \left|\frac{dr}{ds}(s)\right|^2 + \left\|\frac{\partial\sigma}{\partial s}(r(s),s)\right\|^2.$$

Thus,

$$\left\|\frac{d\lambda}{ds}(s)\right\| \geq \left|\frac{dr}{ds}(s)\right| \qquad (21.1.9)$$

for all s in $(0,1]$. For a real number $0 < \varepsilon < 1$,

$$\begin{aligned}
L(\lambda) &\geq \int_\varepsilon^1 \left\|\frac{d\lambda}{ds}(s)\right\| ds \qquad [(19.2.1)] \\
&\geq \int_\varepsilon^1 \left|\frac{dr}{ds}(s)\right| ds \qquad [(21.1.9)] \\
&\geq \left|\int_\varepsilon^1 \frac{dr}{ds}(s)\, ds\right| \\
&= |r(1) - r(\varepsilon)|, \qquad [\text{FTC}]
\end{aligned} \qquad (21.1.10)$$

where FTC stands for the fundamental theorem of calculus. It follows from (21.1.2) and (21.1.10) that

$$L(\lambda) \geq \lim_{\varepsilon \to 0^+} |r(1) - r(\varepsilon)| = |r(1)| = L(\rho_{p,v}). \qquad \square$$

An observation on Theorem 21.1.1 is that, in geometric terms, (21.1.8) says that the latitude curve σ_r and longitude curve σ_s intersect orthogonally along the radial geodesic σ_s, something that is intuitively clear in certain cases, such as the sphere.

21.2   Classical Vector Calculus Theorems

Section 10.6 presents some of the classical theorems of differential vector calculus. At various points in the book, we encountered their modern counterparts, as summarized in the following table:

Classical	Modern
Theorem 10.6.1(a)	Theorem 20.3.3
Theorem 10.6.1(b)	Theorem 20.4.4
Theorem 10.6.1(c)	Example 16.1.7
Theorem 10.6.1(d)	Example 16.1.8
Theorem 10.6.2	Theorems 16.9.2, 16.9.3, 16.9.8

In this section, we provide modern versions of several of the classical theorems of integral vector calculus.

Let $\lambda(t) : [a, b] \longrightarrow \mathbb{R}_0^3$ be a smooth curve, and let $X = (\alpha^1, \alpha^2, \alpha^3)$ be a vector field in $\mathfrak{X}(\mathbb{R}_0^3)$. The **classical line integral of X** is denoted by $\int_C X \cdot ds$ (among other notations) and defined by

$$\int_C X \cdot ds = \int_a^b \left\langle X_{\lambda(t)}, \frac{d\lambda}{dt}(t) \right\rangle dt,$$

where C is the image of λ. The next result shows that, other than a switch from a vector field to an equivalent covector field, the classical line integral and the line integral introduced in Section 16.8 are essentially the same.

Theorem 21.2.1 (Classical Line Integral). *Continuing with the above setup, let ω be the covector field in $\mathfrak{X}^*(\mathbb{R}^3)$ given by*

$$\omega = \alpha^1 d(x) + \alpha^2 d(y) + \alpha^3 d(z).$$

Then

$$\int_C X \cdot ds = \int_\lambda \omega.$$

Proof. Let $\lambda = (\lambda^1, \lambda^2, \lambda^3)$. We have

$$\omega_{\lambda(t)}\left(\frac{d\lambda}{dt}(t)\right) = \alpha^1(\lambda(t)) \, d(x)\left(\frac{d\lambda}{dt}(t)\right) + \alpha^2(\lambda(t)) \, d(y)\left(\frac{d\lambda}{dt}(t)\right)$$
$$+ \alpha^3(\lambda(t)) \, d(z)\left(\frac{d\lambda}{dt}(t)\right)$$
$$= \alpha^1(\lambda(t)) \frac{d\lambda^1}{dt}(t) + \alpha^2(\lambda(t)) \frac{d\lambda^2}{dt}(t) + \alpha^3(\lambda(t)) \frac{d\lambda^3}{dt}(t)$$
$$= \left\langle X_{\lambda(t)}, \frac{d\lambda}{dt}(t) \right\rangle.$$

The result now follows from (16.8.4).   □

Let U be an open set in $\mathbb{R}^2$, let $\mathcal{S}$ be a set in $\mathbb{R}^3$, let $F = (F^1, F^2, F^3) : U \longrightarrow \mathcal{S}$ be a diffeomorphism that has compact support, and let $X = (\alpha^1, \alpha^2, \alpha^3)$ be a vector field in $\mathfrak{X}(\mathbb{R}_0^3)$. The **classical surface integral of X** is denoted by $\iint_\mathcal{S} X \cdot dA$ (among other notations) and defined by

$$\iint_\mathcal{S} X \cdot dA = \iint_U \left\langle X \circ F, \frac{\partial F}{\partial r} \times \frac{\partial F}{\partial s} \right\rangle dr\,ds,$$

where we note that $\mathcal{S}$ is the image of F.

Theorem 21.2.2 (Classical Surface Integral). *Continuing with the above setup, let ω be the form in $\Lambda^2(\mathbb{R}^3)$ given by*

$$\omega = \alpha^1 d(y) \wedge d(z) + \alpha^2 d(z) \wedge d(x) + \alpha^3 d(x) \wedge d(y).$$

Then

$$\iint_\mathcal{S} X \cdot dA = \int_\mathcal{S} \omega.$$

Proof. Let us rewrite ω as

$$\omega = \alpha^3 d(x) \wedge d(y) - \alpha^2 d(x) \wedge d(z) + \alpha^1 d(y) \wedge d(z).$$

Setting

$$\begin{aligned}
(U, (x^i)) &= (U, (r, s)) \\
(V, (y^j)) &= (\mathbb{R}^3, (x, y, z)) \\
(i_1, i_2) &= (1, 2) \\
(j_1, j_2) &= (1, 2), (1, 3), (2, 3) \\
\alpha_{12} &= \alpha^3 \\
\alpha_{13} &= -\alpha^2 \\
\alpha_{23} &= \alpha^1
\end{aligned}$$

and

$$J_F = \begin{bmatrix} \dfrac{\partial(x \circ F)}{\partial r} & \dfrac{\partial(x \circ F)}{\partial s} \\[2ex] \dfrac{\partial(y \circ F)}{\partial r} & \dfrac{\partial(y \circ F)}{\partial s} \\[2ex] \dfrac{\partial(z \circ F)}{\partial r} & \dfrac{\partial(z \circ F)}{\partial s} \end{bmatrix} = \begin{bmatrix} \dfrac{\partial F^1}{\partial r} & \dfrac{\partial F^1}{\partial s} \\[2ex] \dfrac{\partial F^2}{\partial r} & \dfrac{\partial F^2}{\partial s} \\[2ex] \dfrac{\partial F^3}{\partial r} & \dfrac{\partial F^2}{\partial s} \end{bmatrix}$$

in Theorem 15.13.3(a) gives

$$\begin{aligned}
F^*(\omega) &= \left[(\alpha^3 \circ F) \det\left((J_F)^{(1,2)}\right) + (-\alpha^2 \circ F) \det\left((J_F)^{(1,3)}\right) \right. \\
&\qquad \left. + (\alpha^1 \circ F) \det\left((J_F)^{(2,3)}\right) \right] d(r) \wedge d(s) \\
&= \Big\langle (\alpha^1 \circ F, \alpha^2 \circ F, \alpha^3 \circ F), \\
&\qquad \left(\det\left((J_F)^{(2,3)}\right), -\det\left((J_F)^{(1,3)}\right), \det\left((J_F)^{(1,2)}\right) \right) \Big\rangle d(r) \wedge d(s) \\
&= \left\langle X \circ F, \frac{\partial F}{\partial r} \times \frac{\partial F}{\partial s} \right\rangle d(r) \wedge d(s).
\end{aligned}$$

The result now follows from (16.7.2) and (16.7.3). $\square$

The remaining three results of this section all involve applications of Stokes's theorem.

Let $\lambda(t) : [0, 2\pi] \longrightarrow \mathbb{R}^2$ be the smooth curve given by $\lambda(t) = (\cos(t), \sin(t))$, and let $X = (\alpha^1, \alpha^2)$ be a vector field in $\mathfrak{X}(\mathbb{R}^2)$. The **closed unit disk** is defined by $\overline{D} = \{(r, s) \in \mathbb{R}^2 : r^2 + s^2 \leq 1\}$. We view $\overline{D}$ as a Riemannian 2-manifold with boundary and observe that its boundary $\partial \overline{D}$ is the unit circle, which is parametrized by λ in a counterclockwise direction.

Theorem 21.2.3 (Classical Green's Theorem). *With the above setup,*

$$\iint_{\overline{D}} \left(\frac{\partial \alpha^2}{\partial r} - \frac{\partial \alpha^1}{\partial s} \right) dr ds = \int_{\partial \overline{D}} \omega.$$

Proof. Let ω be the covector field in $\mathfrak{X}^*(\mathbb{R}^2)$ given by $\omega = \alpha^1 d(r) + \alpha^2 d(s)$. By Stokes's theorem,

$$\int_{\overline{D}} d(\omega) = \int_{\partial \overline{D}} \omega.$$

A computation similar to that leading to (16.1.2) gives

$$d(\omega) = \left(\frac{\partial \alpha^2}{\partial r} - \frac{\partial \alpha^1}{\partial s} \right) d(r) \wedge d(s),$$

and then (16.7.2) yields

$$\int_{\overline{D}} d(\omega) = \iint_{\overline{D}} \left(\frac{\partial \alpha^2}{\partial r} - \frac{\partial \alpha^1}{\partial s} \right) dr ds.$$

The result follows. $\square$

The setup for the next result is the same as that for Theorem 19.11.4, except that $(M, \mathfrak{e}|_M)$ is assumed to be compact, something needed to ensure certain integrals exist.

Theorem 21.2.4 (Classical Stokes's Theorem). *Let $(M, \mathfrak{e}|_M)$ be a compact orientable Riemannian hypersurface with boundary of $\mathbb{R}^3_0 = (\mathbb{R}^3, \mathfrak{e})$, let V be a unit normal vector field in $\mathfrak{X}_{\mathbb{R}^3_0}(M)$ (as given by Theorem 19.11.1), and let $\mathfrak{O}_M$ be the orientation of M induced by V, so that $(M, \mathfrak{e}|_M, \mathfrak{O}_M)$ is a compact oriented Riemannian hypersurface with boundary of $\mathbb{R}^3_0$. Let Ω_M be its volume form. Let W be the unique outward-pointing unit normal vector field in $\mathfrak{X}_M(\partial M)$ (as given by Theorem 19.11.3), and let $\mathfrak{O}_{\partial M}$ be the (Stokes) orientation induced by W, so that $(\partial M, \mathfrak{e}|_{\partial M}, \mathfrak{O}_{\partial M})$ is an oriented Riemannian hypersurface of $(M, \mathfrak{e}|_M, \mathfrak{O}_M)$. Let $\Omega_{\partial M}$ be its volume form, and let $U = V \times W$. If X is a vector field in $\mathfrak{X}(\mathbb{R}^3_0)$, then*

$$\int_M \langle \mathrm{Curl}(X), V \rangle \, \Omega_M = \int_{\partial M} \langle X, U \rangle \, \Omega_{\partial M}.$$

Proof. By Stokes's theorem,

$$\int_M d(X^{\mathsf{F}}) = \int_{\partial M} X^{\mathsf{F}}. \tag{21.2.1}$$

We seek alternative expressions for the integrands in (21.2.1). For the left-hand side, Theorem 19.11.2(b) and Theorem 20.5.1 give

$$d(X^{\mathsf{F}}) = i_{\mathrm{Curl}(X)}(\Omega_{\mathbb{R}^3_0})|_M = \langle \mathrm{Curl}(X), V \rangle \, \Omega_M, \tag{21.2.2}$$

where $\Omega_{\mathbb{R}^3_0}$ is the volume form on $\mathbb{R}^3_0$. For the right-hand side, by Theorem 19.11.4(e),

$$\Omega_{\partial M}(U) = 1_{\partial M}. \tag{21.2.3}$$

Since $X^{\mathsf{F}}|_{\partial M}$ is a covector field in $\mathfrak{X}^*(\partial M) = \Lambda^1(\partial M)$, it follows from Theorem 3.3.3, Theorem 15.8.5, and Theorem 19.10.2 that there is a uniquely determined function f in $C^\infty(U)$ such that

$$X^{\mathsf{F}}|_{\partial M} = f\,\Omega_{\partial M}. \tag{21.2.4}$$

We have

$$\begin{aligned}
f &= f\,\Omega_{\partial M}(U) &&[(21.2.3)] \\
&= X^{\mathsf{F}}|_{\partial M}(U) &&[(21.2.4)] \\
&= \langle X|_{\partial M}, U\rangle. &&[(19.4.1)]
\end{aligned} \tag{21.2.5}$$

Then (21.2.4) and (21.2.5) give

$$X^{\mathsf{F}}|_{\partial M} = \langle X|_{\partial M}, U\rangle\,\Omega_{\partial M}. \tag{21.2.6}$$

Substituting (21.2.2) and (21.2.6) into (21.2.1) gives the result. $\qquad\square$

Theorem 21.2.5 (Classical Divergence Theorem). *Let $(M, \mathfrak{e}|_M, \mathfrak{D}_M)$ be a compact oriented Riemannian 3-submanifold with boundary of $\mathbb{R}_0^3 = (\mathbb{R}^3, \mathfrak{e})$, so that $(\partial M, \mathfrak{e}|_{\partial M})$ is a Riemannian hypersurface of $(M, \mathfrak{e}|_M, \mathfrak{D}_M)$, and let X be a vector field in $\mathfrak{X}(M)$. Let V be the unique outward-pointing unit normal vector field in $\mathfrak{X}_M(\partial M)$ (as given by Theorem 19.11.3), and let ∂M have the orientation $\mathfrak{D}_{\partial M}$ induced by V, so that $(\partial M, \mathfrak{e}|_{\partial M}, \mathfrak{D}_{\partial M})$ is an oriented Riemannian hypersurface of $(M, \mathfrak{e}|_M, \mathfrak{D}_M)$. Let $\Omega_{\partial M}$ be its volume form. Then*

$$\int_M \mathrm{Div}(X)\,\Omega_M = \int_{\partial M} \langle X, V\rangle\,\Omega_{\partial M}.$$

Proof. We have

$$\begin{aligned}
\int_M \mathrm{Div}(X)\,\Omega_M &= \int_{\partial M} i_X(\Omega_M) &&[\text{Th } 20.4.5] \\
&= \int_{\partial M} \langle X, V\rangle\,\Omega_{\partial M}. &&[\text{Th } 19.11.2(\text{b})] \qquad\square
\end{aligned}$$

Chapter 22

Applications to Physics

22.1 Linear Isometries on Lorentz Vector Spaces

Let $(V, \mathfrak{g}, \mathfrak{O})$ be an oriented Lorentz vector space, and let $\mathcal{E} = (e_1, \ldots, e_m)$ be an orthonormal basis for V that is positively oriented with respect to $\mathfrak{O}$. We denote by L_m the set of linear isometries on V:

$$\mathsf{L}_m = \{A \in \mathrm{Lin}(V, V) : A \text{ a linear isometry}\},$$

where we recall that $\mathrm{Lin}(V, V)$ is the vector space of linear maps from V to V.

Theorem 22.1.1. L_m *is a group (under composition of linear maps), called the* **Lorentz group on** V.

Proof. Let A be a map in L_m. It follows from Theorem 1.1.9 and Theorem 4.4.1(a) that the inverse A^{-1} exists and is in $\mathrm{Lin}(V, V)$. Let v, w be vectors in V. Then

$$\langle v, w \rangle = \langle A(A^{-1}(v)), A(A^{-1}(w)) \rangle = \langle A^{-1}(v), A^{-1}(w) \rangle,$$

so A^{-1} is in L_m. The rest of the verification that L_m is a group is straightforward. $\square$

Let A be a map in $\mathrm{Lin}(V, V)$. For brevity, we denote $[A]_{\mathcal{E}}^{\mathcal{E}}$ by $[A]$, and likewise omit $\mathcal{E}$ from the notation for other matrices. Corresponding to (4.4.1) and (4.4.2), we have

$$[A] = \begin{bmatrix} [A]_1^1 & [A]_2^1 \\ [A]_1^2 & \gamma \end{bmatrix}$$

Semi-Riemannian Geometry, First Edition. Stephen C. Newman.
© 2019 John Wiley & Sons, Inc. Published 2019 by John Wiley & Sons, Inc.

and

$$[A]^{-1} = \begin{bmatrix} \left([A]_1^1\right)^T & -\left([A]_1^2\right)^T \\ -\left([A]_2^1\right)^T & \gamma \end{bmatrix},$$

respectively, where

$$[A]_1^1 = \begin{bmatrix} a_1^1 & \cdots & a_{m-1}^1 \\ \vdots & \ddots & \vdots \\ a_1^{m-1} & \cdots & a_{m-1}^{m-1} \end{bmatrix},$$

$$[A]_2^1 = \begin{bmatrix} a_m^1 \\ \vdots \\ a_m^{m-1} \end{bmatrix},$$

and

$$[A]_1^2 = \begin{bmatrix} a_1^m & \cdots & a_{m-1}^m \end{bmatrix}.$$

The ring isomorphism given by Theorem 2.2.3(b) restricts to a (multiplicative) group isomorphism between L_m and a subgroup of the general linear group $\mathrm{GL}(m) \subset \mathrm{Mult}_{m \times m}$. In what follows, but without further mention, this group isomorphism is used to prove assertions about the group structure of L_m. In particular, we rely on the following: a subset of L_m is a subgroup if and only if the corresponding subset of $\mathrm{GL}(m)$ is a subgroup.

Theorem 22.1.2. *If A is a map in L_m, then:*
(a)

$$[A]_1^1\left([A]_1^1\right)^T = I_{m-1} + [A]_2^1\left([A]_2^1\right)^T = I_{m-1} + \left([A]_1^2\right)^T [A]_1^2.$$

(b)

$$\gamma[A]_2^1 = [A]_1^1\left([A]_1^2\right)^T.$$

(c)

$$\gamma[A]_1^2 = \left([A]_2^1\right)^T [A]_1^1.$$

(d)

$$\gamma^2 = 1 + \left([A]_2^1\right)^T [A]_2^1 = 1 + [A]_1^2\left([A]_1^2\right)^T.$$

Proof. (a): This follows from parts (a) and (d) of Theorem 4.4.6.
    (b): This follows from Theorem 4.4.6(b).
    (c): This follows from Theorem 4.4.6(e).
    (d): This follows from parts (c) and (f) of Theorem 4.4.6.   □

It follows from Theorem 22.1.2(d) that

$$\gamma^2 \geq 1. \tag{22.1.1}$$

Let us define

$$\beta = \frac{\sqrt{\gamma^2 - 1}}{\gamma}, \tag{22.1.2}$$

where we note that, by definition, γ and β have the same sign. Then

$$-1 < \beta < 1 \qquad \text{and} \qquad \gamma^2(1 - \beta^2) = 1. \tag{22.1.3}$$

When $\gamma \geq 1$, which is the case of greatest interest,

$$0 \leq \beta < 1 \qquad \text{and} \qquad \gamma = \frac{1}{\sqrt{1 - \beta^2}}.$$

Given a map A in L_m, we have from Theorem 4.4.3 that $\det(A) = \pm 1$, and from (22.1.1) that $\gamma \geq 1$ or $\gamma \leq -1$. Consider the sets

$$\mathsf{L}_m^+ = \{A \in \mathsf{L}_m : \det(A) = 1\},$$

$$\mathsf{L}_m^\uparrow = \{A \in \mathsf{L}_m : \gamma \geq 1\},$$

and

$$\mathsf{L}_m^{+\uparrow} = \mathsf{L}_m^+ \cap \mathsf{L}_m^\uparrow.$$

In the present context, an orientation-preserving (orientation-reversing) map is said to be **proper (improper)**.

Theorem 22.1.3. *Let A be a map in L_m. Then A is in L_m^+ if and only if A is proper.*

Proof. Since $\det(A) = \pm 1$, the result follows from Theorem 8.2.8. $\qquad \square$

Let us recall the convention adopted in Section 4.9 that for a given orthonormal basis $(e_1, \ldots, e_m)$ for V, the future time cone $\mathcal{T}^+$ is defined to be the one containing the timelike unit vector e_m.

Theorem 22.1.4. *If A is a map in L_m, then the following are equivalent:*
(a) *A is in $\mathsf{L}_m^\uparrow$.*
(b) *$A(e_m)$ is in $\mathcal{T}^+$.*
(c) *A is orthochronous.*

Proof. (a) $\Leftrightarrow$ (b): According to the above convention on e_m, we have from (4.9.8) that $A(e_m)$ is in $\mathcal{T}^+$ if and only if $\langle A(e_m), e_m \rangle < 0$. Since $[e_m] = \begin{bmatrix} 0 & \cdots & 0 & 1 \end{bmatrix}^{\mathrm{T}}$, by Theorem 2.2.4,

$$[A(e_m)] = \begin{bmatrix} [A]\frac{1}{2} \\ \gamma \end{bmatrix},$$

and because $\mathfrak{g}_\varepsilon = \mathrm{diag}(1, \ldots, 1, -1)$, by Theorem 3.1.2(a),

$$\langle A(e_m), e_m \rangle = [A(e_m)]^{\mathrm{T}} \mathfrak{g}_\varepsilon [e_m] = -\gamma.$$

It follows that $\langle A(e_m), e_m \rangle < 0$ if and only if $\gamma > 0$. Since $\gamma^2 \geq 1$, we have $\gamma > 0$ if and only if $\gamma \geq 1$. Thus, $A(e_m)$ is in $\mathcal{T}^+$ if and only if $\gamma \geq 1$.

(b) $\Leftrightarrow$ (c): Given the above convention on e_m, this follows from Theorem 4.9.6. $\qquad \square$

In view of the preceding two results, we can characterize L_m^+, $\mathsf{L}_m^\uparrow$, and $\mathsf{L}_m^{+\uparrow}$ as follows:

$$\mathsf{L}_m^+ = \{A \in \mathsf{L}_m : A \text{ is proper}\},$$

$$\mathsf{L}_m^\uparrow = \{A \in \mathsf{L}_m : A \text{ is orthochronous}\},$$

and

$$\mathsf{L}_m^{+\uparrow} = \{A \in \mathsf{L}_m : A \text{ is proper and orthochronous}\}.$$

Theorem 22.1.5. L_m^+, $\mathsf{L}_m^\uparrow$, *and* $\mathsf{L}_m^{+\uparrow}$ *are subgroups of* L_m.

Proof. It is easily shown that L_m^+ is a subgroup of L_m. Using Theorem 4.9.6, we see that $\mathsf{L}_m^\uparrow$ is a subgroup of L_m. Since the intersection of subgroups is a subgroup, the result for $\mathsf{L}_m^{+\uparrow}$ is immediate. □

In the physics literature, $\mathsf{L}_m^{+\uparrow}$ is called the **proper orthochronous Lorentz group on V**.

Recall from Section 2.4 that the orthogonal group $O(m)$ consists of all matrices P in $\mathrm{Mat}_{m \times m}$ such that $P^\mathrm{T} P = I_m$, and the special orthogonal group $SO(m)$ is the subgroup of $O(m)$ consisting of those matrices P such that $\det(P) = 1$.

We say that a map A in $\mathrm{Lin}(V, V)$ is a **rotation** (with respect to $\mathcal{E}$) if $[A]$ is of the form

$$[A] = \begin{bmatrix} [A]_1^1 & O_{(m-1)\times 1} \\ O_{1\times(m-1)} & 1 \end{bmatrix},$$

where $[A]_1^1$ is in $SO(m-1)$. By definition, $\det\big([A]_1^1\big) = 1$, so $\det(A) = 1$. The set of rotations is

$$\mathsf{R}_m = \{A \in \mathrm{Lin}(V, V) : A \text{ a rotation}\}.$$

Theorem 22.1.6. R_m *is a subgroup of* $\mathsf{L}_m^{+\uparrow}$.

Proof. Since $SO(m-1)$ is a group, it is easily shown that so is R_m. Let A be a map in R_m. We have from $\mathfrak{g}_\mathcal{E} = \mathrm{diag}(1, \ldots, 1, -1)$ that

$$[A]^\mathrm{T} \mathfrak{g}_\mathcal{E} [A] = \begin{bmatrix} \big([A]_1^1\big)^\mathrm{T} [A]_1^1 & O_{(m-1)\times 1} \\ O_{1\times(m-1)} & -1 \end{bmatrix} = \begin{bmatrix} I_{m-1} & O_{(m-1)\times 1} \\ O_{1\times(m-1)} & -1 \end{bmatrix} = \mathfrak{g}_\mathcal{E}.$$

By Theorem 4.4.2, A is a linear isometry, and is therefore in L_m. Since $\det(A) = 1$, A is in L_m^+. By definition, $\gamma = 1$, so A is in $\mathsf{L}_m^\uparrow$. This shows that A is in $\mathsf{L}_m^{+\uparrow}$. Thus, $\mathsf{R}_m \subseteq \mathsf{L}_m^{+\uparrow}$. □

Theorem 22.1.7. *If A is a map in $\mathsf{L}_m^{+\uparrow}$, then the following are equivalent:*
(a) *A is in R_m.*
(b) *$\gamma = 1$.*
(c) *$[A]_2^1 = O_{(m-1)\times 1}$.*
(d) *$[A]_1^2 = O_{1\times(m-1)}$.*

Proof. (b) $\Rightarrow$ (c): We have from Theorem 22.1.2(d) that

$$(a_m^1)^2 + \cdots + (a_m^{m-1})^2 = 0,$$

from which the result follows.

(c) $\Rightarrow$ (b): By Theorem 22.1.2(d), $\gamma^2 = 1$, hence $\gamma = \pm 1$. Since A is in $\mathsf{L}_m^\uparrow$, we have $\gamma = 1$.

(c) $\Leftrightarrow$ (d): We have from Theorem 22.1.2(d) that

$$(a_m^1)^2 + \cdots + (a_m^{m-1})^2 = (a_1^m)^2 + \cdots + (a_{m-1}^m)^2,$$

from which the result follows.

(a) $\Rightarrow$ (b): By definition.

(b) $\Rightarrow$ (a): We have just shown that (b), (c), and (d) are equivalent, so $[A]$ is of the form

$$[A] = \begin{bmatrix} [A]_1^1 & O_{(m-1)\times 1} \\ O_{1\times(m-1)} & 1 \end{bmatrix}$$

for some matrix $[A]_1^1$ in $\mathrm{Mat}_{(m-1)\times(m-1)}$. By Theorem 22.1.2(a), $[A]_1^1([A]_1^1)^\mathsf{T} = I_{m-1}$, hence $[A]_1^1$ is in $\mathrm{O}(m-1)$. Since A is in L_m^+, we have $\det([A]_1^1) = \det([A]) = 1$, so $[A]_1^1$ is in $\mathrm{SO}(m-1)$. Thus, A is in R_m. $\qquad\square$

We say that a map A in $\mathrm{Lin}(V, V)$ is a **boost** (with respect to $\mathcal{E}$ in the $(m-1)$st spacelike direction) if $[A]$ is of the form

$$[A] = \begin{bmatrix} I_{m-2} & O_{(m-2)\times 2} \\ O_{2\times(m-2)} & \begin{bmatrix} \gamma & -\beta\gamma \\ -\beta\gamma & \gamma \end{bmatrix} \end{bmatrix},$$

where $\gamma \geq 1$ and β is computed using (22.1.2). For convenience of notation, we have defined boosts in the $(m-1)$st spacelike direction, but any other spacelike direction would do. For instance, a boost in the first spacelike direction takes the form

$$\begin{bmatrix} \gamma & O_{1\times(m-2)} & -\beta\gamma \\ O_{(m-2)\times 1} & I_{m-2} & O_{(m-2)\times 1} \\ -\beta\gamma & O_{1\times(m-2)} & \gamma \end{bmatrix}.$$

The set of boosts is

$$\mathsf{B}_m = \{A \in \mathrm{Lin}(V, V) : A \text{ a boost}\}.$$

It is clear that B_m contains the identity map. Let A' be another boost, with

$$[A'] = \begin{bmatrix} I_{m-2} & O_{(m-2)\times 2} \\ O_{2\times(m-2)} & \begin{bmatrix} \gamma' & -\beta'\gamma' \\ -\beta'\gamma' & \gamma' \end{bmatrix} \end{bmatrix}.$$

Then

$$[A][A'] = \begin{bmatrix} I_{m-2} & O_{(m-2)\times 2} \\ O_{2\times(m-2)} & \begin{bmatrix} \gamma'' & -\beta''\gamma'' \\ -\beta''\gamma'' & \gamma'' \end{bmatrix} \end{bmatrix},$$

where

$$\gamma'' = \gamma\gamma'(1 + \beta\beta') \qquad \text{and} \qquad \beta'' = \frac{\beta + \beta'}{1 + \beta\beta'}.$$

It is easily shown that $\gamma'' \geq 1$, and that γ'' and β'' satisfy (22.1.2). Thus, $A \circ A'$ is a boost. This demonstrates that B_m is closed under multiplication. However,

$$\begin{bmatrix} I_{m-2} & O_{(m-2)\times 2} \\ O_{2\times(m-2)} & \begin{bmatrix} \gamma & -\beta\gamma \\ -\beta\gamma & \gamma \end{bmatrix} \end{bmatrix}^{-1} = \begin{bmatrix} I_{m-2} & O_{(m-2)\times 2} \\ O_{2\times(m-2)} & \begin{bmatrix} \gamma & \beta\gamma \\ \beta\gamma & \gamma \end{bmatrix} \end{bmatrix}.$$

Since the matrix on the right-hand side is missing the necessary minus signs to make it a boost, B_m does not contain inverses (except in the case of the identity map). Thus, B_m is not a group.

Theorem 22.1.8. B_m *is a subset of* L_m.

Proof. Let A be a map in B_m. We have from $\mathfrak{g}_\varepsilon = \text{diag}(1,\ldots,1,-1)$ and the identity in (22.1.3) that

$$\begin{aligned}
[A]^{\mathrm{T}}\,\mathfrak{g}_\varepsilon\,[A] &= \begin{bmatrix} I_{m-2} & O_{(m-2)\times 2} \\ O_{2\times(m-2)} & \begin{bmatrix} \gamma & \beta\gamma \\ -\beta\gamma & -\gamma \end{bmatrix}\begin{bmatrix} \gamma & -\beta\gamma \\ -\beta\gamma & \gamma \end{bmatrix} \end{bmatrix} \\
&= \begin{bmatrix} I_{m-2} & O_{(m-2)\times 2} \\ O_{2\times(m-2)} & \begin{bmatrix} 1 & 0 \\ 0 & -1 \end{bmatrix} \end{bmatrix} = \mathfrak{g}_\varepsilon.
\end{aligned}$$

By Theorem 4.4.2, A is a linear isometry and is therefore in L_m.   □

Theorem 22.1.9. $\mathsf{R}_m \cap \mathsf{B}_m = \{\text{id}_V\}$.

Proof. If A is in $\mathsf{R}_m \cap \mathsf{B}_m$, then, in the above notation,

$$\begin{bmatrix} [A]^1_1 & O_{(m-1)\times 1} \\ O_{1\times(m-1)} & 1 \end{bmatrix} = [A] = \begin{bmatrix} I_{m-2} & O_{(m-2)\times 2} \\ O_{2\times(m-2)} & \begin{bmatrix} \gamma & -\beta\gamma \\ -\beta\gamma & \gamma \end{bmatrix} \end{bmatrix},$$

where, by definition, $[A]^1_1$ is in $\text{SO}(m-1)$, $\gamma \geq 1$, and β is computed using (22.1.2). It follows that $\gamma = 1$, and then from (22.1.2) that $\beta = 0$. Thus, $[A] = I_m$.   □

Based on Theorem 22.1.5, Theorem 22.1.6, Theorem 22.1.8, and Theorem 22.1.9, Figure 22.1.1 depicts the relationships between R_m, B_m, and $\mathsf{L}^{+\uparrow}_m$ as subsets of L_m, where the dot represents id_V.

Recall from Section 2.4 the vector space $\text{Mat}_{m\times 1}$ of column matrices and its basis $\mathbb{E}_m = (E_1,\ldots,E_m)$. We make $\text{Mat}_{m\times 1}$ into an oriented inner product space by choosing the orientation $\mathbb{O}_m = [\mathbb{E}_m]$ and defining the inner product using matrix multiplication as follows: for matrices P, Q in $\text{Mat}_{m\times 1}$, the inner product is $P^{\mathrm{T}}Q$. Endowed with this structure, we identify $\text{Mat}_{m\times 1}$ in an obvious way with the inner product space $\mathbb{R}^m_0$.

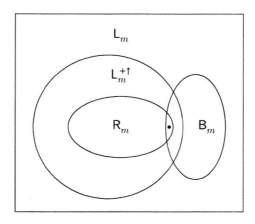

Figure 22.1.1. R_m, B_m, $\mathsf{L}_m^{+\uparrow}$, and L_m

Theorem 22.1.10. *Let $P_1, \ldots, P_{m-1}$ be matrices in $\mathrm{Mat}_{(m-1)\times 1}$, and consider the matrices*

$$\mathbb{P} = (P_1, \ldots, P_{m-1}), \qquad P_1^1 = \begin{bmatrix} P_1 & \cdots & P_{m-1} \end{bmatrix},$$

and

$$P = \begin{bmatrix} P_1^1 & O_{(m-1)\times 1} \\ O_{1\times(m-1)} & 1 \end{bmatrix}.$$

Then:
(a) *$\mathbb{P}$ is an orthonormal basis for $\mathrm{Mat}_{(m-1)\times 1}$ if and only if P_1^1 is in $\mathrm{O}(m-1)$.*
(b) *$\mathbb{P}$ is a positively oriented orthonormal basis for $\mathrm{Mat}_{(m-1)\times 1}$ if and only if P_1^1 is in $\mathrm{SO}(m-1)$.*
(c) *If $\mathbb{P}$ is a positively oriented orthonormal basis for $\mathrm{Mat}_{(m-1)\times 1}$ and A is a map in $\mathrm{Lin}(V, V)$ such that $[A] = P$ or $[A] = P^{\mathrm{T}}$, then A is in R_m.*

Proof. For brevity, let $\mathbb{M} = \mathrm{Mat}_{(m-1)\times 1}$ and $\mathbb{E} = \mathbb{E}_{m-1}$. We have from (2.2.7) that

$$P_1^1 = \begin{bmatrix} [P_1]_{\mathbb{E}} & \cdots & [P_{m-1}]_{\mathbb{E}} \end{bmatrix} = [\mathrm{id}_{\mathbb{M}}]_{\mathbb{P}}^{\mathbb{E}},$$

hence

$$\det(P_1^1) = \det\left([\mathrm{id}_{\mathbb{M}}]_{\mathbb{P}}^{\mathbb{E}}\right). \tag{22.1.4}$$

It follows that if $\mathbb{P}$ is a basis for $\mathbb{M}$, then: $\mathbb{P}$ is consistent with $\mathbb{E}$ (hence positively oriented) if and only if $\det(P_1^1) > 0$.

(a): By definition, P_1^1 is in $\mathrm{O}(m-1)$ if and only if $(P_1^1)^{\mathrm{T}}P_1^1 = I_{m-1}$, which is equivalent to $(P_i)^{\mathrm{T}}P_j = \delta_j^i$ for $i, j = 1, \ldots, m-1$.

(b)($\Rightarrow$): Since $\mathbb{P}$ is an orthonormal basis, we have from part (a) that P_1^1 is in $\mathrm{O}(m-1)$. Then

$$1 = \det(I_{m-1}) = \det\left((P_1^1)^{\mathrm{T}}P_1^1\right) = \det(P_1^1)^2,$$

so $\det(P_1^1) = \pm 1$. Since $\mathbb{P}$ is positively oriented, hence consistent with $\mathbb{E}$, it follows from (22.1.4) that $\det(P_1^1) = 1$. Thus, P_1^1 is in $SO(m-1)$.

(b)($\Leftarrow$): Since P_1^1 is in $SO(m-1)$, which is a subgroup of $O(m-1)$, we have from part (a) that $\mathbb{P}$ is an orthonormal basis for $\mathbb{M}$. Also, because P_1^1 is in $SO(m-1)$, we have $\det(P_1^1) = 1$. It follows from (22.1.4) that $\mathbb{P}$ is consistent with $\mathbb{E}$, hence positively oriented.

(c): If $[A] = P$, the result follows from part (b). Suppose $[A] = P^T$, in which case

$$[A] = \begin{bmatrix} (P_1^1)^T & O_{(m-1)\times 1} \\ O_{1\times(m-1)} & 1 \end{bmatrix}.$$

Let $(P_1^1)^T = \begin{bmatrix} Q_1 & \cdots & Q_{m-1} \end{bmatrix}$, where $Q_1,\ldots,Q_{m-1}$ are matrices in $\mathbb{M}$, and let $\mathbb{Q} = (Q_1,\ldots,Q_{m-1})$. Since $\mathbb{P}$ is a basis for $\mathbb{M}$, we have from Theorem 2.3.2 that so is $\mathbb{Q}$. Furthermore, because $\mathbb{P}$ is orthonormal, $\mathbb{Q}$ is too. Arguing as above gives

$$(P_1^1)^T = \begin{bmatrix} [Q_1]_\mathbb{E} & \cdots & [Q_{m-1}]_\mathbb{E} \end{bmatrix} = \begin{bmatrix} \mathrm{id}_\mathbb{M} \end{bmatrix}_\mathbb{Q}^\mathbb{E},$$

hence $\det(P_1^1) = \det\left(\begin{bmatrix}\mathrm{id}_\mathbb{M}\end{bmatrix}_\mathbb{Q}^\mathbb{E}\right)$. Since $\mathbb{P}$ is positively oriented, it follows from (22.1.4) that so is $\mathbb{Q}$. We have shown that $\mathbb{Q}$ is a positively oriented orthonormal basis for $\mathbb{M}$. The result now follows from part (b). $\square$

We close this section by showing that any map in a proper orthochronous Lorentz group can be expressed as the composition of a rotation, followed by a boost, followed by another rotation.

Theorem 22.1.11. *Any map A in $\mathsf{L}_m^{+\uparrow}$ can be expressed as $A = R_1 \circ B \circ R_2$, where R_1, R_2 are maps in R_m and B is a map in B_m.*

Proof. If $\left([A]\frac{1}{2}\right)^T[A]\frac{1}{2} = 0$, then $[A]\frac{1}{2} = O_{(m-1)\times 1}$. By Theorem 22.1.7, A is in R_m, so in this case we set $R_1 = A$ and $B = R_2 = \mathrm{id}_V$.

Now suppose $\left([A]\frac{1}{2}\right)^T[A]\frac{1}{2} \neq 0$, and let

$$P_{m-1} = -\frac{1}{\beta\gamma}[A]\frac{1}{2}, \tag{22.1.5}$$

where we recall that β is computed from γ using (22.1.2). It follows from Theorem 22.1.2(d) and the identity in (22.1.3) that P_m is a unit vector in $\mathrm{Mat}_{(m-1)\times 1}$. As in the proof of Theorem 4.2.4, we extend P_{m-1} to an orthonormal basis $(P_1,\ldots,P_{m-1})$ for $\mathrm{Mat}_{(m-1)\times 1}$. Reordering $P_1,\ldots,P_{m-2}$ if necessary, suppose $(P_1,\ldots,P_{m-1})$ is positively oriented. Let

$$P_1^1 = \begin{bmatrix} P_1 & \cdots & P_{m-1} \end{bmatrix}^T \tag{22.1.6}$$

and

$$P = \begin{bmatrix} P_1^1 & O_{(m-1)\times 1} \\ O_{1\times(m-1)} & 1 \end{bmatrix}. \tag{22.1.7}$$

Then (22.1.5) gives

$$P_1^1[A]_2^1 = -\beta\gamma \begin{bmatrix} (P_1)^{\mathrm{T}} P_{m-1} \\ \vdots \\ (P_{m-1})^{\mathrm{T}} P_{m-1} \end{bmatrix} = \begin{bmatrix} 0 & \cdots & 0 & -\beta\gamma \end{bmatrix}^{\mathrm{T}}. \qquad (22.1.8)$$

Let

$$P_1^1[A]_1^1 = \begin{bmatrix} Q_1 & \cdots & Q_{m-1} \end{bmatrix}^{\mathrm{T}}, \qquad (22.1.9)$$

where $Q_1, \ldots, Q_{m-1}$ are matrices in $\mathrm{Mat}_{(m-1)\times 1}$. We have

$$
\begin{aligned}
\begin{bmatrix} (Q_1)^{\mathrm{T}} \\ \vdots \\ (Q_{m-1})^{\mathrm{T}} \end{bmatrix} & \begin{bmatrix} Q_1 & \cdots & Q_{m-1} \end{bmatrix} \\
&= \left(P_1^1[A]_1^1 \right) \left(P_1^1[A]_1^1 \right)^{\mathrm{T}} \\
&= P_1^1 \left([A]_1^1 ([A]_1^1)^{\mathrm{T}} \right) (P_1^1)^{\mathrm{T}} \\
&= P_1^1 \left(I_{m-1} + [A]_2^1 ([A]_2^1)^{\mathrm{T}} \right) (P_1^1)^{\mathrm{T}} \qquad \text{[Th 22.1.2(a)]} \\
&= I_{m-1} + \left(P_1^1[A]_2^1 \right) \left(P_1^1[A]_2^1 \right)^{\mathrm{T}} \qquad \text{[Th 22.1.10(a)]} \\
&= I_{m-1} + \mathrm{diag}(0, \ldots, 0, \beta^2\gamma^2) \qquad \text{[(22.1.8)]} \\
&= \begin{bmatrix} I_{m-2} & O_{(m-2)\times 1} \\ O_{1\times(m-2)} & \gamma^2 \end{bmatrix}, \qquad \text{[(22.1.3)]}
\end{aligned}
\qquad (22.1.10)
$$

which shows that $(Q_1, \ldots, Q_{m-1})$ is "close" to being an orthonormal basis for $\mathrm{Mat}_{(m-1)\times 1}$. Since A is in $\mathrm{L}_m^{+\uparrow}$, we have, by definition, that $\gamma \geq 1$. Let $\varepsilon = \pm 1$, and let

$$R_{m-1} = \frac{\varepsilon}{\gamma} Q_{m-1}.$$

We see from (22.1.10) that R_{m-1} is a unit vector and

$$(Q_{m-1})^{\mathrm{T}} R_{m-1} = \varepsilon\gamma. \qquad (22.1.11)$$

For an appropriate choice of ε, $(Q_1, \ldots, Q_{m-2}, R_{m-1})$ is a positively oriented orthonormal basis for $\mathrm{Mat}_{(m-1)\times 1}$. (We will see shortly that, in fact, $\varepsilon = 1$.) Let

$$R_1^1 = \begin{bmatrix} Q_1 & \cdots & Q_{m-2} & R_{m-1} \end{bmatrix}^{\mathrm{T}} \qquad (22.1.12)$$

and

$$R = \begin{bmatrix} R_1^1 & O_{(m-1)\times 1} \\ O_{1\times(m-1)} & 1 \end{bmatrix}. \qquad (22.1.13)$$

By Theorem 2.2.3(a), there are maps F and G in $\mathrm{Lin}(V,V)$ such that

$$[F] = P \quad \text{and} \quad [G] = R^{\mathrm{T}}, \qquad (22.1.14)$$

hence

$$[F \circ A \circ G] = [F][A][G] \qquad\qquad \text{[Th 2.2.2(c)]}$$
$$= P[A]R^{\mathrm{T}} \qquad\qquad \text{[(22.1.14)]}$$
$$= \begin{bmatrix} P_1^1[A]_1^1(R_1^1)^{\mathrm{T}} & P_1^1[A]_2^1 \\ [A]_1^2(R_1^1)^{\mathrm{T}} & \gamma \end{bmatrix}. \qquad \text{[(22.1.7), (22.1.13)]}$$

$(22.1.15)$

We seek alternative expressions for $P_1^1[A]_1^1(R_1^1)^{\mathrm{T}}$ and $[A]_1^2(R_1^1)^{\mathrm{T}}$. For the former,

$$P_1^1[A]_1^1(R_1^1)^{\mathrm{T}}$$
$$= \begin{bmatrix} (Q_1)^{\mathrm{T}} \\ \vdots \\ (Q_{m-2})^{\mathrm{T}} \\ (Q_{m-1})^{\mathrm{T}} \end{bmatrix} \begin{bmatrix} Q_1 & \cdots & Q_{m-2} & R_{m-1} \end{bmatrix} \quad \text{[(22.1.9), (22.1.12)]} \qquad (22.1.16)$$
$$= \begin{bmatrix} I_{m-2} & O_{(m-2)\times 1} \\ O_{1\times(m-2)} & \varepsilon\gamma \end{bmatrix}. \qquad\qquad \text{[(22.1.10), (22.1.11)]}$$

For the latter,

$$\begin{bmatrix} Q_1 & \cdots & Q_{m-1} \end{bmatrix} = \left([A]_1^1\right)^{\mathrm{T}}(P_1^1)^{\mathrm{T}} \qquad\qquad \text{[(22.1.9)]}$$
$$= \left[\left([A]_1^1\right)^{\mathrm{T}}P_1 \quad \cdots \quad \left([A]_1^1\right)^{\mathrm{T}}P_{m-1} \right], \qquad \text{[(22.1.6)]}$$

hence

$$Q_{m-1} = \left([A]_1^1\right)^{\mathrm{T}}P_{m-1}$$
$$= -\frac{1}{\beta\gamma}\left([A]_1^1\right)^{\mathrm{T}}[A]_2^1 \qquad \text{[(22.1.5)]}$$
$$= -\frac{1}{\beta}\left([A]_1^2\right)^{\mathrm{T}}, \qquad\qquad \text{[Th 22.1.2(c)]}$$

so

$$[A]_1^2 = -\beta(Q_{m-1})^{\mathrm{T}}. \qquad\qquad (22.1.17)$$

It follows from (22.1.10) and (22.1.17) that

$$[A]_1^2 Q_i = -\beta(Q_{m-1})^{\mathrm{T}}Q_i = 0$$

for $i = 1, \ldots, m-2$, and from (22.1.11) and (22.1.17) that

$$[A]_1^2 R_{m-1} = -\beta(Q_{m-1})^{\mathrm{T}}R_{m-1} = -\varepsilon\beta\gamma,$$

hence

$$[A]_1^2 (R_1^1)^{\mathrm{T}}$$
$$= \Big[[A]_1^2 Q_1 \quad \cdots \quad [A]_1^2 Q_{m-2} \quad [A]_1^2 R_{m-1} \Big] \qquad [(22.1.12)] \qquad (22.1.18)$$
$$= \big[0 \quad \cdots \quad 0 \quad -\varepsilon\beta\gamma \big].$$

Substituting (22.1.8), (22.1.16), and (22.1.18) into (22.1.15) yields

$$[F \circ A \circ G] =
\begin{bmatrix}
\begin{bmatrix} I_{m-2} & O_{(m-2)\times 1} \\ O_{1\times(m-2)} & \varepsilon\gamma \end{bmatrix} &
\begin{bmatrix} 0 \\ \vdots \\ 0 \\ -\beta\gamma \\ \gamma \end{bmatrix} \\[20pt]
\begin{bmatrix} 0 & \cdots & 0 & -\varepsilon\beta\gamma \end{bmatrix}
\end{bmatrix} \qquad (22.1.19)$$

$$= \begin{bmatrix}
I_{m-2} & O_{(m-2)\times 2} \\
O_{2\times(m-2)} & \begin{bmatrix} \varepsilon\gamma & -\beta\gamma \\ -\varepsilon\beta\gamma & \gamma \end{bmatrix}
\end{bmatrix}.$$

By construction, $(P_1,\dots,P_{m-1})$ and $(Q_1,\dots,Q_{m-2},R_{m-1})$ are positively oriented orthonormal bases for $\mathrm{Mat}_{(m-1)\times 1}$. It follows from (22.1.7), Theorem 22.1.10(b), and (22.1.13) that

$$\det(P) = \det(R) = 1. \qquad (22.1.20)$$

Since A is in $\mathsf{L}_m^{+\uparrow}$,

$$\begin{aligned}
1 &= \det\big([A]\big) \\
&= \det\big(P[A]R^{\mathrm{T}}\big) && [(22.1.20)] \\
&= \det\big([F][A][G]\big) && [(22.1.14)] \\
&= \det\big([F \circ A \circ G]\big) && [\text{Th } 2.2.2(\text{c})] \\
&= \varepsilon(\gamma^2 - \beta^2\gamma^2) && [(22.1.19)] \\
&= \varepsilon, && [(22.1.3)]
\end{aligned}$$

so (22.1.19) becomes

$$[F \circ A \circ G] = \begin{bmatrix}
I_{m-2} & O_{(m-2)\times 2} \\
O_{2\times(m-2)} & \begin{bmatrix} \gamma & -\beta\gamma \\ -\beta\gamma & \gamma \end{bmatrix}
\end{bmatrix}.$$

By definition, β and γ satisfy (22.1.2). Thus, $F \circ A \circ G$ is in B_m.

We have from (22.1.6), (22.1.7), Theorem 22.1.10(c), and (22.1.14) that F is in R_m. Similarly, we have from (22.1.12), (22.1.13), Theorem 22.1.10(c), and (22.1.14) that G is in R_m. It follows from Theorem 22.1.6 that F^{-1} and G^{-1} are in R_m. Setting $B = F \circ A \circ G$, $R_1 = F^{-1}$, and $R_2 = G^{-1}$ gives $A = R_1 \circ B \circ R_2$, which is the desired expression for A in terms of two rotations and a boost. $\square$

22.2   Maxwell's Equations

Let (x, y, z, t) be standard coordinates on $\mathbb{R}_0^3 \times \mathbb{R}$. In what follows, we think of x, y, z as "space" variables and t as a "time" variable. Using the differential operators in Section 10.6, an **electromagnetic field** in $\mathbb{R}_0^3 \times \mathbb{R}$ is characterized by **Maxwell's equations**:

$$[M1] \qquad \operatorname{div}(E) = \frac{1}{\epsilon_0}\rho$$

$$[M2] \qquad \operatorname{div}(B) = 0$$

$$[M3] \qquad \operatorname{curl}(E) = -\frac{\partial B}{\partial t}$$

$$[M4] \qquad \operatorname{curl}(B) = \mu_0 J + \mu_0\epsilon_0\frac{\partial E}{\partial t},$$

where the vector fields E, B, J in $\mathfrak{X}(\mathbb{R}_0^3 \times \mathbb{R})$ are the **electrical field, magnetic field**, and **current density**, respectively; the function ρ in $C^\infty(\mathbb{R}_0^3 \times \mathbb{R})$ is the **charge density**; and the real numbers ϵ_0 and μ_0 are the **permittivity of free space** and the **permeability of free space**, respectively. We interpret the operators div and curl in the above identities as applying only to the space variables, and likewise for other vector calculus operators to follow. Let us assume units have been chosen so that $c = 1$, where $c = \mu_0\epsilon_0$ is the speed of light. Thus, [M4] becomes

$$[M4'] \qquad \operatorname{curl}(B) = \mu_0 J + \frac{\partial E}{\partial t} = \frac{1}{\epsilon_0}J + \frac{\partial E}{\partial t}.$$

We observe that J and ρ are not independent of each other:

$$
\begin{aligned}
0 &= \operatorname{div}\big(\operatorname{curl}(B)\big) && \text{[Th 10.6.1(d)]} \\
&= \operatorname{div}\left(\frac{\partial E}{\partial t} + \mu_0 J\right) && \text{[M4']} \\
&= \frac{\partial\big(\operatorname{div}(E)\big)}{\partial t} + \mu_0\operatorname{div}(J) \\
&= \frac{1}{\epsilon_0}\frac{\partial\rho}{\partial t} + \mu_0\operatorname{div}(J). && \text{[M1]}
\end{aligned}
$$

Then $\mu_0\epsilon_0 = 1$ gives

$$\operatorname{div}(J) + \frac{\partial\rho}{\partial t} = 0, \qquad\qquad (22.2.1)$$

which is called the **continuity equation**.

The rest of this section is devoted to finding alternative expressions for Maxwell's equations, especially in terms of differential forms. This is an opportunity to showcase some of the computational tools developed earlier.

Let us reformulate Maxwell's equations using what are referred to as potentials. It follows from Theorem 10.6.3 and [M2] that there is a vector field A in $\mathfrak{X}(\mathbb{R}_0^3 \times \mathbb{R})$, called the **magnetic potential**, such that

$$B = \operatorname{curl}(A). \qquad\qquad (22.2.2)$$

Then [M3] gives

$$\operatorname{curl}(E) = -\operatorname{curl}\left(\frac{\partial A}{\partial t}\right),$$

hence

$$\operatorname{curl}\left(E + \frac{\partial A}{\partial t}\right) = 0.$$

By Theorem 10.6.2, there is a function ϕ in $C^\infty(\mathbb{R}_0^3 \times \mathbb{R})$, called the **electric potential**, such that

$$E + \frac{\partial A}{\partial t} = -\operatorname{grad}(\phi),$$

so

$$E = -\left(\operatorname{grad}(\phi) + \frac{\partial A}{\partial t}\right). \tag{22.2.3}$$

Equations (22.2.2) and (22.2.3) express E and B in terms of A and ϕ. As it stands, this representation is not unique. For example, let ψ be an arbitrary function in $C^\infty(\mathbb{R}_0^3 \times \mathbb{R})$, and let

$$\widetilde{A} = A + \operatorname{grad}(\psi) \qquad \text{and} \qquad \widetilde{\phi} = \phi - \frac{\partial \psi}{\partial t},$$

which is referred to as a **gauge transformation** of A and ϕ. Then

$$\operatorname{grad}(\widetilde{\phi}) + \frac{\partial \widetilde{A}}{\partial t} = \operatorname{grad}\left(\phi - \frac{\partial \psi}{\partial t}\right) + \frac{\partial}{\partial t}(A + \operatorname{grad}(\psi))$$
$$= \operatorname{grad}(\phi) + \frac{\partial A}{\partial t},$$

and by Theorem 10.6.1(c),

$$\operatorname{curl}(\widetilde{A}) = \operatorname{curl}(A) + \operatorname{curl}\big(\operatorname{grad}(\psi)\big) = \operatorname{curl}(A).$$

Thus, $\widetilde{A}$ and $\widetilde{\phi}$ also satisfy (22.2.2) and (22.2.3). The range of potentials is limited by the constraint

$$\operatorname{div}(A) + \frac{\partial \phi}{\partial t} = 0, \tag{22.2.4}$$

called the **Lorenz gauge**.

Theorem 22.2.1. *In the Lorenz gauge, Maxwell's equations are equivalent to:*

$$\Box(A) = -\mu_0 J$$

$$\Box(\phi) = -\frac{1}{\epsilon_0}\rho,$$

where $\Box$ is the d'Alembert operator, as defined in Example 20.7.3.

Proof. Suppose Maxwell's equations are satisfied. We have from (22.2.4) that

$$\text{div}\left(\frac{\partial A}{\partial t}\right) + \frac{\partial^2 \phi}{\partial t^2} = 0, \tag{22.2.5}$$

hence

$$\frac{1}{\epsilon_0}\rho = \text{div}(E) \qquad\qquad\qquad \text{[M1]}$$

$$= -\text{div}\big(\text{grad}(\phi)\big) - \text{div}\left(\frac{\partial A}{\partial t}\right) \qquad [(22.2.3)]$$

$$= -\text{lap}(\phi) - \text{div}\left(\frac{\partial A}{\partial t}\right) \qquad\quad \text{[Table 10.6.1]}$$

$$= -\left(\text{lap}(\phi) - \frac{\partial^2 \phi}{\partial t^2}\right) \qquad\qquad [(22.2.5)]$$

$$= -\square(\phi).$$

We also have

$$\mu_0 J = \text{curl}(B) - \frac{\partial E}{\partial t} \qquad\qquad\qquad \text{[M4']}$$

$$= \text{curl}\big(\text{curl}(A)\big) + \frac{\partial}{\partial t}\left(\text{grad}(\phi) + \frac{\partial A}{\partial t}\right) \qquad [(22.2.2),\ (22.2.3)]$$

$$= \text{grad}\big(\text{div}(A)\big) - \text{lap}(A) + \text{grad}\left(\frac{\partial \phi}{\partial t}\right) + \frac{\partial^2 A}{\partial t^2} \qquad \text{[Th 10.6.1(e)]}$$

$$= -\text{lap}(A) + \frac{\partial^2 A}{\partial t^2} \qquad\qquad\qquad [(22.2.4)]$$

$$= -\square(A).$$

Conversely, suppose $\square(A) = -\mu_0 J$ and $\square(\phi) = -\rho/\epsilon_0$ are satisfied. The preceding computations "in reverse" show that Maxwell's equations are satisfied. $\square$

We now switch our focus from $\mathbb{R}_0^3 \times \mathbb{R}$ to $\mathbb{R}_1^4$ and recast the preceding discussion in terms of differential forms. Let $J = (J^1, J^2, J^3)$, $E = (E^1, E^2, E^3)$, $A = (A^1, A^2, A^3)$, and $B = (B^1, B^2, B^3)$, and define corresponding covector fields $\mathcal{J}, \mathcal{E}, \mathcal{A}$ in $\mathfrak{X}^*(\mathbb{R}_1^4)$, and forms $\mathcal{B}, \mathcal{F}$ in $\Lambda^2(\mathbb{R}_1^4)$ by

$$\mathcal{J} = \frac{1}{\epsilon_0}[J^1 d(x) + J^2 d(y) + J^3 d(z) - \rho\, d(t)],$$

$$\mathcal{E} = E^1 d(x) + E^2 d(y) + E^3 d(z),$$

$$\mathcal{A} = A^1 d(x) + A^2 d(y) + A^3 d(z) - \phi\, d(t), \tag{22.2.6}$$

$$\mathcal{B} = B^1 d(y) \wedge d(z) + B^2 d(z) \wedge d(x) + B^3 d(x) \wedge d(y),$$

and

$$\mathcal{F} = \mathcal{E} \wedge d(t) + \mathcal{B}$$
$$= E^1 d(x) \wedge d(t) + E^2 d(y) \wedge d(t) + E^3 d(z) \wedge d(t)$$
$$\quad + B^1 d(y) \wedge d(z) + B^2 d(z) \wedge d(x) + B^3 d(x) \wedge d(y).$$

Theorem 22.2.2. *Maxwell's equations are equivalent to:*

$$d(\mathcal{F}) = 0$$

$$\delta(\mathcal{F}) = \mathcal{J},$$

where δ is the codifferential, as defined in Section 20.2.

Proof. Setting $\varphi = \mathcal{F}$, $(\alpha_1, \alpha_2, \alpha_3) = (E^1, E^2, E^3)$, and $(\beta_1, \beta_2, \beta_3) = (B^1, B^2, B^3)$ in Example 16.1.10 and Example 20.2.2 gives

$$
\begin{aligned}
d(\mathcal{F}) = {} & \left(\frac{\partial B^1}{\partial x} + \frac{\partial B^2}{\partial y} + \frac{\partial B^3}{\partial z} \right) d(x) \wedge d(y) \wedge d(z) \\
& + \left(\frac{\partial E^2}{\partial x} - \frac{\partial E^1}{\partial y} + \frac{\partial B^3}{\partial t} \right) d(x) \wedge d(y) \wedge d(t) \\
& - \left(\frac{\partial E^1}{\partial z} - \frac{\partial E^3}{\partial x} + \frac{\partial B^2}{\partial t} \right) d(x) \wedge d(z) \wedge d(t) \\
& + \left(\frac{\partial E^3}{\partial y} - \frac{\partial E^2}{\partial z} + \frac{\partial B^1}{\partial t} \right) d(y) \wedge d(z) \wedge d(t)
\end{aligned}
$$

and

$$
\begin{aligned}
\delta(\mathcal{F}) = {} & \left(\frac{\partial B^3}{\partial y} - \frac{\partial B^2}{\partial z} - \frac{\partial E^1}{\partial t} \right) d(x) + \left(\frac{\partial B^1}{\partial z} - \frac{\partial B^3}{\partial x} - \frac{\partial E^2}{\partial t} \right) d(y) \\
& + \left(\frac{\partial B^2}{\partial x} - \frac{\partial B^1}{\partial y} - \frac{\partial E^3}{\partial t} \right) d(z) - \left(\frac{\partial E^1}{\partial x} + \frac{\partial E^2}{\partial y} + \frac{\partial E^3}{\partial z} \right) d(t).
\end{aligned}
$$

It follows that $d(\mathcal{F}) = 0$ is equivalent to [M2] and [M3]; and $\delta(\mathcal{F}) = \mathcal{J}$ is equivalent to [M1] and [M4']. $\square$

Theorem 22.2.3. *The continuity equation (22.2.1) can be expressed as*

$$d\star(\mathcal{J}) = 0,$$

and the Lorenz gauge (22.2.4) as

$$\delta(\mathcal{A}) = 0.$$

Proof. Setting $\omega = \mathcal{J}$ and $(\alpha_1, \alpha_2, \alpha_3, \alpha_4) = (J^1, J^2, J^3, -\rho)/\epsilon_0$ in Example 20.1.3 yields

$$
\begin{aligned}
d\star(\mathcal{J}) = {} & \frac{1}{\epsilon_0} \left(-\frac{\partial J^1}{\partial x} - \frac{\partial J^2}{\partial y} - \frac{\partial J^3}{\partial z} - \frac{\partial \rho}{\partial t} \right) d(x) \wedge d(y) \wedge d(z) \wedge d(t) \\
= {} & -\frac{1}{\epsilon_0} \left(\text{div}(J) + \frac{\partial \rho}{\partial t} \right) d(x) \wedge d(y) \wedge d(z) \wedge d(t),
\end{aligned}
$$

which proves the identity for the continuity equation. We have from the manifold version of Theorem 4.5.4 and (22.2.6) that $\mathcal{A}^S = (A^1, A^2, A^3, \phi)$, hence

$$-\delta(\mathcal{A}) = \text{Div}(\mathcal{A}^S) \qquad \text{[Th 20.4.3(b)]}$$

$$= \frac{\partial A^1}{\partial x} + \frac{\partial A^2}{\partial y} + \frac{\partial A^3}{\partial z} + \frac{\partial \phi}{\partial t} \qquad \text{[Ex 20.4.6]}$$

$$= \text{div}(A) + \frac{\partial \phi}{\partial t},$$

which proves the identity for the Lorenz gauge.   □

Theorem 22.2.4. *In the Lorenz gauge (22.2.4), Maxwell's equations are equivalent to:*

$$d(\mathcal{A}) = \mathcal{F}$$
$$\Delta(\mathcal{A}) = \mathcal{J},$$

where Δ is the Laplace–de Rham operator, as defined by (20.8.1).

Proof. We need a couple of preliminary observations. Setting $\omega = \mathcal{A}$, hence $(\alpha_1, \alpha_2, \alpha_3, \alpha_4) = (A^1, A^2, A^3, -\phi)$, in Example 16.1.9 gives

$$d(\mathcal{A}) = \left(-\frac{\partial \phi}{\partial x} - \frac{\partial A^1}{\partial t} \right) d(x) \wedge d(t) + \left(-\frac{\partial \phi}{\partial y} - \frac{\partial A^2}{\partial t} \right) d(y) \wedge d(t)$$

$$+ \left(-\frac{\partial \phi}{\partial z} - \frac{\partial A^3}{\partial t} \right) d(z) \wedge d(t)$$

$$+ \left(\frac{\partial A^3}{\partial y} - \frac{\partial A^2}{\partial z} \right) d(y) \wedge d(z) + \left(\frac{\partial A^1}{\partial z} - \frac{\partial A^3}{\partial x} \right) d(z) \wedge d(x)$$

$$+ \left(\frac{\partial A^2}{\partial x} - \frac{\partial A^1}{\partial y} \right) d(x) \wedge d(y).$$

We also have

$$\text{grad}(\phi) + \frac{\partial A}{\partial t} = \left(\frac{\partial \phi}{\partial x} + \frac{\partial A^1}{\partial t}, \frac{\partial \phi}{\partial y} + \frac{\partial A^2}{\partial t}, \frac{\partial \phi}{\partial z} + \frac{\partial A^3}{\partial t} \right)$$

and

$$\text{curl}(A) = \left(\frac{\partial A^3}{\partial y} - \frac{\partial A^2}{\partial z}, \frac{\partial A^1}{\partial z} - \frac{\partial A^3}{\partial x}, \frac{\partial A^2}{\partial x} - \frac{\partial A^1}{\partial y} \right).$$

Thus, $d(\mathcal{A}) = \mathcal{F}$ if and only if (22.2.2) and (22.2.3) are satisfied, which is our first preliminary observation. It follows from (20.8.1) and Theorem 22.2.3 that

$$\Delta(\mathcal{A}) = d\delta(\mathcal{A}) + \delta d(\mathcal{A}) = \delta d(\mathcal{A}), \qquad (22.2.7)$$

which is our second preliminary observation.

Now for the main proof. Suppose Maxwell's equations are satisfied. Then (22.2.2) and (22.2.3) are satisfied, hence $d(\mathcal{A}) = \mathcal{F}$, which proves the first identity. Since $d(\mathcal{A}) = \mathcal{F}$, we have from Theorem 22.2.2 and (22.2.7) that $\Delta(\mathcal{A}) = \delta d(\mathcal{A}) = \delta(\mathcal{F}) = \mathcal{J}$, which proves the second identity. Conversely, suppose $d(\mathcal{A}) = \mathcal{F}$ and $\Delta(\mathcal{A}) = \mathcal{J}$. We have from Theorem 16.1.1(c) that $d(\mathcal{F}) = d^2(\mathcal{A}) = 0$, and from (22.2.7) that $\mathcal{J} = \Delta(\mathcal{A}) = \delta d(\mathcal{A}) = \delta(\mathcal{F})$. By Theorem 22.2.2, Maxwell's equations are satisfied.   □

22.3 Einstein Tensor

Let $(M, \mathfrak{g})$ be a semi-Riemannian manifold, and let $\mathcal{R}$ be the Riemann curvature tensor. The **Ricci curvature tensor (field)** in $T_2^0(M)$ is defined by

$$\mathsf{R} = \mathsf{C}_{14}(\mathcal{R}), \tag{22.3.1}$$

the **scalar curvature (function)** in $C^\infty(M)$ is defined by

$$\mathsf{S} = \mathsf{C}_{12}(\mathsf{R}), \tag{22.3.2}$$

and the **Einstein tensor (field)** in $T_2^0(M)$ is defined by

$$\mathsf{G} = \mathsf{R} - \frac{1}{2}\,\mathsf{S}\mathfrak{g}.$$

In physics, the terms "Ricci curvature tensor", "scalar curvature", and "Einstein tensor" are typically used only when M is a Lorentz 4-manifold, but we will continue with the more general context.

Theorem 22.3.1 (Local Coordinate Expression for Ricci Curvature). *If $(M, \mathfrak{g})$ is a semi-Riemannian manifold, then, in local coordinates:*
(a)

$$\mathsf{R}_{ij} = \sum_k R^k_{kij}$$

$$= \sum_k \left(\frac{\partial \Gamma^k_{ij}}{\partial x^k} - \frac{\partial \Gamma^k_{jk}}{\partial x^i}\right) + \sum_{kl}(\Gamma^k_{kl}\Gamma^l_{ij} - \Gamma^k_{il}\Gamma^l_{jk}). \tag{22.3.3}$$

(b) R *is symmetric; that is,* $\mathsf{R}_{ij} = \mathsf{R}_{ji}$ *for* $i, j = 1, \ldots, m$.

Proof. A preliminary result is

$$\begin{aligned}
\mathsf{R}_{ij} &= \mathsf{C}_{14}(\mathcal{R})_{ij} && [(22.3.1)] \\
&= \mathsf{C}_1^1 \circ \mathsf{S}_4^1(\mathcal{R})_{ij} && [(19.6.2)] \\
&= \sum_{kl} \mathfrak{g}^{kl} \mathcal{R}_{kijl}. && [\text{mv Th } 6.3.5(\text{a})]
\end{aligned} \tag{22.3.4}$$

(a): We have

$$\begin{aligned}
\mathsf{R}_{ij} &= \sum_{kl} \mathfrak{g}^{kl} \mathcal{R}_{kijl} && [(22.3.4)] \\
&= \sum_{kl} \mathfrak{g}^{kl}\left(\sum_n \mathfrak{g}_{ln} R^n_{kij}\right) && [\text{Th } 19.8.2(\text{a})] \\
&= \sum_{kn}\left(\sum_l \mathfrak{g}^{kl}\mathfrak{g}_{ln}\right) R^n_{kij} = \sum_{kn} \delta^k_n R^n_{kij} = \sum_k R^k_{kij},
\end{aligned}$$

which proves the first equality. The second equality follows from Theorem 18.7.2.

(b): Local coordinate versions of the Riemann curvature symmetries in Theorem 19.8.5 yield
$$\mathcal{R}_{kijl} = \mathcal{R}_{jlki} = \mathcal{R}_{ljik}.$$
Then
$$\mathsf{R}_{ij} = \sum_{kl} \mathfrak{g}^{kl} \mathcal{R}_{kijl} \qquad [(22.3.4)]$$
$$= \sum_{kl} \mathfrak{g}^{lk} \mathcal{R}_{ljik}$$
$$= \mathsf{R}_{ji}. \qquad [(22.3.4)]$$

$\square$

Theorem 22.3.2 (Orthonormal Frame Expression for Ricci Curvature). *Let* $(M, \mathfrak{g})$ *be a semi-Riemannian m-manifold with signature* $(\varepsilon_1, \ldots, \varepsilon_m)$, *let* $(E_1, \ldots, E_m)$ *be an orthonormal frame on an open set* U *in* M, *and let* X, Y *be vector fields in* $\mathfrak{X}(M)$. *Then:*
(a)
$$\mathsf{R}(X, Y) = \sum_i \varepsilon_i \mathcal{R}(E_i, X, Y, E_i)$$
on U.
(b) R *is symmetric; that is,* $\mathsf{R}(X, Y) = \mathsf{R}(Y, X)$ *for all vector fields* X, Y *in* $\mathfrak{X}(M)$.

Proof. (a): This follows from Theorem 19.6.2(a) and (22.3.1).
(b): By Theorem 19.8.5,
$$\mathcal{R}(E_i, X, Y, E_i) = \mathcal{R}(Y, E_i, E_i, X) = \mathcal{R}(E_i, Y, X, E_i).$$
The result now follows from part (a). $\square$

Theorem 22.3.3 (Local Coordinate Expression for Scalar Curvature). *If* $(M, \mathfrak{g})$ *is a semi-Riemannian manifold, then, in local coordinates,*
$$\mathsf{S} = \sum_{ij} \mathfrak{g}^{ij} \mathsf{R}_{ij}.$$

Proof. We have from (19.6.3) and (22.3.2) that
$$\mathsf{S} = \mathsf{C}_{12}(\mathsf{R}) = \mathsf{C}_1^1 \circ \mathsf{S}_1^1(\mathsf{R}).$$
The result now follows from the manifold version of Theorem 6.3.5(b). $\square$

Theorem 22.3.4 (Orthonormal Frame Expression for Scalar Curvature). *If* $(M, \mathfrak{g})$ *is a semi-Riemannian manifold with signature* $(\varepsilon_1, \ldots, \varepsilon_m)$, *and* $(E_1, \ldots, E_m)$ *is an orthonormal frame on an open set* U *in* M, *then*
$$\mathsf{S} = \sum_i \varepsilon_i \mathsf{R}(E_i, E_i)$$
on U.

Proof. This follows from Theorem 19.6.2(b) and (22.3.2). □

Theorem 22.3.5 (Orthonormal Frame Expression for Divergence of Ricci Curvature). *Let $(M, \mathfrak{g})$ be a semi-Riemannian m-manifold with signature $(\varepsilon_1, \ldots, \varepsilon_m)$, let $(E_1, \ldots, E_m)$ be an orthonormal frame on an open set U in M, and let X be a vector field in $\mathfrak{X}(M)$. Then*

$$\mathrm{Div}(\mathsf{R})(X) = \sum_i \varepsilon_i \nabla_{E_i}(\mathsf{R})(X, E_i) = \sum_i \varepsilon_i \nabla_{E_i}(\mathsf{R})(E_i, X)$$

on U.

Proof. This follows from Theorem 20.9.1 and Theorem 22.3.2(b). □

Theorem 22.3.6. *If $(M, \mathfrak{g})$ is a semi-Riemannian m-manifold, then:*
(a)
$$\mathrm{Div}(\mathsf{R}) = \frac{1}{2} d(\mathsf{S}) = \frac{1}{2} \mathrm{Div}(\mathsf{S}\mathfrak{g}).$$

(b) $\mathrm{Div}(\mathsf{G}) = 0$.

Proof. (a): For a vector field X in $\mathfrak{X}(M)$, we have

$$
\begin{aligned}
d(\mathsf{S})(X) &= X(\mathsf{S}) && \text{[Th 15.4.5(b)]} \\
&= \nabla_X(\mathsf{S}) && \text{[(18.1.1)]} \\
&= \nabla_X \circ \mathsf{C}_{12} \circ \mathsf{C}_{14}(\mathcal{R}) && \text{[(22.3.1), (22.3.2)]} \\
&= \mathsf{C}_{12} \circ \mathsf{C}_{14} \circ \nabla_X(\mathcal{R}). && \text{[Th 19.6.3]}
\end{aligned}
$$
(22.3.5)

Let p be a point in M. By Theorem 19.3.7, there is an orthonormal frame $(E_1, \ldots, E_m)$ on a neighborhood of p. It follows from Theorem 19.6.2(a) that

$$\mathsf{C}_{14} \circ \nabla_X(\mathcal{R})(Y, Z) = \sum_i \varepsilon_i \nabla_X(\mathcal{R})(E_i, Y, Z, E_i),$$

and then from Theorem 19.6.2(b) that

$$
\begin{aligned}
\mathsf{C}_{12} \circ \mathsf{C}_{14} \circ \nabla_X(\mathcal{R}) &= \sum_j \varepsilon_j \mathsf{C}_{14} \circ \nabla_X(\mathcal{R})(E_j, E_j) \\
&= \sum_j \varepsilon_j \left(\sum_i \varepsilon_i \nabla_X(\mathcal{R})(E_i, E_j, E_j, E_i) \right) \\
&= \sum_{ij} \varepsilon_i \varepsilon_j \nabla_X(\mathcal{R})(E_i, E_j, E_j, E_i).
\end{aligned}
$$
(22.3.6)

By Theorem 19.8.4(b),

$$\nabla_X(\mathcal{R})(E_i, E_j, E_j, E_i) = -\nabla_{E_j}(\mathcal{R})(X, E_i, E_j, E_i) - \nabla_{E_i}(\mathcal{R})(E_j, X, E_j, E_i),$$

hence

$$\sum_{ij} \varepsilon_i \varepsilon_j \nabla_X(\mathcal{R})(E_i, E_j, E_j, E_i)$$

$$= -\sum_{ij} \varepsilon_i \varepsilon_j \nabla_{E_j}(\mathcal{R})(X, E_i, E_j, E_i) - \sum_{ij} \varepsilon_i \varepsilon_j \nabla_{E_i}(\mathcal{R})(E_j, X, E_j, E_i)$$

$$= -\sum_{ij} \varepsilon_i \varepsilon_j \nabla_{E_i}(\mathcal{R})(X, E_j, E_i, E_j) - \sum_{ij} \varepsilon_i \varepsilon_j \nabla_{E_i}(\mathcal{R})(E_j, X, E_j, E_i) \quad (22.3.7)$$

$$= \sum_{ij} \varepsilon_i \varepsilon_j \nabla_{E_i}(\mathcal{R})(E_j, X, E_i, E_j) + \sum_{ij} \varepsilon_i \varepsilon_j \nabla_{E_i}(\mathcal{R})(E_j, X, E_i, E_j)$$

$$= 2\sum_{ij} \varepsilon_i \varepsilon_j \nabla_{E_i}(\mathcal{R})(E_j, X, E_i, E_j),$$

where the third equality follows from Theorem 19.8.5. By Theorem 19.6.2(a),

$$\mathsf{C}_{14} \circ \nabla_{E_i}(\mathcal{R})(X, E_i) = \sum_j \varepsilon_j \nabla_{E_i}(\mathcal{R})(E_j, X, E_i, E_j),$$

hence

$$\sum_i \varepsilon_i \mathsf{C}_{14} \circ \nabla_{E_i}(\mathcal{R})(X, E_i) = \sum_{ij} \varepsilon_i \varepsilon_j \nabla_{E_i}(\mathcal{R})(E_j, X, E_i, E_j). \quad (22.3.8)$$

It follows from Theorem 19.6.3 and (22.3.1) that

$$\mathsf{C}_{14} \circ \nabla_{E_i}(\mathcal{R}) = \nabla_{E_i} \circ \mathsf{C}_{14}(\mathcal{R}) = \nabla_{E_i}(\mathsf{R}),$$

so

$$\sum_i \varepsilon_i \mathsf{C}_{14} \circ \nabla_{E_i}(\mathcal{R})(X, E_i) = \sum_i \varepsilon_i \nabla_{E_i}(\mathsf{R})(X, E_i). \quad (22.3.9)$$

Combining the above identities yields

$$
\begin{aligned}
d(\mathsf{S})(X) &= \mathsf{C}_{12} \circ \mathsf{C}_{14} \circ \nabla_X(\mathcal{R}) && [(22.3.5)] \\
&= \sum_{ij} \varepsilon_i \varepsilon_j \nabla_X(\mathcal{R})(E_i, E_j, E_j, E_i) && [(22.3.6)] \\
&= 2\sum_{ij} \varepsilon_i \varepsilon_j \nabla_{E_i}(\mathcal{R})(E_j, X, E_i, E_j) && [(22.3.7)] \\
&= 2\sum_i \varepsilon_i \mathsf{C}_{14} \circ \nabla_{E_i}(\mathcal{R})(X, E_i) && [(22.3.8)] \\
&= 2\sum_i \varepsilon_i \nabla_{E_i}(\mathsf{R})(X, E_i) && [(22.3.9)] \\
&= 2\,\mathrm{Div}(\mathsf{R})(X). && [\text{Th } 22.3.5]
\end{aligned}
$$

hence

$$\mathrm{Div}(\mathsf{R})(X) = \frac{1}{2}\,d(\mathsf{S})(X).$$

Since p and X were arbitrary, this proves the first equality. By Theorem 20.9.2, $\text{Div}(S\mathfrak{g}) = d(S)$, so the second equality follows from the first.

(b): This follows from part (a).   □

Adopting widely-accepted units, **Einstein's field equations** for a Lorentz 4-manifold are given by the tensor identity

$$\mathsf{G} = \frac{8\pi G}{c^4}\mathsf{T},$$

where G is Newton's gravitational constant, c is the speed of light, and T is the so-called **stress–energy tensor**. These 10 partial differential equations, which form the foundation of the general theory of relativity, describe the way gravity results from spacetime being curved by mass and energy. It is remarkable that the "left-hand sides" of Einstein's equations are expressed entirely in geometric terms, in the sense that the Einstein tensor is completely determined by the metric of a Lorentz 4-manifold.

Part IV

Appendices

Appendix A

Notation and Set Theory

In this appendix, we recall some of the elements of set theory.

Let X be a set, and let S_1 and S_2 be subsets of X. We say that S_1 and S_2 **intersect** or **overlap** if $S_1 \cap S_2 \neq \varnothing$, and that they **do not intersect, do not overlap**, or are **disjoint** if $S_1 \cap S_2 = \varnothing$, where $\varnothing$ denotes the **empty set**. The **difference** between S_1 and S_2 (in that order) is the set

$$S_1 \backslash S_2 = \{x \in S_1 : x \notin S_2\}.$$

Let $\{S_\alpha : \alpha \in A\}$ be a collection of subsets of X, where A is some indexing set. The union and intersection of the S_α are denoted by

$$\bigcup_{\alpha \in A} S_\alpha \quad \text{and} \quad \bigcap_{\alpha \in A} S_\alpha,$$

respectively. The **(Cartesian) product** of sets $X_1, \ldots, X_n$ is defined by

$$X_1 \times \cdots \times X_n = \{(x_1, \ldots, x_n) : x_i \in X_i \text{ for } i = 1, \ldots, n\},$$

and each $(x_1, \ldots, x_n)$ is called an **n-tuple**. For a set X, we denote $X \times \cdots \times X$ [n copies] by X^n.

Theorem A.1. *If X is a set and $\{S_\alpha : \alpha \in A\} \cup \{T\}$ is a collection of subsets of X, then*
(a)

$$\left(\bigcup_{\alpha \in A} S_\alpha \right) \cap T = \bigcup_{\alpha \in A} (S_\alpha \cap T).$$

(b)

$$\left(\bigcap_{\alpha \in A} S_\alpha \right) \cup T = \bigcap_{\alpha \in A} (S_\alpha \cup T).$$

Semi-Riemannian Geometry, First Edition. Stephen C. Newman.
© 2019 John Wiley & Sons, Inc. Published 2019 by John Wiley & Sons, Inc.

(c)

$$X \setminus \left(\bigcup_{\alpha \in A} S_\alpha \right) = \bigcap_{\alpha \in A} (X \setminus S_\alpha).$$

(d)

$$X \setminus \left(\bigcap_{\alpha \in A} S_\alpha \right) = \bigcup_{\alpha \in A} (X \setminus S_\alpha). \qquad \square$$

Theorem A.2. *If X_i is a set, and S_i and T_i are subsets of X_i for $i = 1, \ldots, n$, then*

$$(S_1 \times \cdots \times S_n) \cap (T_1 \times \cdots \times T_n) = (S_1 \cap T_1) \times \cdots \times (S_n \cap T_n). \qquad \square$$

Let $F : X \longrightarrow Y$ be a **map**. (The term "function" is reserved for a special type of map, as discussed below.) We refer to X and Y as the **domain of F** and **codomain of F**, respectively. The **image of F** is denoted by $\operatorname{im}(F)$ or $F(X)$ and defined by

$$\operatorname{im}(F) = F(X) = \{F(x) : x \in X\}.$$

It is implicit in the notation $F : X \longrightarrow Y$ that $F(x)$ is defined for all x in X. On the other hand, $F(X)$ may be a proper subset of Y. Let T be a subset of Y, and define

$$F^{-1}(T) = \{x \in X : F(x) \in T\}.$$

Observe that we do not require T to be a subset of $F(X)$. Thus,

$$F^{-1}(T) = F^{-1}(T \cap F(X)).$$

Theorem A.3. *Let $F : X \longrightarrow Y$ be a map, let $\{S_\alpha : \alpha \in A\}$ be a collection of subsets of X, and let $\{T_\beta : \beta \in B\} \cup \{T, T_1, T_2\}$ be a collection of subsets of Y. Then:*
(a)

$$F\left(\bigcup_{\alpha \in A} S_\alpha \right) = \bigcup_{\alpha \in A} F(S_\alpha).$$

(b)

$$F^{-1}\left(\bigcup_{\beta \in B} T_\beta \right) = \bigcup_{\beta \in B} F^{-1}(T_\beta).$$

(c)

$$F^{-1}\left(\bigcap_{\beta \in B} T_\beta \right) = \bigcap_{\beta \in B} F^{-1}(T_\beta).$$

(d)

$$F\left(F^{-1}(T) \right) = T \cap F(X).$$

(e)

$$F^{-1}(T_1 \setminus T_2) = F^{-1}(T_1) \setminus F^{-1}(T_2). \qquad \square$$

If there is an element y_0 in Y such that $F(x) = y_0$ for all x in X, then F is said to be **constant** or to have **constant value**. This is sometimes expressed as $F = y_0$. We say that F is **injective** if distinct elements of X are mapped to distinct elements of Y; that is, if x_1 and x_2 are elements of X such that $F(x_1) = F(x_2)$, then $x_1 = x_2$. We say that F is **surjective** or **onto** if every element of Y is the image of some element of X; that is, $F(X) = Y$. A map that is both injective and surjective is called **bijective**.

The **restriction** of $F : X \longrightarrow Y$ to a subset S of X is the map denoted by

$$F|_S : S \longrightarrow Y$$

and defined by $F|_S(x) = F(x)$ for all x in S. It is sometimes convenient to use the notation $F|S$ instead of $F|_S$, especially when S is a complicated expression.

Given a map $G : Y \longrightarrow Z$, the **composite** of G and F is the map denoted by

$$G \circ F : X \longrightarrow Z$$

and defined by

$$(G \circ F)(x) = G\big(F(x)\big)$$

for all x in X. Observe that we do not require F to be surjective, so that $G \circ F(X)$ might be a proper subset of $G(Y)$.

The **identity map**

$$\mathrm{id}_X : X \longrightarrow X$$

is defined by $\mathrm{id}_X(x) = x$ for all x in X.

Theorem A.4. *The map $F : X \longrightarrow Y$ is bijective if and only if there is a map*

$$F^{-1} : Y \longrightarrow X,$$

*called the **inverse of F**, such that $F^{-1} \circ F = \mathrm{id}_X$ and $F \circ F^{-1} = \mathrm{id}_Y$.* □

Let $F : X \longrightarrow Y_1 \times \cdots \times Y_n$ be a map, and for each x in X, let us express $F(x)$ as

$$F(x) = \big(F_1(x), \ldots, F_n(x)\big).$$

This defines maps $F_i : X \longrightarrow Y_i$ for $i = 1, \ldots, n$, called the **component maps of F** or simply the **components of F**. We often write

$$F = (F_1, \ldots, F_n).$$

Theorem A.5. *Let $F = (F_1, \ldots, F_n) : X \longrightarrow Y_1 \times \cdots \times Y_n$ be a map, and let T_i be a subset of Y_i for $i = 1, \ldots, n$. Then*

$$F^{-1}(T_1 \times \cdots \times T_n) = \bigcap_{i=1}^{n} F_i^{-1}(T_i).$$ □

The set of real numbers is denoted by $\mathbb{R}$. We adopt the following convention.

Throughout, a map with the codomain $\mathbb{R}$ is called a function.

Let $f : X \longrightarrow \mathbb{R}$ be a function, and let x be an element of X. We say that f **vanishes at x** if $f(x) = 0$, is **nonvanishing at x** if $f(x) \neq 0$, and is **nowhere-vanishing (on X)** if it is nonvanishing at every x in X. We say that f is **strictly positive** if $f(x) > 0$ for all x in X, and **strictly negative** if $f(x) < 0$ for all x in X. The function $f : \mathbb{R} \longrightarrow \mathbb{R}$ is said to be **increasing** if $f(x_1) \leq f(x_2)$ for all $x_1 < x_2$, and **strictly increasing** if $f(x_1) < f(x_2)$ for all $x_1 < x_2$. Analogous definitions are given for **decreasing** and **strictly decreasing**.

Given a real number c, the **constant function** on X with value c, denoted by c_M or simply c, is the function that sends all elements of X to c. In particular, we have 0_X, called the **zero function**, and 1_X.

The **cardinality** of a set X is denoted by $\mathrm{card}(X)$ and defined to be the number of elements in the set, which is either **finite** or **infinite**. Two sets are said to have the same cardinality if there is a bijective map between them. It is known that there are different types of "infinity". For instance, we say that a set is **countable** if it has the same cardinality as the positive integers. It can be shown that the rational numbers are countable, but the real numbers are not.

A binary relation on a set X is a subset $\sim$ of $X \times X$. If (x_1, x_2) is in $\sim$, then x_1 is said to **related to** x_2, written $x_1 \sim x_2$. We say that $\sim$ is an **equivalence relation** on X if:

[**E1**] $x \sim x$ for all x in X.   (reflexivity)
[**E2**] If $x_1 \sim x_2$, then $x_2 \sim x_1$ for all x_1, x_2 in X.   (symmetry)
[**E3**] If $x_1 \sim x_2$ and $x_2 \sim x_3$, then $x_1 \sim x_3$ for all x_1, x_2, x_3 in X. (transitivity)

If $\sim$ is an equivalence relation on X and x is an element of X, the **equivalence class** determined by x is denoted by $[x]$ and defined to be the set of elements of X that are related to x:

$$[x] = \{y \in X : y \sim x\}.$$

In particular, we have from [E1] that $[x]$ contains x.

A **partition** of X is a collection of disjoint nonempty subsets of X whose union equals X.

Theorem A.6. *Given an equivalence relation on a set X, the collection of distinct equivalence classes forms a partition of X.*   □

The **sign** of a nonzero real number x is defined by

$$\mathrm{sgn}(x) = \frac{x}{|x|}.$$

The sign of 0 is not defined.

Kronecker's delta is the function

$$\delta : \{1, \ldots, n\} \times \{1, \ldots, n\} \longrightarrow \{0, 1\}$$

defined by

$$\delta(i,j) = \begin{cases} 1 & \text{if } i = j \\ 0 & \text{if } i \neq j. \end{cases}$$

We henceforth denote $\delta(i,j)$ by either of the equivalent expressions δ_{ij} or δ^i_j, the choice depending on the context.

For an integer $1 \leq k \leq n$, we say that a k-tuple $(i_1, \ldots, i_k)$ of *distinct* elements of $\{1, \ldots, n\}$ is a **multi-index** if $i_1 < \cdots < i_k$. The set of multi-indices on $\{1, \ldots, n\}$ with k elements is denoted by $\mathcal{I}_{k,n}$, and a typical multi-index I in $\mathcal{I}_{k,n}$ is denoted by $I = (i_1, \ldots, i_k)$. For example,

$$\mathcal{I}_{3,4} = \{(1,2,3), (1,2,4), (2,3,4)\}.$$

Evidently, $\mathcal{I}_{k,n}$ has $\binom{n}{k}$ elements, where $\binom{n}{k}$ is a binomial coefficient. The **complement** of the multi-index $I = (i_1, \ldots, i_k)$ in $\mathcal{I}_{k,n}$ is the multi-index $I^c = (i_{k+1}, \ldots, i_n)$ in $\mathcal{I}_{n-k,n}$ where

$$\{i_{k+1}, \ldots, i_n\} = \{1, \ldots, n\} \setminus \{i_1, \ldots, i_k\}.$$

In particular, the complement of the multi-index (i) in $\mathcal{I}_{1,n}$ is the multi-index

$$(i)^c = (1, \ldots, \widehat{i}, \ldots, n) = (1, \ldots, i-1, i+1, \ldots, n)$$

in $\mathcal{I}_{n-1,n}$ for $i = 1, \ldots, n$, where $\widehat{}$ indicates that an expression is omitted. To illustrate, for (2) in $\mathcal{I}_{1,6}$ and $(1,2,3,6)$ in $\mathcal{I}_{4,6}$, we have the complements $(2)^c = (1,3,4,5,6)$ in $\mathcal{I}_{5,6}$ and $(1,2,3,6)^c = (4,5)$ in $\mathcal{I}_{2,6}$.

Appendix B

Abstract Algebra

B.1 Groups

A **group** is a pair $(G, \circ)$, where G is a set and $\circ\colon G \times G \longrightarrow G$ is a map, called **multiplication**, such that for all f, g, h in G:

[**G1**] Multiplication is associative:

$$f \circ (g \circ h) = (f \circ g) \circ h.$$

[**G2**] There is an element id_G in G, called the **identity**, such that

$$\mathrm{id}_G \circ g = g \circ \mathrm{id}_G = g.$$

[**G3**] There is an element g^{-1} in G, called the **inverse** of g, such that

$$g \circ g^{-1} = g^{-1} \circ g = \mathrm{id}_G.$$

Usually $g \circ h$ is abbreviated to gh. It is easily shown that the identity element and the inverse of an element are unique. It is standard practice to refer to G as a group, leaving the multiplication map understood. We say that G is **commutative** if multiplication is commutative; that is, $gh = hg$ for all g, h in G. In that situation, it is usual to employ additive notation, with the group denoted by $(G, +)$, the identity by 0_G or 0, and the inverse of g by $-g$.

A subset G' of G is said to be a **subgroup** of G if: (i) G' is a group in its own right (when multiplication on G is restricted to G'), and (ii) G' has the same identity element as G. We say that a group is **finite** if it is finite as a set.

Theorem B.1.1. *If $\{G_\alpha : \alpha \in A\}$ is a collection of subgroups of G, then $\bigcap_{\alpha \in A} G_\alpha$ is a subgroup of G.*

Proof. Straightforward. $\qquad\square$

Semi-Riemannian Geometry, First Edition. Stephen C. Newman.
© 2019 John Wiley & Sons, Inc. Published 2019 by John Wiley & Sons, Inc.

Let G and H be groups. A map $F : G \longrightarrow H$ is said to be a **group homomorphism** if $F(gh) = F(g)F(h)$ for all g, h in G. We say that F is a **group isomorphism**, and that G and H are **isomorphic**, if F is bijective.

Theorem B.1.2. *If G and H are groups and $F : G \longrightarrow H$ is a group isomorphism, then $F^{-1} : H \longrightarrow G$ is a group isomorphism.*

Proof. The proof is analogous to that of Theorem 1.1.9.   □

Example B.1.3. Clearly, $(\mathbb{R}, +)$ is a commutative group, as is $(\mathbb{R} \setminus \{0\}, \times)$. In Section 2.4, we introduced the general linear group $\mathrm{GL}(m)$ consisting of all invertible real $m \times m$ matrices. For $m \geq 2$, $\mathrm{GL}(m)$ is an infinite noncommutative group, where group multiplication is given by the usual matrix multiplication. In Section B.2, we discuss the permutation group $\mathcal{S}_m$. For $m \geq 3$, $\mathcal{S}_m$ is a finite noncommutative group.   ◇

B.2   Permutation Groups

Let m be a positive integer, and let $\mathcal{S}_m$ be the set of bijective functions on $\{1, 2, \ldots, m\}$. In the present context, each element of $\mathcal{S}_m$ is referred to as a **permutation**. We make $\mathcal{S}_m$ into a group, called the **symmetric group** on $\{1, 2, \ldots, m\}$, by defining group multiplication to be composition of functions. The identity element of $\mathcal{S}_m$ is denoted by id. Let σ and ρ be permutations in $\mathcal{S}_m$. For brevity, we denote $\sigma \circ \rho$ by $\sigma\rho$, and $\sigma \cdots \sigma$ [k copies] by σ^k. Each permutation in $\mathcal{S}_m$ can be displayed as follows:

$$\sigma = \begin{pmatrix} 1 & 2 & \cdots & m-1 & m \\ \sigma(1) & \sigma(2) & \cdots & \sigma(m-1) & \sigma(m) \end{pmatrix}.$$

This makes it clear that $\mathrm{card}(\mathcal{S}_m) = m!$.

Let a be an integer in $\{1, 2, \ldots, m\}$. The **cycle** containing a is the permutation defined by

$$\rho = \begin{pmatrix} a & \rho(a) & \rho^2(a) & \cdots \\ \rho(a) & \rho^2(a) & \rho^3(a) & \cdots \end{pmatrix}, \tag{B.2.1}$$

which sends a to $\rho(a)$, $\rho(a)$ to $\rho^2(a)$, and so on. Since $\{1, 2, \ldots, m\}$ is finite, the sequence $a, \rho(a), \rho^2(a), \ldots$ eventually repeats an earlier element of the sequence. It is easily shown that the first element to repeat is a. Let k be the smallest exponent such that $\rho^k(a) = a$. Then (B.2.1) can be expressed as

$$\rho = \begin{pmatrix} a & \rho(a) & \rho^2(a) & \cdots & \rho^{k-2}(a) & \rho^{k-1}(a) \\ \rho(a) & \rho^2(a) & \rho^3(a) & \cdots & \rho^{k-1}(a) & a \end{pmatrix}.$$

An alternative and more compact notation, called **cycle notation**, is

$$\rho = \begin{pmatrix} a & \rho(a) & \rho^2(a) & \cdots & \rho^{k-1}(a) \end{pmatrix}.$$

We say that the cycle ρ has **length** k. A cycle of length 1 corresponds to an element that is fixed by the permutation. Cycles with no elements in common are said to be **disjoint**. Every permutation in $\mathcal{S}_m$ can be written as the product of disjoint cycles, and in exactly one way (up to order of terms).

A cycle of length 2 is called a **transposition**. Any cycle of length 3 or more can be written in several ways as the product of transpositions. For example,

$$(a_1\ a_2\ a_3\ a_4\ \ldots\ a_{k-1}\ a_k) = (a_1\ a_2)(a_2\ a_3)(a_3\ a_4)\cdots(a_{k-1}\ a_k) \qquad (\text{B.2.2})$$

and

$$(a_1\ a_2\ a_3\ a_4\ \ldots\ a_{k-1}\ a_k) = (a_1\ a_k)(a_1\ a_{k-1})\cdots(a_1\ a_3)(a_1\ a_2). \qquad (\text{B.2.3})$$

The inverse of a cycle is easily obtained:

$$(a_1\ a_2\ a_3\ a_4\ \ldots\ a_{k-1}\ a_k)^{-1} = (a_1\ a_k\ a_{k-1}\ \ldots\ a_4\ a_3\ a_2). \qquad (\text{B.2.4})$$

It needs to be emphasized that since composition of functions proceeds from right to left, so does multiplication of permutations, including multiplication of transpositions. For example, $(1\ 2)(2\ 3)$ sends 1 to 2, and 2 to 3, and 3 to 2 to 1, so that $(1\ 2)(2\ 3) = (1\ 2\ 3)$.

Example B.2.1. Consider the permutation

$$\sigma = \begin{pmatrix} 1 & 2 & 3 & 4 & 5 \\ 4 & 5 & 1 & 3 & 2 \end{pmatrix} \qquad (\text{B.2.5})$$

in $\mathcal{S}_5$, which has the cycle decomposition $\sigma = (1\ 4\ 3)(2\ 5)$. Although there are no cycles of length 1 in this example, it is conventional to suppress such cycles in a cycle decomposition. Since the cycles in a cycle decomposition are disjoint, they commute, so we could equally write $\sigma = (2\ 5)(1\ 4\ 3)$. The inverse of σ is found simply by interchanging the rows of (B.2.5) and reordering columns to obtain

$$\sigma^{-1} = \begin{pmatrix} 1 & 2 & 3 & 4 & 5 \\ 3 & 5 & 4 & 1 & 2 \end{pmatrix},$$

which has the cycle decomposition $\sigma^{-1} = (1\ 3\ 4)(2\ 5)$. The cycle decomposition of σ^{-1} can be obtained directly from the cycle decomposition of σ using (B.2.4):

$$\sigma^{-1} = (1\ 4\ 3)^{-1}(2\ 5)^{-1} = (1\ 3\ 4)(2\ 5).$$

With the aid of (B.2.2) and (B.2.3), we obtain alternative cycle decompositions of σ^{-1}:

$$\sigma^{-1} = (1\ 3)(3\ 4)(2\ 5) = (1\ 4)(1\ 3)(2\ 5).$$

Let $\tau = (2\ 4)$, and consider the permutation

$$\tau\sigma = \begin{pmatrix} 1 & 2 & 3 & 4 & 5 \\ 2 & 5 & 1 & 3 & 4 \end{pmatrix}, \qquad (\text{B.2.6})$$

which has the cycle decomposition $\tau\sigma = (1\ 2\ 5\ 4\ 3)$. We note that (B.2.6) is obtained simply by switching the entries 2 and 4 in the second row of (B.2.5).

The general principle underlying this observation suggests another way of computing the inverse of a permutation. For example, corresponding to σ, consider the 5-tuple $(4, 5, 1, 3, 2)$ representing the second row of (B.2.5). We seek a series of transpositions that reorders $(4, 5, 1, 3, 2)$ to $(1, 2, 3, 4, 5)$. In light of preceding remarks, their product (in the correct order) gives a cycle decomposition of σ^{-1}. In an obvious notation, we have

$$(1\ 5)(4, 5, 1, 3, 2) = (4, 1, 5, 3, 2), \text{ then}$$
$$(3\ 5)(4, 1, 5, 3, 2) = (4, 1, 3, 5, 2), \text{ then}$$
$$(2\ 5)(4, 1, 3, 5, 2) = (4, 1, 3, 2, 5), \text{ then}$$
$$(4\ 1)(4, 1, 3, 2, 5) = (1, 4, 3, 2, 5), \text{ then}$$
$$(3\ 4)(1, 4, 3, 2, 5) = (1, 3, 4, 2, 5), \text{ then}$$
$$(2\ 4)(1, 3, 4, 2, 5) = (1, 3, 2, 4, 5), \text{ then}$$
$$(2\ 3)(1, 3, 2, 4, 5) = (1, 2, 3, 4, 5).$$

Collecting the transpositions together gives

$$(2\ 3)(2\ 4)(3\ 4)(4\ 1)(2\ 5)(3\ 5)(1\ 5) = (1\ 3\ 4)(2\ 5) = \sigma^{-1}. \qquad \diamond$$

Let $x_1, \ldots, x_m$ be indeterminates, and consider the multinomial

$$\Delta = \prod_{1 \leq i < j \leq m} (x_i - x_j).$$

For each σ in $\mathcal{S}_m$, let

$$\sigma(\Delta) = \prod_{1 \leq i < j \leq m} (x_{\sigma(i)} - x_{\sigma(j)}).$$

The **sign** of σ is defined by

$$\text{sgn}(\sigma) = \frac{\sigma(\Delta)}{\Delta} = \prod_{1 \leq i < j \leq m} \frac{x_{\sigma(i)} - x_{\sigma(j)}}{x_i - x_j}.$$

Evidently, $\text{sgn}(\sigma)$ equals 1 or -1.

Theorem B.2.2. *If σ, ρ are permutations in $\mathcal{S}_m$, then:*
(a) $\text{sgn}(\sigma\rho) = \text{sgn}(\sigma)\,\text{sgn}(\rho)$.
(b) $\text{sgn}(\sigma) = \text{sgn}(\sigma^{-1})$.
(c) $\text{sgn}(\sigma\rho\sigma^{-1}) = \text{sgn}(\rho)$.

Proof. (a): Since $\rho(\Delta) = \pm\Delta$, we have

$$\frac{(\sigma\rho)(\Delta)}{\rho(\Delta)} = \frac{\sigma\big(\rho(\Delta)\big)}{\rho(\Delta)} = \frac{\sigma(\pm\Delta)}{\pm\Delta} = \frac{\pm\sigma(\Delta)}{\pm\Delta} = \frac{\sigma(\Delta)}{\Delta} = \text{sgn}(\sigma),$$

hence

$$\text{sgn}(\sigma\rho) = \frac{(\sigma\rho)(\Delta)}{\Delta} = \frac{(\sigma\rho)(\Delta)}{\rho(\Delta)}\,\frac{\rho(\Delta)}{\Delta} = \text{sgn}(\sigma)\,\text{sgn}(\rho).$$

(b): By part (a),

$$1 = \text{sgn}(\text{id}) = \text{sgn}(\sigma\sigma^{-1}) = \text{sgn}(\sigma)\,\text{sgn}(\sigma^{-1}),$$

from which the result follows.

(c): By parts (a) and (b),

$$\text{sgn}(\sigma\rho\sigma^{-1}) = \text{sgn}(\sigma)\,\text{sgn}(\rho)\,\text{sgn}(\sigma^{-1}) = \text{sgn}(\rho). \qquad \square$$

We see from (B.2.2) and (B.2.3) that every permutation in $\mathcal{S}_m$ can be written as (decomposed into) a product of transpositions, but not in a unique manner. As we now show, there is nevertheless something invariant about such decompositions.

Theorem B.2.3. *If a permutation σ in $\mathcal{S}_m$ has a cycle decomposition into k transpositions, then*

$$\text{sgn}(\sigma) = (-1)^k.$$

Thus, the number of transpositions in any cycle decomposition of σ is always an even number or always an odd number.

Proof. Expressing Δ as

$$\Delta = (x_1 - x_2)\left(\prod_{3 \le j \le m}(x_1 - x_j)\right)\left(\prod_{3 \le j \le m}(x_2 - x_j)\right)\left(\prod_{3 \le i < j \le m}(x_i - x_j)\right),$$

we see that $\text{sgn}(1\ 2) = -1$. It follows from Theorem B.2.2(a) that

$$\text{sgn}(1\ 3) = \text{sgn}\big((2\ 3)(1\ 2)(2\ 3)\big) = \text{sgn}(1\ 2) = -1$$
$$\text{sgn}(1\ 4) = \text{sgn}\big((3\ 4)(1\ 3)(3\ 4)\big) = \text{sgn}(1\ 3) = -1,$$

and so on. Elaborating on this argument, it can be shown that the sign of any transposition equals -1. Let $\sigma = \tau_1 \tau_2 \cdots \tau_k$ be a decomposition of σ into transpositions. By Theorem B.2.2(a),

$$\text{sgn}(\sigma) = \text{sgn}(\tau_1)\,\text{sgn}(\tau_2) \cdots \text{sgn}(\tau_k) = (-1)^k. \qquad \square$$

In Example B.2.1, we expressed a certain permutation as the product of 3 transpositions, and also as the product of 7 transpositions. This is consistent with Theorem B.2.3.

Theorem B.2.4. *If $1 \le k < m$ are integers and σ is the permutation in $\mathcal{S}_m$ defined by*

$$\sigma = \begin{pmatrix} 1 & \cdots & k & k+1 & \cdots & m \\ m-k+1 & \cdots & m & 1 & \cdots & m-k \end{pmatrix},$$

then

$$\text{sgn}(\sigma) = (-1)^{k(m-k)}.$$

Proof. We use the algorithm illustrated in Example B.2.1 to compute $\mathrm{sgn}(\sigma)$ based on the number of transpositions required to permute $(m - k + 1, \ldots, m; 1, \ldots, m - k)$ into $(1, \ldots, k; k+1, \ldots, m)$. It takes $m - k$ transpositions to move m to the position after $m - k$; that is, permute $(m - k + 1, \ldots, m; 1, \ldots, m - k)$ into $(m - k + 1, \ldots, m - 1; 1, \ldots, m - k, m)$. It then takes $m - k$ transpositions to move $m - 1$ to the position after $m - k$; that is, permute $(m - k + 1, \ldots, m - 1; 1, \ldots, m - k, m)$ into $(m - k + 1, \ldots, m - 2; 1, \ldots, m - k, m - 1, m)$. This process continues for a total of k steps. Thus, $k(m - k)$ transpositions give the desired permutation.   □

Recall from Appendix A the multi-index $I = (i_1, \ldots, i_k)$ in $\mathcal{I}_{k,m}$ and its complement $I^c = (i_{k+1}, \ldots, i_m)$ in $\mathcal{I}_{m-k,m}$. Let

$$(I, I^c) = (i_1, \ldots, i_k, i_{k+1}, \ldots, i_m) \quad \text{and} \quad (I^c, I) = (i_{k+1}, \ldots, i_m, i_1, \ldots, i_k).$$

Corresponding to (I, I^c) and (I^c, I), we define permutations in $\mathcal{S}_m$ by

$$\sigma_{(I,I^c)} = \begin{pmatrix} 1 & \cdots & k & k+1 & \cdots & m \\ i_1 & \cdots & i_k & i_{k+1} & \cdots & i_m \end{pmatrix}$$

and

$$\sigma_{(I^c,I)} = \begin{pmatrix} 1 & \cdots & m - k & m - k + 1 & \cdots & m \\ i_{k+1} & \cdots & i_m & i_1 & \cdots & i_k \end{pmatrix}.$$

Theorem B.2.5. *With the above setup,*

$$\mathrm{sgn}(\sigma_{(I,I^c)}) \, \mathrm{sgn}(\sigma_{(I^c,I)}) = (-1)^{k(m-k)}.$$

Proof. We have

$$(\sigma_{(I^c,I)})^{-1} = \begin{pmatrix} i_1 & \cdots & i_k & i_{k+1} & \cdots & i_m \\ m - k + 1 & \cdots & m & 1 & \cdots & m - k \end{pmatrix},$$

hence

$$(\sigma_{(I^c,I)})^{-1} \sigma_{(I,I^c)} = \begin{pmatrix} 1 & \cdots & k & k+1 & \cdots & m \\ m - k + 1 & \cdots & m & 1 & \cdots & m - k \end{pmatrix}.$$

The result now follows from Theorem B.2.2 and Theorem B.2.4.   □

Example B.2.6. Let $m = 6$ and $k = 3$, and let $I = (2, 4, 6)$. Then $I^c = (1, 3, 5)$, and

$$(I, I^c) = (2, 4, 6, 1, 3, 5) \quad \text{and} \quad (I^c, I) = (1, 3, 5, 2, 4, 6),$$

hence

$$\sigma_{(I,I^c)} = \begin{pmatrix} 1 & 2 & 3 & 4 & 5 & 6 \\ 2 & 4 & 6 & 1 & 3 & 5 \end{pmatrix} \quad \text{and} \quad \sigma_{(I^c,I)} = \begin{pmatrix} 1 & 2 & 3 & 4 & 5 & 6 \\ 1 & 3 & 5 & 2 & 4 & 6 \end{pmatrix}.$$

In cycle notation,

$$\sigma_{(I,I^c)} = (1\ 2\ 4)(3\ 6\ 5) = (1\ 2)(2\ 4)(3\ 6)(6\ 5)$$

and

$$\sigma_{(I^c,I)} = (2\ 3\ 5\ 4) = (2\ 3)(3\ 5)(5\ 4),$$

so

$$\mathrm{sgn}(\sigma_{(I,I^c)}) = 1 \qquad \text{and} \qquad \mathrm{sgn}(\sigma_{(I^c,I)}) = -1.$$

Thus,

$$\mathrm{sgn}(\sigma_{(I,I^c)})\,\mathrm{sgn}(\sigma_{(I^c,I)}) = -1 = (-1)^{3(6-3)},$$

which agrees with Theorem B.2.5. ◇

B.3   Rings

A **ring** is a triple $(R, +, \circ)$, where R is a set, and $+ : R \times R \longrightarrow R$ and $\circ : R \times R \longrightarrow R$ are maps, called **addition** and **multiplication**, respectively, such that for all a, b, c in R:

[R1] $(R, +)$ is a commutative group.
[R2] Multiplication is **associative**:

$$a \circ (b \circ c) = (a \circ b) \circ c.$$

[R3] Addition and multiplication obey the **distributive laws**:

$$a \circ (b + c) = (a \circ b) + (a \circ c)$$
$$(a + b) \circ c = (a \circ c) + (b \circ c).$$

Usually $a \circ b$ is abbreviated to ab. We say that R is **commutative** if multiplication is commutative; that is, $ab = ba$ for all a, b in R. Also, R is said to be a **ring with identity** if there is an element id_R in R, called the **(multiplicative) identity**, such that

$$\mathrm{id}_R\, a = a\,\mathrm{id}_R = a$$

for all a in R. For example, the set of integers $\{\ldots, -2, -1, 0, 1, 2, \ldots\}$ is a commutative ring with identity.

B.4   Fields

A **field** is a commutative ring with identity $(F, +, \circ)$ such that $(F \backslash \{0\}, \circ)$ is a group. That is, for every nonzero element a in F, there is an element a^{-1} in F, called the **(multiplicative) inverse** of a, such that

$$aa^{-1} = a^{-1}a = \mathrm{id}_F.$$

For example, the set of real numbers is a field.

B.5   Modules

Let R be a commutative ring with identity. A **module over R** (or **R-module**) is a triple $(V, +, \circ)$, where V is a set, and $+: V \times V \longrightarrow V$ and $\circ: R \times V \longrightarrow V$ are maps, called **addition** and **scalar multiplication**, respectively, such that for all v, v_1, v_2 in V and all a, b in R:

[**M1**] $(V, +)$ is a commutative group.

[**M2**] Scalar multiplication is associative:

$$(a \circ b) \circ v = a \circ (b \circ v).$$

[**M3**] Addition and scalar multiplication obey the **distributive laws**:

$$a \circ (v_1 + v_2) = (a \circ v_1) + (a \circ v_2)$$
$$(a + b) \circ v = (a \circ v) + (b \circ v).$$

[**M4**] $\mathrm{id}_R \circ v = v$.

Usually $a \circ v$ is abbreviated to av. A subset V' of V is said to be an **R-submodule** of V if (i) V' is an R-module in its own right (when addition and scalar multiplication in V are restricted to V'), and (ii) V' has the same zero element as V.

Theorem B.5.1. *If $\{V_\alpha : \alpha \in A\}$ is a collection of R-submodules of V, then $\bigcap_{\alpha \in A} V_\alpha$ is an R-submodule of V.*

Proof. Straightforward.   □

Let V and W be R-modules. A map $A : V \longrightarrow W$ is said to be an **R-module homomorphism** if $A(v_1 + v_2) = A(v_1) + A(v_2)$ and $A(cv) = cA(v)$ for all v_1, v_2, v in V and all c in R. The **kernel** of A is defined by

$$\ker(A) = \{v \in V : A(v) = 0\}.$$

We say that A is an **R-module isomorphism**, and that V and W are **isomorphic** (as R-modules), written $V \approx W$, if A is bijective.

Theorem B.5.2. *If V and W are R-modules and $A : V \longrightarrow W$ is an R-module isomorphism, then $A^{-1} : W \longrightarrow V$ is an R-module isomorphism.*

Proof. The proof is virtually identical to that of Theorem 1.1.9.   □

Theorem B.5.3. *Let V and W be R-modules, and let $A : V \longrightarrow W$ be an R-module homomorphism. Then A is injective if and only if $\ker(A) = \{0\}$.*

Proof. The proof is virtually identical to that of (c) $\Leftrightarrow$ (d) in Theorem 1.1.12.
                                                                                                           □

Let $V_1, \ldots, V_m$ be R-modules. We make $V_1 \times \cdots \times V_m$ into an R-module by defining addition and scalar multiplication as follows: for all $(v_1, \ldots, v_m)$, $(w_1, \ldots, w_m)$ in $V_1 \times \cdots \times V_m$ and all c in R, let

$$(v_1, \ldots, v_m) + (w_1, \ldots, w_m) = (v_1 + w_1, \ldots, v_m + w_m)$$

and

$$c(v_1, \ldots, v_m) = (cv_1, \ldots, cv_m).$$

If V and W are R-modules such that $V_1 = \cdots = V_r = V$ and $V_{r+1} = \cdots = V_s = W$, where $r + s = m$, we denote

$$V \times \cdots \times V \times W \times \cdots \times W \qquad \text{by} \qquad V^r \times W^s.$$

For R-modules $V_1, \ldots, V_m, W$, we say that a map

$$\mathcal{A} : V_1 \times \cdots \times V_m \longrightarrow W$$

is **R-multilinear** if it is R-linear in each of its arguments. That is, for all v_i, w_i in V_i and all c in R,

$$\mathcal{A}(v_1, \ldots, cv_i + w_i, \ldots, v_m)$$
$$= c\mathcal{A}(v_1, \ldots, v_i, \ldots, v_m) + \mathcal{A}(v_1, \ldots, w_i, \ldots, v_m)$$

for $i = 1, \ldots, m$. The set of R-multilinear maps from $V_1 \times \cdots \times V_m$ to W is denoted by

$$\text{Mult}_R(V_1 \times \cdots \times V_m, W).$$

We make $\text{Mult}_R(V_1 \times \cdots \times V_m, W)$ into an R-module by defining addition and scalar multiplication as follows: for all maps $\mathcal{A}, \mathcal{B}$ in $\text{Mult}_R(V_1 \times \cdots \times V_m, R)$ and all c in R, let

$$(\mathcal{A} + \mathcal{B})(v_1, \ldots, v_m) = \mathcal{A}(v_1, \ldots, v_m) + \mathcal{B}(v_1, \ldots, v_m)$$

and

$$(c\mathcal{A})(v_1, \ldots, v_m) = c\mathcal{A}(v_1, \ldots, v_m)$$

for all $(v_1, \ldots, v_m)$ in $V_1 \times \cdots \times V_m$. The **zero element** of $\text{Mult}_R(V_1 \times \cdots \times V_m, W)$, denoted by 0, is the zero map, that is, the map that sends all elements of $V_1 \times \cdots \times V_m$ to the zero element of W. When $m = 1$, an R-multilinear map is simply an R-linear map. In that setting, we use the notation $\text{Lin}_R(V, W)$. Since R is an R-module over itself, the above considerations also encompass the R-modules $\text{Mult}_R(V_1 \times \cdots \times V_m, R)$, $\text{Mult}_R(V^m, R)$, and $\text{Lin}_R(V, R)$.

B.6   Vector Spaces

A module V over a field F (viewed as a commutative ring with identity) is called a **vector space over F** (or **F-vector space**). An F-submodule of V is said to be a **subspace** of V, and an F-module homomorphism (isomorphism) is called an **F-linear map (F-linear isomorphism)**.

Let W be a vector space over F. In the notation of Section B.5, we have the F-vector spaces $\mathrm{Mult}_F(V_1 \times \cdots \times V_m, W)$, $\mathrm{Mult}_F(V^m, W)$, and $\mathrm{Lin}_F(V, W)$. For brevity, when $F = \mathbb{R}$, we drop the subscript $\mathbb{R}$ from the notation and write $\mathrm{Mult}(V_1 \times \cdots \times V_m, W)$, $\mathrm{Mult}(V^m, W)$, and $\mathrm{Lin}(V, W)$. In particular, when $F = W = \mathbb{R}$, we have $\mathrm{Mult}(V_1 \times \cdots \times V_m, \mathbb{R})$, $\mathrm{Mult}(V^m, \mathbb{R})$, and $\mathrm{Lin}(V, \mathbb{R})$.

B.7   Lie Algebras

Let R be a commutative ring with identity. A **Lie algebra over R** is a 4-tuple $(V, +, \circ, [\cdot, \cdot])$, where $(V, +, \circ)$ is an R-module and $[\cdot, \cdot] : V \times V \longrightarrow V$ is a map, called **bracket**, such that for all u, v, w in V and all c in R:

[L1] Bracket is **R-bilinear**:

$$[cu + v, w] = c[u, w] + [v, w]$$
$$[u, cv + w] = c[u, v] + [u, w].$$

[L2] Bracket is **alternating**:
$$[v, w] = -[w, v].$$

[L3] Bracket satisfies **Jacobi's identity**:

$$\big[u, [v, w]\big] + \big[v, [w, u]\big] + \big[w, [u, v]\big] = 0.$$

As an example, let R be a (not necessarily commutative) ring with identity and define a map $[\cdot, \cdot] : R \times R \longrightarrow R$ by $[a, b] = ab - ba$ for all a, b in R. Then $(R, +, \circ, [\cdot, \cdot])$ is a Lie algebra over R.

Further Reading

Abate, M. and Tovena, F. (2012). *Curves and Surfaces*. Milan: Springer-Verlag Italia.

Banchoff, T. and Lovett, S. (2010). *Differential Geometry of Curves and Surfaces*. Natick, MA: A K Peters.

Bär, C. (2010). *Elementary Differential Geometry*. Cambridge: Cambridge University Press.

Bishop, R.L. and Goldberg, S.I. (1980). *Tensor Analysis on Manifolds*. New York: Dover.

Boothby, W.M. (2003). *An Introduction to Differentiable Manifolds and Riemannian Geometry, 2nd edn., revised*. Amsterdam: Academic Press.

Conlon, L. (2008). *Differentiable Manifolds, 2nd edn*. Boston, MA: Birkhäuser.

Darling, R.W.R. (1994). *Differential Forms and Connections*. Cambridge: Cambridge University Press.

do Carmo, M.P. (1976). *Differential Geometry of Curves and Surfaces*. Upper Saddle River, NJ: Prentice-Hall.

do Carmo, M.P. (1992). *Riemannian Geometry*. Boston, MA: Birkhäuser.

do Carmo, M.P. (1994). *Differential Forms and Applications*. Berlin: Springer -Verlag.

Flanders, H. (1989). *Differential Forms with Applications to the Physical Sciences*. New York: Dover.

Galbis, A. and Maestre, M. (2012). *Vector Analysis Versus Vector Calculus*. New York: Springer.

Grøn, Ø. and Hervik, S. (2007). *Einstein's General Theory of Relativity: With Modern Applications in Cosmology*. New York: Springer.

Kühnel, W. (2006). *Differential Geometry: Curves–Surfaces–Manifolds, 2nd edn*. Providence, RI: American Mathematical Society.

Lee, J.M. (1997). *Riemannian Manifolds: An Introduction to Curvature*. New York: Springer-Verlag.

Lee, J.M. (2009). *Manifolds and Differential Geometry*. Providence, RI: American Mathematical Society.

Lee, J.M. (2011). *Introduction to Topological Manifolds, 2nd edn*. New York:

Semi-Riemannian Geometry, First Edition. Stephen C. Newman.
© 2019 John Wiley & Sons, Inc. Published 2019 by John Wiley & Sons, Inc.

Springer.

Lee, J.M. (2013). *Introduction to Smooth Manifolds, 2nd edn.* New York: Springer.

Lovett, S. (2010). *Differential Geometry of Manifolds.* Natick, MA: A K Peters.

Naber, G.L. (2012). *The Geometry of Minkowski Spacetime: An Introduction to the Mathematics of the Special Theory of Relativity, 2nd edn.* New York: Springer.

O'Neill, B. (1983). *Semi-Riemannian Geometry: With Applications to Relativity.* New York: Academic Press.

O'Neill, B. (1997). *Elementary Differential Geometry, 2nd edn.* San Diego, CA: Academic Press.

Pressley, A. (2012). *Elementary Differential Geometry, 2nd edn.* London: Springer-Verlag.

Spivak, M. (1965). *Calculus on Manifolds.* Boulder, CO: Westview Press.

Spivak, M. (1999). *A Comprehensive Introduction to Differential Geometry, Volume Two, 3rd edn.* Houston, TX: Publish or Perish.

Spivak, M. (2005). *A Comprehensive Introduction to Differential Geometry, Volume One, 3rd edn., with corrections.* Houston, TX: Publish or Perish.

Sternberg, S. (2012). *Curvature in Mathematics and Physics.* Mineola, NY: Dover.

Torres del Castillo, G.F. (2012). *Differentiable Manifolds: A Theoretical Physics Approach.* Boston, MA: Birkhäuser.

Tu, L.W. (2011). *An Introduction to Manifolds, 2nd edn.* New York: Springer.

Yokonuma, T. (1992). *Tensor Spaces and Exterior Algebra.* Providence, RI: American Mathematical Society.

Index

Semi-Riemannian Geometry, First Edition. Stephen C. Newman.
© 2019 John Wiley & Sons, Inc. Published 2019 by John Wiley & Sons, Inc.